**Springer Textbooks in Earth Sciences,
Geography and Environment**

T0210528

The Springer Textbooks series publishes a broad portfolio of textbooks on Earth Sciences, Geography and Environmental Science. Springer textbooks provide comprehensive introductions as well as in-depth knowledge for advanced studies. A clear, reader-friendly layout and features such as end-of-chapter summaries, work examples, exercises, and glossaries help the reader to access the subject. Springer textbooks are essential for students, researchers and applied scientists.

More information about this series at https://link.springer.com/bookseries/15201

Andrew D. Miall

Stratigraphy: A Modern Synthesis

Second Edition

 Springer

Andrew D. Miall
Earth Sciences
University of Toronto
Toronto, ON, Canada

ISSN 2510-1307   ISSN 2510-1315   (electronic)
Springer Textbooks in Earth Sciences, Geography and Environment
ISBN 978-3-030-87538-1   ISBN 978-3-030-87536-7   (eBook)
https://doi.org/10.1007/978-3-030-87536-7

For Meredith, Henry, Owen and Nora

Preface

The stratigraphic record is the major repository of information about the geological history of Earth, a record stretching back for nearly 4 billion years. Stratigraphic studies fill out our planet's plate-tectonic history with the details of paleogeography, past climates and the record of evolution, and stratigraphy is at the heart of the effort to find and exploit fossil fuel resources.

The exploration of this history has been underway since James Hutton first established the basic idea of uniformitarianism toward the end of the eighteenth century, and William Smith developed the stratigraphic basis for geological mapping a few decades later. Modern stratigraphic methods are now able to provide insights into past geological events and processes on time scales with unprecedented accuracy and precision, and have added much to our understanding of global tectonic and climatic processes. But it has taken 200 years and a modern revolution to bring all the necessary developments together to create the modern, dynamic science that this book sets out to describe. It has been a slow revolution, but stratigraphy now consists of a suite of integrated concepts and methods, several of which have considerable predictive and interpretive power.

It is argued in Chap. 1 of this book that the new, integrated, dynamic science that Stratigraphy has become is now inseparable from what were its component parts, including sedimentology, chronostratigraphy and the broader aspects of basin analysis. In this chapter, the evolution of this modern science is traced from its nineteenth-century beginnings, including the contributions that such special fields as facies analysis, fluid hydraulics, plate tectonics, and the reflection-seismic surveying method have made to its evolution.

The following are just some of the major features of the Stratigraphy of the early twenty-first century: Sequence stratigraphy has become the standard methodology for documentation, mapping and interpretation, replacing the old descriptive practices of lithostratigraphy; reflection-seismic methods, including the use of 3-D seismic and the application of seismic geomorphology, have become steadily more advanced tools for subsurface exploration and development; the Geological Time Scale is being standardized with the universal adoption of the system of Global Stratigraphic Sections and Points (GSSPs) and has become much more precise, with the incorporation of several new methods for evaluating deep time. The stratigraphic record is now able to help generate answers to many complex questions about Earth's past tectonic and climatic history.

The basic field and subsurface observations on which Stratigraphy is based are described in Chap. 2. Facies analysis methods are detailed in Chap. 3, and the recognition of depositional environments by facies methods is described in Chap. 4. Chapter 5 provides a succinct summary of sequence models for siliciclastic and carbonate sediments, and Chap. 6 describes modern mapping methods for use in surface and subsurface studies, including seismic methods. The synthesis of all this material is detailed in Chap. 7, which includes a discussion of the current attempts to standardize sequence-stratigraphic terminology and the Geological Time Scale.

Chapter 8, the concluding chapter of the book, focuses on the new understanding we are acquiring about the processes by which the stratigraphic record preserves elapsed geologic time. Refinements in chronostratigraphic methods are revealing the importance of breaks in the sedimentary record and the ubiquity of missing time, and are revealing an important disconnect between sedimentation rates and preservational processes operating at the present day versus those we interpret from the rock record. This calls for a significant modification in the way that we apply the traditional principles of uniformitarianism to our reconstructions of geologic history. Some examples of modern stratigraphic work based on very detailed data bases, and making use of modern concepts of sedimentation and presentation, are included in this concluding chapter.

The new synthesis that is the subject of this book is offered for advanced undergraduate and graduate training and for use by professionals, particularly those engaged in mapping and subsurface exploration and development.

Toronto, Canada   Andrew D. Miall
July 2021

Acknowledgments

Colleagues who assisted with the earlier editions of *Principles* by critically reading parts or all of the manuscript include Tony Tankard, Andy Baillie, Guy Plint and Ray Ingersoll. I am eternally grateful for their wise advice, and if they choose to look they will find significant portions of the book that describe basic methods, such as the descriptions of field observations methods and the foundations of facies analysis methods, largely unchanged in this book, in which they constitute Chaps. 2–4. Having the photographs in color has been a nice improvement for which thanks are due to the God of Technology (and Springer).

Reviewers for the journal articles from which material in Chaps. 7 and 8 was drawn included Felix Gradstein, Ashton Embry, Bruce Wilkinson, Tony Hallam, Alan Smith, Brian Pratt, Gerald Bryant, John Holbrook, Chris Paola, Pete Sadler, Robin Bailey, Dave Smith, John Howell, and Torbörn Törnqvist. The historical section in Chap. 1 was reviewed by Bill Fisher, Ron Steel, Bob Dalrymple, and Martin Gibling. Chapters 5–7 were critically read by David Morrow. Conversations over the years with Geoff Norris, Nick Eyles, Dale Leckie, Steve Hubbard, Brian Zaitlin, Gerry Reinson, Jim Dixon, Phil Fralick, Jun Cowan, Tobi Payenberg, Bill Galloway, my former colleagues at the Geological Survey of Canada, and many others in Calgary, that amazing center of geological activity, have all helped to shape my understanding of stratigraphy and basin analysis. I extend my thanks to all these individuals. Any remaining errors or omissions remain my responsibility.

For the second edition, I am once again indebted to Guy Plint for his review and comments of the first edition, which have drawn my attention to some new ideas, some errors, and to places where rewriting would improve clarity. David Morrow revealed the mysteries of dolomite to my clastic tin ears. For discussions on the ever-evolving subject of sequence stratigraphy, I benefited from guidance by Octavian Catuneanu. Simon Pattison shared insights from his remarkable era of fieldwork on the Mesaverde Group of the Book Cliffs in Utah. Finally, I have enjoyed many e-mail discussions with Janok Bhattacharya and John Holbrook, which I hope helped us to remain sane during the pandemic, while we discussed the finer points of what came to be called the stratigraphy machine.

Once again, I must acknowledge the enthusiastic support and encouragement of my wife, Charlene Miall. Her insights into the scientific method and the sociology of science have been particularly invaluable. She has enjoyed the fieldwork, too. Our children, Chris and Sarah, have consistently been supportive, and our grandchildren, Meredith, Henry, Owen, and Nora, provide that joy of renewal that makes it all feel worthwhile.

Revision History

My first book, *Principles of Sedimentary Basin Analysis*, went through three editions, published successively in 1984, 1990, and 1999. In 2014, I realized that it might be time for a new edition. In reviewing the changes that had taken place in the whole broad field of sedimentary geology since that last edition, it became clear that stratigraphy is the area that has undergone the most significant changes in the last decades, and that is what I decided would most usefully be treated at length in this book.

Stratigraphy has undergone a revolution that has brought together multiple developments dealing with different themes and concepts in sedimentary geology and basin analysis. The following notes refer to the first edition of the present book.

Chapter 1 includes a new section in which I trace the evolution of these many themes and attempt to show how they have come together during the last few decades (since about 1990). This is based on a review prepared for Geoscience Canada (Miall, 2015a).

The text of Chaps. 2 and 4 from *Principles* was updated and became Chaps. 2–4 in the present book. Chapter 3 of *Principles,* which dealt with dating and correlation, and the formal methods for the definition and naming of units, was substantially rewritten and incorporates much of the material I wrote for "Sophisticated Stratigraphy," a review prepared at the invitation of the Geological Society of America (Miall 2013). It has been moved further along in the present book, appearing as Chap. 7, the point being that stratigraphy should now be seen as a science that synthesizes sedimentary geology, and which therefore requires that the subject is best addressed once the work of sedimentological description and interpretation is underway.

Chapters 5 and 6 of the present book are those that underwent the most complete rewriting, to reflect the major changes in the science since the last edition of *Principles.* Sequence stratigraphy (Chap. 5) has been the standard method for formal description and paleo-geographic interpretation since the 1980s. Mapping methods (Chap. 6) are now dominated, at least in the petroleum industry, by the techniques of the reflection-seismic method, including 3-D seismic and the interpretive methods of seismic geomorphology. The study of detrital zircons has added many new dimensions to stratigraphic studies, and this, also, is touched on in Chap. 6.

The book culminates with Chap. 8, which is intended primarily as a review of current research into the nature of deep time as preserved in the sedimentary record. It is partly based on three research publications (Miall 2014, 2015b, 2016) that focus on modern data dealing with sedimentation and accommodation rates, and the implications of these data for stratigraphic interpretation. The chapter concludes with a review of the current advanced research into cyclostratigraphy and astrochronology.

The *second edition* of this book has been improved with references to many new informative examples of stratigraphic work. Chapters 5 and 7 reflect continuing developments in sequence stratigraphy. New ideas concerning the issue of sedimentation rates, time scales, and the preservability of the stratigraphic record are highlighted in Chap. 8. Research and review in this area (Bhattacharya et al., 2019; Holbrook and Miall, 2020; Miall et al., 2021) have provided many of the new ideas discussed in Chap. 8, which is significantly modified from the first edition.

References

Bhattacharya, J.P., Miall, A.D., Ferron, C., Gabriel, J., Randazzo, N., Kynaston, D., Jicha, B.R., Singer, S., 2019, Balancing sediment budgets in deep time and the nature of the stratigraphic record. Earth-Science Reviews, v. 199, #102985, 25 p.

Holbrook, J.M., and Miall, A.D., 2020, Time in the Rock: A field guide to interpreting past events and processes from siliciclastic stratigraphy. Earth-Science Reviews, v. 203, #103121, 23 p.

Miall, A.D., 2013, Sophisticated stratigraphy, In: Bickford, M.E., ed., The web of geological sciences: Advances, impacts and interactions. Geol Soc Am Spec Pap 500, p. 169–190.

Miall, A.D., 2014, The emptiness of the stratigraphic record: A preliminary evaluation of missing time in the Mesaverde Group, Book Cliffs, Utah. J Sediment Res., v. 84, p. 457–469.

Miall, A.D 2015a, Making stratigraphy respectable: from stamp collecting to astronomical calibration. Geosci Canada, v. 42, p. 271–302.

Miall, A.D., 2015b, Updating uniformitarianism: stratigraphy as just a set of "frozen accidents", In: Smith, D. G., Bailey, R.J., Burgess, P., and Fraser. A., eds., Strata and time: Geological Society, London, Special Publication 404, p. 11–36.

Miall, A.D., 2016, The valuation of unconformities. Earth-Science Reviews, v. 163, p. 22–71.

Miall, A.D., Holbrook, J.M., and Bhattacharya, J.P., 2021, The Stratigraphy Machine. J Sediment Res., v. 91, p. 595–610.

Contents

The Scope of Modern Stratigraphy

1

Contents

Abstract

This chapter explains the important place Stratigraphy holds within geology. It provides a summary of the development of the main ideas in stratigraphic mapping, geophysical methods and sedimentological interpretation that emerged through the twentieth century, up to the emergence of modern-day "sophisticated stratigraphy," which completes the integration of all relevant disciplines into a modern science. A summary is provided of the seventeen orders of magnitude of time scales that need to be considered in the evaluation of Earth history. The chapter concludes with a description of the types of research and exploration projects that stratigraphers may
be engaged in, and a synopsis of recommended workflow and reporting practices.

1.1 The Importance of Stratigraphy

It could be argued that in some respects **Stratigraphy** is the most important component of the science of Geology. Here's why:

McLaren (1978) provided nine reasons why the study of Stratigraphy with, at its center, an accurate geological time scale, is important:

© The Author(s), under exclusive license to Springer Nature Switzerland AG 2022
A. D. Miall, *Stratigraphy: A Modern Synthesis*, Springer Textbooks in Earth Sciences, Geography and Environment,
https://doi.org/10.1007/978-3-030-87536-7_1

[Stratigraphy supplies unique and essential information regarding:] (1) rates of tectonic processes; (2) rates of sedimentation and accurate basin history; (3) correlation of geophysical and geological events; (4) correlation of tectonic and eustatic events; (5) are epeirogenic movements worldwide [?]… (6) have there been simultaneous extinctions of unrelated animal and plant groups [?]; (7) what happened at era boundaries [?]; (8) have there been catastrophes in earth history which have left a simultaneous record over a wide region or worldwide [?]; and (9) are there different kinds of boundaries in the geologic succession [?] (That is, "natural" boundaries marked by a worldwide simultaneous event versus "quiet" boundaries, man-made by definition).[question marks added]

Doyle and Bennett (1998, p. 1) stated that "Stratigraphy is the key to understand the Earth, its materials, structure and past life. It encompasses everything that has happened in the history of the planet." In this statement is the recognition that the stratigraphic history of layered sedimentary rocks preserved on the continents, and on the ocean floors constitutes the documented record of Earth history. No other branch of geology can provide this information.

Berggren et al. (1995, p. v) explained that the "essence of Stratigraphy and its handmaiden Geochronology" is to "understand the dynamic relationship which certainly exists between the evolution of ocean-continental geometries and concomitant changes in the climate and ocean circulation system and the evolution of life itself" by situating "the progression of events in this intricately related system in a precise temporal framework."

Torrens (2002, p. 251) pointed out a unique and essential component of stratigraphy: "The science of geology is all about time. … So stratigraphy must first and foremost concern questions of time. It is the only area of geology that is truly unique, other branches of geology are too often borrowed bits of physics, chemistry or biology."

The foundational basis of stratigraphy is the principle of the superposition of strata—the concept that the layers at the bottom of a pile were laid down first, and the subsequent layers could not have been created until the underlying ones were in place. The formulation of this principle is attributed to Nicolas Steno in the 1660s (Cutler 2004).

The world's first stratigrapher was William Smith, a canal surveyor, who produced the first regional geological map in 1815, covering England, Wales and part of Scotland. The construction and refinement of geological maps, and the documentation of the subsurface for the purposes of petroleum and mineral exploration, have constituted two of the primary activities of practicing stratigraphers worldwide for the past 200 years. Dating and correlating the rocks have formed an integral part of this work, and questions about the nature of the time signal preserved in stratigraphic successions, and the developments of methods to investigate it have constituted a large part of this activity. William Smith's principal theoretical contribution—which is what made

geological mapping possible—was the recognition of the reliability of the fossil record: the same assemblages of fossils always occur in the same order, and thus were born both the method of relative age dating and the first reliable method for correlation on the basis of time.

Faunal succession data provided a necessary support for the theory of evolution (Darwin 1859) and is the basis for our modern concepts of the gradual appearance of all of Earth life forms, culminating (in our egocentric view) in the emergence of the human race.

As I demonstrated in my review paper for the Geological Society of America ("Sophisticated Stratigraphy": Miall 2013), modern stratigraphic methods are now providing extraordinary insights into the history of our Earth. Modern methods of age dating and modern analytical methods have revolutionized the business of **historical geology**. However, in the drive to develop and apply ever more precise laboratory methods to geological samples, whether this be with the aim of age dating or the reconstruction of past climates, it is all too easy to lose sight of exactly where samples come from and what was their field context. How typical and how representative are samples, relative to the variability of their field setting? How were they situated with respect to breaks in sedimentation, the record of rare events, or disturbance induced by bioturbation or syndepositional tectonism? Specialists in quantitative methods, particularly geophysicists, geochemists and those using numerical and statistical methods, including models and simulations, may be particularly susceptible to a neglect of these important questions. As Spychala (2020) has pointed out, sedimentary geologists have so many tools at their disposal, including high-resolution remote sensing and modeling, that there is a tendency by some to downplay the importance of field work. Wright (2019, p. 311) refers to "helicopter science,"

"where researchers 'drop' into a section they barely know, ignore any local expertise, grab some samples which are taken without any consideration of context, depositional or diagenetic, perform some analyses, and propose, for example, a theory of changed global ocean chemistry, when if they had looked at the context of the sample they would have realized that their theory was deeply flawed."

However, one of the key elements of the stratigraphic data base is **field context**. What is the stratigraphic and sedimentologic setting of the rocks that we are using to make these sophisticated interpretations? This is one of the unique characteristics of the science of stratigraphy, of particular relevance to the reconstruction of events in past time.

Until the 1960s, stratigraphy was largely a descriptive science, concerned primarily with the documentation of the lithologic and biostratigraphic successions of sedimentary basins as a basis for locating and exploiting fossil fuel and mineral deposits. Figure 7.3 illustrates an example of a regional cross-section dating from 1882 that illustrates this phase of development. Textbooks on petroleum geology, contain some of the most advanced and detailed treatments of stratigraphy as actually practiced in the field (that by Levorsen 1954 is a classic; a modern example is the text by Chapman 2000), In many respects, William Smith's focusing on basic mapping survived as a central focus through several of the revolutions that were took place in the earth sciences, beginning in the 1960s. However, over the last fifty years a profound change in approach has taken place, initially under the rubric of **Sedimentology,** which took sedimentary geologists away from description and classification into a focus on processes (Seibold and Seibold 2002, provided a detailed history from a European perspective; see also Middleton 2005). Only in recent years have **Stratigraphy**, **Sedimentology, Chronostratigraphy** and **Basin Analysis** come together to provide a dynamic, unified approach to the study of sedimentary basins. It would now be accurate to state that "Stratigraphy IS Sedimentology and Sedimentology IS Stratigraphy."

The evolution of modern methods can be understood as a series of separate developments that partially overlapped in time and which have gradually coalesced to create what I called "Sophisticated Stratigraphy," in my review written for the Geological Society of America (Miall 2013). Middleton (2005, p. 628) suggested that:

> Only after 1950 was it common to find specialists who studied sedimentary rocks, but declined to be called stratigraphers, and since 1977 an increasing number of specialists refuse to make a hard distinction between sedimentology and at least some aspects of paleontology and stratigraphy, which they include together as "sedimentary geology."

A major incentive for the development of ideas about sedimentary rocks has been the drive to explore for and exploit fossil fuels. This became an American national imperative with the mechanization of society and the introduction of mass manufacturing methods early in the twentieth century, followed by the entry of the United States into the First World War. All of this nicely coincided with the discovery of the continent's first giant oil fields in Texas, Oklahoma and California. This was the motivation for the establishment of the first technical association devoted to the science of petroleum, the *American Association of Petroleum Geologists*, in Tulsa, Oklahoma, in 1917 (where the society still maintains its headquarters). Petroleum companies were all very active in the development of the science, until the advent of new business methods in the 1980s led most companies to shed their research operations. But until then, company research arms were important contributors to the theory and methods of stratigraphy. Notably this includes the Shell Development company, which was at the forefront of research in sedimentology through the 1950s and 1960s.

Exxon, Shell and British Petroleum (with Peter Vail at Exxon at the lead, followed by Bert Bally at Shell) were very active in the development of reflection-seismic methods and the advent of modern sequence stratigraphy in the 1970s.

European developments tended to follow the American lead until the birth of the North Sea petroleum province in the 1960s. The Petroleum Exploration Society of Great Britain was established in 1964. The journal Sedimentology, based in Europe, was established in 1962 (although its parent society, the International Association of Sedimentologists was founded in 1952).

Some of the developments in the study of sedimentary rocks were initiated many years ago, but it has only been since sequence stratigraphy matured as a standard descriptive and mapping method during the 1990s that it has become apparent that it has drawn on, exploited, and pulled together these earlier developments that commonly tended to be considered and written about in isolation. The historical evolution of these concepts is summarized in the next section.

1.2 The Evolution of "Sophisticated Stratigraphy"

The roots of modern, dynamic stratigraphy go back to the recognition of the concept of **facies** in the early nineteenth century, but it is argued here that the modern era began with the increased understanding of **fluid hydraulics** and **cyclic sedimentation** and the evolution of the **facies-model** concept in the 1960s. The evolution is broken down below into fourteen steps. A critical fifteenth strand of development concerns the developments of concepts about geologic time and the increasing accuracy and precision with which geologist can now reconstruction the ages of events in the distant geological past (this topic is addressed in Sect. 7.8). These strands of development did not take place in isolation; however, they represent separate concepts or areas of specialization, which took some time to come together into the unified, integrated science that is now practiced. The fifteen components of modern stratigraphy are shown in their historical relationship to each other, together with a few other key developments, including the establishment of key technical societies and journals, in Fig. 1.1. The age ranges shown in brackets in each heading span the period during which the key ideas and publications emerged for the ideas described in that section.

The discovery of the Spindletop field in Texas in 1901 is included in Fig. 1.1 because it has long been regarded as the first oil field to be discovered by the application of scientific methods to the testing of a scientific idea, in this case the "anticlinal theory." As discussed in Sect. 6.3, this is when "oil became an industry."

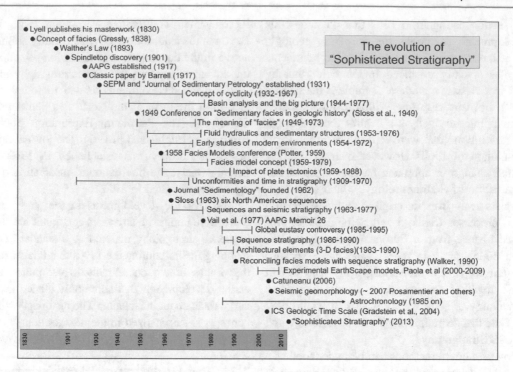

Fig. 1.1 The components of modern sophisticated stratigraphy, shown in their sequential relationships. The brackets beside each component span the age range of the major publications that represent the beginnings of modern developments (as discussed in the text), although in many cases these were preceded by earlier work that contained embryonic beginnings. For example, plate tectonics was preceded by geosyncline theory that began with the work of James Hall and James Dana in the nineteenth century; the geologic time scale is partly based on geochronology, which was founded by Arthur Holmes early in the twentieth century

This summary is intended only to touch on the main highlights. A full historical and analytical account remains to be written. Many of the publications cited here have become classics, with many hundreds of citations.

1.2.1 Beginnings (Nineteenth Century)

Middleton (2005) divided the history of sedimentology into six periods or stages. The first stage ended about 1830 with the publication of Lyell's (1830) masterwork that led to the general acceptance of **uniformitarianism**, or **actualism**, as the basis for geology. What follows in this section falls into his second period. The subsequent discussion does not adhere to his subdivision into "periods" because I focus on specific themes which overlapped in time.

Two key early developments were the recognition of the concept of **facies** (Gressly 1838), and the establishment of **Walther's Law** (Walther 1893–1894). Teichert (1958), Middleton (1973) and Woodford (1973) reviewed the history and use of the concepts in light of contemporary ideas. Note the dates of these papers (1958,1973), in light of the stages of development summarized below, because they help to explain the chronological evolution of modern stratigraphic thought and theory. Cross and Homewood (1997) provided a translation and analysis of Gressly's work and

showed how he anticipated many modern concepts. Walther's law is discussed further in Sect. 3.4.1.

Developments in biostratigraphy were enormously important, in establishing some of the basic ideas about stratal succession, relative ages, and correlation. The evolution of the concepts of zone and stage are discussed in detail elsewhere (Hancock 1977; Miall 2004), topics that are not repeated here. Stratigraphic paleontology was a central theme of stratigraphy until relatively recent times. In fact, the first professional society in the field of sedimentary geology, the *Society of Economic Paleontologists and Mineralogists*, founded in Tulsa in 1931, emphasized this fact in the title of the society. Paleontology and mineralogy were important elements of petroleum geology and basin analysis until the seismic revolution of the 1970s, mainly because of their use in the identification and correlation of rock units in petroleum-bearing basins. Modern biostratigraphic methods are discussed in Sect. 7.5.

Although stratigraphy remained an essentially descriptive science until the 1960s, there was a recurring theme in stratigraphic studies through the nineteenth and early twentieth centuries concerning the mechanisms controlling the location and pattern of stratigraphic accumulation. These included the geosynclinal concepts of Hall and Dana (Aubouin 1965), and what Miall (2004, p. 14) termed "the continual search for a 'pulse of the earth'". As described in that paper, workers such as Ulrich, Chamberlin and Grabau sought to explain repetition and rhythmicity in the stratigraphic record by hypotheses about tectonism and sea-level change that, to some extent, anticipated the ideas emerging from early modern work on sequence stratigraphy in the 1970s (see Miall 2004 for a more extensive discussion of this history).

1.2.2 Cyclic Sedimentation (1932–1968)

Implicit in the early work on facies and on Walther's Law is the concept of recurrence of certain environments and their deposits. Grabau (1906) was one of stratigraphy's early theorists, who recognized the repetition in North American cratonic successions of an onlap–offlap relationship caused by cycles of sea-level change (Fig. 1.2). Gilbert (1895) speculated about astronomical control of cyclic sedimentation in Cretaceous rocks in Colorado. The first modern study of cyclic sedimentation was that by Bradley (1929), who analyzed the varves of the Eocene Green River Formation in Utah and Wyoming. However, the study of Carboniferous deposits of the US Mid-continent in the early 1930s was far more influential. The deposits consist of repetitions of a coal-bearing clastic-carbonate succession. These came to be called **cyclothems,** a term which immediately made its way into the permanent lexicon of geological terminology. Wanless and Weller (1932, p. 1003) are credited with the original definition of this term:

> The word "cyclothem" is therefore proposed to designate a series of beds deposited during a single sedimentary cycle of the type that prevailed during the Pennsylvanian period.

Shepard and Wanless (1935) and Wanless and Shepard (1936) subsequently attributed the cyclicity to cycles of sea-level change, an explanation that has never been seriously challenged.

The beginnings of an understanding of the significance of the lithofacies signatures of common environmental settings are implicit in the paper by Nanz (1954), where coarsening- and fining-upward trends extracted from some modern sedimentary environments in Texas are presented. There is no discussion of repetitiveness or cyclicity in this paper, but the work was clearly foundational for the very important papers by Nanz's Shell colleagues that followed less than a decade later (see Sect. 1.2.6), as attested to by unpublished work by Nanz that this author saw in Shell files while employed with that company in the 1970s.

Duff and Walton (1962) demonstrated that the cyclothem concept had become very popular by the early 1960s. For example, J. R. L. Allen, who is credited as one of the two originators of the meandering-river point-bar model for fluvial deposits, used the term cyclothem for cycles in the Old Red Sandstone in his first papers on these deposits (Allen 1962, 1964). Duff and Walton (1962) addressed the widespread use (and misuse) of the term cyclothem, and discussed such related concepts as modal cycle, ideal cycle, idealized cycle and theoretical cycle, the differences between cyclicity, rhythmicity and repetition, and the possible value of statistical methods for refining cyclic concepts. They speculated about the possibility of repeated delta-lobe migration as a cause of cyclicity, in contrast to the prevailing interpretation of the cycles as the product of sea-level change.

With Carboniferous coal-bearing deposits as the focus, two edited compilations dealing with cyclic sedimentation made essential contributions to the birth of modern sedimentology at about this time. Merriam (1964), based in Kansas, provided a focus on the US Mid-continent deposits, while Duff et al. (1967) dealt at length with European examples. The Kansas publication included a study of cyclic mechanisms by Beerbower (1964) that introduced the concepts of **autocyclic** and **allocyclic** processes. Autocyclic processes refer to the processes that lead to the natural redistribution of energy and sediment within a depositional system (e.g., meander migration, shoreline progradation)— the preference is now to use the term **autogenic** because they are not always truly cyclic—whereas **allocyclic (allogenic)** processes are those generated outside the sedimentary system by changes in discharge, load and slope. Beerbower (1964) was dealing specifically with alluvial deposits in this

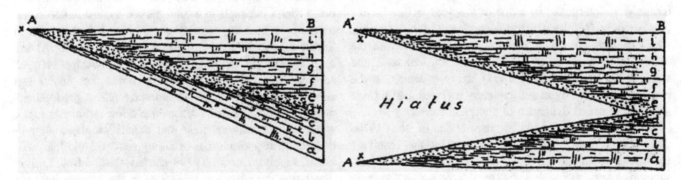

Fig. 1.2 The concept of sedimentary onlap and offlap, as envisaged by Grabau (1906)

paper, but his two terms have subsequently found universal application for other environments and their deposits. The term **allogenic** is now used to refer to processes external to a sedimentary basin, including eustasy, tectonism and climate change.

Another important contribution at this time was that by Visher (1965). The purpose of his paper was to build on the ideas contained in Walther's Law to highlight the importance of the **vertical profile** in environmental interpretation. He provided detailed descriptions of the profiles for six clastic environments, regressive marine, fluvial (channel or valley fill), lacustrine, deltaic, transgressive marine and bathyal–abyssal, drawing on both modern settings and ancient examples. This was, therefore, one of the first comprehensive attempts to apply the principles of actualism (uniformitarianism) to sedimentological interpretations. Interestingly (and this highlights one of the arguments of this chapter that some ideas develop as separate lines of research, which take time to come together), Visher's paper makes no reference to what are now the classic papers on Bouma's turbidite model (Bouma 1962), or Allen's (1964, 1965) work on alluvial deposits, which include his block diagram of a fluvial point bar. However, Beerbower's (1964) description of autocyclicity and Duff and Walton's (1962) speculation about deltaic processes (neither of which are referenced by Visher) indicate the beginnings of what shortly became a flood of new work providing the basis for the facies-model revolution. Early applications of these ideas to the interpretation of the subsurface are exemplified by Berg's (1968) study of an interpreted point-bar complex constituting a reservoir unit in Wyoming.

1.2.3 Basin Analysis and the Big Picture (1948–1977)

Driven in large measure by the needs of the petroleum industry to understand subsurface stratigraphic successions, geologists devised a number of ways to explore the broader origins of a basin fill and understand its paleogeographic evolution. Until the plate tectonics revolution of the 1970s, basins were interpreted in terms of the **geosyncline theory** (see Aubouin 1965 for a historical treatment of this concept), which reached its full expression in this period with the definition of a range of classes based on structural and stratigraphic attributes (Kay 1951), many of which, as the plate-tectonics paradigm subsequently revealed, had little to do with the actual dynamics of continental crust.

Whereas the facies-model revolution of the 1970s (Sect. 1.2.7) dealt with sedimentology on the relatively small scale of individual depositional systems (rivers, deltas, submarine fans, reefs, etc.), paleogeographic reconstruction for industry meant attempting to understand entire basins.

Provenance studies based on detrital petrography were central to this work, hence the title of the first specialized journal in this field, the *Journal of Sedimentary Petrology*, founded in 1931. Isopachs revealed broad subsidence patterns, and (for outcrop work) regional paleocurrent studies confirmed regional transport patterns, even in the absence of the understanding of the hydraulics of sedimentary structures that came later with the development of the flow-regime concept (Sect. 1.2.5). Krumbein (1948) pioneered the generation of lithofacies maps based on such indices as a clastic-carbonate ratios. Examples of his work are shown in Fig. 6.47. Dapples et al. (1948) demonstrated how these maps could be used to deduce tectonic controls in a basin. The subject of stratigraphy meant classical lithostratigraphy. The books and reviews by Pettijohn (1949, 1962; Potter and Pettijohn 1963; Pettijohn et al. 1973) and Krumbein and Sloss (1951, 1963) and Levorsen's (1954) textbook on petroleum geology exemplify this approach.

However, some interesting new ideas that we would now classify under the headings of basin architecture, accommodation and sequence stacking patterns began to emerge, although little of this work was widely used at the time, it being only from the perspective of modern sequence methods that we can look back and see how a few individuals were ahead of their time. Rich (1951) described what we would now term the continental shelf, the continental slope and the deep basin as the **undaform**, **clinoform** and **fondoform**, respectively, and provided descriptions of the processes and resulting sedimentary facies to be expected in each setting (Fig. 1.3). The only one of his terms to survive is **clinoform**, although now it is used as a general term for deposits exhibiting a significant depositional dip, rather than as a term for a depositional environment.

Van Siclen (1958) examined the late Paleozoic cyclothems where they tip over the southern continental margin which, at that time, lay within what is now central Texas. His work includes a diagram of the stratigraphic response of a continental margin to sea-level change and variations in sediment supply that is very similar to present-day sequence models (Fig. 1.4). Oliver and Cowper (1963, 1965) may have been the first to specifically identify "clino" beds in the subsurface using Rich's concepts in a stratigraphic reconstruction based on petrophysical log correlation. This work is discussed in Sect. 6.2.3 (see Fig. 6.14). Curray (1964) was among the first to recognize the importance of the relationships between sea-level and sediment supply. He noted that fluvial and strand plain aggradation and shoreface retreat predominate under conditions of rising sea level and low sediment supply, whereas river entrenchment and deltaic progradation predominate under conditions of falling sea level and high sediment supply. Curtis (1970) carried these ideas further, illustrating the effects of variations in the balance between subsidence and sediment supply as controls on the stacking

Fig. 1.3 The major environments and deposits of the continental shelf, slope and deep basin, as envisaged by Rich (1951)

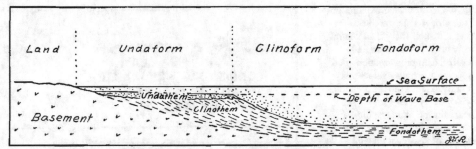

FIGURE 1.—SKETCH ILLUSTRATING DEFINITIONS
Undaform, clinoform, fondoform, undathem, clinothem, fondothem, and wave base and the distribution of muddy water after a storm.
Muddy water shown by stippling; density currents by arrows.
Vertical scale greatly exaggerated.

patterns of deltas, concepts that are now encapsulated by the terms **progradation**, **aggradation** and **retrogradation** (Fig. 5.3). Frazier (1974) subdivided the Mississippi deltaic successions into **transgressive**, **progradational**, and **aggradational** phases, and discussed autogenic (delta switching) and glacioeustatic sedimentary controls.

Perhaps it is because Texas specializes in bigness; this may be the explanation why some critical concepts concerning large-scale sedimentological environments were first developed there. The location of petroleum research laboratories, such as that of Shell Oil in Texas (referred to below) may also have been very influential. I refer to the concept of the **depositional system**, the concept that takes sedimentological analysis beyond the shoreface or the river meander or the reef talus slope to an analysis that encompasses entire systems. Fisk's (1944) work on the lower Mississippi Valley and Delta is an early example of this approach, but it was the later work of William L. Fisher that better exemplifies this next step and was more influential. The work he and his colleagues carried out on the deltas and other depositional systems of the Texas coast (Fisher et al. 1969, 1972) established a whole different scale of operation. Application of current subsurface stratigraphic methods to part of the Eocene section of the Gulf Coast (Fisher and McGowen 1967) demonstrated that existing rivers and deltas along a huge swath of the Gulf Coast had occupied essentially the same map locations for about 40 million years. The depositional systems approach provided the foundation for the **systems tracts** that became a critical part of sequence stratigraphy twenty years later. Lastly, in a paper that appears in the famous memoir that introduced seismic stratigraphy to the geological community (Payton et al. 1977), Brown and Fisher (1977) summarized the ideas of this important group of stratigraphers at the Bureau of Economic Geology (at the University of Texas) and helped to bridge the intellectual next step from large-scale sedimentology to sequence stratigraphy.

1.2.4 The Meaning of "Facies" (1949–1973)

The concept of **facies** and the importance of **Walther's Law** were well understood and used in continental Europe during the nineteenth century, according to Teichert (1958), but did not become widely used in the English-speaking world until the 1930s.

On November 11, 1948 a conference was organized by the Geological Society of America in New York to discuss "Sedimentary facies in geologic history." This was a landmark event, the outcome of which was a Geological Society of America memoir (Longwell 1949) that marked the beginnings of several important developments. The memoir begins with a lengthy paper by Moore (1949) which set the scene by describing and illustrating, with the use of a block diagram, the various facies present within a modern carbonate reef complex in Java, from which he derived this definition:

> Sedimentary facies are areally segregated parts of different nature belonging to any genetically related body of sedimentary deposits.

The paper includes numerous examples of complex stratigraphic relationships from the Phanerozoic record of the United States, illustrating the inter-tonguing of facies of a wide range of environments. Moore's paper also includes an interpretation of the cyclicity exhibited by the cyclothems of the Mid-Continent, accompanied by a diagram showing how different facies develop as a result of repeated transgression and regression. Other papers by E. D. McKee, E. M. Spieker, and others, provide many other examples of complex stratigraphy, indicating that by this time there was a sophisticated understanding of the diachronous nature of facies in the stratigraphic record, and its control by sea-level change. The concluding contribution in this memoir is a lengthy paper by Sloss et al. (1949) in which the concept of the **sequence** is first described (see Sect. 1.2.9).

Fig. 1.4 Subsurface exploration of the Upper Paleozoic section along the shelf margin in central Texas after WW2 generated shelf-to-basin cross-sections that displayed a strong cyclothemic cyclicity. This is the set of models developed by Van Siclen (1958) to explain the stratigraphic architecture in terms of different patterns of sea-level change

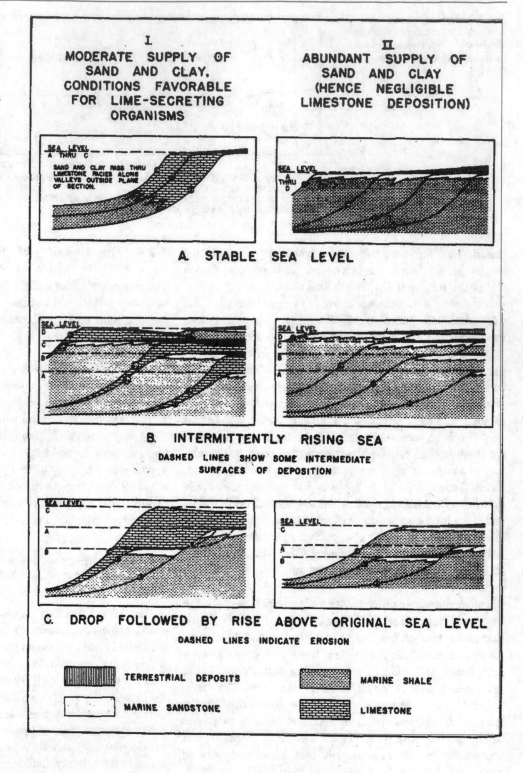

A decade later, Teichert (1958, p. 2719), working from Gressly's original discussion, explained the derivation of the term Facies:

Facies is a Latin word meaning face, figure, appearance, aspect, look, condition. It signifies not so much a concrete thing, as an abstract idea. The word was introduced into geological literature by Nicolaus Steno (1669, pp. 68-75) for the entire aspect of a part of the earth's surface during a certain interval of geologic time.

In his abstract Teichert (1958, p. 2718) provided this succinct definition:

[Facies means] the sum of lithologic and paleontologic characteristics of a sedimentary rock from which its origin and the environment of its formation may be inferred.

Teichert (1958) asserted that the concept of **facies associations** and the importance of **vertical facies successions** were well understood by nineteenth-century European geologists. Cross and Homewood (1997) provided an extended historical discussion of the contributions of Teichert and others to the development of the facies concept.

Interest in the work of the founders of modern sedimentology was renewed in the 1960s, with the new developments in the study of modern sediments, structures and environments. Woodford (1973, p. 3737) translated Gressly (1838) "second law" as follows:

Facies of the same petrographic and geologic nature assume, in different formations, very similar paleontologic characteristics and even succeed each other generally across a more or less numerous series of formations lying one upon the other.

Middleton (1973, p. 981) provided a translation of Walther's methodology from the original German. Walther referred to it as **"ontology"** (**actualism** or **uniformitarianism**, in modern useage) as follows:

It consists in trying to investigate the events of the past through modern phenomena. From being (existence), we explain becoming (genesis).

Middleton (1973, p. 982) translation of Walther's original statement of his Law is as follows:

The various deposits of the same facies-area and similarly the sum of the rocks of different facies-areas are formed beside each other in space, though in a cross-section we see them lying on top of each other. ... it is a basic statement of far-reaching significance that only those facies and facies-areas can be superimposed primarily which can be observed beside each other at the present time.

Middleton (1973, p. 980) suggested that "Walther must be named with Sorby, Gilbert, Grabau, and a few others, as one of the founders of the modern sciences of sedimentology and paleoecology," although he pointed out that whereas Walther's work was cited and acknowledged in much of the pioneer work in the early twentieth century, in the first modern treatment of the subject of facies (Longwell 1949) there was no explicit mention of Walther or his law. He had a much greater influence in Russia, where facies studies were termed **"comparative lithology"**

1.2.5 Fluid Hydraulics and Sedimentary Structures (1953–1976)

A key step in the development of modern sedimentology was the emergence of the idea that sedimentary structures represent measurable and repeatable physical processes, and that they therefore provide information on depositional environments and processes. Early work on the subject included the observations by Sorby (1859, 1908) and Gilbert (1884, 1899) on sedimentary structures, and Gilbert's experimental work (Gilbert 1914). Sorby (1852) was the first to recognize the utility of crossbedding for determining current directions (Sect. 6.6.1). However, as Allen (1993) pointed out, it was not until the appearance of the synthesis by Potter and Pettijohn (1963) that the richness and significance of the preserved record caught the general attention of sedimentary geologists.

A necessary first step toward a modern study of sedimentary structures is accurate description and classification. McKee and Weir (1953) made an important contribution in this direction, with their description of the scales of structures, their internal architecture and bounding surfaces (Fig. 2.10). It is in this paper that the familiar terms **planar**- and **trough-cross-stratification** first appear. A decade later, a comprehensive classification by Allen (1963a) introduced a set of Greek letters for different types of crossbedding, a system that was widely used for some time. Figure 2.11 illustrates the descriptive criteria Allen (1963a) used in his classification. Several illustrated atlases of sedimentary structures also appeared during this period (Pettijohn and Potter 1964; Conybeare and Crook 1968), indicating that sedimentary geologists were coming to grips with the full range of preserved and observable structures.

By the 1950s, sedimentary geologists had become more widely aware of the directional information contained in sedimentary structures, and some pioneering studies of what came to be known as **paleocurrent analysis** were being performed. For example, Reiche (1938) analyzed eolian crossbedding, Stokes (1945) studied primary current lineation in fluvial sandstones, and several authors were dealing with grain and clast orientation (e.g., Krumbein 1939). Pettijohn (1962) provided an overview of the subject, with many examples of the different techniques for analysis and data display that were then in use. Curray (1956) published what became the standard work on the statistical treatment of paleocurrent data.

Meanwhile, several pioneers were attempting to make sedimentary structures in the laboratory, in part as a means to understand the sedimentary record. There was also an interest in understanding fluid hydraulics from an engineering perspective, to aid in the construction of marine facilities, such as bridges and breakwaters. Kuenen and Migliorini (1950), in a classic paper, brought together flume experiments and observations of the ancient record to demonstrate that graded bedding could be generated by turbidity currents (Fig. 1.5). As with many such contributions, it had been preceded by observations and suggestions by many other authors, but this was the paper that brought these observations together into the comprehensive synthesis that made it the benchmark contribution that it became. The term **turbidite** was subsequently coined by Kuenen (1957). McKee (1957), following his many years observing cross-stratification in outcrop, particularly in fluvial and eolian deposits in the Colorado Plateau area, experimented with the formation of cross-stratification by traction currents in a flume.

The critical theoretical development at this time was the series of flume experiments carried out by the US Geological Survey to study sediment transport and the generation of bedforms. The breakthrough work by Simons and Richardson (1961) on the "forms of bed roughness" (Fig. 1.5) led to the definition of the **flow-regime concept**, and the recognition of **lower** and **upper flow regimes** based on flow characteristics (particularly the structure of turbulence), sediment load and resulting bedforms (Harms and Fahnestock 1965; Simons et al 1965. A modern version of the flow-regime concept is provided in Fig. 3.18). At this time,

Allen (1963b) reviewed the observational work of Sorby and made one of the first attempts to interpret sedimentary structures in terms of flow regimes. However, the most important next step was a symposium organized by Middleton (1965), which brought together current field and experimental studies in a set of papers that firmly established flow-regime concepts as a basic tool for understanding the formation of hydraulic sedimentary structures formed by traction currents as preserved in the rock record.

Middleton (1966, 1967) extended the work of Kuenen with further experiments on turbidity currents and the origins of graded bedding, work that was ultimately to lead to a significant new classification of sediment gravity flows, of which it was now apparent that turbidity currents were only one type (Middleton and Hampton 1976). Reference is made in the first of these papers to field observations of turbidites by Roger G. Walker (1965), a reference which marks the beginning of a significant professional collaboration between Walker and Middleton, to which I return later.

Walker's (1967, 1973) field experience with turbidites led to a proposal for the calculation of an index of the proximal-to-distal changes that occur down flow within a turbidite. This marked an attempt at an increasingly quantitative approach to the study of sedimentary structures, although this index was not to survive an increasing knowledge of the complexities of the submarine fan environment within which most turbidites are deposited.

The important new developments in this field were well summarized in a short course, organized by the Society for Sedimentary Geologists, the manual for which provides an excellent review of the state of knowledge at this time

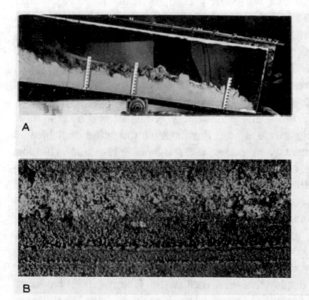

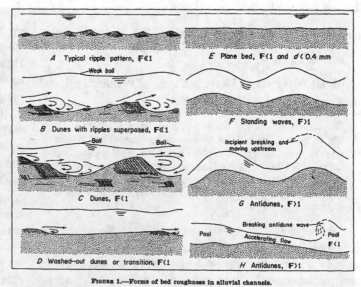

FIGURE 1.—Forms of bed roughness in alluvial channels.

Fig. 1.5 Early work exploring fluid hydraulics and the origins of sedimentary structures. LEFT: An experiment to generate a turbidity current, from Kuenen and Migliorini (1950). RIGHT: The first definition of the flow-regime concept for hydrodynamic structures in sand, from Simons and Richardson (1961)

(Harms et al. 1975). This review contains the first description and definition of hummocky cross-stratification (HCS), and the recognition of this structure as a key indicator of combined-flow (unimodal and oscillatory) storm sedimentation. Additional details regarding the history of development of ideas about flow-regime bedforms, hydrodynamic sedimentary structures and sediment gravity flows are provided in Sects. 3.5.4 and 3.5.5.

1.2.6 Early Studies of Modern Environments (1954–1972)

As noted above, references to modern depositional settings appear in much of the early stratigraphic literature, but in the 1950s studies of "the modern" became more focused. Much of this was due to the recognition by some oil companies of the value of understanding the primary origins of petroleum-bearing rocks. A leader in this field was the research team at Shell Development Company.

Some of the earliest of these studies of modern environment were carried out in carbonate environments, including the work of Illing (1954), and Newell and Rigby (1957) on the Great Bahamas Bank, and Ginsburg and Lowenstam (1958) on the Florida platform. This, and other work on ancient carbonates (referred to below) led to two approaches to the classification of carbonate rocks (Folk 1962; Dunham 1962) that are still used today. In fact, these two papers (which appeared in the same SEPM Special Publication) are among the most important of the "classic" papers mentioned in this chapter, because of their long survival. Later studies of the Bahamas and Florida by Purdy (1963) and Ball (1967) contributed much to the subsequent growth of facies models for carbonate platforms and reefs. Purdy's use of cluster analysis to identify the major lithofacies comprising shelf carbonate sediments has also become a classic (see Fig. 3.7).

The other outstanding set of classic works consists of the research on the Texas coastal plain by Bernard, Leblanc and their colleagues at Shell, building on the preliminary work of Nanz (1954). The first facies model for barrier islands emerged from the work of these individuals on Galveston Island (Bernard et al. 1959; 1962) and the point-bar model for meandering rivers is also attributed to this group, based on their studies of the Brazos River (Bernard and Major 1963; Bernard et al. 1962).

The Mississippi River and Delta is one of the largest of modern fluvial-delta systems, and its location in the center of one of the most important, well-populated, industrial and tourist regions of the United States, in a petroleum province that generates a quarter of the US domestic supply, has led to intensive environmental and geological studies. The stratigraphic significance and complexity of the deposits of this system were first brought to geologists' attention by the detailed work of Fisk (1944). From the point of view of the growth of sedimentology the studies of Frazier (1967) were more significant, providing architectural block diagrams that illustrated the growth of distributaries in a river-dominated delta. Later studies by Fisher et al. (1969, 1972) broadened the scope of delta studies to other regions of the Texas coast and to other deltas worldwide, providing an essential basis for the subsequent development of formal delta facies models. Shepard et al. (1960) edited a collection of broader studies of the Gulf Coast.

Exploration methods for the continental shelf and deep oceans were primitive, until the introduction of side-scan sonar methods and improvements in navigation. The GLORIA sonar system was developed in 1970 but did not receive widespread use for geological purposes until it was adopted by the US Geological Survey in 1984 at the commencement of a program to map the newly established US Exclusive Economic Zone. The Deep Sea Drilling Project (DSDP) began in 1968. Extensive use of seismic stratigraphic techniques had to await the developments taking place in Shell, Exxon and BP, as noted below (in particular, the work of Vail et al. 1977). Sedimentological studies of the continental shelves and slopes and the deep basin were being carried out at this time, but the main breakthroughs in sedimentological analysis came from studies of the ancient sedimentary record, and are referred to below.

1.2.7 Facies-Model Concept (1959–2010)

By the late 1950s a key idea was emerging that environments could be categorized into a limited number of depositional configurations, which are amenable to basic descriptive summaries. The first explicit use of the term "**facies model**" was in a conference report by Potter (1959, p. 1292). He opened the report with the following words:

A discussion concerning sedimentary rocks was held at the Illinois State Geological Survey on 4-5 Nov. 1958, for the purpose of pooling the knowledge and experience of the group concerning three topics: the existence and number of sedimentary associations; the possibility of establishing a model for each association that would emphasize the areal distribution of lithologic units within it; and the exploration of the spatial and sequential relations between the associations.

Later, on the same page, this definition is provided:

A facies model was defined as the distribution pattern or arrangement of lithologic units within any given association. In the early stages of geological exploration, the function of the model is to improve prediction of the distribution of lithologic types.

Note that the essential basis for a facies model is the recognition of a distinctive **facies association**. Much work to identify these associations now ensued.

A mention should be made here of the term **process-response model**. This term has sometimes been used with essentially the same meaning as facies model. Whitten's (1964, p. 455) discussion of this term quoted from Krumbein and Sloss (1963, p. 501), who:

> suggested that in the search for "… generalizing principles it is a useful philosophical device to recognize *models* actual or conceptual frameworks to which observations are referred as an aid in identification and as a basis for prediction."

The journal *Sedimentology* was founded by the International Association of Sedimentologists (IAS) in 1962. The editor was Aart Brouwer from the University of Leyden in the Netherlands, representing what had become a strong Dutch school of sedimentological studies. All the early work on tidal flat sedimentation emerged from this school (e.g., Van Straaten 1954). The then President of the IAS, the American marine geologist Francis Shepard said this, in the Preface on p. 1 of v. 1 of the new journal:

> As this is written, there appear to be several primary purposes in sedimentological studies. One is to relate more completely the present day sediments to ancient sedimentary rocks. Although much has been done in this field recently, there are numerous types of sedimentary rocks for which no equivalent has yet been found in the sediments of today and some correlations need careful reexamination to see if they are correctly interpreted. Another need is for more careful study of sedimentary structures that are often obscured both in old and recent sediments. These structures can be very useful in interpreting paleoclimates and conditions of deposition of ancient sediments. A third important field to investigate is the geochemistry of sediments. Some of the early indications from the chemical nature of sediments have proven misleading and are in need of further study to explain apparent anomalies. Fourth, the rates of sedimentation can be given much more study with all of the new radioactive counting methods.

In an introductory assessment of sedimentary studies immediately following the Preface, editor Brouwer (1962, p. 2–3) reviewed the early history and origins of the separate discipline now called *Sedimentology*:

> Essential parts are derived from sedimentary petrography, others from stratigraphy and still others have a purely palaeontological source. Perhaps stratigraphy takes a more or less central position, and many definitions recently given of stratigraphy (Hedberg 1948; Weller 1960; and others) seem to include nearly all of sedimentology, at least of ancient rocks. This is quite understandable, as sedimentary rocks are the stratigrapher's natural environment. Three modem textbooks, whose scope is mainly sedimentological, have "stratigraphy" in their title (Krumbein and Sloss 1951; Dunbar and Rodgers 1957; Weller 1960).

The reference to sedimentary petrography should be noted here. The first journal to deal specifically with sedimentological topics, the *Journal of Sedimentary Petrology*, was founded in 1931, and initially dealt exclusively with petrographic studies, including studies of detrital composition and provenance, and diagenesis. The scope of the journal gradually widened, and the name was changed to the *Journal of Sedimentary Research* in 1994. According to Gerard V. Middleton (2005) the term **Sedimentology** was coined by A. C. Trowbridge in 1925 and first used in print by Waddell (1933), but did not come into common usage until the 1950s.

Now began a focused program to identify specific lithofacies and lithofacies associations by direct comparison between modern sediments and the preserved record. The comparison went both ways, determined in large measure by the initial interests of the researcher. One of the first of these studies was that by Beales (1958, p. 1846) who proposed the term **Bahamite** for "the granular limestone that closely resembles the present deposits of the interior of the Bahamas Banks described by Illing (1954)." Although this new term did not become part of the sedimentological lexicon, the methods pioneered by Beales and his colleagues were about to become part of the mainstream.

Two classic studies appeared in the early 1960s, Bouma's (1962) turbidite model and Allen's (1964) point-bar model for meandering-river deposits. Both are concerned primarily with interpretation of the rock record, but make extensive reference to deposits and structures forming at the present day.

There appeared a flood of new work during the 1960s and 1970s making use of the new facies-model concepts. Potter (1967) reviewed sandstone environments. He stated (Potter 1967, p. 360):

> The facies-model concept with its emphasis on the existence of relatively few recurring models represents cause-and-effect "deterministic geology"-an approach that attempts to relate distribution and orientation of sand bodies in a basin to measurable, causal factors.

However, much of Potter's discussion dealt with grain size and other petrographic issues, and discussions about the shape and orientation of sand bodies (of importance for stratigraphic-trap prospecting) rather than facies modeling, as this term has come to be understood.

An edited compilation that appeared in the middle of this period (Rigby and Hamblin 1972) provides another good snapshot of the state of sedimentology at this time. It opens with a brief review of the topic of "environmental indicators" by H. R. Gould and this is followed by a classification of sedimentary environments by E. J. Crosby, and by eleven chapters providing details of seven depositional environments (three chapters on alluvial sediments and one discussing the use of trace fossils). There were also several important new textbooks published during this period (e.g., Blatt et al. 1972; Reineck and Singh 1973; Wilson 1975; Freidman and Sanders 1978; Reading 1978). That by Blatt et al. (1972) contains the first summary of depositional environments specifically focused on the concept of the facies model (and using that term in the chapter heading. It was written by co-author Gerard Middleton).

The critical contribution at this time was the development by Walker (1976) of a formal, theoretical description of the concept of the facies model and its value as a summary and a predictor. Central to this work was a new concept that environments could be characterized by a discrete and limited number of specific facies states. Drawing on Middleton's (1973) restatement of Walther's Law, Walker emphasized the importance of the vertical succession of facies, and introduced the **facies relationship diagram** to a broader audience. This is a semi-quantitative expression of the range of vertical transitions revealed by careful vertical measurement of a stratigraphic succession. Reference was made to a detailed study of de Raaf et al (1965), which was the first to employ the concept of facies states and the use of a facies relationship diagram (Fig. 1.6). Another study of vertical facies relationships at this time was that by Miall (1973) using the basic concepts of Markov Chain Analysis.

Walker's (1976, Fig. 4) diagram summarizing the construction of a facies model as a process of "distilling away the local variability" to arrive at the "pure essence of

environmental summary" has been much reproduced (Fig. 3.9 is the current version, from James and Dalrymple 2010).

Walker's (1976) paper appeared first in a new journal, *Geoscience Canada* (founded and edited by his colleague at McMaster University, Gerard Middleton), and was intended as the introductory paper in a series of invited articles written mainly by Canadian sedimentologists dealing with specific environments and facies models. The series was later published as a stand-alone volume (Walker 1979) which became a best-seller and eventually, under changing editorships, went into four editions (Walker and James 1984, 1992; James and Dalrymple 2010). Its success was due in large measure to the concise nature of the descriptions, the elegant diagrams, and the emphasis on the nature of the vertical profile, making this a very practical approach for undergraduate teaching and for work with well-logs and cores. A close competitor was the edited volume compiled by Reading (1978), a book written at a more advanced, graduate to the professional level by him and some of his graduate students at the University of Oxford. This book went into two later editions (1986,1996).

Among the other widely used facies models that appeared in the *Geoscience Canada* series (and subsequently in

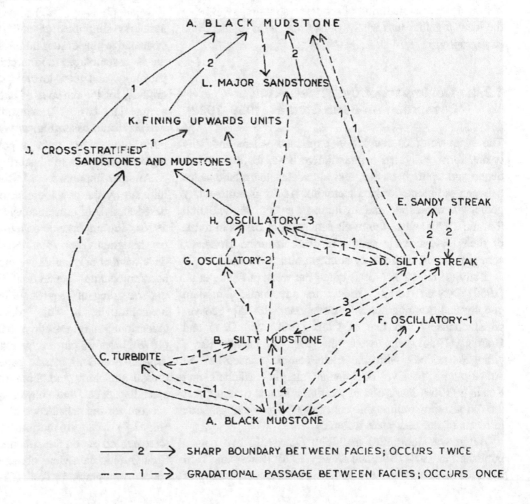

Fig. 1.6 Facies relationship diagram to show the type of boundary between facies and the number of times the facies are in vertical contact with each other. This was the first attempt of its kind to demonstrate quantitatively the vertical relationships between identifiable lithofacies (de Raaf et al. 1965)

Walker 1979) was a treatment of continental shelf sedimentation highlighting the rock record of hummocky cross-stratification, and a simple and elegant model for submarine fans based almost entirely on ancient fan deposits in California and Italy. In this book, carbonate facies models were compiled and co-authored by Noel P. James, who became a co-editor of later editions. **Ichnology**, the study of trace fossils, evolved into an enormously valuable subsurface facies analysis tool, allowing detailed analysis of sedimentary environments in drill core, as well as throwing much useful light on the significance of stratal surfaces, with the preservation of evidence of non-deposition and early lithification (Frey and Pemberton 1984; McEachern et al. 2010).

By the mid-1970s the stage was set for Sedimentology to flourish. The Walker (1979) *Facies Models* volume, and Reading's (1978) textbook were enormously influential. However, through the 1980s it remained largely isolated from the "big picture" concepts that were emerging from the plate-tectonics revolution, and developments in seismic stratigraphy. These I discuss below. Textbooks that appeared during this period (e.g., Miall 1984; Matthews 1984; Boggs 1987) deal with all these topics essentially in isolation, as separate chapters with little cross-referencing. As I argue below, it took the maturing of sequence stratigraphy to bring the ideas together into what we may now term sophisticated stratigraphy.

1.2.8 The Impact of the Plate-Tectonics Revolution on Basin Studies (1959–1988)

The plate-tectonics revolution explained where and why basins form, provided a quantitative basis for their subsidence and uplift behavior, and elucidated the relationships between sedimentation and tectonics. As far as sedimentary geology is concerned, the revolution was not complete until the mid-1970s, when the re-classification of basins in terms of their plate-tectonic setting reached maturity. However, some important preliminary studies pointed the way.

Bally (1989, p. 397–398) noted the work of Drake et al. (1959) "who first tried to reconcile modern geophysical-oceanographic observations with the geosynclinal concept" and that of Dietz (1963) and Dietz and Holden (1974) who were the first to equate Kay's "miogeosyncline" with the plate-tectonic concept of an Atlantic-type passive continental margin. Mitchell and Reading (1969) made one of the first attempts to reinterpret the old tectono-stratigraphic concepts of **flysch** and **molasse** in terms of the new plate tectonics.

But it was John Bird and John Dewey, in two papers published in 1970 (Bird and Dewey 1970; Dewey and Bird 1970), who completely revolutionized our understanding of the origins of sedimentary basins (and much of the rest of geology) with reference to the geology of the Appalachian orogen, in particular, that portion of it exposed throughout the island of Newfoundland. Dickinson (1971) made reference to all of this work in his own first pass at relating sedimentary basins to plate tectonics.

These breakthroughs of the 1970s initiated a worldwide explosion of studies of basins and tectonic belts exploring the new plate-tectonic concepts. Through the 1970s, a series of books and papers was published containing the results (Dickinson 1974a; Dott and Shaver 1974; Burk and Drake 1974; Strangway 1980; Miall 1980, 1984). One of the more important of these contributions was a paper by Dickinson (1974b) which constituted the first comprehensive attempt to classify sedimentary basins of all types in terms of their plate-tectonic setting. This paper was particularly notable for the extensive treatment of arc-related basins and was followed up by a more detailed paper on this subject (Dickinson and Seely 1979) that remained the standard work on the subject for many years. This latter work was based in part on the recognition of a series of arc-related sedimentary basins within the Cordillera (Dickinson 1976), especially the Great Valley basin of California, which has long served as a type example of a forearc basin (e.g., Ingersoll 1978a, b, 1979).

Miall (1984, p. 367) argued that by the application of judicious simplification and by skillful synthesis we can systematize the descriptions of depositional systems (their facies assemblages and architecture), structural geology, petrology, and plate-tectonic setting into a series of **basin models**, for the purpose of interpreting modern and ancient sedimentary basins. Dickinson (1980, 1981) used the term **petrotectonic assemblages** with the same meaning. These basin models are then a powerful tool for interpreting regional plate-tectonic history.

Another important era in the field of basin analysis was initiated by the development of quantitative, geophysically based models of crustal subsidence, commencing in the late 1970s. The importance of these models to the development of stratigraphy was that they provided the basis for the development of quantitative models of basin subsidence and accommodation generation that greatly improved our understanding of large-scale basin architectures. The main breakthrough in the development of a modern extensional-margin basin model was made by McKenzie (1978), based in part on his studies of the subsidence of the Aegean Sea. This classic paper introduced the concept of crustal stretching and thinning during the initial sea-floor spreading event, and showed quantitatively how this could account for the subsidence history of Atlantic-type margins (Fig. 1.7). Many of the important early tests of this model were carried out on the Atlantic margin of the United States. Stratigraphic data were obtained from ten Continental Offshore Stratigraphic Test (COST) wells drilled on the

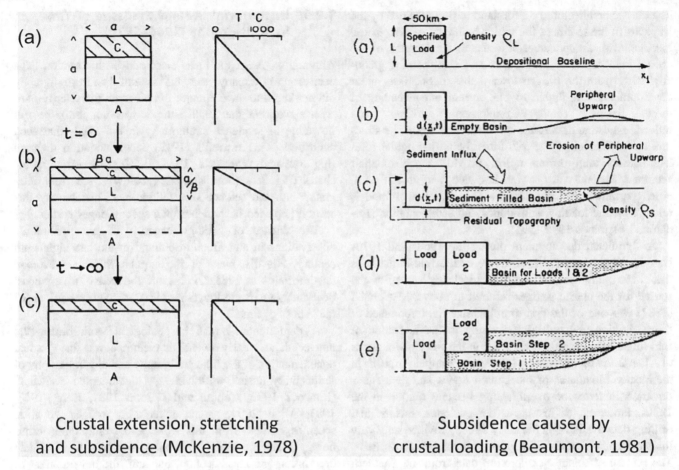

Crustal extension, stretching and subsidence (McKenzie, 1978)

Subsidence caused by crustal loading (Beaumont, 1981)

Fig. 1.7 Key diagrams from the first two modern quantitative models for sedimentary basins. At LEFT, the extensional-basin model shows three schematic stages of the evolution of a cube of crust, with the geothermal gradient at right: **a** initial state, **b** stretching, thinning, and steepening of geothermal gradient, **c** gradual cooling of the thickened lithosphere leads to subsidence. RIGHT: Loading of a level baseline leads to subsidence and filling with sediment. These models are more fully explained by Miall (1999) and Allen and Allen (2013)

continental shelf off New England between 1976 and 1982, and led to the development of formal backstripping procedures (Watts and Ryan 1976; Steckler and Watts 1978; Watts 1989) and to simple computer graphic models of subsiding margins (Watts 1981) that were very useful in illustrating the development of the basic architecture of Atlantic-type margins. Dewey (1982) emphasized their simple two-stage development: the early phase of rifting, typically capped by a regional unconformity, followed by a thermal relaxation phase which generates a distinctive pattern of long-term onlap of the basement.

An important modification of the McKenzie model was to recognize the importance of simple shear during continental extension, as expressed by through-going extensional crustal detachment faults (Wernicke 1985). This style of crustal extension was first recognized in the Basin and Range Province of Nevada, and was suggested by preliminary seismic data from the facing continental margins of Iberia and the Grand Banks of Newfoundland (Tankard and Welsink 1987).

The North Sea basin is the best studied rift basin and has provided many insights regarding subsidence styles and structural geology (White and McKenzie 1988).

Turning to the other major class of sedimentary basins, those formed by flexural loading of the crust, it was Barrell (1917) who was the first to realize that "the thick nonmarine strata of the Gangetic plains accumulated in space made available by subsidence of the Indian crust beneath the mass of thrust plates of the Himalayan Range" (Jordan 1995, p. 334). Price (1973) revived the concept of regional isostatic subsidence beneath the supracrustal load of a fold-thrust belt that generates the marginal moat we now term a foreland basin (a term introduced by Dickinson 1974b), based on his work in the Southern Canadian Cordillera. Beaumont (1981) and Jordan (1981) were the first to propose quantitative flexural models for foreland basins, constraining the models with detailed knowledge of the structure and stratigraphy of the studied basins (Fig. 1.7). It is clear that the crust must have mechanical strength for a wide foredeep, such as the Alberta Basin or the Himalayan foreland basin, to be created.

The classic architecture of a foreland basin is defined by the isopachs of the sediment fill, which is that of an asymmetric lozenge, with a depocenter adjacent to the location of the crustal load, tapering along strike and also thinning gradually away from the orogen toward the craton. Two major developments contributed to our current understanding of these basins. Firstly, exploration drilling and reflection-seismic data led to an understanding of the structure and dynamics of the fold-thrust belts that border foreland basins and, during uplift, provide much of their sediment. Secondly, a growing knowledge of crustal properties permitted the development of quantitative models relating crustal loading, subsidence and sedimentation (Jordan and Flemings 1990,1991).

A significant development during the 1960s and 1970s was the elucidation of the structure of the fold-thrust belts that flank many orogenic uplifts and clearly served as the source for the clastic wedges referred to above. McConnell (1887) was one of the first to emphasize the importance of thrust faulting and crustal shortening in the formation of fold-thrust belts, based on his work in the Rocky Mountains of Alberta. As noted by Berg (1962), the mapping of faults in the Rocky Mountains of the United States and their interpretation in terms of overthrusting became routine in the 1930s. However, as his paper demonstrates, seismic and drilling data available in the early 1960s provided only very limited information about the deep structure of thrust belts. The release of seismic exploration data from the Southern Rocky Mountains of Canada by Shell Canada led to a landmark study by Bally et al. (1966) and set the stage for modern structural analyses of fold-thrust belts. A series of papers by Chevron geologist Clinton Dahlstrom, concluding with a major work in 1970 (Dahlstrom 1970), laid out the major theoretical principles for the understanding of the thrust faulting mechanism.

The final piece of the puzzle was to explain accommodation generation and the occurrence of regional tilts and gentle angular unconformities on cratons hundreds of kilometers from plate margins—the phenomenon termed **epeirogeny**. Modeling of mantle processes indicated the presence of convection currents that caused heating and uplift or cooling and subsidence of the crust. Gurnis (1988, 1990, 1992) termed this **dynamic topography**. Cloetingh (1988) described the process of **intraplate stress** (also termed **in-plane stress**) whereby horizontal stresses exerted on plates, such as the outward-directed compressive stress from sea-floor spreading centers ("ridge push"), and the downward pull of subsiding oceanic plates, may be expressed as intraplate earthquakes that cumulatively develop faults and long-wavelength folds. The best modern treatment of basin dynamics and models is contained in the book by Allen and Allen (2013).

1.2.9 Unconformities and the Issue of Time in Stratigraphy (1909–1970)

Although some of the ideas discussed in this section have been around for many years, the issue of time in stratigraphy did not begin to have a major influence on the science until Ager's work in the 1970s, and it was not until the full flowering of sequence stratigraphy in the 1990s that such contributions as Barrell's (1917) accommodation diagram (Fig. 5.2) and Wheeler's (1958, 1959) chronostratigraphic charts (Fig. 8.1) (both discussed below) were fully integrated into the science of stratigraphy. This is why this section is placed here, rather than several pages earlier.

The science of geology began with James Hutton's observations in and around Scotland in the late eighteenth century. His discovery of the angular Silurian–Devonian unconformity at Siccar Point on the coast of southeast Scotland gave rise to Playfair's (1802) famous remark about the "abyss of time."

A predominant strand in geological work during the nineteenth and early twentieth centuries was the gradual documentation of the lithostratigraphy and biostratigraphy of sedimentary basins worldwide. As documented elsewhere (Hancock 1977; Conkin and Conkin 1984; Berry 1968, 1987; Miall 2004), some remarkably refined zonation schemes resulted from this work, and stratigraphic terminology and methods gradually evolved to facilitate description and classification, but until the development of radioisotopic dating by Ernest Rutherford and Arthur Holmes (Holmes first book on the geological time scale was published in 1913) the development of a quantitative understanding of earth processes was limited.

Geological mapping and research in North America during the "frontier" period is usefully summarized by Blackwelder (1909), who discussed the various types of sedimentary break (angular versus structurally conformable) and the duration of the missing time that they represented. His paper contained what is probably the first chronostratigraphic chart for the interior (cratonic) stratigraphy of North America, showing what was then known about the extent of the major Phanerozoic stratigraphic units on this continent and the unconformities that separate them.

In a paper that was remarkably ahead of its time, written shortly after the discovery of the concept of radioisotopic dating, Barrell (1917, p. 747–748) set out what we now refer to as the concept of **accommodation**:

In all stratigraphic measures of time, so far as the writer is aware, the rate of deposition of a sedimentary series has been previously regarded as dependent on the type of sediment, whether sandstone, shale, or limestone, combined with the present rate of supply of such sediment to regions of deposition. Here is developed an opposite view: that the deposition of nearly all sediments occurs just below the local baselevel, represented by wave base or river flood level, and is dependent on upward oscillations of baselevel or downward oscillations of the bottom, either of which makes room for sediments below baselevel. According to this control, the rate of vertical thickening is something less that the rate of supply, and the balance is carried farther by the agents of transportation.

Barrell (1917) was probably the first to understand the relationships among sedimentation, preservation, and accommodation. He constructed a diagram (Fig. 5.2) showing the "Sedimentary Record made by Harmonic Oscillation in Baselevel" (Barrell 1917, p. 796) that is remarkably similar to diagrams that have appeared in some of the Exxon sequence model publications since the 1980s (e.g., Van Wagoner et al. 1990, Fig. 39; Fig. 5.5 of this book). It shows that when long-term and short-term curves of sea-level change are combined, the oscillations of base level provide only limited time periods when base level is rising and sediments can accumulate. In his diagram "Only one-sixth of time is recorded" by sediments (Barrell 1917, p. 797). This remarkable diagram (1) anticipated Jervey (1988) ideas about sedimentary accommodation that became fundamental to models of sequence stratigraphy, (2) it also anticipated Ager's (1981, 1993) point that the sedimentary record is "more gap than record;" and (3) it constitutes the first systematic exploration of the problem of preservation potential.

During the early part of the twentieth century there was much theorizing about the forces at work within the Earth to form mountain ranges and sedimentary basins. This is summarized elsewhere (e.g., Miall 2004) and not dealt with here, because ultimately it did not contribute much to the development of modern stratigraphy. However, the practical work of petroleum exploration did make a difference. The distinguished petroleum geologist A. I. Levorsen was one of the first to describe in detail some examples of the "natural groupings of strata on the North American craton":

A second principle of geology which has a wide application to petroleum geology is the concept of successive layers of geology in the earth, each separated by an unconformity. They are present in most of the sedimentary regions of the United States and will probably be found to prevail the world over (Levorsen 1943, p. 907).

This principle appears to have been arrived at on the basis of practical experience in the field rather than on the basis of theoretical model building. These unconformity-bounded successions, which are now commonly called "**Sloss sequences**," for reasons which we mention below, are tens to hundreds of meters thick and, we now know, represent tens to hundreds of millions of years of geologic time (Fig. 1.8). They are therefore of a larger order of magnitude than the cyclothems. Levorsen did not directly credit Grabau, Ulrich, or any of the other contemporary theorists who were at work during this period (see Miall 2004), nor did he

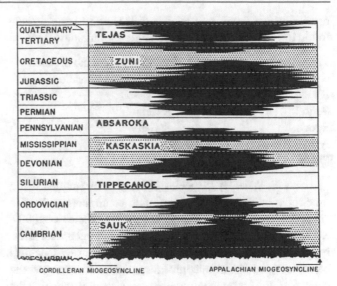

Fig. 1.8 The classic six sequences established by Sloss (1963) for North American cratonic stratigraphy

cite the description of unconformity-bounded "rock systems" by Blackwelder (1909). Knowledge of these seems to have been simply taken for granted.

The symposium on "Sedimentary facies in geologic history" referred to above in Sect. 1.2.4 contained a lengthy treatment of facies variability in the Paleozoic rocks of the cratonic interior of the United States by Sloss et al. (1949). In this paper much use is made of isopachs and lithofacies maps using Krumbein's (1948) concepts of clastic ratios and sand-shale ratios. The work revealed to the authors the contradictions inherent in current classifications of rock units in North America according to standard geologic time units. The use of the standard time scale (Cambrian, Ordovician, etc.) as a basis for mapping, obscured the fact that the major sedimentary breaks within the succession commonly did not occur at the divisions provided by the time scale, and so they set out to establish "operational units" for mapping purposes. Thus were born the first **sequences** for the North American interior: the Sauk, Tippecanoe, Kaskaskia and Absaroka.

The Sloss et al. (1949) paper in the symposium volume (Longwell 1949) is followed by nearly 50 pages of published discussion by many of the leading American geologists of the day, in which the issues raised by detailed mapping and the concepts and classifications available at the time for their systematization were fully discussed. This broader discussion is dealt with at length elsewhere (Miall 2004; 2010, Chap. 1). For the purpose of this book, the importance of the Sloss et al. (1949) paper and the wider discussion of sedimentary facies contained in the other papers (see Sect. 1.2.4) is that it clearly confirmed, at the time of publication, the need for a systematic differentiation of descriptive terminologies for "time" and for the "rocks." This had been provided by Schenk and Muller (1941), who proposed the following codification of stratigraphic terminology:

Time division (for abstract concept of time)	Time-stratigraphic division (for rock classification)
Era	–
Period	System
Epoch	Series
Age	Stage
Phase	Zone

Harry E. Wheeler (Wheeler 1958, p. 1050) argued that a time-rock (chronostratigraphic) unit could not be both a "material rock unit" and one whose boundaries could be extended from the type section as isochronous surfaces, because such isochronous surfaces would in many localities be represented by an unconformity. Wheeler developed the concept of the chronostratigraphic cross-section, in which the vertical dimension in a stratigraphic cross-section is drawn with a time scale instead of a thickness scale (Fig. 8.1). In this way, time gaps (unconformities) become readily apparent, and the nature of time correlation may be accurately indicated. Such diagrams have come to be termed "**Wheeler plots.**" Wheeler cited with approval the early work of Sloss and his colleagues, referred to in more detail below:

> As a tangible framework on which to hang pertinent faunal and lithic data, the *sequence* of Sloss, Krumbein and Dapples (1949, pp. 110-11) generally fulfills these requirements. Paraphrasing these authors' discussion, a *sequence comprises an assemblage of strata exhibiting similar responses to similar tectonic environments over wide areas, separated by objective horizons without specific time significance* (Wheeler 1958, p. 1050; italics as in original).

Sequences came later to be called simply "**unconformity-bounded units.**" However, a brief mention should be made of the concept of the **format**, a term suggested by Forgotson (1957) for laterally equivalent formations enclosed by widely mappable marker beds above and below. Wheeler's (1958) methods are now universally accepted, although in practice they are still rarely applied (see Sects. 8.10 and 8.11). Ager (1973) is famous for his remark that "the sedimentary record is more gap than record." In a later book he expanded on the theme of gaps. Following a description of the major unconformities in the record at the Grand Canyon, he said, (Ager 1993, p. 14):

> We talk about such obvious breaks, but there are also gaps on a much smaller scale, which may add up to vastly more unrecorded time. Every bedding plane is, in effect, an unconformity. It may seem paradoxical, but to me the gaps probably cover most of earth history, not the dirt that happened to accumulate in the moments between. It was during the breaks that most events probably occurred.

Dott (1983, 1996) similarly warned about the episodic nature of sedimentation. However, as discussed in Chaps. 7

and 8, stratigraphers are still not dealing fully with the issue of time and its representation in the rock record.

Modern biostratigraphic methods are discussed in Sect. 7.5. The evolution of chronostratigraphic methods and the increasing accuracy and precision with which sedimentary rocks can be dated is discussed in detail elsewhere (Miall 2004, 2010, Chap. 14) and is the focus of Sect. 7.8 and Chap. 8 of this book. A landmark in the development of modern stratigraphy was the adoption in the 1970s of the GSSP principal for the fixing of major chronostratigraphic boundaries. GSSP stands for **Global Stratigraphic Sections and Points** and is a system for identifying outcrop sections that are accepted by the international community as marking the boundaries of stages and series (McLaren 1970).

1.2.10 Sequences and Seismic Stratigraphy (1963–1977)

Building on his earlier work (Sloss et al. 1949), further analysis by Sloss (1963) added two more sequences of Mesozoic–Cenozoic age to the North American suite (Zuni, Tejas) and firmly established the concept of the large-scale control of cratonic stratigraphy by cycles of sea-level change lasting tens of millions of years (Fig. 1.8). In later work Sloss (1972) demonstrated a crude correlation of these sequences with a similar stratigraphy on the Russian Platform, thereby confirming that global sea-level cycles constituted a major sedimentary control. However, Sloss, unlike his student Peter Vail, was never convinced that global eustasy told the entire story (Sloss 1988, 1991). In his 1963 paper Sloss included a pair of diagrammatic cross-sections of the Sauk and Tippecanoe sequence across the cratonic interior of North America that clearly indicated an angular unconformity between the two sequences, a relationship that could only have been developed as a result of broad warping of the craton before deposition of the Tippecanoe sediments.

Ross (1991) pointed out that all the essential ideas that form the basis for modern sequence stratigraphy were in place by the 1960s. The concept of repetitive episodes of deposition separated by regional unconformities was developed by Wheeler and Sloss in the 1940s and 1950s (Sect. 1.2.9). The concept of the "ideal" or "model" sequence had been developed for the mid-continent cyclothems in the 1930s (Sect. 1.2.2). The hypothesis of glacioeustasy was also widely discussed at that time. Van Siclen (1958) provided a diagram of the stratigraphic response of a continental margin to sea-level change and variations in sediment supply that is very similar to present-day sequence models (Sect. 1.2.3; Fig. 1.4). An important symposium on cyclic sedimentation convened by the Kansas Geological Survey marks a major milestone in the progress of research in this area (Merriam 1964); yet the subject did not "catch on." There are probably two main reasons for this. Firstly, during the 1960s and 1970s sedimentologists were preoccupied mainly by autogenic processes and the process-response model, and by the implications of plate tectonics for large-scale basin architecture (Sect. 1.2.7). Secondly, geologists lacked the right kind of data. It was not until the advent of high-quality seismic-reflection data in the 1970s, and the development of the interpretive skills required to exploit these data, that the value and importance of sequence concepts became widely appreciated. Shell, British Petroleum and Exxon were all actively developing these skills in their research and development laboratories in the 1970s. The first published use of the term "**seismic stratigraphy**" was in a paper by Fisher et al. (1973) describing a subsurface succession in Brazil (the term appeared in the Portuguese language as estratigrafia sismica). Peter Vail, working with Exxon, was the first to present his ideas in the English-speaking world, at the 1974 annual meeting of the Geological Society of America, but it was his presentation the following year at the American Association of Petroleum Geologists (Vail 1975) that caught the attention of the petroleum geology community. This was the beginning of the modern revolution in the science of stratigraphy.

The key idea that Vail and his colleagues proposed was that large-scale stratigraphic architecture could be reconstructed from reflection-seismic records. Their publication of Memoir 26 of the American Association of Petroleum Geologists (Vail et al. 1977) was one of the major landmark events in the development of modern stratigraphy. Vail had learned about sequences from his graduate supervisor, Larry Sloss, and added to these his own ideas about global sea-level change (eustasy) as the major allogenic control of sequence development. The debate about global eustasy was long and controversial and has been amply aired elsewhere (see Miall 2010). However, what emerged from the debate was the critical importance of the "big-picture" in stratigraphic reconstruction, and the predictive value of sequence models. Having once seen a seismic record interpreted in terms of seismic stratigraphy, with its emphasis on seismic terminations and regional unconformities, and the common occurrence of clinoform architectures, old concepts of "layer-cake" stratigraphy were dead forever. The section reproduced here as Fig. 1.15 appeared in Memoir 26, and was one of the first to bring this point home to the geological community.

It also seems likely that, working in the Gulf Coast, Vail learned from the "big-picture" stratigraphers at the Bureau of Economic Geology. The regional view exemplified by work such as the Texas atlas (Fisher et al. 1972) and the seismic interpretation that these individuals were already working on, and which eventually appeared in the same AAPG memoir (Brown and Fisher 1977) were very influential in

helping sedimentary geologists understand the large-scale setting and tectonic influences on sedimentary basins at the very time that geophysical basin models were providing the quantitative basis for the plate-tectonic interpretations of these basins (Sect. 1.2.8).

Peter Vail has come to be called the "Father" of sequence stratigraphy, while his graduate supervisor, Larry Sloss (1913–1996), has posthumously earned the title of the "Grandfather" of sequence stratigraphy.

1.2.11 Architectural Elements: Sedimentology in Two and Three Dimensions (1983–1990)

Lithofacies maps and isopachs and the reconstruction of regional paleocurrent patterns had become standard tools of the sedimentary geologist (or basin analyst) by the 1970s (the second edition of the Potter and Pettijohn book "*Paleocurrents and basin analysis*" was published in 1977), but they often failed to capture the fine detail of sedimentary processes that were by now emerging from facies studies. As Miall (1984, Sect. 5.3) pointed out, these mapping methods tended to produce generalizations that did not always reflect the rapidly shifting patterns of depositional systems that could now be reconstructed from detailed sedimentological study of outcrops, well records and cores.

There was also a scale mismatch. Lithofacies maps deal with large map areas (tens to hundreds of kilometers across) and are essentially two dimensional. Facies studies at this time (the 1970s to early 1980s) were one dimensional, focusing on the vertical profile in drill records or outcrops (typically a few meters to tens of meters high). What were clearly needed were the tools to put the observations together. Three-dimensional sedimentological studies provided part of the answer, particularly for outcrop analysis, and sequence studies focused on the larger picture (Sect. 1.2.12).

Work on fluvial systems by Allen (1983) and by Ramos and his colleagues (Ramos and Sopeña, 1983; Ramos et al. 1986) led the way. These papers focused on large two-dimensional outcrops of complex fluvial deposits and offered classifications of the lithofacies units that described them in two or three dimensions (Fig. 1.9). Picking up on this early work, Miall (1988a; b; 1985) offered a systematized approach that re-stated the lithofacies classification idea in terms of a limited suite of **architectural elements** that, it was proposed, constitute the basic building blocks of fluvial assemblages. One of the strengths of the approach is the ability to relate paleocurrent observations to the fine detail of the channel and bar complexes, revealing whole new insights into the bar construction and preservation processes. Comparable approaches have subsequently been adopted for other depositional environments (in 2017 the

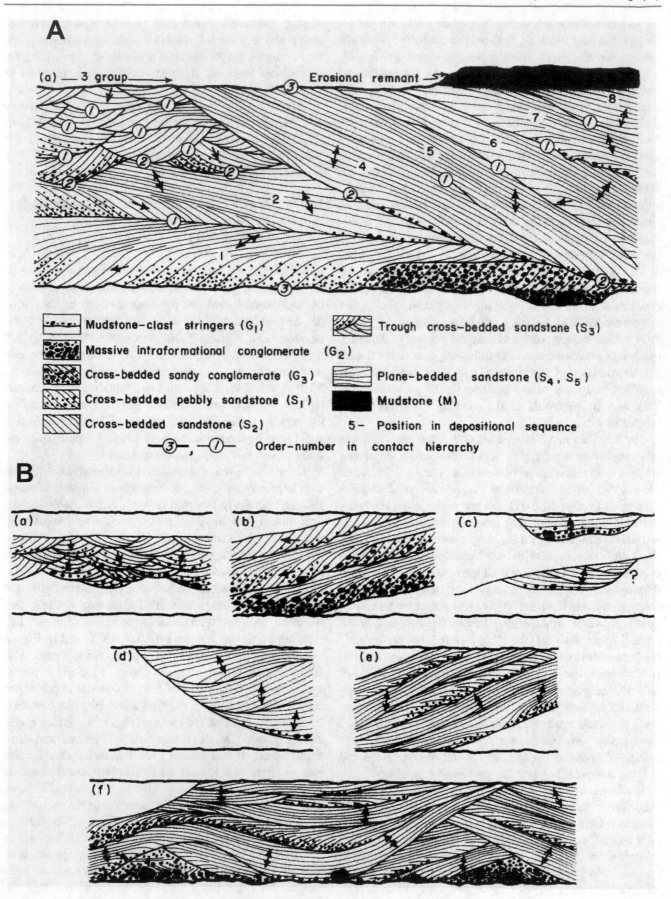

Fig. 1.9 A Schematic summary of the features of the sheet sandstones present in the Brownstones (Devonian, Welsh borders) showing the hierarchy of bedding contacts, sedimentary facies and position of sedimentation units in depositional sequence; **B** Summary of the main kinds of depositional feature: **a** Tabular layers of dune cross-bedded (trough cross-bedded) sandstone; **b** assemblages of down-climbing (forward-accreting) bar units; **c** minor channel forms and fills; **d** major channel form and fill; **e** groups of laterally accreted bar units; **f** symmetrical complexes (sand shoals) of laterally accreted bar units with gravel cores. From Allen (1983)

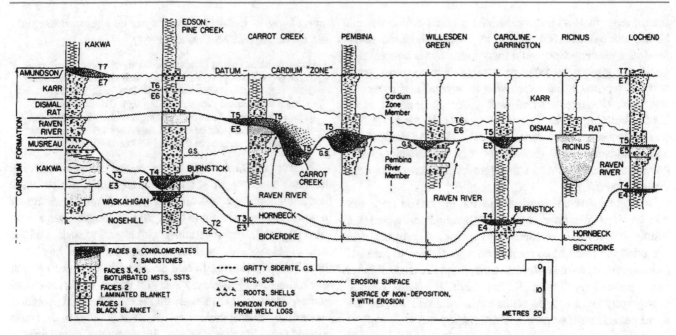

Fig. 1.10 The first sequence model for part of the Cretaceous succession of Alberta—the Cardium Sandstone (Turonian-Coniacian) (Plint et al. 1986)

APPG determined that the 1985 paper had been one of the most influential papers in siliciclastic sedimentology during the previous decade). The use of photomosaics as base maps for analyzing large outcrops has become standard, and there have been technological developments, such as the use of LIDAR methods for outcrop documentation, facilitating the digitization of observations, corrections for scale problems and perspective effects in ground observations and so on. This topic is dealt with at greater length in Sect. 3.5.11.

On a broader scale, the increasing use of seismic-reflection data in basin analysis has now revealed the three-dimensional complexity of many stratigraphic successions. Sequence analysis has highlighted the importance of onlap and offlap relationships, and the ubiquity of clinoform architectures in continental margin strata has been demonstrated repeatedly (Sect. 6.3).

1.2.12 Sequence Stratigraphy (1986–1990)

In the decade following the publication of AAPG Memoir 26 (Payton 1977) a wholesale re-evaluation of regional stratigraphy was underway. The significance of this revolution can be exemplified by the first publication that applied the new sequence concepts to an important swath of regional geology, the Cardium Sandstone of Alberta. This loosely defined unit is host to the largest oil field hosted in a clastic reservoir in Canada, the Pembina field, and stratigraphic and sedimentologic studies of the unit had been underway since it was discovered in 1953. The Pembina reservoir was

difficult to understand. It consists of locally as much as 9 m of wave- and tide-deposited conglomerate accumulated some 200 km from the assumed contemporary shoreline. How did it get there? There was much discussion of "offshore bars" and other concepts in the Canadian literature through the 1970s and 1980s; however, none of the ideas were fully satisfactory. But then arrived a new interpretation by Plint et al. (1986), who reconstructed from well-logs a set of seven basin-wide surfaces of erosion and transgression, that tied together all the complex local stratigraphies and set out a depositional model implying cycles of base-level change lasting about 125 ka (Fig. 1.10). In this model, the gravel was delivered to the middle of the basin by fluvial systems during sea-level lowstands, and then reworked into shoreface deposits during subsequent transgressions. The interpretation was controversial and was subject to intense discussion at the time (Rine et al. 1987), but the interpretation has stood the test of time and has led to a complete remapping of Alberta Basin stratigraphy using the new sequence concepts (Mossop and Shetsen 1994).

Meanwhile, researchers working with seismic data, particularly in the research laboratories of Shell, BP and Exxon, were applying sequence concepts to basins around the world, yielding many insights into stratigraphic architecture and regional basin controls, particularly the importance of tectonism, even though the global-eustasy paradigm remained dominant throughout the 1980s and early 1990s. Several atlases were published at this time, taking advantage of the large atlas format to display reflection-seismic cross-sections at large scales (Bally 1987). Even more importantly, in 1988 a second major product by the team at Exxon was published (Wilgus, et al. 1988), showing in detail how sequence concepts could incorporate facies analysis and could be applied to outcrop studies. The **systems tract** concept reached a full expression in several key papers in this book (Posamentier and Vail 1988; Posamentier et al. 1988), building on experimental models of Jervey (1988) that essentially reinvented Barrell's (1917) ideas about accommodation and its control on sedimentation and developed them further in the light of modern facies concepts (Sect. 5.2.1).

Another important publication from the Exxon team was that by Van Wagoner et al. (1990) which presented the results of several detailed field mapping projects and extended the reach of sequence concepts further, to regional outcrop and subsurface studies. Largely on the basis of these two publications by the Exxon team, the term seismic stratigraphy began to be replaced in common use by the more general term **sequence stratigraphy**.

1.2.13 Reconciling Facies Models with Sequence Stratigraphy (1990)

By the year 1990 a moment of tension had arrived in the evolution of sophisticated stratigraphy. The enormously successful facies-model approach, focusing on very detailed local studies, including meticulous analysis of drill cores, had resulted in a proliferation of sedimentological studies and numerous refinements of ideas about how to classify and subdivide sedimentary environments in an ever expanding range of tectonic and climatic settings. Most interpretations dwelt at length on autogenic sedimentary processes. Meanwhile, sequence stratigraphy had introduced an entirely different scale of research, encompassing whole basins, and focusing on allogenic controls, particularly sea-level change. In addition, the architectural-element approach to facies studies departed from the clean simplicity of the vertical profile by suggesting two- and three-dimensional assemblages of sedimentary building blocks in patterns difficult to pin down and classify.

The problems may be exemplified by an examination of a paper by Walker (1990), who was attempting to reconcile his facies-model approach to the new concepts and methods. He (Walker 1990, p. 779) complained that the architectural-element approach, which treated elements as building blocks that could be assembled in multiple ways (Miall 1985), constituted "sedimentological anarchy." Walker (1990) conceded that the proliferation of information about environments and facies associations that had resulted from the explosion of facies studies rendered the simple facies-model approach for complex depositional systems (such as submarine fans) inadequate. He referred

approvingly to the depositional systems approach exemplified by Fisher and McGowen (1967).

> Future facies modeling must emphasize these contemporaneous, linked depositional environments, and their response to tectonics and changes of relative sea level. This will combine the strengths of classical facies modeling with the recognition that widely spaced and "distinct" geographic environments (summarized as models) can be rapidly superimposed as part of one transgressive or regressive system (Walker 1990, p. 780).

Walker (1990, p. 781) also expressed concern regarding the new concepts of sequence stratigraphy, which were becoming popular at this time. He pointed out the ambiguity in some of the definitions (e.g., that of the **parasequence**; see Sect. 5.2.2), the uncertainty with regard to scale, and the lack of clarity in such expressions as "relatively conformable." The issue of scale arises with reference to such expressions as "genetically related" strata. In facies-model studies, genetically related implies gradational contacts between lithofacies that are related to each other in the sense implied by Walther's Law. In sequence stratigraphy, genetically related means the deposits formed during a full cycle of base-level change, although, as Walker (1990, p. 784) pointed out, using Galloway's (1989) **genetic stratigraphic sequence model** implies that strata above and below a subaerial erosion surface (the E/T surfaces of Plint et al. 1986) are genetically related, which they are certainly not.

While reluctant to fully embrace the new methods and terminology of sequence stratigraphy, Walker (1990) conceded that the regional patterns and the emphasis on large-scale sedimentary controls that were being revealed by sequence studies were valuable. As a compromise he suggested the adoption of the new system of **allostratigraphy** that had been proposed in 1983 by the North American Commission on Stratigraphic Nomenclature. Allostratigraphy is based on the recognition, mapping and subdivision of unconformity-bounded units. For example, a typical sequence, in the sense implied by Vail et al. (1977) constitutes an alloformation.

1.2.14 The Full Flowering of Modern Sequence-Stratigraphic Methods

When sequence stratigraphy was introduced to the geological community through the landmark publications of the Exxon Group (Payton 1977; Wilgus et al. 1988; Van Wagoner et al. 1990) it came with an overriding hypothesis that eustatic sea-level change was the main driver of changes in accommodation, and hence of sequence architecture. Doubts about the universal applicability of this hypothesis began to emerge in the 1980s, and by the mid-1990s most earth scientists had accepted that other factors, including climate change and regional tectonism may play a key role (Miall 1995). The controversy is described in detail elsewhere (Miall and Miall 2001, 2002; Miall 2010, Chap. 12).

The realization that many allogenic processes are at work during the accumulation of a basin fill gave renewed impetus to stratigraphic studies, because it became clear that sequence methods, combining the large scale of reflection-seismic surveying with the facies scale of the outcrop or drill core, could be very powerful tools for the reconstruction of geologic history, as well as provide much more useful predictive stratigraphic models for petroleum exploration and development. Some examples of this are discussed and illustrated in Chap. 6.

The evolution of seismic records from analog (paper records of wiggle traces) to digital, facilitated an enormous development in computer-processing and display techniques. One of the most important outcomes was the emerging ability to develop horizontal "**seiscrop sections**" from three-dimensional data volumes (Brown 1985; Fig. 6.34 illustrates an early example). Automatic tracking procedures enabled the geologist to follow stratigraphic surfaces through structural disturbances, and processing could flatten the result to restore an original horizontal depositional surface. Soon this led to the development of an entirely new discipline, **seismic geomorphology**, which deals with the analysis of ancient depositional systems based on their preserved landscape architecture and three-dimensional construction (Davies et al. 2007; Hart 2013). Furthermore, the debate about global eustasy placed renewed emphasis on the need for accurate global chronostratigraphic correlations in order to test regional and global correlations, and this also encouraged new work in this field.

The flourishing of sequence stratigraphy as a research topic inevitably led to differences of interpretation and even to differences in the methods for defining sequences. For example, Hunt and Tucker (1992) showed how the Exxon sequence model was quite inadequate in dealing with the falling stage of a base-level cycle. Galloway (1989) proposed defining sequence boundaries at the maximum flooding surface rather than the subaerial erosion surface and its basinward correlative conformity. This and other controversies (discussed in Sect. 7.7) hindered the development of a uniform methodology and common language for dealing with sequences on a formal basis.

Catuneanu (2006), in what has become the standard textbook on sequence stratigraphy, addressed the controversies, and showed how different approaches could be reconciled if care is taken with descriptions and definitions. In a series of papers culminating in a review for Newsletters in Stratigraphy he and selected colleagues have been leading the way in the work to gain acceptance for sequence stratigraphy as the appropriate formal basis for modern stratigraphic work (Catuneanu et al. 2009,2010,2011). More

recently, Steel and Milliken (2013) have provided a very useful documentation of the many incremental additions to our knowledge of siliciclastic facies associations and models and their incorporation into sequence-stratigraphic interpretations.

Modern theoretical and experimental work is making substantial contributions to our understanding of processes of sedimentation and sequence generation. The specially designed experimental facility (eXperimental EarthScape Facility, or XES) described by Paola (2000) and Paola et al. (2001, 2009) is particularly well-equipped to explore what Sheets et al. (2002) termed the stratigraphic "mesoscale," the time scale of years to thousands of years. Within this time frame, "the depositional pattern shifts from reflecting the short-term flow pattern to reflecting long-term basinal accommodation. Individual events are averaged to produce large-scale stratal patterns" (Sheets et al. 2002, p. 288). At this scale, autogenic processes grade into, or are affected by and modified by allogenic forcing. Muto and Steel (2004) demonstrated that, given steady conditions of discharge and sediment supply, prograding deltas will eventually start to "autoincise" over the mesoscale time scale. Strong and Paola (2008) explored the evolving nature of valley incision, terrace formation and valley fill, and demonstrated that the valley-floor surface that ultimately is preserved in the geological record during a cycle of base-level change is an erosion surface that never actually existed in its entirety as a topographic surface in its preserved form, because it undergoes continuous modification by erosion or sedimentation until final burial. Kim and Paola (2007) demonstrated that the autogenic process of delta and channel switching may, under the influence of fault movement, develop cyclothem-like cycles over time periods of 10^5 years.

Meanwhile, the research theme centered on facies analysis is by no means complete. As discussed in Chap. 4, advances in the understanding of processes and environments continue, aided by the experimental work touched on above and by improved observational methods. Three topics merit note: (1) the increasing recognition of the importance of cool-water environments for carbonate sedimentation, (2) an improved understanding of the development of deep-water turbidite deposits relative to the cycle of sea-level change and sediment delivery patterns, together with a much expanded understanding of the variability and complexity of turbidite systems, due in large measure to developments in marine geology, three-dimensional seismic surveying, and large-scale outcrop work and (3) the increasing realization that in natural systems mud forms silt- and sand-sized floccules, and most mud is transported and deposited by currents of all kinds. Pelagic settling may be of minor importance as a source of mud deposits.

International work on sedimentary and stratigraphic geology has become of increasing importance in recent

Fig. 1.11 The main intellectual themes that have now merged into the multidisciplinary subject of modern "sophisticated" stratigraphy. Shown are references to the key foundational papers that led to major shifts in earth science theory and methods

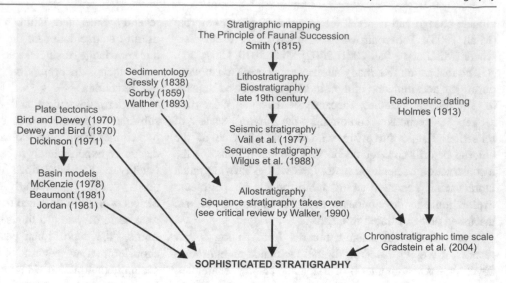

years. Chen et al. (2019) provided a useful overview and introduction to Chinese research in the areas of chronostratigraphy and sedimentology.

As discussed in Chap. 8, the use of "big data" in stratigraphy (hundreds to thousands of well sections, or many kilometers of measured section), exploiting the new sequence methods, is providing literally revolutionary new insights into the processes of basin evolution.

1.2.15 Stratigraphy: The Modern Synthesis

The full flowering of modern stratigraphy represents the amalgamation of the concepts and methods encompassed in all of the separate developments described in the preceding sections. The power of the modern science could not possibly have evolved without the contributions from all of these strands of development. However, for the purpose of education and training, the basic components of modern stratigraphy can be broken down into the following list of the major topics (Fig. 1.11). Key references are provided here to some of the main recent reviews and textbooks:

1. Facies analysis methods and facies models (James and Dalrymple 2010)
2. Sequence stratigraphy, concepts, definitions and methods (Catuneanu 2006)
3. Interpretations of the origins of sequences in terms of basin processes (tectonism, eustasy, climate change, etc.) (Miall 2010; Allen 2013)
4. Modern seismic methods, including seismic geomorphology (Veeken 2007; Davies et al. 2007; Hart 2013)
5. Chronostratigraphy and the Geologic Time Scale (Gradstein et al. 2004, 2012, 2020; see also https://www.stratigraphy.org)

6. Basin geodynamics: the origins of basins in terms of plate tectonics and crustal behavior (Miall 1999; Busby and Azor 2012)
7. Modern formal stratigraphic methods (Salvador 1994). Updated methods at https://www.stratigraphy.org
8. Advanced field methods, documentation, analysis, interpretation: Holbrook and Miall (2020).

A specialized branch of stratigraphy deals with the Quaternary record. Specialists include archeologists and anthropologists. Age dating reaches levels of accuracy and precision in the 10^3–10^4-year range, based on dating methods designed specifically for the Recent, including ^{14}C and U-Th radiometric methods, optically stimulated luminescence, cosmogenic radionuclides, and amino-acid geochronometry (http://www.inqua-saccom.org/stratigraphic-guide/geochronometry/).

1.3 Time in Stratigraphy

Sedimentologists and stratigraphers study sedimentary events over time periods extending through 17 orders of magnitude, from the long-term global changes in plate distribution, atmospheric composition, and oceanic geochemistry that take place over billions of years (10^9 yr), to the entrainment and displacement of individual sand grains in traction carpets, events that take but a few seconds (3 s = 10^{-7} yr) (Miall 1991, 2014, 2015; Holbrook and Miall 2020). Different parts of this time spectrum (shown schematically in Fig. 1.12) have become the special interest of different scientific groups (and are considered in different parts of this book). Consider this wide variety of observations:

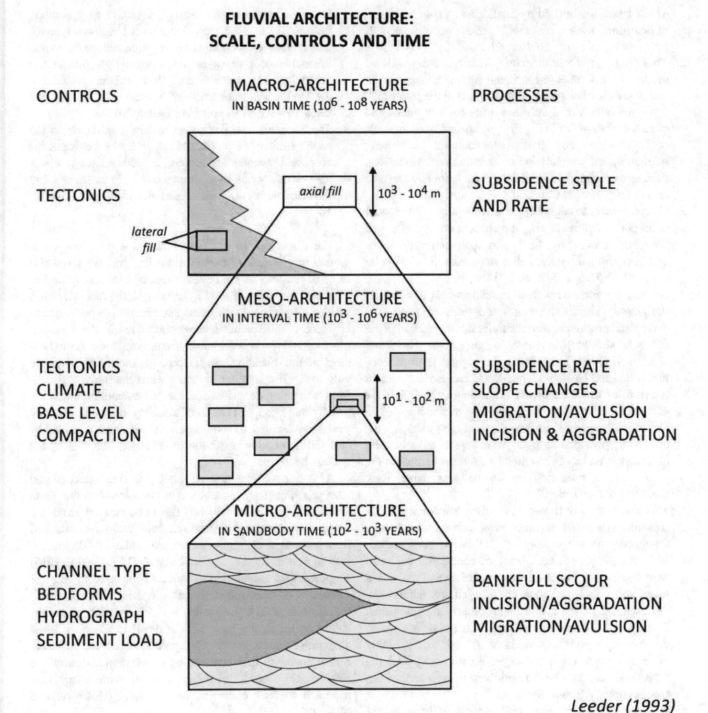

Fig. 1.12 The hierarchy of scales in fluvial systems (from Leeder 1993)

1. Those interested in fluid hydraulics and the development of bedforms are concerned with periods of a few seconds to a few hours (10^{-7}–10^{-3} yr; Chap. 2).
2. Those who study sedimentary facies and depositional models have to take into account diurnal changes in air and water circulation and tidal effects at one extreme (10^{-3} yr), the effects of **dynamic events** such as floods and storms (durations of 10^{-3}–10^{-2} yr, spaced over intervals of 10^{-1}–10^1 yr), and, at the other extreme, the growth, migration, and progradation of various subenvironments, such as river channels, tidal flats, delta lobes and crevasse splays (10^1–10^4 yr; Chap. 3).
3. Conventional stratigraphy—the mapping and formal definition of formations, members and groups, and sequence stratigraphy, deal with sedimentary packages and depositional systems that may take 10^2–10^7 yr to accumulate (Chaps. 5, 6, and 7).
4. For those concerned with the establishment of chronostratigraphic relationships and the rates of geological events and processes, chronostratigraphic precision in the dating of stratigraphic events is rarely better than about ± 0.1 m.y. (10^5 yr), this being the highest level of resolution that the Geologic Time Scale can currently attain for most of the Phanerozoic (Gradstein et al. 2004; see Chap. 7). For the Neogene the availability of a chemostratigraphic scale based on oxygen isotope variations has provided an even finer degree of resolution, and work is underway to extend a cyclostratigraphic time scale, with a resolution in the 10^5-year range, back through the Cenozoic (Sect. 7.8.2).
5. There is increasing recognition of the importance of **astronomic (orbital) forcing** as a control on climate. Commonly, in the geologic past, this has been responsible for the growth and decay of continental ice caps, with consequent major effects on climate and global sea level, and hence, on stratigraphic cyclicity, notably the development of the cyclic deposits termed cyclothems. These are the so-called **Milankovitch processes**, which deal mainly with time periods of 10^4–10^5 yr. The study of these cycles constitutes **cyclostratigraphy** and the time scale that may be derived from them is termed **astrochronology** (Chap. 8).
6. A major task for modern stratigraphers is the establishment and correlation of the major sequence framework for the world's sedimentary basins. These have durations ranging from 10^4 to 10^7 yr. The driving force for these sequences includes eustasy, driven by changes in the global average rate of sea-floor spreading, regional tectonics, and climate change (Miall 2010).
7. Geophysical modeling of sedimentary basins has, in recent years, focused on the flexural and rheological behavior of the earth's crust in response to stretching, heating and loading events, and on the regional (epeirogenic) changes in continental elevation caused by mantle thermal processes (dynamic topography). Subsidence driven by these mechanisms is slow, and the evolution of the large-scale architecture of a basin is a process that takes 10^6–10^7 yr to complete (Allen 2013).
8. The assembly and dispersion of supercontinents on the earth's surface is a product of the thermal behavior of the core and mantle. Evidence is accumulating for a long-term cyclicity throughout Proterozoic and Phanerozoic time, with a periodicity in the order of 10^8 yr.

The disparities in the time scales that form the background to the work of these various groups have given rise to some tensions in the development of basin analysis. For example, the study of fluid hydraulics in the laboratory has yielded elegant phase diagrams that permit interpretations to be made of instantaneous flow conditions in the development of sequences of hydrodynamic structures (crossbedding, parting lineation, etc.). However, such analyses relate only to the few minutes or hours when the individual bedform was being deposited and can tell us nothing about the usually much longer periods of time (up to several or many years) represented by non-deposition or erosion, and they also tell us nothing about the preservation potential of the various bedforms.

As data have accumulated relating to time intervals and rates of sedimentary processes it has become clear that there are major disparities between the rates derived from the study of modern sedimentary processes and those obtained from analysis of the geological record. Sadler (1981) was the first to document this in detail. Miall (1991, 2014, 2015) suggested that a hierarchical approach be used to manage discussion of rates and time scales in the geological record (Miall et al. 2021). This issue, and the question of preservation potential, are addressed in detail in Chap. 8. Ideas emerging from the new stratigraphic synthesis are requiring a new approach to the recording and interpretation of the stratigraphic record, involving a much more meticulous approach to data collection and analysis (Holbrook and Miall 2020).

1.4 Types of Project and Data Problems

Data collection and analysis procedures are, of course, determined by the nature of the project. The following are some typical basin-analysis problems, with a brief discussion of the data collection potential.

1.4.1 Regional Surface Stratigraphic Mapping Project

Work of this nature is one of the primary functions of government surveys, intent on providing complete map-sheet coverage of their area of responsibility, both as a service to industry and as a basis for expert advice to government economic planners. Similar regional surveys are commonly undertaken by industry as a preliminary to detailed surface or subsurface exploration, although their studies are rarely as thorough. Many academic theses are also of this type.

Many government surveys are carried out by individuals who are not specifically trained in the analysis of sedimentary basins, the idea being that members of the survey should be generalists, capable of mapping anything. This was an old British tradition, and it is an unfortunate one because it means that the individual survey officer cannot possibly be aware of all the skills that are now available for mapping work in sedimentary rocks, nor are they encouraged to take the time for the specialized observations which would make their work so much more effective. The argument that this is "left for the academics to do later" is not always satisfactory, for it is commonly the case that the stratigraphy of a succession can only be clarified by those who thoroughly understand its sedimentology. Many fruitless arguments about stratigraphic terminology can be avoided if this is realized at the beginning of a mapping endeavor. A team approach to a project is often an ideal solution. For instance, the research synthesis edited by Besly and Kelling (1988) is an excellent example of an integrated basin-analysis approach to a broad and complex research problem, in this case the origin of a major series of sedimentary basins in northwest Europe.

Another argument that is sometimes heard about geological mapping is that once it is done it does not need to be done again. Political pressures to defund or disband state geological surveys have been based on this argument, which, of course, ignores the fact that maps constantly need to be updated to incorporate new data, or redrawn to encompass new concepts or to take advantage of new technologies, such as Geographical Information Systems (GIS: digital mapping). It is likely that eventually most standard surface geological maps will be redrawn using sequence concepts instead of lithostratigraphic terminology.

The basis of all surface basin-analysis projects is the careful compilation of vertical stratigraphic sections. These are described by the geologist in the field, who also collects samples for subsequent laboratory analysis, taking care to label each sample according to its position in the section. Where should such sections be located? The choice depends on a variety of factors. First, they should be typical of the area in which they are found and should be as free as possible of structural deformation. Obviously, they should also be well exposed and, ideally, free of chemical or organic weathering, which disturbs or obscures textures and structures. Methods for documenting surface sections are discussed in detail in Sect. 2.2.

The best sections for regional correlation are those that include several stratigraphic intervals, but it is rare to find more than a few of these except in exceptionally well-exposed areas. Short, partial sections of a stratigraphic unit can provide much valuable sedimentological data, such as facies and paleocurrent information, although it may not be possible to locate them precisely within a stratigraphic framework unless they can be correlated by a marker bed or structural interpretation (Chap. 6). Exposures of this type tend to be ignored by the regional mapper, but they should not be ignored by those intent on producing an integrated basin analysis, because they add to the data base and may provide many useful sedimentological clues.

The great advantage of surface over subsurface studies is the potential, given adequate clean outcrop, to see medium- to large-scale sedimentary features, such as crossbedding, and larger scale architectural elements, including channels, large bar deposits, bioherms, etc., that may be difficult or impossible to identify in a drill hole. Large outcrops can be used to construct lateral profiles. These, of course, add immeasurably to any basin interpretation, particularly paleogeographic aspects, such as the size, geometry, and orientation of depositional elements (e.g., reefs, channels). The geologist should also always be searching for lateral variations in lithologies, fossil content, or sedimentary structures, as these changes may provide crucial control for paleogeographic interpretations.

The disadvantage of studies carried out exclusively at the surface is that most of the rocks in any given basin are buried and may be inaccessible to observation over very large areas. Many basins are depositional basins, in the sense that they preserve at the present day essentially the same outline as during sedimentation. The rocks exposed at the surface, especially around the margins of the basin, may have a quite different thickness and facies to those preserved at the center and may show erosion surfaces and unconformities not present in the center because of the tendency of basin margins to be affected by a greater degree of tectonic instability. This is illustrated in Fig. 1.13. A basin analysis carried out under such circumstances might therefore produce very incomplete or misleading results.

1.4.2 Local Stratigraphic-Sedimentologic Mapping Project

A common incentive for a detailed local study is the occurrence of some highly localized economic deposit, such

Fig. 1.13 Contrasts between stratigraphic thicknesses and facies at basin center and basin margin. The geology of the deep, hydrocarbon-producing regions of a basin might be quite different from that at the margins, so that surface geology gives little useful information on what lies below. The North Sea Basin is an excellent example. This example is of the Messinian evaporite basin, Sicily (Schreiber et al. 1976)

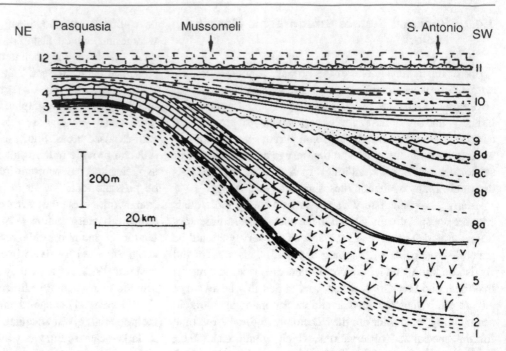

as an ore body or coal seam. A geologist will examine every available outcrop within a few square kilometers of the deposit, and may also supplement the analysis with logs of diamond drill holes. Another type of local study is an academic thesis project, particularly at the master's level. A small-scale project may be chosen because of the time and cost limits imposed by academic requirements. As discussed in Chap. 8, the modern synthesis of stratigraphic and sedimentologic methods requires extremely detailed field documentation and advanced methods of analysis and interpretation in order for us to fully understand the stratigraphic record (Holbrook and Miall 2020).

Many of the comments given in the preceding section apply to local studies, but there are some additional complications that often arise because of the nature of such projects. A common fault is the emphasis on local features to

the exclusion of any real consideration of regional implications. The geologist will erect a detailed local stratigraphy and fail to show clearly how it relates to any regional framework that may have been established, or there may be an overemphasis on certain selected parameters chosen, perhaps, as a training exercise, to the exclusion of others.

To set against these problems is the advantage a detailed local project may offer for carrying out very complete, sophisticated paleogeographic reconstruction. Rather than relying on selected stratigraphic sections, the geologist may be able to trace out units on foot or study their variation in a close network of diamond drill holes (Fig. 1.14). In this way, the detailed variations within, for example, an individual reef or channel network or a coal swamp, can be reconstructed. Typically, there will be more time to spend developing architectural data from large outcrops. Such reconstructions

Fig. 1.14 A typical diamond drill hole (DDH) network across a mining property, a gold prospect in Precambrian metasediments, northern Ontario

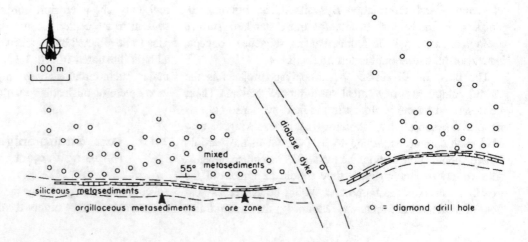

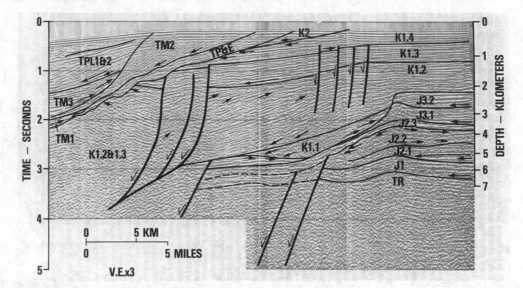

Fig. 1.15 Example of an interpreted seismic line across the continental margin of North Africa (Mitchum et al. 1977a). This is a famous section, having been reproduced in many textbooks as one of the first examples to be published showing the wealth of stratigraphic detail that could be extracted from the data. Note the presence of several angular unconformities, buried paleotopography, the internal architectural details of individual stratigraphic units, and the system of standard coded stratigraphic units employed by the authors. AAPG © 1977, reprinted by permission of the AAPG whose permission is required for further use

are particularly valuable because they provide a mass of three-dimensional data against which to test theoretical facies models and sequence concepts. Some of the most fascinating information now being obtained from the sedimentary record is based on hundreds, even thousands of data points, with numerous outcrop horizons literally "walked out" to ensure accurate correlation. Examples are described in Chap. 8.

Diamond drill holes are produced by a process of continuous coring. They therefore provide a complete lithologic and stratigraphic sample through the units of interest. The small diameter of the core (2–5 cm) does not permit recognition of any but the smallest sedimentary structures, but the close drill hole spacing of a few hundred meters or less means that very detailed stratigraphic correlation is usually possible.

Unfortunately, a tradition of retention and curation of diamond drill-hole core has rarely emerged in government or industrial organizations except, perhaps, for a few key holes in areas of particular interest. The core is normally discarded once the immediate interest in the area has subsided, and much valuable material is lost to future workers. The mining industry commonly seems to prefer it this way, but what is good for corporate competitiveness is not necessarily the best method for developing a national stratigraphic data base.

1.4.3 Regional Subsurface Mapping Project

Exploration activity in the petroleum industry is now mainly of this type. Companies may initially send surface geological

mapping parties into the field, but ultimately the most thorough field studies in a given field area are likely to be done by government survey organizations. Much industry activity is now located in offshore regions for which surface geological information is, in any case, sparse, unobtainable or irrelevant. As noted previously, the beds exposed at the edge of a basin may bear little relation to those buried near its center (Fig. 1.13).

Regional subsurface work is based initially on geophysical data and subsequently on test drilling. Gravity and aeromagnetic information may provide much useful information on broad structural features, particularly deep crustal structures. Refraction seismic lines may be shot for the same purpose. More detailed structural and stratigraphic data are obtained from reflection-seismic surveys (Fig. 1.15; Chap. 6), and these provide the basis for all exploration drilling in the early phases of basin development. Deep reflection surveys have profoundly changed our ideas about deformed belts during the last decade. Seismic shooting and processing is now a highly sophisticated process, and its practitioners like to talk about seismic stratigraphy as if it can provide virtually all the answers, not only about structure, but about the stratigraphic subdivision of a basin, regional correlation, and even lithofacies. This is particularly the case where, as is now common, selected areas of a basin are explored using three-dimensional seismic methods. However, the seismic method is only one exploration tool, and its results must be tested against those derived in other ways. For example, test drilling may show that stratigraphic correlations predicted from seismic interpretation are incorrect (Chap. 6).

Problems of stratigraphic velocity resolution, the presence of low-angle unconformities, and the obscuring effects of local structure can all introduce errors into seismic interpretation. Resolution of bedding units is still relatively crude, as illustrated in the comparison between typical seismic wave forms and the scale of actual bedding units shown in Fig. 7.1. Dating and correlation of seismic sequences depend heavily on biostratigraphic and geophysical log information from exploration wells that are used to constrain and calibrate stratigraphic reconstructions made using sequence principles.

Exploration wells, especially the first ones to be drilled in a frontier basin, are as valuable in their own way as space probes sent out to study the planets. As noted above (Sect. 1.2.8), early developments in the evolution of geotectonic models for extensional continental margins depended on the results from research wells drilled off the US Atlantic margins (the COST series). However, it is unfortunate that in many countries well data are not treated with the same respect afforded space information, but are regarded as the private property of the organization that paid for the drilling. In a competitive world, obviously, a company has the right to benefit from its own expenditures but, in the long term, knowledge of deep basin structure and stratigraphy belongs to the people and should eventually be made available to them. In Canadian frontier areas, well samples and logs must be deposited with the federal government and released for public inspection two years after well completion. Two years competitive advantage is quite long enough for any company in the fast-moving world of the oil industry, and after this time period the well records become part of a national data repository that anybody can use—with obvious national benefits. Seismic records are released after 5 years.

The nature of the stratigraphic information derived from a well is both better and worse than that derived from surface outcrops. It is better in the sense that there are no covered intervals in a well section, and such sections are generally much longer than anything that can be measured at the surface (perhaps exceeding 6000 m), so that the stratigraphic record is much more complete. The disadvantage of the well section is the very scrappy nature of the actual rock record available for inspection. Three types of sample are normally available:

1. Cuttings. These are produced by the grinding action of the rotary drill bit. These generally are less than 1 cm in length (Fig. 1.16) and can therefore only provide information on lithology, texture and microfossil content. The North American practice is to collect from the mud stream and bag for examination samples every 10 ft (3 m) of drilling depth. When drilling in soft lithologies,

Fig. 1.16 Typical well cuttings. Photo courtesy of J. Dixon

Fig. 1.17 Examples of typical core from a petroleum exploration hole

rock may cave into the mud stream from the side of the hole many meters above the drill bit, so that the samples become contaminated. Also, cuttings of different density may rise in the mud stream at different rates, which is another cause of mixing. It is thus necessary to observe the first appearance rule, which states that only the first (highest) appearance of a lithology or fossil type can be plotted with some confidence. Even then, depth distortions can be severe.

2. Full-hole core (Fig. 1.17). The rotary drill bit may be replaced by a coring tool when the well is drilling through an interval of interest, such as a potential or

actual reservoir bed. However, this type of core is expensive to obtain, and it is rare to find that more than a few tens of meters, perhaps only a few meters, of core are available for any given hole.

The advantage of a core is that because of the large diameter (usually on the order of 10 cm) it permits a detailed examination of small- to medium-sized sedimentary structures. Macrofossils may be present, and trace fossils are usually particularly well seen. The amount of sedimentological detail that can be obtained from a core is thus several orders of magnitude greater than is provided by chip samples. However, it is frustrating not to be able to assess the scale significance of a feature seen in a core. An erosion surface, for example, may be the product of a local scour, or it may be a major regional disconformity, but both could look the same in a core. Hydrodynamic sedimentary structures might be present, but it may not be possible to interpret their geometry and, except in rare instances where a core has been oriented in the hole, they provide no paleocurrent information. Orientation can sometimes be deduced if the core shows a structural dip that can be determined from regional structural data or dipmeter (formation microscanner) logs.

3. Side-wall cores. These are small plugs extracted by a special tool from the wall of a hole after drilling has been completed. These cores are rarely available to the geologist because they are used in porosity-permeability tests and are disaggregated for caving-free analysis by biostratigraphers. In any case, their small size limits the amount of sedimentological information that they can yield.

In addition to the samples and core, the analysis of which is described in Sect. 2.3, each exploration hole nowadays is subjected to an extensive series of petrophysical logging methods, which provide records by direct analog tracing or digitization. The description and interpretation of such logs has been the subject of several textbooks and cannot be treated exhaustively here. Log information is discussed further in Sect. 2.4. The following are a few preliminary remarks discussing the utility of logs in a subsurface data collection scheme.

Most geophysical tools measure a single physical property of the rock, such as its electrical resistivity, sonic velocity, and gamma radioactivity. These properties reflect lithology, and can therefore be used, singly or in combination, to interpret lithology. Because measurements with modern tools have depth resolution of a few centimeters they are of great potential value in deriving accurate, depth-controlled lithologic logs free of the problems of sample caving (Fig. 2.28). The response of a single tool is not unique to each rock type; for example, many different formations will contain rocks with the same electrical resistivity, and so it is not possible to interpret lithology directly from a single log type. However, such interpretations may be possible from a combination of two or more logs, and attempts have been made to automate such interpretations based on computerized calculation routines from digitized log records. Unfortunately, the physical properties of rocks and their formation fluids vary so widely that such automated interpretation procedures can only be successful if they are adapted to the specific conditions of each basin. They thus lose much of their exploration value, but become of considerably greater importance once an initial scatter of exploration wells has become available to calibrate the logs.

A particularly common use to which geophysical logs are put is in stratigraphic correlation (Sect. 6.2). Log records through a given unit may have a distinctive shape, which a skilled geologist can recognize in adjacent holes. Correlation may therefore be possible even if details of lithology are unknown and, indeed, the establishment of correlations in this way is standard practice in subsurface work. Similarly, log shape is a useful tool in environmental interpretation, though it must be used with considerable caution.

Logs are also of considerable importance in calibrating seismic records—seismic velocities are routinely derived from sonic logs and used to improve seismic correlations (e.g., see Fig. 6.19 and discussion in text).

Regional subsurface work may lead to the development of a list of petroleum **plays** for an area. These are conceptual models to explain the local **petroleum system**, including the history of generation, migration and trapping for petroleum pools. For example, a Devonian reef play, which might be based on the occurrence in Devonian strata of porous, dolomitized reef masses enclosed in mudrock, would provide an ideal series of stratigraphic traps. The development of a petroleum play leads to the evolution of an exploration methodology, which summarizes exploration experience in choosing the best combination of exploratory techniques and the type of data required to locate individual pools within the play area. At this stage, exploration moves from the play stage to the **prospect** stage. The first wells drilled in an area might have been drilled on specific prospects, with a specific play in mind, but it is rare for early wells to be successful. Prospect development is the next phase in an exploration program.

It is now common practice in petroleum exploration programs to carry out studies of burial history and thermal maturation of the stratigraphic section in a basin. This is based on what are now called backstripping procedures. Subsidence history is reconstructed by computer programs that reverse the process of sedimentation, stripping off each layer in turn, decompacting the remaining section, and calculating the isostatic balance. In this way, the component of subsidence due to tectonic effects can be isolated from that caused by the weight of sediment, and this, in turn, can be related to the rheological behavior of the crust underlying the basin. Burial history can be related to thermal history in order to document the organic maturation of the sediments and, in this way, the timing of petroleum generation and release can be evaluated.

The recognition and correlation of stratigraphic sequences (Chaps. 5 and 6) now forms the core of regional stratigraphic work. This framework is then interpreted to assess regional paleogeographic histories and cycles of local to regional sea-level changes. The latter may be compared to regional and global standards, although this is of less importance to the exploration geologist than the production of a useable regional stratigraphic framework.

The reader is advised to turn to modern textbooks on petroleum geology (e.g., Chapman 2000) for further discussion of methods of subsurface exploration and development.

1.4.4 Local Subsurface Mapping Project

A network of closely spaced holes a few kilometers (or less) apart may be drilled to develop a particular petroleum or mineral prospect. These may be in virtually virgin territory or they may be step-out wells from a known pool or deposit in a mature petroleum basin or mine. A three-dimensional seismic survey may be carried out over selected areas of interest. It is at this stage in exploration work that geological skills come most strongly into play. In the early exploration phases, virtually the only information available is geophysical; almost all early petroleum exploration wells are drilled on potential structural traps. Once a hydrocarbon-bearing reservoir is located, the next problem is to understand why it is there, how large and how porous it is, and in what direction it is likely to extend. Interpretations of depositional environments, applications of facies models, and reconstructions of facies architecture and paleogeographic evolution may become crucial in choosing new drilling sites. Similar procedures must be followed in order to develop stratabound ore bodies. In the case of offshore petroleum exploration the exploitation of a new petroleum pool may be achieved by directional drilling of many wells from a central platform, and great pressure may be placed upon the development geologists to design a production strategy based on extremely limited data.

The types of data available for local subsurface mapping are discussed in Sect. 1.4.3. Three-dimensional seismic surveying is becoming an increasingly valuable tool for local structural and stratigraphic mapping, particularly for areas characterized by complex depositional systems, because of the ability to use the techniques of **seismic geomorphology**

to map depositional units and their respective lithofacies in great detail (Sect. 6.3.4). At this point, it is pertinent to add a few cautionary notes on the uses of subsurface data. Because of its inherent limitations it is natural that exploration geologists would wish to exploit the data to the full extent but, paradoxically, this can lead to a very limited approach to an exploration problem. Geologists may be so impressed with a particular interpretive technique that they may tend to use it to the exclusion of all others. This has led to many false interpretations. It cannot be overemphasized that, in basin analysis as, no doubt, in every area of geology, every available tool must be brought to bear on a given problem. Examples of techniques in the area of sedimentology that have been overutilized include the use of vertical profile studies, grain-size analysis, and grain-surface-texture analysis to interpret depositional environments.

1.5 Summary of Research and Reporting Procedures

Vail et al. (1991) suggested a working method for the application of sequence methods to the exploration of a new basin. Following a discussion of the application of Exxon sequence analysis methods and concepts to two basin examples, Miall (1997, Sect. 17.3.3) proposed a modified set of research procedures (A simplified version of this is provided in Fig. 1.18):

1. Develop an allostratigraphic framework based on detailed lithostratigraphic well correlation and/or seismic facies analysis.
2. Develop a suite of possible stratigraphic models that conform with available stratigraphic and facies data.
3. Establish a regional sequence framework with the use of all available chronostratigraphic information.
4. Determine the relationships between sequence boundaries and tectonic events by tracing sequence boundaries into areas of structural deformation, and documenting the architecture of onlap/offlap relationships, fault offsets, unconformable discordancies, etc.
5. Establish the relationship between sequence boundaries and regional tectonic history, based on plate-kinematic reconstructions.
6. Refine the sequence-stratigraphic model. Subdivide sequences into depositional systems tracts and interpret facies.
7. Construct regional structural, isopach, and facies maps, interpret paleogeographic evolution, and develop plays and prospects based on this analysis.
8. Develop detailed subsidence and thermal-maturation history by backstripping/geohistory analysis.

Fig. 1.18 A summary of
workflow suggestions

So how do we get a grasp of modern stratigraphy?

- It starts with the mapping of lithofacies
  - classification and coding of lithofacies categories for speedy logging

- We extend our ground observations by methods of correlation
  - lithostratigraphy
  - biostratigraphy
- In the subsurface using petrophysical logs
  - wiggle tracing and the erection of regional units

- On the regional scale: seismic reflection

- Development of sequence stratigraphic framework

- We try to understand what we're looking at on the local scale:
  - Walther's Law and facies analysis
  - Cycles, cyclothems, vertical profiles
  - generalized facies models tied to systems tracts

- The bigger picture
  - Krumbein's facies mapping techniques (isopachs, isopleths)
  - seismic stratigraphy
  - allostratigraphy and regional sequence stratigraphy

- Interpretation
  - Paleogeography
  - Rates of processes
  - How did the sequences form?
  (mechanisms provide correlation models)

It was suggested that subsidence and maturation analysis (Step 8) be carried out at the conclusion of the detailed analysis, rather than near the beginning, as proposed in the Exxon approach. The reason is that a complete, thorough analysis requires the input of a considerable amount of stratigraphic data. Corrections for changing water depths and for porosity/lithification characteristics, which are an integral part of such analysis, all require a detailed knowledge of the stratigraphic and paleogeographic evolution of the basin.

Catuneanu (2006, pp. 63–71) provided a detailed set of suggestions for completing a regional sequence analysis. His "workflow," which is summarized here is based on "a general understanding [that] the larger scale tectonic and depositional setting must be achieved first, before the smaller scale details can be tackled in the most efficient way and in the right geological context." His workflow therefore proceeds from the large-scale through a decreasing scale of observation and an increasing level of detail. These are the basic components of the workflow:

1. Interpret the tectonic setting. This determines the type of sedimentary basin, and therefore controls the subsidence pattern and the general style of stratigraphic architecture.

Ideally, the analysis should start from regional seismic lines.
2. Determine the broad regional paleographic setting, including the orientation of regional depositional dip and strike, from which broad predictions may be made about the position and orientation of coastlines, and regional facies architecture.
3. Determine paleodepositional environments, making use of well-log ("well-log motifs"), core data, and 3-D seismic data, as available.
4. From the regional concepts developed by steps 1–3, depositional trends may be predicted, and this provides essential diagnostic clues for the interpretation of sequence architecture. The focus should now be on the recognition and mapping of seismic terminations and major bounding surfaces. For example, coastal onlap may indicate transgression (retrogradation) and downlap indicates regression (progradation). Only after the depositional trends are constrained, can the sequence-stratigraphic surfaces that mark changes in such trends be mapped and labeled accordingly.
5. The last step of a sequence analysis is to identify the systems tracts.

Catuneanu (2006) was not primarily concerned in his book with driving mechanisms, and so his workflow ends with the construction of a detailed sequence stratigraphy. It is only once this has been completed that geohistory (backstripping) analysis can be carried out, a reconstruction of the history of relative sea-level change constructed, and a search for sequence-generating mechanisms be undertaken.

The most recent addition to the geologists' tool kit is three-dimensional reflection-seismic data. The concept of extracting seismic-reflection data horizontally, or along gently dipping bedding planes, has been around since at least the mid-1980s (Brown 1985), but it has only been with the steady increase in computer power that 3-D seismic has become a routine exploration tool. Alistair Brown "wrote the book" on this topic in 1986, and his memoir has gone into multiple editions (Brown 2011). At depths of up to a few hundred meters, the resolution of 3-D seismic is adequate to image depositional systems in all their complex detail, providing an essential complement to well data in any project to map potential stratigraphic traps. The ability to map ancient landscapes in this way has given rise to a whole new special research topic, **seismic geomorphology**, pioneered by Henry Posamentier (2000). The subject is explored in detail by Davies et al. (2007) (see Chap. 6).

Formal stratigraphic methods are described at the website of the International Commission on Stratigraphy (http://www.stratigraphy.org). They do not yet include formal definitions and procedures for the application of sequence-stratigraphic methods to the subdivision and naming of units in a basin fill (see Sect. 7.7).

The structure of a final report should follow a logical order. A summary of the typical contents of a regional report is given below. Elements of this structure commonly are missing in the first drafts of graduate theses and reports submitted for publications in journals, or are not presented in a logical order. Not all components may be required for every individual study.

1. Introduction to the project. Why has this research been undertaken? What regional problem or scientific question has been addressed?
2. A description of the area covered by the report. This is normally illustrated by the use of a map. A project area may be defined geographically (mapping quadrant, state boundary) or geologically (tectonic province, sedimentary basin).
3. The geological setting of the area. The reader must be able to understand the purpose of the report within the context of the regional geological history. If appropriate, a description of the geological setting may include a brief outline of the plate-tectonic history of the region.

Significant earlier work on the problem at hand or on the regional geology of the area may be referenced.
4. The stratigraphy of the area, typically including a stratigraphic table. The basis for regional correlations and age designations may be provided. If the purpose of the report is to describe the sequence stratigraphy of the region, earlier lithostratigraphic analyses should be briefly summarized, with references to this earlier work
5. Location of study sections, or wells used, or samples collected. These should be located by annotated points on a map, and other details may be provided if pertinent, such as GPS data or national map grid references for key sections. If highly detailed location information is required this may be placed in an appendix. Annotated aerial photographs/satellite images/digital elevation models may be used to locate key outcrop sections.
6. Research methods should be described. This may include, where relevant, sample collection and preparation procedures, field mapping and measurement methods, laboratory procedures, data reduction techniques. Be sure to reference the source of the methods used. Lengthy descriptions of methodology, such as laboratory standardization procedures, may be placed in an appendix (see also point 10, below). Methods that have been used before do not have to be described in detail, but the reader should be able to understand what you have done without needing to read a body of earlier literature.
7. Presentation of observations and/or laboratory results. As far as possible this presentation should be separate from any interpretation. Data should be illustrated without interpretation (e.g., uninterpreted seismic sections, raw stratigraphic sections). This may be difficult, for example, where observations have been guided by a new interpretive hypothesis. The use of systems tract designations in the description of stratigraphic sections is a good example of a supposedly descriptive approach that has generated many problems.
8. Discussion of results, interpretations, and new hypotheses or theories arising from the project. Suggestions for further work.
9. Summary and conclusions. The major results. What new insights does this research provide? What new research or new problems does the project highlight? Do not introduce new references at this point.

10. Many mainstream journals now offer the opportunity to supplement the information in a published paper with additional documentation accessible on a dedicated journal website. This could include details of research methodologies, calculations, statistical procedures, data tables, etc.

There are also some simple rules to follow regarding illustrations.

1. All locations and geological regions referenced in the text must be named on a map, preferably near the beginning of the report.
2. All symbols on all figures must be explained, preferably by legends on the figures themselves. The old European and Russian practice of providing symbols in numbered boxes that are explained by numbered notes in the caption is cumbersome and not to be recommended (this practice facilitated translations into different languages while avoiding the need to redraft figures, but now that science is almost universally published in the English language, this need no longer exists).
3. Lettering on figures should not be smaller than 1 mm high when published.
4. Photographs, whether of outcrops, samples, thin-sections, SEM images, etc., must include a scale, and orientation information may also be critical.
5. Figures, tables, and plates must be numbered in the order by which they are referenced in the text.

References

Ager, D. V., 1973, The nature of the stratigraphical record: New York, John Wiley, 114 p.

Ager, D. V., 1981, The nature of the stratigraphical record (second edition): John Wiley, New York, 122 p.

Ager, D. V., 1993, The new catastrophism, Cambridge University Press, 231 p.

Allen, J. R. L., 1962, Petrology, origin and deposition of the highest Lower Old Red Sandstone of Shropshire, England: Journal of Sedimentary Petrology, v. v. 32, p. 657–697.

Allen, J. R. L., 1963a, The classification of cross-stratified units, with notes on their origin: Sedimentology, v. 2, p. 93–114.

Allen, J. R. L., 1963b, Henry Clifton Sorby and the sedimentary structures of sands and sandstones in relation to flow conditions: Geologie en Mijnbouw, v. 42, p. 223–228.

Allen, J. R. L., 1964, Studies in fluviatile sedimentation: six cyclothems from the Lower Old Red Sandstone, Anglo-Welsh basin: Sedimentology, v. 3, p. 163-198.

Allen, J. R. L., 1965, A review of the origin and characteristics of recent alluvial sediments: Sedimentology, v. 5, p. 89–191.

Allen, J. R. L., 1983, Studies in fluviatile sedimentation: bars, bar complexes and sandstone sheets (low-sinuosity braided streams) in the Brownstones (L. Devonian), Welsh Borders: Sedimentary Geology, v. 33, p. 237-293.

Allen, J. R. L., 1993, Sedimentary structures: Sorby and the last decade: Journal of the Geological Society, London, v. 150, p. 417-425.

Allen, P. A., and Allen, J. R., 2013, Basin analysis: Principles and application to petroleum play assessment, third edition: Chichester: Wiley-Blackwell, 619 p.

Aubouin, J., 1965, Geosynclines: Developments in Geotectonics, Elsevier, Amsterdam, v.1, 352 p.

Ball, M. M., 1967, Carbonate sand bodies of Florida and the Bahamas: Journal of Sedimentary Petrology, v. 37, p. 556–591.

Bally, A. W., ed., 1987, Atlas of seismic stratigraphy: American Association of Petroleum Geologists Studies in Geology 27, in 3 vols.

Bally, A. W., 1989, Phanerozoic basins of North America, in Bally, A. W., and Palmer, A. R., eds., The geology of North America—an overview: The geology of North America, Geological Society of America, v. A, p. 397–446.

Bally, A. W., Gordy, P. L., and Stewart, G. A., 1966, Structure, seismic data and orogenic evolution of southern Canadian Rockies: Bulletin of Canadian Petroleum Geology, v. 14, p. 337–381.

Barrell, Joseph, 1917, Rhythms and the measurement of geologic time: Geological Society of America Bulletin, v. 28, p. 745–904.

Beales, F. W., 1958, Ancient sediments of Bahaman type: American Association of Petroleum Geologists, v. 42, p. 1845–1880.

Beaumont, C, 1981, Foreland basins: Geophysical Journal of the Royal Astronomical Society; v. 65, p. 291–329.

Beerbower, J. R., 1964, Cyclothems and cyclic depositional mechanisms in alluvial plain sedimentation: Geological Survey of Kansas Bulletin 169, v. 1, p. 31–42.

Berg, R. R., 1962, Mountain flank thrusting in Rocky Mountain foreland, Wyoming and Colorado: American Association of Petroleum Geologists Bulletin, v. 46, p. 2019-2032.

Berg, R. R., 1968, point-bar origin of Fall River Sandstone reservoirs, northeastern Wyoming: American Association of Petroleum Geologists Bulletin, v. 52, p. 2116–2122.

Berggren, W. A., Kent, D. V., Aubry, M.-P., and Hardenbol, J., eds., 1995, Geochronology, time scales and global stratigraphic correlation: Society for Sedimentary Geology Special Publication 54, 386 p.

Bernard, H. A., and Major, C. J., 1963, Recent meander belt deposits of the Brazos River; an alluvial "sand" model (abs): American Association of Petroleum Geologists Bulletin, v. 47, p. 350.

Bernard, H. A., Leblanc, R. J., and Major, C. J., 1962, Recent and Pleistocene geology of southeast Texas, in Rainwater, E. H., and Zingula, R. P., eds., Geology of the Gulf Coast and central Texas: Geological Society of America, Guidebook for 1962 Annual Meeting., p. 175–224.

Bernard, H. A., Major, C. F. Jr., and Parrott, B. S., 1959, The Galveston Barrier Island and environs - a model for predicting reservoir occurrence and trend: Transactions of the Gulf Coast Association of Geological Societies, v. 9, p. 221–224.

Berry, W. B. N., 1968, Growth of prehistoric time scale, based on organic evolution: W.H. Freeman and Co., San Francisco, 158 p.

Berry, W. B. N., 1987, Growth of prehistoric time scale based on organic evolution, revised edition: Oxford: Blackwell Science, 202 p.

Besly, B. M., and Kelling, G., eds., 1988, Sedimentation in a synorogenic basin complex: the Upper Carboniferous of northwest Europe: Blackie, Glasgow, 276 p.

Bird, J. M., and Dewey, J. F., 1970, Lithosphere plate-continental margin tectonics and the evolution of the Appalachian Orogen: Geological Society of America Bulletin, v. 81, p. 1031–1060.

Blackwelder, E., 1909, The valuation of unconformities: Journal of Geology, v. 17, p. 289–299.

Blatt, H., Middleton, G. V., and Murray, R. C., 1972, Origin of sedimentary rocks: Englewood Cliffs, New Jersey: Prentice-Hall, 634 p.

Boggs, S., Jr., 1987, Principles of sedimentology and stratigraphy: Prentice Hall, Englewood Cliffs, New Jersey, 784 p.

Bouma, A. H., 1962, Sedimentology of some flysch deposits: Elsevier, Amsterdam, 168 p.

Bradley, W. H., 1929, The varves and climate of the Green River epoch: US Geological Survey Professional Paper 158, p. 87-110.

Brouwer, A., 1962, Past and present in Sedimentology: Sedimentology, v. 1, p. 2–6.

Brown, A. R., 1985, The role of horizontal seismic sections in stratigraphic interpretation, in Berg, O. R., and Woolverton, D. G., eds., Seismic stratigraphy II: American Association of Petroleum Geologists Memoir 39, p. 37–47.

Brown, A. R., 2011, Interpretation of three-dimensional seismic data, seventh edition, American Association of Petroleum Geologists Memoir 42, 646 p.

Brown, L. F., Jr., and Fisher, W. L., 1977, Seismic-stratigraphic interpretation of depositional systems: examples from Brazilian rift and pull-apart basins, in Payton, C. E., ed., Seismic stratigraphy — applications to hydrocarbon exploration: American Association of Petroleum Geologists Memoir 26, p. 213–248.

Burk, C. A., and Drake, C. L., eds., 1974, The geology of continental margins, Springer-Verlag, New York, 1009 p.

Busby, C., and Azor, A., eds., 2012, Tectonics of sedimentary basins: Recent Advances: Wiley-Blackwell, Chichester, 647 p.

Catuneanu, O., 2006, Principles of sequence stratigraphy: Elsevier, Amsterdam, 375 p.

Catuneanu, O., Abreu, V., Bhattacharya, J. P., Blum, M. D., Dalrymple, R. W., Eriksson, P. G., Fielding, C. R., Fisher, W. L., Galloway, W. E., Gibling, M. R., Giles, K. A., Holbrook, J. M., Jordan, R., Kendall, C. G. St. C., Macurda, B., Martinsen, O. J., Miall, A. D., Neal, J. E., Nummedal, D., Pomar, L., Posamentier, H. W., Pratt, B. R,. Sarg, J. F., Shanley, K. W., Steel, R. J., Strasser, A., Tucker, M. E., and Winker, C., 2009, Toward the Standardization of Sequence Stratigraphy: Earth Science Reviews, v. 92, p. 1–33.

Catuneanu, O., Bhattacharya, J. P., Blum, M. D., Dalrymple, R. W., Eriksson, P. G., Fielding, C. R., Fisher, W. L., Galloway, W. E., Gianolla, P., Gibling, M. R., Giles, K. A., Holbrook, J. M., Jordan, R., Kendall, C. G. St. C., Macurda, B., Martinsen, O. J., Miall, A. D., Nummedal, D., Posamentier, H. W., Pratt, B. R,. Shanley, K. W., Steel, R. J., Strasser, A., and Tucker, M. E., 2010, Sequence stratigraphy: common ground after three decades of development: First Break, v. 28, p. 21–34.

Catuneanu, O., Galloway, W.E., Kendall, C.G.St.C., Miall, A.D., Posamentier, H.W., Strasser A., and Tucker M.E., 2011, Sequence Stratigraphy: Methodology and Nomenclature: Report to ISSC: Newsletters on Stratigraphy, v. 4 (3), p. 173–245.

Chapman, R. E., 2000, Petroleum Geology, Elsevier, Amsterdam, 414 p.

Chen, Z-Q., Hu, X., Montañez, I. P., and Ogg, J. G., 2019, Sedimentology as a key to understanding earth and life processes: Earth Science reviews, v. 189, p. 1–5.

Cloetingh, S., 1988, Intraplate stress: a new element in basin analysis, in Kleinspehn, K., and Paola, C., eds., New Perspectives in basin analysis: Springer-Verlag, New York, p. 205–230.

Conkin, B. M., and Conkin, J. E., eds., 1984, Stratigraphy: foundations and concepts: Benchmark Papers in Geology, New York: Van Nostrand Reinhold, 363 p.

Conybeare, C. E. B., and Crook, K. A. W., 1968, Manual of sedimentary structures: Australian Bureau of Mineral Resources, Geology and Geophysics, Bulletin 102, 327 p.

Cross, T. A., and Homewood, P. W., 1997, Amanz Gressly's role in founding modern stratigraphy: Geological Society of America Bulletin, v. 109, p. 1617–1630.

Curray, J. R., 1956, The analysis of two-dimensional orientation data: Journal of Geology, v. 64, p. 117–131.

Curray, J. R., 1964, Transgressions and regressions, in Miller, R. L., ed., Papers in marine geology, Shepard Commemorative volume: MacMillan Press, New York, p. 175–203.

Curtis, D. M., 1970, Miocene deltaic sedimentation, Louisiana Gulf Coast, in Morgan, J. P., ed., Deltaic sedimentation modern and ancient: Society of Economic Paleontologists and Mineralogists Special Publication 15, p. 293-308.

Cutler, A., 2004, The seashell on the mountaintop: Plume Books, Penguin, New York, 228 p.

Dahlstrom, C. D. A., 1970, Structural geology in the eastern margin of the Canadian Rocky Mountains: Bulletin of Canadian Petroleum Geology, v. 18, p. 332–406.

Dapples, E. C., Krumbein, W. C., and Sloss, L. L., 1948, Tectonic control of lithologic associations: American Association of Petroleum Geologists Bulletin, v. 32, p. 1924–1947.

Darwin, C., 1859, On the origin of species by means of natural selection, or the reservation of favoured races in the struggle for life: John Murray, London, 502 p.

Davies, R. J., Posamentier, H. W., Wood, L. J., and Cartwright, J. A., eds., 2007, Seismic geomorphology: applications to hydrocarbon exploration and production: Geological Society, London, Special Publication 277, 274 p.

De Raaf, J. F. M., Reading, H. G., and Walker, R. G., 1965, Cyclic sedimentation in the Lower Westphalian of North Devon, England: Sedimentology, v. 4, p. 1–52.

Dewey, J. F., 1982, Plate tectonics and the evolution of the British Isles: Journal of the Geological Society, London, v. 139, p. 371-412.

Dewey, J. F., and Bird, J. M., 1970, Plate tectonics and geosynclines, Tectonophysics, v. 10, p. 625-638.

Dickinson, W. R., 1971, Plate tectonic models of geosynclines: Earth and Planetary Science Letters, v. 10, p. 165–174.

Dickinson, W. R., ed., 1974a, Tectonics and sedimentation: Society of Economic Paleontologists and Mineralogists Special Publication 22.

Dickinson, W. R., 1974b, Plate tectonics and sedimentation, in Dickinson, W. R., ed., Tectonics and sedimentation: Society of Economic Paleontologists and Mineralogists Special Publication 22, p. 1-27.

Dickinson, W. R., 1976, Sedimentary basins developed during evolution of Mesozoic-Cenozoic arc-trench system in western North America: Canadian Journal of Earth Sciences, v. 13, p. 1268–1287.

Dickinson, W. R., 1980, Plate tectonics and key petrologic associations, in Strangway, D. W., ed., The continental crust and its mineral deposits: Geological Association of Canada Special Paper 20, p. 341-360.

Dickinson, W. R., 1981, Plate tectonics and the continental margin of California, in Ernst, W. G., ed., The geotectonic development of California: Prentice-Hall Inc., Englewood Cliffs, New Jersey, p. 1-28.

Dickinson, W. R., and Seely, D. R., 1979, Structure and stratigraphy of forearc regions: American Association of Petroleum Geologists Bulletin, v. 63, p. 2–31.

Dietz, R. S., 1963, Collapsing continental rises; an actualistic concept of geosyncline mountain building: Journal of Geology, v. 71, 314–333.

Dietz, R. S., and Holden, J. C., 1974, Collapsing continental rises; actualistic concept of geosynclines—a review, in Dott, R. H., Jr., and Shaver, R. H., eds., Modern and ancient geosynclinal sedimentation: Society of Economic Paleontologists and Mineralogists, Special Publication 19, p. 14–27.

Dott, R. H., Jr., 1983, Episodic sedimentation—how normal is average? How rare is rare? Does it matter? Journal of Sedimentary Petrology, v. 53, p. 5-23.

Dott, R. H., Jr., 1996, Episodic event deposits versus stratigraphic sequences—shall the twain never meet? Sedimentary Geology, v. 104, p. 243-247.

Dott, R. H., Jr., and Shaver, R. H., eds., 1974, Modern and ancient geosynclinal sedimentation: Society of Economic Paleontologists and Mineralogists Special Publication 19, 380 p,

Doyle, P. and Bennett, M. R., eds., 1998, Unlocking the stratigraphical record: John Wiley and Sons, Chichester, 532 p.

Drake, C. L., Ewing, M., and Sutton, G. H., 1959, Continental margins and geosynclines—the east coast of North America north of Cape Hatteras, in Physics and Chemistry of the Earth: Pergamon Press, London, v. 3, p. 110-198.

Duff, P. M. D., Hallam, A., and Walton, E. K., 1967. Cyclic sedimentation: Developments in Sedimentology, v. 10, Elsevier, Amsterdam, 280 p.

Duff, P. McL. D., and Walton, E. K., 1962, Statistical basis for cyclothems: a quantitative study of the sedimentary succession in the east Pennine Coalfield: Sedimentology, v. 1, p. 235–255.

Dunbar, C. O., and Rodgers, J., 1957. Principles of Stratigraphy: Wiley, New York, 356 p.

Dunham, R. J., 1962, Classification of carbonate rocks according to depositional texture, in Ham, W. E., ed., Classification of carbonate rocks: American Association of Petroleum Geologists Memoir 1, p. 108-121.

Fisher, W. L., and McGowen, J. H., 1967, Depositional systems in the Wilcox Group of Texas and their relationship to occurrence of oil and gas: Transactions of the Gulf Coast Association of Geological Societies, v. 17, p. 105–125.

Fisher, W. L., Brown, L. F., Scott, A. J., and McGowen, J. H., 1969, Delta systems in the exploration for oil and gas: University of Texas Bureau of Economic Geology, 78 p.

Fisher, W. L., Gama, E., and Ojeda, H. A., 1973, Estratigrafia sismica e sistemas deposicionais da Formação Piaçabuçu, Sociedade Brasileira de Geologia, Anai do XXVII Congresso, v. 3, p. 123-133.

Fisher, W. L., McGowen, J. H., Brown, L. F., Jr., and Groat, C. G., 1972, Environmental geologic atlas of the Texas coastal zone — Galveston-Houston area: Bureau of Economic Geology, Texas.

Fisk, H. N., 1944, Geological investigations of the alluvial valley of the lower Mississippi River: U. S. Army Corps of Engineers, Mississippi River Commission, Vicksburg, Mississippi, 78 p.

Folk, R. L., 1962, Spectral subdivision of limestone types, in Ham, W. E., ed., Classification of carbonate rocks: American Association of Petroleum Geologists Memoir 1, p. 62-84.

Forgotson, J. M., 1957, nature, usage and definition of marker-defined vertically segregated rocks units: American Association of Petroleum Geologists Bulletin, v. 41, p. 2108–2113.

Frazier, D. E., 1967; Recent deltaic deposits of the Mississippi River— their development and chronology: Transactions of the Gulf Coast Association of Geological Societies, v. 17, p. 287–315.

Frazier, D. E., 1974, Depositional episodes: their relationship to the Quaternary stratigraphic framework in the northwestern portion of the Gulf Basin: Bureau of Economic Geology, University of Texas, Geological Circular 74–1, 26 p.

Frey, R. W., and Pemberton, S. G., 1984, Trace fossil facies models, in Walker, R. G., ed., Facies models, 2nd edition, Geoscience Canada Reprint Series 1, p. 189–207.

Friedman, G. M., and Sanders, J. E., 1978, Principles of Sedimentology: new York: John Wiley and Sons, 791 p.

Galloway, W. E., 1989, Genetic stratigraphic sequences in basin analysis I: Architecture and genesis of flooding-surface bounded depositional units: American Association of Petroleum Geologists Bulletin, v. 73, p. 125–142.

Gilbert, G. K., 1884, Ripple marks: Science, v. 3, p. 375–376.

Gilbert, G. K., 1895, Sedimentary measurement of geologic time: Journal of Geology, v. 3, p. 121–127.

Gilbert, G. K., 1899, Ripple marks and cross-bedding: Geological Society of America Bulletin, v. 10, p. 135–140.

Gilbert, G. K., 1914, The transportation of debris by running water: U. S. Geological Survey Professional Paper 86.

Ginsburg, R. N., and Lowenstam, H. A., 1958, The influence of marine bottom communities on the depositional environment of sediments: Journal of Geology, v. 66, p. 310–318.

Grabau, A. W., 1906, Types of sedimentary overlap. Geological Society of America Bulletin, v. 17, p. 567-636.

Gradstein, F. M., Ogg, J. G., and Smith, A. G., eds., 2004, A geologic time scale: Cambridge University Press, Cambridge, 610 p.

Gradstein, F. M., Ogg, J. G., Schmitz, M. D., and Ogg, G. M., 2012, The Geologic time scale 2012: Elsevier, Amsterdam, 2 vols., 1176 p.

Gradstein, F. M., Ogg, J. G., Schmitz, M. D., and Ogg, G. M., eds., 2020, Geologic time scale 2020: Elsevier, Amsterdam, 1357 p.

Gressly, A., 1838, Observations geologiques sur le Jura Soleurois: nouv. Mem. Soc. Helv. Sci. Nat. 2: 1-112.

Gurnis, M., 1988, Large-scale mantle convection and the aggregation and dispersal of supercontinents: Nature, v. 332, p. 695–699.

Gurnis, M., 1990, Bounds on global dynamic topography from Phanerozoic flooding of continental platforms: Nature, v. 344, p. 754–756.

Gurnis, M., 1992, Long-term controls on eustatic and epeirogenic motions by mantle convection: GSA Today, v. 2, p. 141–157.

Hancock, J. M., 1977, The historic development of biostratigraphic correlation, in Kauffman, E. G. and Hazel, J. E., eds., Concepts and methods of biostratigraphy: Dowden, Hutchinson and Ross Inc., Stroudsburg, Pennsylvania, p. 3-22.

Harms, J. C., and Fahnestock, R. K., 1965, Stratification, bed forms, and flow phenomena (with an example from the Rio Grande): Society for Sedimentary Geology (SEPM) Special Publication 12, p. 84–115.

Harms, J. C., Southard, J. B., Spearing, D. R., and Walker, R. G., 1975, Depositional environments as interpreted from primary sedimentary structures and stratification sequences: Society of Economic Paleontologists and Mineralogists Short Course 2, 161 p.

Hart, B. S., 2013, Whither seismic stratigraphy: Interpretation: A journal of subsurface characterization: Society of Exploration Geophysicists and American Association of Petroleum Geologists, v. 1, p. SA3-SA20.

Hedberg, H. D., 1948, Time stratigraphic classification of sedimentary rocks: Bulletin of the Geological Society of America, v. 59, p. 447–462.

Holbrook, J., and Miall, A. D., 2020, Time in the Rock: A field guide to interpreting past events and processes from siliciclastic stratigraphy: Earth Science Reviews, v. 203, 103121, 23 p.

Hunt, D., and Tucker, M. E., 1992, Stranded parasequences and the forced regressive wedge systems tract: deposition during base-level fall: Sedimentary Geology, v. 81, p. 1–9.

Illing, L. V., 1954, Bahaman calcareous sands: American Association of Petroleum Geologists, v. 38, p. 1–95.

Ingersoll, R. V., 1978a, Submarine fan facies of the Upper Cretaceous Great Valley Sequence, northern and central California: Sedimentary Geology, v. 21, p. 205–230.

Ingersoll, R. V., 1978b, Petrofacies and petrologic evolution of the late Cretaceous fore-arc basin, northern and central California: Journal of Geology, v. 86, p. 335–352.

Ingersoll, R. V., 1979, Evolution of the Late Cretaceous forearc basin, northern and central California; Geological Society of America Bulletin, v. 90, Pt. 1, p. 813–826.

James, N. P., and Dalrymple, R. W., eds., 2010, Facies Modesl 4: GEOtext 6, Geological Association of Canada, St. John's, Newfoundland, 586 p.

Jervey, M. T., 1988, Quantitative geological modeling of siliciclastic rock sequences and their seismic expression, in Wilgus, C. K., Hastings, B. S., Kendall, C. G. St. C., Posamentier, H. W., Ross, C. A., and Van Wagoner, J. C., eds., Sea level Changes - an integrated approach: Society of Economic Paleontologists and Mineralogists Special Publication 42, p. 47–69.

Jordan, T. E., 1981, Thrust loads and foreland basin evolution, Cretaceous, western United States: American Association of Petroleum Geologists Bulletin, v. 65, p. 2506-2520.

Jordan, T. E., 1995, Retroarc foreland and related basins, in Busby, C. J., and Ingersoll, R. V., eds., Tectonics of sedimentary basins: Blackwell Science, Oxford, p. 331–362.

Jordan, T. E., and Flemings, P. B., 1990, From geodynamic models to basin fill—a stratigraphic perspective, in Cross, T. A., ed., Quantitative dynamic stratigraphy: Prentice-Hall, Englewood Cliffs, p. 149–163.

Jordan, T. E., and Flemings, P. B., 1991, Large-scale stratigraphic architecture, eustatic variation, and unsteady tectonism: a theoretical evaluation: Journal of Geophysical Research, v. 96B, p. 6681–6699.

Kay, M., 1951, North American geosynclines: Geological Society of America Memoir 48.

Kim, W., and Paola, C., 2007, Long-period cyclic sedimentation with constant tectonic forcing in an experimental relay ramp: Geology, v. 35, p. 331–334.

Krumbein, W. C., 1939, Preferred orientation of pebbles in sedimentary deposits: Journal of Geology, v. 47, p. 673–706.

Krumbein, W. C., 1948, Lithofacies maps and regional sedimentary-stratigraphic analysis: American Association of Petroleum Geologists, v. 32, p. 1909–1923.

Krumbein, W. C. and Sloss, L. L., 1951, Stratigraphy and Sedimentation: Freeman, San Francisco, 497 p.

Krumbein, W. C., and Sloss, L. L., 1963, Stratigraphy and sedimentation (2nd. Edition): San Francisco, W. H. Freeman and Co., 660 p.

Kuenen, Ph. H., 1957, Review of Marine Sand-Transporting Mechanisms: Journal of the Alberta Society of Petroleum Geologists, v. 5, No. 4, p. 59–62

Kuenen, Ph. H., and Migliorini, C. I., 1950, Turbidity currents as a cause of graded bedding: Journal of Geology, v. 58, p. 91–127.

Leeder, M. R., 1993, Tectonic controls upon drainage basin development, river channel migration and alluvial architecture: implications for hydrocarbon reservoir development and characterization, in North, C. P. and Prosser, D. J., eds., Characterization of fluvial and aeolian reservoirs: Geological Society, London, Special Publication 73, p. 7–22.

Levorsen, A. I., 1943, Discovery thinking: American Association of Petroleum Geologists Bulletin, v. 27, p. 887–928.

Levorsen, A. I., 1954, Geology of petroleum: San Francisco, W. H. Freeman, 703 p.

Longwell, C. R., ed., 1949, Sedimentary facies in geologic history: Geological Society of America Memoir 39, 171 p.

Matthews, R. K., 1984, Dynamic Stratigraphy, second edition, Englewood Cliffs, New Jersey: Prentice-Hall, 489 p.

McConnell, R. G., 1887, Report of the geological structure of a portion of the Rocky Mountains, accompanied by a section measured near the 51st parallel: Annual Report, 1886, Geological Survey of Canada, p. 1D-41D.

McEachern, J. A., Pemberton, S. G., Gingras, M. K., and Bann, K. L., 2010, Ichnology and facies models, in James, N. P., and Dalrymple, R. W., eds., Facies Models 4: Geotext 6, Geological Association of Canada, p. 19–58.

McKee, E. D., 1957, Flume experiments on the production of stratification and cross-stratification: Journal of Sedimentary Petrology, v. 27, p. 129–134.

McKee, E. D., and Weir, G. W., 1953, Terminology for stratification in sediments: Geological Society of America Bulletin, v. 64, p. 381–389.

McKenzie, D. P., 1978, Some remarks on the development of sedimentary basins: Earth and Planetary Science Letters, v. 40, p. 25–32.

McLaren, D. J., 1970, Presidential address: time, life and boundaries: Journal of Paleontology, v. 44, p. 801–813.

McLaren, D. J., 1978, Dating and correlation: A review, in Cohee, G. V., Glaessner, M. F., and Hedberg, H. D., eds., Contributions to the geologic time scale: American Association of Petroleum Geologists, Studies in Geology 6, p. 1–7.

Merriam, D. F., ed., 1964, Symposium on cyclic sedimentation: Kansas Geological Survey Bulletin 169, 636 p.

Miall, A. D., 1973, Markov chain analysis applied to an ancient alluvial plain succession: Sedimentology, v. 20, p. 347–364.

Miall, A. D., ed., 1980, Facts and Principles of World Petroleum Occurrence: Canadian Society of Petroleum Geologists Memoir 6, 1003p.

Miall, A. D., 1984, Principles of sedimentary basin analysis: Berlin: Springer-Verlag, 490 p.

Miall, A. D., 1985, Architectural-element analysis: A new method of facies analysis applied to fluvial deposits: Earth Science Reviews, v. 22, p. 261–308.

Miall, A. D., 1988a, Reservoir heterogeneities in fluvial sandstones: lessons from outcrop studies: American Association of Petroleum Geologists Bulletin, v. 72, p. 682–697.

Miall, A. D., 1988b, Facies architecture in clastic sedimentary basins, in Kleinspehn, K., and Paola, C., eds., New perspectives in basin analysis: Springer-Verlag Inc., New York, p. 67–81.

Miall, A. D., 1991, Hierarchies of architectural units in terrigenous clastic rocks, and their relationship to sedimentation rate, in A. D. Miall and N. Tyler, eds., The three-dimensional facies architecture of terrigenous clastic sediments and its implications for hydrocarbon discovery and recovery: Society of Economic Paleontologists and Mineralogists, Concepts in Sedimentology and Paleontology, v. 3, p. 6–12.

Miall, A. D., 1995, Whither stratigraphy? Sedimentary Geology, v. 100, p. 5-20.

Miall, A. D., 1997, The geology of stratigraphic sequences, First edition: Springer-Verlag, Berlin, 433 p.

Miall, A. D., 1999, Principles of sedimentary basin analysis, Third edition: New York, N. Y.: Springer-Verlag Inc., 616 p.

Miall, A. D., 2004, Empiricism and model building in stratigraphy: the historical roots of present-day practices. Stratigraphy: American Museum of Natural History, v. 1, p. 3–25.

Miall, A. D., 2010, The geology of stratigraphic sequences, second edition: Springer-Verlag, Berlin, 522 p.

Miall, A. D., 2013, Sophisticated stratigraphy, in Bickford, M. E., ed., The web of geological sciences: Advances, impacts and interactions: Geological Society of America Special Paper 500, p. 169-190.

Miall, A. D., 2014, The emptiness of the stratigraphic record: A preliminary evaluation of missing time in the Mesaverde Group, Book Cliffs, Utah: Journal of Sedimentary Research, v. 84, p. 457-469.

Miall, A. D., 2015, Updating uniformitarianism: stratigraphy as just a set of "frozen accidents", in Smith, D. G., Bailey, R., J., Burgess, P., and Fraser, A., eds., Strata and time: Geological Society, London, Special Publication 404, p. 11–36.

Miall, A. D., Holbrook, J. M., and Bhattacharya, J. P., 2021, The Stratigraphy Machine: Journal of Sedimentary Research, in press.

Miall, A. D., and Miall, C. E., 2001, Sequence stratigraphy as a scientific enterprise: the evolution and persistence of conflicting paradigms: Earth Science Reviews, v. 54, #4, p. 321–348.

Miall, C. E., and Miall, A. D., 2002, The Exxon factor: the roles of academic and corporate science in the emergence and legitimation of a new global model of sequence stratigraphy: Sociological Quarterly, v. 43, p. 307–334.

Middleton, G. V., ed., 1965, Primary sedimentary structures and their hydrodynamic interpretation: Society of Economic Paleontologists and Mineralogists Special Publication 12, 265 p.

Middleton, G. V., 1966, 1967, Experiments on density and turbidity currents, I, II, and III, Canadian Journal of Earth Sciences, v. 3, p. 523–546, p. 627–637, v. 4, p. 475–505.

Middleton, G. V., 1973, "Johannes Walther's law of the correlation of facies": Geological Society of America Bulletin, v. 84, p. 979–988.

Middleton, G. V., 2005, Sedimentology, History, in Middleton, G. V., ed., Encyclopedia of sediments and sedimentary rocks, Berlin: Springer-Verlag, p. 628-635.

Middleton, G. V., 1966, 1967, Experiments on density and turbidity currents, I, II, and III, Canadian Journal of Earth Sciences, v.3, p. 523–546, p. 627–637, v. 4, p. 475–505.

Middleton, G. V., and Hampton, M. A., 1976, Subaqueous sediment transport and deposition by sediment gravity flows, in Stanley, D. J., and Swift, D. J. P., eds., Marine sediment transport and environmental management; Wiley, New York, p. 197-218.

Mitchell, A. H., and Reading, H. G., 1969, Continental margins, geosynclines and ocean floor spreading: Journal of Geology, v. 77, p. 629–646.

Mitchum, R. M., Jr., Vail, P. R., and Thompson, S. III, 1977a, Seismic stratigraphy and global changes of sea level, Part 2, The depositional sequence as a basic unit for stratigraphic analysis, in Payton, C. E., ed., Seismic stratigraphy—applications to hydrocarbon exploration: American Association of Petroleum Geologists Memoir 26, p. 53–62.

Moore, R. C., 1949, Meaning of facies, in Longwell, C., ed., Sedimentary facies in geologic history, Geological Society of America Memoir 39, p. 1-34.

Mossop, G. D., and Shetsen, I., compilers, 1994, Geological Atlas of the Western Canada Sedimentary Basin: Canadian Society of Petroleum Geologists, 510 p.

Muto, T., and Steel, R.J., 2004, Autogenic response of fluvial deltas to steady sea-level fall: Implications from flume-tank experiments: Geology, v. 32, p. 401–404.

Nanz, R. H., Jr., 1954, Genesis of Oligocene sandstone reservoir, Seeligson field, Jim Wells and Kleberg Counties, Texas: American Association of Petroleum Geologists Bulletin, v. 38, p. 96-117.

Newell, N. D., and Rigby, J. K., 1957, Geological studies on the Great Bahama Bank: regional aspects of carbonate deposition: Society of Economic Paleontologists and Mineralogists Special Publication 5, p. 15-72.

North American Commission on Stratigraphic Nomenclature, 1983, North American Stratigraphic Code: American Association of Petroleum Geologists Bulletin, v. 67, p. 841–875.

Oliver, T. A., and Cowper, N. W., 1963, Depositional environments of the Ireton Formation, central Alberta: Bulletin of Canadian Petroleum Geology, v. 11, p. 183–202.

Oliver, T. A., and Cowper, N. W., 1965, Depositional environments of the Ireton Formation, central Alberta: Bulletin of the American Association of Petroleum Geologists, v. 49, p. 1410–1425.

Paola, C., 2000, Quantitative models of sedimentary basin filling: Sedimentology, v. 47 (supplement 1) p. 121–178.

Paola, C., Mullin, J., Ellis, C., Mohrig, D. C., Swenson, J. B., Parker, G., Hickson, T., Heller, P. L., Pratson, L., Syvitski, J., Sheets, B., and Strong, N., 2001 Experimental stratigraphy: GSA Today, v. 11, #7, p. 4–9

Paola, C., Straub, K., Mohrig, D., and Reinhardt, L., 2009, The unreasonable effectiveness of stratigraphic and geomorphic experiments: Earth Science Reviews, v. 97, p. 1–43.

Payton, C. E., ed., 1977, Seismic stratigraphy—applications to hydrocarbon exploration: American Association of Petroleum Geologists Memoir 26, 516 p.

Pettijohn, F. J. 1949, Sedimentary rocks: New York, Harper and Bros, 513 p.

Pettijohn, F. J., 1962, Paleocurrents and paleogeography: American Association of Petroleum Geologists Bulletin, v. 46, p. 1468–1493.

Pettijohn, F. J., and Potter, P. E., 1964, Atlas and glossary of primary sedimentary structures: Springer-Verlag, New York, 370 p.

Pettijohn, F. J., Potter, P. E., and Siever, R., 1973, Sand and sandstone: Springer-Verlag, New York, 618 p.

Playfair, J., 1802, Illustrations of the Huttonian theory of the Earth, Dover Publications, New York 528 p (ed. G.W. White, 1956).

Plint, A. G., Walker, R. G., and Bergman, K. M., 1986, Cardium Formation 6. Stratigraphic framework of the Cardium in subsurface: Bulletin of Canadian Petroleum Geology, v. 34, p. 213-225.

Posamentier, H. W., 2000, Seismic stratigraphy into the next millennium: a focus on 3-D seismic data: American Association of Petroleum Geologists, Annual Conference, New Orleans, p. 16–19.

Posamentier, H. W., and Vail, P. R., 1988, Eustatic controls on clastic deposition II—sequence and systems tract models, in Wilgus, C. K., Hastings, B. S., Kendall, C. G. St. C., Posamentier, H. W., Ross, C. A., and Van Wagoner, J. C., eds., Sea level Changes - an integrated approach: Society of Economic Paleontologists and Mineralogists Special Publication 42, p. 125–154.

Posamentier, H. W., Jervey, M. T., and Vail, P. R., 1988, Eustatic controls on clastic deposition I—Conceptual framework, in Wilgus, C. K., Hastings, B. S., Kendall, C. G. St. C., Posamentier, H. W., Ross, C. A., and Van Wagoner, J. C., eds., Sea level Changes - an integrated approach: Society of Economic Paleontologists and Mineralogists Special Publication 42, p. 109–124.

Potter, P. E., 1959, Facies models conference: Science, v. 129, p. 1292–1294.

Potter, P. E., 1967, Sand bodies and sedimentary environments: A review: American Association of Petroleum Geologists Bulletin, v. 51, p. 337–365.

Potter, P. E., and Pettijohn, F. J., 1963, Paleocurrents and basin analysis: Springer-Verlag, Berlin, 296 p.

Potter, P. E., and Pettijohn, F. J., 1977, Paleocurrents and basin analysis, 2nd edition: Academic Press, San Diego, California, 296 p.

Price, R. A., 1973, Large-scale gravitational flow of supracrustal rocks, southern Canadian Rockies, in DeJong, K. A., and Scholten, R. A., eds., Gravity and tectonics: John Wiley, New York, p. 491–502.

Purdy, E. G., 1963, Recent calcium carbonate facies of the Great Bahama Bank: Journal of Geology, v. 71, p. 334–355, p. 472–497.

Ramos, A., and Sopeña, A., 1983, Gravel bars in low-sinuosity streams (Permian and Triassic, central Spain), in Collinson, J. D., and Lewin, J., eds., Modern and ancient fluvial systems: International Association of Sedimentologists Special Publication 6, p. 301–312.

Ramos, A., Sopeña, A., and Perez-Arlucea, M., 1986, Evolution of Buntsandstein fluvial sedimentation in the northwest Iberian Ranges (Central Spain): Journal of Sedimentary Petrology, v. 56, p. 862–875.

Reading, H. G., ed., 1978, Sedimentary environments and facies: Oxford: Blackwell, 557 p.

Reading, H. G., ed., 1986, Sedimentary environments and facies, second ed.: Blackwell, Oxford, 615 p.

Reading, H. G., ed., 1996, Sedimentary environments: processes, facies and stratigraphy, third edition: Blackwell Science, Oxford, 688 p.

Reiche, P., 1938, An analysis of cross-lamination: The Coconino Sandstone: Journal of Geology, v. 46, p. 905–932.

Reineck, H. E., and Singh, I. B., 1973, Depositional sedimentary environments—with reference to terrigenous clastics: Berlin: Springer-Verlag, 439 p.

Rich, J. L., 1951, Three critical environments of deposition and criteria for recognition of rocks deposited in each of them: Geological Society of America Bulletin, v. 62, p. 1–20.

Rigby, J. K., and Hamblin, W. K., eds., 1972, recognition of ancient sedimentary environments: Society of Economic Paleontologists and Mineralogists Special Publication 16, 340 p.

Rine, J. M., Helmold, K. P., Bartlett, G. A., Hayes, B. J. R., Smith, D. G., Plint, A. G., Walker, R. G., and Bergman, K. M., 1987, Cardium

Formation 6. Stratigraphic framework of the Cardium in subsurface: Discussions and reply: Bulletin of Canadian Petroleum Geology, v. 35, p. 362-374.

Ross, W. C., 1991, Cyclic stratigraphy, sequence stratigraphy, and stratigraphic modeling from 1964 to 1989: twenty-five years of progress?, in Franseen, E. K., Watney, W. L., and Kendall, C. G.. St.C., Sedimentary modeling: computer simulations and methods for improved parameter definition: Kansas Geological Survey Bulletin 233, p. 3-8.

Sadler, P. M., 1981, Sedimentation rates and the completeness of stratigraphic sections: Journal of Geology, v. 89, p. 569–584.

Salvador, A., ed., 1994, International Stratigraphic Guide, Second edition: International Union of Geological Sciences, Trondheim, Norway, and Geological Society of America, Boulder, Colorado, 214 p.

Schenk, H. G., and Muller, S. W., 1941, Stratigraphic terminology: Geological Society of America Bulletin, v. 52, p. 1419–1426.

Schreiber, B. C., Friedman, G. M., Decima, A., and Schreiber, E., 1976, Depositional environments of Upper Miocene (Messinian) evaporite deposits of the Sicilian Basin: Sedimentology, v. 23, p. 729–760.

Seibold, E and Seibold, I., 2002, Sedimentology: from single grains to recent and past environments: some trends in sedimentology in the twentieth century: in Oldroyd, D. R., ed., The Earth Inside and Out: Some major contributions to Geology in the twentieth century: Geological Society, London, Special Publication 192, p. 241-250.

Sheets, B. A., Hickson, T. A., and Paola, C., 2002, Assembling the stratigraphic record: depositional patterns and time-scales in an experimental alluvial basin: Basin Research, v. 14, p. 287–301.

Shepard, F. P., and Wanless, H. R., 1935, Permo-Carboniferous coal series related to southern hemisphere glaciation: Science, v. 81, p. 521–522.

Shepard, F. P., Phleger, F. B., and van Andel, T. H., eds., 1960, Recent sediments, northwest Gulf of Mexico, AmericanAssociation of Petroleum Geologists, 394 p.

Simons, D. B., and Richardson, E. V., 1961, Forms of bed roughness in alluvial channels: American Society of Civil Engineers Proceedings, v. 87, No. HY3, p. 87-105.

Simons, D. B., Richardson, E. V., and Nordin, C. F., 1965, Sedimentary structures generated by flow in alluvial channels, in Middleton, G. V., ed., Primary sedimentary structures and their hydrodynamic interpretation: Society of Economic Paleontologists and Mineralogists Special Publication 12, p. 34–52.

Sloss, L. L., 1963, Sequences in the cratonic interior of North America: Geological Society of America Bulletin, v. 74, p. 93–113.

Sloss, L. L., 1972, Synchrony of Phanerozoic sedimentary-tectonic events of the North American craton and the Russian platform: 24th International Geological Congress, Montreal, Section 6, p. 24-32.

Sloss, L. L., 1988, Tectonic evolution of the craton in Phanerozoic time, in Sloss, L. L., ed., Sedimentary cover—North American Craton: U.S.: The Geology of North America, Boulder, Colorado, Geological Society of America, v. D-2, p. 25–51.

Sloss, L. L., 1991, The tectonic factor in sea level change: a countervailing view: Journal of Geophysical Research, v. 96B, p. 6609–6617.

Sloss, L. L., Krumbein, W. C., and Dapples, E. C., 1949, Integrated facies analysis; in Longwell, C. R., ed., Sedimentary facies in geologic history: Geological Society of America Memoir 39, p. 91-124.

Sorby, H. C., 1852, On the oscillation of the currents drifting the sandstone beds of the southeast of Northumberland, and on their general direction in the coalfield in the neighbourhood of Edinburgh: Proceedings of the West Yorkshire Geological Society, v. 3, p. 232–240.

Sorby, H. C., 1859, On the structures produced by the currents present during the deposition of stratified rocks: The Geologist, v. II, p. 137–147.

Sorby, H. C., 1908, On the application of quantitative methods to the study of the structure and history of rocks. Quarterly Journal of the Geological Society, London, v. 64, p. 171-232.

Spychala, Y. T., 2020, About the (in)value of field work: Journal of Sedimentary Research, v. 90, p. 102–103.

Steckler, M. S., and Watts, A. B., 1978, Subsidence of the Atlantic-type continental margin off New York: Earth and Planetary Science Letters, v. 41, p. 1–13.

Steel, R. J., and Milliken, K. L., 2013, Major advances in siliciclastic sedimentary geology, 1960-2012, in Bickford, M. E., ed., The web of geological sciences: Advances, impacts and interactions: Geological Society of America Special Paper 500, p. 121-167.

Steno, Nicolaus, 1669, De Solido intra Solidum naturaliter contento dissertationis prodromus, 78 p. Florence.

Stokes, W. L., 1945, Primary lineation in fluvial sandstones: a criterion of current direction: Journal of Geology, v. 45, p. 52–54.

Strangway, D. W., ed., 1980, The continental crust and its mineral deposits: Geological Association of Canada Special Paper 20.

Strong, N., and Paola, C., 2008, Valleys that never were: time surfaces versus stratigraphic surfaces: Journal of Sedimentary Research, v. 78, p. 579–593.

Tankard, A. J., and Welsink, H. J., 1987, Extensional tectonics and stratigraphy of Hibernia oil field, Grand Banks, Newfoundland: American Association of Petroleum Geologists Bulletin, v. 71, p. 1210-1232.

Teichert, C., 1958, Concepts of facies: Bulletin of the American Association of Petroleum Geologists, v. 42, p. 2718–2744.

Torrens, H. S., 2002, Some personal thoughts on stratigraphic precision in the twentieth century, in Oldroyd, D. R., ed., The Earth inside and out: some major contributions to geology in the twentieth century: Geological Society, London, Special Publication 192, p. 251-272.

Vail, P. R., 1975, Eustatic cycles from seismic data for global stratigraphic analysis (abstract): American Association of Petroleum Geologists Bulletin, v. 59, p. 2198–2199.

Vail, P. R., Audemard, F., Bowman, S. A., Eisner, P. N., and Perez-Crus, C., 1991, The stratigraphic signatures of tectonics, eustasy and sedimentology—an overview, in Einsele, G., Ricken, W., and Seilacher, A., eds., Cycles and events in stratigraphy: Springer-Verlag, Berlin, p. 617–659.

Vail, P. R., Mitchum, R. M., Jr., Todd, R. G., Widmier, J. M., Thompson, S., III, Sangree, J. B., Bubb, J. N., and Hatlelid, W. G., 1977, Seismic stratigraphy and global changes of sea-level, in Payton, C. E., ed., Seismic stratigraphy - applications to hydrocarbon exploration: American Association of Petroleum Geologists Memoir 26, p. 49-212.

Van Siclen, D. C., 1958, Depositional topography—examples and theory: American Association of Petroleum Geologists Bulletin, v. 42, p. 1897–1913.

Van Straaten, L.M. J. U., 1954, Composition and structure of recent marine sediments in the Netherlands: Leidse Geol. Mededel., v. 19, p. 1-110.

Van Wagoner, J. C., Mitchum, R. M., Campion, K. M. and Rahmanian, V. D. 1990, Siliciclastic sequence stratigraphy in well logs, cores, and outcrops: American Association of Petroleum Geologists Methods in Exploration Series 7, 55 p.

Veeken, 2007, Seismic stratigraphy, basin analysis and reservoir characterization: Elsevier, Amsterdam, Seismic Exploration, v. 37, 509 p.

Visher, G. S., 1965, Use of vertical profile in environmental reconstruction; American Association of Petroleum Geologists Bulletin: v. 49, p. 41-61.

Waddell, H., 1933, Sedimentation and sedimentology: Science, v. 77, p. 536–537.

Walker, R. G., 1965, The origin and significance of the internal sedimentary structures of turbidites: Proceedings of the Yorkshire Geological Society, v. 35, p. 1–29.

Walker, R. G., 1967, Turbidite sedimentary structures and their relationship to proximal and distal depositional environments: Journal of Sedimentary Petrology, v. 37, p. 25–43.

Walker, R. G., 1973, Mopping up the turbidite mess, in Ginsburg, R. N., ed., Evolving concepts in sedimentology: Johns Hopkins University Press, Baltimore, p. 1–37.

Walker, R. G., 1976, Facies models 1. General Introduction: Geoscience Canada, v. 3, p. 21-24.

Walker, R. G., ed., 1979, Facies models: Geoscience Canada Reprint Series No. 1, 211 p.

Walker, R. G., 1990, Facies modeling and sequence stratigraphy: Journal of Sedimentary Petrology, v. 60, p. 777–786.

Walker, R. G., and James, N. P., ed., 1984, Facies Models, second edition: Geoscience Canada Reprint Series 1, 317 p.

Walker, R. G. and James, N. P., eds., 1992, Facies models: response to sea-level change: Geological Association of Canada, 409 p.

Walther, Johannes, 1893–1894, Einleisung in die Geologie alshistorische Wissemchaft: Jena, Verlag von Gustav Fischer, 3 vols., 1055 p.

Wanless, H. R., and Shepard, E. P., 1936, Sea level and climatic changes related to Late Paleozoic cycles: Geological Society of America Bulletin, v. 47, p. 1177–1206.

Wanless, H. R., and Weller, J. M., 1932, Correlation and extent of Pennsylvanian cyclothems: Geological Society of America Bulletin, v. 43, p. 1003–1016.

Watts, A. B., 1981, The U. S. Atlantic margin: subsidence history, crustal structure and thermal evolution: American Association of Petroleum Geologists, Education Course Notes Series #19, Chap. 2, 75 p.

Watts, A. B., 1989, Lithospheric flexure due to prograding sediment loads: implications for the origin of offlap/onlap patterns in sedimentary basins: Basin Research, v. 2, p. 133–144.

Watts, A. B., and Ryan, W. B. F., 1976, Flexure of the lithosphere and continental margin basins: Tectonophysics, v. 36, p. 24–44.

Weller, J. M., 1960, Stratigraphic Principles and Practice: Harper, New York, 725 p.

Wernicke, B., 1985, Uniform-sense normal simple shear of the continental lithosphere: Canadian Journal of Earth Sciences, v. 22, p. 108–125.

Wheeler, H. E., 1958, Time-stratigraphy: American Association of Petroleum Geologists Bulletin, v. 42, p. 1047–1063.

Wheeler, H. E., 1959, Stratigraphic units in time and space: American Journal of Science, v. 257, p. 692–706.

White, N., and McKenzie, D., 1988, Formation of the "steers' head" geometry of sedimentary basins by differential stretching of the crust and mantle: Geology, v. 16, p. 250–253.

Whitten, E. H. T., 1964, Process-response models in geology: Geological Society of America Bulletin, v. 75, p. 455–464.

Wilgus, C. K., Hastings, B. S., Kendall, C. G. St. C., Posamentier, H. W., Ross, C. A., and Van Wagoner, J. C., eds., 1988, Sea level changes - an integrated approach: Society of Economic Paleontologists and Mineralogists Special Publication 42, 407 p

Wilson, J. L., 1975, Carbonate facies in geologic history: Springer-Verlag, New York, 471 p.

Woodford, A. O., 1973, Johannes Walther's Law of the Correlation of Facies: Discussion: Geological Society of America Bulletin, v. 84, p. 3737–3740.

Wright, V. P., 2019, Memes, false news, and the death of empiricism: Journal of Sedimentary Research, v. 89, p. 310–311.

The Stratigraphic-Sedimentologic Data Base

2

Contents

Abstract

This chapter details the types of observation that are made on sedimentary rocks in the field, and in the subsurface, including drill cuttings and cores. Field observations include lithology, petrology, bedding types, sedimentary structures, and trace and body fossils. Petrophysical (wireline) logging techniques are also described. Numerous photographs are used to illustrate the field characteristics of sedimentary rocks.

2.1 Introduction

A successful basin analysis requires the collection and integration of several, perhaps many, different kinds of data. Direct observation of the rocks may or may not be fundamental to the study. In the case of a surface geological project, they will be preeminent, though perhaps supplemented by geochemical and geophysical information, plus laboratory analysis of collected samples. For subsurface petroleum studies, actual rock material available for examination may be very limited, consisting of well cuttings from rotary drilling, plus a few short cores. Petrophysical well logs and regional seismic lines may provide at least as important a part of the total data base. Investigations for stratabound ores and minerals typically employ networks of diamond drill holes from which a continuous core normally is available. This provides a wealth of material for analysis, although certain types of observation, such as analysis of sedimentary structures, may be difficult or impossible in such small-diameter cores.

In this chapter, we discuss the collection and description of stratigraphic and sedimentologic data from outcrops and drill holes. Advanced workers should supplement this chapter with a review of the methods of documentation and analysis discussed by Holbrook and Miall (2020).

The stratigraphic record is characterized by three-dimensional complexity, on all scales from the thin section to the basin fill. Depending on the objective of the project, from local to regional, it will be necessary to measure and describe several or many sections and document the stratigraphic links between them. On a regional or reconnaissance scale there are many mapping methods with which to accomplish this, as described in Chap. 6. Regional mapping methods, such as the use of air-photo reconnaissance, biostratigraphy, reflection-seismic data, or sequence mapping, necessarily introduce some possibility of imprecision or error, and wherever possible, it is advantageous to literally "walk out" correlations between sections by following individual horizons from outcrop to outcrop or, in the subsurface, make effective use of electrofacies and "wiggle tracing" of petrophysical logs (Sects. 2.4.2, 6.2). The recognition and tracing of key marker beds may facilitate this work. The kinds of detailed modern studies described in Chap. 8 were all carried out in areas where extensive outcrop or a dense data base of subsurface sections permitted highly detailed correlations between sections. As described in that chapter, very detailed stratigraphic studies, informed by recent ideas about rates and styles of sedimentary preservation are leading to important new breakthroughs about the nature of the stratigraphic record.

2.2 Describing Surface Stratigraphic Sections

Vertical stratigraphic sections, whether measured at the surface or derived from subsurface records, constitute the single most important data set that the basin analyst should assemble. Lithostratigraphic or sequence-stratigraphic classification and correlation, and many sedimentological interpretations, depend on the documentation of vertical relationships within and between lithological units.

2.2.1 Methods of Measuring and Recording the Data

2.2.1.1 Vertical Stratigraphic Sections

The simplest way to record the details of a surface outcrop is by measuring and describing a vertical stratigraphic section. Ideally, the location of the section should be chosen to include important stratigraphic features, such as formation contacts, but, in practice, the location is commonly determined by accessibility, e.g., the presence of bars or beaches allowing us to walk along a river cut, or a negotiable gully cutting through a cliff section. In some cases, especially where units demonstrate considerable local variability in thickness or facies, the collection of the necessary data for correlation may require rock-climbing skills to work across steep or vertical outcrop faces.

In reconnaissance work, rapid measurement and description techniques are acceptable. For example, a hand-held altimeter (aneroid barometer) may be used in conjunction with dip measurements to reconstruct stratigraphic thicknesses using simple trigonometry. Another method that is commonly described in field handbooks is the pace-and-compass technique, suitable for estimating thicknesses across relatively level ground, given accurate stratigraphic dip. The same distances may be measured from maps or air photographs. Long experience with these methods has shown that they are not very reliable; errors of up to 50% can be expected.

By far the simplest and most accurate method for measuring a section is the use of a Jacob's staff or "pogo stick." The stick is constructed of a 1.5 m wooden rod, with a clinometer and sighting bar (Fig. 2.1). The clinometer is preset at the measured structural dip and can then be used to measure stratigraphic thickness as fast as the geologist can write down descriptive notes. The best technique is to use two persons. The senior geologist observes the rocks and makes notes, while the junior (who can be an inexperienced student) "pogo's" his or her way up the section recording increments of 1.5 m on a tally counter and collecting samples. The length 1.5 m is convenient for all but the tallest or

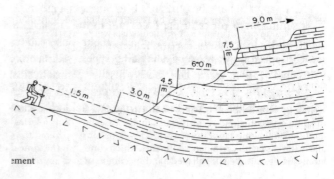

Fig. 2.1 Use of pogo stick in section measurement

shortest persons, although it can be awkward to manipulate on steep slopes. The only skill required by the pogo operator is the ability to visualize the dip of the strata in three dimensions across whatever terrain the geologist may wish to traverse. This is important so that the pogo can at all times be positioned perpendicular to bedding with the line of sight extending from the sighting bar parallel to bedding.

Another field method, to be used for large cliff outcrops, is to work from photomosaics or LIDAR output. The geologist works across the face using climbing equipment, making observations and measurements and conveying these by radio to an assistant who sits at a distance, where he/she can see the geologist, marking the observations on the photograph. Stratigraphic thicknesses and other spatial data may be calculated from a scaled photograph or using LIDAR (light detection and ranging) data.

It is far preferable to measure up a stratigraphic section rather than down, even though this often means an arduous climb up steep slopes. Many geologists working by helicopter in rugged terrain have made their traverses physically easier by working downhill wherever possible. But not only is it difficult to manipulate a pogo stick downward in a section, it makes it more difficult for the geologist to comprehend the order of events he or she is observing in the outcrops. It is often convenient to measure a section in separate increments, making use of the most accessible talus slopes or gullies, and traversing laterally along a prominent stratigraphic surface to an adjacent accessible area wherever necessary (Fig. 2.2).

The geologist should search for the cleanest face on which to make observations. Normally, this should be weathered and free of vegetation, talus, or rain wash. Most sedimentary features show up best where they have been etched out by wind or water erosion, or where a face is kept continuously clean and polished by running water, as in a river bed or an intertidal outcrop. Such features rarely show up better on fresh fracture surfaces, so a hammer should only be used for taking samples. Carbonates may benefit from etching with dilute acid. The geologist should methodically examine both vertical cuts and the topside and underside of bedding planes; all may have something to reveal, as described later in this chapter. It is also useful, on larger outcrops, to walk back and examine them from a distance, even from a low-flying helicopter or from a boat offshore, or photograph them with a drone, as this may reveal large-scale channels, facies changes, and many other features of interest. A different technique may be used to document such large outcrops, which may conveniently be termed lateral profiling to distinguish it from vertical profiling. We discuss this in the next section.

In the interests of maximum efficiency, the geologist obviously should ensure that all the necessary measurements, observations and sample collections are made during the first visit to a section. It may be useful to carry along a checklist, for reference each time a lithologic change in the section requires a new bed description. Many geologists have attempted to carry this process one step further by designing the checklist in the form of a computer processible data card, or they record the data in the form of a computer code (e.g., Alexander-Marrack et al. 1970; Friend et al. 1976). There may be two problems with this:

Fig. 2.2 Measuring a vertical stratigraphic section in rugged terrain. The section may be offset along a selected stratigraphic surface in order to facilitate ease of access

formation/sequence boundaries

—— segment of measured vertical section

·········· along-strike traverse

1. If attempts are made to record every piece of field information on the computer file, the resulting file is likely to be very large and cumbersome. Storage and retrieval programming may consume far more time than the original fieldwork, unless the geologist can draw on some preexisting program package. This leads to the second problem.

2. If data are to be coded in the field or if a preexisting program system is to be used, it means that decisions will already have been made about what data are to be recorded and how they are to be recorded before the geologist goes into the field. If the geologist knows in advance what is likely to be found, as a result of some previous descriptions or reconnaissance work, this may be satisfactory, but in the case of isolated field areas such foreknowledge may not be available. In this case, there is a certain risk involved in having the observation system designed in advance.

Individual geologists vary in their interests and in the observations they make and, of course, the rocks are highly variable, so that it is not possible to design a single, all-purpose, section-measuring software package. Specialized systems have to be designed for specific projects, and although this leads to expense in programming and debugging, it means that the program can be designed for the specific type of output required. It should not be forgotten, though, that programming, as such, is a technical, not a scientific, skill. Students and other workers who write programs for their research get few points for producing a workable program, only for the interesting scientific ideas their programming allows them to test.

2.2.1.2 The Construction of Lateral Profiles

Some stratigraphic units are essentially tabular at the scale of the outcrop, and can be quickly and accurately documented using vertical profiles, in the way outlined in the previous section. However, some types of sedimentary assemblage contain complex facies changes, which may be at a small enough scale to observe in individual outcrops, especially in large outcrops. For example, a reef core, with its reef-front talus slope and back-reef lagoonal deposits, or a large fluvial or submarine-fan channel, with its fill of complex bar deposits, may be spectacularly displayed in a road cut or mountainside. The measurement of a few vertical sections across such an outcrop is a quite inadequate way to document the wealth of facies detail that may be available. Petroleum companies have developed a considerable interest in such large outcrops because of their use as potential analogues of subsurface reservoir units. Internal heterogeneities of these reservoir analogues can be studied and their porosity and permeability characteristics studied, for example, by the use of minipermeameters (e.g., Miall and Tyler 1991; Doyle and Sweet 1995).

In order to document the details in a large outcrop it may be necessary to construct a lateral profile, a long section that encompasses the full vertical stratigraphic height of the outcrop, and also extends along strike as far as possible, to illustrate the facies changes. This may be constructed by careful surveying, but a much quicker method is to make use of photographic mosaics of the outcrop. The geologist moves well back from the exposed face, perhaps as much as several hundred meters in the case of a very large outcrop, and carries out a traverse parallel to the face, taking a series of overlapping frames until he or she has covered the entire outcrop. By taking care to remain at the same distance from the face, each frame will be at approximately the same scale. The same end can be achieved by taking photographs from a boat, or even from an aircraft flying low over the outcrop. Unmanned drones are ideal platforms for cameras to take carefully positioned and oriented images.

The images are carefully overlapped in order to construct a mosaic, which can then be used in the field as a kind of topographic map base on which to enter stratigraphic and sedimentologic detail (Fig. 2.3). Some scale distortions inevitably arise. Outcrops are rarely flat, and projections and gullies will not fit together precisely in the mosaic because of the differing perspectives of adjacent frames. Such distortions may be trivial relative to the immense amount of detail that can be shown on such profiles, and if accurate measurements of individual features are required, they should, in any case, be made in the field and not from the photograph.

Modern digital methods, such as the use of LIDAR (light detection and ranging), may make the preparation of the outcrop image for interpretation easier (Hodgetts 2013). This method makes use of laser surveying to record the image, and because the range of each sample point is also recorded, the image can be processed to reduce or eliminate parallax problems, and may then also be rotated to position the image, for example, to remove a structural tilt or to simulate a view parallel to stratigraphic dip.

Techniques for documenting and interpreting lateral profiles are discussed further in Sect. 3.3.4.

2.2.2 Types of Field Observation

2.2.2.1 Subdivision of the Section into Descriptive Units

This is a subjective operation based on the rock types present, the quality and accessibility of the exposure, and the amount of detail required in the description. Very detailed descriptions may require subdivision into units containing (for example) a single mudstone lens or crossbed set, and will therefore be on the order of a few centimeters or tens of centimeters thick. Thicker units can be defined by grouping similar rock types, but sedimentologically useful detail may

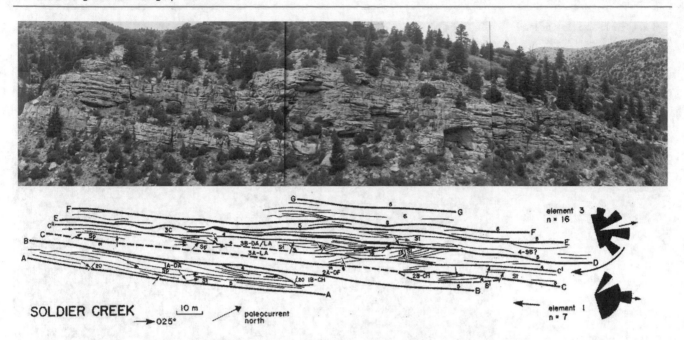

Fig. 2.3 An example of a large outcrop (Upper Cretaceous fluvial Castlegate Sandstone, Utah), with annotation, below. Shown are major bedding surfaces, paleocurrent measurements and letter codes for major bedding units and bounding surfaces

be lost thereby. For each unit, the kinds of observations listed in the succeeding paragraphs are made where appropriate.

In the case of lateral profiles (e.g., Fig. 2.3), among the most valuable kinds of observation that can be made is the documentation and classification of the various kinds of bounding surface that separate stratigraphic units. These range from the simple bedding-plane surfaces that separate individual crossbed sets through the surfaces that bound channels and bars to the major (usually horizontal) surfaces that delimit mappable stratigraphic units (formations, members, stratigraphic sequences, etc.). A discussion and classification of bounding surfaces is presented in Sect. 3.5.11.

2.2.2.2 Lithology and Grain Size

Lithologic classification of clastic rocks can usually be done satisfactorily by visual observation in the field, without the necessity of follow-up laboratory work. Classification is based on grain size (Fig. 2.4), which is easily measured on the outcrop. For sand-grade rocks, it is useful to take into the field a grain-size chart (Fig. 2.5) or a set of sand samples each representing one phi class interval through the sand size ranges. These are used for comparison purposes and permit recognition of the main sand-grade subdivisions: very fine, fine, medium, coarse, and very coarse. Many tests by the author and others have shown that such observations provide adequate, accurate information on the modal size range of the sandstones.

Sorting is also an important descriptive criterion. The description should be modified by appropriate adjectives if sorting characteristics require it, for example, pebbly coarse-grained sandstone, silty mudstone, etc. For the purpose of regional facies analysis, this is usually the only kind of grain-size information required. Some examples of the use of sorting adjectives are shown in Fig. 2.6. The distinction between **well sorted** and **poorly sorted** refers to the distribution of grain sizes within an individual bed. For example, eolian dune sands (Fig. 2.6a) and many beach sands exhibit a limited range of grains sizes because of current winnowing, whereas many fluvial deposits are rapidly deposited and may consist of mixtures of a range of grain sizes (Fig. 2.6b, d) although this is not invariably the case, as shown by the well-sorted sandstone in Fig. 2.6d. Likewise, debris-flow conglomerates are typically very poorly sorted (Fig. 2.6c, e, f).

Siltstone and mudstone can be distinguished in a hand specimen by the presence or absence of a gritty texture, as felt by the fingers or the tongue. This is, of course, a crude method, and should be checked by making thin sections of selected samples. However, field identifications of this type commonly are adequate for the purpose of facies analysis.

Fine-grained rocks, including those consisting of a mixture of sandstone, siltstone, and mudstone, are difficult to classify and describe. Dean et al. (1985) discussed the methods used in the Deep Sea Drilling Project based on smear slides of soft sediments made on board ship.

For conglomerates, maximum clast size is often a useful parameter to measure. Typically, this is estimated by taking the average of the 10 largest clasts visible within a specified region of an outcrop, such as a given area of a certain bedding plane. In thick conglomerate units, it may be useful

PARTICLE DIAMETER
m mm Φ
CLASTICS
CARBONATES
allochems matrix
10^0 2048 -11 v. large boulders v. coarse calcirudite extremely coarsely crystalline
1024 -10 large
512 -9 medium
256 -8 small
10^{-1} 128 -7 large cobbles
64 -6 small
32 -5 v. coarse pebbles coarse calcirudite
10^{-2} 16 -4 coarse
8 -3 medium m. calcirudite
4 -2 fine
2 -1 v. fine f. calcirudite v.c.crystalline
gravel
10^{-3} 1 0 v. coarse sand
0.5 +1 500 coarse c. calcarenite c. crystalline
0.25 +2 250 medium m. calcarenite
10^{-4} 0.125 +3 125 fine f. calcarenite m. crystalline
0.0625 +4 62 v. fine v. f. calcarenite
0.0312 +5 31 v. coarse silt c. calcilutite f. crystalline
10^{-5} 0.0156 +6 16 coarse m. calcilutite
0.0078 +7 8 medium f calcilutite v. f. crystalline
0.00390 +8 4 fine v. f. calcilutite
0.00195 +9 2 v. fine clay/mud aphano-crystalline
micrometres (μm)

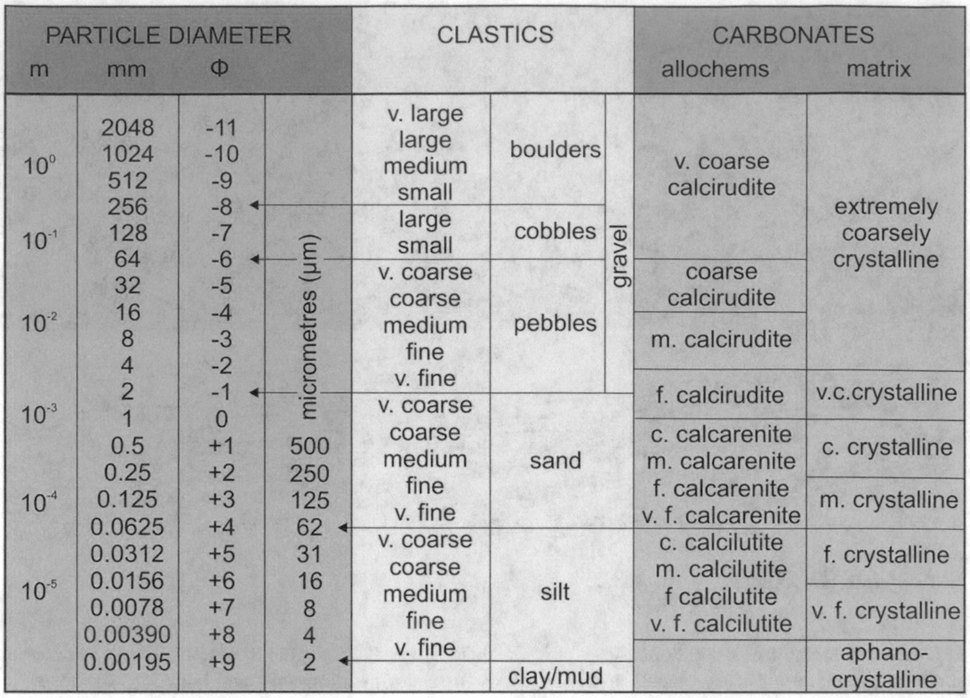

Fig. 2.4 Grain-size classification of clastic and carbonate rocks

mm 1 2 3 4 5 6 7 8
Imperial Oil
Esso
vcU = 1410 – 2000 u = – 0.5 – – 1.0ø
vcL = 1000 – 1410 u = 0.0 – – 0.5ø
cU = 710 – 1000 u = 0.5 – 0.0ø
cL = 500 – 710 u = 1.0 – 0.5ø
mU = 350 – 500 u = 1.5 – 1.0ø
mL = 250 – 350 u = 2.0 – 1.5ø
fU = 177 – 250 u = 2.5 – 2.0ø
fL = 125 – 177 u = 3.0 – 2.5ø
vfU = 88 – 125 u = 3.5 – 3.0ø
vfL = 62 – 88 u = 4.0 – 3.5ø
Angular
Subangular
Subrounded
Rounded
Well Rounded

Fig. 2.5 A chart for estimating grain size in the field

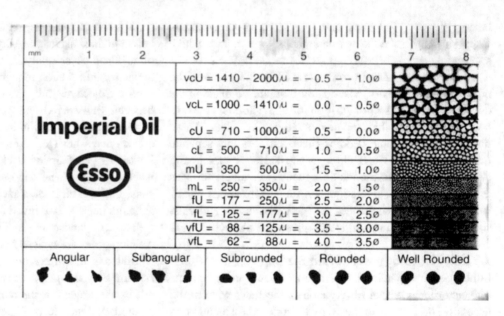

to repeat such measurements over regular vertical intervals of the section. It is also important to note the degree of sorting, clast shape and roundness, matrix content, and fabric of conglomerate beds. For example, does the conglomerate consist predominantly of very well-rounded clasts of approximately the same size, or is it composed of angular fragments of varying size and shape (breccia)? Do the clasts "float" in abundant matrix, a rock type termed **matrix-supported conglomerate**, or do the clasts rest on each other with minor amounts of matrix filling the interstices—a **clast-supported conglomerate**? These features are discussed at length in Chap. 4.

Carbonate rocks commonly cannot be described adequately or accurately in outcrops, and require description from thin sections or polished sections observed under a low-power microscope. Among the reasons for this are the ready susceptibility of carbonate rocks to fine-scale diagenetic change, and the fact that weathering behavior in many cases obscures rather than amplifies such changes, as seen in outcrops. Another important reason for not relying on outcrop observation is that some of the types of information required for carbonate facies analysis are simply too small to be seen properly with the naked eye. These include mud content, certain sedimentary textures and biogenic features.

Fig. 2.6 Sorting and grading in sandstones and conglomerates

Field geologists traditionally take a dropper bottle of 10% hydrochloric acid with them to test for carbonate content and to aid in distinguishing limestone from dolomite (on the basis of "fizziness"). However, for research purposes, the test is quite unsatisfactory, and the geologist is advised to abandon the acid bottle (and stop worrying about leakage corroding equipment and clothing). Dolomite commonly can be distinguished from limestone by its yellowish weathering color in the field, but a better field test is to use alizarin red-S

in weak acid solution. This reagent stains calcite bright pink but leaves dolomite unstained. In both hand specimens and thin sections use of this reagent can reveal patterns of dolomitization on a microscopic scale.

Because of the problem with carbonate rocks discussed previously, the geologist is advised not to rely on field notes for facies analysis of these rocks, but to carry out a rigorous sampling program and supplement (and correct) the field notes using observations made on polished slabs or thin sections. Sampling plans are discussed later in this section. Laboratory techniques for studying carbonates are described by Wilson (1975, Chap. 3).

Evaporates are difficult to study in surface outcrops. They are soft and recessive and commonly poorly exposed, except in arid environments. Like carbonates, they are highly susceptible to diagenetic change, so that field observations must be supplemented by careful laboratory analysis.

Mixed carbonate-clastic sediments are common and are typically dealt with as if they were carbonate or clastic, which may not be the most effective way to emphasize subtle lithologic characteristics. Mount (1985) discussed the problems of classifying these rocks and suggested some methodological approaches.

2.2.2.3 Porosity

Porosity and permeability are of particular interest if the rocks are being studied for their petroleum potential. Observations in surface outcrops may be of questionable value because of the effects of surface weathering on texture and composition, but the geologist should always break off a fresh piece of the rock and examine the fracture surface because such observations commonly constitute the only ones made. The geologist should distinguish the various types of porosity, such as intergranular (in detrital rocks), intercrystalline (in chemical rocks), and larger pores, such as vugs, bird's-eye texture, molds of allochems, such as oolites or pellets, fossil molds, fracture porosity, etc. (e.g., Fig. 4.53). Porosity types should be reported in terms of the estimated percentage they occupy in the bulk volume of the sample.

More accurate observations may be made from thin sections, and samples may be submitted to a commercial laboratory for flow tests if required. Measurements of relative permeability are now routinely made on outcrop profiles of reservoir analogues using minipermeametry equipment.

2.2.2.4 Color

Color may or may not be an important parameter in basin analysis. Individual lithologic units may display a very distinctive color, which aids in recognition and mapping. Sometimes it even permits a formation to be mapped almost entirely using helicopter observations from the air, with a minimum of ground checking. However, the sedimentological meaning and interpretation of color may be difficult to resolve.

Some colors are easily interpreted—sandstones and conglomerates commonly take on the combined color of their detrital components, pale grays and white for quartzose sediments, pinks for feldspathic sandstones and darker colors for lithic rocks. As noted, limestones and dolomites may also be distinguished using color variations. However, color is strongly affected by depositional conditions and diagenesis, particularly the oxidation–reduction balance. Reduced sediments may contain organically derived carbon and Fe^{2+} compounds, such as sulfides, imparting green or drab gray colors. Oxidized sediments may be stained various shades of red, yellow, or brown by the presence of Fe^{3+} compounds such as hematite and limonite. However, local reducing environments, such as those created around decaying organisms, may create localized areas or spots of reduction color. Color can change shortly after deposition, as shown for example by T. R. Walker (1967) and Folk (1976). Moberly and Klein (1976) found that oxidation and bacterial action cause permanent color changes when fresh sediments, such as deep-sea cores, are exposed to the air. Leaching by groundwaters can also drastically change formation colors.

Thus, the problem is to decide how much time to devote to recording color in the field. Ideally, each descriptive unit in the stratigraphic section should be studied for color using a fresh rock-fracture surface and comparisons to some standard color scheme, such as the U.S. National Research Council Rock-Color Chart (Goddard et al. 1948). In practice, for the purpose of facies and basin analysis, such precision is not required. Simple verbal descriptions, such as pale gray, dark red-brown, etc., are adequate. More precise descriptions may be useful if detailed studies of diagenetic changes are to be undertaken, but recent work has shown that such studies may give misleading results if carried out exclusively on surface exposures because of the effects of recent weathering (Taylor 1978).

2.2.2.5 Bedding

An important type of observation, particularly in clastic rocks, is the thickness of bedding units. Thickness relates to rate of environmental change and to depositional energy. In some cases, bed thickness and maximum grain size are correlated, indicating that both are controlled by the capacity and competency of single depositional events. Bed-thickness changes may be an important indicator of cyclic changes in the environment, and sedimentologists frequently refer to **thinning-upward** and **fining-upward** or **coarsening-and-thickening-upward cycles**. It is important to distinguish bedding from weathering characteristics. For example, a unit may split into large blocks or slabs upon weathering, but close examination may reveal faint internal bedding or lamination not emphasized by weathering. Bedding can be measured and recorded numerically, or it can be described in field notes semi-quantitatively (Fig. 2.7) using the descriptive classification given in Table 2.1.

2.2.2.6 Inorganic Sedimentary Structures

Sedimentary structures include a wide variety of primary and post-depositional features (Table 2.2). All individually yield useful information regarding depositional or diagenetic events in the rocks, and all should be meticulously recorded and described in the context of the lithology and grain size of

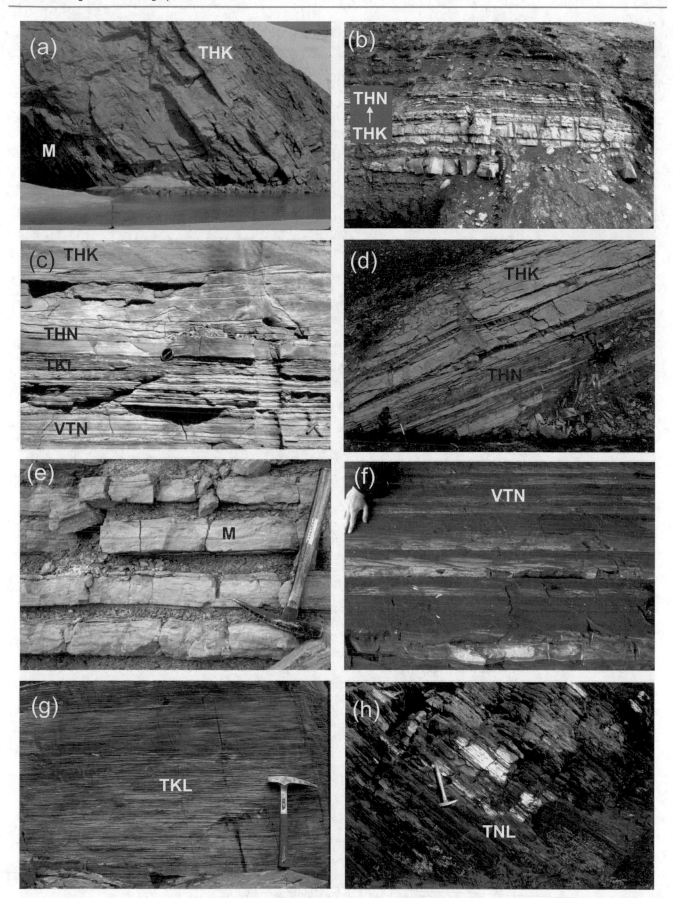

Fig. 2.7 Bedding and lamination. See Table 2.1 for key to abbreviations

Table 2.1 Table of stratification thicknesses

Descriptor	Range	Fig. 2.7
Very thickly bedded	>1 m	
Thickly bedded	30–100 cm	THK
Medium bedded	10–30 cm	M
Thinly bedded	3–10 cm	THN
Very thinly bedded	1–3 cm	VTN
Thickly laminated	0.3–1 cm	TKL
Thickly laminated	<0.3 cm	TNL

Table 2.2 Classification of inorganic sedimentary structures

I. Hydrodynamic structures
  A. By mass gravity transport ·
  Graded bedding
      a. Normal
      b. Reversed
  B. By noncohesive flow
    1. Lamination
    2. Cross-lamination (amplitude < 5 cm).
    3. Crossbedding (amplitude > 5 cm)
    4. Clast imbrication
    5. Primary current lineation
    6. Fossil orientation
II. Hydrodynamic erosion of the bed
  A. Macroscopic
    1. Scours
    2. Channels
    3. Low-relief erosion surfaces
  B. Mesoscopic
    1. Intraformational breccias
    2. Hardgrounds
    3. Lag concentrates
    4. Flutes
    5. Tool markings
    6. Rain prints
III. Liquefaction, load, and fluid loss structures
  A. Load cast
  B. Flame structures
  C. Ball, pillow, or pseudonodule structures
  D. Convolute bedding
  E. Syndepositional faults and slumps
  F. Growth faults
  G. Deformed crossbedding
  H. · Dish and pillar structures
  I. Sand volcanoes
  J. Injection features
    1. Dikes
    2. Mud lumps
    3. Diapirs
  K. Synaeresis features
  L. Desiccation cracks
  M. Ice and evaporite crystal casts
  N. Gas bubble escape marks
  O. Teepee structures
  P. Ptygmatic/enterolithic/chicken wire gypsum/anhydrite

the bed in which they occur. The assemblage of structures and, in some cases, their orientation can yield vital paleogeographic information.

Inorganic sedimentary structures can be divided into three main genetic classes, as shown in Table 2.2. These are described briefly in order to aid their recognition in the field, but a discussion of their origin and interpretation is deferred to later chapters. Useful texts on this subject include Potter and Pettijohn (1977), Allen (1982), Harms et al. (1982), and Collinson and Thompson (1982). Ashley (1990) carried out an important overview of bedform types and their names, and proposed a unification of concepts and terminology (referred to briefly below) that has received universal acceptance.

Sediment carried in turbulent suspension by mass gravity-transport processes, such as debris flows and turbidity currents, is subjected to internal sorting processes. When the flow slows and ceases, the sorting may be preserved as a distinct texture termed **graded bedding**. Grading commonly consists of an upward decrease in grain size, as illustrated in Fig. 2.6; this is termed **normal grading**. However, certain sedimentary processes result in an upward increase in grain size, termed **inverse grading**.

Clastic grains can be divided into two classes on the basis of their interactive behavior. Cohesive grains are those that are small enough that they tend to be bound by electrostatic forces and thus resist erosion once deposited on a bed. This includes the clay minerals and fine silt particles. A range of erosional sedimentary structures is present in such rocks (Table 2.2), as discussed later. Larger clastic grains, including siliciclastic, evaporite and carbonate fragments, of silt to cobble size, are **noncohesive**. They are moved by flowing water or wind as a traction carpet along the bed, or by intermittent suspension. The dynamics of movement causes the grains to be mounded into a variety of **bedforms,** which are preserved as **crossbedding** within the rock (Figs. 2.8, 2.9).

There are three main classes of bedforms and crossbedding found in ancient rocks:

1. Those formed from unidirectional water currents such as are found in rivers and deltas, and oceanic circulation currents in marine shelves and the deep sea.
2. Those formed by oscillatory water currents, include both wave- and tide-generated features. Although the time scale of current reversal is, of course, quite different, there are comparable features between the structures generated in these different ways.
3. Those formed by air currents. Such currents may be highly variable, and the structure of the resulting deposits will be correspondingly complex. However, examination of ancient wind-formed (eolian) rocks indicates some consistent and surprisingly simple patterns (Chaps. 4, 5).

Recognition in an outcrop or in a core of the diagnostic features of these crossbedding classes is an invaluable aid to environmental interpretation (as discussed at length in Chap. 4), and therefore crossbedding structures must be examined and described with great care wherever they are found.

Attributes of crossbedding and examples are illustrated in Figs. 2.8, 2.9, 2.10 and 2.11. A **foreset** represents an avalanche face, down which grains roll or slump or are swept down by air or water currents. Continuous deposition produces repeated foreset bedding or lamination as the bedform accretes laterally, resulting in a crossbed **set** (Fig. 2.10; McKee and Weir 1953). A **coset** is defined as a sedimentary unit made up of two or more sets of strata or crossbedding separated from other cosets by surfaces of erosion, nondeposition or abrupt change in character (McKee and Weir 1953). Note that a coset can contain more than one type of bedding.

When describing crossbedding, attention must be paid to seven attributes (Fig. 2.11). All but the last of these were first described in detail in an important paper on crossbedding classification by Allen (1963a). In the field, crossbeds are classified first according to whether they are solitary or grouped. Solitary sets are bounded by other types of bedding or crossbedding, grouped sets are cosets consisting entirely of one crossbed type. Scale is the next important attribute. In water-laid strata, it is found that a bedform amplitude of about 5 cm is of hydrodynamic significance (Allen 1982; Ashley 1990) and, accordingly, this amplitude is used to subdivide crossbeds into small- and large-scale forms. An assumption is made that little or none of the top of a bedform is lost to erosion prior to burial; generally, the amount lost seems to increase in approximate proportion to the scale of the bedform or the thickness of the crossbed structure. Forms thinner than 5 cm are termed **ripples**, whereas forms larger than 5 cm have been given a variety of names, reflecting in part a diversity of hydrodynamic causes and in part a considerable terminological confusion. Ashley (1990) proposed several recommendations for a simplification of the terminology that have become widely adopted. Most forms larger than ripples are now termed **dunes**. This will be discussed further in Chap. 4.

Most crossbed sets contain foresets that terminate at the base of the set, in which case the foresets are said to be discordant. In rare cases where the crossbeds are parallel to the lower bounding surface, as occurs in some sets with curved lower surfaces, the crossbeds are described as concordant.

The crossbeds may show either homogeneous or heterogeneous lithology. Homogeneous crossbeds are those composed of foresets whose mean grain size varies by less than two phi classes. Heterogeneous (or heterolithic) crossbeds

Fig. 2.8 Examples of cross-stratification in outcrop and drill core. **a** Wave ripples in shallow-water dolomite; **b** Ripples and climbing ripples in glaciofluvial outwash; **c** Planar crossbedding in drill core; **d** Trough crossbedding; flow direction toward the right; **e** Low-angle crossbedding; flow direction toward the left

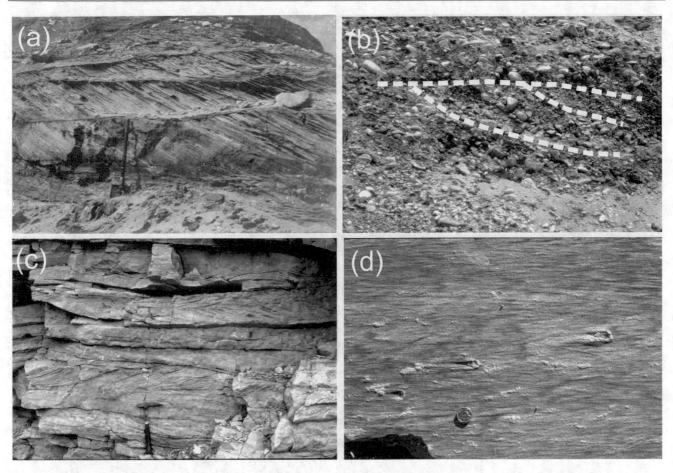

Fig. 2.9 Examples of types of crossbedding. **a** Planar crossbedding in sandstone; **b** Planar crossbedding in modern river gravel; **c** Herringbone crossbedding (crossbed dip direction reverses 180° from set to set); **d** view of the underside of a bed showing parting lineation with scour hollows around small pebbles. Flow direction was from right to left

Fig. 2.10 Terminology for stratified and cross-stratified (crossbedded) units (McKee and Weir, 1953)

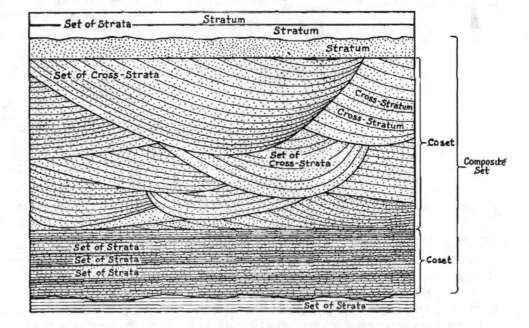

Fig. 2.11 Criteria used in the description and definition of crossbedding types (Allen, 1963a)

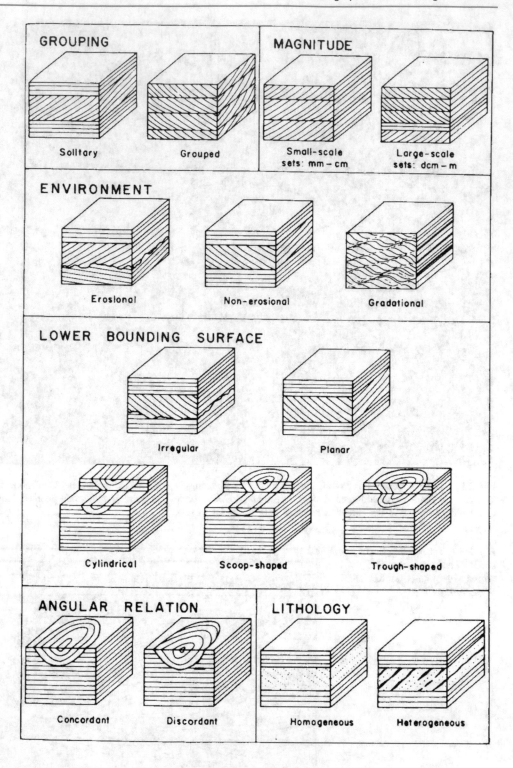

may contain laminae of widely varying grain size, including interbedded sand and mud or sand and gravel, possibly even including carbonaceous lenses.

The minor internal structures within crossbeds are highly diagnostic of their origin. The dip angle of the foreset relative to the bounding surface is of considerable dynamic significance. Are the foresets curved, linear or irregular in sections parallel to the dip? Is the direction of dip constant, or are there wide variations or reversals of dip within a set or between sets? Do the sets contain smaller scale hydrodynamic sedimentary structures on the foresets, and if so, what is the dip orientation of their foresets relative to that of the larger structure? What is the small-scale internal geometry of the foresets—are they tabular, lens or wedge shaped? Do

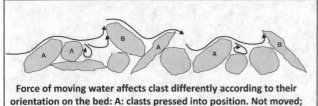

Force of moving water affects clast differently according to their orientation on the bed: A: clasts pressed into position. Not moved; B: clasts unstable, liable to be flipped over.

Fig. 2.12 Class imbrication. Top: imbrication fabric in a modern river. Currents flowed from left to right. Bottom: how turbulent currents generate the imbricated fabric

they display other kinds of sedimentary structures, such as trace fossils, synsedimentary faults or slumps? Are **reactivation surfaces** present? These are minor erosion surfaces on bedforms that were abandoned by a decrease in flow strength and then reactivated at some later time.

These attributes can be used to classify crossbed sets in the field. It is time-consuming to observe every attribute of every set, but it is usually possible to define a limited range of crossbed types that occur repeatedly within a given stratigraphic unit. These can then be assigned some kind of local unique descriptor, enabling repeated observations to be recorded rapidly in the field notebook. Modern methods of facies classification that encompass variations in crossbed type are discussed in Sect. 2.3.3.

A vital component of basin analysis is an investigation of sedimentological trends, such as determining the shape and orientation of porous rock units. **Paleocurrent analysis** is one of several techniques for investigating sedimentary trends based, among other attributes, on studying the size, orientation and relative arrangement of crossbedding structures. Therefore, when describing outcrop sections, it is essential to record the orientation of crossbed sets. The procedures for doing this and the methods of interpretation are described in Sect. 6.7.

Crossbedding represents a macroscopic orientation feature, but each clastic grain is individually affected by a flow system and may take up a specific orientation within a

deposit in response to flow dynamics. The longest dimension of elongated particles tends to assume a preferred position parallel or perpendicular to the direction of movement and is commonly inclined upflow, producing an imbricated or shingled fabric. This fabric may be present in sand-sized grains and can be measured optically, in thin section (Martini, 1971) or using bulk properties such as dielectric or acoustic anisotropy (Sippel, 1971). In recent years, paleomagnetic data have also been recognized to contain much useful information relating to primary sedimentary fabrics. Eyles et al. (1987) discussed magnetic orientation and anisotropy data with reference to the depositional processes of till and till-like diamict deposits. Oriented specimens must be collected in the field for such an analysis (see Sect. 2.2.3.2). In conglomerates, an **imbrication** fabric commonly is visible in an outcrop and can be readily measured by a visual approximation of average orientation or by laborious individual measurements of clasts. Figure 2.12 illustrates imbrication in a modern river bed. As discussed in Sect. 6.7.2, it has been found that in nonmarine deposits, in which imbrication is most common, the structure is one of the most accurate of paleocurrent indicators (Rust 1975).

Grain sorting is responsible for generating another type of fabric in sand-grade material, which is also an excellent paleocurrent indicator. This is **primary current lineation**, also termed **parting lineation** because it occurs on bedding-plane surfaces of sandstones that are flat bedded and usually readily split along bedding planes. An example is illustrated in Fig. 2.9d. Primary current lineation is the product of a specific style of water turbulence above a bed of cohesionless grains, as are the various bedforms that give rise to crossbedding (Sect. 3.5.4). It therefore has a specific hydrodynamic meaning and is useful in facies analysis as well as paleocurrent analysis.

Rather than the bed itself, objects such as plant fragments, bones or shells may be oriented on a bedding plane. This should be observed if possible, but interpretation commonly is not easy, as discussed in Sect. 6.7.

2.2.2.7 Sedimentary Structures Produced by Hydrodynamic Erosion of the Bed

A wide variety of erosional features is produced by water erosion of newly deposited sediment. These result from changes in water level or water energy in response to floods, storms, tides or wind-driven waves and currents. They can also result from evolutionary change in a system under steady equilibrium conditions. These processes result in the development of various types of **bounding surfaces** in the rocks (Figs. 2.13). Recognition and plotting of these features in outcrop sections are important components of facies analysis, and with adequate exposure, orientation studies may contribute significantly to the analysis of depositional trends.

Fig. 2.13 Examples of bounding surfaces exposed in outcrop

These features range in size up to major **river** and **tidal channels**, **submarine canyons**, and **distributary channels** several kilometers across and tens or hundreds of meters deep, but large features such as these can rarely be detected in the average small outcrop. They may be visible in large outcrop sections, where they can be documented using lateral profiles (Sect. 3.5.11) and on seismic sections (Chap. 6), and it may be possible to reconstruct them by careful lithostratigraphic correlation and facies analysis of scattered outcrops, but this is beyond the scope of our immediate discussion. At the outcrop scale, there are two types of small-scale erosional features to discuss, those that truncate one or more bedding units and those that scour or pit the bedding plane without significantly disrupting it.

The first type includes **channels**, **scours**, low-relief **erosion surfaces** and **rill markings**, in decreasing order of scale. These may be classified as macroscopic erosion features (Table 2.2). Channels and scours are usually filled by sediment that is distinctly different in grain size and bedding characteristics from that into which the channel is cut. Almost invariably the channel fill is coarser than the eroded strata indicating, as might be expected, that the generation of the channel was caused by a local increase in energy level. Exceptions are where a channel was abandoned and subsequently filled by fine sediment and coal under low-energy conditions.

It is a common error to confuse trough crossbedding with channels. Troughs are formed by the migration of trains of dunes or sinuous-crested megaripples (Sect. 3.5.4). They rest on curved scour surfaces, but these are not channels. The scours are formed by vortex erosion in front of the advancing dunes and are filled with sediment almost immediately. Channels, on the other hand, may not be filled with sediment for periods ranging from hours to thousands of years after the erosion surface is cut, and so the cutting and filling of the channel are quite separate events.

Erosion surfaces may exhibit little erosional relief, which may belie their importance. In the nonmarine environment, sheet erosion, wind deflation and pedimentation can generate virtually planar erosion surfaces. In subaqueous environments, oceanic currents in sediment-starved areas, particularly in abyssal depths, can have the same result. Exposed carbonate terrains may develop **karst** surfaces, with the formation of extensive cave systems. At the outcrop scale, careful examination of erosion surfaces may reveal a small-scale relief and the presence of features such as infilled **desiccation cracks**, basal intraformational or extrabasinal **lag gravels**, fissures filled with sediment from the overlying bed, zones of bioclastic debris, etc. In some subaerial environments, soil or weathering profiles may have developed, including the development of **caliche** or **calcrete**, and the presence of surfaces of nondeposition. In carbonate environments, surfaces of nondeposition commonly develop subaqueously. **Hardgrounds** are organically bored surfaces that may be encrusted with fossils in growth positions (Fig. 2.14). Alternatively, they may be discolored by oxidation, giving a red stain, or blackened by decayed algal matter. Long-continued winnowing of a surface of nondeposition may leave **lag concentrates** or **condensed sections** consisting of larger particles, blackened by algal decay, and possibly including abundant phosphatized fossil material (Wilson, 1975, pp. 80–81). Condensed sections may be generated by rapid transgression, which forces coastal sediment sources to undergo retreat (Chap. 5). In continental-slope deposits, giant slumps and slides are common and are particularly well exposed as **intraformational truncation surfaces** in deep water carbonate sediments.

It may be difficult to assess the length of time missing at erosion surfaces. Some may even represent major time breaks detectable by biostratigraphic zonation. In any case, a careful search for and description of such features in the field is an important part of section description. The recognition and mapping of erosion surfaces is a critical element in the study of sequence stratigraphy, as discussed at greater length in Chap. 6. As further discussed in Chaps. 7 and 8, modern chronostratigraphic documentation of sedimentation rates is indicating that a significant proportion of the elapsed time during which a stratigraphic succession accumulated may be represented by hiatuses and surfaces of nondeposition.

Mesoscopic erosional features fall mainly into a class of structure termed **sole markings**. These are features seen on the underside of bedding planes, usually in sandstones, and they represent the natural casts of erosional features cut into the bed below, which is typically siltstone or mudstone. They attest to the erosive power of the depositional event that formed the sandstone bed, but beyond this, most have little facies or environmental significance. However, they can be invaluable paleocurrent indicators. **Tool markings** are a class of sole structure formed by erosional impact of large objects entrained in the flow, including pebbles, plant fragments, bone or shell material. The many varieties that have been observed have been assigned names that indicate the interpreted mode of origin. They include groove, drag, bounce, prod, skip, brush and **roll markings**. A few examples are illustrated in Fig. 2.15, which show the strongly linear pattern on the bed, providing excellent paleocurrent indicators.

Flute markings are formed by vortex erosion, typically at the base of turbidity currents, although they have also been observed at the base of fluvial channels (Fig. 2.16). Erosion is deepest at the up-current end of the scour and decreases down current, so that in a flute cast the high-relief nose of the cast points up-current. In rare examples, vortex flow lines may be perceived in the walls of the flute. Flutes generally are in the order of a few centimeters deep.

Fig. 2.14 Structures formed during periods of intraformational nondeposition or erosion

Fig. 2.15 Sole marks on the underside of a sandstone bed. These are groove casts formed at the base of a turbidity current

2.2.2.8 Liquefaction, Load and Fluid Loss Structures

Clay deposits saturated with water are characterized by a property termed **thixotropy**: when subjected to a sudden vibration, such as that generated by an earthquake, they tend to liquify and lose all internal strength. This behavior is responsible for generating a variety of structures in clastic rocks. Clay beds commonly are interbedded with sand or silt and, when liquefied, the coarser beds have a higher density than the clay and tend to founder under gravity. This may or may not result in the disruption of the sand units. Where complete disruption does not take place, the sand forms bulbous shapes projecting into the underlying clay, termed **load structures** (Fig. 2.17). These may be well seen in ancient rocks by examining the underside of a sandstone bed. They are therefore a class of sole structure, though one

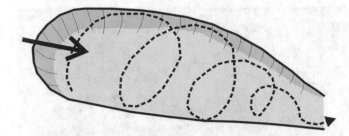

Fig. 2.16 Flute marks. Diagram at top illustrates the formation of flutes by vortex erosion at the base of a high-energy turbid flow. Middle: view of underside of bed showing flutes and scour around a pebble. Bottom: flutes at the base of a fluvial channel sandstone

produced without current movement. Clay wisps squeezed up between the load masses form pointed shapes termed **flame structures** because of a resemblance to the shape of flames. These are best seen in cross-section. Occasionally, loading may take place under a moving current, such as a turbidity flow, and load structures may then be stretched, possibly by shear effects, into linear shapes paralleling the direction of movement.

Commonly, load masses become completely disrupted. The sand bed may break up into a series of ovate or spherical masses that sink into the underlying bed and become surrounded by mud. Lamination in the sand is usually preserved in the form of concave-up folds truncated at the sides or tops of each sand mass, attesting to the fragmentation and sinking

of the original layer. Various names have been given to these features, including **ball**, **pillow** or **pseudonodule structures** (Fig. 2.17). These structures rarely have a preferred shape or orientation and are not to be confused with slump structures, which are produced primarily by lateral rather than vertical movement.

In many environments, sediment is deposited on a sloping rather than a flat surface (large-scale examples of this are visible on seismic-reflection surveys and are called **clinoforms**—see Chaps. 5 and 6), for example, the subaqueous front of a delta, which is something like a very large-scale foreset built into a standing body of water. The difference is that once deposited such material usually does not move again as individual grains because the angle of the slope is too low. Typically, large-scale submarine fans and deltas exhibit slopes of less than 2°. However, the sediment in such environments is water saturated and has little cohesive strength. Slopes may therefore become oversteepened, and masses of material may be induced to slump and slide downslope by shock-induced failure. Undoubtedly, the thixotropic effects described previously facilitate this process. The result is the production of internal shear or glide surfaces and deformed masses of sediment, termed **slump structures**. Failure surfaces may be preserved as **syndepositional faults**. Some examples of **convolute bedding** may be produced this way, although others are the result of water escape, as discussed below.

Structures produced by failure and lateral movement commonly retain an internal orientation with a simple geometric relationship to the orientation of the depositional slope. This could include the elongation of slump masses and the orientation (strike) of slide surfaces, both parallel to depositional strike, or the asymmetry, even overturning, of folds in convolute beds. Recognition of these geometric properties in an outcrop is important because it helps distinguish the structures from those of different origin, and orientation characteristics obviously have potential as paleoslope indicators.

Very large-scale slumps and syndepositional faults are developed on major deltas. The latter are termed **growth faults. Olisthostromes** are giant slumps developed on tectonically active continental slopes.

Deformed or **overturned crossbedding** (Fig. 2.18) is developed in saturated sand beds by the shearing action of water or turbid flow across the top of the bedform. The upper few centimeters of the crossbedded unit move down-current by a process of intragranular shear, and foreset lamination is overturned as a result, producing an up-current dip. Obviously, to produce this structure the shearing current must have a similar orientation to that of the current that generated the crossbedding. Deformed crossbedding is common in fluvial and deltaic environments. As additional sediment is laid on top of saturated deposits, grains within the substrate

Fig. 2.17 Load structures, in the form of balls, pillows and flame structures of mud trapped between them. Note the deformed lamination. The example (below) is from a turbidite succession

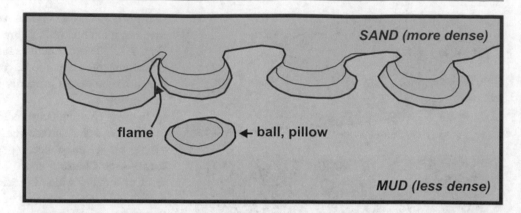

begin to settle and pack more tightly. Pore waters are expelled in this process and move upward or laterally to regions of lower hydrostatic pressure. Eventually, they may escape to the surface. This process may take place slowly if sediment is being deposited grain by grain and the fluid movement leaves little or no impression on the sediments.

If loading is rapid, a much more energetic process of fluid loss takes place, and the sediment itself will be moved around in the process. The result, in sand-grade deposits, is a group of features called **dish-and-pillar structures** (Fig. 2.19c, d, e). Dishes are produced by escaping water breaking upward through a lamination and turning up the edges; pillars record the vertical path of water flow moving to the surface. Dishes may be up to 50 cm in diameter. These structures are particularly common in the deposits of sediment-gravity flows, such as fluidized flows, in which sediment emplacement is rapid. They are produced by water escape as a flow ceases movement, the loss of the lubricating effect of water itself being one of the main reasons why the

flow stops. However, dish-and-pillar structures have also been observed in fluvial and other deposits and are therefore not environmentally diagnostic. Obviously, they can only be seen in deposits containing lamination and will not be present if the sand is uniform in texture.

Fluid movement within a bed is an additional cause of convolute lamination. In this case, the laminations are folded by internal shear and may occasionally be broken through by pipes (Fig. 2.19f). At the sediment–water interface, escaping water may bubble out as a small spring, building up a miniature **sand volcano** (Fig. 2.19g).

The emplacement of relatively more dense material over a lower density layer was discussed previously as the cause of load casts. In addition to the downward movement of the denser material, this situation can be the cause of upward movement of lighter sediment, which is injected, together with contained pore fluids, into the overlying rocks. This can occur on a large scale, producing **diapirs** of evaporite or mud, both of which flow readily under the overburden

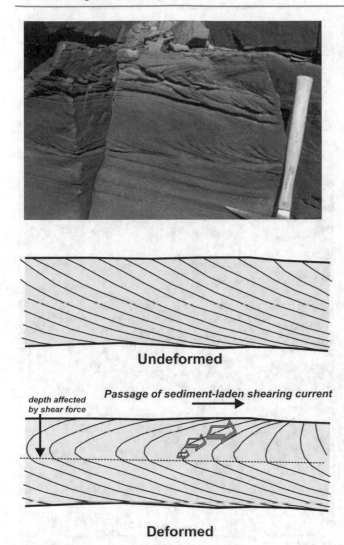

Fig. 2.18 The development of deformed or overturned crossbedding

weight of a few hundred meters of sediment. These diapirs may be several kilometers across and may extend up for several kilometers through overlying deposits. Evaporite diapirs commonly develop on continental margins. Mud diapirs are a characteristic feature of deltas, where coarser deltaic sediment is dumped rapidly on marine mud deposits. Pore fluids may be sealed in the sand beds by this process, with the resulting high overburden pressures leading to overpressuring. Exploration drilling into such deposits must include the use of blow-out preventers to guard against the explosive release of pore fluids and gases when an over-pressured bed is penetrated by the drill.

On a smaller scale, the same injection process can generate **clastic dikes**, consisting of sheets of sandstone or conglomerate (siliciclastic or carbonate) cutting through overlying or underlying beds (Fig. 2.20). The host rocks usually are sharply truncated and not internally deformed, indicating that they were at least partially lithified prior to

intrusion. Some dikes intrude along fault planes. Some are intensely folded, suggesting deformation by compaction and further dewatering after injection.

Desiccation cracks are readily recognized by even the untrained eye (Fig. 2.21). They are one of the best and most common indicators of subaerial exposure in the rock record (note the spelling of "desiccation," a word that is almost invariably misspelled by students). They may penetrate as deep as a few meters into the underlying rocks (although a few centimeters is more typical) and are normally filled by sediment from the overlying bed. **Teepee structures** are a variety of large desiccation cracks caused by limestone or evaporite expansion on tidal flats. Desiccated beds on tidal flats may peel or curl as they dry, and disrupted fragments commonly are redeposited nearby as an **intraclast breccia** (Fig. 2.14). A subtly different kind of shrinkage feature, termed as **synaeresis structure,** may be distinguished from desiccation cracks by two principal differences: unlike desiccation cracks they do not normally form continuous polygonal networks across bedding planes, but may appear as a loose assemblage of small worm-like relief markings on a bedding surface; second, they do not show deep penetration into the substrate, but appear to rest on the bedding plane in which they are found. Synaeresis structures are common in such environments as lakes, lagoons and tidal pools, where salinity changes may be frequent. As noted in Sect. 3.5.7, these features were thought to be subaqueous shrinkage cracks, but have now been reinterpreted as gypsum pseudomorphs.

Evaporation and freezing may cause the development of crystals of evaporite salts and water, respectively, on the depositional surface, particularly on alluvial floodplains, supratidal flats, and lake margins. Evaporite crystals and nodules may be preserved in the rock record and, of course, major evaporite deposits are common, but individual crystals commonly are replaced by pseudomorphs, or are dissolved and the cavity filled with silt or sand, forming a **crystal cast**. Such structures are a useful indicator of subaerial exposure and desiccation, but do not necessarily imply long-term aridity. Gypsum and halite are the two commonest minerals to leave such traces. Gypsum forms blade-shaped casts and halite characteristic cubic or "hopper-shaped" structures. Ice casts may be formed in soft sediments during periods of freezing, but have a low preservation potential.

Particularly distinctive evaporite structures on supratidal (sabkha) flats are termed **ptygmatic, enterolithic**, and **chicken-wire structures**. These are caused by in-place crystal growth, expansion and consequent lateral compression of evaporite nodules, possibly aided by slight overburden pressures. Enterolithic structure is so named for a resemblance to intestines; chicken-wire structure is caused by squeezing of carbonate films between the nodules;

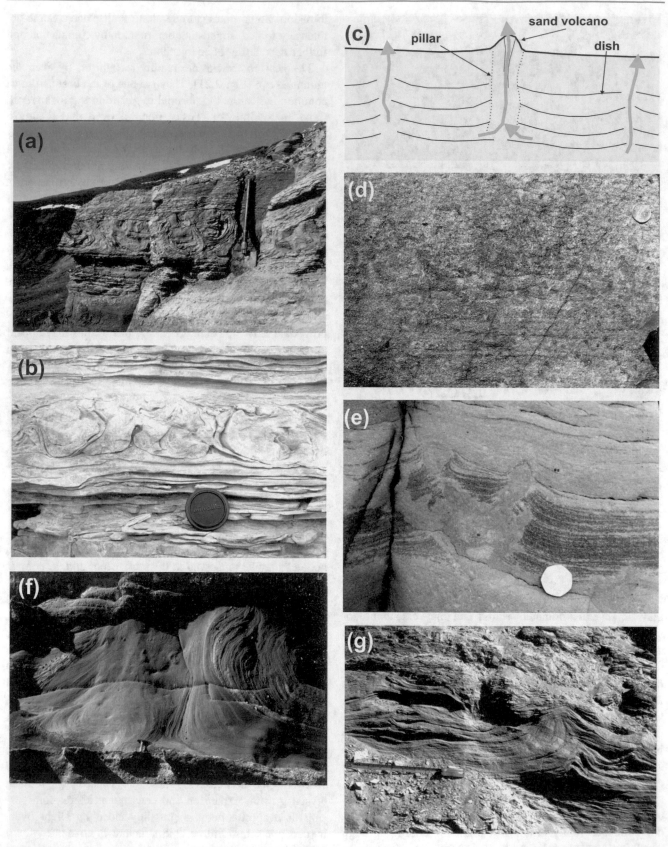

Fig. 2.19 a, b convolute bedding generated by syndepositional collapse or gravity sliding; **c**: mechanism of formation of dish-and-pillar structures; **d, e** examples of dish-and-pillar structures; **f** water escape structure formed by sediment loading; **g**: sand volcano formed by water escape at the depositional surface

Fig. 2.20 Clastic dykes

Fig. 2.21 Desiccation cracks, in bedding plane view (left) and cross-section view (right)

ptygmatic folds may be caused largely by overburden pressures (Maiklem et al. 1969).

Lastly, **gas bubble escape structures** should be mentioned. These are produced by carbon dioxide, hydrogen sulfide or methane escaping from buried, decaying organic matter. Gas passes up through wet, unconsolidated sediment and forms bubbles at the sediment–water interface, leaving small pits on the bedding surface. These structures have been confused with **rain-drop imprints**, but form subaqueously, as may be apparent from associated features preserved in the rocks. True rain-drop imprints probably are very rare.

Mills (1983) provided a discussion and review of this class of structures.

2.2.2.9 Fossils

Body fossils are obviously among the most powerful environmental indicators to be found in sedimentary rocks and should be observed and identified with care. Paleontology and paleoecology are specialized subjects, a detailed discussion of which is beyond the scope of this book. However, those engaged in describing outcrop sections for basin-analysis purposes should be able to make use of such information as they can gather. A complete and thorough paleontological-paleoecological examination of an outcrop section may take several hours, days or even weeks of work, involving the systematic examination of loose talus and breaking open fresh material or sieving unconsolidated sediment in the search for a complete suite of fossil types. Extensive suites of palynomorphs or microfossils may be extracted by laboratory processing of field samples. Many apparently unfossiliferous or sparsely fossiliferous stratigraphic intervals have been found to contain a rich and varied fauna or flora by work of this kind, but it is the kind of research for which few basin analysts have the time or inclination.

We are concerned in this book with reconstructing depositional environments and paleogeography. Fossils can be preserved in three different ways that yield useful environmental information.

1. In-place life assemblages include invertebrate forms attached to the sea bottom, such as corals, archaeocyathids, rudists, some brachiopods and pelecypods in growth positions, some bryozoa, stromatoporoids (Fig. 2.22), stromatolites and trees. In-place preservation usually is easy to recognize by the upright position of the fossil and presence of roots, if originally part of the organism. This type of preservation is the easiest to interpret because it permits the drawing of close analogies with similar modern forms, in the knowledge that the fauna or flora almost certainly is an accurate indicator of the environment in which the rocks now enclosing it were formed.

2. Environmental indicators almost as good as in-place life assemblages are examples of soft-bodied or delicately articulated body fossils preserved intact. These indicate very little transport or agitation after death, and preservation in quiet waters, such as shallow lakes, lagoons, abandoned river channels or deep oceans. The Cambrian Burgess Shale in the Rocky Mountains near Field, British Columbia, contains one of the most famous examples of such a fossil assemblage, including impressions of soft, nonskeletal parts of many organisms, such as sponges, that are not found anywhere else (Fig. 2.22b). The Jurassic Solnhofen Limestone of Germany is another good example. These are both examples of what are termed **lagerstätte**, sedimentary units with exceptional, *in situ* preservation (Sect. 3.5.8). The bones of vertebrate animals tend to disarticulate after death, because of the decay of muscle and cartilage

and the destructive effects of predators (Fig. 2.22c). Nevertheless, entire skeletons of bony fish, reptiles and mammals are commonly found in certain rock units, indicating rapid burial under quiet conditions. The Solnhofen Limestone is a well-known example, containing, among other fossils, the entire skeleton and feather impressions of a primitive bird. Such fossil assemblages must, nevertheless, be interpreted with care because a limited amount of transportation is possible from the life environment to the site of eventual burial. Presumably the bird of the Solnhofen Limestone lived in the air, not at the sediment–water interface where it was deposited!

3. Much more common than either of the foregoing are death assemblages of fossils that may have been transported a significant distance, perhaps many kilometers from their life environment. These commonly occur as lag concentrates of shelly debris such as gastropod, pelecypod, brachiopod or trilobite fragments, and fish bones or scales (Fig. 2.22). Such concentrations may be abundant enough to be locally rock forming, for example, the famous Silurian Ludlow Bone Bed of the Welsh borderlands. They normally indicate a channel-floor lag concentration or the product of wave winnowing, and can usually be readily recognized by the fact that fragment grain size tends to be relatively uniform.

Transported body fossils may not occur as concentrations but as scattered, individual occurrences, in which case each find must be interpreted with care. Did it live where it is now found or was it transported a significant distance following death? The environmental deduction resulting from such an analysis may be quite different depending on which interpretation is chosen. Evidence of transportation may be obvious and should be sought, for example, broken or abraded fossils may have traveled significant distances. Overturned corals, rolled stromatolites (including oncolites) and uprooted tree trunks are all obviously transported.

These problems are particularly acute in the case of microfossils and palynomorphs that, on account of their size, are particularly susceptible to being transported long distances from their life environment. For example, modern foraminiferal tests are blown tens of kilometers across the supratidal desert flats of India and Arabia, and shallow-water marine forms are commonly carried into the deep sea by sediment-gravity flows, such as turbidity currents. Detritus eroded from earlier stratigraphic units may also include derived fossil material. Environmental interpretations based on such fossil types may therefore be very difficult, though it may still be possible if the analysis is carried out in conjunction with the examination of other sedimentary features.

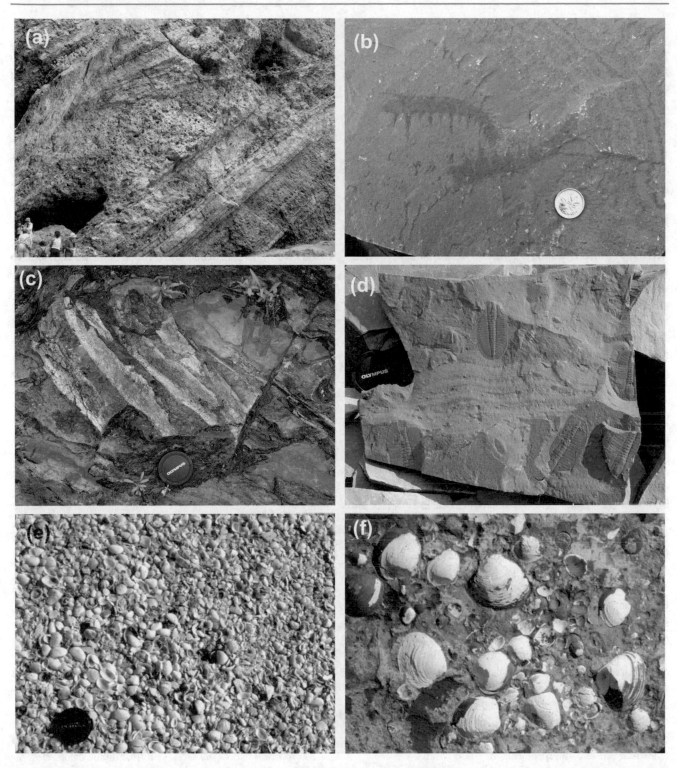

Fig. 2.22 Examples of fossils in outcrop. **a.** Stromatoporoid "reef," Devonian, Alberta; **b** soft-bodied preservation of arthropod jaws, Burgess Shale, British Columbia; **c** marine reptile bones, Cretaceous, northeastern British Columbia; **d.** Trilobites, Burgess Shale, British Columbia; **e** Modern shell beach, Florida; **f** Pelecypod shell bed, Cretaceous, Arctic Canada

In fact, it was the occurrence of sandstones containing shallow-water foraminifera interbedded with mudstones containing deep-water forms in the Cenozoic of the Ventura Basin, California, that was one of the principal clues leading to the development of the turbidity-current theory for the origin of deep-water sandstones.

Because of the great variety of life forms preserved as fossils, the subject of paleoecology is a large and complex one. Detailed studies are for the specialist and a complete treatment is beyond the scope of this book. The reader is referred to such texts as Hallam (1973), McKerrow (1978), and Dodd and Stanton (1981). A few examples are discussed in Chap. 4.

2.2.2.10 Biogenic Sedimentary Structures

Footprints, burrows, resting, crawling or grazing trails and escape burrows are examples of what are termed **trace fossils**. They are abundant in some rock units, particularly shallow-marine deposits; all may yield useful environmental information, including water depth, rate of sedimentation and degree of agitation (Figs. 2.23, 2.24). The study of trace fossils is termed **ichnology**. Distinctive assemblages of trace fossils, even the nondescript structure **bioturbation**, which is ubiquitous in many shallow-marine rocks, can be interpreted usefully by the sedimentologist. Footprints are, of course, best seen on bedding-plane surfaces, as are many types of feeding trails and crawling traces (Fig. 2.25). Burrows are better examined in vertical cross-section and are most visible either in very clean, fresh, wetted rock surfaces or in wind- or water-etched weathered outcrops (Fig. 2.23b, d).

Stromatolites are a distinctive component of many carbonate successions, particularly those of Precambrian age, and have been the subject of much detailed study.

Studies of trace fossils in modern settings and in ancient rocks have led to the recognition of distinctive assemblages, or **ichnofacies**, that are environmentally dependent, reflecting such conditions as the water depths, energy levels and nutrient availability preferred by the organisms that leave the traces. An ecological zonation is shown in Fig. 2.26. Here, for example, can be seen the *Skolithos* assemblage of traces, one which is characteristic of high-energy shoreline environments. Here, a burrowing mode of life permits organisms to live in protected environments while benefiting from the oxygenated and nutrient-rich waters of wave-influenced coastal settings. The *Nereites* assemblage is one composed of organisms that systematically mine the nutrient-poor sediments of deep-water settings. *Glossifungites* organisms like semi-consolidated substrates, and are commonly to be found at major stratigraphic bounding surfaces and so on.

Figure 2.26 is from MacEachern et al. (2010), which is by far the best modern text dealing with the description, classification and interpretation of trace fossils.

2.2.3 Sampling Plan

The amount of sampling to be carried out in outcrop section depends on the nature of the problem in hand, as discussed in Sect. 1.4. We are not concerned here with sampling of ore, hydrocarbon source beds, or coal to be analyzed for economic purposes, but the sampling required to perform a satisfactory basin analysis. Sampling is carried out for three basic purposes: (1) to provide a suite of typical lithologic samples illustrating textures, structures or distinctive fossils on a hand-specimen scale; (2) to provide a set for laboratory analysis of petrography and maturation using polished slabs, the optical microscope and possibly other tools, such as x-ray diffraction analysis and the scanning electron microscope and (3) to gather macrofossil and lithologic samples for microfossil or palynological examination to be used for studying biostratigraphy.

2.2.3.1 Illustrative Samples

The choice of such samples is usually simple and can be based on a trade-off between how much it would be useful to take and how much the geologist can physically carry. Large samples showing sedimentary structures or suites of fossils may be an invaluable aid in illustrating the geology of the project area to the geologist's supervisor or for practical demonstration at a seminar or a poster display at a conference. Unless fieldwork is done close to a road or is being continuously supported by helicopter, it is rarely possible to collect as much as one would like.

2.2.3.2 Petrographic Samples

Before carrying out a detailed sampling program for petrographic work, the geologist should think very carefully about what the samples are intended to demonstrate. Here are some typical research objectives:

1. Grain-size analysis of siliciclastic and carbonate-clastic rocks, as a descriptive parameter and as an aid to interpreting the depositional environment.
2. Petrography of detrital grains, including heavy minerals, as an aid to determining sediment sources. Studies of detrital zircons have become very useful in recent years, as discussed in Sect. 6.6.2.
3. Petrography of carbonate grains as an aid to determining the depositional environment.
4. Studies of grain interactions, matrix, and cement of both carbonate and clastic grains in order to investigate diagenetic history.
5. Studies of detrital grain fabric in order to determine paleocurrent patterns or, in certain cases, as an aid to interpreting the depositional environment; oriented samples are required.
6. Studies of thermal basin maturity using clay mineral characteristics, vitrinite reflectance, microfossil color or fluid inclusions.
7. Oriented samples for paleomagnetic study, for use in developing a reversal stratigraphy, a paleopole, or for studying diagenesis.

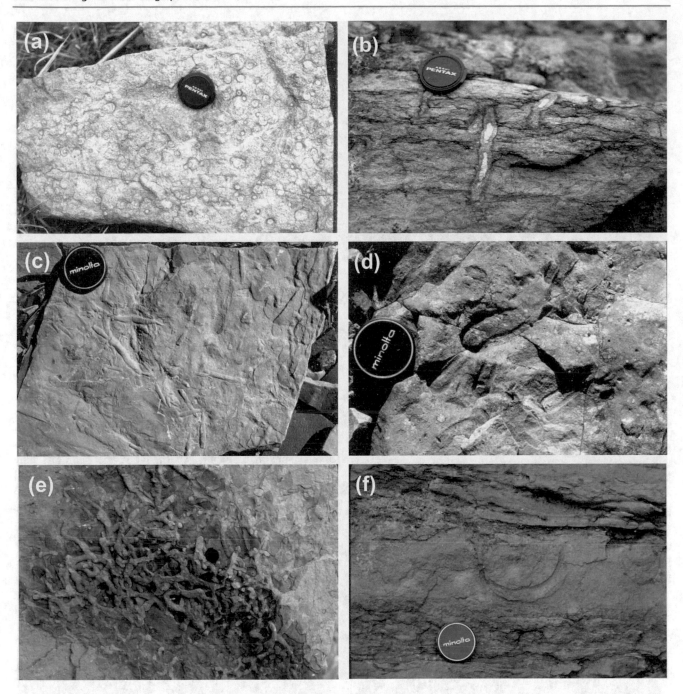

Fig. 2.23 Trace fossils. A, B. *Skolithos* (Jurassic) in bedding plane **a** and vertical **b** exposure; **c** Feeding and crawling trails on a shallow-marine bedding-plane surface (Ordovician); **d, e, f** shallow-marine burrows, all Cretaceous, including *Diplocraterion* (**d**) *Thalassinoides* (**e**) and *Arenicolites* (**f**)

The geologist must define the scope of the problem before collecting any samples, otherwise it may later be discovered that the collection is unsatisfactory. Work in remote regions is excellent training in such planning exercises, because only rarely is there a chance to return to an outcrop a second time. For the purpose of most regional studies, it is useful to examine petrographic variations vertically through a sequence and areally within a single stratigraphic unit. For example, the composition of detrital grains in a sequence may show progressive vertical changes, recording erosional unroofing of a source area or the switching of source areas. Sampling should be adequate to document this statistically. Samples taken every 10 to 50 m through a sequence will normally suffice. For diagenetic and fabric studies and for paleomagnetic work,

Fig. 2.24 Example of trace fossils in a drill core, showing how well these features can be seen. They include *Skolithos* (top) and *Rosellia* (bottom)

position in space can be reconstructed in the laboratory. For paleomagnetic and fabric studies, this need is obvious, but it may also be useful for petrographic purposes, for example, where it is necessary to examine cavity-filling detrital matrix to determine the time of filling relative to tectonic deformation, as an aid to determining structural top, or for studying microscopic grain-size changes related to bedding (e.g., graded bedding). To collect an oriented sample, the geologist selects a projecting piece of the outcrop that is still in place, not having been moved by frost heave, exfoliation or other processes, and yet is still removable by hammer and chisel. A flat face on this piece is measured for orientation and marked by felt pen before removal. A more detailed discussion of the methods and purpose of sample orientation is given by Prior et al. (1987).

How much should be collected at each sample station? A few hundred grams is adequate for most purposes. Thin sections can be made from blocks with sides less than 2 cm. Grain mounts of unconsolidated sediment can be made from less than 20 g. Where a particular component is sought, such as disseminated carbonaceous fragments for vitrinite-reflectance measurements, it may be necessary to take a larger sample to ensure that enough fragments are included for a statistical study at each sample station. Samples for paleomagnetic study are collected in a variety of ways, including the use of portable drills, which collect cores about 2 cm in diameter. Alternatively, the geologist may wish to take oriented blocks about 4 × 8 × 15 cm, from which several cores can be drilled in the laboratory. Oriented specimens of unconsolidated sediment are collected by means of small plastic core boxes or tubes pushed into an unweathered face by hand. Textures may be preserved by on-site infiltration with resin (see also Prior et al., 1987).

How do we ensure that samples are truly representative? There is a conflict here between statistically valid experimental design and what is practically possible. Statistical

more detailed sampling may be required. Paleomagnetic research normally requires several samples from a single locality in order to permit checks for accuracy.

The most detailed sampling program is required for carbonate sequences, particularly those of shallow-marine origin, which show the most facies variation. Laboratory work on polished slabs or thin sections is required for a reliable facies description of most carbonate rocks, and this may call for sampling every meter, or less, through a section.

For certain purposes, it is necessary that the sample be oriented, that is, it should be marked in the field so that its

Fig. 2.25 Examples of vertebrate footprints. Left: view of underside of bed, showing a projecting dinosaur footprint; right: bird footprints

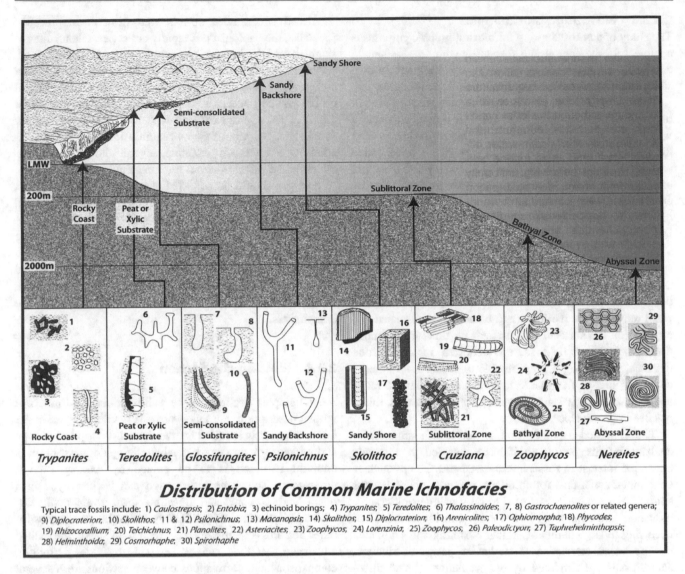

Distribution of Common Marine Ichnofacies

Typical trace fossils include: 1) *Caulostrepsis*; 2) *Entobia*; 3) echinoid borings; 4) *Trypanites*; 5) *Teredolites*; 6) *Thalassinoides*; 7, 8) *Gastrochaenolites* or related genera; 9) *Diplocraterion*; 10) *Skolithos*; 11 & 12) *Psilonichnus*; 13) *Macanopsis*; 14) *Skolithos*; 15) *Diplocraterion*; 16) *Arenicolites*; 17) *Ophiomorpha*; 18) *Phycodes*; 19) *Rhizocorallium*; 20) *Teichichnus*; 21) *Planolites*; 22) *Asteriacites*; 23) *Zoophycos*; 24) *Lorenzinia*; 25) *Zoophycos*; 26) *Paleodictyon*; 27) *Tuphrehelminthopsis*; 28) *Helminthoida*; 29) *Cosmorhaphe*; 30) *Spirorhaphe*

Fig. 2.26 Ecological zonation of trace-fossil assemblages (McEachern et al. 2010)

theory requires that we take samples according to some specific plan, such as once every 10 or 30 m (for example) through a vertical section, or by dividing a map area into a square grid and taking one sample from somewhere within each cell of the grid. By these methods we can satisfy the assumptions of statistical theory that our samples are truly representative of the total population of all possible samples. In practice, we can never fully satisfy such assumptions. Parts of any given rock body are eroded or too deeply buried for sampling. Exposures may not be available where sampling design might require them, or a particular interval might be covered by talus. An additional consideration is that the very existence of exposures of a geological unit might be governed by weathering factors related to the parameter the geologist hopes to measure. For example, imagine a sandstone bed formed at the confluence of two

river systems, one draining a quartz-rich terrain and one a quartz-poor terrain. The quartzose sandstone intervals may be quartz cemented, much more resistant to erosion, and therefore much more likely to crop out at the surface. Sampling of such a unit might give very biased petrographic results.

In carrying out a basin analysis, we often deal with large areas and considerable thicknesses of strata. Petrographic, textural and maturity trends are usually strong enough to show through any imperfections in our sampling program. We collect what we can, taking care that our measurements are controlled by the appropriate geological variables, for example, that counts of detrital components are all made on the same grain-size range. For a further discussion of sampling issues, the reader is referred to the review of field methods by Holbrook and Miall (2020).

2.2.3.3 Biostratigraphic Samples

The study of any fossil group for biostratigraphic purposes is a subject for the appropriate specialists who, ideally, should collect their own material. However, this may not be possible, and the geologist often is required to do the collecting.

Unless a unit is particularly fossiliferous, the collecting of macrofossils can rarely be performed in a fully satisfactory way by the geologist, who is also measuring and describing the section. The search for fossils may take a considerable amount of time, far more than is necessary for the other aspects of the work. In practice, what the geologist usually ends up with are scattered bits and pieces and spot samples of more obviously fossiliferous units, in which it may be fortuitous whether or not any species of biostratigraphic value are present. In arid regions, where vegetation cover is sparse, some types of fossils, particularly vertebrate remains, are best located by carefully scanning the sloping surfaces of poorly exposed fine-grained sediments. In such cases, rain wash removes fine particles, and bones are left isolated and exposed on the surface. Given adequate time, for example, the two or three field seasons required for dissertation research, the geologist may be able to spend more time on collecting and to familiarize him or herself with the fauna and/or flora, but in reconnaissance mapping exercises, this is usually impracticable.

The increased sophistication of subsurface stratigraphic analysis by the petroleum industry has led to a greatly expanded interest in fossil microorganisms. Groups such as conodonts, acritarchs, foraminifera, palynomorphs, diatoms and radiolaria have been found to be sensitive stratigraphic indicators and commonly have the inestimable advantage of occurring in large numbers, so that biostratigraphic zonation can be based on statistical studies of taxon distribution. Microfossils are extracted by deflocculation or acid dissolution of suitable host rocks. Most useful fossil forms are pelagic, or are distributed by wind (palynomorphs), so that potentially they may be found in a wide variety of rock types. However, their occurrence is affected by questions of hydrodynamics in the depositional setting and post-depositional preservation. Palynomorphs may be rare in sandstones because they cannot settle out in the turbulent environments in which sand is deposited, but they are abundant in associated silts, mudrocks and coal. Radiolarians and other siliceous organisms may be entirely absent in mudstones but abundantly preserved in silts, cherts and volcanic tuffs, because they are dissolved in the waters of relatively high pH commonly associated with the formation of mud rocks. Conodonts are most commonly preserved in limestones and calcareous mudstones and may be sparse in dolomites, because dolomitization commonly occurs penecontemporaneously in environments inimical to conodonts, such as sabkha flats.

Armed with advice of this kind from the appropriate specialist, the geologist can rapidly collect excellent suites of samples for later biostratigraphic analysis and can cut down on the amount of barren material carried home at great effort and expense, only to be discarded. Advice should be sought on how much material to collect. Normally, a few hundred grams of the appropriate rock type will yield a satisfactory fossil suite, but more may be required for more sparse fossiliferous intervals, for example, several kilograms for conodonts to be extracted from unpromising carbonate units.

Samples should be collected at regular vertical intervals through a section, preferably every 10 to 50 m. This will vary with rock type, in order to permit more detailed sampling from condensed units or particularly favorable lithologies. Such sample suites permit the biostratigrapher to plot range charts of each taxon and may allow detailed zonation. Scattered or spot samples have to be examined out of context and may not permit very satisfactory age assignments.

2.2.4 Plotting the Section

A stratigraphic section can be published in the form of a written description, but this is not a very effective use of the information. It is required for the formal description of type and reference sections of named stratigraphic units, but it is doubtful if much is to be gained by reading such a description. The same information can be conveyed in a much more compact and digestible form as a graphic log, with a central column for lithology and adjacent columns for other features of importance. Such logs have the added advantage that they can be laid side by side, permitting comparisons and correlations between sections from several locations.

It is possible to devise logging techniques with columns and symbols to convey every scrap of lithologic and petrographic information, plus details of fossils, sedimentary structures, paleocurrent measurements, even chemical composition. Many companies and government organizations print blank logging forms of this type for use by their geologists to facilitate the logging process and to standardize the results of different workers. Johnson (1992) developed a comprehensive, all-purpose logging form that he offered as a standard for fieldworkers.

However, such logs have the disadvantage that significant information may be lost in a welter of detail. If a main purpose of a log is to permit visual comparisons between different sections, it is advisable to simplify the logs so that the critical features stand out from the page. For the purpose of basin analysis, the most important data are those that carry paleogeographic information, such as lithology, grain size, sedimentary structures, and fossil content. Paleocurrent data can be added to these, but commonly are treated separately (Sect. 6.7). As we shall see in Chap. 4, the vertical succession of facies is often of crucial importance in interpreting the depositional environment, and so it is helpful to emphasize this in the logs.

For clastic rocks, one of the most useful techniques is to vary the width of the central lithology column according to grain size, the wider the column, the coarser the rocks. Examples are illustrated in Fig. 2.27. Many other examples are given by Reading (1996) and James and Dalrymple (2010). This method imitates the weathering profile of most clastic

Fig. 2.27 Typical representation of surface stratigraphic sections. Grain-size scale at base of column: m = mud, s = silt, vf to vc = very fine to very coarse sand, c = conglomerate

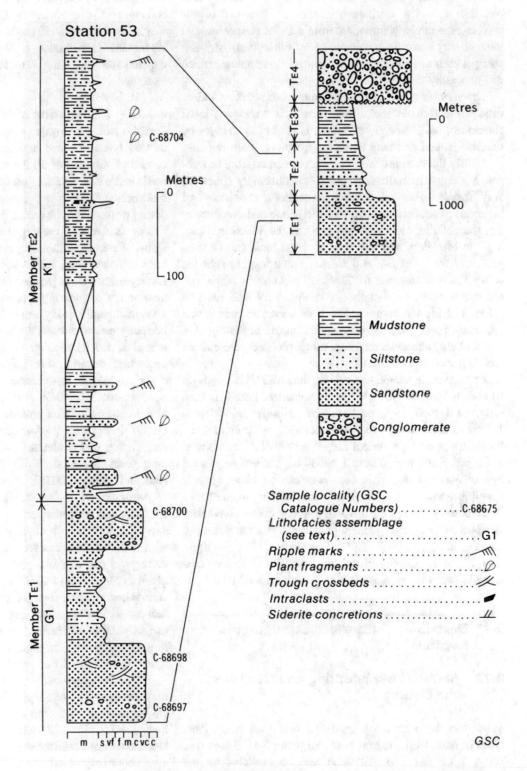

rocks, as muddy units tend to be less resistant and to form recessive intervals, in contrast to the projecting buttresses and cliffs of coarser sandstones and conglomerates. Drawing the column in this way enables rapid visual comparisons between sections and also permits instant recognition of gradual trends, such as upward fining, and sharp breaks in lithology, as at an erosion surface. Subtleties, such as changes in sorting or a bimodal grain-size distribution, cannot be displayed in this way, but are rarely as important from a facies perspective and may, in any case, be accompanied by other kinds of facies change, such as in a sedimentary structure assemblage, which can be readily displayed in visual logs.

The variable column-width technique has not been widely used for carbonate rocks, but there is no reason why it should not be. However, grain size is subject to changes by diagenesis, and this may make interpretation more difficult.

Within the column, various patterns can be used to indicate lithology, including symbols for sedimentary structures and fossils. For siliciclastic sequences consisting of interbedded mudstone, siltstone, sandstone and conglomerate, the column width conveys most of the necessary lithologic information, and the body of the column can be used primarily for structures and fossils. Some loggers split the column in two, one side for lithology and one for structures, but this is rarely as visually successful. Very little need be placed outside the column, which preserves an uncluttered appearance and increases the graphic impact of the log.

Symbols, abbreviations and other plotting conventions are discussed in Sect. 2.3.4.

At what scale should the logs be drawn? This depends on what it is they are intended to demonstrate. Detailed, local sedimentological logs may require a scale of between 1 cm:1 m and 1 cm:10 m. Regional stratigraphic studies can be illustrated in large foldout diagrams or wall charts at scales in the order of 1 cm:10 m to 1 cm:50 m, whereas page-sized logs of major stratigraphic sequences can be drawn (grossly simplified) at scales as small as 1 cm:1000 m. In the petroleum industry, the scale 1 in:100 ft has long been a convenient standard for subsurface stratigraphic work. This translates to a convenient approximate metric equivalent of 1 cm:10 m, although it is actually closer to 1 cm:12 m. Metric units should always be used, preferably in multiples of 10.

2.3 Describing Subsurface Stratigraphic Sections

2.3.1 Methods of Measuring and Recording the Data

Subsurface sections are logged and described using three types of data, well cuttings, core and petrophysical logs (see Sects. 2.3.2 and 2.4). All three may be available for the

large-diameter holes drilled by petroleum exploration companies. Diamond-drill holes (DDH) provide a continuous core but nothing else. The logging techniques and the results obtainable are therefore different.

To reduce costs, corporate practice may make use of outside consultant services to provide wireline-log and core descriptions, or they may limit staff to using photographs of core instead of encouraging them to view the actual core. Key observational detail could be lost as a result, particularly where the investigator is developing a hypothesis that depends on recording critical depositional or diagenetic features.

2.3.1.1 Examination of Well Cuttings

Samples stored in company and government laboratories are of two types, washed and unwashed. Unwashed cuttings consist of samples of all the material that settles out of the mud stream into a settling pit at the drill site. Unconsolidated mudrocks may disperse completely into the mud stream during drilling, in which case little of them will be preserved except as coatings on larger fragments or occasional soft chips. Washed cuttings are those from which all mud has been removed (Fig. 1.16). The washing process makes the cutting examination process easier, but it further biases the distribution of rock types present if the drill penetrated any unconsolidated muddy units. Stratigraphic well logging is normally carried out on the washed cuttings, whereas palynological and micropaleontological sampling is done on the unwashed material. Stratigraphic logging techniques are described later in this chapter. A detailed guide and manual was published by Low (1951) and is well worth reading. Many companies also provide their own manuals. McNeal (1959) has some useful comments, and the American Association of Petroleum Geologists has also issued a logging guide (Swanson, 1981). A more recent text is that by Ellis and Singer (2007).

As described in Sect. 1.4.3, samples are collected at the well site and bagged every 10 ft (3 m). The bag is labeled according to the depth of recovery by the well-site geologist, who makes allowances for the time taken for the mud and cuttings to rise to the surface. Measurements are normally given as "depth below K.B"; K.B. stands for kelly bushing, a convenient measurement location on the drilling platform a few meters above ground level. The altitude above sea level of this point is determined by surveying, so that these drilling depths can be converted to "depths subsea." On offshore rigs K.B. is 25 to 30 m above sea level.

For various reasons, not all the cuttings in any given bag may be derived from the depth shown. The problems of caving and variable chip density have been referred to in Sect. 1.4.3. (Fig. 2.28). These problems are particularly acute in soft or unconsolidated rocks, and samples from such a sequence may consist of a heterogeneous mixture bearing

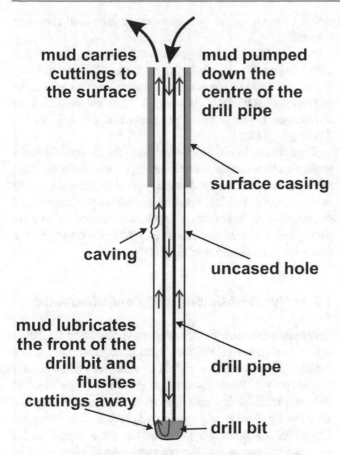

mud carries cuttings to the surface

mud pumped down the centre of the drill pipe

surface casing

caving

uncased hole

mud lubricates the front of the drill bit and flushes cuttings away

drill pipe

drill bit

Fig. 2.28 How cavings enter the mud stream in a drill hole

little relation to such stratigraphic detail as thinly interbedded units of contrasting lithology. The loss of the muds from the cutting suite compounds the problem. There are several ways in which these problems can be at least partially resolved. Caved material may be obvious by the large size of the fragments or by its exotic lithology or fossil content. For example, in one well in the Canadian Arctic, I logged Jurassic pelecypods and foraminifera from Devonian cuttings, 270 m below the unconformable contact of the Jurassic with the Devonian. The geologist will gradually become familiar with the formations under study and will then readily recognize such obvious contamination.

A more powerful tool is available to the logger and that is to study the cuttings in conjunction with the suite of petrophysical (wireline) logs from the hole. These logs record various physical properties as a measuring tool is slowly run the length of the hole. Modern petrophysical logging methods are capable of resolving lithologic variations over vertical intervals of a few decimeters or less and can therefore be used, with practice, to interpret lithologies in conjunction with the well cuttings. A description of the common petrophysical methods and their uses is given later in this chapter. Once the geologist is familiar with the petrophysical response of the various lithologies in the hole under

examination, it is possible to use the logs to adjust or correct the sample description. A lithologic log may therefore be drawn up that bears only a loose relationship to the material actually present in the sample bags. Such a log is an interpretive log and should be clearly labeled as such. It is likely to be more useful in basin interpretation than a log which simply records the cuttings dogmatically, particularly in the case of poorly consolidated beds or those with rapid vertical lithologic variations. Soft muds will be entirely unrecorded in a straight sample log, which may give the geologist a very inaccurate picture of the subsurface stratigraphy. All this can be allowed for in an interpretive log, but interpretations can be wrong, and the geologist must be aware of it when using this type of record.

Cuttings are observed under a low-power, reflected-light binocular microscope. Immersing the samples in water may aid observation, particularly as dust adhering to the chips can be washed off. The most useful magnification range for such a microscope is from about X5 to X50. The critical petrophysical logs should be unfolded to the appropriate depth interval and placed at one side of the microscope. It is advisable to scan rapidly the samples from several tens of meters of section before beginning the detailed description. Like standing back from an outcrop section, this gives the geologist the opportunity to perceive major lithologic variations. These may be correlated with changes in the petrophysical log response, permitting precise depth control.

As the samples are described, the information may be recorded directly on a preprinted logging form or written out in note form. If the log is to be an interpretive one, several tens of meters of section should be examined before plotting the graphic log, so as to give the geologist time to digest what is being observed. It may be necessary for publication or other purposes to produce a written sample description, but, as discussed under surface sections, these are difficult for a reader to absorb and are likely to be seldom used. The *Geological Survey of Canada* now publishes them in microfiche form to save paper and space.

Many petroleum and mining exploration companies and service companies (such as the *American Stratigraphic Company, Canadian Stratigraphic Service Ltd.* and *International Geosystems Corporation*) now use computer processible logging forms. The data are then stored in digital form in data banks. Retrieval programs may be available that can use these data for automated log plotting, and the same data bank can be exploited for automated plotting of maps and sections, as discussed in Chap. 6.

2.3.1.2 Examination of Core

The large-diameter core produced by petroleum drilling (Fig. 1.17) is stored either in a company laboratory or an official government repository. North American practice is to divide the core into 2.5 ft (0.8 m) lengths, which are stored side by side, two to a box. Top and bottom should be marked on the box. The core usually consists of a series of short pieces, broken from each other by torque during the drilling process. The well-site geologist should number or label each piece with respect to its position in the box and its orientation because, unless this is done, once a piece is moved it may be very difficult to restore it to its correct position, and serious errors may be introduced in reconstructing the vertical lithologic succession. Diamond-drill-hole (DDH) cores are normally stored in 5 ft (1.6 m) lengths, five to a box. The same remarks apply with regard to the position and orientation of core pieces. DDH cores are rarely brought from the field back into the office, except for crucial holes. They are examined and logged in the field, and then are commonly abandoned. They may even be tipped out of the box to prevent competitors from taking a look. This is a great waste of research material, but the practice seems unlikely to change.

Cores are most conveniently examined in a laboratory specifically designed for this purpose (Fig. 2.29). Core boxes are laid on roller tables, so that they can be readily loaded and unloaded using a fork-lift vehicle. A movable platform is positioned above the core boxes, on which are placed the microscope, petrophysical logs, notebook, etc.

The grinding action of the core barrel during drilling may smear the core surface and obscure lithologic features; sometimes this can be rectified by washing the core with water or even dilute hydrochloric acid. An even more useful technique is to have the core cut longitudinally with a rock saw, creating a flat section. This should always be wetted with water or etched lightly with dilute acid before examination. The etching technique is particularly useful when examining carbonate rocks, as it tends to generate a fine

Fig. 2.29 A core laboratory. Energy Resources Conservation Board, Calgary, Alberta

relief between grains and cement or different carbonate minerals.

Where petrophysical logs are available, it is important to correlate core lithology with log response. This exercise may reveal that parts of the core are missing, perhaps as a result of fragmentation of soft lithologies. Such information is of importance in attempts to reconstruct a detailed vertical lithologic profile.

For a discussion of description and plotting routines, refer to the previous section on well cuttings. Essentially the same methods are used, except that many more features are visible in core, such as bedding features, sedimentary structures and macroscopic trace and body fossils. Orientation of core for geological purposes was discussed by Davison and Haszeldine (1984) and Nelson et al. (1987).

2.3.2 Types of Cutting and Core Observation

Large-scale features, including most sedimentary structures and the subtleties of bedding, cannot be identified in well cuttings. They are partly visible in cores, but cores usually provide only frustratingly small snapshots of major features, and core research is rather like trying to describe elephant anatomy by examining a piece of skin with a microscope. The following notes are given in the same format as for surface sections, so that the contrasts with the latter can be emphasized. The description of field observation techniques should be referred to where appropriate.

2.3.2.1 Subdivision of the Section into Descriptive Units

This is best carried out by core and sample examination in conjunction with petrophysical logs, in the case of petroleum exploration wells. The combination is a powerful one and yields good generalized stratigraphic subdivisions with precise depth control. For DDH cores, the absence of petrophysical logs is compensated by the availability of a continuous core, and stratigraphic subdivision is simple. Because so many features, such as sedimentary structures, cannot be observed in cuttings, descriptive units in the subsurface tend to be thicker and more generalized than those observed in the outcrop. However, for core, the focus on what are really very small samples and the attempt to maximize the use of limited amounts of information tends to lead to very detailed descriptions. Examination of surface sections, sections based on cuttings and those on core, particularly short petroleum cores, require very different concepts of scale. These should be borne in mind when a basin-analysis exercise calls for the correlation of all three types of data.

2.3.2.2 Lithology and Grain Size

These can be observed satisfactorily in cuttings for all rock types except conglomerates, using the same techniques as described under surface sections. Conglomerates cannot be adequately studied in cuttings where clast size is larger than cutting size. It may not be possible to ascertain which cuttings represent clast fragments and which matrix, and no observations of clast grain size can be made. Remember, also, that unconsolidated silts, muds and evaporites may not be represented in well cuttings. These may require identification using petrophysical logs.

Well cuttings commonly contain contaminants that the logger should discard or ignore. Many are easy to recognize, such as metal pipe shavings or bit fragments. Oily substances, such as pipe dope or grease, may coat some fragments, but can usually be distinguished from natural oil stains by the fact that they coat the cuttings and do not penetrate them. Drilling mud may also coat the cuttings, particularly poorly washed samples. Casing cement may appear as a flood of cuttings at certain levels, where the hole was reentered following the setting of the casing. Cement can be easily mistaken for sandy, silty or chalky carbonate. Finally, foreign materials, such as feathers, sacking, seeds, cellophane, perlite or coarse mica flakes, may be present. These are used in the drilling mud to clog large pores and prevent loss of mud circulation.

Usually there are few restrictions on lithology and grain-size determinations in core, except where particularly coarse conglomerates are present.

2.3.2.3 Porosity

See the discussion in Sect. 2.2.2.3. Observations and measurements from subsurface rocks are more reliable than are those from outcrops because of the complications of surface weathering.

2.3.2.4 Color

See the discussion under the heading of surface sections (Sect. 2.2.2.4).

2.3.2.5 Bedding

For core, see the discussion under the heading of surface sections (Sect. 2.2.2.5). Bedding cannot be seen in well cuttings except for fine lamination, whereas core provides good information on bedding variation. Caution should be exercised in attempts to extract any quantitative information about bed thickness from cores, because of the possibility of core loss, as discussed earlier.

2.3.2.6 Sedimentary Structures

Very few, if any structures can be observed in well cuttings. However, a wide range of structures is visible in core (e.g., Figs. 2.8c, 2.24). Large structures, such as major erosion features and giant crossbedding, may be difficult to discern because the small sample of the structure visible in the core may easily be confused with something else. For example, thick crossbed sets could be mistaken for structural dip if the upper or lower termination of the set against a horizontal bedding plane cannot be determined. Dipmeter interpretations and formation microscanner observations may help here, as discussed in Sect. 6.7. Erosion surfaces are practically impossible to interpret in small outcrops or cores. A break may indicate anything from a storm-induced scour surface to an unconformity representing several hundred million years of nondeposition. The presence of soils or regoliths below the erosion surface or lag gravels above the surface are good indicators of a major sedimentary break, if present (except, of course, where there is structural discordance).

Ripple marks and crossbed set up to a few decimeters thick are more readily recognizable in core, and the reader should turn to the discussion of these in the section on surface exposures for methods of study. The large-scale geometry of crossbed sets is difficult to interpret in core. For example, the difference between the flat shape of a foreset in planar crossbedding and the curved surface of a trough crossbed is practically impossible to detect in core, even the large-diameter petroleum core. Curvature of a typical trough crossbed across such a core amounts to about 2°. Sensitive dipmeter and microscanner logs have considerable potential for interpreting crossbedding in the subsurface (Sect. 6.7), but the method is not widely used. Paleocurrent determinations could be made from oriented core, but the availability of the latter is practically zero.

Small-scale erosion features such as flutes, tool markings and rain prints, and other bedding features such as desiccation cracks and synaeresis markings are difficult to find in core because bedding plane sections are rare and the geologist is discouraged from creating additional sections by breaking up the core.

Liquefaction, load, and fluid-loss structures are commonly visible in core and, except for the larger features, should be readily interpretable.

2.3.2.7 Fossils

Fossil fragments are commonly visible in well cuttings, but are difficult to recognize and interpret. The best solution is to examine them in thin section, when distinctive features of internal structure may be apparent. A program of routine thin-section examination of fossiliferous sequences may be desirable and, particularly if the rocks are carbonates, this can be combined with the lithologic analysis. An excellent textbook by Horowitz and Potter (1971) discusses the petrography of fossils in detail.

Macrofossils can rarely be satisfactorily studied in core, except in the case of highly fossiliferous sections, such as reefs and bioherms. The reason is that the chance sections afforded by core surfaces and longitudinal cuts do not necessarily provide exposures of a representative suite of the forms present, and unlike sparsely fossiliferous surface outcrops, there is no opportunity to break up more rock in a search for additional specimens. Fragments may be studied in thin section, as described previously, but the limitations on the quantity of material still apply.

Both cores and cuttings may be used by biostratigraphic specialists, who extract palynomorphs and microfossils from them by processes of deflocculation or acid dissolution. Much useful ecological information may be obtainable from these suites of fossils. For example, Mesozoic and Cenozoic foraminifera were sensitive to water depth (as are modern forms), and documentation of foram assemblages through a succession can permit a detailed reconstruction of the varying depths of marine depositional environments (Sect. 3.5.8).

2.3.2.8 Biogenic Sedimentary Structures

As in the case of most other mesoscopic features, very little can be seen in cuttings, whereas cores commonly contain particularly well-displayed biogenic sedimentary structures. Those confined to bedding planes may not be particularly easy to find, whereas burrows usually are easy to study and can provide invaluable environmental interpretation (Fig. 2.24). Refer to the notes and references in Sect. 2.2 and to Sect. 2.2.10. An ecological zonation that has been developed for trace fossils has proved particularly useful for subsurface work (Fig. 2.26).

2.3.3 Sampling Plan

As in the case of surface sections, we are concerned here with sampling for basin-analysis purposes in three main categories: illustrative lithologic samples; samples for laboratory petrographic analysis using thin and polished sections, X-ray diffraction, etc. and samples for biostratigraphic purposes.

Very limited quantities of material are available for any kind of sampling in subsurface sections. The cuttings and core stored in the laboratory are all that will ever be available and once used cannot be replaced. Thus, they should be used with care, and permission should always be sought from the appropriate company or government agency before removing any material for research purposes.

A core may provide excellent illustrative material for demonstrating lithology, sedimentary structures and facies sequences. The *Canadian Society of Petroleum Geologists* has established a tradition of holding a "core conference" every year in one of the government core laboratories in Calgary (other societies, such as the *Society for Sedimentary*

Geology now hold their own workshops). Each contributor to the conference provides a display of a selected core from a producing unit and uses this as a basis for presenting an interpretation of its geology, with emphasis on depositional environments, diagenesis and petroleum migration history. The educational value of these conferences is inestimable, and they are always well attended. Some have resulted in the publication of well-illustrated proceedings volumes, for example, Shawa (1974), Lerand (1976), and McIlreath and Harrison (1977). As of Spring 2014, the *Society for Sedimentary Geology (SEPM)* had published 22 "Core Workshop" notes.

The use of well cuttings as illustrative material of this type is clearly limited. However, cuttings may be sampled routinely for petrographic studies, using etching and staining techniques on the raw, unwashed cuttings or preparing polished or thin sections. Fortunately, very small samples are adequate for this kind of work.

Sampling for biostratigraphic purposes should always be carried out on the unwashed rather than the washed cuttings. Depth control is, of course, better for core, but in petroleum wells, there is rarely adequate core for routine biostratigraphic sampling. Very rarely a petroleum exploration well may be drilled by continuous coring for stratigraphic research purposes, and these provide ample sample material free of the problems of sample caving and depth lag. The same is true, of course, for DDH cores. The quantity of material required for biostratigraphic purposes depends on the fossil type under investigation. A few hundred grams is usually adequate for palynological purposes, whereas to extract a representative conodont suite from sparsely fossiliferous carbonate sediments may require several kilograms of material. In the latter case, samples from several depth intervals may have to be combined.

2.3.4 Plotting the Section

Well logging is a routine procedure, and most organizations provide standard forms for plotting graphic logs, with a set of standard symbols and abbreviations. That used by the American Stratigraphic Company and Canadian Stratigraphic Services is typical. An example is illustrated in Fig. 2.30. Lithology is shown by color in a column near the center of the log, and accessories, cements, fossil types and certain structures are shown by symbols. To the left of this column are columns for formation tops, porosity type and porosity grade (amount). Formation tops are shown by a formation code. They are interpreted and may be subject to revision. Porosity type is given in a standard code or symbol and porosity grade by a crude graph based on visual estimates. The depth column may be used for symbols to indicate hydrocarbon shows or stains. To the right of the

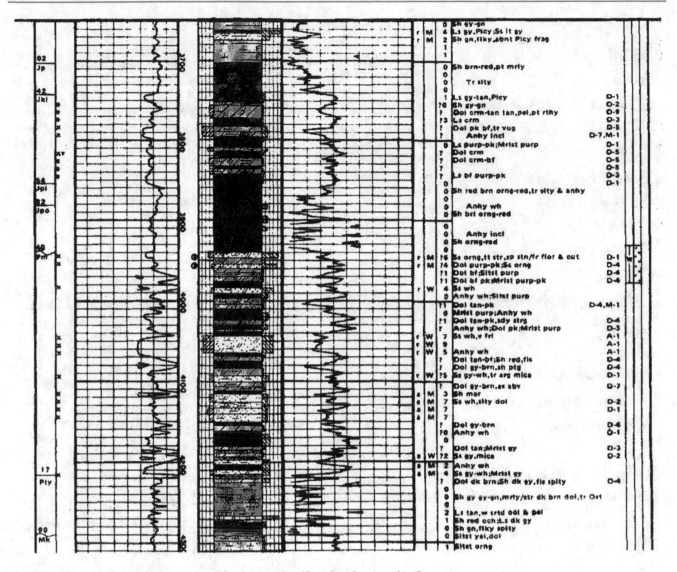

Fig. 2.30 Example of part of a log produced by the *American/Canadian Stratigraphic Company*

lithology column is a column for crystal or detrital grain size, based on visual estimates or measurements against a grain-size comparison set (Fig. 2.5). Both to the left and right of the lithology column are spaces for selected petrophysical logs. On the right-hand side of the log there is a space for typing in an abbreviated description of each lithologic interval. The American/Canadian Stratigraphic Company system uses a list of more than 450 abbreviations, covering almost every conceivable descriptive parameter. Remaining columns are used for grain rounding and sorting and for engineering data.

Much of this detail is not necessary for basin-analysis purposes. Figure 2.31 illustrates the more limited range of symbols and codes used by Tassonyi (1969) in a study of the subsurface stratigraphy of the Mackenzie Basin in northern Canada. Figure 2.32 illustrates one of his graphic logs. This style emphasizes lithology and other important stratigraphic variables.

2.4   Petrophysical Logs

A wide range of physical parameters can be measured using tools lowered down a petroleum exploration hole. Because of the method of data acquisition these are commonly called by the alternative name **wireline logs**. These tools provide information on lithology, porosity, and oil and water saturation (Table 2.3), and may be used in studies of thermal maturation, chemostratigraphy and the calibration of seismic data. In many cases, the measurements are not direct, but require interpretation by analogy or by correlating values between two or more logs run in the same hole. The subject of petrophysics is a highly advanced one and is beyond the scope of this book. Some approaches are described by Pirson (1977); others are given in the various interpretation manuals issued by the logging companies. Some excellent texts that

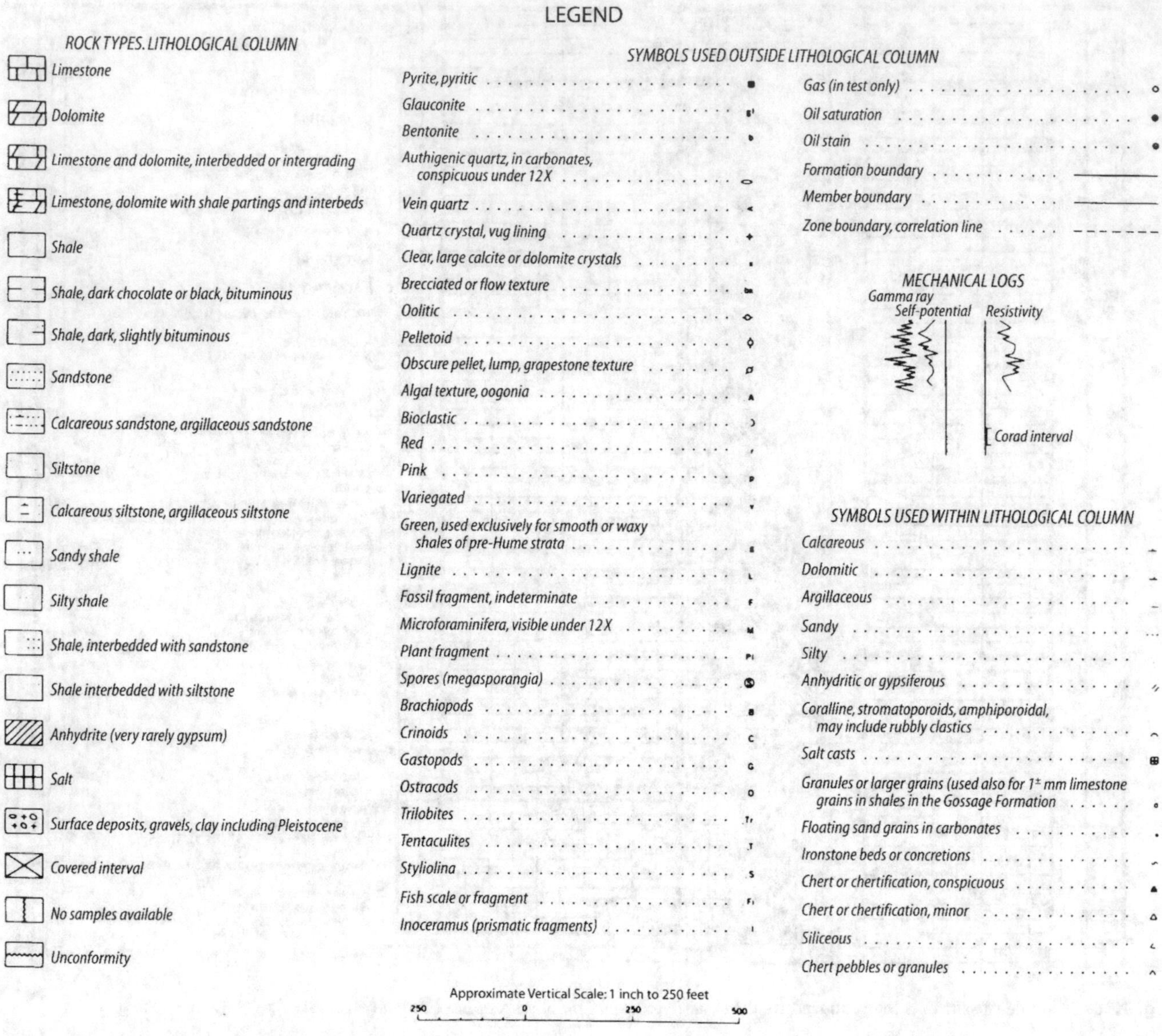

LEGEND

ROCK TYPES. LITHOLOGICAL COLUMN

- Limestone
- Dolomite
- Limestone and dolomite, interbedded or intergrading
- Limestone, dolomite with shale partings and interbeds
- Shale
- Shale, dark chocolate or black, bituminous
- Shale, dark, slightly bituminous
- Sandstone
- Calcareous sandstone, argillaceous sandstone
- Siltstone
- Calcareous siltstone, argillaceous siltstone
- Sandy shale
- Silty shale
- Shale, interbedded with sandstone
- Shale interbedded with siltstone
- Anhydrite (very rarely gypsum)
- Salt
- Surface deposits, gravels, clay including Pleistocene
- Covered interval
- No samples available
- Unconformity

Pyrite, pyritic . . . . ●
Glauconite . . . . ᵍ'
Bentonite . . . . ᵇ
Authigenic quartz, in carbonates, conspicuous under 12X . . . . ○
Vein quartz . . . . ◄
Quartz crystal, vug lining . . . . +
Clear, large calcite or dolomite crystals . . . . . ·
Brecciated or flow texture . . . . ▬
Oolitic . . . . ◇
Pelletoid . . . . ◊
Obscure pellet, lump, grapestone texture . . . ▱
Algal texture, oogonia . . . . ▲
Bioclastic . . . .)
Red . . . .
Pink . . . . ▼
Variegated . . . .
Green, used exclusively for smooth or waxy shales of pre-Hume strata . . . . ᵍ
Lignite L
Fossil fragment, indeterminate . . . . f
Microforaminifera, visible under 12X . . . . м
Plant fragment . . . . Pₗ
Spores (megasporangia) . . . . ☺
Brachiopods . . . . ᵇ
Crinoids c
Gastopods . . . . ɢ
Ostracods . . . . o
Trilobites . . . . Tᵣ
Tentaculites . . . . T
Styliolina . . . . s
Fish scale or fragment . . . . Fₗ
Inoceramus (prismatic fragments) . . . . ᴵ

SYMBOLS USED OUTSIDE LITHOLOGICAL COLUMN

Gas (in test only) . . . . ○
Oil saturation . . . . ●
Oil stain . . . . ●
Formation boundary . . . . _____
Member boundary . . . . _____
Zone boundary, correlation line . . . . . - - - - - -

MECHANICAL LOGS
Gamma ray
Self-potential   Resistivity

Corad interval

SYMBOLS USED WITHIN LITHOLOGICAL COLUMN

Calcareous . . . .
Dolomitic . . . .
Argillaceous . . . .
Sandy . . . .
Silty . . . .
Anhydritic or gypsiferous . . . . . . //
Coralline, stromatoporoids, amphiporoidal, may include rubbly clastics . . . . ^
Salt casts . . . . ⊞
Granules or larger grains (used also for 1± mm limestone grains in shales in the Gossage Formation . . . . . o
Floating sand grains in carbonates . . . .
Ironstone beds or concretions . . . . . .
Chert or chertification, conspicuous . . . ▲
Chert or chertification, minor . . . . . △
Siliceous . . . .
Chert pebbles or granules . . . . ^

Approximate Vertical Scale: 1 inch to 250 feet
250   0   250   500

Fig. 2.31 Codes and symbols used in a regional subsurface study of Paleozoic and Mesozoic stratigraphy in the Mackenzie Valley Region, Northwest Territories (Tassonyi, 1969)

discuss geological interpretations are those by Cant (1992) and Doveton (1994). Whittaker (1998) and Ellis and Singer (2007) provided more recent reviews of petrophysical logging methods.

In this section, some of the principal log types are described briefly, and some demonstrations of their utility are given. Petrophysical logs are used routinely by stratigraphers and basin analysts to provide information on lithology and to aid facies analysis, as discussed in Sects. 1.4.3 and 3.5.10. The use of the gamma-ray and other log types in stratigraphic correlation is discussed in Sect. 6.2

2.4.1 Gamma-Ray Log (GR)

This log measures the natural radioactivity of the formation, and therefore finds economic application in the evaluation of radioactive minerals, such as potash or uranium deposits. In sedimentary rocks, radioactive elements (potassium, thorium) tend to concentrate in clay minerals, and therefore the log provides a measurement of the muddiness of a unit. Texturally and mineralogically mature clastic lithologies, such as quartz arenites and clean carbonate sediments, give a low log response, whereas mudstones and certain special

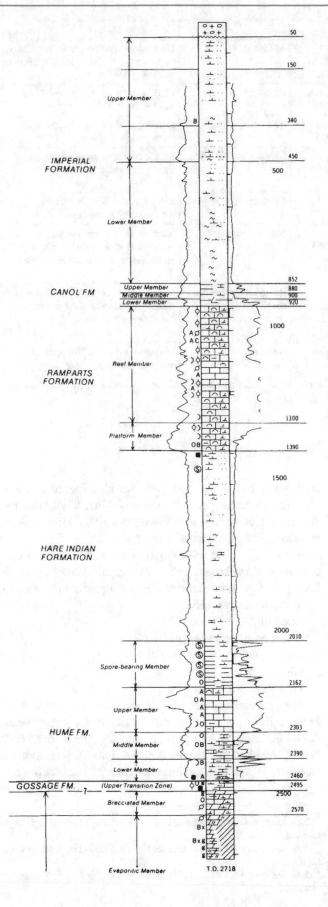

Fig. 2.32 A subsurface log drawn to emphasize stratigraphically important details, using the codes and symbols given in Fig. 2.31. Devonian, Mackenzie Valley, Northwest Territories. SP and resistivity logs have been redrawn beside the lithologic column (Tassonyi 1969)

sediment types, such as volcanic ash and granite wash (which has a high feldspar content), give a high log response. Absolute values and quantitative calculations of radioactivity are not necessary for the stratigrapher. The shape of the log trace is a sensitive lithostratigraphic indicator, and the gamma-ray log is commonly used in correlation and facies studies ("electrofacies," see Sect. 6.2). The log has the advantage that gamma radiation penetrates steel, and so the log can be run in cased holes. GR tools are also available for logging outcrops. In certain research settings, the availability of outcrop GR logs aids considerably in correlating sections against subsurface sections where lithologic information may be sparse.

Low GR readings (deflection to the left) correspond to cleaner, sandy parts of the succession. For clastic rocks, the variable-width column defined by the two log traces for each well provides a graphical portrayal of grain-size variations analogous to the variable-width log-plotting methods recommended for drawing surface sections (Sect. 2.2.4). The gamma-ray log is therefore widely used for plotting interpretive sample logs of subsurface clastic sections, and for log-shape studies. An example is illustrated in Fig. 2.33, showing how "upward-fining" and "upward-coarsening" cycles commonly yield a distinctive signature on gamma-ray logs (see also Fig. 3.61).

2.4.2 Spontaneous Potential Log (SP)

The curve is a recording of the potential difference between a movable electrode in the borehole and the fixed potential of a surface electrode. SP readings mainly record currents of electrochemical origin, in millivolts, and are either positive or negative. There are two separate electrochemical effects, as illustrated in Fig. 2.34. When the movable electrode is opposite a muddy unit, a positive current is recorded as a result of the membrane potential of the mudstone. The latter is permeable to most cations, particularly the Na + of saline formation waters, as a result of ion exchange processes, but is impermeable to anions. A flow of cations therefore proceeds toward the least saturated fluid, which in most cases is the drilling mud in the hole (log deflection to the right).

Opposite permeable units such as porous limestones and sandstones there is a liquid junction potential that generates a negative potential in the movable electrode. Both anions and cations are free to diffuse into the drilling mud from the more concentrated saline formation waters, but Cl- ions have the greater mobility, and the net effect is a negative charge (deflection to the left).

Table 2.3 The major petrophysical log types and their uses

Log	Property Measured	Unit	Geological Uses
Spontaneous Potential	Natural electric potential(compared to drilling mud)	Millivolts	Lithology (in some cases), correlation, curve shape analysis, identification of porous zones
Resistivity	Resistance to electric current flow	Ohm-metres	Identification of coals, bentonites, fluid evaluation
Gamma-ray	Natural radioactivity —related to K,Th, U	API units	Lithology (shaliness), correlation, curve shape analysis
Sonic	Velocity of compressional sound wave	Microseconds/metre	Identification of porous zones, coal, tightly cemented zones
Caliper	Size of hole	Centimetres	Evaluate hole conditions and reliability of other logs
Neutron	Concentrations of hydrogen (water and hydrocarbons) in pores	Per cent porosity	Identification of porous zones, cross-plots with sonic, density logs for empirical separation of lithologies
Density	Bulk density (electron density) includes pore fluid in measurement	Kilograms per cubic metre (gm/cm3)	Identification of some lithologies such as anhydrite, halite, non-porous carbonates
Dipmeter	Orientation of dipping surfaces by resistivity changes	Degrees (and direction)	Structural analysis, stratigraphic analysis

The SP log trace, particularly in clastic sequences, is very similar in shape to that of the GR log (examples are given in Figs. 2.34, 2.35, 2.36, 3.62 and 3.63) and the two may be used alternatively for correlation purposes, as illustrated in Fig. 6.3. The SP curve is normally smoother, and this log type does not offer the same facility for identifying thin beds. SP deflections are small where the salinity and resistivities of the formation fluids and the drilling mud are similar.

2.4.3 Resistivity Logs

Most rock types, in a dry state, do not transmit electric currents and are therefore highly resistive. The main exception consists of those rocks with abundant clay minerals. These transmit electricity by ion exchange of the cations in the clay lattice. In the natural state, rocks in the subsurface are saturated with water or hydrocarbons in pore spaces. Formation waters are normally saline and thus act as electrolytes. The resistivity therefore depends on the salinity and the continuity of the formation waters. The latter depends, in turn, on porosity and permeability, so that resistivity is lowest in units such as clean sandstone and vuggy dolomite and highest in impermeable rocks, for example, poorly sorted, dirty, silty sandstones and tight

carbonates. Evaporites and coal are also highly resistive. Metallic ores have very low resistivity. Oil is highly resistive and so, under certain conditions, resistivity tools may be used to detect oil-saturated intervals.

A wide variety of resistivity measurement tools have been devised, as listed below. Published logs are commonly identified by the appropriate abbreviation, such as shown in Fig. 6.3.

Electrical survey (ES).

Laterolog[1] (LL).

Induction-electrical survey (IES).

Dual induction laterolog (DIL).

Microlog[2].

Microlaterolog[3].

The conventional electrical survey was, together with the SP log, the only logging tool available for many years, during the early days of petrophysical logging before World War II. An electrical current is passed into the formation via an electrode, and this sets up spherical equipotential surfaces

---

[1] These are registered trade names for tools developed by Schlumberger Limited.

[2] These are registered trade names for tools developed by Schlumberger Limited.

[3] These are registered trade names for tools developed by Schlumberger Limited.

GR count increases

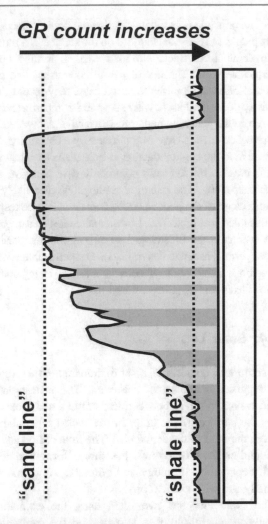

"sand line"

"shale line"

Fig. 2.33 The gamma-ray log. Radioactive traces of uranium, radium, thorium and potassium are concentrated in clay minerals. Gamma-ray readings are therefore a measure of the "clayeyness" of the sediments. The gamma-ray log is therefore an excellent recorder of sand-shale variations, as in this "upward-coarsening" or "funnel-shaped" profile, characteristic of a deltaic succession. The outcrop example is a deltaic profile in the Tertiary Eureka Sound Group of the Canadian Arctic Islands

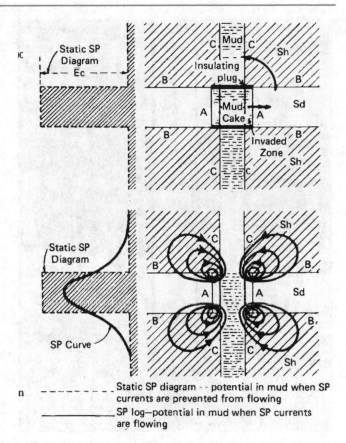

Static SP diagram -- potential in mud when SP currents are prevented from flowing

SP log—potential in mud when SP currents are flowing

Fig. 2.34 The principal of the SP log

centered on the electrode–rock contact. Three additional electrodes are positioned on the tool to intersect these surfaces at set distances from the first electrode. The standard spacing is

short-normal electrode: 40.6 cm (16 in).
medium-normal electrode: 1.63 m (64 in).
lateral electrode: 5.69 m (18 ft 8 in).

The wider the spacing the deeper the penetration of the current into the formation. This permits comparisons between zones close to the hole, permeated by drilling mud, and uninvaded zones further out (Fig. 2.35). The wide spacing of the electrodes also means that the ES tool is not sensitive to thin beds, and so it is not a very satisfactory device for stratigraphic studies.

The laterolog uses an arrangement of several electrodes designed to force an electrical current to flow horizontally out from the borehole as a thin sheet. A monitoring electrode measures a variable current that is automatically adjusted to maintain this pattern as the tool passes through lithologies of variable resistivity. This device is much more responsive to thin beds. Examples, run together with a SP survey and plotted on a logarithmic scale, are shown in Fig. 2.36.

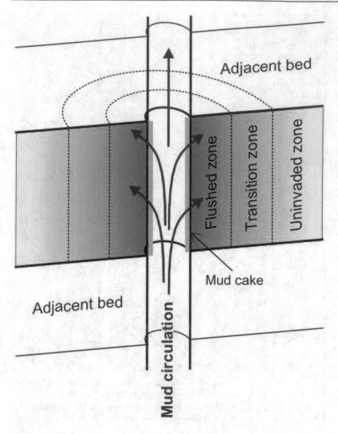

Fig. 2.35 Porous beds adjacent to a drill hole are flushed and filled with the drilling mud. This also leaves a "mud cake" adhering to the wall of the hole

Induction logs were developed for use with oil-based drilling muds which, because they are nonconductive, make the use of electrodes unsatisfactory. A high-frequency alternating current is passed through a transmitting coil, creating a magnetic field that induces a secondary current to flow in the surrounding rocks. This, in turn, creates a magnetic field that induces a current in a receiver coil. The strength of the induced current is proportional to formation conductivity.

The DIL survey is a combination of two induction devices and a laterolog device, with different formation penetration characteristics. It is normally run with an SP tool. An example is illustrated in Fig. 2.36. Separation between the three resistivity curves occurs opposite permeable units, where the low resistivity of the saline formation waters contrasts with the higher resistivity of the zone close to the borehole, which has been invaded by low-salinity drilling mud (Fig. 2.35). The deep-penetration induction log (ILd) therefore gives the lowest resistivity reading and the shallow penetration laterolog (LL8) the highest. The presence of these permeable zones is confirmed by the SP curve, which has a pattern very similar to that of the induction logs in the example shown.

The Microlog is the most sensitive device for studying lithologic variations in thin-bedded sequences. Its primary petrophysical purpose is to measure the resistivity of the invaded zone. The principle is as follows: during the drilling process mud enters permeable beds and hardens on the surface as a mud cake up to about 1 cm thick. No mud cake is formed opposite impermeable units. The microlog tool consists of two closely spaced electrodes. Opposite the mud cake, they record the low resistivity of the mud itself, whereas opposite impermeable units, they record the generally higher resistivities of uninvaded rocks. The presence of mud cake is confirmed by a caliper log, which is sensitive to the slight reduction in hole diameter when a mud cake is present. Muddy units commonly cave and give a very erratic caliper log. The microlog readings are also likely to be erratic because of the poor electrode contact. All these responses are illustrated in Fig. 2.37. The Microlaterolog is a more sensitive version of the Laterolog. Use of both the microresistivity devices permits accurate determinations of permeable sandstone and carbonate thickness, of considerable use in regional subsurface facies studies (Chaps. 5 and 6).

2.4.4 Sonic Log

The sonic tool consists of a set of transmitters for emitting sound pulses and a set of receivers. The fastest path for sound waves to travel between transmitters and receivers is along the surface of the hole, in the rock itself rather than through the mud or the actual tool. The time of first arrival of the sound pulses is therefore a measure of formation density, which depends on lithology and porosity. The sonic log is normally run with a GR tool.

The sonic tool has two main uses, the estimation of porosity where lithology is known, and the calibration of regional seismic data (Sect. 6.3.1). For the latter purpose, a computer in the recording truck at the wellhead integrates the travel time over each increment of depth as the survey is run. Every time this amounts to 1 ms a small pip appears on a track down the center of the log.

2.4.5 Formation Density Log

The tool contains a radioactive source emitting gamma rays. These penetrate the formation and collide with it, in a process known as Compton scattering. The deflected gamma rays are recorded at a detector on the tool. The rate of scattering is dependent on the density of the electrons in the formation with which the gamma rays collide. This, in turn, depends on rock density, porosity and composition of the formation fluids.

2.4.6 Neutron Log

For this log, a radioactive source emitting neutrons is used. These collide with the nuclei of the formation material, with

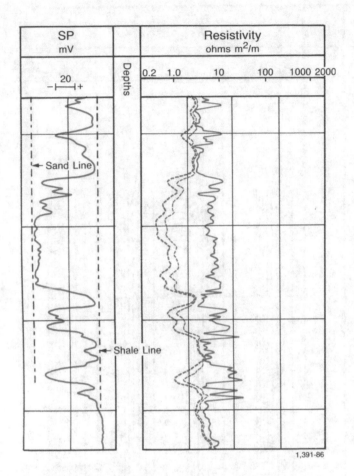

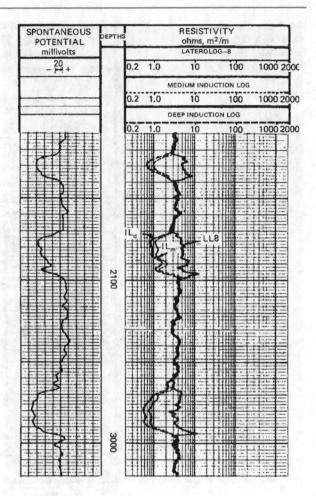

Fig. 2.36 Examples of SP and resistivity logs. Impermeable beds, such as shales, are indicated where the Laterolog curves combine. This is confirmed by the deflection of the SP curve toward the shale line. Permeable beds are indicated where the Laterolog curves diverge, indicating the different resistivity response between the invaded and uninvaded zone

a consequent loss of energy. The greatest loss of energy occurs when the neutron collides with a hydrogen nucleus, and so the total loss of energy depends mainly on the amount of hydrogen present, either in formation waters, hydrocarbons, or bound water in clay minerals, gypsum, etc. The detector measures either the amount of scattered low-energy neutrons or the gamma rays emitted when these neutrons are captured by other nuclei.

2.4.7 Crossplots

Where the formation is known to consist of only two or three rock types, such as sandstone–siltstone–mudstone or limestone–dolomite, combinations of two or more logs can be used to determine lithology, porosity and hydrocarbon content. These crossplots are therefore of considerable stratigraphic use where only generalized lithologic information is available from well cuttings.

For clean, non-muddy formations, combinations of the sonic, formation density and neutron logs are the most useful. For example, Fig. 2.38 shows the neutron-density combination. These logs are commonly calibrated in terms of apparent limestone porosity, that is, if the rock is indeed limestone, its porosity has the value indicated. For other lithologies, the porosity estimate will be in error, but by reading values for both logs, it is possible to determine both lithology and correct porosity. The curves in Fig. 2.38 give ranges of actual porosity readings for four principal rock types. To show how this graph can be used, compare it to the neutron-density overlay given in Fig. 2.38. Such overlays may be provided on a routine basis by the logging company or can be redrawn on request. The thick sandstone interval at the top can be recognized by the distinctly higher readings on the density curve. Values range from about 4 to 12, whereas those on the neutron curve are mostly close to zero. Examination of Fig. 2.38 shows that only sandstone can give this combination. The thick limestone at the bottom of

Fig. 2.37 Example of a
Microlog and Caliper log
(reproduced courtesy of
Schlumberger Inc.)

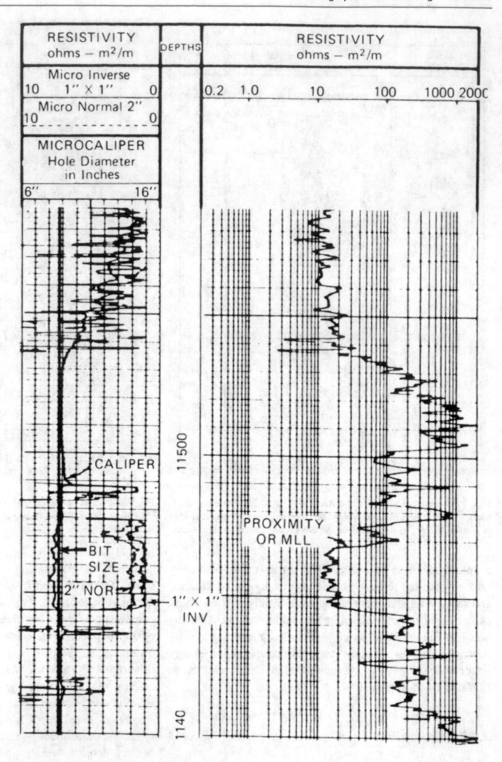

the overlay in Fig. 2.38 is indicated by the near coincidence of the two curves. Dolomite or anhydrite would be suggested by relatively higher neutron readings. Mudstones can commonly be recognized by very high neutron readings relative to density, because of the abundant water in the clay mineral lattice. This would be confirmed by the gamma-ray response or the SP or caliper log, if available.

2.4.8 Integrating Cores and Wireline Logs

The combination of core and petrophysical log data is a powerful one. An essential step is to locate the core on the petrophysical strip log by referring to the core depth information on the core box, as shown in Fig. 2.39. In this example, an important regional bounding surface and

Fig. 2.38 *Left:* A neutron (ON):density (OD) crossplot, showing how readings of the two logs can be used to determine lithology and porosity (reproduced courtesy of Schlumberger Inc.). *Right:* Example of a neutron: density overlay, illustrating how curve separation and deflections can be used to determine lithology (reproduced courtesy of Schlumberger Inc.)

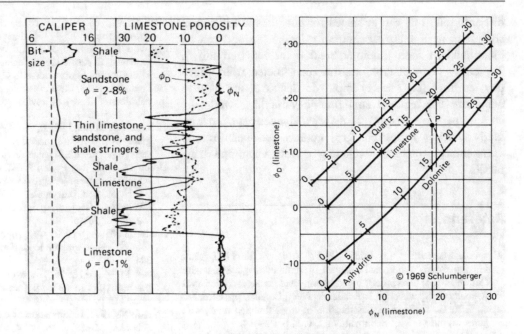

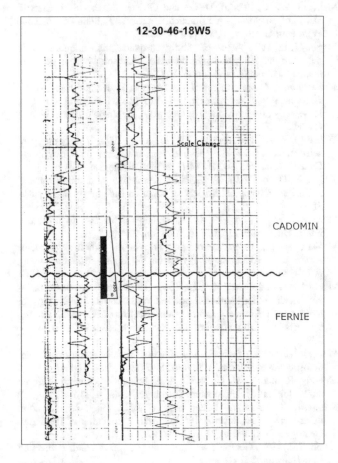

Fig. 2.39 Integrating core and log data is a very important and useful step in the documentation of subsurface stratigraphy and sedimentology. This example is form the Alberta Basin. The cored interval in the well is shown by the black bar down the center of the wireline log. The contact between Fernie shales and siltstones and Cadomin conglomerates and sandstones is indicated on the core by a red arrow

disconformity that can be traced for hundreds of kilometers shows up as a sharp discontinuity in the log and can be identified by a sharp lithologic break in the core (red arrow). Typically, regional cross-sections constructed from petrophysical logs (e.g., see examples in Sect. 6.2) are used to document the regional stratigraphic variability, while the core is used to highlight local facies characteristics, particularly vertical facies changes, contact relationships and sedimentary structures, including crossbedding and trace fossils.

References

Alexander-Marrack, P. D., Friend, P. F., Yeats, A. K. 1970, Mark sensing for recording and analysis of sedimentological data. In: Cutbill JL (ed) Data processing in biology and geology: Systematics Association Special, vol 3. Academic Press, London, pp 1–16.

Allen, J. R. L., 1963a, The classification of cross-stratified units, with notes on their origin: Sedimentology, v. 2, p. 93–114.

Allen, J. R. L., 1982, Sedimentary structures: their character and physical basis: Elsevier Scientific Publications, Developments in Sedimentology 30, 663 p.

Ashley, G. M., 1990, Classification of large-scale subaqueous bedforms: a new look at an old problem: Journal of Sedimentary Petrology, v. 60, p. 160–172.

Cant, D. J., 1992, Subsurface facies analysis, in Walker, R. G., and James, N. P., eds., Facies models: response to sea level change: Geological Association of Canada, St. John's, Newfoundland, p. 27–45.

Collinson, J. D., and Thompson, D. B., 1982, Sedimentary structures: George Allen and Unwin, London, 194 p.

Davison, I., and Haszeldine, R. S., 1984, Orienting conventional cores for geological purposes: a review of methods: Journal of Petroleum Geology, v. 7, p. 461–466.

Dean, W. E., Leinen, M., and Stow, D. A. V., 1985, Classification of deep-sea, fine-grained sediments: Journal of Sedimentary Petrology, v. 55, p. 250–256.

Dodd, J. R., Stanton, R. J. Jr. 1981, Paleoecology, concepts and applications. Wiley, New York

Doveton, J. H., 1994, Geologic log interpretation: Society for Sedimentary Geology (SEPM) Short Course Notes, #29, 184 p.

Doyle, J. D., and Sweet, M. L., 1995, Three-dimensional distribution of lithofacies, bounding surfaces, porosity, and permeability in a fluvial sandstone: Gypsy Sandstone of northern Oklahoma: American Association of Petroleum Geologists Bulletin, v. 79, p. 70–96.

Ellis V, Singer JM (2007) Well logging for earth scientists. Springer-Verlag, Berlin

Eyles, N., Day, T. E., and Gavican, A., 1987, Depositional controls on the magnetic characteristics of lodgement tills and other glacial diamict facies: Canadian Journal of Earth Sciences, v. 24, p. 2436–2458.

Folk, R. L., 1976, Reddening of desert sands: Simpson Desert, N.T, Australia: Journal of Sedimentary Petrology, v. 46, p. 604–615.

Friend, P. F., Alexander-Marrack, P. D., Nicholson, J., and Yeats, A. K., 1976, Devonian sediments of east Greenland I: introduction, classification of sequences, petrographic notes: Meddelelser om Gronland Bd. 206, nr. 1.

Goddard, E. N., Trask, P. D., De Ford, R. K., Rove, O. N., Singewald, J. T., Jr., and Overbeck, R. M., 1948, Rock-color chart: Geological Society of America, 16 p.

Hallam, A., ed., 1973, Atlas of palaeobiogeography: Elsevier, Amsterdam.

Harms, J. C., Southard, J. B., and Walker, R. G., 1982, Structures and sequences in clastic rocks: Society of Economic Paleontologists and Mineralogists Short Course 9, Calgary.

Hodgetts, D., 2013, Laser scanning and digital outcrop geology in the petroleum industry: A review: Marine and Petroleum Geology, v. 46, p. 335–354.

Holbrook, J. M., and Miall, A. D., 2020, Time in the Rock: A field guide to interpreting past events and processes from siliciclastic stratigraphy: Earth Science Reviews, v. 203, 103121, 23 p.

Horowitz, A. S., and Potter, P. E., 1971, Introductory Petrography of Fossils: Springer-Verlag, New York, 302 p.

James, N. P., and Dalrymple, R. W., eds., 2010, Facies Modesl 4: GEOtext 6, Geological Association of Canada, St. John's, Newfoundland, 586 p.

Johnson, M. R., 1992, A proposed format for general-purpose comprehensive graphic logs: Sedimentary Geology, v. 81, p. 289–298.

Lerand, M. M., ed., 1976, The sedimentology of selected clastic oil and gas reservoirs in Alberta: Canadian Society of Petroleum Geologists, 125 p.

Low, J. W., 1951, Examination of well cuttings: Quarterly of Colorado School of Mines, v. 46, no. 4, 47 p.

Maiklem, W. R., Bebout, D. G., and Glaister, R. P., 1969, Classification of anhydrite — a practical approach: Bulletin of Canadian Petroleum Geology, v. 17, p. 194–233.

Martini, I. P., 1971, A test of validity of quartz grain orientation as a paleocurrent and paleoenvironmental indicator: Journal of Sedimentary Petrology, v. 41, p. 60–68.

McEachern, J. A., Pemberton, S. G., Gingras, M. K., and Bann, K. L., 2010, Ichnology and facies models, in James, N. P., and Dalrymple, R. W., eds., Facies Models 4: Geotext 6, Geological Association of Canada, p. 19–58.

McIlreath, I. A., and Harrison, R. D., eds., 1977, The geology of selected carbonate oil, gas and lead-zinc reservoirs in Western Canada: Canadian Society of Petroleum Geologists, 124 p.

McKee, E. D., and Weir, G. W., 1953, Terminology for stratification in sediments: Geological Society of America Bulletin, v. 64, p. 381–389.

McKerrow, W. S., ed., 1978, The ecology of fossils, an illustrated guide: Duckworth & Co. Ltd., London, 384 p.

McNeal, R. P., 1959, Lithologic analysis of sedimentary rocks: American Association of Petroleum Geologists Bulletin, v. 43, p. 854–879.

Miall, A. D., and Tyler, N., eds., 1991, The three-dimensional facies architecture of terrigenous clastic sediments, and its implications for hydrocarbon discovery and recovery: Society of Economic Paleontologists and Mineralogists Concepts and Models Series, v. 3, 309 p.

Mills, P. C., 1983, Genesis and diagnostic value of soft-sediment deformation structures - a review. Sed Geol 35:83–104

Moberly, R., and Klein, G., deV., 1976, Ephemeral color in deep-sea cores: Journal of Sedimentary Petrology, v. 46, p. 216–225.

Nelson, R. A., Lenox, L. C., and Ward, B. J., Jr., 1987, Oriented core: its use, error, and uncertainty: American Association of Petroleum Geologists Bulletin, v. 71, p. 357–367.

Pirson, S. J., 1977, Geologic well log analysis, 2nd ed.: Gulf Publishing Co., Houston, 377 p.

Potter, P. E., and Pettijohn, F. J., 1977, Paleocurrents and basin analysis, 2nd edition: Academic Press, San Diego, California, 296 p.

Prior, D. J., Knipe, R. J., Bates, M. P., Grant, N. T., Law, R. D., Lloyd, G. E., Welbon, A., Agar, S. M., Brodie, K. H., Maddock, R. H., Rutter, E. H., White, S. H., Bell, T. H., Ferguson, C. C., and

Wheeler, J., 1987, Orientation of specimens: essential data for all fields of geology: Geology, v. 15, p. 829–831.

Reading, H. G., ed., 1996, Sedimentary environments: processes, facies and stratigraphy, third edition: Blackwell Science, Oxford, 688 p.

Rust, B. R., 1975, Fabric and structure in glaciofluvial gravels, in Jopling, A. V., and McDonald, B. C., eds., Glaciofluvial and glaciolacustrine sedimentation: Society of Economic Paleontologists and Mineralogists Special Publication 23, p. 238–248.

Shawa, M. S., ed., 1974, Use of sedimentary structures for recognition of clastic environments: Canadian Society of Petroleum Geologists, 66 p.

Sippel, R. F., 1971, Quartz grain orientation - 1 (the photometric method): Journal of Sedimentary Petrology, v. 41, p. 38–59.

Swanson, R. G., 1981, Sample examination manual: American Association of Petroleum Geologists.

Tassonyi, E. J., 1969, Subsurface geology, Lower Mackenzie River and Anderson River area, District of Mackenzie: Geological Survey of Canada Paper 68–25.

Taylor, J. C. M., 1978, Introduction to state of the art meeting, 1977, on sandstone diagenesis: Journal of the Geological Society. London v. 135, p. 3-6.

Walker, T. R., 1967, Formation of redbeds in modern and ancient deserts: Geological Society of America Bulletin, v. 78, p. 353–368.

Whittaker, A., 1998, Borehole data and geophysical log stratigraphy, in Doyle, P., and Bennett, M. R., eds., Unlocking the stratigraphic record: advances in modern stratigraphy: John Wiley and Sons, Chichester, p. 243–273.

Wilson, J. L., 1975, Carbonate facies in geologic history: Springer-Verlag, New York, 471 p.

Facies Analysis

Contents

Abstract

The study and interpretation of the textures, sedimentary structures, fossils and lithologic associations of sedimentary rocks on the scale of an outcrop, well section or small segment of a basin comprise the subject of facies analysis. Characteristic associations of these features constitute distinct lithofacies, biofacies or ichnofacies, the specific features of which provide information on depositional processes, local environmental characteristics and, in some cases, immediate post-depositional processes, such as water-loss during compaction. The concept of the facies model is explained, as a local "summary of the environment," and issues of sedimentary preservation are explored. The importance and limitations of "uniformitarianism" are discussed, with reference to whether or not the "present" is a suitable analogue of past conditions, and the reverse (is the past a suitable basis for interpreting the present?). A detailed review of environmental criteria is provided, which is intended to enable geologists to focus on the key features of the rocks that provide the most information on processes and environments.

3.1   Introduction

The study and interpretation of the textures, sedimentary structures, fossils and lithologic associations of sedimentary rocks on the scale of an outcrop, well section or small segment of a basin (the subject of Chap. 2) comprise the subject of facies analysis. A very large literature has grown since the 1960s that synthesizes this information in the form of facies models of varying complexity and sophistication. Several excellent texts deal extensively with facies analysis methods and process-response facies models (Reading 1996; Boggs 2012; James and Dalrymple 2010). The purpose of this chapter is to focus on a discussion of analytical methods and reviews of the kinds of practical information relating to depositional processes and sedimentary environments that can be obtained from sediments, based on the observations described in Chap. 2. It is hoped that this material will provide students with an introduction to modern facies modeling methods. Practice on specific ancient examples can then be carried out using the summaries of facies models set out in Chapter 4 (and perhaps also including one of the advanced texts mentioned above).

As explained in Sect. 1.2, since the 1990s, facies analysis methods have become an integral component of the broader study of sequence stratigraphy (Sect. 5.2.3). Some of the issues associated with the methodological transition from the local-scale focus of facies modeling to the basin-scale approach of sequence stratigraphy were discussed in the review paper of Walker (1990), which is examined in Sect. 1.2.13. Sequence analysis is aided by various mapping methods (Chap. 6) and by the separate and independent skills involved in correlation and dating (Chap. 7).

3.2   The Meaning of Facies

The term **facies** refers to those attributes of a sedimentary rock that provide information about depositional processes or the depositional environment. Beyond that basic point, the meaning of the word facies has been much debated in geology (e.g., Longwell 1949; Teichert 1958; Krumbein and

Sloss 1963; Middleton 2005; see Sect. 1.2.4). It is widely used in sedimentary geology, and also in metamorphic petrology, where it has a different meaning (Fawcett 1982). Anderton (1985), Reading (1996, Chap. 2), Pirrie (1998) and Dalrymple (2010a, b) provided excellent discussions of the modern sedimentological uses of the term and the methods of interpreting individual facies and facies relationships.

The word facies is now used in both a descriptive and an interpretive sense, and the word itself may have either a singular or plural meaning. Descriptive facies include **lithofacies** and **biofacies**, both of which are terms used to refer to certain observable attributes of sedimentary rock bodies that can be interpreted in terms of depositional or biological processes. (When used without a prefix in this book, the word facies is intended to mean either lithofacies or biofacies.) An individual lithofacies is a rock unit defined on the basis of its distinctive lithologic features, including composition, grain size, bedding characteristics and sedimentary structures. Each lithofacies represents an individual depositional event. Lithofacies may be grouped into **lithofacies associations** or **assemblages**, which are characteristic of particular depositional environments. These assemblages form the basis for defining **lithofacies models**; they commonly are repetitive and may be fully cyclic. A biofacies is defined on the basis of fossil components, including either body fossils or trace fossils. The term biofacies is normally used in the sense of an assemblage of such components. For the purpose of sedimentological study, a deposit may be divided into a series of lithofacies units, each of which displays a distinctive assemblage of lithologic or biologic features. These units may be single beds a few millimeters thick or a succession of beds tens to hundreds of meters thick. For example, a river deposit may consist of decimeter thick beds of a conglomerate lithofacies interbedded with a crossbedded sandstone facies. Contrast this with the biofacies terms used to describe the fill of many major early Paleozoic basins. Commonly, this may be divided into units hundreds of meters thick comprising a shelly biofacies, containing such fossils as brachiopods and trilobites, and a graptolitic biofacies. At the other extreme, J. L. Wilson (1975) recommended the use of microfacies in studying thin sections of carbonate rocks and defined 24 standard types.

The scale of an individual lithofacies or biofacies unit depends on the level of detail incorporated in its definition. It is determined by the variability of the succession, by the nature of the research undertaken (basin-wide reconnaissance versus a detailed local study), or by the availability of rock material for examination. Facies units defined on the basis of outcrop, core, well-cutting or geophysical criteria tend to refer to quite different scales and levels of detail. Geophysicists in the petroleum industry refer to seismic facies (Sects. 5.2.2, 6.3.3), but this is not comparable to the small-scale type of facies discussed in this chapter. Modern, high-resolution shallow seismic surveying coupled with side-scan sonar imaging is providing a powerful tool for the analysis of facies compositions and geometries in modern environments and has had a major impact on the understanding of shelf, slope and basin sedimentary environments (Sect. 6.3). Considerable attention is now being paid to the three-dimensional geometry of facies units in outcrop studies, and in subsurface studies involving reservoir development, three-dimensional seismic-reflection methods are increasingly being employed. To a large extent, the scales at which facies units are defined reflect criteria of convenience. The term is thus a very flexible and convenient one for descriptive purposes.

The term facies can also be used, usually for lithofacies assemblages, in an interpretive sense for groups of rocks that are thought to have been formed under similar conditions. This usage may emphasize specific depositional processes, such as till facies or turbidite facies. Alternatively, it may refer to a particular depositional environment, such as shelf-carbonate facies or fluvial facies, encompassing a wide range of depositional processes (see also Dalrymple 2010a for additional discussion of the meaning and usage of the term facies).

At one time, widespread use was made of two nineteenth-century Swiss stratigraphic terms that had acquired a generalized facies meaning encompassing lithologic characteristics, depositional environment and tectonic setting. The first of these is the *flysch facies*, comprising marine sediments, typically turbidites and other sediment-gravity flow deposits, formed on tectonically active continental margins. The *molasse facies* consists of nonmarine and shallow-marine sediments, mainly sandstones and conglomerates, formed within and flanking fold belts during and following their elevation into mountain ranges. Both of these facies types may make up major stratigraphic units hundreds or thousands of meters thick and extending for hundreds or thousands of kilometers. Continued use of the terms flysch and molasse is not recommended because of ambiguities about their implications for tectonic setting. The vague association with orogenic belts is inadequate now that we have the theory of plate tectonics to assist us in interpreting the relationship between sedimentation and tectonics (Allen and Allen 2013).

Lithostratigraphy and lithofacies analyses are two contrasting approaches to the study of sedimentary rocks. The first is the traditional descriptive approach. The second is based on detailed facies descriptions, which provide the basis for the genetic study of sediments using facies models. Lithofacies and biofacies analyses must be used to assist in stratigraphic studies, because by understanding the depositional environments and paleogeography existing at the time a rock unit was formed we are much better placed to make predictions and extrapolations about lateral changes in thickness and composition. Obviously, this will be invaluable for correlation purposes, and can make for a much more logical definition of

formal lithostratigraphic units. Biofacies analysis is crucial in the definition and comprehension of biostratigraphic units. Lithostratigraphy and facies analyses are essential components of the study of **sequence stratigraphy** (Sect. 5.2.3).

3.3 Recognition and Definition of Facies Types

3.3.1 Philosophy and Methods

In order to make sense of the lithologic variability present in most sedimentary basins, it is necessary to generalize, categorize and simplify what we see in well sections and outcrops. In sedimentology, we find that much of the variability disguises a limited range of basic lithofacies and biofacies types, and that variations between these types represent minor random environmental fluctuations or are even the result of accidental exposure or the position of a thin-section cut. The existence of this natural pattern is what makes facies studies, facies modeling and paleogeographic reconstruction possible. For example, for many years sedimentologists categorized most deep-sea deposits in terms of only five basic lithofacies. These were the T_a, T_b, T_c, T_d and T_e divisions of the classic Bouma turbidite sequence (Bouma 1962). This Bouma model served sedimentologists well, and was used as the type example in Walker's (1976) classic definition of facies models. But it has always been known that some beds do not fit the pattern, as discussed further below. The standard submarine-fan model that evolved in the 1970s (Mutti and Ricci-Lucchi 1972; Walker 1978) interpreted Bouma sequences as occurring mainly on the outer, non-channelized part of a fan. But we now know that the lithofacies spectrum for submarine fans and for other types of deep-sea sand deposit is rather more complex than was hypothesized in these early models (see Sect. 4.2.9).

The deposits of many other depositional environments have proved to be amenable to some simple, empirical classifications. For example, Miall (1977, 1978, 1996) showed that most river deposits could be described using about 20 lithofacies types. J. L. Wilson (1975), in a major review of Phanerozoic carbonate rocks, concluded that most could be described satisfactorily by drawing from a list of only 24 standard microfacies. Eyles et al. (1983) proposed a simple lithofacies classification scheme for the description of glacial-marine and glacial-lacustrine deposits. Farrell et al. (2012) provided the basis for a universal set of codes for describing clastic sediments, which they suggest (p. 377) "supports process-based interpretations of stratigraphic sequences." Many fossil groups await detailed biofacies analysis.

It is recommended that every basin analyst should study each basin with a fresh eye and, at least in the preliminary stages of the research, erect a local facies scheme without too much dependence on previously published work. Slavish adherence to such readily available research keys may result in minor but critical lithofacies types and lithofacies relationships being missed or forced to fit an inappropriate mold. New facies models or a better definition of old ones will never come about if geologists are content merely with such replication studies. Bridge (1993) expressed this concern forcefully with respect to the facies scheme now in widespread use for fluvial deposits (see Sect. 4.2.1), and Shanmugam (1997) argued strongly that the Bouma sequence concept had been seriously misused in the study of deep-sea sandstones. Others, such as Walker (1990, 1992), Reading and Levell (1996), and Dalrymple (2010a, b) are strongly in favor of the simplifying and categorizing methods of facies analysis. The problem that concerned Bridge (1993) and Shanmugam (1997) is that hasty or lazy field observation can too easily lead to incorrect facies classification. For example, Shanmugam (1997) argued that for too long sedimentologists have been applying the Bouma sequence concept and its fivefold subdivisions to deposits that exhibit few of the key characteristics of the original sequence. Many successions contain no graded basal unit, or they contain internal breaks in sedimentation indicating that they are not the product of single depositional events, or they contained evidence of deposition from other types of sediment-gravity flow, such as rafted clasts, that would indicate debris-flow deposition, and so on. In fact, such practices have become so common, in Shanmugam's (1997) view, that he refers to a phenomenon that he termed the "turbidite mind set."

Hummocky cross-stratification is an excellent example of a structure that sedimentologists had been looking at without seeing for many years, until focused upon and given a name by Harms et al. (1975). Suddenly, it was realized that this distinctive style of crossbedding is characteristic of many storm-generated ancient deposits. Its recognition led to the reappraisal of numerous shallow-marine assemblages, to the extent that the term became something of a cliché in the early 1980s (e.g., see Byers and Dott 1981). This is a good example of critical advances in sedimentological interpretation being missed because of the widespread lack of critical, independent observation.

How, then, does the basin analyst perform a facies analysis of an undescribed rock sequence or succession of fossils? The method for measuring and describing stratigraphic sections is discussed in detail in Chap. 2. For the purpose of lithofacies analysis, the focus must be on recognizing associations of attributes that are repeated through the section (or parts of the entire basin). Lithofacies may be distinguished by the presence of bedding units with a characteristic sedimentary structure or structures, a limited grain-size range, a certain bed thickness and perhaps a distinctive texture or color (color is subject to diagenetic change and should not be used as a primary criterion in definition;

see Sect. 2.2.2.4). Biofacies represent associations within the same stratigraphic interval of a limited suite of genera or species. Biofacies studies may be carried out on single taxonomic groups, because many of these, if examined by specialists, can yield highly detailed paleoecological (and hence depositional) information. The definition of biofacies is an exercise in defining assemblages and will be discussed in the next section.

In order to recognize these associations of lithologic attributes or fossil types, it may be useful to set up a checklist and tabulate their occurrence. Statistical analysis is commonly used by paleoecologists to establish biofacies from such tabulated data. It is now less used for erecting individual lithofacies because many of the common associations of lithologic attributes are well enough known to be recognized without the aid of statistics. But the key to success is very careful, empirical observation.

3.3.2 Field Examples of Facies Schemes

Two examples of lithofacies schemes that have been erected from field data are given in this section to illustrate the methods. A simple biofacies association is also described.

Cant and Walker (1976) described a Devonian fluvial section in eastern Quebec and subdivided it into eight lithofacies. The section is shown in Fig. 3.1, in which lithologic symbols are keyed to the lithofacies scheme. Note the use in this illustration of the variable-width column plotting technique for drawing stratigraphic sections that is discussed in Sect. 2.2.4. Here are some examples of their lithofacies descriptions, edited in order to focus on salient features.

Well defined trough crossbedded facies (B): This facies is composed of well-defined sets of trough crossbedding..., with trough depths averaging 15 to 20 cm (range 10 to 45 cm). The troughs are regularly stacked on top of each other, but in some individual occurrences of the facies, trough depths decrease upward... The sets are composed of well-sorted medium sand... A few of the coarser sets have granules and pebbles concentrated at their bases.

Asymmetrical scour facies (E): This facies consists of large, asymmetrical scours and scour fillings, up to 45 cm deep and 3 m wide... The scours cut into each other and into underlying troughed facies (A and B), and occurrences of the asymmetrical scour facies have a flat, erosionally truncated top... The main difference between the scour fillings and the two troughed facies (A and B) lies in the geometry of the infilling strata. In the asymmetrical scours, the layers are not at the angle of repose, but are parallel to the lower bounding surface.

Rippled sandstone and mudstone Facies (F): This facies includes cross-laminated sandstones..., and alternating cross-laminated sandstones and mudstones. [An example of the latter] is 1.5 m thick and consists of three coarsening-upward sequences, which grade from basal mudstones into trough cross-laminated fine sandstone and finally into granule sandstone. The sandstones capping each coarsening-upward sequence have sharp, bioturbated tops.

A variety of criteria has been used in defining these lithofacies, but the authors refrained, at this stage, from making any environmental interpretation.

A carbonate example was described by R.C.L. Wilson (1975), who studied the Upper Jurassic Oolite Series (actually a formation, but an incorrect, older nomenclature is still in use) of Dorset, England. He recognized four lithofacies, as shown in Fig. 3.2. Note in this illustration the relationship between lithostratigraphic and lithofacies units. The four lithofacies are as follows (described using the carbonate classification of Folk 1962):

1. Coarsening-upward units shown by two beds (1) the *Chlamys qualicosta* bed of intramicrite–oomicrite–oosparite poorly washed biosparite and (2) pisolite, consisting of quartz sands and phyllosilicate clay-intramicrite–oomicrite–oosparite–oncolites.
2. Crossbedded sets of oosparite showing 20 to 25° dips and sharp contacts either with phyllosilicate clays with nodular micrites or bioturbated oolite. Some minor flaser bedding and clay drapes over current ripples also occur.
3. Association of *Rhaxella* biomicrites.
4. Sheet deposits (5 to 10 cm) and large accretion sets (30 cm) of oomicrite and biomicrite with subsidiary oosparite and biosparite. Shell debris often shows imbricate structure, and the oomicrites are texturally inverted sediments, being a mixture of extremely well-sorted oolites in a micrite matrix. Some sets showing alternating current directions occur.

The careful reader will note two points about these Oolite lithofaces. First, a mixture of lithofacies and biofacies criteria is used. This commonly is desirable in fossiliferous successions where the fossils are distinctive rock-forming components. Second, several of the lithofacies consist of different carbonate rock types interbedded on a small scale. Each of these could be described as a lithofacies in its own right, in which case Wilson's four facies become facies assemblages. It is a question of the scale of description that is the most suitable for the purpose at hand. Those engaged in analyzing entire basins or major stratigraphic intervals in a basin cannot afford to spend too much time on fine detail. Decisions must be made with the core or outcrop in front of the analyst about how thick the descriptive units are going to be and how to combine thin beds for the purpose of lithofacies description. Generally, lithofacies thicknesses on the order of a few decimeters to a few meters have been found to be the most useful. Smaller scale subdivisions may be erected for selected examples of well-exposed sections or continuous cores if desired, to illustrate particular points in a facies description.

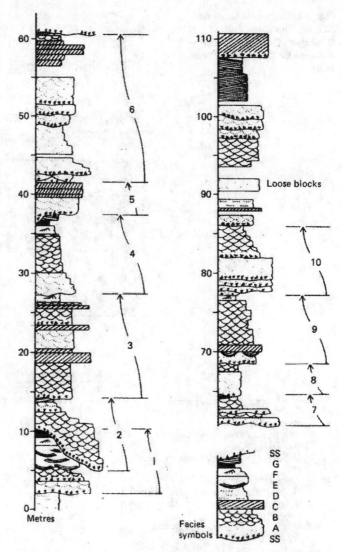

Fig. 3.1 A stratigraphic section plotted using a standardized facies scheme and the variable-width column technique. Fluvial cycles of the Battery Point Formation (Devonian), Quebec (Cant and Walker 1976)

Wilson suspected that his Oolite Series beds were tidal in origin, and he tabulated various features of each of the four facies using a range of criteria suggested by Klein (1971) for recognizing tidal deposits (Fig. 3.3). Many of these make specialized use of paleocurrent evidence, which is discussed in Sect. 2.2.4. Figure 3.3 illustrates the use of a checklist for lithofacies description. This can be a useful technique for field observation if the geologist suspects in advance what it is he/she is about to study. It is then possible to incorporate all the special observational detail and environmental criteria available from studies of the appropriate modern environment or other ancient rock units. In doing so there is, of course, the danger that other features might be missed, and then interpretation becomes a self-fulfilling prophecy. Tables of this kind are not recommended for illustration of written or

Fig. 3.2 Carbonate facies and their distribution in the Osmington Oolite (Jurassic), Southern England (R.C.L. Wilson 1975)

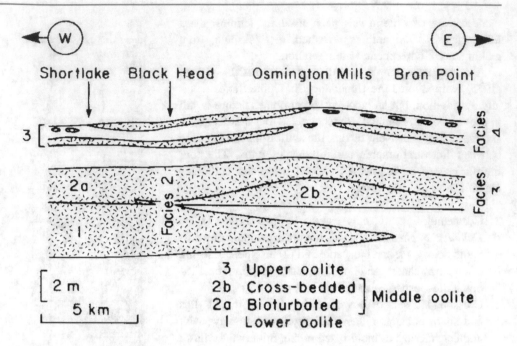

	Facies of Osmington Oolites			
	1	2	3	4
A. 1. Cross stratification with sharp set boundaries			■	■
2. Herringbone cross-stratification			■	■
4. Parallel laminae	■			
5. Complex internal organization of dunes, etc.			■	■
B. 6. Reactivation surfaces			■	■
7. Bimodal distribution of set thicknesses			■	
9. Unimodal distribution of dip direction of cross strata			■	
10. Orientation of cross-strata parallels sand-body trend			?	
C. 13. Small current ripples to larger current ripples			■	
19. Symmetrical ripples			■	
D. Flaser bedding			■	
F. 35 & 36 channel log deposits			■	
H. Burrowing, etc.	■	■	■	

Fig. 3.3 Distribution of selected lithofacies attributes in the Osmington Oolite (R.C.L. Wilson 1975)

oral presentation of a lithofacies study except, perhaps, as an appendix. They are visually difficult to absorb, and the geologist does better to provide summary descriptions, such as those given previously, plus photographs and graphic logs.

3.3.3 Establishing a Facies Scheme

The examples in the previous section serve to illustrate the methods used in three different kinds of basin problems to define facies types. The objective of the basin analyst should be to erect a facies scheme that can encompass all the rock types present in his or her project area. If the project encompasses several or many stratigraphic units deposited under widely varying environments, it may make more sense to erect a separate scheme for each lithostratigraphic unit, because the lithofacies and, most likely, the biofacies assemblages will be different for each such unit. However, this can lead to repetition and overlap, so some judgment is required.

Facies schemes should be kept as simple as possible, otherwise they defeat the whole purpose of carrying out a facies analysis. Some authors have subdivided their rocks into 20 or 30 lithofacies and erected subclasses of some of these. This gives the appearance of meticulous research and great analytical precision, but such schemes are difficult to understand or absorb, and make it more difficult to identify common associations. Remember that the purpose of a facies analysis is to aid in interpretation, and in basin analysis, this is often best accomplished by judicious simplification. It is rare that many more than half a dozen distinct lithofacies occur together in intimate stratigraphic association. Commonly, two to six lithofacies occur in repeated, monotonous succession through tens or hundreds of meters of section. In defining only a few facies to cover thick and varied successions, the problem of what to group together and what to separate is always present. What should be done about thin beds? Should units that show gradational contacts be split, and if so, how? The only answer is to do what works. One approach is to compile a very detailed description of the first well or section of a new project and to use the resulting notes to erect a preliminary facies scheme; this is then employed for describing each new section. It has the advantage that, instead of writing down basic descriptions for every bedding unit, it may be adequate simply to assign it to the appropriate facies and record in one's notes any additional observations that seem necessary, such as grain-size differences or additional sedimentary structures relative to the original descriptive scheme. The scheme itself may thus be modified as one goes along. This approach has worked well for the author for many years of studying clastic sedimentary sequences in outcrops and cores, in the Canadian Arctic, the southwest United States and elsewhere. It is more difficult to apply this method to carbonate rocks, because of their finer scale variability and the difficulty of seeing all the necessary features in weathered outcrops. One approach is to take into the field a rock saw and other necessary equipment for constructing etched slabs or peels. These can then be examined and described as fieldwork proceeds (J. L. Wilson 1975, pp. 56–60). Friend et al. (1976) avoided these problems entirely by recording only attributes of a sequence, not lithofacies as such. They then used statistical techniques to group these attributes into lithofacies assemblages.

Anderton (1985) suggested that facies criteria be ranked in order of priority, permitting an increasingly refined subdivision of the facies units as the work proceeds. His suggested ranking for channelized clastic deposits (as an example) is

1. bases of major channels or scoured surfaces;
2. tops and bases of nonclastic lithologies (e.g., coal, limestone);
3. abrupt changes from section composed predominantly of one grain size to another (e.g., sandstone to shale, fine sandstone to coarse) and
4. changes of internal structure within units of similar grain size.

This approach, if extended regionally, can form the basis for a practical means of recognizing and defining sequences, of which the facies assemblages are distinct components.

With continual study of the same kind of rocks, a standard set of facies classes will gradually emerge and can be used for rapid field or laboratory description. For the writer, this led to the erection of a generalized lithofacies scheme for describing the deposits of braided rivers based on field studies of various ancient deposits and a review of literature on both ancient deposits and modern rivers. The initial scheme contained 10 lithofacies types (Miall 1977). Further research by the writer and use by other workers led to the addition of about 10 more lithofacies types (Miall 1978), but the original 10 include most of the common ones. This scheme has now been applied to a wide variety of fluvial deposits and to the fluvial component of deltaic successions by numerous workers around the world. Table 3.1 lists these lithofacies, showing the codes used for note taking and a sedimentological interpretation of each. The lithofacies codes consist of two parts, a capital letter for modal grain size (G, gravel; S, sand; F, fines) and a lowercase letter or letters chosen as a mnemonic of a distinctive texture or structure of each lithofacies. The three lithofacies B, E and F of Cant and Walker (1976), discussed in the previous section, are St, Ss and Fl in this scheme. Le Blanc Smith (1980, 1980and Uba et al. (2005) developed this fluvial facies scheme still further by incorporating additional structures and information on grain size (see also Miall 1996, for further discussion of code schemes). However, as Dalrymple (2010, p. 6) noted, this scheme does not contain categories

Table 3.1 Facies classification (from Miall 1996)

Facies code	Facies	Sedimentary structures	Interpretation
Gmm	Matrix-supported, massive gravel	Weak grading	Plastic debris flow (high-strength, viscous)
Gmg	Matrix-supported gravel	Inverse to normal grading	Pseudoplastic debris flow(low strength, viscous)
Gci	Clast-supported gravel	Inverse grading	Clast-rich debris flow (high strength), or pseudoplastic debris flow (low strength)
Gcm	Clast-supported massive gravel	–	Pseudoplastic debris flow (inertial bedload, turbulent flow)
Gh	Cast-supported, crudely bedded gravel	Horizontal bedding, imbrication	Longitudinal bedforms,lag deposits,sieve deposits
Gt	Gravel, stratified	Trough crossbeds	Minor channel fills
Gp	Gravel, stratified	Planar crossbeds	Transverse bedforms, deltaic growths from older bar remnants
St	Sand, fine to v. coarse may be pebbly	Solitary or grouped Trough crossbeds	Sinuous-crested and linguoid 3-D dunes
Sp	Sand, fine to v. coarse may be pebbly	Solitary or grouped Planar crossbeds	Transverse and linguoid bedforms (2-D dunes)
Sr	Sand, very fine to coarse	Ripple cross-lamination	Ripples (lower flow regime)

(continued)

Table 3.1 (continued)

Facies code	Facies	Sedimentary structures	Interpretation
Sh	Sand, very fine to coarse may be pebbly	Horizontal lamination parting or streaming lineation	Plane-bed flow (critical flow)
Sl	Sand, very fine to coarse may be pebbly	Low-angle (<15°) crossbeds	Scour fills, humpback or washed-out dunes, antidunes
Ss	Sand, very fine to coarse may be pebbly	Broad, shallow scours	Scour fill
Sm	Sand, very fine to coarse	Massive, or faint lamination	Sediment-gravity flow deposits
Fl	Sand, silt, mud	Fine lamination, v. small ripples	Overbank, abandoned channel, or waning flood deposits
Fsm	Silt, mud	Massive	Backswamp or abandoned channel deposits
Fm	Mud, silt	Massive, desiccation cracks	Overbank, abandoned channel, or drape deposits
Fr	Mud, silt	Massive, roots, bioturbation	Root bed, incipient soil
C	Coal, carbon-aceous mud	Plant, mud films	Vegetated swamp deposits
P[1]	Paleosol carbonate (calcite, siderite)	Pedogenic features: nodules, filaments	Soil with chemical precipation

and codes for the tidal features that are common in some coastal (especially estuarine) fluvial systems.

J. L. Wilson's microfacies scheme for carbonates contains 24 types. Figure 3.4 illustrates his standard legend, facies numbers and abbreviated description. Figures 3.5 and 3.6 illustrate the use of these two schemes in drawing stratigraphic sections.

3.3.4 Facies Architecture

Many of the facies-analysis techniques discussed in this chapter focus on the use of vertical facies relationships, as derived from the measurement of stratigraphic sections (Sects. 2.2, 2.3, 3.5.10). Although section measuring has evolved into a sophisticated art, it is essentially a

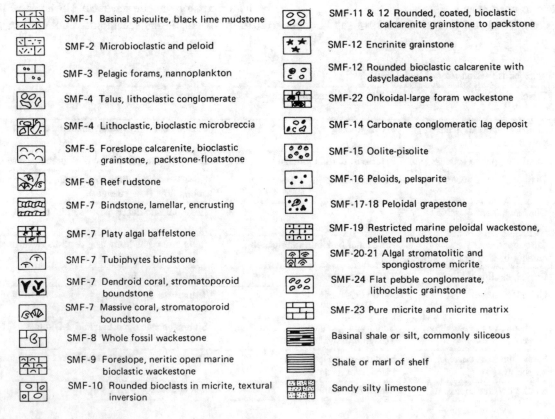

Fig. 3.4 The standard carbonate microfacies scheme of Wilson (1975)

Fig. 3.5 Use of the fluvial litho-facies scheme of Table 3.1 for plotting a section in braided fluvial deposits (Miall 1978)

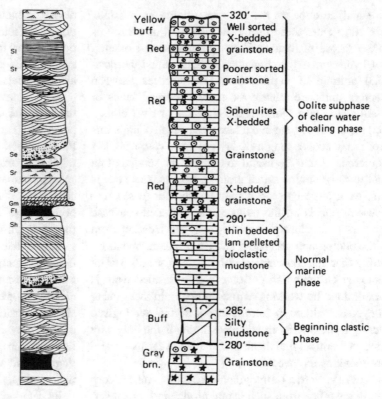

Fig. 3.6 Use of the carbonate microfacies scheme for plotting a limestone section (Wilson 1975)

one-dimensional approach to the study of what commonly are very complex three-dimensional rock bodies. Increasingly, it has been realized in recent years that vertical sections are inadequate for the description of facies variations in many environments, and techniques are evolving for the description and classification of facies units in three dimensions. At the surface, this can be accomplished by the search for and the description of large, laterally extensive outcrops. In the subsurface, lithostratigraphic correlation of closely spaced cores, coupled with high-resolution seismic data (especially 3-D seismic, if available), may serve the purpose. The results are important not only to facilitate the development of more quantitatively accurate reconstructions of depositional environments, but to provide a usable data base for development geologists and engineers to model fluid flow in reservoir units.

Outcrop techniques were described by Anderton (1985) and Miall (1985, 1988a, b, 1996), and field examples were given by Allen (1983), Miall and Tyler (1991), and Miall

(1988b, 1996). Ideally, the geologist should search for long natural or artificial cliff faces in areas of minimal structural disturbance. The entire outcrop should be photographed to permit the construction of a lateral profile based on a photomosaic or LIDAR profile, using the techniques described in Sect. 2.1.2.2. Hodgetts (2013) provided a review of LIDAR techniques. Nowadays, unmanned drones provide ideal platforms for photographic inaccessible outcrops.

Once printed and mounted on card or board, the mosaic, with a transparent overlay, can be taken back to the field and used as a base map on which to plot detailed sedimentological observations. Alternatively, hand-held digital devices, such as tablets, are now flexible enough that they may be taken into the field and used for the direct entry of field data at the time of observation. The geologist should exploit the detail available in locations of excellent outcrop to measure and describe as many facies criteria as possible, taking care to enter these on the mosaic with as precise as possible a record of their locations. Locations and orientations of hydrodynamic sedimentary structures and the position and attitude of various bounding surfaces between sedimentary units are among the most important kinds of observations to make. For average roadcuts or river cliffs a hundred meters or so long and a few tens of meters high, this fieldwork may take as long as a day or two to complete. For large outcrops, it may be useful to employ the service of an assistant to carry out the actual plotting, because once one is standing on a large outcrop it may be difficult to see precisely where one is on the mosaic. The assistant can base him or herself as closely as possible at the location from which the photograph was taken and enter measurements on the profile that are communicated by voice (or radio) from the geologist on the outcrop. Some highly detailed work is now being done by working across vertical cliff faces using climbing gear. Although introducing a significant logistic complexity to fieldwork, in some cases this is the only way to ensure that critical fine detail is collected at the exact spot that observations are needed.

If necessary, vertical stratigraphic sections can be constructed in the office from such lateral profile data. However, the wealth of detail yielded by a good quality lateral profile may make such section drawing superfluous. As noted by Anderton (1985), "Photographic techniques are sufficiently important in facies description that if one ever has to choose between taking a camera and a hammer into the field, take the camera."

Figure 2.3 is an example of a lateral profile constructed using these techniques; a brief discussion of the kinds of facies observations that can be made from profiles is presented in Sect. 3.5.11. When combined with paleocurrent data, detailed architectural reconstructions may be made from good outcrop data (Fig. 6.65).

3.4 Facies Associations and Models

3.4.1 The Association and Ordering of Facies

The term **facies association** was defined by Potter (1959) as "a collection of commonly associated sedimentary attributes" including "gross geometry (thickness and areal extent); continuity and shape of lithologic units; rock types ..., sedimentary structures, and fauna (types and abundances)" (Sect. 1.2.7). A facies association (or assemblage) is therefore based on observation, perhaps with some simplification. It is expressed in the form of a table, a statistical summary or a diagram of typical stratigraphic occurrences (e.g., a vertical profile). A **facies model** is an interpretive

device, which is erected by a geologist to explain the observed facies association. A facies model may be developed at first to explain only a single stratigraphic unit, and similar units may then be studied in order to derive generalized models.

Lithofacies group together into assemblages because they represent various types of depositional events that frequently occur together in the same overall depositional environment. For example, a submarine-fan environment typically contains canyon, channel, levee, overbank, proximal slope subenvironments and distal outer fan lobes, each of which is characterized by a distinctive lithofacies. These lithofacies become stacked into stratigraphic units because the environments shift through time, permitting different lithofacies to accumulate along any given vertical axis. The nature of these environmental shifts is often predictable, which means that the resulting lithofacies successions are equally predictable. For example, on the submarine fan, channels may shift in position and subenvironments within a turbidity current may move (very rapidly) down slope. Deltas and tidal flats prograde seaward, river channels and tidal inlets migrate, and so on. These processes may repeat themselves many times. This is the basis for the principle of cyclic sedimentation. It was clearly understood by one of the founders of modern sedimentary geology, Johannes Walther. He enunciated what has come to be known as Walther's law, the "rule of succession of facies," which states

> The various deposits of the same facies area and, similarly, the sum of the rocks of different facies areas were formed beside each other in space, but in a crustal profile we see them lying on top of each other... it is a basic statement of far-reaching significance that only those facies and facies-areas can be superimposed, primarily, that can be observed beside each other at the present time (Walther 1893–1894, as translated from the German by Middleton 1973).

The value of this rule to basin analysts is that it means lateral facies relationships can be predicted by studying them in vertical succession. Stratigraphic sections, such as outcrops measured up cliffs or drill hole sections, comprise most of our data base. Lateral facies relationships are much more difficult to study because they tend to be gradual and stretched out over considerable distances. Few outcrops are long enough to reveal such features in their entirety. Paleogeographic reconstruction therefore relies heavily on the examination of vertical profiles.

It must be stressed that Walther's law only applies to continuous sections without significant stratigraphic breaks. An erosion surface may mark the removal of the evidence of several or many subenvironments. It is therefore important to observe the nature of contacts between lithofacies units when describing sections. Contacts may be sharp or gradational. Sharp contacts may or may not imply significant sedimentary breaks. Periods of erosion or nondeposition may be revealed by truncated sedimentary structures, extensive boring, penecontemporaneous deformation, hardgrounds or submarine diagenesis. Sedimentary breaks may represent erosional episodes of a few seconds or minutes as a turbulent scour rolls by on a channel floor, or the shifting of sedimentary environments by autogenic processes, or they may represent the product of structural disturbances lasting hundreds of thousands to millions of years. The interpretation of the significance of sedimentary breaks should form an essential part of an overall stratigraphic analysis, as discussed in Sect. 7.6.

Walther's Law does not apply in some settings. For example, shallow-water carbonate platforms, lakes and evaporite environments may undergo rapid, basin-wide changes in water chemistry and physical conditions (temperature, water clarity) in response to climate change or modest sea-level change, with the sedimentary response being basin-wide. The vertical succession of facies is then not a reflection of a lateral shift in environments.

Given these constraints, the study of cyclic sedimentation has been one of the most popular and fruitful in sedimentology because, as Reading (1986, p. 5) stated, "it enabled geologists to bring order out of apparent chaos, and to describe concisely a thick pile of complexly interbedded sedimentary rocks." Interpretations of cyclic sedimentation

were one of the earliest approaches used to analyze sedimentary succession (Sect. 1.2.7) and form the basis of most modern facies models.

Statistical methods have been used for defining and describing lithofacies and biofacies associations. Cluster analysis methods were used by Imbrie and Purdy (1962) and Purdy (1963) to document and describe the associations of facies components on the modern Bahama Platform (Fig. 3.7). Multivariate statistical methods have also been used in the erection of biofacies associations (Mello and Buzas 1968; Ludvigsen 1978). These methods were described at some length in earlier treatments of stratigraphy by this author (Miall 1984), but this material has been omitted here, because such techniques, having served their purpose, have now fallen into disuse. After more than 40 years of intensive research, the major facies associations in the most important sedimentary environments have now been recognized (Reading 1996; Galloway and Hobday 1996; James and Dalrymple 2010), and such exploratory techniques are no longer needed. Attention has shifted to the investigation of the three-dimensional attributes of sedimentary successions, including architectural-element analysis, at the outcrop scale (Sect. 3.5.11), and seismic stratigraphy, at the basin scale (Sect. 6.3). The following is a brief summary of two statistical methods that were popular in the 1960s and 1970s.

Fig. 3.7 Cluster analysis (R-mode) of facies components on the modern Bahama Platform (Purdy 1963)

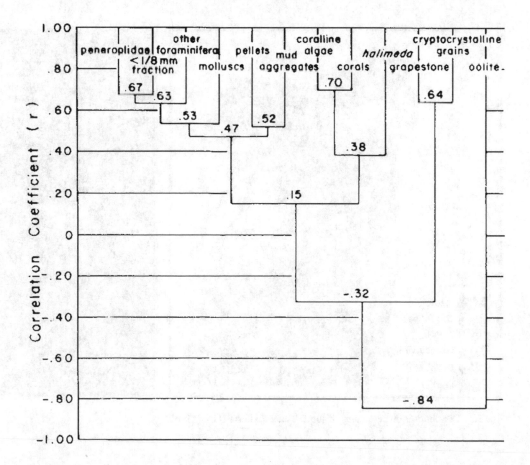

Imbrie and Purdy (1962) and Purdy (1963) carried out a landmark study of carbonate facies that involved a statistical analysis of data from modern sediments of the Great Bahama Bank. Purdy performed point count analyses under the microscope of 218 samples from the northern part of this bank. He recognized 12 major organic and inorganic components: coralline algae, *Halimeda* (calcareous alga), Peneroplidae (a family of Foraminifera), other foraminifera, corals, molluscs, fecal pellets, mud aggregates, grapestone, oolite, cryptocrystalline grains and silt grains < 1/8 mm in size. Cluster analysis was then used to determine which components tended to occur together; the results are shown in Fig. 3.7. From this information, five major lithofacies were defined, and by referring back to the sample data, these could be mapped as shown in Fig. 3.8. Purdy (1963) interpreted the lithofacies data in terms of physical and chemical processes operating over the bank. The work has been widely used and quoted as an example of a lithofacies study in carbonate rocks.

There has been considerable controversy over the methods used to define sedimentary cycles. Early attempts were subjective and qualitative. The culmination of the analysis was commonly the erection of a single model or ideal cycle (e.g., Visher 1965). This approach was criticized by Duff and Walton (1962) on the grounds that, in fact, this ideal

cycle may rarely be represented in the section. Random environmental changes, local reversals of the sequence or the effect of some external event, such as a flash flood or tectonic pulse, commonly intercede to modify the succession. The many analyses of the Carboniferous Coal Measure cyclothems prior to about 1960 illustrate this difficulty well, but cannot be entered into here for reasons of space (modern ideas about these cycles are discussed in Miall 2010).

As noted in Sect. 1.2.7, a step forward in the analysis of cyclicity was introduced by de Raaf et al. (1965) in the form of the *facies relationship diagram*, which is a semi-quantitative illustration of all possible vertical facies transitions, highlighting those that are more common (Fig. 1.6). The technique of Markov chain analysis, which was popular in the 1970s, added a certain statistical rigor to this type of analysis. Like the multivariate techniques described above, it results in a grouping of lithofacies into an assemblage, but it also has the additional property of revealing the order in which the lithofacies tend to occur. It can reveal not only a single-model cycle, but also the several or many statistically most probable variations on this theme. A Markov processes is one "in which the probability of the process being in a given state at a particular time may be deduced from knowledge of the immediately preceding state" (Harbaugh and Bonham-Carter 1970, p. 98). The methods

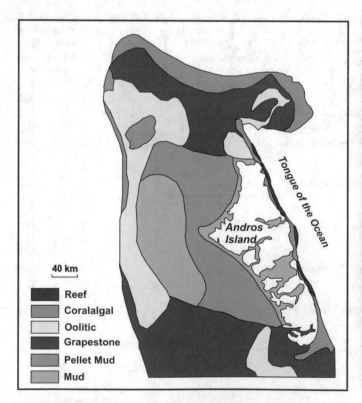
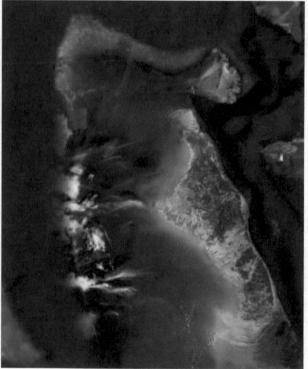

Fig. 3.8 Distribution of lithofacies of the Bahama Platform (Purdy 1963)

were described by Miall (1973) and Harper (1984), but are no longer used, because the extensive body of empirical research that has accumulated since the 1970s has thoroughly documented cyclicity in the sedimentary record, and attention has turned to architectural analysis and sequence stratigraphy.

3.4.2 The Theory of Facies Models

The concept of a facies model as a summary of a depositional environment and its products is a recent one in sedimentology. The term was first discussed at a conference reported by Potter (1959), but was used in the sense we now imply by the term facies assemblage (Sect. 1.2.7). The difference is critical. A facies assemblage is essentially descriptive, whereas a facies model attempts to provide an interpretation of a particular type of facies assemblage in terms of depositional environments. Much use is usually made of comparisons with actual modern environments, and

so it is now common practice to refer to these mental constructs as *actualistic facies models* (although Shea 1982 has criticized the use of the word "actualistic," partly on etymological grounds).

As noted in Sect. 1.2.7, the modern usage of the term facies model commenced with the classic paper by Walker (1976). The most recent expression of this concept is given by Dalrymple (2010a), drawing on Walker (1992), who described the process of constructing a facies model, using submarine fans as an example. Data from various sources (modern sediments, ancient deposits) are collected, and the wealth of information sifted to determine the important features that they have in common. The local variability is "distilled" away (Fig. 3.9). Various statistical procedures may be used to achieve this, such as those discussed in the previous section. The objective is to produce a general summary of submarine fans. But, as Walker (1992) stated, "what constitutes local detail, and what is general? Which aspects do we dismiss, and which do we extract and consider important?"

Fig. 3.9 The development of a facies model, using submarine-fan deposits as an example (Dalrymple 2010a; Modified from Walker 1976)

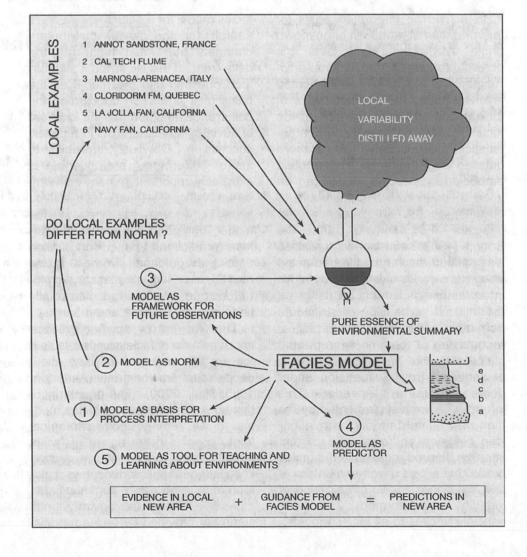

Walker (1976) stated that a facies model should fulfill four functions:

1. it must act as a **norm**, for purposes of comparison;
2. it must act as a **framework** and **guide** for future observations;
3. it must act as a **predictor** in new geological situations and.
4. it must act as a basis for **environmental interpretation**.

This is illustrated in Fig. 3.9. Facies models are a powerful tool for the interpretation of poorly exposed sediments because they suggest certain critical observations or clues for interpreting the sedimentary record. As with all such summaries and models, however, uncritical use may lead to loss of information or misinterpretation, because it is tempting to observe strata in terms of a preconceived model. If a facies model is being used properly in new field situations, each use may generate a refinement of the model or it may lead to the recognition of situations where the particular model is inappropriate. The development of a new model could then follow.

What does a facies model look like? The answer depends on how the model was constructed. Paleogeographic sketches, block diagrams and vertical profile logs are typical components of a published facies model. Some models are based primarily on geomorphology (e.g., deltas, alluvial fans), others on transport processes (e.g., turbidite, glacial till and eolian dune models) and others on organic processes (e.g., reefs). Many are governed by climatic and tectonic variables, which have not, in all cases, been exhaustively explored.

Many models are based primarily on the study of modern environments, for example, the threefold subdivision of deltas erected by Galloway (1975) used maps of typical deltas to convey information on lithofacies geometry. For continental environments, there is no problem seeing and describing a wealth of detail in modern settings, and this has led to the recognition that for fluvial deposits at least 16 distinct fluvial styles can be described (Miall 1996). Rapid evolution in eolian facies models came about by improved understanding of eolian bedforms (Sect. 3.5.4.7).

Facies models for subaqueous environments are less dependent on studies of modern settings, because the techniques of marine geology (geophysical profiling, shallow coring, sonar, satellite navigation) only began to enter general use in the early 1980s and did not, at first, permit very precise descriptions of lithofacies geometry and relationships. The first comprehensive submarine-fan model of Mutti and Ricci-Lucchi (1972) was based mainly on well-exposed Cenozoic fan deposits in Italy, and interpretations of other ancient deposits figure prominently in the first detailed discussions of this environment (e.g., Walker 1978). Models for storm-dominated shelf deposits incorporate hummocky cross-stratification (e.g., Hamblin and Walker 1979), a structure not reliably documented from modern environments until the 1980s. Howard (in Byers and Dott 1981, p. 342) explained how it may be obscured in cores of recent sediments by our methods of sampling the sea bottom and offered what may be the first observation from a modern offshore region (Howard and Reineck 1981). Application of side-scan sonar techniques to the study of submarine depositional systems brought about a revolution in our knowledge of shelf and deep-marine environments, yielding a knowledge of channels, bedforms and other depositional features that had a major impact on facies models for these environments (e.g., Bouma et al. 1985).

Many of these points are reviewed at greater length in Chap. 4, but this summary should serve to illustrate the varied state of the art of facies modeling at the present day. It is very much a case of work in progress, and basin analysts can potentially add much to our fund of knowledge by careful studies of newly explored stratigraphic units.

3.4.3 The Present as the Key to the Past, and Vice Versa

We have all heard the cliche "the present is the key to the past." It was one of the great generalizations to emerge from James Hutton's enunciation of the principle of uniformitarianism toward the end of the eighteenth century. Charles Lyell's work a half century later seemed to nail it down forever as a cornerstone of the still unborn science of sedimentology. Yet it is only true in a limited sense. Undoubtedly, the study of modern depositional environments provides the essential basis for modern facies studies, but there are at least a dozen major problems that emerge to confuse this work. It turns out that the past is also a very important key to the present that many geomorphological processes can best be understood by adopting a geological perspective and looking at the ancient record. The art of facies modeling is therefore a two-way process.

A fundamental problem with uniformitarianist interpretations of the sedimentary record is its fragmentary nature. It can be demonstrated that in clastic successions as little as 10% of the elapsed time which the succession represents is actually recorded by sediment (Miall 2014b). Where it can be documented by detailed sedimentary and chronostratigraphic work it can be demonstrated that sections typically consist of numerous short sections separated by minor sedimentary breaks. This does not materially affect work on an outcrop scale, because the time scales of sedimentary accumulation and erosional processes can be assessed from the field evidence and accommodated into interpretations. On a broader scale, that of whole depositional systems, sequences and regional studies, the issues of rates and time scales become more significant. We touch on this later in this section, and turn to the modern research on this topic in Chap. 8.

The greatest advantage in studying modern environments is that we can observe and measure sedimentary processes in action. We can measure current strength in the rivers and oceans; we can observe at least the smaller bedforms moving and evolving; we can measure temperatures and salinities and study the physics and chemistry of carbonate sedimentation and the critical effects of organic activity in the photic zone (shallow depth zone in the sea affected by light penetration). We can sample evaporite brines in inland lakes and deep oceans, such as the Red Sea, and deduce the processes of concentration and precipitation. We can put down shallow core holes and study the evolution of the sedimentary environment through the recent to the present day. Many physical, chemical, thermodynamic and hydrodynamic sedimentary models have resulted from this type of work since the 1960s. As noted in the previous section, for some sedimentary environments, work of this type has constituted the major advance in our understanding. Examples of early facies models that depended largely on studies of modern environments include those for deltas (Coleman and Wright 1975; Galloway 1975), barrier islands (Bernard et al. 1959; Dickinson et al. 1972), tidal inlets (Kumar and Sanders 1974), clastic tidal flats (Evans 1965; Van Straaten 1951, 1954; Reineck and Singh 1980), tidal deltas (Hayes 1976) and sabkhas (Shearman 1966; Kinsman 1966, 1969).

Some of the most difficult problems to resolve when using actualistic models stem from the inadequate time scale available to us for observation purposes. A very persistent geologist may be stubborn enough to pursue the same field project for 10, perhaps even 20, years. Aerial photographs might push observations of surface form as far back as about 1920 at the earliest. Old maps may go back for another 100 years or more, but become increasingly unreliable. Weather records may be available back into the nineteenth century; stream gauge data have been collected for only a few decades. It is difficult to assess the relevance of a 100-year record to a geological unit that may have taken a million or more years to accumulate. Have the last 100 or so years been typical? Were the same sorts of processes occurring at the same rates in the distant geological past? As discussed in the following paragraphs the answer is a qualified maybe. The reader is also referred forward to Chap. 8, where the issue of assessing elapsed time in the geologic record is discussed at length.

The most important aspects of this question are the difficulty of assessing the importance of ephemeral events and judging the preservation potential of deposits we can see forming at the present day. Dott (1983) and Reading and Levell (1996) have called attention to the need to distinguish between normal and catastrophic sedimentation. Normal processes persist for the greater proportion of time. Net sedimentation is usually slow. It may be nil or even negative if erosion predominates. Normal processes include pelagic settling, organic growth, diagenesis, tidal and fluvial currents (Reading and Levell 1996). Reading and Levell (1996) distinguished catastrophic processes as those that "occur almost instantaneously. They frequently involve 'energy' levels several orders of magnitude greater than those operating during normal sedimentation." They may deposit a small proportion of the total rock and give rise to only an occasional bed, or they may deposit a large proportion of the total rock and so become the dominant process of deposition. Examples of catastrophic processes include flash floods in rivers, earthquakes, hurricanes, tsunamis and sediment-gravity flows. Although geologists have studied many modern flash flood deposits and the effects of several recent hurricanes (e.g., Hayes 1967), we cannot be sure that their magnitude and frequency at the present day are the same as that in some past period of interest without attempting to obtain some geological perspective from studying the ancient record. The most violent and geologically important event may be one that only occurs every 200 or 500 years and has not yet been seen. Sediment gravity flows are thought to be the chief agent of erosion of submarine canyons, and have now been observed or monitored on several continental margins (Piper and Normark 2009), but major events are rare. Jobe et al. (2018) examined the recurrence interval and volume of turbidite events on a number of Quaternary fan systems. Costa and Andrade (2020) provided an introduction to a special journal issue on tsunami deposits.

Research by B. R. Pratt has suggested that the sedimentary products of several important sedimentary processes are underrepresented in the published record of the geological past. This is probably because there have been insufficient observations of the results of modern events to provide adequate descriptive data for routine observation and analysis. Among such processes Pratt suggests are earthquakes (generating seismites) and tsunamis. He has carried out many research projects, particularly in Precambrian and early Paleozoic successions to explore the nature of the evidence (e.g., Pratt 2001, 2002; Pratt and Bordonaro 2007). Tsunamis triggered by earthquakes along the Cascadia subduction zone of western North America have left a sedimentary record in coastal marshes along the west coast, providing some analogue data for interpreting the ancient record. Clague et al. (2000) and Witter et al. (2012) discussed specific examples, including major events in 1700 and in 1964.

Ager (1973, 1981) and Dott (1983) argued that many stratigraphic sequences contain more gaps than record, and that significant proportions of the sedimentary record are deposited in a very short time by particularly violent dynamic events. Such events (hurricanes, sediment-gravity flows, flash floods, etc.) are rare and difficult to study in action. Most of our energy as sedimentologists is expended in studying the less violent processes that occupy most of geological time but may contribute volumetrically far less sediment to the total record. Ager (1986) re-examined a well-known basal conglomerate in the Jurassic Lias of Wales, and concluded that it was a storm deposit formed very rapidly. In fact, he facetiously suggested that (p. 35) "it all happened one Tuesday afternoon," a remark that was deliberately designed to draw attention to the difference between "normal" and "event" sedimentation. In Chap. 8, we discuss the current state of knowledge of sedimentation rates, and it is argued that the stratigraphic record represents an amalgam of the products of numerous sedimentary processes acting over a wide range of rates and time scales (Miall 2015, 2016).

Studies of modern environments also suffer from the fact that many of the deposits we see forming at the present day are lost to erosion and never preserved. Thus, the geological record may be biased. The bias may be in favor of more deeply buried sediments, which the geologist, scratching the surface with trenches and box cores, never sees. For example, Picard and High (1973) published a detailed study of the sedimentary structures of modern ephemeral streams, but many of the structures are sufficiently ephemeral to be rarely, if ever, found in the ancient record. Fluvial flash flood deposits commonly are preferentially preserved because they infill deep scours below the normal level of fluvial erosion (Miall 1996, Chap. 8). Many of our fluvial facies models are based on studies of modern rivers in upland regions undergoing net degradation. How relevant are they to research in some of the great ancient alluvial basin that by the thickness of preserved deposits demonstrate a long history of aggradation? Facies models for barrier and shoreface environments are subject to similar constraints. For years, the classic Galveston Island (Texas) model of coarsening-upward beach accretion cycles dominated geological thinking (Bernard et al. 1959), but more recently it was realized that the barrier sediments may be removed by lateral migration of deep tidal inlets and that many barriers consist of superimposed inlet deposits with a superficial skin of wave-formed shoreline sediments (Kumar and Sanders 1974). Hunter et al. (1979) argued that many shallow-subtidal deposits are systematically removed by rip currents and are never preserved in the geological record.

Some of the best studies of modern environments are those that use shallow drill cores to penetrate into pre-recent deposits. Such sediments can be said to have "made it" into the geological record, and yet they can be placed in the context of a still extant and presumably little modified modern environment. Some work on modern turbidites (Bennetts and Pilkey 1976), anastomosed rivers (Smith and Smith 1980) and reefs (Adey 1975; Adey et al. 1977; Shinn et al. 1979) is of this type.

In two important ways, the late Cenozoic has been quite unlike most of the more distant past, and is therefore a misleading laboratory for reconstructing some sedimentary environments. The Pleistocene ice age generated several geologically rapid changes of sea-level culminating in a major rise and transgression since about 12,000 years ago. Second, the present configuration of continents and oceans is a unique pattern, different from any in the past because of the long history of sea-floor spreading, rifting, subduction and suturing. This plate movement has had an important effect on some of the broader aspects of facies models. For example, the Tethyan Ocean dominated tropical regions until it was closed by the breakup of Pangea during the Mesozoic. The extreme cold of the Antarctic continent and the vigor of the oceanic currents and weather systems surrounding it did not become fully established until the separation of the continent from the other Gondwana continents around 30 Ma.

Because of the recent sea-level rise, modern continental shelf, shallow-marine and coastal plain deposits around the world have been formed under transgressive conditions. Sea level was approximately 150 m lower during the Plio-Pleistocene glacial phases, rivers graded their profile to mouths located near the edge of present continental shelves and carbonate platforms, such as the Bahamas, were exposed to subaerial erosion and may have developed extensive karst systems. Submarine canyons were deeply entrenched by active subaerial and submarine erosion (Shepard 1981; McGregor 1981). During the rapid transgressions that followed in interglacial phases, shoreline sands were continually reworked. On the Atlantic shelf of North America, extensive barrier islands were formed and receded into their present position (Swift 1975). River valleys were drowned and filled with estuarine and deltaic deposits. Submarine canyons were commonly deprived of their abundant supply of river-borne detritus, as this was now deposited on the landward side of a widening and deepening continental shelf. This had a drastic effect on the rate of growth of some submarine fans, such as the Mississippi fan, which has become essentially dormant. Carbonate sedimentation began afresh in warm, detritus-free waters, over resubmerged platforms, but the local water depths, circulation patterns and facies distribution may have been partly controlled by erosional topography (Purdy 1974; although this is disputed by Adey 1978). We therefore have excellent modern analogues for studies of rapid transgressions in the geological record, but few for rapid regressions or for periods of stillstand. The

Mississippi and other large deltas are good examples of regressive deposits built since the last ice age, but they represent an environment that may be characterized by unusually rapid progradation. Some have maintained that most modern continental shelves are covered by relict sediments, implying that their study may not be of much geological relevance (Emery 1968), but more recent work has shown that the dynamic effects of tidal currents and storms do in fact result in continual change (Swift et al. 1971).

Miall (2014a, b; 2015) has argued that basing stratigraphic interpretations on comparisons with the post-glacial record is a mis-application of the principal of uniformitarianism for two important reasons: (1) firstly, the issue of preservation—the present post-glacial record exemplifies the unfinished nature of the geological preservation machine. A future glacioeustatic fall in sea level would remove much of the sedimentary record along continental margins and within estuaries, and subsequent (future) cycles of rise and fall would generate a long-term geological record by the superimposition of the lowermost fragments of successive cycles. (2) The rates of processes calculated from study of the post-glacial record (e.g., fluvial and deltaic channel filling and avulsion; lobe switching on deltas and submarine fans) are orders of magnitude greater that the rates that can be calculated from apparently similar processes in the geological past, suggesting that uniformitarianist comparisons of processes may be incorrect. For example, Miall (2014a) suggested that changes in fluvial channel stacking patterns in the ancient record are not reflections of changing rates of sedimentary accommodation but record proximal–distal shifts in fluvial styles in response to allogenic forcing. We explore these points further in Chap. 8. Holbrook and Miall (2020) offered an in-depth discussion of the implications of missing time and the hierarchies of scale in sedimentation and preservation, to which advanced readers might wish to turn.

Earth's climates and paleogeography have changed dramatically during its 4.5-Ga history. Ancient environments and stratigraphic configurations have been described from the stratigraphic record that have no modern counterparts, which complicates the work of facies analysis. We discuss some of these issues below, and refer to a few of the unique periods in Earth's stratigraphic history in Sect. 7.9.

Rapid transgressions may have occurred commonly during other ice ages in the geological past (Miall 2010), but most other changes in the geological past were somewhat slower. For two lengthy periods in the Phanerozoic, our suite of modern analogues is particularly inadequate. These were the times of high global sea-level stand during the Ordovician–Silurian and again in the Cretaceous, when vast areas of the world's continents were covered by shelf seas. We simply do not have modern equivalents of these huge inland seas, and many of the large shelf seas that do exist, such as the Bering Sea and Yellow Sea, have yet to be studied in detail. Likewise, there are no modern analogues for the very large evaporite basins that exist in the geological record (James et al. 2010).

Ginsburg (in Byers and Dott 1981) discussed the problem of interpreting Cambro-Ordovician carbonate banks of the North American craton. These are up to hundreds of kilometers long and thousands of kilometers wide. "The dilemma is how the vast extent of the banks, most of which suggest carbonate production and deposition in but a few meters or less of water, could all be bathed in normal marine or slightly restricted water. Would not such banks form major circulation barriers?" (Ginsburg, as reported by L. C. Pray, in Byers and Dott 1981). It has been suggested that the banks are actually diachronous, and developed by seaward progradation, or they may have been crossed by "irregular to channelized deeper water areas facilitating water circulation." Tidal currents certainly would have been able to assist with the latter. Careful biostratigraphic work has now demonstrated the diachronous nature of these tidal deposits. Runkel et al. (2007) showed that the Sauk sequence in Wisconsin consists of suites of superimposed, extremely low-angle, clinoform sets.

Climatic patterns and the network of oceanic currents are controlled by global plate configurations. In many respects, our present geography is unique. Therefore, we have modern analogues for situations that may not have existed in the past and, conversely, we cannot replicate certain conditions that did exist. For example, the Mesozoic Tethyan Ocean and the Pangea supercontinent had profound effects on climate and water circulation and hence on sedimentation patterns, and we have only generalized models for interpreting them. The study of paleoclimates is beyond the scope of this book, but is considered in an excellent review by Frakes et al. (1992).

The last group of problems with the actualistic modeling method arises from the important effects organisms have on sedimentary processes. Plants and animals have, of course, both evolved profoundly since Archean time. Therefore, in many environments, sedimentation is controlled or modified by sediment–organism relations that did not exist, or were different, in the geological past (James et al. 2010). Many authors, beginning with Schumm (1968), have speculated on the implications of the evolution of land plants on fluvial patterns. Vegetation has a crucial effect on stabilizing channel banks, colonizing bars and islands, as controlling chemical weathering, sediment yield and discharge fluctuations. Our typical braided and meandering fluvial facies models reflect these effects as they have been studied in temperate, humid, and hot and cold arid climates. Going back in time, the evolution of grasses in Mesozoic time must have changed geomorphic patterns profoundly. Abundant land vegetation is thought to have first appeared in the Devonian, and prior to that time the majority of river channels may have been unconfined, ephemeral and braided, as in modern arid regions (Davies and Gibling 2010a, b). Modern deserts are commonly used as analogues for pre-Devonian rivers, but there is no reason why they had to be arid. However, we have no modern analogue for a humid, vegetation-free environment with which to study the pre-Devonian except, possibly, the south coast of Iceland.

Turning to shallow-marine environments, the same kinds of difficulties apply. H. E. Clifton (in Byers and Dott 1981) pointed out that salt marsh vegetation, so important on modern tidal flats, did not exist prior to the Cenozoic. Similarly, foraminifera, which provide sensitive bathymetric indicators for younger sediments, particularly those of Cenozoic age, did not exist in the Paleozoic. James et al. (2010) discussed the implications of these various evolutionary developments on changes in carbonate depositional environments.

The ecology of forms that are now extinct may be difficult to interpret, which makes them less useful in facies studies (Sect. 3.5.7). Instead of providing independent, unambiguous environmental information, as do still living forms, it may be necessary to interpret them with reference to their sedimentary context, which itself may be of uncertain origin. Functional morphology is studied to determine probable habits, but many uncertainties may remain. In the Precambrian, most sedimentary units are entirely devoid of fossils, and here problems of lithofacies interpretation without supporting fossil evidence may become acute. Many Precambrian units have been reinterpreted several times for this reason, as different environmental criteria are brought to bear on particular problems. Long (1978) discussed many of these difficulties with reference to the recognition of fluvial deposits in the Proterozoic.

Reviews of Precambrian clastic sedimentation systems by Eriksson et al. (1998, 2013) documented several major differences in the preserved Precambrian record relative to that of the Phanerozoic (see Sect. 7.9). Marine shelf deposits are characterized by very uniform suites of sediments lacking distinctive vertical trends. Fluvial deposits are dominated by those of braided style. Foreshore deposits are rare, and eolian dune deposits appear to be absent in rocks older than about 1.8 Ga.

Meanwhile, the arrival of sophisticated rover vehicles on Mars is now allowing close-up photographic observation and stratigraphic and sedimentologic interpretations of rock outcrops, such as the exposure of fluvial deposits described by Salese et al. (2020). The vastly different and largely unknown atmospheric and climatic conditions of Mars make such studies particularly informative, and it seems likely that stratigraphy and sedimentology will soon move to the forefront of research into former environmental conditions on the planet.

3.4.4 To Classify and Codify, or Not?

We now know, more than two decades into the twenty-first century, a great deal about the origins of sedimentary rocks, and have amassed a considerable published record of case studies documenting sedimentary processes, facies and facies assemblages in a wide variety of modern settings, plus a wealth of detailed descriptions and interpretations of ancient units (e.g., James and Dalrymple 2010). These data are in danger of overwhelming the practitioner, and summaries and classifications are more needed than ever.

Stephen Jay Gould (1989) stated: "Classifications are theories about the basis of natural order, not dull catalogues complied only to avoid chaos." In a descriptive science like geology, there will always be a need for classifications. These serve two purposes, they attempt to create order out of apparent chaos, as aids to memory, and they are an attempt to understand genesis. The best classifications are those that highlight differences in origins of geological features, and are therefore particularly useful for purposes of prediction and extrapolation, for example, in geological exploration.

There will always be a creative tension between those who focus on the minute differences among geological attributes and those who like to simplify and generalize. This is the tension between those, the purists, who seek constantly to perfect our understanding of descriptive detail and of processes, and those, the synthesizers, who think they know enough details for the time being, and want to move onto larger problems. It is the same as the debate among taxonomists between the "lumpers" and the "splitters." Facies classification and facies models offer plenty of fuel for debate between the advocates of both sides of this argument. Anderton (1985) provided an excellent discussion of the various levels of complexity that are embraced by the modeling process (see also Pirrie 1998).

The purists like to point to classifications that have failed, as proof of the failure of the idea of classification. For example, Bridge (1993) used Mike Leeder's (pers. com.) example of the geochemical classification of igneous rocks proposed by Shand (1947), a complex system that did not catch on and was not used, no doubt because it was not based on any understanding of genetic reality that withstood the test of time. Likewise, Kay's (1951) classification of geosynclines was widely used until the 1960s, at which time it virtually vanished, to be replaced by classifications based on plate tectonics. Ingersoll and Busby (1995) proposed 27 varieties of sedimentary basin, based mainly on plate-tectonic principles—more types than there were geosynclines. Is this being objected to? No, in fact the contrary (e.g., see Dickinson 1997), because it is now recognized that this is a classification genetically based on plate-tectonic concepts, and therefore has deep meaning, much like a phylogenetic classification of organisms.

Replacement and abandonment are always likely to be the fate of classifications. But does this mean that we should avoid the attempt to classify? At the very least, classifications represent way stations on the road to perfect understanding. At best, they offer a common language for description and interpretation. This was why I proposed my classification and code scheme for fluvial deposits in 1977. Numerous descriptions of modern and ancient deposits were accumulating in the literature at that time, but most were described using unique systems of nomenclature, rendering comparative studies almost impossible. Reinterpretation of the descriptions in terms of a standard classification revealed a number of common themes in terms of facies assemblages and vertical profiles (Miall 1977), and this began the work of recognizing and systematizing the variety of processes in braided fluvial systems that is still continuing. The classification of fluvial sandstone facies is based in part on fluvial hydraulics. It has been demonstrated that common bedforms are predictable in form and are generated under predictable conditions (Ashley 1990; Sect. 3.5.4). This justifies the erection of a simple classification of those lithofacies that represent the preserved deposits of these bedforms. Revisions and improvements to this fluvial facies classification have not ceased. An updated version was offered by Miall (1996), but other workers have developed their own, and in fact the system has been adopted and adapted several times by others without reference to the original work (e.g., Uba et al. 2005), indicating that the ideas have, to a considerable extent, developed a life of their own.

Turning to fluvial facies models: the explosion of descriptive work in fluvial sedimentology has led to the discovery of numerous variants on the original simple suite of models, and this has led to a variety of attacks on the concept of the model itself, rather than a recognition that the models were only intended as mental concepts, as teaching aids and as temporary fixed points to aid in the comprehension of nature. The proliferation of facies studies led Dott and Bourgeois (1983) to remark that by the early 1980s fluvial facies models had "multiplied like rabbits so that every real-world example now seems to require a new model. Such proliferation defeats the whole purpose of the conceptual model by encouraging excessive pigeon-holing, which obscures rather than reveals whatever unity may exist among the variants." It was partly to manage this problem that I began to explore the concept of the architectural element as an object capable of simplification and classification at a sub-facies model level (Miall 1985).

Some textbooks detailing simple classifications have been very successful. The Geological Association of Canada "Facies Models" volume, now in its fourth edition (James and Dalrymple 2010), has sold more than 70,000 copies.

This work is popular because it is useful in teaching and preliminary interpretation, but nobody ever claimed that the book represents the last word in description and interpretation.

Warnings about the dangers of oversimplification in the use of classifications and models are common in the literature describing them (e.g., see Miall 1980, 1985, 1996, on my own fluvial models, and the general remarks of Bridge 1993), but this does not prevent such misuse. Don Gorsline (personal communication 1984) has told the story of a graduate student of his who came to him deeply upset one day and about to quit the program, because he could not make his rocks fit any of the standard models, and felt that he had failed as a sedimentologist. Gorsline had to gently point out to him the purpose of research, which is to do original things.

The debate continues. In Shanmugam's (1997) fascinating and provocative review of the use of the Bouma sequence model, he states: "Miall (1995, p. 379) asks, ' ... who would now object to the use of Bouma's (1962) five divisions (A-E) as a framework for the field description of turbidites?' I, for one, would." And then, in the Geotimes (February 1988, p. 32) review of advances in clastic sedimentology Neil Wells wrote, with reference to the book "The geology of fluvial deposits" (Miall 1996) the opinion that the "codification of facies and architectural elements [in this book] risks promoting rigid classification, superficial observation, and simplistic interpretation."

The use of predetermined classifications in clastic sedimentology was debated at length by Bridge (1993, 1995) and Miall (1995, 1996, p. 78, 89) with regard to the specific application of classification techniques to fluvial deposits. Such classifications are widely used, but their existence causes unease to some researchers. Shanmugam's discussion of the Bouma sequence concept and the turbidite mindset shows why this is so. He has demonstrated rather thoroughly the misuse of the original concept by subsequent workers. The misuse of the turbidite model by workers on deep-marine sandstones had even led to the description of something called the "non-turbulent turbidity current," which Shanmugam rightly called the "ultimate example of an oxymoron."

What has happened here, and is there a systemic problem with certain sedimentological methods? I do not think so. Shanmugam's critique of the Bouma model does not mean we should abandon the model, only use it more critically. Shanmugam's work is, in fact, an application of Walker's approach that we use a model as a guide to future observations in order to determine whether they fit or not. He also refers to Anderton's (1985) critique of modeling techniques and methods. But it is precisely because we have in our minds a good concept of what a "true" Bouma turbidite should look like, that we can readily appreciate how far off the track many sedimentological descriptions and interpretations have strayed, when someone like Shanmugam comes along and brings us up short with careful, detailed observation. It turns out that many deep-marine sands may not be turbidites at all (modern work on sediment-gravity flows is summarized in Sect. 3.5.5). But this does not prove that Bouma was wrong or that the concept of the facies model is wrong, only that these elegant concepts have been applied too carelessly.

Sequence stratigraphy, both as a body of concepts for the teaching of stratigraphy and as a research tool, has been heavily dependent on models, especially the by-now virtually classic papers of Posamentier et al. (1988), Posamentier and Vail (1988), and Van Wagoner et al. (1990). Yet this area, too, has been bedeviled by the misunderstanding and misuse of models, to the extent that the original proponents of the models have had to become apologists for them, pointing out in such overview papers as Posamentier and James (1993) and Weimer and Posamentier (1993) that the models were never intended to be of universal application. These controversies have now largely been resolved by the comparative, integrative work of Catuneanu (2006; see Sect. 7.7).

There is no question but that classifications and models in sedimentology have been used in too facile a manner as research tools by many workers, but this does not mean that the approach is wrong, only that all due caution is required in the application of the models. Classifications and models that genuinely reflect variations in the genesis of sedimentary units are likely to survive, thrive and be used, possibly with numerous revisions and accretions. Those that do not will be ignored.

3.4.5 Facies Analysis and Sequence Stratigraphy

As noted by Nummedal (in Nummedal et al. 1987, p. iii),

> The interpretation of the sedimentary geological record has been greatly stimulated over the past few years by rapid conceptual advances in "sequence stratigraphy", i.e., the attempt to analyze stratigraphic successions in terms of genetically related packages of strata. The value of the concept of a "depositional sequence" lies both in the recognition of a consistent three-dimensional arrangement of facies within the sequence, the facies architecture, and the regional (and inter-regional) correlation of the sequence boundaries. It has also been argued that many sequence boundaries are correlatable globally, and that they reflect periods of sea-level lowstand, i.e., sequence boundaries are subaerial erosion surfaces.

Sequence stratigraphy is not a radically new concept (Sect. 1.2.10). For example, many of the sedimentological ideas were discussed by Frazier (1974), and the first assemblage of sequences, in the modern sense, was established by Sloss et al. (1949). But the study of sequences has received considerable impetus since developments in seismic stratigraphy in the late 1970s reminded geologists of the ubiquity of sea-level changes throughout the geologic past.

Sequence stratigraphy is a combination of lithostratigraphy, allostratigraphy and facies analysis (Fig. 7.30). The vertical arrangement of facies is interpreted in terms of Walther's law, leading to reconstructions of the lateral movement through time of depositional environments (Chap. 5, Sect. 7.7). Sequences typically consist of transgressive and regressive half-sequences that together comprise repeated cyclic successions. Up to this point the study of sequences does not differ greatly from the analysis of vertical profiles for the purpose of documenting autogenic depositional events, for example, the progradation and lobe switching of a deltaic coastal plain complex. However, as noted by Nummedal, careful regional lithostratigraphic, allostratigraphic and chronostratigraphic correlation demonstrates that, typically, the sequences extend laterally for much greater distances than could be attributed to autogenic controls. The influence of a regional or global allogenic control, such as changes in climate or relative sea level, may then be suspected (Miall 1995, 2010).

3.5 Review of Environmental Criteria

In Chap. 2, a summary of major environmental indicators is given under the heading of what to look for in outcrop and subsurface sections. This list is by no means exhaustive, for example, it concentrates almost exclusively on what can be seen with the naked eye. However, it includes many of the sedimentary features vital to a generalized environmental interpretation and hence to an effective basin analysis. In this section, some indications are given of how to use this information. Beginners at the art of basin analysis, or those who wish to understand what their specialist colleagues are up to, may find this section both enlightening and confusing. The numerical precision or well-defined statistical error of laboratory-oriented geological studies is not to be found in the area of facies interpretation. Statistical methods may be used to aid in analyzing the composition of facies assemblages, but the business of interpreting their meaning depends heavily on qualitative study. The sedimentologist must be aware of the meaning and limitations of all the facies criteria visible and must be able to weigh the evidence of all these against each other. **Field context** is key. A good knowledge of published facies models and modern analogues is, of course, essential.

A common basin-analysis problem is that the geologist is faced with a new outcrop or core showing certain assemblages of lithologies, textures, structures and, perhaps, fossils, and may have no idea which facies model to turn to for assistance in making an environmental interpretation. The intent of this section is to review very briefly the types of interpretation that can be made from the principal kinds of sedimentological observation in order to provide some environmental clues and an entry into some of the crucial literature. Discussions of the physical and chemical conditions of formation of most sediment types are beyond the scope of this book.

In some of the notes that follow, siliciclastics and carbonates are treated separately because they require a different approach. This is particularly the case with interpretations of grain size and texture. For others, such as hydrodynamic sedimentary structures, there is no reason not to consider all rock types under the same heading. Structures such as crossbedding and ripple marks are commonly regarded as the domain of the sandstone specialists, but they also occur in carbonates and evaporites. Chemical sediments are often studied by geologists whose first interest is in the chemistry of their formation and diagenesis, but could usually benefit from the approach taken to sedimentology by clastic sedimentologists.

The emphasis throughout is on features that are most useful for constructing depositional environments and paleogeography. Diagenesis and geochemistry are not dealt with in this book.

It cannot be overemphasized that very few sedimentological criteria have an unambiguous environmental interpretation. Many years ago it was thought that ripple marks only occurred in shallow water, until oceanographers started taking photographs of the bottom of the oceans. More recently dish structures were thought to be indicators of submarine grain flows, but have now been found in fluvial and other deposits (Nilsen et al. 1977). The common trace fossil *Ophiomorpha* is a good indicator of shallow-marine environments, but J. Coleman (personal communication, 1976) reported finding it many miles inland in the deposits of the modern Mekong River. Marine it certainly is, but the Mekong River has an extensive marine salt wedge that flows far upstream during high tide. *Ophiomorpha* has also been found in outer shelf environments (Weimer and Hoyt 1964). Reliance should never be placed on a single structure or feature of the rocks or on a single method of analysis for making environmental interpretations. The geologist must review and assess all the evidence available.

3.5.1 Grain Size and Texture

In a general sense, the grain size of a clastic sediment indicates the relative amount of energy required to emplace the grains in their final resting place. This energy might have been derived from the force transmitted by air or water movement or it may represent downward movement under gravity. Most clastic sediments represent a combination of both of these processes.

Grain-size interpretation in siliciclastic rocks (Fig. 2.4) tends to be much simpler than it does in carbonates and evaporites because diagenesis in these chemical sediments frequently obscures original grain relationships. The grain size of carbonate sediments may be related entirely to in-place primary or diagenetic crystal growth and not at all to transport processes. The size of allochems in carbonate sediments is commonly determined by their organic origins. However, some chemical sediments show evidence of having behaved as clastic detritus at some stage in their formation, so that considerations of grain dynamics might provide useful environmental information.

There are two ways in which general grain-size data yield environmental information in siliciclastic sediments. Local vertical variations in mean or maximum grain size are frequently cyclic or rhythmic, and these, coupled with variations in sedimentary structures, are powerfully diagnostic. Analysis of vertical profiles for cyclic patterns has been discussed earlier in this chapter (Sect. 3.4.1) and is dealt with in greater detail in Sect. 3.5.10. The second aspect of grain-size information is the environmental information contained in the size distribution of individual samples. This subject has received an enormous amount of attention from sedimentologists, who have proposed a wide variety of statistical techniques for distinguishing the deposits of various depositional environments on the basis of some supposed environmental signature retained in the sample. Most of this attention has been directed toward the study of sandstones. It has long been realized that the hydraulic sorting effects of waves, wind and unidirectionally flowing water result in the movement of different populations of grains. Most sandstones are mixtures of several populations, but it has long been the hope that the right analytical methods would infallibly recognize these subpopulations based on some size or sorting criteria, enabling the depositional environment to be recognized. All such techniques depend on careful laboratory size analyses, preferably carried out using sieves or a settling tube or, failing this in the event of lithified samples, counts of grains in thin section. Various methods were discussed by Visher (1969), Solohub and Klovan (1970), Glaister and Nelson (1974), Friedman and Sanders (1978), and Friedman (1979). None of these methods is fully reliable because of the problem of inherited size distributions in the case of second or multicycle sands, and the effects of diagenesis (cementation, secondary porosity development) on lithified sandstones. Extensive laboratory work and data analysis are required to complete a rigorous grain-size analysis. Even then the results are usually ambiguous and, in this writer's opinion, do not justify the great effort expended. Accordingly, the use of this tool is not normally recommended. The method does, perhaps, have some uses in the analysis of well cuttings or sidewall cores, where these are the only samples available, because the technique can be applied to small samples, whereas most of the observational techniques discussed here cannot. Even here, visual inspection of petrophysical logs may provide equally reliable environmental information in a fraction of the time. A new, comprehensive, descriptive classification of clastic sediments, including those consisting of very mixed grain-size populations, was offered by Farrell et al. (2012).

A useful technique in the study of conglomerates is the measurement of maximum clast size. Normally, this is determined by averaging the intermediate diameter of the 10 largest clasts present at the sample level. Such measurements have been used in the study of grading and cyclic changes in subaerial alluvial fans (Gloppen and Steel 1981; Steel 1974) and in subaqueous resedimented conglomerates (Nemec et al. 1980). It is found that there is a direct relationship between maximum particle size and bed thickness for deposits formed by subaerial or subaqueous debris flows, indicating that the beds were formed by single depositional events without subsequent reworking (Bluck 1967; Nemec et al. 1980).

Grain size, fabric and texture are, together with sedimentary structures and vertical profiles, important criteria for distinguishing flow type in the study of sediment-gravity flows and their deposits. The recognition of particular grain-size populations in a deposit, the internal sorting and grading of each population, and the vertical and lateral relationships of the populations to each other have been interpreted in terms of grain-support mechanisms and their evolution during the passage of flow events (Middleton and Hampton 1976; Lowe 1979, 1982; Postma 1986; See Sect. 3.5.5). Magnetic fabrics have been used to help distinguish sediment-gravity flow deposits from other types of

deposits. For example, Shor et al. (1984) used this technique in an attempt to discriminate turbidites from contourites. Gravenor and Wong (1987) and Eyles et al. (1987) employed magnetic-fabric data in studies of glacigenic deposits.

Grain-size and textural studies in conglomerates were used by Walker (1975) to suggest models of deposition on submarine fans (Fig. 3.10). The disorganized-bed model is thought to represent rapid, clastic deposition on steep slopes, perhaps in the feeder submarine canyon at the head of a fan. Debris flows passing to the inner fan are thought to exhibit first an inverse to normally graded texture, passing down-current into a graded-stratified type. Inverse grading develops as a result of dispersive pressure, possibly including a "kinetic sieve" mechanism whereby smaller clasts fall down between the larger ones. Hein (1982) extended this work to pebbly and massive sandstones.

Similar studies of texture and fabric in glacigenic deposits have helped to distinguish the various mechanisms of formation of poorly sorted conglomerates. Early workers referred to all glacigenic conglomerates as **till** (tillite is the lithified variety), a deposit formed by the accumulation of unsorted debris below or in front of grounded glacial ice. However, careful facies analysis of many Quaternary and more ancient glacial deposits has shown that most are water laid (Eyles and Eyles 2010). Analyses of clast fabric and the magnetic fabric of the matrix have been used as a supplement to outcrop facies studies in order to determine depositional processes (Eyles et al. 1987). Glacial conglomerates may have originated as tills, but have moved down local depositional slopes as slumps or sediment-gravity flows or by the accumulation of material moved into the basin by ice rafting. Fabric studies can help distinguish these processes. Eyles et al. (1983) suggested referring to all poorly sorted conglomerates by the nongenetic term **diamictite** (unlithified

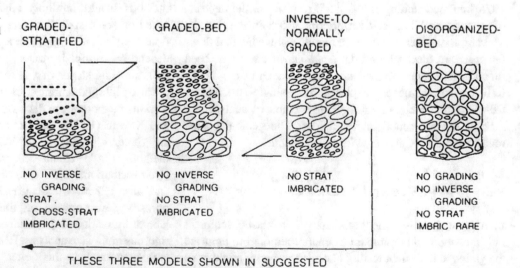

Fig. 3.10 Stratigraphic-textural-fabric models for resedimented conglomerates (Walker 1975)

variety: diamict or diamicton), and they erected a lithofacies classification for diamictites that places primary emphasis on the presence or absence of grading and stratification, and the density of clast packing (Sect. 4.2.4). Schultz (1984) applied this same classification to the analysis of diamictites formed in a subaerial alluvial fan setting.

In the conglomerate deposits of alluvial fans and other gravelly rivers, grain-size variations commonly reveal crude stratification and large-scale crossbedding (Fig. 2.9b). The stratification is the product of longitudinal-bar growth or superimposition of debris flows. Individual bar deposits may show an upward decrease in grain size (Miall 1996). True grading is rare in fluvial gravels formed by traction transport.

Many resedimented conglomerates and pebbly sandstones contain a clast fabric with the long (a) axis dipping upstream, indicating that they were deposited from a dispersed sediment mass without bedload rolling (Walker 1975; Hein 1982). This contrasts with the common imbrication of flat clasts in conglomerates deposited by traction transport (Fig. 2.12), such as in gravelly rivers (Rust 1972). The fabric is therefore a useful indicator of transport mode.

For most carbonate sediments, a different approach must be taken to grain size and texture. Most grains, both fine and coarse, are locally produced at least in part by organic

activity. The mean grain size or the size of the largest grains may mean little in terms of local hydraulics. Dunham (1962) has pointed out that it is more useful to focus on the fine material in the rock, because this is an accurate measure of the strength of winnowing currents. Dunham (1962) devised a system for classifying the texture of carbonate rocks based on whether the sediment is a self-supporting grain framework or whether the grains are enclosed in a micrite matrix. The spectrum of packing categories erected by Dunham is shown in Fig. 3.11. The terms **mudstone**, **wackestone**, **packstone**, **grainstone** and **boundstone** are widely used for describing the texture of carbonates. In general, the amount of current winnowing energy implied by the rock type name increases from left to right. A similar carbonate textural spectrum was devised by Folk (1962), who used a two-part terminological system (Fig. 3.11). The suffixes **micrite** and **sparite** are used for the predominant matrix type. Sparry rocks are packstones and grainstones from which the micrite matrix has been removed by winnowing and replaced by coarse calcite cement. Documenting textural relations in this way is an important first step in subdividing the various subenvironments of carbonate platforms. Wilson (1975, Chap. 1) discussed the subject at greater length.

Fig. 3.11 The textural spectrum in limestones and the two basic classifications used for limestones (Dunham 1962; Folk 1962)

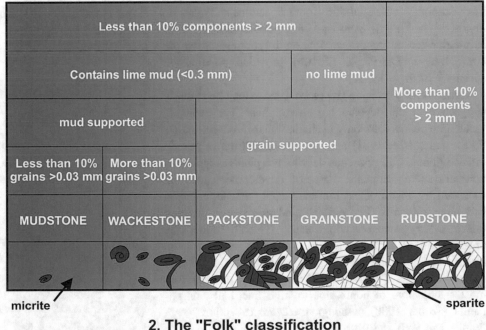

LIMESTONE CLASSIFICATION
1. The "Dunham" classification

Less than 10% components > 2 mm				More than 10% components > 2 mm
Contains lime mud (<0.3 mm)			no lime mud	
mud supported		grain supported		
Less than 10% grains >0.03 mm	More than 10% grains >0.03 mm			
MUDSTONE	WACKESTONE	PACKSTONE	GRAINSTONE	RUDSTONE

micrite   sparite

2. The "Folk" classification

micrite	sparse- to packed biomicrite	poorly washed biosparite	biosparite	coarse biosparite

Schreiber et al. (1976) and Schreiber (1981, 1986) have shown that many evaporite deposits, particularly those formed in marginal-marine environments, can be treated as clastic deposits in that they show similar bedding and sedimentary structures (Fig. 4.66). The same is true for some carbonate rocks, particularly oolite sand shoals (Ball 1967) and carbonate turbidites and debris flows (Mountjoy et al. 1972), which are formed by clastic sorting and redistribution processes, although consisting of carbonate particles (Fig. 4.61). Considerations of grain size (cyclic or lateral changes, grading) and texture (packing) might provide useful environmental information in these cases.

3.5.2 Petrology

The composition of the major rock-forming constituents of siliciclastic sediments (including mudrocks) is not directly diagnostic of environment. Krynine (1942) suggested that many sandstone types were characteristic of particular tectonic-sedimentary environments, for example, **arkoses** (feldspathic sandstones) supposedly represent nonmarine sediments derived from granitic orogenic complexes and **greywackes** (lithic sandstones) represent early "geosynclinal" sedimentation. These interpretations have been discarded, although Dickinson and Suczek (1979) and Schwab (1981) have shown that sandstone composition may closely reflect the plate-tectonic setting of the basin. However, this does not necessarily translate into depositional environment. For example, volcaniclastic forearc sediments may be deposited in fluvial, lacustrine, marginal-marine, shelf or submarine-fan settings, depending on the continental-margin configuration. Dalrymple (2010b) provided an updated summary of these ideas.

Certain minor components of sandstones may be strongly suggestive of the depositional environment. For example, glauconite pellets form only in shallow-marine environments (Odin and Matter 1981) and are rare as detrital or resedimented grains. Carbonaceous debris from plants is typical of nonmarine environments. Paleosol development may be indicated by lenses of calcium carbonate (caliche, calcrete) and other minerals (Fig. 4.6). Abundant red iron staining indicates oxygenated environments, typically either the preservation of oxidized states in detrital particles (Van Houten 1973) or the production of oxidized colors during early diagenesis (Walker 1967). Red beds are therefore mostly indicative of nonmarine or high intertidal environments (Turner 1980), although there are exceptions (e.g., Franke and Paul 1980).

Within a given basin, detrital composition may reflect variations in source area or depositional environment. These data may therefore be used as a paleocurrent indicator (Sect. 6.5.1) and may assist in the stratigraphic correlation of units formed under the same hydraulic conditions. Davies and Ethridge (1975) showed that detrital composition varied between fluvial, deltaic, beach and shallow-marine environments on the Gulf Coast and in various ancient rock units as a result of hydraulic sorting and winnowing processes and chemical destruction. Thus, although composition is not environmentally diagnostic, it may be useful in extending interpretations from areas of good outcrop or core control into areas where only well cuttings are available. Detrital composition, including clastic petrography and chemistry, is commonly a useful aid to stratigraphic mapping and the reconstruction of paleogeography (Sect. 6.6).

For carbonate sediments, in contrast to siliciclastics, petrographic composition is one of the most powerful environmental indicators (James et al. 2010). Carbonate facies studies, therefore, require routine thin-section petrographic analysis, nowadays supplemented by the use of the scanning electron microscope and cathodoluminescence (Wilson 1975; Scholle 1978; Flügel 1982; Machel 1985). Most carbonate grains are at least partly organic in origin, including micrite mud derived from organic decay or mechanical attrition, sand-sized and larger particles consisting of organic fragments, fecal pellets, grapestone and ooliths, all of which are produced in part by organic cementation processes, and boundstones or biolithites, formed by framework-building organisms. Most carbonate particles are autochthonous; therefore, an examination of the composition of a carbonate sediment is of crucial importance.

These and other differences between carbonate and siliciclastic sediments are summarized in Fig. 3.12. Textural classifications of carbonate sediments were discussed in the

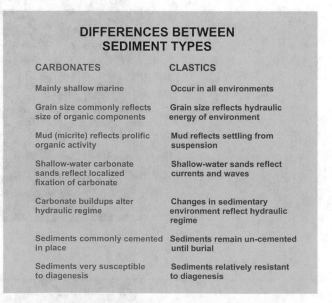

DIFFERENCES BETWEEN SEDIMENT TYPES

CARBONATES	CLASTICS
Mainly shallow marine	Occur in all environments
Grain size commonly reflects size of organic components	Grain size reflects hydraulic energy of environment
Mud (micrite) reflects prolific organic activity	Mud reflects settling from suspension
Shallow-water carbonate sands reflect localized fixation of carbonate	Shallow-water sands reflect currents and waves
Carbonate buildups alter hydraulic regime	Changes in sedimentary environment reflect hydraulic regime
Sediments commonly cemented in place	Sediments remain un-cemented until burial
Sediments very susceptible to diagenesis	Sediments relatively resistant to diagenesis

Fig. 3.12 The differences between carbonate and clastic sediments (adapted from James 1984a)

previous section (Fig. 3.11) and have been expanded to include compositional details of framework-building organisms by Embry and Klovan (1971) and Cuffey (1985). Interpretation of ancient carbonate sediments is complicated by the fact that the organisms that generate carbonate particles have changed with time (James et al. 2010, Fig. 12). However, there are many similarities in form and behavior between modern and extinct groups, so that actualistic modeling can usually be carried out with caution (see also Sect. 4.3). Another problem is that diagenetic change is almost ubiquitous in carbonate rocks and may obscure primary petrographic features (this is briefly discussed in Sect. 4.3.1). Ginsburg and Schroeder (1973) showed that some carbonates are converted contemporaneously from reef boundstones or grainstones to wackestones by continual boring, followed by infill of fine-grained sediment and cement. Mountjoy (1980) suggested that many carbonate mud mounds may owe their texture to this process.

The Chalk of northwest Europe is a unique, distinctive carbonate deposit consisting primarily of coccoliths, with minor constituents of calcareous microfossils and a specialized fauna of brachiopods, bryozoans and low-Mg shells. The absence of an aragonite fauna has long been attributed to early dissolution. Tagliavento et al. (2021) have now demonstrated through isotopic analysis that the smallest size fraction of chalk (1 to 5 μm) consists of microcarbonate precipitates generated from the early dissolution of aragonite fossils.

Most environmental interpretation of carbonates is based on thin-section examination using a microfacies description system such as that erected by Wilson (1975; Fig. 3.4 of this book). Wilson was able to define a set of standard facies belts for the subenvironments of a carbonate platform and slope (Fig. 3.13), each characterized by a limited suite of microfacies reflecting the variations in water depth, water movement, oxygenation and light penetration. A classic

example of a carbonate petrology study, the analysis of modern sediments of the Bahama Platform, is described briefly in Sect. 4.3 (see Figs. 3.7 and 3.8). Each facies assemblage can readily be related to environmental variables, such as the quiet-water pelletoidal facies and the high-energy skeletal sand and oolite lithofacies.

Many carbonates consist of dolomite rather than limestone. Dolomite was early suggested to be largely a replacement of preexisting unconsolidated lime sediment by the addition of magnesium from seawater to the sea-floor calcite accumulations before burial and subsequent lithification could form beds of limestone (van Tuyl 1916). Even at that time, however, there were also other suggestions concerning the origin of dolomite beds, such as dolomite accumulation as a primary precipitate, or as replacements of limestone during burial lithification after deposition (van Tuyl 1916). In most occurrences, however, dolomite is clearly a replacement, as indicated by the presence of dolomite rhombs penetrating allochemical particles such as shell fragments or ooliths. For the following paragraphs concerning the origin of dolomite, I am indebted to D. W. Morrow (Pers. Com., 2021).

In dolomitized strata, dolomite may be grouped according to average crystal size into one of three groups (Morrow 1982), microcrystalline (up to ∼ 20 μm), macrocrystalline (∼ 20 μm to ∼ 2.0 mm, i.e., sucrosic dolomite) and megacrystalline (> ∼ 2.0 mm). The primary controls on dolomite crystal size are the texture and average crystal size of the precursor limestone or sediment, the degree of dolomite supersaturation, and the dolomite precipitational temperature (Sibley and Gregg 1987; Lucia 1995; Huang et al. 2014). Dolomititization that occurs in low temperature earth surface or shallow burial settings tend to be more fabric retentive, whereas late diagenetic high-temperature dolomitization, such as the megacrystalline white dolospars of "hydrothermal dolomites" are fabric destructive (Davies and

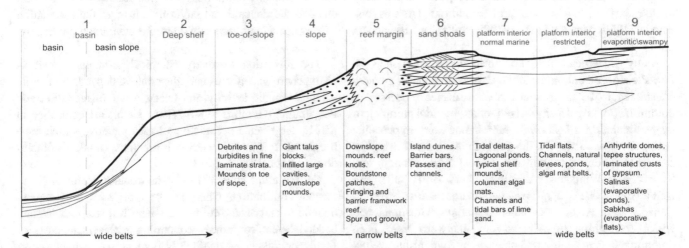

Fig. 3.13 The standard facies belts of Wilson (1975), as modified by Schlager (2005, Fig. 4.3)

Smith 2006; Morrow 2014), although even these late-post-lithification dolomites retain aspects of some megascopic elements, such as larger fossil fauna and early pre-burial fractures, such as in the Presqu'ile and Manetoe dolomites of Western Canada (Morrow et al. 2002; Morrow 2014).

Fabric preservative penecontemporaneous dolomite replacement of lime sediments (aragonite) within present-day sabkhas bordering the Persian Gulf (Patterson and Kinsman 1982) occurs within a very distinctive peritidal and supratidal facies succession several meters thick of dolomitized facies capped by an interval of displacive anhydrite nodules (Wood and Wolfe 1969; McKenzie et al. 1980). Supratidal dolomite crusts form above intertidal carbonates on Modern (Holocene) tidal flat settings of the Caribbean Islands and have been found within dolomitized Plio-Pleistocene strata beneath these islands (Budd 1997). Aragonite and calcite sediments are also actively dolomitized in shallow island and shelf lagoonal settings in areas where seawater is evaporated at high rates, such as Bonaire (Lucia and Major 1994) where "seepage refluxion" (Adams and Rhodes 1960) has dolomitized subjacent sediments, and in Belize where subtidal coastal lagoon sediments have been partly dolomitized in seawater of near normal salinity but elevated alkalinity (Mazzullo et al. 1995). Microcrystalline primary dolomite may be forming in highly evaporitic settings as in Deep Springs Lake, California (Meister et al. 2011) and the Coorong Lagoon, Australia (Warren 2000). The dolomite-evaporite association, together with the evidence of evaporites or desiccation textures, is strongly environmentally diagnostic. A largely dolomitic succession several meters thick with a laminated subaqueous unit overlain by a massive unit capped by "tepee" structures, intraclast breccias and stromatolites is typical for Coorong Lagoon-type deposits. Dolomitization and/or dolomite precipitation in all these settings depends on subsurface "seepage refluxion" or "evaporative pumping" of sea water to the surface, under hot, arid conditions (Adams and Rhodes 1960; McKenzie et al. 1980; Saller and Henderson 2001).

Many widespread ancient platform dolomites, such as the Plio-Pleistocene dolomites of the Caribbean Islands, are now considered to be the product of near-surface processes of dolomitization, by density-driven reflux, by tidal pumping or by recirculation of normal, mesohaline and hypersaline seawater or connate marine pore fluids beneath shallow surface meteoric groundwater lenses (Vahrenkamp and Swart 1994; Whitaker et al. 1994; Budd 1997; Saller and Henderson 2001). Platform carbonates across western Canada, such as the Devonian Wabamun, Grosmont and Nisku formations of Alberta, are likely to have been dolomitized penecontemporaneously by hypersaline reflux (Saller and Yaremko 1994; Jones et al. 2003) and the microcrystalline and macrocrystalline dolomites of the Devonian Presqu'ile Barrier (Keg River equivalent) complex beneath the Alberta-NWT boundary (Qing 1998; Fu and Qing 2011).

Dolomitized carbonate platform facies successions, such as shoaling-upward bar complexes and tidal flat capped cycles, commonly exhibit a strong control on dolomite crystal size, from a microcrystalline to macrocrytalline size range, by the precursor limestone sediment grain size (Lucia 1995). Earlier interpretations invoking late, post-lithification, dolomitization of Western Canadian carbonate platforms, such as for the Cooking Lake and Leduc formations, involving deep topographically driven or burial compaction-driven subsurface fluid circulation (e.g., Amthor et al. 1993) have significant drawbacks (Morrow 1998) and have been superseded by interpretations of early diagenesis involving reflux-driven evaporated marine fluids supplemented by geothermal convection in the shallow subsurface (Jones et al. 2003) to form the extensive dolomitized Western Canada platform dolomites (e.g., Fig. 4.54). The recognition of evaporated marine fluids as primary agents of dolomitization is a belated affirmation of the inferences of van Tuyl (1916), made at a time when the many examples of modern-day dolomite precipitation were not known.

The composition of evaporite minerals is not a good guide to the depositional environment of evaporites. A sample of normal seawater if evaporated to dryness yields a sequence of precipitates in the following order: calcite, gypsum, halite, epsomite, sylvite and bischofite. However, the composition of the final deposit in nature may vary considerably because of the effects of temperature, the availability of earlier formed components for later reaction, and the rate and nature of replenishment of the water supply. Evaporites may form in a variety of marine and nonmarine environments (Schreiber et al. 1976; Schreiber 1981; Kendall 2010), and it is not their composition as much as their internal structure and lithologic associations that are the best clues to the depositional environment (e.g., the association with penecontemporaneous dolomite mentioned previously; Sect. 4.4).

The formation of many chemical sediments, such as chalk, chert, phosphates and glauconites, depends on factors such as organic activity and ocean water oxygenation and temperature (Gorsline 1984). These factors are controlled in part by large-scale oceanic circulation patterns, which have been studied in order to develop predictive models of facies development (Parrish 1983).

Certain other chemical deposits contain useful environmental information. Chert is common as a replacement mineral in carbonate sediments, where it forms nodules and bedded layers commonly containing replacement casts of fossils, ooliths, etc. Knauth (1979) suggested that such chert was formed by the mixing of fresh and marine waters in a shallow subsurface, marginal-marine setting. Chert also occurs in abyssal oceanic sediments in association with mafic and ultramafic igneous rocks. Radiolarians, sponge spicules and diatoms are common. These are some of the typical components of **ophiolites**, which are remnants of oceanic crust and indicate a former deep-water environment (Grunau 1965; Barrett 1982).

Iron-rich rocks occur in a variety of settings. Their chemistry is controlled partly by Eh and pH conditions. Pyrite and siderite are common as early-diagenetic crystals and nodules in the reduced environment of organic-rich muds, particularly in fluvial or coastal (deltaic, lagoonal) swamps. Occasionally, such deposits may be present in

rock-forming abundance, as in the pisolitic bog iron ores. Plant remains, impressions and replaced (petrified) wood are commonly associated with these forms of iron. Pyrite is also associated with unoxidized, disseminated organic particles in the black muds of anoxic lake and ocean basins.

Organic-carbon-rich black shales are of considerable importance as petroleum source beds. Arthur et al. (1984) reviewed their origins and significance in terms of oceanic organic productivity, sedimentation rate and oxygenation of ocean bottom waters (Fig. 3.14). Ultimate controls on sedimentation are plate-tectonic configurations, which determine global sea level and climate, and which control the nature of oceanic circulation patterns. Such shales cannot, therefore,

Fig. 3.14 Oxygen content (redox state) at the sediment–water interface and generalized relationships between types of benthic organisms, sedimentary structures, water chemistry, mineralogy, organic content and sediment color. Notes: 1*, certain polychaete organisms and foraminifera may inhabit surficial environments where benthic metazoans are excluded; 2*, degree of lamination depends in part on seasonal variations in clastic input and productivity. (Arthur et al. 1984, reproduced by permission of the Geological Society, London)

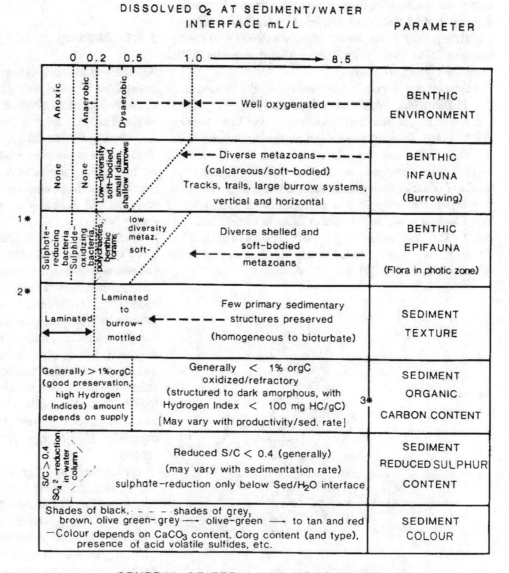

GENERAL CRITERIA FOR RECOGNITION OF
BOTTOM-WATER DISSOLVED OXYGEN
LEVELS IN MUDDY MARINE ENVIRONMENTS

be interpreted strictly on a local basis, but must be considered in a regional or even a global context.

The increasing importance of shale gas as an energy source has focused considerable attention on the facies analysis and petrology of mudrocks. Contrary to common assumptions, most mudrocks contain significant proportions of non-mud components, particularly quartz and calcium carbonate, with clay minerals comprising less than 50% by volume, based on thin-section point counts (Shaw and Weaver 1965; Macquaker and Adams 2003; Fig. 3.15). When examined in detail, shales are not homogeneous, but vary significantly in petrographic composition and internal structures, which, in turn, significantly affects their reservoir properties. It is increasingly being recognized that mud deposits are in many cases not simply passive pelagic accumulations in quiet, deep-water environments, but are the deposits of mud floccules transported by traction currents or sediment-gravity flows, and commonly exhibiting bedforms comparable to those formed in sand-bed deposits (Plint 2010; Shieber et al. 2013).

Hematite and chamosite iron ores are locally important in the Phanerozoic. Most are oolitic and display typical shallow-water sedimentary structures. As Van Houten (1985) noted, these ironstones are particularly abundant in the Ordovician and Jurassic record, but display widely varying paleogeographic and paleoclimatic settings, and so their use as facies indicators is uncertain at present. The iron is probably an early-diagenetic replacement. Precambrian iron formations are widespread in rocks between about 1.8 and 2.6 Ga. Their mineralogy is varied and unusual (Eichler 1976); it may reflect formation in an anoxic environment (Cloud 1973; Pufhal 2010).

Manganese- and phosphate-rich rocks are locally important. Both commonly occur as crusts and replacements on disconformity and hardground surfaces, where they are taken as indicators of nondeposition or very slow sedimentation. Phosphates are particularly common on continental margins in regions of upwelling oceanic currents. More detailed discussions of these sediments are given by Blatt et al. (1980) and Pufhal (2010).

Coal always indicates subaerial swamp conditions, usually on a delta plain, river floodplain or raised swamp. However, lacustrine coals and coal formed in barrier–lagoon settings have also been described. Calcrete (or caliche) occurs in alluvial and coastal environments and is an excellent indicator of subaerial exposure (Bown and Kraus 1981; James 1972; Wright 1986).

3.5.3 Bedding

Are the various rock types present in a stratigraphic unit interbedded in major packages several or many meters thick, or are they interlaminated on a scale of a few centimeters or millimeters (Fig. 2.7)? Is the bedding fine or coarse? Is it flat, disturbed, undulatory, distinct or indistinct (gradational)? These questions, while rarely providing answers that are uniquely diagnostic of a depositional environment, may provide important supplementary information. In a general way, the bed thickness is proportional to the depositional energy level.

Finely laminated sediments are mostly formed in quiet-water environments. X-radiography and scanning electron microscopy are commonly used to define lamination and very small-scale sedimentary structures in fine-grained rocks (e.g., see several papers in Stow and Piper 1984; O'Brien 1990). Such sediments may include laminated pelagic mudrocks, prodelta deposits, deep-water evaporites, thin-bedded basin-plain turbidites, and delta plain lagoonal and fluvial overbank muds. Laminated fluvial sheetflood sandstones are an exception to this pattern (see next section). Thicker beds form in a variety of high-energy wave- or current-dominated environments. Reef rocks, formed in extremely high-energy conditions, may lack bedding entirely.

An interbedding of contrasting lithofacies may be environmentally indicative. For example, wavy, flaser and lenticular bedding (**tidal bedding**: Reineck and Wunderlich 1968; Fig. 3.16) record the alternation of quiet-water mud sedimentation and higher energy flow conditions under which rippled sand is deposited. This can occur during tidal reversals on exposed mudflats, on fluvial floodplains or below normal wave base on the shelf, at depths affected by infrequent storm waves. Another common bedding association is that produced by the alternation of storm- and

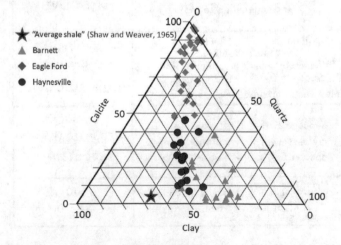

Fig. 3.15 Composition of typical mudrocks. Note that clay mineral composition is rarely more than 50% (Hart et al. 2013, after Shaw and Weaver 1965). AAPG © 2013, reprinted by permission of the AAPG whose permission is required for further use

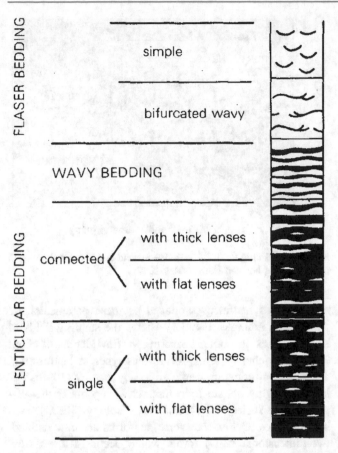

Fig. 3.16 Flaser, wavy and lenticular bedding Reineck and Singh (1973)

fair-weather processes on the shorcface. A sequence, basal gravel → laminated sand → bioturbated or rippled sand, indicates storm suspension followed by decreasing energy levels and then a return to low-energy wave activity and bioturbation during periods of fair weather (Kumar and Sanders 1976).

As noted in Sect. 3.5.1, maximum grain size and bed thickness are commonly correlated in deposits formed by individual sediment-gravity flow events. This can provide invaluable interpretive data.

Bedding surfaces that bound depositional units have a particular importance in facies analysis, as discussed in Sect. 3.5.11.

3.5.4 Hydrodynamic Sedimentary Structures

Many environmental deductions can be made from the details of internal structure of hydrodynamic sedimentary structures and from orientation (paleocurrent) information. Three general groups of structures can be distinguished:

1. structures formed by unimodal water currents in gravel, sand and mud in rivers, deltas, parts of ebb and flood tidal deltas in inlets, submarine fans and continental slopes (contour currents);
2. structures formed by reversing (bimodal) water currents, such as tides and wave oscillation in sand and mud in shelf and marginal-marine environments and in lakes and
3. structures formed by eolian currents in coastal dune complexes, inland sand seas and some alluvial-lacustrine environments.

Unimodal currents are readily recognized from unimodal foreset orientations (e.g., Fig. 2.8), but such patterns are not necessarily environmentally diagnostic. For example, as discussed below, it has been found that in areas of strongly reversing currents such as tidal inlets and their associated deltas, ebb and flood currents are segregated into different parts of the system. Structures in a single outcrop of a tidal delta may therefore be misinterpreted as fluvial in origin, based on structure type and paleocurrent patterns (Fig. 3.27). Simple paleocurrent models, such as those of Selley (1968), should therefore be used with caution (paleocurrent analysis is discussed in detail in Sect. 6.7). Other evidence, such as fauna, might yield clues as to the correct interpretation. Crossbedding structures may contain evidence of stage fluctuation, in the form of **reactivation surfaces** (Collinson 1970; Fig. 3.26 of this book). These are erosion surfaces formed during a fall in the water level, but again, they are not environmentally diagnostic as water levels rise and fall in rivers, deltas and tidal environments.

Reversing currents can be recognized from such structures as herringbone crossbedding (Figs. 2.9c; 3.27) or wave-ripple cross-lamination, in which foreset dip directions are at angles of up to 180° to each other. Herringbone crossbedding is a classic indicator of reversing tidal currents, but it can also form under oscillatory wave-generated flow conditions (Clifton et al. 1971), and even in fluvial environments, where bars migrate toward each other across a channel. Because of the segregation of ebb and flood currents in estuaries and inlets, herringbone crossbedding is, in fact, not common in many marginal-marine deposits. Some examples of this structure may result from the deposition from the dominant and subordinate currents that generate tidal dunes (Fig. 3.27; see Allen 1980; Dalrymple 2010c, p. 210). Some examples of herringbone crossbedding as observed in flat outcrops may, in fact, be oblique exposures of unimodally oriented, cross-cutting trough cross-sets.

The reversing ripples, chevron ripples, lenticular foresets, and variable symmetry and orientation of wave-formed ripple cross-lamination, as discussed below, are strongly diagnostic of a low-energy wave environment, such as a gently shelving marine beach or a lake margin (Fig. 3.29;

see de Raaf et al. 1977). Similar structures could also form in abandoned meanders or floodplain ponds in an alluvial environment, but would comprise a less conspicuous part of the overall succession. In many marginal-marine environments, crossbedding will be formed by both waves and tides, resulting in very complex paleocurrent patterns (e.g., Klein 1970). Careful documentation of structure types and their orientations may be necessary to distinguish the precise environment and mode of origin, but such work may also yield invaluable information on sand-body geometry, shoreline orientation, beach and barrier configuration, etc. (see further discussion on Sect. 3.5.4.4, below).

Work is underway to document and classify the hydrodynamic structures that are increasingly being recognized in mudrocks as a result of the almost ubiquitous tendency of mud to flocculate and then respond to shear stress much as a bed of loose sand (Schieber et al. 2013). It has been realized that most mud rocks are probably deposited from moving water, not by pelagic settling. The lack of recognition of this important interpretive point is partly explained by the fact that bedforms are flattened out and almost disappear during burial, generating what appears to be simple lamination. Newly deposited mud is highly porous (70–80%) and loses most of this porosity during burial compaction, with almost complete loss of the vertical dimension in preserved sedimentary structures. This can readily be recovered by directional stretching of digital images of clean mudrock surfaces, such as core slabs.

3.5.4.1   The Flow-Regime Concept

The interpretation of hydrodynamic sedimentary structures is one of the most important components of facies analysis, particularly in siliciclastic sediments. Certain carbonate and evaporite deposits contain similar structures and can be studied in the same way.

The basis for interpreting structures formed in aqueous environments is the **flow-regime concept**. This fundamental theory states that the flow of a given depth and velocity over a given bed of noncohesive grains will always produce the same type of bed configuration and therefore the same internal stratification. If such structures are predictable, their presence can be used to interpret flow conditions. These fundamental ideas were first enunciated following an extensive series of flume experiments by Simons and Richardson (1961) and were developed further for use by geologists by Simons et al. (1965), Harms and Fahnestock (1965), Southard (1971), and Harms et al. (1975, 1982) (Sect. 1.2.5). Valuable syntheses of hydrodynamic sedimentary structures were provided by Collinson and Thompson (1982) and Allen (1982), and reviews of later developments were provided by Leeder (1983), Ashley (1990) and Allen (1993).

It is now realized that bedforms are controlled mainly by three parameters, sediment grain size, flow depth and flow velocity, and a series of experiments has demonstrated the

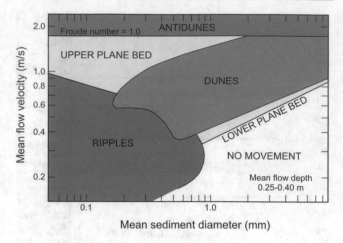

Fig. 3.17 Stability fields of bedforms in sand and silt in a flow depth of 25–40 cm (adapted from Ashley 1990)

sequences of bedforms produced as these parameters are varied. For example, Fig. 3.17 shows the stability fields of ripples, dunes and other bedforms for flow depths of about 20 cm. A welter of terms for various types of bedforms in different environments was clarified by Ashley (1990), following lengthy debate by a research group under the auspices of the Society for Sedimentary Geology (SEPM). Two broad classes of flow-transverse bedforms are now defined, two-dimensional (2-D) forms which occur at lower flow velocities, and three-dimensional (3-D) forms, which occur at higher velocities for a given grain size. There is a continuum of sizes of these bedforms from small-scale ripples to very large dunes more than 10 m in height. A natural discontinuity in size occurs at a spacing (wavelength) of 0.5 to 1.0 m, which reflects the structure of the turbulence that generates these bedforms (Leeder 1983). Forms smaller than this are termed ripples; Ashley (1990) recommended that those larger should all be termed **dunes**. Small-scale ripples, dunes and sand waves are forms out of phase with surface water movement; indeed, their form may bear no relation to surface water patterns at all. These bedforms have traditionally been classified as **lower flow-regime** forms (Fig. 3.18).

Increasing the flow velocity over a field of dunes causes changes in the turbulence in the boundary layer, and this has important consequences for the geometry of the bedform that results. In particular, the separation eddies become flatter, and skewed more and more downstream. Grains are swept over the tops of dunes from which the crests are eroded in the strong currents, and distinct avalanche faces may not form. Bedforms assume a "humpback" or "washed-out" shape (Fig. 3.18). The crossbed deposit that is preserved may be characterized by the low dip of the foresets (<10°), as shown in the example of low-angle crossbedding illustrated in Fig. 2.8e.

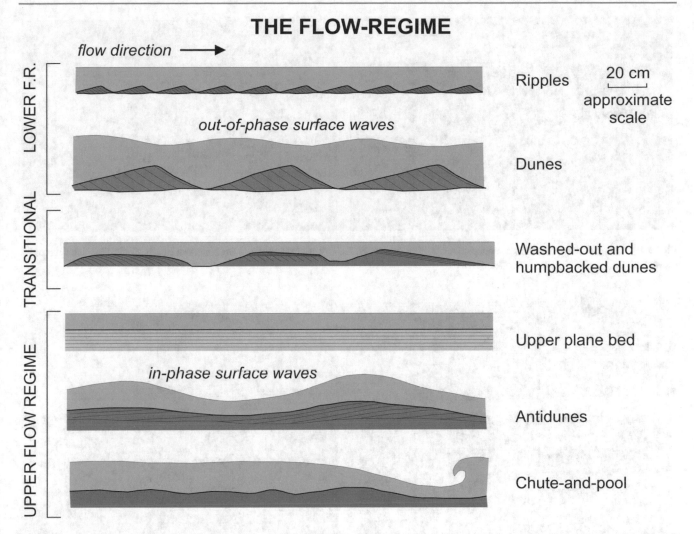

Fig. 3.18 The flow-regime concept, illustrating the general succession of bedforms that develops with increasing flow velocity. Note internal stratification (adapted from Simons et al. 1965; Blatt et al. 1980)

The **upper flow regime** is characterized by **antidunes** and standing waves (Fig. 3.19), which are in phase with surface water waves. The transition from lower to upper flow regime is marked by a streaming out of transverse turbulent eddies into longitudinal eddies. An intermediate **upper flat-bed condition** is marked by streaming flow, which aligns the sand grains and produces primary current lineation (parting lineation; Fig. 2.9d).

How can these flume data be used to interpret ancient sediments? First, Allen (1968) and Harms et al. (1975) demonstrated the relationships between bedforms and sedimentary structures. For example, planar-tabular crossbedding is produced by the migration of straight-crested megaripples, such as sand waves (what are now termed 2-D dunes; Fig. 3.20a), whereas trough crossbedding develops from the migration of 3-D dunes (Fig. 3.20b). Allen (1968) demonstrated the dependence of dune and ripple shape on water depth (Fig. 3.21). Second, the flow-regime concept

may be used to interpret ordered sequences of sedimentary structures in terms of gradations in flow conditions. Examples of applications to fluvial point-bar deposits, Bouma turbidite sequences and wave-formed sedimentary structures are discussed below. This is by no means an exhaustive listing. For example, the concepts have been adapted by Dott and Bourgeois (1982) to the interpretation of hummocky crossbedding, a product of storm wave activity (Fig. 3.32; see below and Sect. 4.2.8).

The size of a bedform depends on the depth of the system in which it forms and the amount of sediment available. In large, energetic channels with an abundant sand supply, very large dunes may form, possibly exceeding 10 m in height (e.g., Fig. 3.22). Large dunes also occur on the continental shelf, in areas where there are strong tidal or oceanic currents. Numerous observations made in a range of channel types shows that there is a continuum in sizes of dunes, from those about 10 cm high, up to forms two orders of

Fig. 3.19 Left: Standing waves in a modern river, Alaska (photo: N. Smith); Right: Antidunes on a modern beach, South Carolina (photo: M. Hayes)

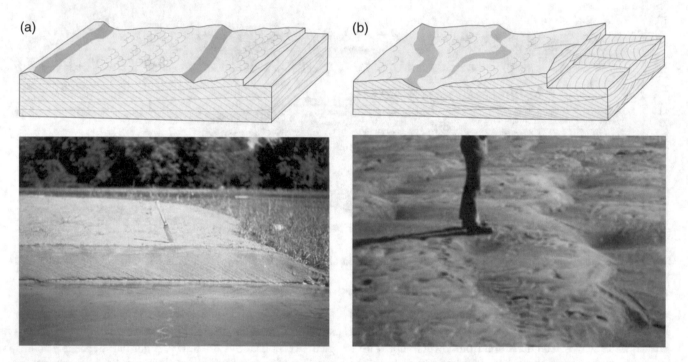

Fig. 3.20 Relationships between bedforms and sedimentary structures. **a** linguoid (3-D) dunes and trough crossbedding; **b** sand waves (2-D dunes) and planar crossbedding (diagrams adapted from Harms et al. 1975; photographs courtesy: N. D. Smith)

magnitude larger (Fig. 3.23). Note, however, the gap in this diagram corresponding to heights of about 4 to 8 cm. Bedforms of this scale are rare, because of the change in turbulence structure that takes place as flow velocity increases from the ripple to the dune stage.

There is a fairly consistent relationship between dune heights and wavelength with flow depth. This may be useful in estimating water depths or channel size from preserved dune deposits in the ancient record.

3.5.4.2 Bedform Preservation

In nature, bedform conditions change continually in response to changing flow conditions. What finally gets preserved into the rock record is an important consideration, because it is the final preserved result that we need to be able to understand in order to interpret the sedimentary record. For example, the gradual fill of a river or tidal channel is commonly recorded by a succession showing an upward decrease in grain size and an upward change in bedform type

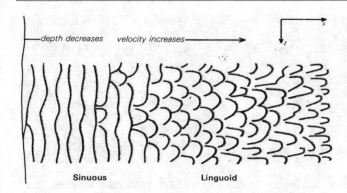

Fig. 3.21 Variations in bedform morphology with depth and velocity. (Allen 1968)

Fig. 3.22 An example of a large, simple, planar-crossbed set formed in a deep channel. Devonian Old Red Sandstone, Scotland

from forms higher to ones lower in the flow-regime suite, reflecting the gradual reduction in flow velocity and depth in the channel. It is common for the base of a channel fill to be characterized by dune deposits, with the upper part of the fill showing abundant ripples. The changing suite of crossbed structures is part of the description of the standard fining-upward cycle of channel-fill deposits (Sect. 3.5.10; Fig. 4.3), and similar suites occur in other cycle types.

The actual act of preservation may come about because of a sudden local switch in flow directions, which leaves the last-formed deposit abandoned. Antidune and chute-and-pool structures are rare in the rock record, because typically they are modified back into a lower energy structure, such as the plane bed condition, as flow velocities drop. Their preservation may, therefore, indicate a sudden change of this type. Consider, for a moment, how bedforms advance along a bed. Given a space on the floor of a channel or an area of the continental shelf, so long as the sediment supply is constant and current strength remains the same, each bedform will migrate in the direction of the current, to be replaced by the

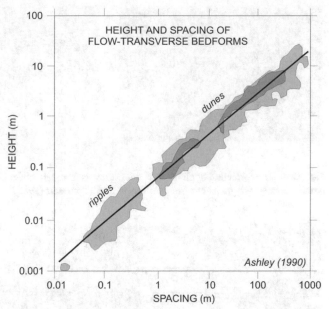

Fig. 3.23 The range of ripple and dune sizes observed in all depositional setting. The two darker areas indicate clusters of particularly common dune sizes (adapted from Ashley 1990)

one behind it. Under these conditions, how do bedforms ever stack upon each other to form successions that may consist of several or many individual crossbed sets one above the other? Usually, the scour hollow in the lee of the bedform crests is slightly erosive. As each bedform advances, the top of the one in front is stripped away. Complete bedforms are rarely preserved. Stacking to form multiple crossbed sets may be simply a case of successive trains of bedforms filling available accommodation space.

Under certain conditions, successive bedforms climb up the backs of each other without intervening erosion. This occurs when sediment is settling from suspension in addition to the sediment rolled or bounced along the bed by traction. The structure so produced is called **climbing ripples** (Fig. 2.8b). The angle of climb increases with the amount of suspended sediment in the mix (Fig. 3.24). The right-hand image in Fig. 3.24 shows deposition and preservation of the up-current (stoss) sides of the ripples, because of the addition of suspended sediment to each ripple as it forms. As a result, the angle of climb is high.

Examples of planar and trough crossbedding are illustrated in Fig. 3.25. Illustration (a) shows five virtually identical sets of planar-tabular crossbedding, with almost identical foreset orientations, indicating a consistency in channel orientation. Diagram (b) shows several large-scale crossbed sets more than 1 m thick. Diagram (c) shows intersecting trough sets in cross-section view, while Diagram (d) is a bedding-plane view of trough crossbedding.

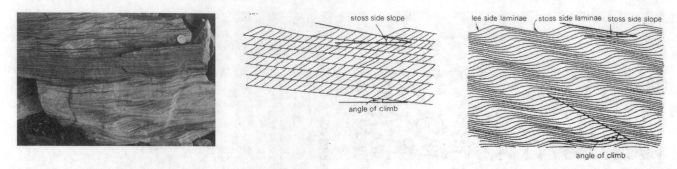

Fig. 3.24 The formation of climbing ripples. Flow is from left to right in all three images. Photo at left is of a Paleogene unit in Ellesmere Island, Arctic Canada; Diagrams are from Collinson and Thompson (1982)

Fig. 3.25 Examples of planar (**a**, **b**) and trough (**c**, **d**) crossbedding

3.5.4.3 Bedforms and Crossbedding in Gravels

Most of what we have been describing up to this point concerns sand systems. However, gravel also forms bedforms, and these are preserved as crossbedding in many conglomerate deposits formed in rivers (Fig. 2.9b), on beaches and in some submarine deposits that are otherwise dominated by sediment-gravity flows.

Individual clasts are oriented by flow (Fig. 2.12). We have not discussed this in the case of sand grains, because it requires specially oriented samples and careful laboratory examination to measure and document grain orientation. Nevertheless, such orientation measurements have been made on a number of sandstone bodies, and provide a useful supplementary form of paleocurrent measurement. In the case of gravels, the orientation of individual particles is much more obvious, and more readily measured in outcrop (Fig. 2.12). The reasons for imbricated orientations are explained in Fig. 2.12, which makes it clear that this is simply an issue of clast stability under the force of moving water. Once a clast assumes a position dipping upstream it is more likely to remain in this position, and become buried and preserved into the sedimentary record holding such an orientation.

3.5.4.4 Structures Formed by Reversing (Tidal) Currents

Flow-regime concepts apply to tidal settings, with the difference that current directions reverse on a diurnal (daily) or semi-diurnal (12-h) timetable, as tides flood and ebb. The details of where and how this occurs, and the relationship to other processes, including river influx, wave action and so on, are discussed in in Sect. 4.2.5. Here we focus on the specifics of how reversing tidal currents generate distinctive bedforms and crossbed structures.

The key to understanding the distinctiveness of tidal structures is to realize the importance of the process of current reversal. During the flood or the ebb, currents in tidal channels are typically comparable in strength to those occurring in rivers of comparable channel dimensions. Bedforms and crossbed sets of similar size and geometry are formed. However, fluvial currents are unidirectional, and vary in strength mainly on a seasonal basis. Tidal currents slow and stop altogether every 12 h (diurnal currents) or every 6 h (semi-diurnal currents). During the slackwater phase, mud may be deposited, and then current strength picks up again, but in the reverse direction. These processes are shown in diagrammatic form in Fig. 3.26. The "first mud drape" formed during a slackwater phase may be partially re-eroded when the tidal current resumes in the opposite direction. Commonly, this current will plane off the top of the earlier dune deposit, forming a curved **reactivation surface**, while ripples may form at the base of the set, where current strength is lower. What commonly characterize tidal dune deposits are the rhythmically spaced reactivation surfaces denoting regular intervals of current cessation and/or erosion as the dune forms.

Sand waves that develop in tidal environments range from 1 to 15 m in thickness and have characteristic internal structures, as described by Allen (1980, 1982). In detail, the geometry of tidal bedforms depends on the relative strengths of the flood and ebb currents. As we show in Sect. 4.2.5, these are usually not equal locally, with one being dominant over the other in terms of current strength, sediment transport and water depths during the most active phase of sedimentation. The result can be a complex construction of reversing crossbedding, the relative size of the sets corresponding to each phase of the tide providing some indication of the relative strengths of the tidal phases (Fig. 3.27). It is not at all uncommon for some areas of tidal deposition, particularly in the middle of tidal deltas, at the inner or outer mouths of tidal channels, to be dominated by one phase of the tidal current, with the result that, locally, the crossbed structures formed and preserved into the rock record appear

THE DEVELOPMENT OF TIDAL MUD-SAND COUPLETS

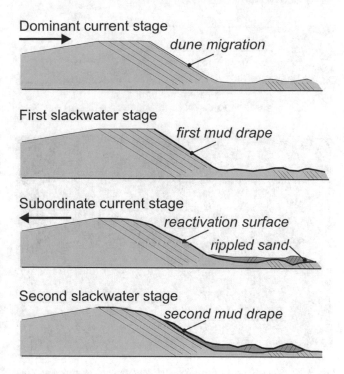

Fig. 3.26 The succession of events as tidal currents pass through a slackwater condition and then are renewed in the opposite direction (adapted from Visser 1980)

unimodal and quite fluvial in character (Fig. 3.27). It may be necessary to carefully search for other evidence of marine influence, such as fossil evidence, to arrive at the correct environmental interpretation.

Another distinctive character of many tidal deposits is the variation in current strength and resulting deposit thickness that reflects the monthly evolution from spring to neap tides and back again. Spring tides are the stronger tidal currents and higher tidal range that occur when the earth and moon are acting together, whereas the weaker, neap tides occur when the sun and the moon exert their gravitational force in divergent directions, reducing the scale of the global tidal circulation. The result may be rhythmic variations in the thickness of tidal bed sets (Fig. 3.28). Meticulous work on modern tidal flats in the Netherlands and some ancient deposits has shown how the thickness variations may be related to the specifics of the sun–moon system at the time of deposition (Visser 1980; Williams 1989). Many other subtle indicators of tidal influence were described by Reineck and Singh (1980) and Nio and Yang (1991).

The alternation of strong and weak currents is shown at a smaller scale, within the ripple deposits that form on most tidal flats. A terminology for these various forms of so-called **tidal bedding** was evolved by Dutch workers, and is shown

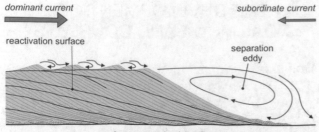

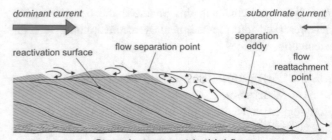

Fig. 3.27 Tidal deposits formed under the influence of reversing flow. Left: ebb and flood have comparable strengths. The result may be herringbone crossbedding. Right: The more unimodal appearance of crossbedding formed when one phase of the tide is strongly dominant, resulting in affectively unimodal current strengths (diagrams adapted from Allen 1980)

in Fig. 3.16, and examples are illustrated in Fig. 3.29. **Flasers** are the small mud drapes that develop over ripples during slackwater phases. **Wavy** and **lenticular bedding** occur when there is a limited sand supply on a tidal flat, and ripples may not join together to form continuous beds, but occur as isolated ripple crests.

A word of caution: tidal bedding is a very distinctive lithofacies in tidal settings, but its presence is not definitive proof of a tidal environment. The alternation of sand and mud indicates fluctuating current strengths, but this can also occur on river floodplains on a seasonal or more irregular basis, and can occur on the deeper parts of continental shelves, where occasional storm waves may stir the sediment on a sea floor that is otherwise a tranquil environment of mud deposition. In making any kind of environmental interpretation from the rock record, the geologist should use all the facies indicators that he or she can observe.

3.5.4.5 Structures Formed by Oscillating Currents (Waves)

Clifton et al. (1971) carried out one of the first detailed studies of the sedimentary structures that form on coastlines under breaking waves. They recognized a direct relationship between wave type, resulting water motion and structure type (Fig. 3.30). Waves are generated by wind shear across a water surface. The greater the distance over which this takes place (called wave "fetch") and the stronger the winds, the greater the amplitude of the waves that are generated. The water motion that is expressed as surface waves is circular, with the diameter of the circle increasing with wave height. The circular motion is transmitted to the water column below (Fig. 3.30). Under average conditions this motion extends downward for 5–15 m, but under storm conditions, it exceeds 20 m, and may reach as much as 200 m. Where the waves touch bottom, the circular motion is expressed as oscillation back and forth, the flow velocity usually being adequate to move sediment of sand grade. Waves are, therefore, a significant factor in the generation of sediments on continental shelves and the bottom of lakes. Their effects are even more pronounced at shorelines, where the waves shoal, steepen and break, with the familiar rush of water up the beach slope and then back (breaker, surf bore and swash zones in Fig. 3.30).

The nature of the deposits formed beneath shoaling and breaking waves was first studied by SCUBA divers on the Pacific coast of Oregon. The divers anchored themselves to the sea bottom so that they could make observations in this very energetic environment. The results of their studies are

NEAP-SPRING BUNDLING OF TIDAL DUNE CROSSBEDDING

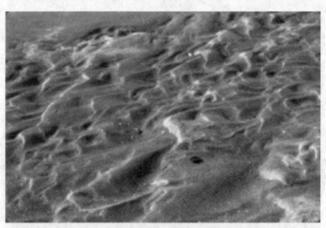

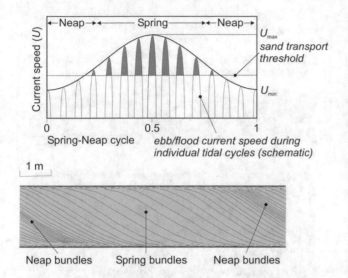

Neap bundles   Spring bundles   Neap bundles

Fig. 3.28 Current speed during individual flood-ebb cycles is shown by the successive peaks in the upper diagram. The peaks are joined by the smooth curve at the top, which traces out the rise and fall in maximum current strength from neap to spring and back to the neap condition. Areas colored in brown indicate those periods when sand transport is possible. These show a maximum duration around the time of the spring tides, as reflected by the increased bundle thicknesses formed at such times (lower diagram). Adapted from Yang and Nio (1989)

Fig. 3.29 Examples of tidal bedding. Top: ripples on a modern tidal flat, near Seattle; Bottom: wavy bedding in a Cretaceous unit, Utah

Fig. 3.30 The formation of bedforms beneath breaking waves. Adapted from Clifton et al. (1971)

WAVE BUILD-UP AND BEDFORM ZONES ON A SANDY COASTLINE

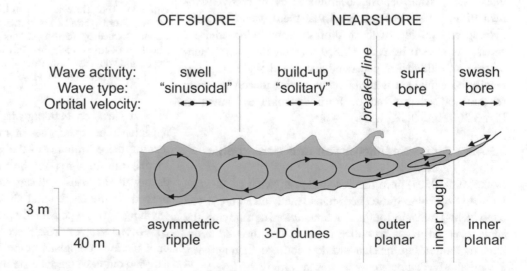

shown in Fig. 3.30. It was found that there are distinct zones of bedforms in the nearshore to offshore, corresponding to the evolving state of the waves, as they approach the beach. Flow-regime concepts can be adapted to understand the formation of these zones.

In offshore areas, as the oscillatory motion of the waves begin to touch bottom, sand may be moved into ripples. With continued approach of the wave toward the beach, the velocity of the oscillation increases as frictional shear at the base of wave increases. A zone of three-dimensional dunes is formed. The zone of highest energy is that beneath the breaking wave, and extending immediately to the landward of that. Here, the characteristic bedform is a plane bed. These three zones (Fig. 3.30) correspond to the progression through the lower flow regime to the beginning of the upper flow regime, as flow velocity is increased in a flume.

The "inner rough zone" is an area of rapid change in flow direction and energy. Lower flow-regime bedforms are the result. Finally, the swash zone, where the water from the breaking wave rushes up onto the beach and then flows back or is infiltrated, is a zone where flow velocity may be high for a few seconds, but water depths are small. This is called the "inner planar zone," and is an area where flat-bed conditions may develop, but on gently shelving beaches this is commonly an area of wave-ripple formation (Fig. 3.31).

Wave ripples are typically symmetrical in external form, and have distinctive inner structures, reflecting the rapidly alternating flow directions. The points noted in the upper diagram of Fig. 3.31 may be used to rapidly identify wave ripples in outcrop and drill core. These are small structures, and close examination is necessary.

In the rock record it is not expected that the zones identified in Fig. 3.30 would be perfectly preserved exactly in their original configuration, because the zones normally move up and down the shore with the rise and fall of the tides. As a beach deposit accumulates by regressive sedimentation, it could be predicted that there would be some overlap and mixing of the bedforms representing adjacent zones. As shown in Sect. 4.2.5, however, the overall succession of lithofacies generated by beach regression can readily be explained with reference to the general zonation of transport energy and bedform assemblages shown in Fig. 3.30 (Figs. 3.58, 4.15).

3.5.4.6 Storm Sedimentation and Geostrophic Flow

In the mid-1970s, once the description of what came to be known as **flow-regime bedforms** (and their preserved equivalents) became routine, sedimentologists studying the ancient record began to notice types of crossbedded structures that did *not* fit the standard descriptions. Among these, a particularly distinctive type was a variety of low-angle crossbedding with both concave- and convex-up curvature.

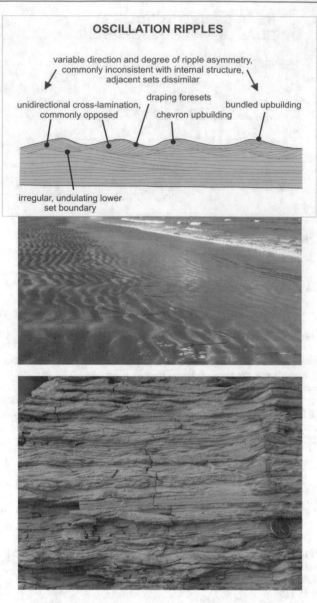

Fig. 3.31 Top: The fine-scale structure of wave-formed (oscillation) ripples. These typically are symmetrical and consist of bundles of laminae formed by short-lived bursts of flow in opposing directions. Middle: modern wave-formed beach, South Carolina; Bottom: wave ripples in Ordovician sediments, near Ottawa (diagram: de Raaf et al. 1977; middle photo: M. O. Hayes)

In plan view, on bedding planes, these structures had the appearance of low domes or mounds (Fig. 3.32). The context of these structures—the presence of shallow-marine fossils, the stratigraphic bracketing between coastal and deep-water deposits—all suggested that these bedforms were formed on the continental shelf, but nobody had observed anything of this type actually present, or forming, on modern continental shelves. This presented something of a problem, until marine geologists, going back over their records, and carrying out new, targeted surveys of the sea floor, were able to identify the structure and confirm that, indeed, this

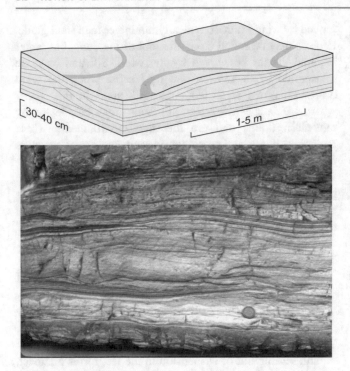

Fig. 3.32 Hummocky cross-stratification. The outcrop example is from the Jurassic of Yorkshire, UK (diagram adapted from Harms et al. 1975)

structure is a characteristic feature of the continental shelf. The descriptive name **hummocky cross-stratification** was coined for this new structure (Byers and Dott 1981; Dott and Bourgeois 1982). The acronym HCS is often used as a convenient shorthand (Harms et al. 1975).

It is now known that HCS is the product of what is called **combined flow**, that is, both waves and ocean currents are involved. This combination occurs during storms, when wind pressures lead to the generation of temporary **geostrophic currents**. These, combined with the effects of deep

scour by storm waves, cause a considerable amount of sand transport across the shelf (Swift et al. 1983; Nottvedt and Kreisa 1987; Allen 1993; Plint 2010).

The principle of the geostrophic current is illustrated in Fig. 3.33. Storm winds approaching the shore push coastal waters onland. This coastal **setup** may raise local water levels by as much as 5 m. The setup is part of a temporary overturning of water, with a return flow at depth, across the continental shelf. The return flow is influenced by the Coriolis effect, and may be diverted to flow obliquely across the continental shelf, as shown in Fig. 3.33. At the same time, the oscillatory sediment motion induced by large storm waves reaches deep into the water column, and adds a back-and-forth component to the transport of individual sand grains. The overall result of this complex sand movement pattern is the development of the mounds, or hummocks, that we now recognize as HCS. The local pattern of growth of the hummocks is indicated by the progradation and truncation of the crossbedding within the hummocks.

3.5.4.7 Eolian Bedforms

Eolian sand dunes are photogenic. They are often very large, ranging up to several hundred meters high, and forming dune fields many kilometers across. Images taken from space and from the ground of Earth's various deserts, and now, also, of the dunes on Mars, are commonplace. In certain parts of the stratigraphic record, eolian dunes make up significant thicknesses, particularly in the Mesozoic of the American southwest and the Permian of the North Sea Basin (where they form a very important gas reservoir) (Brookfield 2010).

Two important developments since the mid-1970s led to a revolution in the study of eolian strata in the rock record. First Brookfield (1977) and Gradzinski et al. (1979) examined the mechanics of dune construction and migration and

Fig. 3.33 Storm-driven return flow across the continental shelf is diverted by the Coriolis effects, resulting in sediment transport obliquely across the shelf (adapted from Walker, in Walker and James 1992)

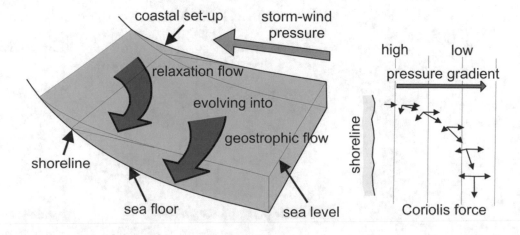

presented some useful ideas on the nature of large-scale crossbed bounding surfaces. Second, Hunter (1977a, b, 1980, 1981) and Kocurek and Dott (1981) studied the details of sand movement by wind on modern and ancient dunes and showed that several distinctive crossbedding and lamination patterns are invariably produced. In addition, Walker and Harms (1972) and Steidtmann (1974) carried out useful detailed facies studies of ancient eolian units. All this work has brought us to the point where eolian crossbedding should now be relatively simple to recognize, even in small outcrops.

Figure 3.34 is adapted from a global study of sand seas carried out in the late 1970s that utilized early "Landsat" images, and shows four of the main types of large sand dune. Barchans (Fig. 3.35d) tend to occur in areas of limited sand supply. They are commonly seen migrating across stony, sand-free, desert surfaces. Transverse dunes are comparable to water-lain planar-tabular crossbedding in terms of their internal structure. The other dune types (and only a few are shown in Figs. 3.34, 3.35) reflect the product of complex wind patterns.

The pattern of air flow over a bed of loose sand is superficially similar to that of aqueous traction currents. Sand grains are moved by traction, saltation (bouncing) and intermittent suspension; flow separates at the dune crest and there is turbulence in the lee of the dune. In detail, of course, flow patterns are distinctly different. For example, turbulence patterns commonly result in air flowing horizontally *across* the dune foreset. Air currents are less "competent" than aqueous traction currents, in the sense that the lower density of air means that the size range of particles that can be

moved by wind is small, typically not larger than sand grade. Pebbles may be rolled along the foot of the dune, but larger material cannot be moved by the wind. Silt and clay are transported in suspension and may settle when the wind drops. Interdune areas (e.g., the dark areas in Fig. 3.35c, which are dark because they are damp) may be affected by rare flash floods and are therefore environments where temporary ponds and erosional river channels may occur, as discussed in more detail in Sect. 4.2.2.

Sand is transported and deposited on dunes by three processes, which generate distinctive textures and structures that readily enable the processes to be identified in small outcrops or drill core. These processes are:

1. Formation of **wind ripples**. Long-wavelength, low-amplitude ripples commonly form as a result of airflow along the active front face of a dune (Fig. 3.36a). Because of their small size, ripple foresets are rarely visible. An example is shown in Fig. 3.36b.
2. **Grain-fall lamination**. Sand blown over a dune crest and settling out of suspension in the lee, forms compact, commonly laminated units on the dune face (Fig. 3.36e). The carpet of sand moving close to the dune surface as it passes over the crest causes the "fuzzy" appearance of the dune crest behind the observer in Fig. 3.36c.
3. **Grain-flow lobes**. These are small avalanches of sand triggered whenever the depositional surface becomes too steep (Fig. 3.36c, d). They are most easily identified in strike-parallel cross-section, where their lenticular shape is quite distinctive (Fig. 3.36f).

Fig. 3.34 The main types of large-scale eolian dune (adapted from McKee 1979)

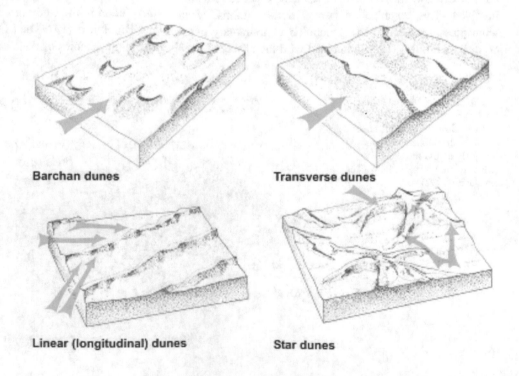

Barchan dunes

Transverse dunes

Linear (longitudinal) dunes

Star dunes

Fig. 3.35 Modern eolian dunes. **a** Great Sand Dunes National Monument, Colorado; **b, c**, Aerial views of the Gobi Desert, Western China; **d**. A barchan in the Gobi Desert

Figure 3.37 shows an exposed dune foreset in the Permian Canyon de Chelly Sandstone, Arizona (at the point of the arrow). The enlarged view of this face (at right) shows the linear pattern of wind-ripple crests oriented down the dune face, exactly as in the photograph of the modern dune surface shown in Fig. 3.36a. Figure 3.38 illustrates the three "orders" of bounding surface that may be identified in large eolian crossbed sets.

3.5.5 Sediment Gravity Flows

Sediment gravity flow deposits include at least five distinct types of lithofacies assemblage (Fig. 3.39). They are initiated by a failure of some kind on a sloping basin margin (Fig. 3.40). Sedimentary oversteepening on a delta slope may cause this, or an unstable slope may undergo collapse as a result of an earthquake. Sediment gravity flows can also occur where a sediment-laden water body flows into a standing body of water, either a lake or the sea.

Turbidity currents were the first type of sediment-gravity flow to be identified. After the construction of the Hoover Dam on the Colorado River in the 1930s, it was noticed that as the turbid water of the river entered Lake Mead (the water body created by damming the river) it descended beneath the surface as underflows, because of the greater density of the turbid water relative to the clear water of the lake (Grover and Howard 1938). A Canadian example is illustrated in Fig. 3.41. Engineers noticed that sediment was being transported well into the lake before being deposited. At about the same time, marine geologists suggested that submarine currents might be responsible for eroding out submarine canyons (Daly 1936). Then, in the 1950s, turbidity currents were modeled in laboratory flumes by Kuenen and Migliorini (1950; see Fig. 1.5). This classic paper clearly established subaqueous sediment-gravity flows as the cause of graded bedding and other features in the rock record.

The ability of turbidity currents to carry sediment from shallow into deep water explained a puzzle that had bothered petroleum geologists working in the Ventura Basin of California for some time. There, sandstones containing a shallow-water fauna of foraminifera are interbedded with shales containing a deep-water fauna (Natland and Kuenen 1951). How could this be? We now know that turbidity currents can carry large quantities of sediment for thousands of kilometers under water (see Sect. 4.2.9). In the early 1970s, data relating to a major earthquake that had occurred on the Grand Banks of Newfoundland in 1929 was interpreted in terms of turbidity-current activity (Heezen and Hollister 1971), and became a classic object lesson in the power and scale of turbidity-current events (Figs. 3.42, 3.43). Middleton carried out an important set of experiments on turbidity currents in the 1960s (Middleton 1966, 1967;

Fig. 3.36 Types of eolian stratification. **a** Wind ripples formed by air currents passing across the front of a dune; **b** Example of wind ripples in a Permian eolian sandstone, Arizona; **c, d.** Grain-fall deposits (beneath the observer's feet) and grain-flow deposits (the sand lobes oriented down the dune slope); **e, f** Permian sandstone in Northern England showing grain-fall (dark, laminated beds) and grain-flow lenses (lenticular units seen in strike-parallel cross-section in **f**)

see Sect. 1.2.5 for additional information concerning the origins of our knowledge about sediment-gravity flows).

In order to understand the several different kinds of sediment-gravity flow and the variety of textures and fabrics that they display, it is necessary to make a short diversion into some elements of fluid hydraulics (Middleton and Hampton 1976; Lowe 1979). Fluids such as water, including

water that contains a dilute suspension of dispersed particles, have low viscosity, and are said to be **Newtonian fluids**, because they obey Newton's laws of viscosity. Fluids that contain a modest concentration of sediment show increased viscosity, a viscosity that may vary with the rate of shear. Such fluids are called **non-Newtonian**, and are characterized by concentrations of greater than 30% sand, or somewhat

Fig. 3.37 Exposure of active dune face (at point of arrow). Close up (right) shows wind ripples oriented down the dune face. Permian, Arizona

Fig. 3.38 Structure of typical ancient eolian dune deposit. MBS, main bounding surface; SS, second-order bounding surface; TS, third-order bounding surface; ER, eolian ripples; FC, fluvial channel or sheetflood deposits (after Gradzinski et al. 1979). Example at right is from the Schnebly Hills Formation, near Flagstaff, Arizona

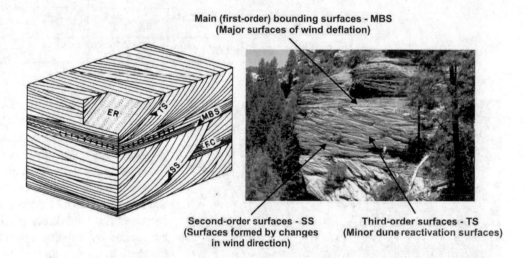

Fig. 3.39 A classification of sediment-gravity flows (adapted from Lowe 1979)

Flow behaviour	Flow type		Sediment support mechanism
Fluid	Fluidal flow	Turbidity current	Fluid turbulence
		Fluidized flow	Escaping pore fluid (full support)
		Liquefied flow	Escaping pore fluid (Partial support)
Plastic (Bingham)	Debris flow	Grain flow	Dispersive pressure
		Mudflow or cohesive debris flow	Matrix strength Matrix density

Fig. 3.40 The evolution of flow types with time, showing the changes in sediment-support mechanisms, and the resulting flow types (Pickering et al. 1986)

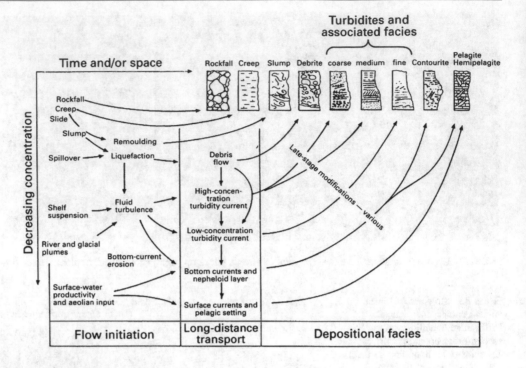

Fig. 3.41 Turbid stream waters draining into Lake Louise, Alberta, plunge beneath the clear lake waters as a turbidity current

less if the sediment consists of cohesive clay particles. If the concentration of sediment is large, viscosity is also large, and the flow then has a certain mechanical strength. In fact, its strength may hold it in place on a sloping basin margin until an applied shear force exceeds a certain critical strength. Such a substance has the properties of a plastic. If the viscosity remains constant after the failing shear stress is exceeded, the flow is called a **Bingham plastic**. If the viscosity varies, it is called a pseudoplastic. Actual sediment-gravity flows, at different times and under different conditions, display all of these conditions. Study and experimentation led Middleton and Hampton (1976) to the

classification shown in Fig. 3.44. Figure 3.39 is the modern version of this classification. Column three in Fig. 3.39 lists the five main types of sediment-gravity flow. The boxes to the left indicate how these correspond to the types of flow discussed in the preceding paragraph, and the column on the right lists the specific processes that mobilize the sediment and keep it mobilized during transportation. Some of these processes are remarkably persistent. Notably, the sediment transported by turbidity currents may remain in suspension for days to weeks, and major flows can travel across the ocean floor for hundreds to several thousands of kilometers. All these types of flow are common on the ocean floor. The two "plastic" types of flow, grain flows and mudflows/debris flows may also occur on land, under subaerial conditions.

Flow types can also evolve with time (Pickering et al. 1986). As a flow passes down slope, a loss of sediment from the base of the flow, or turbulence causing incorporation of additional fluid or sediment, may cause a change in the flow dynamics. The evolution from one flow type to another is shown in Fig. 3.40.

3.5.5.1 Debris Flow

Debris flows consist of poorly sorted mixtures of clay, silt, sand and larger clasts, possibly including large boulders (examples are illustrated in Fig. 3.45). *Mudflows* are a type of debris flow dominated by a mud matrix, with floating clasts. Debris flows may occur subaerially or subaqueously. They evolve from slumps or slides that become lubricated by infiltrated water. Subaerial flows are characteristic of arid areas, where weathered debris may accumulate on hillsides

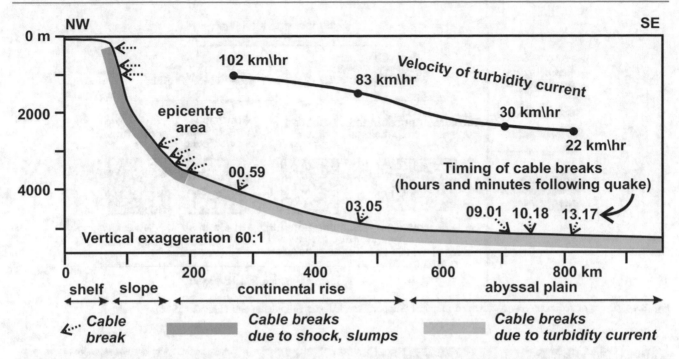

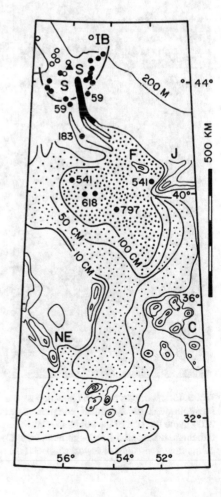

Fig. 3.42 Cross-section of the continental margin of the Grand Banks of Newfoundland, showing data relating to a major earthquake that occurred there in 1929. Slumps triggered by the quake generated a turbidity current that flowed down the slope and continued for more than 1000 km out across the Atlantic Ocean. The times at which successive submarine cables were broken by the current enabled geologists to calculate the velocity of the flow (graph at top) (adaped from Heezen and Hollister 1971)

Fig. 3.43 The extent of the turbidite deposited following the 1929 quake. Laurentian Channel at top left (Walker, in Walker and James 1992)

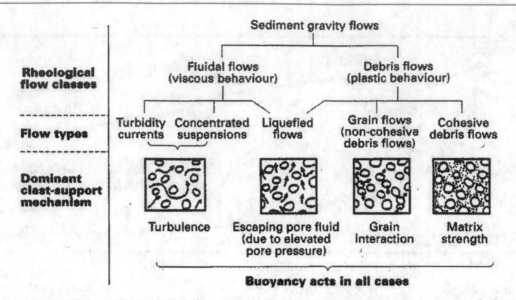

Fig. 3.44 Types of sediment-gravity flow and principal "support" mechanisms (Middleton and Hampton 1976)

Fig. 3.45 Examples of debris flow deposits. **a** Modern debris flow that swept through an apartment complex, coastal Venezuela, February 2000. Note the size of the boulders relative to the size of the apartment buildings. **b** Subaerial flow, with large logs; Cenozoic, Japan. **c** Marine debris flow consisting of reworked glacial debris, Proterozoic, Ontario. **d, e** Miocene deposits, Turkey. **f** Cenozoic deposit, California. **g, h** Carbonate debris flows, Cambrian–Ordovician, Newfoundland. Note deformed clast in **h**. **i** Distant view of carbonate debris flow lenses, Cretaceous, France (*photos*: **a**: O. Colmenares, **d, e** W. Nemec)

Fig. 3.46 A large slump on the continental margin of NW Africa triggered a muddy debris flow that traveled for several hundred kilometers across the floor of the Atlantic Ocean (after Embley 1976)

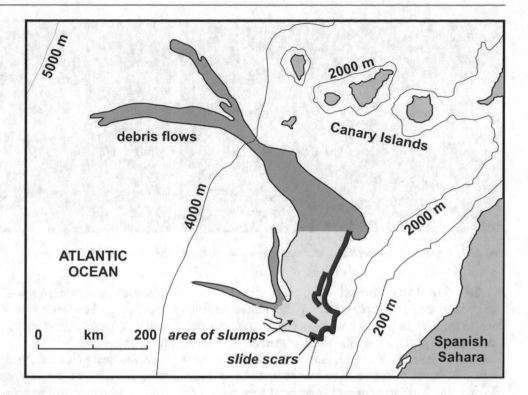

for many years, until mobilized by a sudden downpour. They flow at speeds of 1–3 m/s. Witnesses report that flows are very noisy, sounding like the onrush of several large freight trains. Debris flows may also occur in humid, vegetated areas when exceptionally heavy rains occur after periods of several or many decades (Fig. 3.45a). They are common on the slopes of arc volcanoes, where loose ash is easily mobilized by the rain. Under-water debris flows occur on continental slopes, as a result of sedimentary oversteepening. They may be triggered by an earthquake. Subaerial flows may travel for 10–20 km out across a flat basin floor. Subaqueous flows may travel for several hundred kilometers under water (Fig. 3.46).

The concentrated sediment–water mixture of a debris flow gives it a certain mechanical strength, which enables the flow to suspend and transport large boulders for significant distances (Fig. 3.45a). Loss of water as a flow moves down slope may lead to the development of a "rigid plug" of sediment, whereas within the remainder of the flow, sediment and water continue to interact with each other, until enough water is lost or gravitational energy is dissipated on a flat surface. At this point, the internal frictional forces absorb the flow momentum and it ceases to move. Internal shear may develop a recognizable fabric, with elongate grains all tending to be rotated into vertical or imbricated fabrics (e.g., Fig. 3.45c). Large clasts may sink to the base of the flow and settle out earlier than smaller clasts, resulting in a **graded fabric** (Fig. 3.45e). Alternatively, agitation within the flow may cause smaller clasts to move downward between the

larger ones (a process called a kinetic sieve), resulting in a **reverse-graded** fabric, in which the large clasts are at the top (Fig. 3.45d).

3.5.5.2   Grain Flow

Grain flows are a rather specialized form of sediment-gravity flow; one we already encountered in the discussion of eolian bedforms. The grain-flow lobes that form as small avalanches on the front of active dunes are grain flows (Fig. 3.36c, d, f). They are categorized this way because they constitute a form of mass sediment movement. In this case, the "support" mechanism that keeps the grains in suspension so that gravity overcomes friction is what is called **dispersive pressure**. This is the force of grains "bouncing" against each other as a result of their elasticity (Fig. 3.44). This process only works effectively where the sediment mass consists of well-sorted grains, and flow requires a slope in the range of 30° to provide the necessary gravitational energy.

3.5.5.3   Liquified/fluidized Flow

These flows are kept in suspension by the movement of pore fluids between the grains, which thereby overcomes the internal grain-on-grain friction. This mechanism is important mainly in the last stages of movement of the flow, as the grains begin to settle and pore fluids are extruded. The upward movement of the pore fluids typically forms dish-and-pillar structures (Fig. 2.19c, d, e), which constitutes the most distinctive criterion for recognizing this category of sediment-gravity flow. Examples are shown in Fig. 3.47.

Fig. 3.47 Fluidized/liquefied flows. **a** Cretaceous, south of France; **b** Cenozoic, Riverdale, California

3.5.5.4 Turbidity Current

The first systematic exploitation of the turbidity-current concept for the interpretation of the ancient record was by a young Dutch geologist, by the name of Arnold H. Bouma, who was completing his Ph.D. thesis work on some Cretaceous sandstones in the south of France in the early 1960s (Bouma 1962). He recognized the repeated occurrence of a type of sedimentary cycle a few decimeters to a few meters in thickness, and attributed these cycles to the action of turbidity currents. The publication containing his results generated widespread interest, and his name rapidly became attached to the cycle model he had proposed. Nowadays, it is commonplace to refer to the typical or classical turbidity-current deposit as a **Bouma sequence**. An example of one of these sequences, observed in a drill core, is shown in Fig. 3.48, together with a modern interpretation of the cycle in terms of the flow-regime concept.

The succession of five divisions constituting the standard Bouma sequence provides a good example of how sediment support and transport mechanisms change with time within an individual flow event (Fig. 3.39). Initially, the sediment is maintained in suspension by the turbulence within the flow. The suspended sediment is what provides the flow with its increased density and hence the gravitational momentum to keep flowing, which in turn generates more turbulence. This process, known as "autosuspension," is what allows the flows to keep traveling under water, across the floor of an ocean, for hundreds of kilometers. Sediment gradually drops out of suspension as the momentum of the flow decreases, the coarser particles being deposited first, resulting in a graded bed (Bouma division A). This basal unit usually rests on a bed marked by sole markings (flutes, tool marks; Figs. 2.25, 2.16), which are evidence of the erosive power of the flow.

The deposition of the A division corresponds to a "quick bed" phase of the upper flow regime. A decrease in flow power configures the flow as a plane bed, from which is

deposited the laminated sands of the B division (Fig. 3.48). The flow is here acting as a traction current, with loose sediment particles moved along the bed by the frictional force of the flow.

In the final phases of the flow, at its most distal end, the frictional force of the flow on the bed corresponds to the lower flow regime, forming the ripples or small dunes of the C division. The remaining very fine sands and silts being transported by the flow then settle out as the D division.

The A to D divisions of the flow deposit are formed rapidly, the A division perhaps within a few minutes, the remainder of the deposit within hours. The E division is not a product of turbidity-current deposition, but represents the background **pelagic settling** of fine muds that continues at all the time on a quiet ocean floor. This division represents the tens to thousands of years between flow events.

The composition of each turbidity-current deposit changes in a down-current direction, as shown in Fig. 3.49. The wedge-shaped cross-section of a flow shown at the center of this diagram represents an individual flow unit, perhaps hundreds of kilometers long, and varying from a couple of meters thick, at most, near source, to a few centimeters, at its most distant end.

3.5.6 Sedimentary Structures Produced by Hydrodynamic Erosion of the Bed

Few of these are environmentally diagnostic, although their presence may add weight to an interpretation made from other evidence.

As discussed in Chap. 2, there are two main classes of erosional structure at the outcrop scale, macroscopic and mesoscopic. Macroscopic structures comprise channels, scours and low-relief erosion surfaces. Examples are illustrated in Fig. 2.13. They can occur in practically any

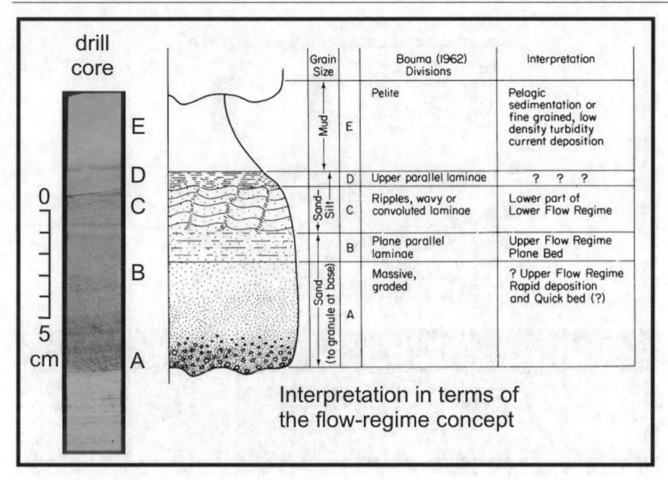

Fig. 3.48 The Bouma sequence model. An example of a complete sequence, from the Precambrian of northern Ontario, is shown at left. The five divisions (A–E) of the standard sequence are interpreted at right (diagram from Blatt et al. 1980. Core photo: Gowganda Formation, Northern Ontario)

environment as a result of current activity. More diagnostic than the channel itself is the channel fill, which can be analyzed in terms of vertical profile, lithology, sedimentary structures, etc. Subordinate features on an erosion surface may also provide some environmental information. Lag concentrates of pebbles, bioclastic debris or phosphatized fossils may be interpreted in terms of such processes as condensed sedimentation or wind deflation, storm erosion and deposition, or submarine nondeposition (hardgrounds). Desiccated surfaces may break up, yielding breccias of partly lithified mud, silt or carbonate clasts, which can then be interpreted as the result of subaerial exposure.

Another term for macroscopic erosion surfaces is **bounding surfaces**. Their physical extent, geometry and facies associations have become important components of the architectural subdivision and classification of clastic deposits (Sects. 2.2.2, 3.5.11), and also feature in the definition and subdivision of stratigraphic sequences (Chap. 5).

Mesoscopic erosional features include the variety of sole markings described and illustrated in Chap. 2 (Fig. 2.15). These are abundant in submarine-fan deposits and submarine basin plain sediments, particularly in sandy turbidites. Flute marks, in particular, indicate a style of vortex turbulence that is common at the base of turbidity currents (Fig. 2.16). However, they have been observed in a wide range of other settings, including fluvial traction current deposits. Turbidity currents are not restricted to the deep oceans, but are common in lakes and subaqueous ice-margin environments.

A very detailed study by Davies and Shillito (2021) has demonstrated how bedding surfaces may be preserved as "true substrates." These are surfaces that accumulate significant evidence of biogenic activity, and other processes that modify, but add or subtract little from a bedding surface, but reveal the passage of significant intervals of time, thus corresponding to what Tipper (2015) referred to as surfaces of "stasis."

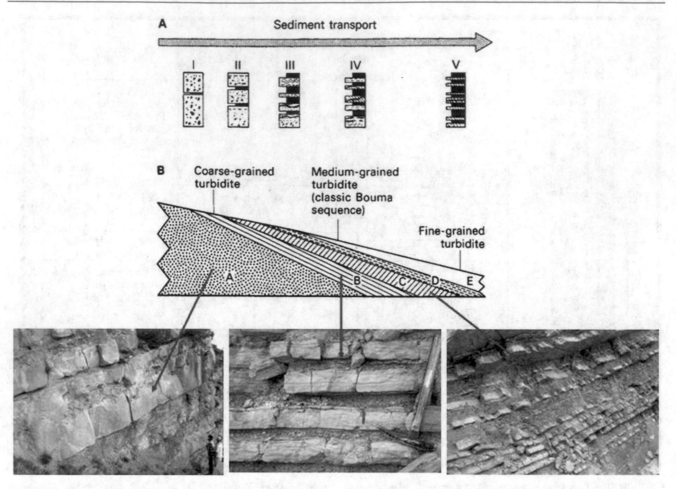

Fig. 3.49 The down-current evolution of turbidity-current deposits. As shown by the much-shortened cross-section of an individual flow (at center of diagram), near the source of the flow, the deposit may consist mainly of the A division, whereas at the distal end of the flow, turbidites may begin with the B or C division (diagram from Reading 1986)

3.5.7 Liquefaction, Load and Fluid Loss Structures

A few of these are environmentally diagnostic, others are practically ubiquitous in occurrence. Load and slump structures, convolute bedding and other products of thixotropy can occur wherever sediments are saturated, which is to say practically anywhere.

Water escape features, such as dish-and-pillar structures (Fig. 2.19c–f), are also found in the deposits of many environments (Nilsen et al. 1977), although they are particularly abundant in fluidized sediment-gravity flows.

As noted in Chap. 2, subtly different sedimentary structures have been attributed to desiccation and synaeresis. Desiccation cracks are an indicator of subaerial exposure, synaeresis has been attributed to volume changes resulting from salinity fluctuations in tidal, lagoonal or lacustrine settings (e.g., Donovan 1973), but a reexamination of one of the classic locations of synaeresis cracks, the Orcadian Basin in Scotland, suggested that the "cracks" are gypsum pseudomorphs (Astin and Rogers 1991). Desiccation may be accompanied by the break-up of the uppermost beds and

resedimentation as an intraformational breccia. Shinn (1983) provided a useful review of the terminology of some of the small-scale structures that develop on tidal flats and the kinds of interpretations that can be made from them.

Displacive evaporite growth in sabkhas produces a variety of distinctive structures that are useful diagnostic indicators.

3.5.8 Paleoecology of Body Fossils

In Sect. 2.2.2.9, we discuss briefly the three main types of body fossil preservation, those in growth positions, soft-bodied organisms that would not survive significant transportation and the far more common death assemblage. While fossils are invaluable as environmental indicators, there may be complications involved in the interpretation of reworked fossils and death assemblages. Difficulties may also arise in the case of faunas and floras limited in their distribution by factors of ecology or biogeography (not necessarily the same thing). All these points are relevant when we focus on the use of body fossils as indicators of depositional environment.

Paleoecology is the study of the relationships between fossil organisms and their environment. In most paleoecological research and in the resulting books and papers, the primary focus is on the animals, with the use of such criteria as species interrelationships and fossil-sediment characteristics in the analysis of the ecology of some fossil species or community. Such research is typically carried out by specialists whose background is usually paleontology or biology, and these individuals may have little or no interest in regional stratigraphy or basin history. Here we wish to reverse the process, to use fossils as paleogeographic indicators. In the Phanerozoic record, fossils are among the most powerful environmental indicators available, but, of course, their distribution and usefulness in the Precambrian are very limited. Useful texts on this subject have been provided by Ager (1963), Boucot (1981), and Dodd and Stanton (1981).

Even the simplest paleoecological observations can be invaluable. The presence of abundant bioturbation or (for example) a brachiopod fauna may be the key evidence in distinguishing the deposits of tidal channels and their point bars, tidal inlets and deltas from certain fluvial and delta plain facies, which can be sedimentologically very similar (Barwis 1978). At the other extreme are sophisticated statistical studies of particular communities or fossil groups, such as the trilobite population analysis of Ludvigsen (1978). Most paleoecological studies by stratigraphers and basin analysts fall somewhere between these extremes.

Fossil data are subject to three biases, which may distort our observations. First, it has been determined that in any depositional or ecological setting, a large proportion of the species present have a low probability of becoming fossils. Shaw et al. (2021) found that taxon fossilization potential of marine genera varies between environments, from 34% in shallow and deep water to 44% in coral reefs, 51% on seamounts and 15% in pelagic assemblages. Within-environment fossilization potential, in contrast, does not exceed 32% (in shallow water).

Second, there is the bias of unskilled or incomplete collection, which is particularly likely to arise when paleoecological analysis is attempted by the nonspecialist or when it is rushed through, as in a reconnaissance survey, by a prospecting team. Ager (1963) demonstrated this with data assembled by B.W. Sparks, who used two methods to collect molluscs from Quaternary deposits in southern Britain (Fig. 3.50). A bulk sample was analyzed by sieving, and the distribution was compared with that of a collection made by hand picking from the outcrop surface. The bias in the second collection in favor of larger or more brightly colored species is startling.

The third bias is that introduced by the geological obstacle courses organisms are put through before they end up under the geologist's hammer. The various hazards and possible paths through them are summarized in Fig. 3.51. The study of this flow diagram is itself the subject of a

Fig. 3.50 Distribution of species in two collections made from the same outcrop by two different methods. (From D.V. Ager, Principles of Palcoecology, 1963, McGraw-Hill, New York. Reproduced with permission of the publisher)

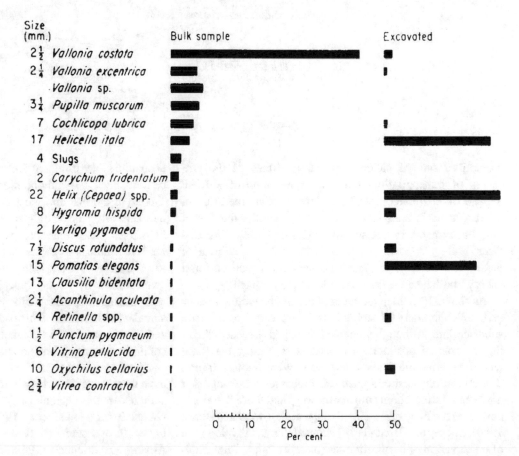

Fig. 3.51 The various
possibilities for the preservation
or destruction of a living animal
community and its eventual
collection as a fossil assemblage.
(From D.V. Ager, Principles of
Paleoecology, 1963,
McGraw-Hill, New York.
Reproduced with permission of
the publisher)

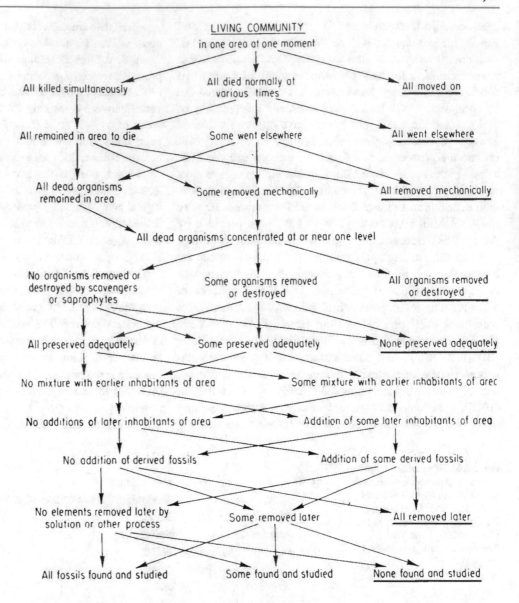

specialized science called **taphonomy**. Many of the processes of transportation, removal or break-up of a fossil depend on sedimentological processes and are therefore of interest to basin analysts. Figure 3.52 illustrates the fate of dinosaur remains in Cretaceous fluvial deposits of the Red Deer Valley, Alberta (from Dodson 1971). The state of preservation may yield much information on local transport energy and hence on the depositional environment.

As shown by a thematic set of papers in the journal Palaios (v.1, #3), the study of taphonomy is growing into a useful subdiscipline. Although valuable in a negative sense, revealing the role of postmortem processes in biasing the fossil record, taphonomic studies can also provide useful information on the physical and chemical processes occurring after death and during burial (**biostratinomy**) and those that occur during early diagenesis (fossil diagenesis)(Brett and Baird 1986). Speyer and Brett (1986) showed that the varying state of preservation of fossil trilobites (disarticulation, fragmentation) records useful information regarding energy levels in the

depositional environment and proposed that these variations be formalized into taphonomic facies or **taphofacies**.

One of the first and most obvious observations that must be made in a paleoecological study is to relate biofacies to lithofacies. The relationships may be obvious, including marked contrasts in facies, such as that shown in Fig. 3.53 (from Hecker 1965). Here, there is a red-bed lithofacies characterized by fish, a brackish lithofacies with Lingula, and marine deposits with corals, stromatoporoids and abundant bioturbation. This pattern of interbedded sediments and fossil types is characteristic of Early Paleozoic continental-marine transitions around the world.

Many excellent paleoecological-sedimentological studies have been made of carbonate reefs, such as the Permian reef complex of the Guadalupe Mountains in the southwestern United States (Newell et al. 1953), Silurian reefs of the Great Lakes (Lowenstam 1950, 1957) and Devonian reefs of Alberta (Andrichuck 1958; Klovan 1964). Figure 3.54 illustrates the lithofacies and biofacies subdivisions of some

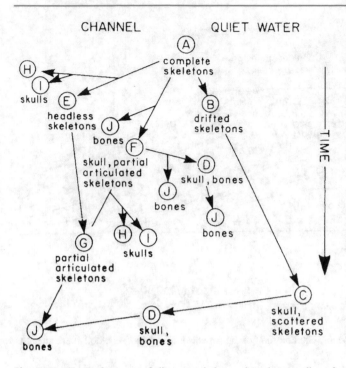

Fig. 3.52 The taphonomy of dinosaur skeletons based on studies of the Cretaceous rocks of Red Deer Valley, Alberta (Dodson 1971)

Silurian carbonate banks (J. L. Wilson 1975; Sellwood 1986). More subtle fossil–sediment relationships may only be apparent after careful statistical studies, such as the brachiopod analysis carried out by Jones (1977). In some cases, biofacies are not obviously related to lithofacies, as in the Welsh Paleozoic brachiopod assemblages discussed in Sect. 7.5.1. (see Fig. 7.8).

How are these observations on biofacies actually used to indicate the depositional environment? In four main ways: (1) by comparison with the ecology of living relatives, (2) by deductions about functional morphology, (3) by deductions from position and preservation and (4) by deductions from assemblages and associations. These lines of reasoning are not intended to be mutually exclusive, and indeed, in many ancient deposits, there is an overlap in their application.

To start with the first point, comparisons with modern relatives become more difficult the further back in time we recede, as we trace evolutionary lines back to more and more distant ancestors, for whom there is no guarantee of unchanged habits. This is where examination of functional morphology, position and associations might provide invaluable supportive evidence. The corals are a classic example of a group of organisms that has undergone marked evolutionary change

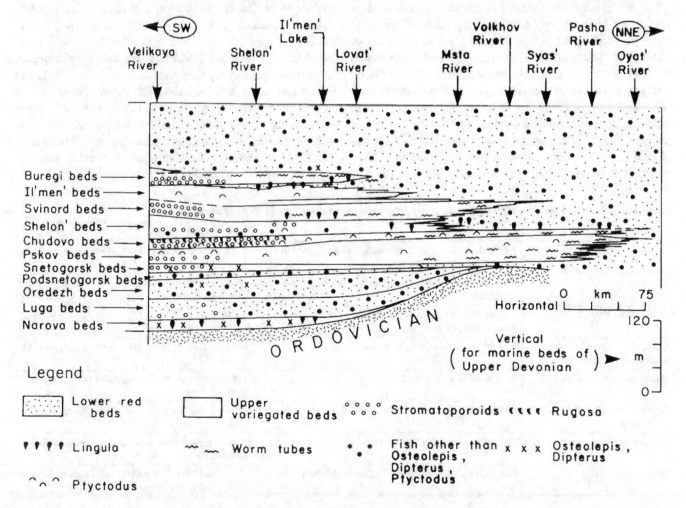

Fig. 3.53 Lithofacies–biofacies relationships for some Upper Devonian rocks in Russia (Hecker 1965)

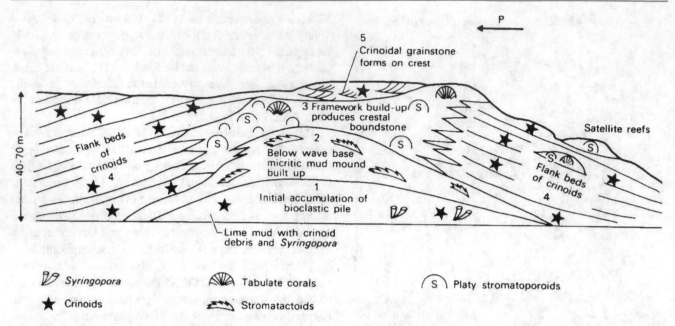

Fig. 3.54 Lithofacies–biofacies composition of carbonate buildups in the mid-Silurian shelf of the American Mid-west (Sellwood 1986; after J.L. Wilson 1975. Reproduced with permission from J.C. Wilson 1975)

during the Phanerozoic, yet seems to have continued to inhabit rather similar ecological niches throughout this time. Modern hermatypic (colonial, reef forming) corals inhabit depths of water down to about 100 m, but are only abundant above about 15 m. They are best adapted to grow in waters between 18° and 36° C, with an optimum of 25° to 29° C. Strong sunlight, a continuous nutrient supply, minimal salinity variations and clean, sediment-free water are also necessary. These requirements restrict colonial corals to a present-day latitudinal range of 30°S to 30°N, with variations depending on the extent of cold or warm oceanic currents, sediment input from large rivers, etc. (Schlager 2005; James and Wood 2010). For

Mesozoic and Cenozoic corals, which are closely related to modern forms (orders Scleractinia and Alcyonaria), environmental deductions based on modern ecology seem safe. Paleozoic corals belong to two different orders, Rugosa and Tabulata, yet their gross appearance, lithofacies and biofacies associations (particularly the symbiotic association with algae) remain similar, as is evident from the various Paleozoic reef studies quoted previously. Heckel (1972a) suggested, however, that they may have occupied less agitated environments, with the now extinct stromatoporoids filling part of this niche. Table 3.2 summarizes some other modern–ancient comparisons for carbonate-producing organisms.

Table 3.2 Modern and ancient carbonate-producing organisms and their sedimentary products (James 1984a)

Modern organism	Ancient counterpart	Sedimentary aspect
Corals	Archaeocyathids, corals, stromatoporoids, Bryozoan, rudistid bivalves, hydrozoans	The large components (often in place) of reefs and mounds
Bivalves	Bivalves, brachiopods, cephalopods, Trilobites, and other arthropods	Remain whole or break apart in several pieces to form sand- and
Gastropods, benthic	Gastropods, tintinitids, tentactulitids, salterelids, benthic foraminifera, brachiopods	Whole skeletons that form sand-and gravel-sized particles
Codiacian algae Halimeda, sponges	Crinoids and other pelmatozoans, sponges	Spontaneously disintegrate upon death into many sand-sized
Planktonic foraminifers	Planktonic foraminifera, coccoliths	Medium sand-sized and smaller particles in basinal deposits
Encrusting foraminifers and coralline algae	Coralline algae, phylloid algae, renalcids,	Encrust on or inside hard substrates, build up thick deposits,
Codiacean algae	Codiacean algae, Penicillus-like forms	Spontaneously disintegrate upon death to form lime mud
Blue-green algae	Blue-green algae (especially in pre-Ordovician)	Trap and bind fine-grained sediments to form mats and stromatolites

A group of organisms that has become extremely important in recent years for unraveling the history of continental
margins is the Foraminifera. Comparisons of ancient and
modern assemblages yield invaluable paleobathymetric data
for Cenozoic and later Mesozoic time, as discussed below.

Many ancient groups, such as the trilobites, graptolites
and many vertebrates, including early fish, amphibians and
reptiles, are now extinct, as are numerous genera and species
of still thriving phyla. Therefore, comparisons with the
ecology of modern relatives becomes tenuous or impossible.
Ecological and environmental interpretations may then
depend partly or wholly on deductions made from functional
morphology. For example, Nichols (1959) was able to
demonstrate the change in burrowing habit of the Cretaceous
echinoid *Micraster* by careful examination of the functional
morphology and comparison with similar features in modern
echinoids. Micraster tests show an increase with time in the
number of pores required for respiratory tube feet, an
obvious adaptation to deeper burial.

There has been much debate about the habitats of trilobites,
which show considerable variation in morphology. Some
have large eyes, suggesting dim light (deep water?), others are
blind and may have lived below the photic zone. Long horizontal spines may have been a device to inhibit sinking in a
soft substrate, and the same effect may have been achieved by
the broad, flat exoskeletons of some species. Others were
burrowers, for example, Bathyrus, which has a concave-up,
wedge shape adapted to wriggling into the substrate. Ludvigsen (1978) discussed these and other morphological features and related them to trilobite biofacies zones and their
position on the Ordovician platform of Western Canada.

A widely used climatic indicator is the shape of tree
leaves. It is known from present-day distributions that in
tropical regions the leaves of dicotyledonous trees are entire
(smooth), whereas in temperate regions they are more
commonly dentate. Edwards (1936) showed that this characteristic could be used to interpret fossil floras, even for
extinct or unidentifiable forms. Modern work confirms this
early discovery (Seyfullah 2012; Royer and Wilf 2006).
Modern and ancient examples are illustrated in Fig. 3.55.

Factors of abundance, state of preservation and position
may yield important environmental information. Fossil
forms that normally grew attached to a substrate may occur
as disoriented, broken and abraded individuals or as
hydraulically accumulated shell beds, suggesting storm
wave activity. Isolated individuals may be found out of
context, for example, I have found rare abraded orthocone
cephalopods and compound corals in laminated stromatolitic
dolomites formed on a Silurian tidal flat in the Canadian
Arctic. Ballance et al. (1981) reported finding coconuts in
Miocene turbidites and suggested that they may have been
transported offshore by tsunamis, the same events that
probably triggered the turbidity currents. Fossils found in

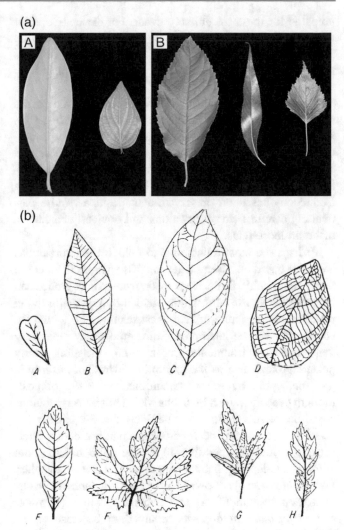

Fig. 3.55 Leaf shape as a reflection of climate. **a** Modern examples:
(A) Leaves of modern plants with smooth ("entire") margins, from
warm climates. (B) Leaves of modern plants with toothed margins,
from cool climates (from Seyfullah, 2006). **b** Ancient examples: **A–D**,
tropical plants typically have entire margins; assemblage from Lower
Eocene, Southern England; **E–H**, temperate plants typically have
dentate margins; assemblage from Miocene, Southern France.
(Edwards 1936; From D.V. Ager, Principles of Paleoecology, 1963.
McGraw-Hill, New York, Reproduced with permission from the
publisher.)

place, such as coral colonies, pelecypods in their burrows
and upright trees attached to their roots, are particularly
diagnostic of depositional environment (so long as their
habits are known) because such occurrences are usually
devoid of any ambiguity about possible transportation into
an unrelated environment. However, Jeletzky (1975) and
Cameron (1975) have engaged in heated controversy over
the significance of shallow-water invertebrate fossils found
in nodules in a turbidite succession on Vancouver Island.
Are the nodules still in place where they were formed around
the fossils, or is there evidence of abrasion indicating
transportation to deeper water? The geologist should be

prepared for the most unlikely event. For example, upright
trees, weighed down by boulders in their roots, have been
found in the flood deposits of Mt. St. Helens and offshore
near islands in the ocean, where they were emplaced by
storms.

Preferred orientations of attached or transported fossils
may have paleocurrent significance. Corals, crinoids, stromatolites and other immobile organisms may grow in particular directions in response to current conditions, and
accumulations of elongated fossils such as graptolites, gastropods, logs or belemnites may be aligned hydraulically.
Deductions based on preservation or position require particularly careful field observations and obviously cannot be
made on loose talus.

Perhaps the most reliable paleoecological interpretations
are those based on assemblages and associations in an entire
fossil community. Deductions made from extinct groups can
be checked against information yielded by those with living
relatives, and a picture of ecological relationships within the
community can be attempted. Some forms may yield very
generalized environmental information, including many
pelagic groups such as the graptolites, whose occurrence is
governed more by energy conditions at the site of postmortem deposition than by habitat during life. Interpretations
based on tolerant groups may be finely tuned if other, more
selective species can be found. Particularly accurate interpretations may be possible if species with different but
slightly overlapping habitat ranges can be found together.
For example, in the Lower Cretaceous of northern Texas,
arenaceous foraminifera are associated with the oyster
Ostrea carinata in deposits assumed to be close to the
shoreline, but the association only occurs in limestones and
marls, not in mudstones. Modern arenaceous foraminifera
are known to be tolerant of brackish water. *O. carinata*
therefore appears to be tolerant of reduced salinities, but not
of turbid water. By contrast, the oyster *Gryphaea
washitaensis* occurs in all rock types, but never in association with arenaceous foraminifera. It therefore seems to have
preferred open marine conditions, but had no preference for
either turbid or non-turbid water (Laughbaum 1960; Ager
1963). Another example is the association of ostracodes,
gastropods and the brachiopod *Lingula* in some Silurian
dolomite beds in the Canadian Arctic. This is interpreted as a
brackish-water tidal flat assemblage and contrasts strongly
with the rich coral, brachiopod, crinoid and trilobite
assemblage of interbedded open marine limestones (Miall
and Gibling 1978). Heckel (1972b) provided tables relating
the modern distribution of fossilizable invertebrate groups to
water salinity, turbidity and depth (e.g., Fig. 3.56), and
within each group individual genera or species may have
much more restricted tolerance ranges, as in the various
examples given above.

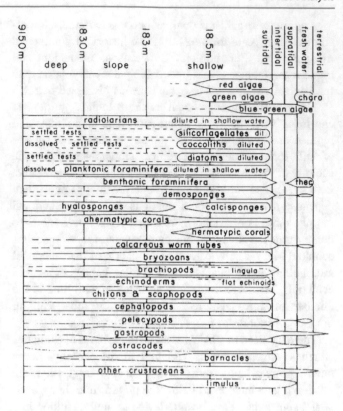

Fig. 3.56 Depth distribution of common types of fossil invertebrates
(Heckel 1972b)

Some of the most detailed assemblage studies carried out
for basin-interpretation purposes are those on ancient reefs
of all ages and on Cenozoic foraminifera. For example,
Fig. 3.57 shows the vertical changes in assemblage through
Niagaran reefs of the Great Lakes (Lowenstam et al. 1956).
The reefs were built on a soft substratum in quiet water and
gradually extended up into the zone of breaking waves.

Lagerstätte (plural **Lagerstätten**) is a German term that
refers to a sedimentary layer with unusual occurrences of
relatively well-preserved organic remains, typically found in
their original life setting. They include two broad types,
complete bodies or parts with soft tissues (Konservat-
Lagerstätten) or densely concentrated skeletal remains
(Konzentrat-**Lagerstätten**) (Seilacher et al. 1985; Allison
1988). The Burgess Shale of British Columbia and the
Solenhofen Limestone of Germany are two well-known
examples.

Studies of foraminifera paleoecology began in the
Cenozoic of the Gulf Coast and California in the 1930s
(Natland 1933). The initial stimulus for the work was the
need to correlate petroleum-bearing sediments, and this
remains one of its most important applications. However, the
same data are now finding increasing application in the
interpretation of the subsidence history of continental margins, a subject of considerable relevance to our attempts at

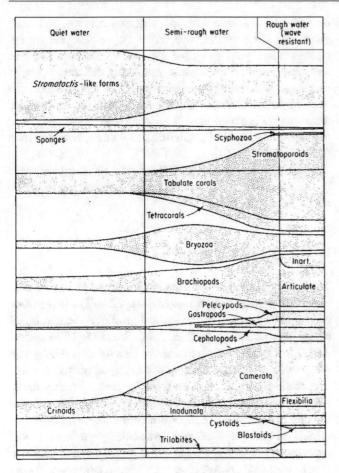

Fig. 3.57 Vertical variations in fossil assemblages of Niagaran reefs, Great Lakes (Lowenstam et al. 1956)

understanding the mechanisms of plate tectonics. An excellent example of this is the work on subsurface Cenozoic strata of the continental margin of Labrador and Baffin Island by Gradstein and Srivastava (1980). They distinguished four biofacies and interpreted their environments based on associated faunas and comparisons with other modern and ancient foraminifera assemblages. Considerable use was made of DSDP data. The biofacies are as follows:

1. Nonmarine: spores and pollen, no foraminifera.
2. Shallow neritic: marginal marine to inner shelf (<100 m deep). Diagnostic foraminiferal assemblages are of very low generic and specific diversity, with rare or no planktonics. In the Late Neogene sections, *Cibicidiodes*, *Elphidium*, *Cassidulina*, *Bulimina*, *Melonis*, gastropods and bryozoans occur in low numbers.
3. Deep neritic: 100–200 m water depths. Foraminiferal generic and specific diversity varies, planktonics occur locally. Cenozoic benthonic genera include *Uvigerina*, *Pullenia*, *Gyroidina* and *Alabamina*. Paleogene assemblages are impoverished.

4. Bathyal: upper slope, water depths 200–1000 m. Foraminiferal generic and specific diversity are high, and the assemblages are dominated by coarse, often large-sized agglutinated taxa. Over 50 species, including rich *Cyclammina* spp. Locally, planktonics are abundant (mostly turborotaliid and globigerinid forms). In a few wells, Paleogene calcareous benthonics occur in low numbers, including *Pleurostomella*, *Osangularia*, *Stilostomella* and *Nuttalides*.

3.5.9 Ichnology

Trace fossils are widely used environmental indicators in marine sediments, where they are most abundant. They have one advantage over body fossils in that they are almost invariably in place; derived and transported trace fossils are very rare. They are well preserved and readily studied in drill core, and have therefore made a substantial contribution to subsurface environmental analysis. Trace fossils may be used in two ways: (1) to indicate local sedimentation and erosion patterns and (2) to indicate the depositional environment. Trace-fossil assemblages have become valuable tools in the analysis of discontinuities and their allostratigraphic significance. The study of **ichnofacies** has become a specialized form of paleoecology. It is widely used in subsurface studies because of the ease with which trace fossils may be studied in cores. Many recent review articles and textbooks have been written on this subject, among the most useful of which is that by MacEachern et al. (2010).

There are various ways in which trace fossils may be used to evaluate local erosion and sedimentation. The density of bioturbation in a bed varies inversely with the rate of sedimentation, so that where sediment supply is low and invertebrate life abundant primary structures such as bedding may be completely destroyed MacEachern et al. (2010, Fig. 3) illustrated a **bioturbation index**, which is an attempt to quantify the intensity of alteration. Beds formed by rapid sedimentation, such as the passage of a turbidity current or a storm, may contain little or no penetration by organisms, except perhaps for escape burrows made vertically through the bed. Types of bioturbation may vary from one lithology to the next, because of different behavior patterns of the organisms in response to different sediment types. Ichnofacies are useful in the examination of cores and outcrops for the analysis of discontinuities, including the various bounding surfaces used in sequence-stratigraphic analysis (MacEachern et al. 2010, p. 45).

Important and widely quoted work by Seilacher (1967) showed how trace fossils could be used to interpret paleobathymetry. A set of five biofacies (now termed ichnofacies) were erected. These have now been expanded to nine (MacEachern et al. 2010; see Fig. 2.26 in this book). In the shallowest waters, waves and currents keep nutrients in suspension. Animals are subject to violent conditions and build deep vertical burrows such as *Skolithus*. In less turbulent waters faunal diversity increases, sediment feeding becomes more important and a variety of crawling and grazing feeding trails appear (*Cruziana* ichnofacies). Below wave base, sedimentation is slower, the oxygen content of the sediments is lower and nutrient supply more sparse. Sediment mining organisms of the *Zoophycos* and *Nereites* ichnofacies are characteristic. The latter, with its complex but highly systematic grazing patterns, is particularly typical of abyssal submarine-fan deposits. Other assemblages include a non-marine *Scoyenia* ichnofacies, a "softground" *Psilonichnus* assemblage, in addition to the *Skolithus* and *Cruziana* assemblages, and the *Glossifungites* ichnofacies, which develops on firm but unlithified substrates. The details of the ichnofacies distributions vary from basin to basin and depend on sediment type, water temperature, salinity and circulation patterns. Frey and Pemberton (1984) pointed out that

> local sets of environmental factors are most important in controlling the distribution of tracemakers, whether or not these parameters occur at specific water depths. For instance, many of the estuarine point bars ... exhibit a high-energy, channelward side typified by a *Skolithus* association and a low-energy bankward side typified by a *Cruziana* association... The respective associations occur in close proximity, at the same stratigraphic or bathymetric datum.

Also, as Howard (in Basan 1978) pointed out, different organisms may make very similar structures, and conversely, the same species may make different structures when engaged in different activities or when interacting with substrates of different composition. Interpretations must therefore always be made in conjunction with other facies studies, and the geologist would be wise to consult published work on trace fossil assemblages in other basins of a similar age. Given these caveats, "ichnofacies are facies models that address animal-sediment responses in the depositional environment" (MacEachern et al. 2010, p. 29). The term facies model is used here with explicit reference to the term as defined by Walker (1976).

In carbonate environments, algal mats and stromatolites generally are excellent indicators of intertidal deposition, although recent discoveries of stromatolites in subtidal settings have indicated that an environment hostile to browsing organisms may be the critical control. According to James (1984), their "upper limit is controlled by climate; in arid areas they cannot grow above the high intertidal into the supratidal zone, whereas in areas of high rainfall where the supratidal zone is moist or flooded for days at a time, mats are prolific. The lower limit is more variable and appears to be controlled by the presence of gastropods that eat algae."

In hypersaline zones, gastropods cannot survive, and mats grow into the subtidal zone. After deposition and burial, mats commonly rot away, but they leave voids as a result of the disappearance of organic materials, entrapped gas or shrinkage. The resulting distinctive type of porosity has been referred to as laminoid fenestrate, loferite or bird's eye. Stromatolites are present well back into the Precambrian and may have lived in similar environments throughout this time (see also Pratt 2010).

MacEachern et al. (2010, p. 31) report on recent advances in the analysis and understanding of ichnofacies in continental environments.

3.5.10 Vertical Profiles

The history of cyclic sedimentation and the early significance of vertical profiles is discussed in Sect. 1.2.2. The importance of Walther's law and of the vertical profile in modern facies analysis is discussed in Sect. 3.4.1. Methods of statistical analysis of cyclic sedimentation are also described briefly in this section. The recognition of cyclic sequences has become one of the most widely used tools for reconstructing depositional environments in the subsurface. One of the reasons for this is that cyclic changes are commonly readily recognizable in petrophysical logs (particularly gamma ray and spontaneous logs) and are often interpreted without access to core or well cuttings. Whether such interpretations are always correct is another question. The sedimentological literature is full of shorthand references to **fining-upward** or **coarsening-upward** cycles or to **fining-and-thinning** or **coarsening-and-thickening upward**. Grain size, bed thickness and scale of sedimentary structures are commonly correlated in clastic rocks, so that the cyclicity may be apparent from several types of observation.

An important warning: cyclicity may be in the eye of the beholder. Many facies interpretations in sedimentology place considerable emphasis on the recognition of repeated, ordered vertical change in facies characteristics, and many well-known facies models have been built upon such interpretations. But, as Hiscott (1981) demonstrated, with respect to submarine-fan deposits, the presence of cyclicity may be based on very subjective interpretations. Harper (1998) developed tests of significance for perceived thickening and thinning trends in stratal successions. Anderton (1995) argued that most submarine-fan deposits are non-cyclic. He stated "the human brain is very good at pattern recognition. It seldom misses a real pattern and can often see patterns in random noise. Thus, objective correlation tends to be over-correlation." Zeller (1964) carried out an experiment to demonstrate that geologists were capable of constructing and correlating apparently meaningful stratigraphic sections from coded data that was actually assembled from sequences of random numbers. Wilkinson et al. (1998) took issue with the interpretations of cyclicity in certain shelf-carbonate successions, descriptions of which have formed the basis for some important studies in sequence stratigraphy. They argued that other natural, random processes can be demonstrated to generate very similar patterns. The solution may be to apply rigorous statistical tests to the data.

There are two common basic types of cycle, those indicating an increase in transport energy upward and those demonstrating a decrease. Both types can be caused by a variety of sedimentary, climatic and tectonic mechanisms. Beerbower (1964) divided these into **autocyclic** and **allocyclic** controls. The terms **autogenic** and **allogenic** are now preferred, in order to avoid the connotation of cyclicity. Autogenic mechanisms are those that result in the natural redistribution of energy within a depositional system. Examples include the meandering of a channel in a river, tidal creek or submarine fan; subaerial flood events; subaqueous sediment-gravity flows; channel switching on a subaerial or submarine fan or a delta (avulsion); storms and tidal ebb and flood. All of these can potentially produce cyclic sequences. Allogenic mechanisms are those in which change in the sedimentary system is generated by some external cause. Tectonic control of basin subsidence, sediment supply, paleoslope tilt, eustatic sea-level change and climatic change are the principal types of allogenic mechanisms. These are large-scale basinal sedimentary controls. A sedimentary basin may be affected by several of these processes at the same time, so that it is not uncommon to find that there are two or three scales of cyclicity nested in a vertical profile. As Holbrook and Miall (2020, p. 12) noted, "allogenic processes do not generate stratigraphy, and are recorded instead by the changes they cause to autogenic processes." Allogenic cycles tend to be thicker and more widespread in their distribution than autogenic cycles. The latter are generally formed only within the confines of the subenvironment affected by the particular autogenic process. This assists the geologist in distinguishing and interpreting sedimentary cycles, a matter of some importance in the definition of various scales of sequences and parasequences, but such interpretations may be far from easy.

Vertical profiles formed a crucial component of the first facies models, including the point-bar model of Bernard and Major (1963) and Allen (1964), the barrier island model of Bernard et al. (1959) and the Bouma (1962) turbidite sequence (Sects. 1.2.2, 1.2.4, 1.2.7). They were the focus of a classic paper on facies models by Visher (1965), and their recognition in the subsurface has been covered in many articles and textbooks (e.g., Pirson 1977; Fisher et al. 1969; Sneider et al. 1978; Selley 1979). They still provide an essential basis of many modern facies models (Reading 1996; James and Dalrymple 2010), because they provide a simple way of synthesizing many different types of observation, including all the environmental criteria described in this chapter.

Vertical profiles have found an important use in recent years in the definition and documentation of stratigraphic sequences (Sect. 5.2; Fig. 5.12). As sea level rises and falls, coastal and shallow-marine depositional environments may migrate landward and seaward, respectively. The result is a corresponding vertical succession of facies assemblages recording the drowning and reestablishment of the various coastal environments. Coarsening-upward and fining-upward cycles may demonstrate the presence of deltaic, estuarine or beach–barrier systems at the shoreline, depending on the local paleogeography. Shoaling-upward successions bounded by marine flooding surfaces have been termed **parasequences** by sequence stratigraphers, and the nature of their vertical stacking is an important component of sequence analysis (Van Wagoner et al. 1987; Catuneanu 2006). However, modern work has resulted in significant confusions over the use of the term parasequence and, as explained in Sect. 5.4, it is now recommended that the term be abandoned. Where younger successions extend successively further into the basin, they are said to exhibit a **progradational stacking pattern**, and this is interpreted to indicate a rate of sediment input more rapid than the rate at which depositional space (**accommodation**) is being generated by basin subsidence or sea-level rise. **Aggradational** (vertically stacked) and **retrogradational** (**backstepping**) patterns indicate a balance between sediment delivery and accommodation, and a deficit in sediment delivery, respectively (Fig. 5.3). The latter condition is particularly common during transgression (Sect. 5.2.3). Lithostratigraphic and chronostratigraphic techniques may be used to correlate sequences and their component facies successions across a basin in order to determine the regional history of sea-level change (Sect. 6.3.2).

Coarsening-and-thickening-upward cycles are the most varied in their origins. Several distinct types are produced by coastal regression and progradation (lateral accretion), where there is a gradation from low-energy environments offshore to higher energy in the shoaling wave and intertidal zones. Examples are illustrated in Fig. 3.58. Other types are formed where there is a steep slope and abundant sediment supply, and the flow system that develops gradually grades itself to a balance by filling in the basin margin with a wedge of sediment. Coarsening-upward cycles are formed under these circumstances by prograding submarine fans and alluvial fans, particularly where the relief is maintained or even increased by active tectonic uplift. Other examples of coarsening-upward cycles (not illustrated) are those produced by crevasse splays in fluvial and deltaic settings, washover fans building from a barrier landward into a lagoon, and fluvioglacial sequences formed in front of an advancing continental ice sheet.

Fining-and-thinning-upward cycles commonly occur in fluvial environments as a result of lateral channel migration (point-bar succession) or vertical channel aggradation.

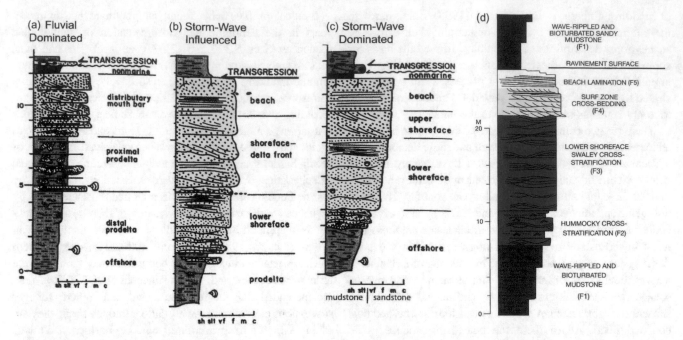

Fig. 3.58 Typical examples of thickening-and-coarsening-upward profiles from Dalrymple and James (2010). **a, b, c** Examples of delta-front successions (Bhattacharya 2010, Fig. 19, p. 246). **d** Shoreface-offshore profile (Dalrymple 2010a, Fig. 5, p. 9)

Alluvial fans may also show fining-upward cycles where they form under conditions of tectonic stability. These three types are shown in Fig. 3.59. Other examples are the tidal-creek point-bar and intertidal beach progradation sequences. Sediments deposited by catastrophic runoff events, including fluvial flash floods and debris flows and many types of subaqueous sediment-gravity flows, also show a fining-upward character. Major marine storms also generate fining-upward successions. A storm peak is usually erosional, and so a storm succession typically records the waning phase of the storm, resting on a regional erosion surface (Fig. 4.32).

Many of these cycles are superficially similar, and it may require careful facies and paleocurrent studies to distinguish them. Information on lateral variability may be crucial but this may be difficult to obtain in the case of subsurface studies.

Fig. 3.59 Typical examples of thinning-and-fining-upward profiles from Dalrymple and James (2010); Vertical scales approximately the same;
a Tidal-inlet channel-fill succession (Boyd 2010, Fig. 23, p. 277). **b** Tidal channel or tidal bar succession (Dalrymple 2010c, Fig. 20, p. 213).
c Meandering-river point bar or channel fill (Dalrymple, 2010b, Fig. 6, p. 64)

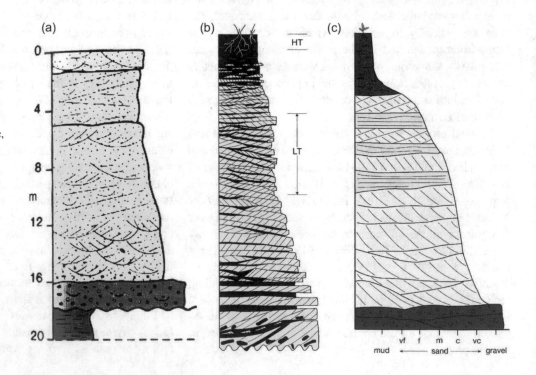

For carbonate environments, less emphasis has traditionally been placed on the vertical facies succession or profile, and more on the grain type, fauna and structures of individual beds. Assemblages of such attributes are commonly environmentally diagnostic, whereas in the case of siliciclastic sediments, much ambiguity may be attached to their interpretation, and such additional features as vertical profile and lateral facies relationships assume a greater importance. The range of environments in which carbonates are formed is much narrower than that of siliciclastics; they are mainly confined to shallow continental shelves, platforms or banks and adjacent shorelines and continental-margin environments. Yet enormous variability is apparent in these various settings, particularly in shallow-water and coastal regions, and this is another reason why standard vertical profile models have not become as popular as they have with clastic sedimentologists.

Ginsburg (1975), James (1984) and Pratt (2010) discussed shoaling-upward succession formed in shallow-subtidal to supratidal settings. These are common in the ancient record, reflecting the fact that the rate of carbonate sedimentation is generally much greater than the rate of subsidence. Shallowing-up sequences therefore repeatedly build up to sea level and prograde seaward. Lateral shifts in the various subenvironments are common. Pratt (2010) offered three generalized sequences as models of vertical profiles that could develop under different climatic and energy conditions (Fig. 3.60). Ginsburg and Hardie (1975) and Ginsburg et al. (1977) developed an exposure index representing the percent of the year an environmental zone is exposed by low tides. By studying tide gauges and careful surveying of part of the modern Andros Island tidal flat, they were able to demonstrate that a variety of physical and organic sedimentary structures is each present over a

Shallowing-upward Cycles

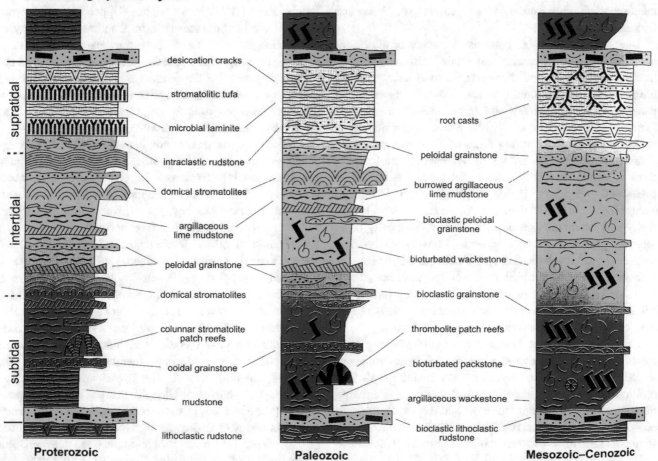

Fig. 3.60 Examples of meter-scale shoaling-upward carbonate succession formed in shallow subtidal to intertidal environments (Pratt 2010, Fig. 18, p. 416)

surprisingly narrow tidal exposure zone. This idea has considerable potential for interpreting shoaling-upward sequences, as demonstrated by Smosna and Warshauer (1981).

Carbonate buildups or reefs may contain an internal cyclicity that is the result of upward reef growth. James and Bourque (1992) suggested that the vertical profile may show an upward transition from an initial pioneer or stabilization phase to colonization, diversification and domination phases, characterized by distinctive textures and faunas. James and Wood (2010) described the response of reefs to sea-level change in terms of **keep-up**, **catch-up** or **give-up** modes, reflecting the rate of accumulation relative to the rate of sea-level change. In practice, most ancient reefs are the product of numerous sedimentation episodes separated by diastems or disconformities, attesting to fluctuating water levels (e.g., Upper Devonian reefs of Alberta: Mountjoy 1980). Analysis of vertical profiles of repeated cyclic patterns may therefore not be very helpful for basin-analysis purposes, although such work may be useful for documenting small-scale patterns of reef growth (e.g., Wong and Oldershaw 1980).

Deep-water carbonates comprise a variety of allochthonous, shelf-derived breccias and graded calcarenites, contourite calcarenites and hemipelagic mudstones, cut by numerous intraformational truncation (slide) surfaces (Cook and Enos 1977; McIlreath and James 1984; Coniglio and Dix 1992; Playton et al. 2010). Slope sedimentation is commonly most rapid during times of high sea level, when abundant carbonate material is being generated on the platform and shed from the margins, the process is termed as **highstand shedding** (Schlager 1991). The lithofacies assemblages are distinctive, but variations in slope topography and the random occurrence of sediment-gravity flows seem to preclude the development of any typical vertical profile. Organic stabilization and submarine cementation of carbonate particles probably prevent the development of carbonate submarine fans comparable to those formed by siliciclastic sediments, with their distinctive channel and lobe morphology and characteristic vertical profile.

Cyclic sequences are common in evaporite-bearing sediments, reflecting a sensitive response of evaporite environments to climatic change, brine level or water chemistry. Vertical profile models are therefore of considerable use in environmental interpretation. One of the most well known of these is the coastal sabkha, based on studies of modern arid intertidal to supratidal flats on the south coast of the Persian Gulf (Shearman 1966; Kinsman 1969; Kendall 1992, 2010). Coastal progradation and growth of displacive nodular anhydrite results in a distinctive vertical profile that has been widely applied (indeed, overapplied) to ancient evaporite-bearing rocks. Kendall (1992) discussed variations in this profile model, reflecting differences in climate and water chemistry that arise in other coastal and playa lake-margin settings.

As noted elsewhere, evaporites can occur in a variety of other lacustrine and hypersaline marine settings. They mimic many kinds of shallow-to-deep marine carbonate and siliciclastic facies, and a range of sedimentary criteria is required to demonstrate origin. The vertical profile is only one of these, but may be useful particularly when examining subsurface deposits in cores. For example, sulfates that accumulate below wave base commonly display a millimeter-scale lamination interbedded with carbonate and organic matter and possibly including evaporitic sediment-gravity flow deposits (Kendall 1992). The latter may even display Bouma sequences (Schreiber et al. 1976). Shoaling-upward intertidal to supratidal cycles have been described by Schreiber et al. (1976) and Vai and Ricci-Lucchi (1977) in Messinian (Upper Miocene) evaporites of the Mediterranean Basin. Caution is necessary in interpreting these cycles because they may not indicate a buildup or progradation under stable water levels, but instead they may be the product of brine evaporation and falling water levels. Many cycles of recharge and evaporation have been proposed for major evaporite basins such as the Mediterranean (Hsü et al. 1973).

Lacustrine environments are, in general, characterized by a wide variety of vertical profiles, reflecting many cyclic processes involving changes in water level and water chemistry. Many of these contain a chemical-sediment component. Lakes are highly sensitive to climate change, and their sediments have, therefore, become important in the investigation of orbital forcing mechanisms (Miall 2010). Van Houten (1964) described a shoaling-upward, coarsening-upward type of cycle in the Lockatong Formation (Triassic) of New Jersey. The cycles are about 5 m thick and consist, in upward order, of black, pyritic mudstone, laminated dolomitic mudstone and massive dolomitic mudstone with desiccation cracks and bioturbation. Chemical cycles which have an upper member of analcime-rich mudstone are also present. The cycles are interpreted as the product of short-term climatic change, with differences between the two types of cycle reflecting a greater tendency toward humidity or aridity, respectively. Eugster and Hardie (1975) described transgressive–regressive playa-margin cycles in the oil-shale-bearing Green River Formation of

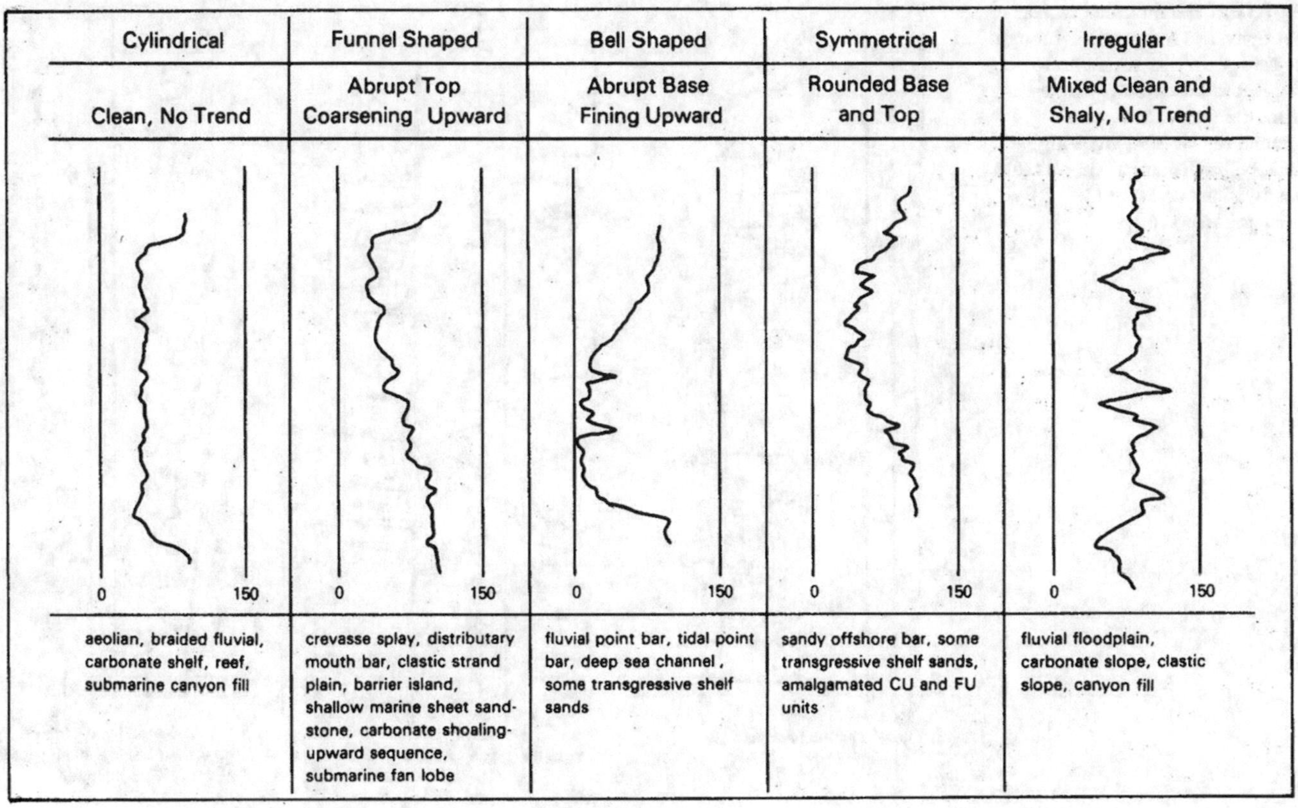

Cylindrical	Funnel Shaped	Bell Shaped	Symmetrical	Irregular
Clean, No Trend	Abrupt Top Coarsening Upward	Abrupt Base Fining Upward	Rounded Base and Top	Mixed Clean and Shaly, No Trend
aeolian, braided fluvial, carbonate shelf, reef, submarine canyon fill	crevasse splay, distributary mouth bar, clastic strand plain, barrier island, shallow marine sheet sandstone, carbonate shoaling-upward sequence, submarine fan lobe	fluvial point bar, tidal point bar, deep sea channel, some transgressive shelf sands	sandy offshore bar, some transgressive shelf sands, amalgamated CU and FU units	fluvial floodplain, carbonate slope, clastic slope, canyon fill

Fig. 3.61 "Log-shape analysis." Certain petrophysical logs, particularly gamma-ray and SP logs, when working with clastic successions, may be very useful indicators of cyclic processes in vertical profiles (Cant 1992)

the Rocky Mountains. Donovan (1975) erected five profile models for cycles that occur as a result of changes in lake level and fluvial-deltaic lake-margin progradation in the Devonian Orcadian Basin of Northern Scotland. Numerous other examples could be quoted.

Much has been written on the recognition of cyclic sedimentation from petrophysical logs (Fisher et al. 1969; Skipper 1976; Pirson 1977; Cant 1984, 1992). At present, the technique is best suited to the study of clastic cycles. As explained in Sect. 2.4, the gamma ray, spontaneous potential and resistivity logs are sensitive indicators of sand-mud variations and so are ideally suited to the identification of fining- and coarsening-upward cycles. These appear as **bell-shaped** and **funnel-shaped** log curves, respectively, and various subtleties of environmental change may be detected by observing the convexity or concavity of the

curves, smooth versus serrated curves, the presence of nested cycles of different thicknesses and so on. The technique has been referred to as **log-shape analysis** (Fig. 3.61). Fisher et al. (1969) published a series of typical profiles for coastal and marginal-marine clastic environments based on examples from the Gulf Coast (Fig. 3.62). Other examples from the Beaufort-Mackenzie Basin (from Young et al. 1976) are illustrated in Fig. 3.63. These curves are commonly interpreted in the absence of cores or cuttings. As should be apparent from the preceding pages similar cycles may be produced in different environments, and so this is a risky procedure. However, by paying close attention to appropriate facies models and scale considerations (cycle thickness, well spacing) good paleogeographic reconstructions can be attempted. The availability of cores in a few crucial holes may make all the difference.

Fig. 3.62 Examples of typical petrophysical log profiles through coastal plain and shelf clastic sequences based on examples from the Gulf Coast. Left log, gamma ray or S.P.; right log, resistivity. Center bar shows scale subdivisions of 30 m (100 ft) (Fisher et al. 1969)

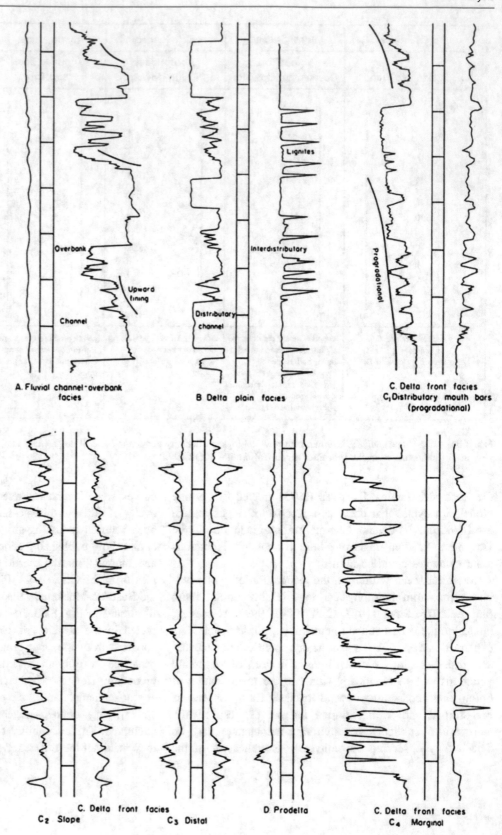

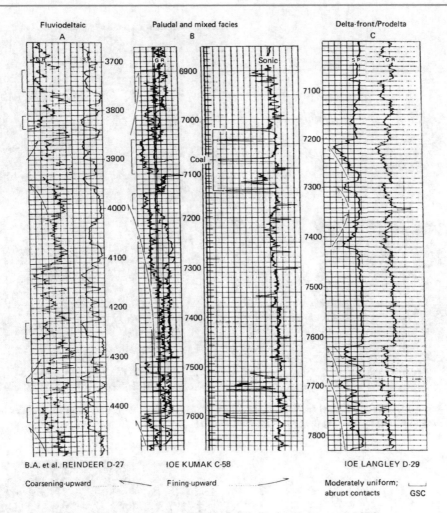

Fig. 3.63 Some examples of actual log profiles from the Beaufort-Mackenzie Basin (Young et al. 1976)

Figures 3.64, 3.65 and 3.66 illustrate outcrop examples of some typical vertical facies associations, together with the type of wireline-log responses that would be generated by running spontaneous potential and sonic logs through or across them. Funnel-shaped log responses are shown in Fig. 3.64. Figure 3.65 illustrates wireline logs showing the typical bell-shaped log profile, together with two examples.

These are both fluvial examples. Figure 3.66a illustrates a variation on the channel-fill log shape, the so-called **blocky** log response, indicating a sand-fill with little vertical variation in mud content or grain size (Fig. 3.66b). Finally, Fig. 3.66c, d illustrates a log response, and example, of a clastic succession showing no cyclicity, in this case a suite of turbidite sandstones.

Fig. 3.64 Wireline log showing the typical "funnel-shaped" profile, with three examples: **b, d** Tertiary delta-front cycles, Arctic Canada; **c** deltaic bay-fill succession, Carboniferous, Kentucky (Log diagram from Pirson 1977)

Fig. 3.65 **a, c** Wireline logs showing the typical "bell-shaped" profile of a fining-upward succession. **b** Point-bar, Carboniferous, Alabama; **d** Fluvial channel-fill, Tertiary, Arctic Canada (log diagram from Pirson 1977)

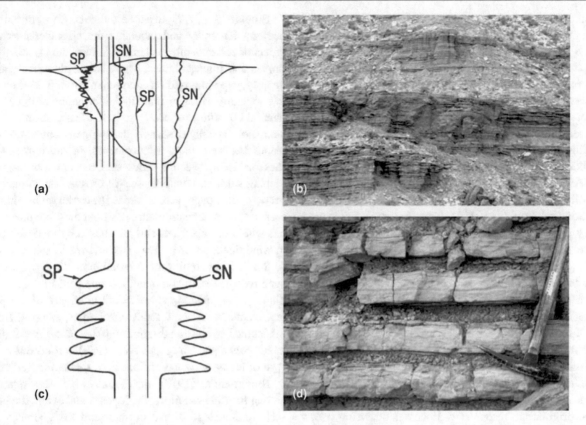

Fig. 3.66 a Blocky log response characteristic of some channel fills, with an example **b** of a channel in the Cutler Group (Permian), New Mexico. **c** Serrated log (from Pirson 1977) and an example (**d**), turbidite sandstones, France

3.5.11 Architectural Elements and Bounding Surfaces

One of the fundamental objectives of stratigraphy is to describe and interpret the three-dimensional architecture of the sediments filling sedimentary basins. On the largest scale, this involves the disciplines of lithostratigraphy, biostratigraphy, allostratigraphy and chronostratigraphy (Chap. 7), the reconstruction of major depositional sequences and the sequence-stratigraphic architecture, perhaps with the aid of regional seismic sections (Chaps. 5 and 6), an assessment of the roles of sea-level change, and the various mechanisms of basin subsidence (Miall 2010). At an intermediate to small scale, basin architecture depends on the interplay among subsidence rates, sea-level and the autogenic processes governing the distribution and accumulation of sediments within a particular range of depositional systems. Exploration for stratigraphic petroleum traps and for many types of stratabound ore bodies require close attention to basin architecture at the intermediate scale. For example, the search for and the interpretation of subsurface trends in potential reservoir rocks is a major preoccupation of petroleum geologists. Since the late 1950s many improvements have been made in techniques of basin analysis that facilitate this task.

For some time, we have been quite good at predicting the location and trend of such major depositional entities as alluvial fans, barrier reefs, various types of delta and barrier island sand bodies (to take some examples at random). But one of the outcomes of modern sedimentological research on depositional processes has been a much improved understanding of the composition and architecture of the various facies assemblages formed *within* each type of depositional system. The predictive value of sequence models has also proved its usefulness. We now have the ability, therefore, to construct much better models of the heterogeneities that exist within clastic reservoirs. Conventional techniques of sedimentological analysis and facies modeling place primary emphasis on the use of Walther's law and the interpretation of vertical profiles, particularly for the study of clastic sediments (Reading 1996; James and Dalrymple 2010). Meanwhile, the study of sequence stratigraphy has reached a level of maturity that terminologies and techniques have largely stabilized (Catuneanu 2006; Catuneanu et al. 2011; See Sect. 7.7).

A good understanding of the types of depositional units or elements that comprise stratigraphic sequences facilitates quantitative modeling of clastic depositional systems. Necessary data include the description in three dimensions of their scales, lithofacies compositions, length of time required for accumulation, preservation potential and their spacing in time and space. A more immediate practical application of these data is to improve the information used by reservoir engineers and geologists in their models of fluid flow in clastic reservoirs. It has been estimated that complex internal architectures are responsible for intra-reservoir stratigraphic entrapment of an average of 30% of the original oil in place, amounting to as much as 100 billion barrels of movable, non-residual oil in the United States alone (Tyler et al. 1984). A better understanding of architectural complexities of petroleum reservoirs would facilitate improved primary production and would increase the success rate of enhanced recovery projects.

There are two important interrelated ideas (Miall 1988a, b, c):

1. The first is the concept of **architectural scale**. Deposits consist of assemblages of lithofacies and structures over a wide range of physical scales, from the individual small-scale ripple mark to the assemblage produced by an entire depositional system. Recent work, particularly in eolian, fluvial, tidal and turbidite environments, suggests that it is possible to formalize a hierarchy of scales (Figs. 1.12, 3.67, 8.9). Depositional units at each size scale originate in response to processes occurring over a particular time scale and are physically separable from each other by a hierarchy of internal **bounding surfaces**.
2. The second is the concept of the **architectural element**. An architectural element is a lithosome characterized by its geometry, facies composition and scale, and is the depositional product of a particular process or suite of processes occurring within a depositional system.

3.5.11.1 Architectural Scale and Bounding Surface Hierarchies

As noted by Allen (1983, p. 249):

> The idea that sandstone bodies are divisible internally into "packets" of genetically related strata by an hierarchically ordered set of bedding contacts has been exploited sedimentologically for many years, although not always in an explicit manner. For example, McKee and Weir (1953) distinguished the hierarchy of the stratum, the set of strata, and the coset of sets of strata, bedding contacts being used implicitly to separate these entities.

Allen (1966) showed that flow fields in such environments as rivers and deltas could be classified into a hierarchical order. His hierarchy was designed as an aid to the interpretation of variance in paleocurrent data collected over various areal scales, from the individual bed to large outcrops or outcrop groups. The hierarchy consists of five categories, small-scale ripples, large-scale ripples, dunes, channels and the integrated system, meaning the sum of the variances over the four scales. Miall (1974) added the scale of the entire river system to this idea (Fig. 6.62).

Brookfield (1977) discussed the concept of an eolian bedform hierarchy and tabulated the characteristics of four orders of eolian bedform elements: draas, dunes, aerodynamic ripples and impact ripples. These four orders occur simultaneously, superimposed on each other. Brookfield showed that this superimposition resulted in the formation of three types of internal bounding surface (Fig. 3.38). His first-order surfaces are major, laterally extensive, flat-lying or convex-up bedding planes between draas (**macroforms**, in the terminology of Jackson 1975). Second-order surfaces are low-to-moderate dipping surfaces bounding sets of cross-strata formed by the passage of dunes across draas (**mesoforms**). Third-order surfaces are reactivation surfaces bounding bundles of laminae within crossbed sets and are caused by localized changes in wind direction or velocity (mesoforms to **microforms**).

Similar hierarchies of internal bounding surfaces have been recognized in some subaqueous bedforms (Table 3.3). Shurr (1984) developed a fivefold hierarchy of morphological elements of shelf sandstone bodies, ranging from the lithosome (widespread stratigraphic units) to individual facies packages. Yang and Nio (1989) carried out a similar type of study on some ancient tidal sandstone bodies.

Brookfield's (1977) development of the relationship among the time duration of a depositional event, the physical scale of the depositional product and the geometry of the resulting lithosome was a major step forward that has been of considerable use in the analysis of eolian deposits. Brookfield (1977), Gradzinski et al. (1979) and Kocurek (1981) showed how these ideas could be applied to the interpretation of ancient eolian deposits. Kocurek (1988) has found that first-order surfaces include two types of surfaces, the most laterally extensive of which he terms **supersurfaces**. Characterization of eolian deposits therefore now requires a fourfold hierarchy of bounding surfaces.

Several workers have attempted to develop a breakdown of the range of physical scales present in fluvial deposits. Allen's (1983) study of the Devonian Brownstones of the Welsh Borders represents the first explicit attempt to formalize the concept of a hierarchy of bounding surfaces in fluvial deposits and makes reference to Brookfield's work in eolian strata as a point of comparison. Allen described three types of bounding surfaces. He reversed the order of numbering from that used by Brookfield (1977), such that the surfaces with the highest number are the most laterally extensive. No reason was offered for this reversal, but the result is an open-ended numbering scheme that can readily accommodate developments in our understanding of larger scale depositional units, as discussed below. First-order contacts, in Allen's scheme, are set boundaries, in the sense of McKee and Weir (1953). Second-order contacts "bound clusters of sedimentation units of the kinds delineated by first-order contacts." They are comparable to the coset boundaries of McKee and Weir (1953), except that more

Table 3.3 Hierarchy of depositional units

SRS	Time scale (years)	Inst. Sed. Rate (m/ka)	Bound Surf.	Sedimentary process	Eolian architecture (Brookfield)	Coastal-estuarine architecture (Allen)	Shelf architecture (Dott and Bourgeois, Shurr)	Submarine fan (Mutti and Normark)
1	10^{-6} [30 s]	10^6	0	Burst-sweep cycle				
2	$10^{-5}-10^{-4}$ [5–50 min]	10^5	0	Ripple migration		Ripple [E3 surface]	3-surface in HCS	
3	10^{-3} [8 hr]	10^5	1	Dune migration, foreset bundles		Tidal bundle [E2 surface]	2 surface in HCS	
4	$10^{-2}-10^{-1}$ [3–36 days]	10^4	2	Diurnal variability to normal meteorological floods (dynamic events)		Neap-Spring bundle [E2] storm layer	HCS sequence [1 surface]	
5	10^0-10^1	10^2-10^3	3	Seasonal to 10-yr flood: macroform growth increment	Reactivation [3rd order surface]			
6	10^2-10^3	10^2-10^3	3,4	Long-term (100-yr) flood: macroform, point bar, splay	Dune [2nd order surface]	Sand-wave field, washover fan	Facies package (V)	Macroform [5]
7	10^3-10^4	10^0-10^1	5	Long-term geomorphic process: channel, delta lobe, coal seam	Draa [1st order surface]	Sand ridge, tidal channel, barrier island	Elongate lens (IV)	Minor lobe, channel-levee [4]
8	10^4-10^5	10^{-1}	6	Channel belt, delta, orbital cycle [5th order]	Erg [supersurface]		Regional lentil (III)	Major lobe [turbidite stage 3]
9	10^5-10^6	$10^{-2}-10^{-1}$	7	Depositional system, alluvial fan, major delta complex, orbital cycle [4th order]	Erg [supersurface]		Sandstone sheet (II)	Depositional system
10	10^6	$10^{-1}-10^0$	7	Rapid subsidence of convergent-margin basins, syntectonic clastic progradation, growth strata			Lithosome (I)	
11	10^6-10^7	$10^{-2}-10^{-1}$		Basin-fill complex, tectonic cyclothem (e.g., "clastic wedge")		Coastal-plain complex	Lithosome (I)	Fan complex
12	10^6-10^7	$10^{-3}-10^{-2}$		Very low-accommodation cratons				

SRS = Sedimentary Rate Scale; HCS=Hummocky cross-stratification
Information in square brackets in columns 6–9 refers to classifications used by the authors named at the head of the column

than one type of lithofacies may comprise a cluster. Allen (1983) stated that "these groupings, here termed complexes, comprise sedimentation units that are genetically related by facies and/or paleocurrent direction." Many of the complexes in the Brownstones are macroforms, in the sense defined by Jackson (1975). Third-order surfaces are comparable to the major surfaces of Bridge and Diemer (1983). No direct relationship is implied between Allen's three orders of surfaces and those of Brookfield because of the different hydraulic behavior and depositional patterns of eolian and aqueous currents.

Miall (1988a, b, 1996) found it useful to expand Allen's classification to a sevenfold hierarchy to facilitate the definition of fluvial macroform architecture and to include the largest, basin-scale heterogeneities in the classification (Fig. 3.67). This hierarchy is summarized in Table 3.3 and is compared with other examples of architectural hierarchies erected for eolian deposits (Brookfield 1977; Kocurek 1988), coastal-estuarine sand waves (Allen 1980), hummocky cross-stratification cycles (Dott and Bourgeois 1982), shelf deposits (Shurr 1984) and turbidite depositional systems (Mutti and Normark 1987). Cullis et al. (2018) carried out a comparative analysis of many different approaches to the hierarchical classification of deep-marine depositional systems. These architectural techniques are now being applied to outcrop study and analysis on Mars, by making use of the sophisticated image capture and processing technology carried on the latest ground rover vehicles (Salese et al. 2020).

Fig. 3.67 A sixfold hierarchy of depositional elements and bounding surfaces for fluvial deposits. Diagrams **a** to **e** represent successive enlargements of part of a fluvial unit in which ranks of bounding surfaces are indicated by circled numbers. A similar hierarchy could probably be developed for channelized deposits in other environments (Miall 1988a)

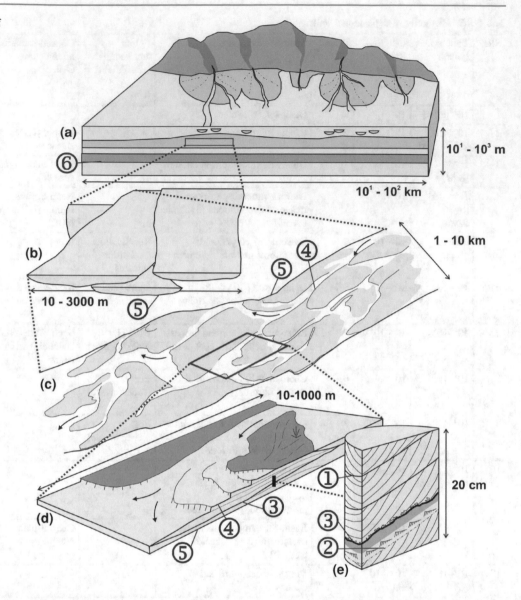

The various rankings in Table 3.4 are correlated mainly on the basis of the time duration represented by the deposits or the length of time stored in the bounding surfaces between them (column 2 in Table 3.3). The numbers in the first column refer to a hierarchy of **Sedimentation Rate Scales** that is discussed in Sect. 8.4. The formal bounding surface classification shown in Fig. 3.67 is provided in column 4 of Table 3.3. An example of an outcrop lateral profile showing the application of the bounding surface classification is shown in Fig. 2.3. Note that the depositional units listed in Table 3.3 represent a range of 12 orders of magnitude in time scale. They also represent at least 14 orders of magnitude in size (area). Note also that although this hierarchy has, at the time of writing, only been applied to fluvial deposits (several examples are given in Miall and Tyler 1991; Miall 1996),

there is considerable potential for similar subdivision of all clastic deposits, including an adaptation of the other hierarchical systems illustrated in the table. Miall (1988c) noted the similarities of fluvial deposits to some coarse-grained submarine-fan deposits, particularly the channel fills, and suggested ways of applying the architectural concepts described here to these rocks. These ideas were picked up by Clark and Pickering (1996), who provided numerous illustrations of modern and ancient deep-marine channel systems.

First- and second-order surfaces in the sixfold fluvial hierarchy record boundaries within microform and mesoform deposits (Fig. 3.67). The definition of *first-order surfaces* is unchanged from Allen (1983). *Second-order surfaces* are simple coset bounding surfaces, in the sense of McKee and Weir (1953). Third- and fourth-order surfaces are defined when architectural reconstruction indicates the presence of macroforms, such as point bars or sand flats (e.g., the large mid-channel bars in Fig. 4.4). *Third-order surfaces* represent growth increments of macroforms, such as the reactivation surfaces of Jones and McCabe (1980). Individual depositional units (storeys or architectural elements) are bounded by surfaces of *fourth-order* or higher rank. *Fifth-order surfaces* are surfaces bounding major sand sheets, such as channel-fill complexes (Fig. 3.67). They are generally flat to slightly concave upward, but may be marked by local cut-and-fill relief and by basal lag gravels (these were termed third-order surfaces by Allen 1983, and are the major surfaces of Bridge and Diemer 1983). *Sixth-order surfaces* are surfaces defining groups of channels or paleovalleys. Mappable stratigraphic units, such as members or submembers and sequence boundaries, are bounded by surfaces of sixth order and higher (Fig. 3.67; not defined by Allen 1983). Miall (1997, Table 15.3) suggested that sixth-order surfaces, as defined here, are equivalent to the "F" surfaces of Nio and Yang (1991) and to the bounding unconformities of allomembers or submembers, in the sense used by some workers in the area of sequence stratigraphy (Chap. 6). *Seventh-order surfaces* define individual depositional systems. *Eighth-order surfaces* bound major basin-fill complexes. Stratigraphic sequences are bounded by surfaces of sixth- and higher order rank.

Surfaces of fifth and higher order are potentially the easiest to map in the subsurface because of their wide lateral extent and essentially simple, flat or gently curved, channelized geometry. Many examples of the mapping of channel and paleovalley surfaces have, in fact, been reported in the literature (e.g., Busch 1974). Doyle and Sweet (1995) reported on an attempt to apply Miall's hierarchical system of bounding surfaces to a subsurface study of a fluvial sandstone succession. Considerable potential now exists for mapping these surfaces with three-dimensional seismic data, at least at shallow depths where resolution is adequate (Brown 2011; see Chap. 6). All higher order surfaces may appear very similar in cores. They are best differentiated by careful stratigraphic correlation between closely spaced cores, an objective that is best achieved in the most intensively developed fields, where well spacing may be a few hundred meters, or less. Third- and fourth-order surfaces may be recognized by gentle depositional dips (usually <10°).

Identification and correlation of these various bounding surfaces can clearly make a major contribution to the unraveling of the complexities of a channelized depositional system. They are likely to be particularly useful in the recognition and documentation of macroforms, about which much remains to be learned. Further discussion and illustration of the fluvial bounding surface hierarchy as applied to outcrop examples of fluvial deposits are given by Miall (1996). A broader approach to bounding surfaces and sedimentary breaks in general was taken by Miall (2016), who provided a new classification of unconformities and sedimentary breaks based on the time scale which they represent and the autogenic and allogenic processes that generate them. This is discussed in greater detail in Sect. 7.6.

The basis for a universal scheme with which to subdivide complex three-dimensional rock bodies is beginning to emerge, as suggested by Table 3.3, and the development of a common terminology may become necessary. Thus, the eolian supersurfaces of Kocurek (1988) are equivalent to Miall's (1988a) sixth-order surfaces, whereas Brookfield's (1977) first-order eolian surfaces could perhaps be equated with fifth-order surfaces. Similarly, in the shelf environment, tidal sand waves and sand banks also contain a hierarchy of facies units and internal bounding surfaces (Allen 1980; Shurr 1984; Berné et al. 1988; Harris 1988). The basal E1 surfaces of tidal sand waves (Allen 1980) are equivalent to Miall's third-order surfaces in terms of their time significance, and E2 surfaces are of first or second order, depending on their origin as diurnal or spring-neap erosion surfaces. The basal erosion surfaces of sand banks (Harris 1988) are of fifth order, whereas the internal master bedding planes are equivalent to fourth-order surfaces. Note that these rankings are based on time significance, not on geometry. E2 surfaces of sand waves and master bedding surfaces of sand banks both show depositional dips and appear superficially similar to Miall's third-order surfaces (e.g., compare Fig. 3.67 with Fig. 4.31), but represent different processes acting over different time periods. These differences make the erection of a common terminology difficult.

3.5.11.2 Architectural Elements

Application of the bounding surface concept permits the subdivision of a clastic succession into a hierarchy of three-dimensional rock units. This facilitates description and also makes it easier to estimate the appropriate physical extent and time duration of the processes that controlled sedimentation at each level of the hierarchy. Rock description involves the definition of lithofacies and the recognition of facies assemblages.

Most deposits may be subdivided into several or many types of three-dimensional bodies characterized by

Fig. 3.68 Principal sandstone and/or conglomerate architectural elements in fluvial deposits. (Modified from Miall 1985)

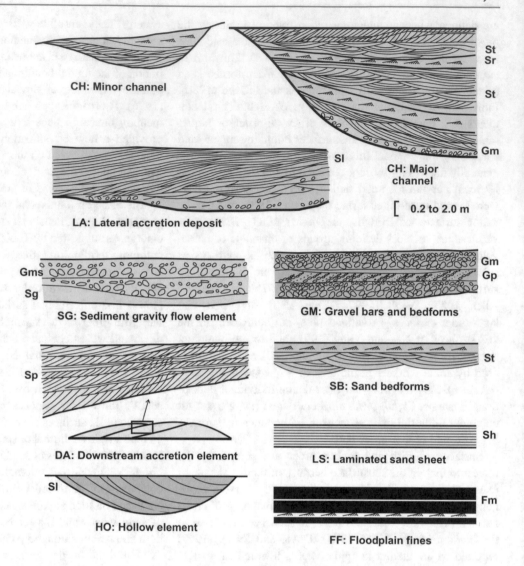

distinctive lithofacies assemblages, external geometries and orientations (many of which are macroforms). Allen (1983) coined the term **architectural element** for these depositional units, and Miall (1985, 1996) attempted a summary and classification of the current state of knowledge of these elements as they occur in fluvial deposits, suggesting that there about eight basic architectural elements comprising the coarse units in fluvial depositional systems (Fig. 3.68). A comparable number of fine-grained deposits make up the overbank or floodplain suite (Fig. 3.69).

Two interpretive processes are involved simultaneously in the analysis of outcrops that contain a range of scales of depositional units and bounding surfaces: (1) the definition of the various types and scales of bounding surfaces and (2) the subdivision of the succession into its constituent lithofacies assemblages, with the recognition and definition

of macroforms and any other large features that may be present. Figure 3.70 illustrates a hypothetical pair of cross-sections through an assemblage of architectural elements, showing how a system of labeling and nomenclature may be used to clarify the sedimentology of the deposit.

In general, the most distinctive characteristic of a macroform is that it consists of genetically related lithofacies, with sedimentary structures showing similar orientations and internal minor bounding surfaces (first to third order of the classification given previously) that extend from the top to the bottom of the element, indicating that it developed by long-term lateral, oblique or downstream accretion. A macroform is comparable in height to the depth of the channel in which it formed and in width and length is of similar order of magnitude to the width of the channel. However, independent confirmation of these dimensions is

Fig. 3.69 Principal architectural elements in the overbank or floodplain of river systems, based on an architectural diagram of a floodplain succession in the Lower Freshwater Molasse of Switzerland (Platt and Keller 1992), showing the range of elements to be expected in a floodplain setting (Miall 1996, Fig. 7.3). Element codes: CH: channel, CR: crevasse channel, CS: crevasse splay, FF: floodplain fines, LA: lateral accretion element, LV: levee. P: pedogenic (paleosoil) unit

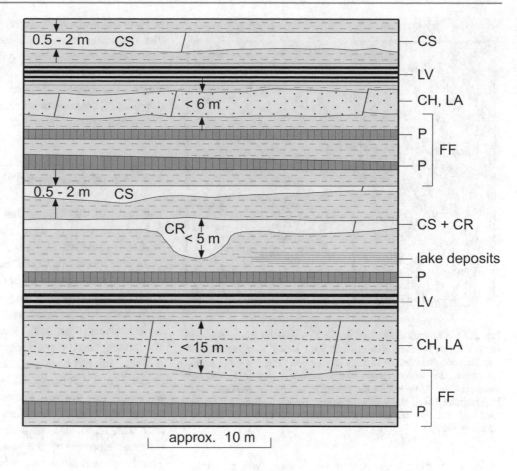

Fig. 3.70 A diagram to explain a system for the documentation of architectural elements and bounding surfaces in a fluvial deposit. The architecture of a simple, hypothetical mid-channel macroform that shows lateral accretion (LA), oblique accretion and downstream accretion (DA) in different parts of the unit. Arbitrary cutoff between these accretion geometries is indicated by dashed line. Numerals in brackets refer to ranks of bounding surfaces. Capital letters indicate numbering system for bounding surfaces as they would be labeled in outcrop. Based on the sand-flat model of Allen (1983, Fig. 19A) and the Brahmaputra bar mapped by Bristow (1987, Fig. 12)

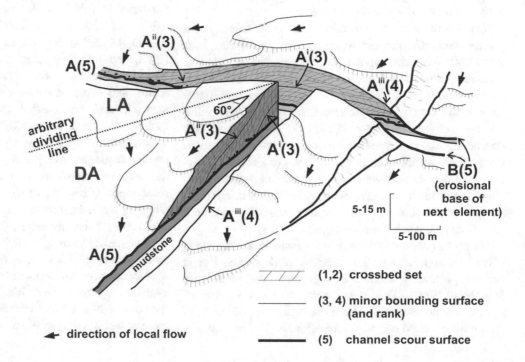

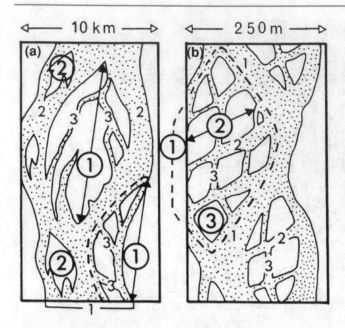

Fig. 3.71 Channel hierarchies in the Brahmaputra River, **a** and the Donjek River, **b** (after Williams and Rust 1969). Numbers in circles refer to bars, other numbers refer to channels. The first-order channel comprises the whole river, which includes several second-order channels. Bars scale within the channels in which they occur. In the Brahmaputra River, third-order channels modify higher order bars but still have bars within them, which cannot be shown at this scale (Bristow 1987)

difficult in multistory sandstone bodies, where channel margins are rarely preserved and the storeys commonly have erosional relationships with each other.

The definition of a macroform in any given outcrop is in part an interpretive process. Some types of macroforms, such as the lateral accretion deposits that comprise the typical point bar, are by now so well known that their recognition in outcrops would be classified by some workers as a descriptive rather than an interpretive exercise. Other types of macroform are less well known. The recognition of a macroform may depend in part on the type of bounding surface that encloses it. Conversely, the appropriate classification of a bounding surface may depend on a description of the lithofacies assemblage and geometry of the beds above and below it. For these reasons, description, classification and interpretation cannot always be completely separate exercises.

It may be possible to document the range in scales and the potential mappability of the hierarchy of depositional units and their bounding surfaces in any given stratigraphic unit (e.g., Miall 1988b). This type of subdivision is of importance in reservoir development. However, hierarchies of this type do not have a fixed range of scales (Fig. 3.71). For example,

in parts of the Donjek River, Yukon, large compound bars (the first-order bars of Williams and Rust 1969, using an earlier classification; fourth-order macroforms in the present scheme) are in the order of 200–400 m in downstream length. Comparable features in the Brahmaputra River, Bangladesh, are several kilometers in length (Coleman 1969) —an order of magnitude larger (Bristow and Best 1993).

3.6 Conclusions and Scale Considerations

The focus of this chapter has been on analytical methods. A brief overview of depositional environments and facies models is provided in the next chapter.

At this point, it would be useful to draw together a number of ideas appearing in different parts of this book that touch on stratigraphic scale.

In Sect. 1.4, we discuss different types and scales of basin-analysis projects, the kinds of data collection typically undertaken, and the problems and opportunities each offers. A distinction has been made here between the terms "facies models" and "depositional systems." The distinction is essentially one of scale. Facies models are the focus of local surface and subsurface studies, whereas depositional systems deal with regional systems and their study now forms an essential element of the study of sequence stratigraphy. These represent different levels of the stratigraphic hierarchy and corresponding analytical complexity. This point is illustrated in Table 3.4. Selected analytical methods are listed in this table, with an indication of the kinds of information obtained at the smaller, facies-analysis scale and the larger, depositional-systems scale. Similarly, each depositional environment can be analyzed at the two different scales, and some examples are given to demonstrate this idea. The table is not exhaustive, but is offered as an illustration of these scale considerations. To tackle the larger scales, including the mapping of systems tracts and stratigraphic sequences, many different basin mapping techniques are required, which is why the discussion of sequence stratigraphy (Chap. 5) needs to be complemented by the mapping methods described in Chap. 6.

To some extent, the difference between facies models and depositional systems is artificial; the boundary between the two is vague, and for some environments, such as deltas, the distinction is all but impossible to define. Yet, it is a useful approach to take, because it helps to distinguish and clarify the purpose of a number of different procedures we perform more or less simultaneously as a basin analysis is carried out.

Table 3.4 Scales of sedimentological analysis

1. Scales	Facies Analysis	Depositional Systems Analysis Sequence stratigraphy
2. General definitions	Outcrop or local-scale facies assemblages and environmental interpretations, within-basin (autogenic) cyclic mechanisms, facies models	Basin-wide paleogeography and stratigraphic architecture, allogenic process: contemporaneous tectonics, sea-level change, climate change, sequence stratigraphy
3. Selected analytical methods	Facies Analysis scale	Depositional Systems scale
Sedimentary structures and paleo currents	Bar and bedform geometry and relative arrangement, local hydraulics, identification of river, wave and tide influence	Basin dispersal patterns and relationship to tectonic elements
Vertical profile studies	Environmentally diagnostic autogenic mechanisms	Allogenic mechanisms
Petrology (siliciclastic)	Authigenic components may be environmentally diagnostic	Basin dispersal patterns, plate- tectonic setting of source rocks
Petrology (carbonate)	Basis of most lithofacies identifications and paleogeographic reconstructions at all scales	
Paleoecology	Local environmental interpretations	Regional paleobathymetry, gross salinity and temperature changes
4. Selected environments	Facies-Analysis scale	Depositional-systems scale
Fluvial	Bar type, channel geometry local fluvial style	Areal and stratigraphic variations of fluvial style, relation of dispersal patterns to contemporaneous tectonic elements
Eolian	Dune type and geometry, inter-dune environments	Erg paleogeography, relation to playa and fluvial systems
Deltaic	Discrimination of river, wave and tide influence is main objective at all levels of analysis as these are main controls on all scales of cyclicity and facies geometry	
Shelf	Types of HCS and storm s equence, tidal sand wave evolution	Relative dominance of tides versus storms
Continental Rise	Fan channel geometry and facies, channel migration and lobe progradation	Canyon-fan-basin plain relations, paleogeography of fan complexes
Carbonate Platform	local environments, tidal exposure, reef paleoecology	Discrimination of windward and leeward shelf margins, reef distribution
Evaporite Basins	Recognition of deep or shallow water origin	Large scale cyclicity and relation to water-level fluctuations

References

Adams J. E., and Rhodes M. L., 1960 Dolomitization by seepage refluction: American Association of Petroleum Geologists Bulletin, v. 44, p. 1912–1920

Adey, W. H., 1975, The algal ridges and coral reefs of St. Croix; their structure and Holocene development: Atoll Research Bulletin, No. 187, 67 p

Adey, W. H., 1978, Coral reef morphologenesis: a multidimensional model: Science, v. 202, p. 831–837.

Adey, W. H., Macintyre, R., and Stuckenrath, R. D., 1977, Relict barrier reef system off St. Croix: its implications with respect to late Cenozoic coral reef development in the Western Atlantic: Proceedings of the 3rd International Coral Reef Symposium, v. 2, p. 15–21.

Ager, D. V., 1963, Principles of palaeoecology: McGraw-Hill, New York, 371 p.

Ager, D. V., 1973, The nature of the stratigraphical record: New York, John Wiley, 114 p.

Ager, D. V., 1981, The nature of the stratigraphical record (second edition): John Wiley, New York, 122 p.

Ager, D., 1986, A reinterpretation of the basal 'Littoral Lias' of the Vale of Glamorgan. Proceedings of the Geologists Association, v. 97, p. 29–35.

Allen, J. R. L., 1964, Studies in fluviatile sedimentation: six cyclothems from the Lower Old Red Sandstone. Anglo-Welsh Basin: Sedimentology v.3, p. 163–198

Allen, J. R. L., 1966, On bed forms and paleocurrents: Sedimentology, v. 6, p. 153–190.

Allen, J. R. L., 1968, Current ripples: North Holland Publishing Company, Amsterdam, 433 p.

Allen, J. R. L., 1980, Sand waves: a model of origin and internal structure: Sedimentary Geology, v. 26, p. 281–328.

Allen, J. R. L., 1982, Sedimentary structures: their character and physical basis: Elsevier Scientific Publications, Developments in Sedimentology 30, 663 p.

Allen, J. R. L., 1983, Studies in fluviatile sedimentation: bars, bar complexes and sandstone sheets (low-sinuosity braided streams) in the Brownstones (L. Devonian). Welsh Borders: Sedimentary Geology, v. 33, p. 237–293

Allen, J. R. L., 1993, Sedimentary structures: Sorby and the last decade: Journal of the Geological Society. London, v. 150, p. 417–425.

Allen, P. A., and Allen, J. R., 2013, Basin analysis: Principles and application to petroleum play assessment: Chichester: Wiley-Blackwell, 619 p.

Allison, P. A., 1988, Konservat-Lagerstatten: cause and classification: Paleobiology, v. 14, p. 331–344.

Amthor, J. E., Mountjoy, E. W., and Machel, H. G., 1993, Subsurface dolomites in Upper Devonian Leduc buildups, central part of Rimbey-Meadowbrook reef trend, Alberta, Canada. Bull Can Pet Geol., v. 41, p. 164–185.

Anderton, R., 1985, Clastic facies models and facies analysis, in Brenchley P. J., and Williams, B. P. J., eds., Sedimentology: Recent developments and applied aspects: Blackwell Scientific Publications, Oxford, p. 31–47.

Andrichuk, J .M., 1958, Stratigraphy and facies analysis of Upper Devonian reefs in Leduc, Stettler and Redwater area, Alberta. Am Asso Petrol Geol Bull., v. 42, p. 1–93.

Arthur, M. A., Dean, W. E., and Stow, D. A. V., 1984, Models for the deposition of Mesozoic-Cenozoic fine-grained organic-carbon-rich sediment in the deep sea, in Stow, D. A. V., and Piper, D. J. W., eds., Fine-grained sediments: deep-water processes and facies: Geological Society of London Special Publication 15, p. 527–560.

Ashley, G. M., 1990, Classification of large-scale subaqueous bedforms: a new look at an old problem: Journal of Sedimentary Petrology, v. 60, p. 160–172.

Astin, T. R., and Rogers, D. A., 1991, "Subaqueous shrinkage cracks" in the Devonian of Scotland reinterpreted: Journal of Sedimentary Petrology, v. 61, p. 850–859.

Ball, M. M., 1967, Carbonate sand bodies of Florida and the Bahamas: Journal of Sedimentary Petrology, v. 37, p. 556–591.

Ballance, P. F., Gregory, M. R., and Gibson, G. W., 1981, Coconuts in Miocene turbidites in New Zealand: possible evidence for tsunami origin of some turbidity currents: Geology, v. 9, p. 592–595.

Barrett, T. J., 1982, Stratigraphy and sedimentology of Jurassic bedded chert overlying ophiolites in the North Apennines, Italy: Sedimentolgy, v. 29, p. 353–373.

Barwis, J. H., 1978, Sedimentology of some South Carolina tidal-creek point bars, and a comparison with their fluvial counterparts, in Miall, A. D., ed. Fluvial Sedimentology: Canadian Society of Petroleum Geologists Memoir 5, p. 29–160.

Basan, P. B., ed., 1978, Trace fossil concepts: Society of Economic Paleontologists and Mineralogists Short Course No. 5, 201 p.

Beerbower, J. R., 1964, Cyclothems and cyclic depositional mechanisms in alluvial plain sedimentation: Geological Survey of Kansas Bulletin 169, v. 1, p. 31–42.

Bernard, H. A., and Major, C. J., 1963, Recent meander belt deposits of the Brazos River; an alluvial "sand" model (abs): American Association of Petroleum Geologists Bulletin, v. 47, p. 350.

Bernard, H. A., Major, C. F. Jr., and Parrott, B. S., 1959, The Galveston Barrier Island and environs - a model for predicting reservoir occurrence and trend: Transactions of the Gulf Coast Association of Geological Societies, v. 9, p. 221–224.

Berné, S., Auffret, J.-P., and Walker, P., 1988, Internal structure of subtidal sandwaves revealed by high-resolution seismic reflection: Sedimentology, v. 35, p. 5–20.

Bhattacharya, J., 2010, Deltas, in James, N. P., and Dalrymple, R. W., eds., Facies Models 4: GEOtext 6, Geological Association of Canada, St. John's, Newfoundland, p. 233–264.

Blatt, H., Middleton, G. V., and Murray, R., 1980, Origin of sedimentary rocks; 2nd edition: Prentice-Hall Inc., Englewood Cliffs, New Jersey, 782 p.

Bluck, B. J., 1967, Deposition of some Upper Old Red Sandstone conglomerates in the Clyde area: a study in the significance of bedding: Scottish Journal of Geology, v. 3, p. 139–167.

Boggs, S. Jr., 2012, Principles of sedimentology and stratigraphy, 5th edn. Prentice Hall, Englewood Cliffs, New Jersey, 585 p.

Boucot, A. J., 1981, Principles of marine benthic paleoecology: Academic Press, New York,463 p.

Bouma, A. H., 1962, Sedimentology of some flysch deposits: Elsevier, Amsterdam, 168 p.

Bouma, A. H., Normark, W. R., and Barnes, N. E., eds., 1985, Submarine fans and related turbidite systems: Springer-Verlag Inc., Berlin and New York, 351 p.

Bown, T. M., and Kraus, M. J., 1981, Lower Eocene alluvial paleosols (Willwood Formation, northwest Wyoming, U.S.A.) and their significance for paleoecology, paleoclimatology, and basin analysis: Palaeogeography, Palaeoclimatology, Palaeoecology, v. 34, p. 1–30.

Boyd, R., 2010, Transgressive wave-dominated coasts, in James, N. P., and Dalrymple, R. W., eds., Facies Models 4: GEOtext 6, Geological Association of Canada, St. John's, Newfoundland, p. 265–294.

Brett, C. E., and Baird, G. C., 1986, Comparative taphonomy: a key to paleoenvironmental interpretation based on fossil preservation: Palaios, v. 1, p. 207–227.

Bridge, J. S., 1993, Description and interpretation of fluvial deposits: a critical perspective: Sedimentology, v. 40, p. 801–810.

Bridge, J. S., 1995, Description and interpretation of fluvial deposits: a critical perspective: Reply to Discussion: Sedimentology, v. 42, p. 384–389.

Bridge, J. S., and Diemer, J. A., 1983: Quantitative interpretation of an evolving ancient river system: Sedimentology, v. 30, p. 599–623.

Bristow, C. S., 1987, Brahmaputra River: Channel migration and deposition, in Ethridge, F. G., Flores, R. M., and Harvey, M. D., eds., Recent developments in fluvial sedimentology: Society of Economic Paleontologists and Mineralogists Special Publication 39, p. 63–74.

Bristow, C. S., and Best, J. L., 1993, Braided rivers: perspectives and problems, in Best, J. L., and Bristow, C. S., eds., Braided rivers: Geological Society, London, Special Publication 75, p. 1–11.

Brookfield, M. E., 1977, The origin of bounding surfaces in ancient aeolian sandstones: Sedimentology, v. 24, p. 303–332.

Brown, A. R., 2011, Interpretation of three-dimensional seismic data, seventh edition, American Association of Petroleum Geologists Memoir 42, 646 p.

Budd, D.A., 1997, Cenozoic dolomites of carbonate islands: their attributes and origin. Earth Sci Rev., v. 42, p. 1–47.

Busch, D. A., 1974, Stratigraphic traps in sandstones—exploration techniques: American Association of Petroleum Geologists Memoir 21.

Byers, C. W., and Dott, R. J., Jr., 1981, SEPM Research Conference on modern shelf and ancient cratonic sedimentation—the orthoquartzite-carbonate suite revisited: Journal of Sedimentary Petrology, v. 51, p. 329–347.

Cameron, B. E. B., 1975, Geology of the Tertiary rocks north of Latitude 49°, west coast of Vancouver Island: Geological Survey of Canada Paper 75–1A, p. 17–19.

Cant, D. J., 1984, Subsurface facies analysis, in Walker, R. G., ed., Facies models, 2nd edition: Geoscience Canada Reprint Series 1, p. 297–310.

Cant, D. J., 1992, Subsurface facies analysis, in Walker, R. G., and James, N. P., eds., Facies models: response to sea level change: Geological Association of Canada, St. John's, Newfoundland, p. 27–45.

Cant, D. J., and Walker, R. G., 1976, Development of a braided-fluvial facies model for the Devonian Battery Point Sandstone. Quebec; Canadian Journal of Earth Sciences, v. 13, p. 102–119.

Catuneanu, O., 2006, Principles of sequence stratigraphy: Elsevier, Amsterdam, 375 p.

Catuneanu, O., Galloway, W. E., Kendall, C. G. St. C., Miall, A .D., Posamentier, H. W., Strasser A., and Tucker M .E., 2011, Sequence Stratigraphy: Methodology and Nomenclature: Report to ISSC: Newsletters on Stratigraphy, v. 4 (3), p. 173–245.

Clague, J. J., Bobrowsky, P. T., and Hutchinson, I., 2000, A review of geological records of large tsunamis at Vancouver Island, British Columbia, and implications for hazard. Quatern Sci Rev., v. 19, p. 849–863.

Clark, J. D., and Pickering, K. T., 1996, Architectural elements and growth patterns of submarine channels: application to hydrocarbon exploration: American Association of Petroleum Geologists Bulletin, v. 80, p. 194–221.

Clifton, H. E., Hunter., R. E., and Phillips, R. L., 1971, Depositional structures and processes in the non-barred, high energy nearshore: Journal of Sedimentary Petrology, v. 41, p. 651–670.

Cloud, P., 1973, Peleoecological significance of the banded iron formation: Economic Geology, v. 68, p. 1135–1143.

Coleman, J. M., 1969, Brahmaputra River: Channel processes and sedimentation: Sedimentary Geology, v. 3, p. 129–239.

Coleman, J. M., and Wright, L. D., 1975, Modern river deltas: variability of processes and sand bodies, in Broussard, M. L., ed., Deltas, models for exploration: Houston Geological Society, Houston, p. 99–149.

Collinson, J. D., 1970, Bedforms of the Tana River, Norway: Geografisker Annaler, v. 52A, p. 31–55.

Collinson, J. D., and Thompson, D. B., 1982, Sedimentary structures: George Allen and Unwin, London, 194 p.

Coniglio, M., and Dix, G. R., 1992, Carbonate slopes, in Walker, R. G. and James, N. P., eds., Facics models: response to sea-level change: Geological Association of Canada, Geotext 1, p. 349–373.

Cook, H. E., and Enos, P., eds., 1977, Deep-water carbonate environments: Society of Economic Paleontologists and Mineralogists Special Publication 25, 336 p.

Costa, P. J. M., and Andrade, C., 2020, Tsunami deposits: present knowledge and future challenges: Sedimentology, v. 67, p. 1189–1206.

Cuffey, R. J., 1985, Expanded reef-rock textural classification and the geologic history of bryozoan reefs: Geology, v. 13, p. 307–310.

Cullis, S., Colombera, L., Patacci, M., and McCaffrey, W. D., 2018, Hierarchical classifications of the sedimentary architecture of deep-marine depositional systems: Earth Science Reviews, v. 179, p. 38–71.

Dalrymple, R. W., 2010a, Interpreting sedimentary successions: facies, facies analysis and facies models, in James, N. P., and Dalrymple, R. W., eds., Facies Models 4: GEOtext 6, Geological Association of Canada, St. John's, Newfoundland, p. 3–18.

Dalrymple, R. W., 2010b, Introduction to siliciclastic facies models, in James, N. P., and Dalrymple, R. W., eds., Facies Models 4: GEOtext 6, Geological Association of Canada, St. John's, Newfoundland, p. 59–72.

Dalrymple, R. W., 2010c, Tidal depositional systems, in James, N. P., and Dalrymple, R. W., eds., Facies Models 4: GEOtext 6, Geological Association of Canada, St. John's, Newfoundland, p. 201–231.

Daly, R. A., 1936, Origin of submarine canyons: American Journal of Science, v. 31, p. 401–420.

Davies, D. K., and Ethridge, F. G., 1975, Sandstone composition and depositional environments: American Association of Petroleum Geologists Bulletin, v. 59, p. 239–264.

Davies, G. R., and Smith, L. B., Jr., 2006, Structurally controlled hydrothermal dolomite reservoir facies: an overview. Am Asso Petrol Geol Bull., v. 90, p. 1641–1690.

Davies, N. S. and Gibling, M. R., 2010a, Cambrian to Devonian evolution of alluvial systems: The sedimentological impact of the earliest land plants: Earth-Science Reviews, v. 98, p. 171–200.

Davies, N. S., and Gibling, M. R., 2010b, Paleozoic vegetation and the Siluro-Devonian rise of fluvial lateral accretion sets: Geology, v. 38, p. 51–54.

Davies, N. S., and Shillito, A. P., 2021, True substrates: the exceptional resolution and unexceptional preservation of deep-time snapshots on bedding surfaces: Sedimentology, in press.

De Raaf, J. F. M., Boersma, J. R., and Van Gelder, A. 1977, Wave generated structures and sequences from a shallow marine succession. Lower Carboniferous, County Cork, Ireland: Sedimentology, v. 24, p. 451–483.

De Raaf, J. F. M., Reading, H. G., and Walker, R. G., 1965, Cyclic sedimentation in the Lower Westphalian of North Devon, England: Sedimentology, v. 4, p. 1–52.

Dickinson, K. A., Berryhill, H. L., Jr., and Holmes, C. W., 1972, Criteria for recognizing ancient barrier coastlines, in Rigby, J. K., and Hamblin, W. K., eds., Recognition of ancient sedimentary environments: Society of Economic Paleontologists and Mineralogists Special Publication 16, p. 192–214.

Dickinson, W. R. (Chair, U.S. Geodynamics Committee), 1997, The dynamics of sedimentary basins: National Research Council, National Academy Press, 43 p.

Dickinson, W. R., and Suczek, C. A., 1979, Plate tectonics and sandstone compositions: American Association of Petroleum Geologists Bulletin, v. 63, p. 2164–2182.

Dodd, J. R., and Stanton, R. J. Jr., 1981, Paleoecology, concepts and applications. Wiley, New York.

Dodson, P., 1971, Sedimentology and taphonomy of the Oldman Formation (Campanian), Dinosaur Provincial Park, Alberta (Canada): Palaeogeography, Palaeoclimatology, Palaeoecology, v. 10, p. 21–74.

Donovan, R. N., 1973, Basin margin deposits of the Middle Old Red Sandstone at Dirlot, Caithness: Scottish Journal of Geology, v. 9, p. 203–212.

Donovan, R. N., 1975, Devonian lacustrine limestones at the margin of the Orcadian Basin, Scotland: Journal of the Geological Society of London, v. 131, p. 489–510.

Dott, R. H. Jr., 1983, Episodic sedimentation—how normal is average? How rare is rare? Does it matter? J Sediment Petrol, v. 53, p. 5–23.

Dott, R. H., Jr., and Bourgeois, J., 1982, Hummocky stratification: significance of its variable bedding sequences: Geological Society of America Bulletin, v. 93, p. 663–680.

Dott, R. H., Jr., and Bourgeois, J., 1983, Hummocky stratification: significance of its variable bedding sequences. Reply to Discussion by R. G. Waker et al.: Geological Society of America Bulletin, v. 94, p. 1245–1251.

Doyle, J. D., and Sweet, M. L., 1995, Three-dimensional distribution of lithofacies, bounding surfaces, porosity, and permeability in a fluvial sandstone: Gypsy Sandstone of northern Oklahoma: American Association of Petroleum Geologists Bulletin, v. 79, p. 70–96.

Duff, P. McL. D., and Walton, E. K., 1962, Statistical basis for cyclothems: a quantitative study of the sedimentary succession in the east Pennine Coalfield: Sedimentology, v. 1, p. 235–255.

Dunham, R. J., 1962, Classification of carbonate rocks according to depositional texture, in Ham, W. E., ed. Classification of Carbonate Rocks: American Association of Petroleum Geologists Memoir 1, p. 108–121.

Edwards, W. N., 1936, The flora of the London Clay: Proceedings of the Geologists' Association of London, v. 47, p. 22–31.

Eichler, J., 1976, Origin of the Precambrian banded iron formations, in Wolf, K. H., ed. Handbook of Strata-Bound and Strataform Ore Deposits: Elsevier, Amsterdam, v. 7, p. 157–201.

Embley, R. W., 1976, New evidence for occurrence of debris flow deposits in the deep sea: Geology, v. 4, 371–374.

Embry, A. F., and Klovan, J. E., 1971, A Late Devonian reef tract on northeastern Banks Island, N.W.T.: Bulletin of Canadian Petroleum Geology, v. 19, p. 730–781.

Emery, K. O., 1968, Relict sediments on continental shelves of the world: American Association of Petroleum Geologists Bulletin, v. 52, p. 445–464.

Eriksson, P. G., Banerjee, S., Catuneanu, O, Corcoran, P. L., Eriksson, K. A., Hiatt, E. E, Laflamme, M., Lenhardt, N., Long, D. G. F., Miall, A. D., Mints, M. V., Pufahl, P. K., Sarkar, S., Simpson, E. L., Williams, G. E., 2013, Secular changes in sedimentation systems and sequence stratigraphy, in Kusky, T., Stern, R., and Dewey, J., eds., Secular Changes in Geologic and Tectonic Processes: Gondwana Research, Special issue, v. 24, issue #2, p. 468–489.

Eriksson, P. G., Condie, K. C., Tirsgaard, H., Mueller, W. U., Alterman, W., Miall, A. D., Aspler, L. B., Catuneanu, O., Chiarenzelli, J. R., 1998, Precambrian clastic sedimentation systems, in Eriksson, P. G., ed. Precambrian Clastic Sedimentation Systems: Sedimentary Geology, Special Issue 120, p. 5–53.

Eugster, H. P., and Hardie, L. A., 1975, Sedimentation in an ancient playa-lake complex: The Wilkins Peak Member of the Green River Formation of Wyoming: Geological Society of America Bulletin, v. 86, p. 319–334.

Evans, G., 1965, Intertidal flat sediments and their environments of deposition in the Wash: Quarterly Journal of the Geological Society of London, v. 121, p. 209–245.

Eyles, C. H., and Eyles, N., 2010, Glacial deposits, in James, N. P., and Dalrymple, R. W., eds., Facies Models 4: GEOtext 6, Geological Association of Canada, St. John's, Newfoundland, p. 73–104.

Eyles, N., Day, T. E., and Gavican, A., 1987, Depositional controls on the magnetic characteristics of lodgement tills and other glacial diamict facies: Canadian Journal of Earth Sciences, v. 24, p. 2436–2458.

Eyles, N., Eyles, C. H., and Miall, A. D., 1983, Lithofacies types and vertical profile models; an alternative approach to the description and environmental interpretation of glacial diamict sequences: Sedimentology, v. 30, p. 393–410.

Farrell, K. M., Harris, W. B., Mallinson, D. J., Culver, S. J., Riggs, S. R., Pierson, J., Self-Trail, J. M., and Lautier, J. C., 2012, Standardizing texture and facies codes for a process-based classification of clastic sediment and rock: Journal of Sedimentary Research, v. 82, p. 364–378.

Fawcett, J. J., 1982, Facies (metamorphic facies), in McGraw-Hill Encyclopedia of Science and Technology, 5th ed.: McGraw-Hill, New York, p. 303.

Fisher, W. L., Brown, L. F., Scott, A. J., and McGowen, J. H., 1969, Delta systems in the exploration for oil and gas: University of Texas Bureau of Economic Geology, 78 p.

Flügel, E., 1982, Microfacies analysis of limestones: Springer-Verlag, Berlin and New York, 633 p.

Folk, R. L. 1962, Spectral subdivision of limestone types, in Ham, W. E., ed., Classification of carbonate rocks. Am Assoc Pet Geol Mem 1, p. 62–84.

Frakes, L. A., Francis, J. E., and Syktus, J. I., 1992, Climate modes of the Phanerozoic: Cambridge University Press, Cambridge, 274 p.

Franke, W., and Paul, J., 1980, Pelagic redbeds in the Devonian of Germany—deposition and diagenesis: Sedimentary Geology, v. 25, p. 231–256.

Frazier, D. E., 1974, Depositional episodes: their relationship to the Quaternary stratigraphic framework in the northwestern portion of the Gulf Basin: Bureau of Economic Geology, University of Texas, Geological Circular 74–1, 26 p.

Frey, R. W., and Pemberton, S. G., 1984, Trace fossil facies models, in Walker, R. G., ed., Facies models, 2nd edition, Geoscience Canada Reprint Series 1, p. 189–207.

Friedman, G. M., 1979, Address of the retiring President of the International Association of Sedimentologists: Differences in size distributions of populations of particles among sands of various origins: Sedimentology, v. 26, p. 3–32.

Friedman, G. M., 1980, Dolomite is an evaporite mineral: evidence from the rock record and from sea-marginal ponds of the Red Sea: Society of Economic Paleontologists and Mineralogists Special. Publication 28, p. 69–80.

Friedman, G. M., and Sanders, J. E., 1978, Principles of Sedimentology: new York: John Wiley and Sons, 791 p.

Friend, P. F., Alexander-Marrack, P. D., Nicholson, J., and Yeats, A. K., 1976, Devonian sediments of east Greenland I: introduction, classification of sequences, petrographic notes: Meddelelser om Gronland Bd. 206, nr. 1.

Fu Q, Qing H., 2011, Medium and coarsely crystalline dolomites in the Middle Devonian Ratner Formation, southern Saskatchewan, Canada: origin and pore evolution. Carbonates Evaporites, v. 26, p. 111–125.

Galloway, W. E., 1975, Process framework for describing the morphologic and stratigraphic evolution of the deltaic depositional systems, in Broussard, M. L., ed., Deltas, models for exploration: Houston Geological Society, Houston, p. 87–98.

Galloway, W. E., and Hobday, D. K., 1996, Terrigenous clastic depositional systems, second edition: Springer-Verlag, Berlin, 489 p.

Ginsburg, R. N., ed., 1975, Tidal deposits: a casebook of recent examples and fossil counterparts: Springer-Verlag, Berlin and New York, 428 p.

Ginsburg, R. N. and Hardie, L. A., 1975, Tidal and storm deposits, northeastern Andros Island, Bahamas, in Ginsburg, R. N., ed., Tidal deposits: a casebook of recent examples and fossil counterparts: Springer-Verlag, Berlin and New York, p. 201–208.

Ginsburg, R. N., Hardie, L. A., Bricker, O. P., Garrett, P., and Wanless, H. R., 1977, Exposure index: a quantitative approach to defining position within the tidal zone, in Hardie, L. A., ed. Sedimentation on the Modern Carbonate Tidal Flats of Northwestern Andros Island, Bahamas: Johns Hopkins University Studies in Geology 22, p. 7–11.

Ginsburg, R. N. and Schroeder, J. H., 1973, Growth and submarine fossilization of algal cup reefs, Bermuda: Sedimentology, v. 20, p. 575–614.

Glaister, R. P., and Nelson, H. W., 1974, Grain-size distributions, an aid in facies identification: Bulletin of Canadian Petroleum Geology, v. 22, p. 203–240.

Gloppen, T. G., and Steel, R. J., 1981, The deposits, internal structure and geometry in six alluvial fan-fan delta bodies (Devonian-Norway)—A study in the significance of bedding sequences in conglomerates, in Ethridge, F. G., and Flores, R. M., eds., Recent and ancient nonmarine depositional environments: models for exploration: Society of Economic Paleontologists and Mineralogists Special Publication 31, p. 49–70.

Gorsline, D. S., 1984, A review of fine-grained sediment origins, characteristics, transport and deposition, in Stow, D. A. V., and Piper, D. J. W., eds., Fine-grained sediments: deep-water processes and facies: Geological Society of London Special Publication 15, p. 17–34.

Gould, S. J., 1965, Is uniformitarianism necessary? Am J Sci, v. 263, p. 223–228.

Gradstein, F. M., and Srivastava, S. P., 1980, Aspects of Cenozoic stratigraphy and paleoceanography of the Labrador Sea and Baffin Bay: Palaeogeography, Palaeoclimatology, Palaeoecology, v. 30, p. 261–295.

Gradzinski, R., Gagol, J., and Slaczka, A., 1979, The Tumlin Sandstone (Holy Cross Mts., central Poland): Lower Triassic deposits of aeolian dunes and interdune areas: Acta Geologica Polonica, v. 29, p. 151–175.

Gravenor, C. P., and Wong, T., 1987, Magnetic and pebble fabrics and origin of the Sunnybrook Till, Scarborough, Ontario, Canada. Can J Earth Sci, v. 24, p. 2038–2046.

Grover, N. C., and Howard, C. S., 1938, The passage of turbid water through Lake Mead: American Society of Civil Engineers, v. 103, p. 720–732.

Grunau, H. R., 1965, Radiolarian cherts and associated rocks in space and time: Eclogae Geologica Helvetica, v. 58, p. 157–208.

Harbaugh, J. W., and Bonham-Carter, G., 1970, Computer simulation in geology: Wiley-Interscience, New York, 98 p.

Harms, J. C., and Fahnestock, R. K., 1965, Stratification, bed forms, and flow phenomena (with an example from the Rio Grande), in Middleton, G. V., ed., Primary sedimentary structures and their hydrodynamic interpretation: Society of Economic Paleontologists and Mineralogists Special. Publication 12, p. 84–115.

Harms, J. C., Southard, J. B., Spearing, D. R., and Walker, R. G., 1975, Depositional environments as interpreted from primary sedimentary structures and stratification sequences: Society of Economic Paleontologists and Mineralogists Short Course 2, 161 p.

Harms, J. C., Southard, J. B., and Walker, R. G., 1982, Structures and sequences in clastic rocks: Society of Economic Paleontologists and Mineralogists Short Course 9, Calgary.

Harper, C. W., Jr., 1984, Improved methods of facies sequence analysis, in Walker, R. G., ed., Facies models, 2nd ed.: Geoscience Canada Reprint Series 1, p. 11–13.

Harper, C. W., Jr., 1998, Thickening and/or thinning upward patterns in sequences of strata: tests of significance: Sedimentology, v. 45, p. 657–696.

Harris, P. T., 1988, Large-scale bedforms as indicators of mutually evasive sand transport and the sequential infilling of wide-mouthed estuaries: Sedimentary Geology, v. 273–298.

Hart, B. S., Macquaker, J. H. S., and Taylor, K. G. 2013, Mudstone ("shale") depositional and diagenetic processes: Implications for seismic analysis of source-rock reservoirs: Interpretation. A Journal of Subsurface Characterization: Society of Exploration Geophysicists and American Association of Petroleum Geologists, v. 1, p. B7–B26.

Hayes, M. O., 1967, Hurricanes as geological agents: Case studies of Hurricanes Carla, 1961, and Cindy, 1963: University of Texas, Bureau of Economic Geology, Austin, Texas, Report of Investigations No. 61.

Hayes, M. O., 1976, Morphology of sand accumulation in estuaries; an introduction to the symposium, in Cronin, L. E., ed., Estuarine Research, v. 2, Geology and Engineering: Academic Press, London, p. 3–22.

Heckel, P. H., 1972a, Pennsylvanian stratigraphic reefs in Kansas, some modern comparisons and implications: Geologische Rundschau, v. 61, p. 584–598.

Heckel, P. H., 1972b, Recognition of ancient shallow marine environments, in Rigby, J. K., and Hamblin, W. K., eds., Recognition of ancient sedimentary environments: Society of Economic Paleontologists and Mineralogists Special Publication 16, p. 226–296.

Hecker, R. F., 1965, Introduction to paleoecology (translated from Russian): American Elsevier Publishing Company, New York, 166 p.

Heezen, B. C. and Hollister, C. D., 1971, The face of the deep: Oxford University Press, New York, 659 p.

Hein, F.J., 1982, Depositional mechanisms of deep-sea coarse clastic sediments. Cap Enragé Formation, Quebec: Canadian Journal of Earth Sciences, v. 19, p. 267–287.

Hiscott, R. N., 1981, Deep-sea fan deposits in the Macigno Formation (Middle-Upper Oligocene) of the Gordana valley, northern Apennines, Italy — Discussion: Journal of Sedimentary Petrology, v. 51, p. 1015–1033.

Hodgetts, D., 2013, Laser scanning and digital outcrop geology in the petroleum industry: A review: Marine and Petroleum Geology, v. 46, p. 335–354.

Holbrook, J., and Miall, A. D., 2020, Time in the Rock: A field guide to interpreting past events and processes from siliciclastic stratigraphy: Earth Science Reviews, v. 203, 103121, 23 p.

Howard, J. D., and Reineck, R.-E., 1981, Depositional facies of high-energy beach-to-offshore sequence: comparison with low energy sequence: American Association of Petroleum Geologists Bulletin, v. 65, p. 807–830.

Hsü, K. J., Cita, M. B., and Ryan, W. B. F., 1973, The origin of the Mediterranean evaporites, in Ryan, W. B. F., Hsü, K. J., et al. Init Rep Deep Sea Drilling Proj, v. 13, p. 1203–1231.

Hunter, R. E., 1977a, Basic types of stratification in small eolian dunes: Sedimentology, v. 24, p. 361–388.

Hunter, R. E., 1977b, Terminology of cross-stratified sedimentary layers and climbing-ripple structures: Journal of Sedimentary Petrology, v. 47, p. 697–706.

Hunter, R. E., 1980, Depositional environments of some Pleistocene coastal terrace deposits, southwestern Oregon—case history of a progradational beach and dune sequence: Sedimentary Geology, v. 27, p. 241–262.

Hunter, R. E., 1981, Stratification styles in eolian sandstones: some Pennsylvanian to Jurassic examples from the Western Interior U.S.A., in Ethridge, F. G., and Flores, R., eds., Recent and ancient nonmarine depositional environments: models for exploration: Society of Economic Paleontologists and Mineralogists Special Publication 31, p. 315–329.

Hunter, R. E., Clifton, H. E., And Phillips, R. L., 1979, Depositional processes, sedimentary structures and predicted vertical sequences in barred nearshore systems, southern Oregon Coast: Journal of Sedimentary Petrology, v. 49, p. 711–726.

Imbrie, J., and Purdy, E. G., 1962, Classification of modern Bahamian carbonate sediments, in Ham, W. E., ed., Classification of carbonate rocks. Am Assoc Pet Geol Mem 1, p. 253–272.

Ingersoll, R. V., and Busby, C. J., 1995, Tectonics of sedimentary basins, in Busby, C. J., and Ingersoll, R. V., eds, Tectonics of sedimentary basins: Blackwell Science, p. 1–51.

Jackson, R. G., II, 1975, Hierarchical attributes and a unifying model of bed forms composed of cohesionless material and produced by shearing flow: Geological Society of America Bulletin, v. 86, p. 1523–1533.

James, N. P., 1972, Holocene and Pleistocene calcareous crust (caliche) profiles: criteria for subaerial exposure: Journal of Sedimentary Petrology, v. 42, p. 817–836.

James, N. P., 1984, Shallowing-upward sequences in carbonates, in Walker, R. G., ed. Facies Models, Second Edition, Geoscience Canada Reprint Series 1, p. 213–228.

James, N. P., and Bourque, P.-A., 1992, Reefs and mounds, in Walker, R. G. and James, N. P., eds., Facies models: response to sea-level change: Geological Association of Canada, Geotext 1, p. 323–347.

James, N. P., and Dalrymple, R. W., eds., 2010, Facies Models 4: GEOtext 6, Geological Association of Canada, St. John's, Newfoundland, 586 p.

James, N. P., Kendall, A. C., and Pufahl, P. K., 2010, Introduction to biological and biochemical facies models, in James, N. P., and Dalrymple, R. W., eds., Facies Models 4: GEOtext 6, Geological Association of Canada, St. John's, Newfoundland, p. 323–340.

James, N. P., and Wood, R., 2010, Reefs, in James, N. P., and Dalrymple, R. W., eds., Facies Models 4: GEOtext 6, Geological Association of Canada, St. John's, Newfoundland, p. 421–447.

Jeletzky, J. A., 1975, Hesquiat Formation (new): a neritic channel and interchannel deposit of Oligocene age, western Vancouver Island, British Columbia (92E): Geological Survey of Canada Paper 75–32.

Jobe, Z. R., Howes, N., Romans, B. W., and Covault, J. A., 2018, Volume and recurrence of submarine-fan-building turbidity currents: The Depositional Record, v. 4, p. 160–176.

Jones, B., 1977, Variations in the Upper Silurian brachiopod Atrypella phoca (Salter) from Somerset and Prince of Wales Islands. Arctic Canada: Journal of Paleontology v. 51, p. 459–479.

Jones, C. M., and McCabe, P. J., 1980, Erosion surfaces within giant fluvial cross-beds of the Carboniferous in northern England: Journal of Sedimentary Petrology, v. 50, p. 613–620.

Jones, G. D., Smart, P. L., Whitaker, F. F., Rostron, B. J., and Machel, H. G., 2003, Numerical modeling of reflux dolomitization in the Grosmont platform complex (Upper Devonian), Western Canada sedimentary basin. Am Asso Petrol Geol Bull, v. 87(8), p. 1273–1298.

Kay, M., 1951, North American geosynclines: Geological Society of America Memoir 48.

Kendall, A. C., 1992, Evaporites, in Walker, R. G. and James, N. P., eds., Facies models: response to sea-level change: Geological Association of Canada, Geotext 1, p. 375–409.

Kendall, A. C., 2010, Marine evaporites, in James, N. P., and Dalrymple, R. W., eds., Facies Models 4: GEOtext 6, Geological Association of Canada, St. John's, Newfoundland, p. 505–539.

Kinsman, D. J. J., 1966, Gypsum and anhydrite of recent age, Trucial Coast, Persian Gulf, in Rau, J. L., ed. Second Symposium on Salt: Northern Ohio Geological Society, Cleveland, Ohio 1, p. 302–326.

Kinsman, D. J. J., 1969, Modes of formation, sedimentary associations, and diagenetic features of shallow-water supratidal evaporites: American Association of Petroleum Geologists Bulletin, v. 53, p. 830–840.

Klein, G. deV., 1970, Depositional and dispersal dynamics of intertidal sand bars: Journal of Sedimentary Petrology, v. 40, p. 1095–1127.

Klein, G. deV., 1971, A sedimentary model for determining paleotidal range: Geological Society of America Bulletin, v. 82, p. 2585–2592.

Klovan, J. E., 1964, Facies analysis of the Redwater Reef complex, Alberta, Canada. Bull Can Pet Geol v. 12, p. 1–100.

Knauth, L. P., 1979, A model for the origin of chert in limestone: Geology, v. 7, p. 274–277.

Kocurek, G. A., 1981, Significance of interdune deposits and bounding surfaces in aeolian dune sands: Sedimentology, v. 28, p. 753–780.

Kocurek, G. A., 1988, First-order and super bounding surfaces in eolian sequences - bounding surfaces revisited: Sedimentary Geology, v. 56, p. 193–206.

Kocurek, G., and Dott, R. H., Jr., 1981, Distinctions and uses of stratification types in the interpretation of eolian sand: Journal of Sedimentary Petrology, v. 51, p. 579–595.

Krumbein, W. C., and Sloss, L. L., 1963, Stratigraphy and sedimentation (2nd. ed., San Francisco, W. H. Freeman and Co., Edition), 660p.

Krynine, P. D., 1942, Differential sedimentation and its products during one complete geosynclinal cycle: Proceedings of the 1st. Pan American Congress on Mining and Engineering Geology, Pt. 1, v. 2, p. 537–560.

Kuenen, Ph. H., and Migliorini, C. I., 1950, Turbidity currents as a cause of graded bedding: Journal of Geology, v. 58, p. 91–127.

Kumar, N., and Sanders, J. E., 1974, Inlet sequence: a vertical succession of sedimentary structures and textures created by the lateral migration of tidal inlets: Sedimentology, v. 21, p. 491–532.

Kumar, N., and Sanders, J. E., 1976, Characteristics of shoreface deposits; modern and ancient: Journal of Sedimentary Petrology, v. 46, p. 145–162.

Laughbaum, L. R., 1960, A paleoecologic study of the Upper Denton Formation, Tarrant, Denton, and Cooke Counties, Texas. J Paleontol., v. 34, p. 1183–1197.

Le Blanc Smith, G., 1980, Logical-letter-coding system for facies nomenclature: Witbank Coalfield. Transactions of the Geological Society of South Africa, v. 83, p. 301–312.

Leeder, M. R., 1983, On the interactions between turbulent flow, sediment transport and bedform mechanics in channelized flows, in Collinson, J. D., and Lewin, J., eds., Modern and ancient fluvial systems: International Association of Sedimentologists Special Publication 6, p. 5–18.

Long, D. G. F., 1978, Proterozoic stream deposits: some problems of recognition and interpretation of ancient sandy fluvial systems, in Miall, A. D., ed. Fluvial Sedimentology: Canadian Society of Petroleum Geologists Memoir 5, p. 313–341.

Longwell, C. R., ed., 1949, Sedimentary facies in geologic history: Geological Society of America Memoir 39, 171 p.

Lowe, D. R., 1979, Sediment gravity flows: their classification and some problems of application to natural flows, in Doyle, L. J., and Pilkey, O. H., Jr., eds., Geology of Continental Slopes: Society of Economic Paleontologists and Mineralogists Special Publication 27, p. 75–82.

Lowe, D. R., 1982, Sediment gravity flows II. Depositional Models with Special Reference to the Deposits of High-Density Turbidity Currents: Journal of Sedimentary Petrology, v. 52, p. 279–297.

Lowenstam, H. A., 1950, Niagaran reefs in the Great Lakes area: Journal of Geology, v. 58, p. 430–487.

Lowenstam, H. A., 1957, Niagaran reefs in the Great Lakes area: Geological Society of America Memoir 67, v. 2, p. 215–248.

Lucia, F. J., 1995, Rock-fabric/petrophysical classification of carbonate pore space for reservoir characterization. Am Asso Petrol Geol Bull., v. 79(9), p. 1275–1300.

Lucia, F. J. and Major, R. P., 1994, Porosity evolution through hypersaline reflux dolomitization. In: B. Purser, M. Tucker and D. Zenger (Editors), Dolomites: a Volume in Honor of Dolomieu. International Association of Sedimentologists Special Publication., v. 21, p.325–341.

Ludvigsen, R., 1978, Middle Ordovician trilobite biofacies, southern Mackenzie Mountains, in Stelck, C. R., and Chatterton, B. D. E., eds., Western and Arctic Canadian biostratigraphy: Geological Association of Canada Special Paper 18, p. 1–37.

MacEachern, J. A., Pemberton, S. G., Gingras, M. K., and Bann, K. L., 2010, Ichnology and facies models, in James, N. P., and Dalrymple, R. W., eds., Facies Models 4: Geotext 6, Geological Association of Canada, p. 19–58.

Machel, H.-G., 1985, Cathodoluminescence in calcite and dolomite and its chemical interpretation: Geoscience Canada, v. 12, p. 139–147.

Mackenzie, J., Hsü, K. J., and Schneider, J. F., 1980, Movement of subsurface waters under the sabkha, Abu Dhabi, UAE and its relation to evaporative dolomite genesis. SEPM Special Publication 28, p. 11–30.

Macquaker, J. H. S., and Adams, A. E., 2003, Maximizing information from fine-grained sedimentary rocks: An inclusive nomenclature for mudstones: Journal of Sedimentary Research, v. 73, p. 735–744.

Mazzullo, S. J., Bischoff, W. D., and Teal, C. S., 1995, Holocene shallow-subtidal dolomitization by near-normal seawater, northern Belize. Geology, v. 23, p. 341–344.

McGregor, B. A., 1981, Ancestral head of Wilmington Canyon: Geology, v. 9, p. 254–257.

McIlreath, I. A., and James, N. P., 1984, Carbonate slopes, in Walker, ed., R. G., Facies Models, 2nd edition: Geoscience Canada Reprint Series 1, p. 245–257.

McKee, E. D., ed., 1979, A study of global sand seas: U.S. Geological Survey Professional Paper 1052.

error

and Rahmani, R. A., eds., Clastic tidal sedimentology: Canadian Society of Petroleum Geologists memoir 16, p. 3–27.

Nottvedt, A., and Kreisa, R. D., 1987, model for the combined-flow origin of hummocky cross-stratification: Geology, v. 15, p. 357–361.

Nummedal, D., and Swift, D. J. P., 1987, Transgressive stratigraphy at sequence-bounding unconformities: some principles derived from Holocene and Cretaceous examples, in Nummedal, D., Pilkey, O. H., and Howard, J. D., eds., Sea-level fluctuation and coastal evolution; Society of Economic Paleontologists and Mineralogists Special Publication 41, p. 241–260.

O'Brien, N. R., 1990, Significance of lamination in Toarcian (Lower Jurassic) shales from Yorkshire. Great Britain: Sedimentary Geology, v. 67, p. 25–34.

Odin, G. S., and Matter, A., 1981, De glauconiarum origine: Sedimentology, v. 28, p. 611–642.

Parrish, J. T., 1983, Upwelling deposits: nature of association of organic-rich rock, chert, chalk, phosphorite and glauconite: American Association of Petroleum Geologists Bulletin, v. 67, p. 529.

Patterson, R. J., and Kinsman, D. J. J., 1982, Formation of diagenetic dolomite in coastal sabkha along Arabian (Persian) Gulf. AAPG Bulletin, v. 66, p. 28–43.

Picard, M. D., and High, L. R., Jr., 1973, Sedimentary structures of ephemeral streams: Developments in Sedimentology 17. Elsevier, Amsterdam, 223 p.

Pickering, K. T., Stow, D. A. V., Watson, M. P., and Hiscott, R. N., 1986, Deep-water facies, processes and models: a review and classification scheme for modern and ancient sediments: Earth Science Reviews, v. 23, p. 75–174.

Piper, D. J. W., and Normark, W. R., 2009, Processes that initiate turbidity currents and their influence on turbidites: a marine geology perspective: Journal of Sedimentary Research, v. 79, p. 347–362.

Pirrie, D., 1998, Interpreting the record, facies analysis, in Blundell, D. J., and Scott, A. C., eds., Lyell: the past is the key to the present: Geological Society of London, Special Publication 143, p. 395–420.

Pirson, S. J., 1977, Geologic well log analysis, 2nd ed.: Gulf Publishing Co., Houston, 377 p.

Platt, N. H., and Keller, B., 1992, Distal alluvial deposits in a foreland basin setting — the Lower Freshwater Molasse (Lower Miocene), Switzerland: sedimentology, architecture and palaeosols: Sedimentology, v. 39, p. 545–565.

Playton, T. E., Janson, X., and Kerans, C., 2010, Carbonate slopes, in James, N. P., and Dalrymple, R. W., eds., Facies Models 4: GEOtext 6, Geological Association of Canada, St. John's, Newfoundland, p. 449–476.

Plint, A. G., 2010, Wave- and storm-dominated shoreline and shallow-marine systems, in James, N. P., and Dalrymple, R. W., eds., Facies Models 4: GEOtext 6, Geological Association of Canada, St. John's, Newfoundland, p. 167–199.

Posamentier, H. W., and James, D. P., 1993, An overview of sequence stratigraphic concepts: uses and abuses, in Posamentier, H. W., Summerhayes, C. P., Haq, B. U., and Allen, G. P., eds., Sequence stratigraphy and facies associations: International Association of Sedimentologists Special Publication 18, p. 3–18.

Posamentier, H. W., Jervey, M. T., and Vail, P. R., 1988, Eustatic controls on clastic deposition I—Conceptual framework, in Wilgus, C. K., Hastings, B. S., Kendall, C. G. St. C., Posamentier, H. W., Ross, C. A., and Van Wagoner, J. C., eds., Sea level Changes - an integrated approach: Society of Economic Paleontologists and Mineralogists Special Publication 42, p. 109–124.

Posamentier, H. W., and Vail, P. R., 1988, Eustatic controls on clastic deposition II—sequence and systems tract models, in Wilgus, C. K., Hastings, B. S., Kendall, C. G. St. C., Posamentier, H. W., Ross, C. A., and Van Wagoner, J. C., eds., Sea level Changes - an integrated approach: Society of Economic Paleontologists and Mineralogists Special Publication 42, p. 125–154.

Postma, G., 1986, Classification for sediment gravity-flow deposits based on flow conditions during sedimentation: Geology, v. 14, p. 291–294.

Potter, P. E., 1959, Facies models conference: Science, v. 129, p. 1292–1294.

Pratt, B. R., 2001, Oceanography, bathymetry and syndepositional tectonics of a Precambrian intracratonic basin: integrating sediments, storms, earthquakes and tsunamis in the Belt Supergroup (Helena Formation, ca. 1.45 Ga), western North America: Sedimentary Geology, v. 141–142, p. 371–394.

Pratt, B. R., 2002, Storm versus tsunamis: dynamic interplay of sedimentary, diagenetic, and tectonic processes in the Cambrian of Montana: Geology, v. 30, p. 423–426.

Pratt, B. R., 2010, Peritidal carbonates, in James, N. P., and Dalrymple, R. W., eds., Facies Models 4: GEOtext 6, Geological Association of Canada, St. John's, Newfoundland, p. 401–420.

Pratt, B. R., and Bordonaro, O. L., 2007, Tsunamis in a stormy sea: Middle Cambrian inner-shelf limestones of western Argentina: Journal of Sedimentary Research, v. 77, p. 256–262.

Pufhal, P. K., 2010, Bioelemental sediments, in James, N. P., and Dalrymple, R. W., eds., Facies Models 4: GEOtext 6, Geological Association of Canada, St. John's, Newfoundland, p. 477–503.

Purdy, E. G., 1963, Recent calcium carbonate facies of the Great Bahama Bank: Journal of Geology, v. 71, p. 334–355, p. 472–497.

Purdy, E. G., 1974, Karst-determined facies patterns in British Honduras: Holocene carbonate sedimentation model: American Association of Petroleum Geologists Bulletin, v. 58, p. 825–855.

Qing, H., 1998, Petrography and geochemistry of early-stage, fine- and medium-crystalline dolomites in the Middle Devonian Presqu'ile Barrier at Pine point, Canada. Sedimentology, v. 45, p. 433–446.

Reading, H. G., ed., 1986, Sedimentary environments and facies, second ed.: Blackwell, Oxford, 615 p.

Reading, H. G., ed., 1996, Sedimentary environments: processes, facies and stratigraphy, third edition: Blackwell Science, Oxford, 688 p.

Reading, H. G., and Levell, B. K., 1996, Controls on the sedimentary rock record, in Reading, H. G., ed., Sedimentary environments: processes, facies and stratigraphy, third edition: Blackwell Science, Oxford, p. 5–36.

Reineck, H. E., and Singh, I. B., 1973, Depositional sedimentary environments—with reference to terrigenous clastics: Berlin: Springer-Verlag, 439 p.

Reineck, H. E., and Singh, I. B., 1980, Depositional sedimentary environments - with reference to terrigenous clastics, 2nd edn. Springer-Verlag, New York, 549p.

Reineck, H. E., and Wunderlich, R., 1968, Classification and origin of flaser and lenticular bedding: Sedimentology, v. 11, p. 99–104.

Royer, D., Wilf, P., 2006, Why do toothed leaves correlate with cold climates? Gas Exchange at Leaf Margins Provides New Insights into a Classic Paleotemperature Proxy: International Journal of Plant Science, v. 167(1), p. 11–18.

Runkel, A. C., Miller, J. F., McKay, R. M., Palmer, A. R., and Taylor, J. F., 2007, High-resolution sequence stratigraphy of lower Paleozoic sheet sandstones in central North America: the role of special conditions of cratonic interiors in development of stratal architecture: Geological Society of America Bulletin, v. 119, p. 860–881.

Rust, B. R., 1972, Pebble orientation in fluviatile sediments: Journal of Sedimentary Petrology, v. 42, p. 384–388.

Salese, F., McMahon, W. J., Balme, M. R., Ansan, V., Davis, J. M., and Kleinhans, M. G., 2020, Sustained fluvial deposition recorded in Mars' Noachian stratigraphic record. Nature Communications, v. 11, article 2067.

Saller, A. H., and Henderson, N., 2001, Insight from the Permian of west Texas: Reply. Am Asso Petrol Geol Bull., v. 85, p. 530–532.

Saller, A. H., Yaremko, K. 1994, Dolomitization and Porosity Development in the Middle and Upper Wabamun Group, Southeast

Peace River Arch, Alberta, Canada. Am Asso Petrol Geol Bull., v. 78, p. 1406–1430.

Schieber, J., Southard, J. B., Kissling, P., Rossman, B., and Ginsburg, R., 2013, Experimental deposition of carbonate mud from moving suspensions: importance of flocculation and implications for modern and ancient carbonate mud deposition: Journal of Sedimentary Research, v. 83, p. 1025–1031.

Schlager, W., 1991, Depositional bias and environmental change— important factors in sequence stratigraphy: Sedimentary Geology, v. 70, p. 109–130.

Schlager, W., 2005, Carbonate sedimentology and sequence stratigraphy: SEPM Concepts in Sedimentology and Paleontology #8, 200p.

Scholle, P. A., 1978, A color illustrated guide to carbonate rock constituents, textures, cements and porosities: American Association of Petroleum Geologists Memoir 27, 241 p.

Schreiber, B. C., 1981, Marine evaporites: facies development and relation to hydrocarbons and mineral genesis: American Association of Petroleum Geologists Fall Education Conference, 44 p.

Schreiber, B. C., 1986, Arid shorelines and evaporites. In: Reading, H. G., ed., Sedimentary environments and facies, 2nd ed., Blackwell Scientific Publications, Oxford, p. 189–228.

Schreiber, B. C., Friedman, G. M., Decima, A., and Schreiber, E., 1976, Depositional environments of Upper Miocene (Messinian) evaporite deposits of the Sicilian Basin: Sedimentology, v. 23, p. 729–760.

Schultz, A. W., 1984, Subaerial debris flow deposition in the Upper Paleozoic Cutler Formation, western Colorado: Journal of Sedimentary Petrology, v. 54, p. 759–772.

Schumm, S. A., 1968: Speculations concerning paleohydrologic control of terrestrial sedimentation: Geological Society of America Bulletin, v. 79, p. 1573–1588.

Schwab, F. L., 1981, Evolution of the western continental margin, French-Italian Alps: sandstone mineralogy as an index of plate tectonic setting. J Geol., v. 89, p. 349–368.

Seilacher, A., Reif, W.-E., Westphal, F., Riding, R., Clarkson, E. N. K., and Whittington, H. B., 1985, Sedimentological, ecological and temporal patters of fossil Lagerstratten: Philosophical Transactions of the Royal Society, Series B, Biological Sciences, v.311, #1148, p. 5–24.

Selley, R. C., 1968, A classification of paleocurrent models: Journal of Geology, v. 76, p. 99–110.

Selley, R. C., 1979, Dipmeter and log motifs in North Sea submarine-fan sands: American Association of Petroleum Geologists Bulletin, v. 63, p. 905–917.

Sellwood, B. W., 1986, Shallow-marine carbonate environments, in Reading, H. G., ed., Sedimentary environments and facies, 2nd ed.: Blackwell, Oxford, p. 283–342.

Seyfullah, L., 2012, Fossil focus: Using plant fossils to understand past climates and environments: Palaeontology online, v. 2(7), p. 1–8.

Shand, S. J., 1947, Eruptive rocks, 3rd ed., John Wiley and Sons, New York.

Shanmugam, G., 1997, The Bouma sequence and the turbidite mind set: Earth Science Reviews, v. 42, p. 201–229.

Shaw, D. B., and Weaver, C. E., 1965, The mineralogical composition of shales: Journal of Sedimentary Petrology, v. 35, p. 213–222.

Shaw, J. O., Briggs, D. E. G., and Hull, P. M., 2021, Fossilization potential of marine assemblages and environments: Geology, v. 49, p. 258–262.

Shea, J. H., 1982, Twelve fallacies of uniformitarianism: Geology, v. 10, p. 449–496.

Shearman, D. J., 1966, Origin of marine evaporites by diagenesis: Transactions of the Institute of Mining and Metallurgy Bulletin, v. 75, p. 208–215.

Shepard, F. P., 1981, Submarine canyons: multiple causes and long-time persistence: American Association of Petroleum Geologists Bulletin, v. 65, p. 1062–1077.

Shinn, E. A., 1983, Birdseyes, fenestrae, shrinkage pores, and loferites: a reevaluation: Journal of Sedimentary Petrology, v. 53, p. 619–628.

Shinn, E. A., Halley, R. B., Hudson, J. H., Lidz, B., and Robbin, D. M., 1979, Three dimensional aspects of Belize patch reefs: American Association of Petroleum Geologists Bulletin, v. 63, p. 528.

Shor, A. N., Kent, D. V., and Flood, R. D., 1984, Contourite or turbidite?: magnetic fabric of fine-grained Quaternary sediments, Nova Scotia continental rise, in Stow, D. A. V., and., Piper, D J. W, eds., Fine-grained sediments: deep-water processes and facies: Geological Society of London Special Publication 15, p. 257–273.

Shurr, G. W., 1984, Geometry of shelf-sandstone bodies in the Shannon Sandstone of southeastern Montana, in Tillman, R. W., and Siemers, C. T., eds., Siliciclastic shelf sediments: Society of Economic Paleontologists and Mineralogists Special Publication 34, p. 63–83.

Sibley, D. F., and Gregg, J. M., 1987, Classification of dolomite rock textures. J Sediment Petrol., v. 57, 967–975.

Simons, D. B., and Richardson, E. V., 1961, Forms of Bed Roughness in Alluvial Channels: American Society of Civil Engineers Proceedings, v. 87(HY3), p. 87–105.

Simons, D. B., Richardson, E. V., and Nordin, C. F., 1965, Sedimentary structures generated by flow in alluvial channels, in Middleton, G. V., ed., Primary sedimentary structures and their hydrodynamic interpretation: Society of Economic Paleontologists and Mineralogists Special. Publication 12, p. 34–52.

Skipper, K., 1976, Use of geophysical wireline logs for interpreting depositional processes: Geoscience Canada, v. 3, p. 279–280.

Sloss, L. L., Krumbein, W. C., and Dapples, E. C., 1949, Integrated facies analysis; in Longwell, C. R., ed., Sedimentary facies in geologic history: Geological Society of America. Memoir 39, p. 91–124.

Smith, D. G., and Smith, N. D., 1980, Sedimentation in anastomosed river systems: examples from alluvial valleys near Banff, Alberta: Journal of Sedimentary Petrology, v. 50, p. 157–164.

Smosna, R., and Warshauer, S. M., 1981, Rank exposure index on a Silurian carbonate tidal flat: Sedimentology, v. 28, p. 723–731.

Sneider, R. M., Tinker, C. N., and Meckel, L. D., 1978, Deltaic environment reservoir types and their characteristics: Journal of Petroleum Technology, v. 20, p. 1538–1546.

Solohub, J. T., and Klovan, J. E., 1970, Evaluation of grain-size parameters in lacustrine sediments: Journal of Sedimentary Petrology, v. 40, p. 81–101.

Southard, J. B., 1971, Representation of bed configurations in depth-velocity-size diagrams: Journal of Sedimentary Petrology, v. 41, p. 903–915.

Speyer, S. E., and Brett, C. E., 1986, Trilobite taphonomy and Middle Devonian taphofacies: Palaios, v. 1, p. 312–327.

Steel, R. J., 1974, New Red Sandstone floodplain and piedmont sedimentation in the Hebridean Province: Journal of Sedimentary Petrology, v. 44, p. 336–357.

Steidtmann, J. R., 1974, Evidence for eolian origin of cross-stratification in sandstone of the Casper Formation, southernmost Laramie Basin, Wyoming: Geological Society of America Bulletin, v. 85, p. 1835–1842.

Stow, D. A. V., and Piper, D. J. W., eds., 1984, Fine-grained sediments: deep-water processes and facies: Geological Society of London Special Publication 15, Blackwell Scientific Publications, Oxford, 659 p.

Swift, D. J. P., 1975, Barrier island genesis: evidence from the Middle Atlantic shelf of North America: Sedimentary Geology, v. 14, p. 1–43.

Swift, D. J. P., Figueiredo, A. G. Jr., Freeland, G. L., and Oertel, G. F., 1983, Hummocky cross-stratification and megaripples: a geological double standard? J Sediment Petrol., v. 53, p. 1295–1318.

Swift, D. J. P., Stanley, D. J., and Curray, J. R., 1971, Relict sediments on continental shelves: a reconsideration: Journal of Geology, v. 79, p. 322–346.

Tagliavento, M., John, C. M., Anderskouv, K., and Stemmerik, L., 2021, Towards a new understanding of the genesis of chalk: diagenetic origin of microcarbs confirmed by clumped isotope analysis: Sedimentology, v. 68, p. 513–530.

Teichert, C., 1958, Concepts of facies: Bulletin of the American Association of Petroleum Geologists, v. 42, p. 2718–2744.

Tipper, J. C., 2015, The importance of doing nothing: stasis in sedimentation systems and its stratigraphic effects: in Smith, D. G., Bailey, R., J., Burgess, P., and Fraser, A., eds., Strata and time: Probing the Gaps in Our Understanding: Geological Society, London, Special Publication 404, p. 105–122.

Turner, P., 1980, Continental red beds: Developments in Sedimentology 29, Elsevier, Amsterdam, 562p.

Tyler, N., Galloway, W. E., Garrett, C. M., Jr., and Ewing, T. E., 1984, Oil accumulation, production characteristics, and targets for additional recovery in major oil reservoirs of Texas: Bureau of Economic Geology, Texas, Geologic Circular 84–2, 31 p.

Uba, E., Heubeck, C., and Hulka, C., 2005, Facies analysis and basin architecture of the Neogene Subandean synorogenic wedge, southern Bolivia: Sedimentary Geology, v. 180, p. 91–123.

Vai, G. B., and Ricci-Lucchi, F., 1977, Algal crusts, autochthonous and clastic gypsum in a cannibalistic evaporite basin: a case history from the Messinian of Northern Apennines: Sedimentology, v. 24, p. 211–244.

Van Houten, F. B., 1964, Cyclic lacustrine sedimentation, Upper Triassic Lockatong Formation, central New Jersey and adjacent Pennsylvania, in Merriam, D. F., ed. Symposium on Cyclic Sedimentation: Geological Survey of Kansas Bulletin 169, p. 495–531.

Van Houten, F. B., 1973, Origin of red beds: a review—1961–1972: Annual Review of Earth and Planetary Sciences, v. 1, p. 39–62.

Van Houten, F. B., 1985, Oolitic ironstones and contrasting Ordovician and Jurassic paleogeography: Geology, v. 13, p. 722–724.

Van Straaten, L. M. J. U., 1951, Texture and genesis of Dutch Wadden Sea sediments: Proceedings of the 3rd International Congress on Sedimentology, Netherlands, p. 225–255.

Van Straaten, L .M. J. U., 1954, Composition and structure of recent marine sediments in the Netherlands: Leidse Geol. Mededel., v. 19, p. 1–110.

Van Tuyl, F. M., 1916, New points on the origin of dolomite. Am J Sci., v. 42, p. 249–260.

Van Wagoner, J. C., Mitchum, R. M., Jr., Posamentier, H. W., and Vail, P. R., 1987, Seismic stratigraphy interpretation using sequence stratigraphy, Part 2: key definitions of sequence stratigraphy, in Bally, A. W., ed., Atlas of seismic stratigraphy: American Association of Petroleum Geologists Studies in Geology 27, v. 1, p. 11–14.

Visher, G. S., 1965, Use of vertical profile in environmental reconstruction. Am Assoc Petrol Geol Bull., v. 49, p. 41–61.

Visher, G. S., 1969, Grain size distributions and depositional processes: Journal of Sedimentary Petrology, v. 39, p. 1074–1106.

Visser, M. J., 1980, Neap-spring cycles reflected in Holocene subtidal large-scale bedform deposits: a preliminary note: Geology, v. 8, p. 543–546.

Walker, R. G., 1975, Generalized facies models for resedimented conglomerates of turbidite association: Geological Society of America Bulletin, v. 86, p. 737–748.

Walker, R. G., 1976, Facies models 1. General Introduction: Geoscience Canada v. 3, p. 21–24.

Walker, R. G., 1978, Deep-water sandstone facies and ancient submarine fans: models for exploration for stratigraphic traps: American Association of Petroleum Geologists Bulletin, v. 62, p. 932–966.

Walker, R. G., 1990, Facies modeling and sequence stratigraphy: Journal of Sedimentary Petrology, v. 60, p. 777–786.

Walker, R. G., 1992, Facies, facies models and modern stratigraphic concepts, in Walker, R. G. and James, N. P., eds., Facies models: response to sea-level change: Geological Association of Canada, p. 1–14.

Walker, T. R., 1967, Formation of redbeds in modern and ancient deserts: Geological Society of America Bulletin, v. 78, p. 353–368.

Walker, T. R., and Harms, J. C., 1972, Eolian origin of Flagstone beds, Lyons Sanstone (Permian), type area. Boulder County, Colorado: Mountain Geologist, v. 9, p. 279–288.

Walther, Johannes, 1893–1894, Einleisung in die Geologie alshistorische Wissemchaft: Jena, Verlag von Gustav Fischer, 3 vols., 1055 p.

Weimer, P., and Posamentier, H. W., 1993, Recent developments and applications in siliciclastic sequence stratigraphy, in Weimer, P., and Posamentier, H. W., eds., Siliciclastic sequence stratigraphy: American Association of Petroleum Geologists Memoir 58, p. 3–12.

Weimer, R. J., and Hoyt, H. J., 1964, Burrows of *Callianassa major* Say, geologic indicators of littoral and shallow neritic environments: Journal of Paleontology v. 38, p. 761–767.

Wilkinson, B. H., Diedrich, N. W, Drummond, C. N., and Rothman, E. D., 1998, Michigan hockey, meteoric precipitation, and rhythmicity of accumulation on peritidal carbonate platforms: Geological Society of America Bulletin, v. 110, p. 1075–1093.

Williams, G. E., 1989, Late Precambrian tidal rhythmites in South Australia and the history of the Earth's rotation: Journal of the Geological Society. London, v. 146, p. 97–111.

Williams, P. F. and Rust, B. R., 1969, The sedimentology of a braided river: Journal of Sedimentary Petrology, v. 39, p. 649–679.

Wilson, J. L., 1975, Carbonate facies in geologic history: Springer-Verlag, New York, 471 p.

Wilson, R. C. L., 1975, Upper Jurassic Oolite Shoals, Dorset Coast, England. In: Ginsburg, R.N., ed., Tidal deposits. Springer-Verlag, New York, p. 355–362.

Witter, R. C., Jaffe, B., Zhang, Y., and Priest, G., 2012, Reconstructing hydrodynamic flow parameters of the 1700 tsunami at Cannon beach. Oregon, USA: Natural Hazards, v. 63, p. 223–240.

Wong, P. K., Oldershaw, A. E., 1980, Causes of cyclicity in reef interior sediments, Kaybob Reef, Alberta. Bull Can Pet Geol., v. 28, p. 411–424.

Wright, V. P., ed. 1986, Paleosols: their recognition and interpretation: Blackwell Scientific Publications, Oxford, 315 p.

Yang Chang-shu, and Nio, S.-D., 1989, An ebb-tide delta depositional model - a comparison between the modern Eastern Scheldt tidal basin (southwest Netherlands) and the Lower Eocene Roda Sandstone in the southern Pyrenees (Spain): Sedimentary Geology, v. 64, p. 175–196.

Young, F. G., Myhr, D. W., and Yorath, C. J., 1976, Geology of the Beaufort-Mackenzie Basin: Geological Survey of Canada Paper 76–11.

Zeller, E. J., 1964, Cycles and psychology: Geological Survey of Kansas. Bulletin 169, p. 631–636.

Facies Models

4

Contents

Abstract

Depositional environments may be classified into three broad groups, siliciclastic, carbonate and evaporite. Siliciclastic environments range from alluvial fan through fluvial, various coastal environments, the continental shelf, slope and deep-marine settings. The most important carbonate environments are shallow platform, tidal flat, reef and slope settings. Evaporite environments include shallow tidal flat and deep basin settings. This chapter provides a succinct "summary of the environment" for each of the major depositional settings, focusing on key facies indicators and sedimentary processes.

4.1 Introduction

As we describe in Chaps. 2 and 3, there are two broad classes of sedimentary rock, clastic sediments and chemical sediments. **Clastic sediments** are generated by the transport of detrital particles by currents of water or air; **chemical sediments** are created in place by chemical and biochemical processes, although current reworking may be an important part of the final set of processes that develops a given facies assemblage. Most clastic detritus is derived from the erosion and weathering, or recycling, of siliceous igneous, metamorphic or sedimentary rocks, and is referred to as

A. D. Miall, *Stratigraphy: A Modern Synthesis*, Springer Textbooks in Earth Sciences, Geography and Environment,
https://doi.org/10.1007/978-3-030-87536-7_4

siliciclastic detritus. Chemical sediments, particularly carbonate sediments, may be reworked as clastic detritus by marine currents or in sediment-gravity flows.

The observations that form the basis of facies analysis are described in Chap. 2. The interpretation of these observations in terms of depositional processes is the subject of facies analysis (Chap. 3). Somewhat different techniques are needed to study clastic and chemical sediments (Chap. 3). The importance of currents of all types in the formation of clastic sediments—unimodal, bimodal and oscillating traction currents, eolian currents and sediment-gravity flows—means that identifying the lithofacies and structures associated with each type of current process represents a substantial part of the work carried out to interpret clastic depositional environments. The processes of chemical sedimentation are identified by the petrological products of various chemical and biochemical processes, including the allochems of carbonate sediments and the precipitates of evaporate sediments. Current reworking is, of course, common in these environments, too.

In this chapter, we show how the **summaries of the environment** that can be assembled from the facies data may be categorized into some 14 broad classes of **facies models**, for use as aids in teaching and research. Each of the models described and discussed in this chapter may be taken as a typical example of its kind, but in nature there is enormous variability in depositional environments, reflecting differences in geography, climate, regional weather and current regimes, sediment supply and so on. No field example is therefore likely to compare exactly with any published model.

Modern techniques to identify sedimentary environments based on sedimentary successions, fossils and structures date back to the early 1960s (see Chap. 1). The development of the flow-regime concept was a critical step forward (Sect. 1.2.5). Another was the realization of the significance of vertical profiles (Sects. 1.2.2, 1.2.7), a concept that began to be exploited by several geologists more or less simultaneously in Britain and the southern United States. In both cases, much of this early work was concerned with river (fluvial) deposits. This is no accident; it obviously reflects the accessibility of rivers for easy examination. Facies models for nonmarine and shallow-marine environments began to mature in the 1970s. The documentation of continental shelf and deeper marine deposits took longer, especially those of the deep-marine environment, notably submarine fans and shelf deposits. The major breakthroughs in the evolution of facies studies for these environments came in the early 1980s with the development of side-scan sonar methods for imaging the sea floor (Sect. 1.2.6). Prior to that time, interpretations of shelf and deep-marine sediments relied on bathymetric charts of variable accuracy, coupled with limited bottom coring.

Nowadays, we can not only confidently identify the products of a wide range of environments, but we can also detect and describe small differences between them. For example, in the 1960s, fluvial deposits were identified as either of the "meandering" or the "braided" variety. A modern synthesis of fluvial environments (Miall 1996) identified at least 16 distinct varieties of fluvial environment, and the point was made that there exist intermediate conditions between most of any pair of the 16 types. Similarly, so many actual modern submarine fans have now been studied and documented in detail that sedimentologists are no longer able to agree on how to classify them—there are simply too many different kinds (Bouma et al. 1985).

Among the most important developments in facies analysis in recent years has been the recognition of the dynamic nature of mud deposition. The long-standing assumption that most mud is deposited in static water by pelagic settling has been shown to be largely incorrect. Mud typically flocculates in most natural environments, and is subject to distribution by ambient currents, just as is loose sand. This opens a rich field for the reinterpretation of muddy deposits, some of which is touched on in this chapter. Readers may also wish to turn directly to some of the research and review literature that focuses on this topic, e.g., Macquaker et al. (2007), Plint (2010), Schieber et al. (2013), Trabucho-Alexandre (2015).

The intent of this chapter is to provide a brief overview of what is, in fact, a very complex topic. Facies models, or "summaries of the environment" (Sect. 3.4.2), are essential mental constructs that stratigraphers must build as they develop sequence-stratigraphic and chronostratigraphic interpretations of their subject rocks. This chapter is designed to provide a flavor of the type of work that needs to be done. For a more complete understanding of depositional systems and facies models, the reader should turn to James and Dalrymple (2010), which is the fourth edition of the famous Canadian "Facies Models volume." Steel and Milliken (2013) provided a review of major advances in siliciclastic facies studies since 1960. Schlager (2005) published a comprehensive overview of carbonate facies and sequence stratigraphy.

4.2 Clastic Environments

4.2.1 Fluvial Environments

All siliciclastic detritus is transported at some point, by a river system. First-cycle debris eroded from a mountainous source area enters torrential mountain streams and is carried into mountain basins, or lakes, deposited at the edges of a basin in coarse alluvial fans, or transported out onto a coastal plain. The work of the sedimentologist involves recognizing various depositional settings (subenvironments) within a river (Fig. 4.1), and accumulating enough information about these to be able to reconstruct the local fluvial style (Fig. 4.2). Three examples of fluvial styles are illustrated in Figs. 4.3, 4.4 and 4.5.

The facies classification and coding system introduced by Miall (1977, 1978) and updated by Miall (1996; see Table 3.1) is now virtually in standard use for fluvial studies, although as noted in Sect. 3.3.1, not all workers agree with the standardized approach to classification implicit in this system (e.g., Bridge 1993). Conversely, Farrell et al. (2012) have developed what they offer as a universal code system for use with all clastic rocks. An increasing number of workers are employing architectural methods for the three-dimensional documentation and classification of fluvial deposits in the surface and subsurface, particularly for channel-fill sandstone and conglomerate bodies (e.g., Miall and Tyler 1991; Miall 1996, Chap. 6). The classification of bounding surfaces proposed by Miall (1988a, b; see Sect. 3.5.1.11) has also found considerable use, although it has, as expected, proved difficult to make much use of the classification in subsurface work (Doyle and Sweet 1995). A simplified classification of channel style, showing their dependency on sediment grain size and other factors, and the range of typical architectural elements present in each type is shown in Fig. 4.2.

Detailed work on modern floodplains, such as the study of the Mississippi by Farrell (1987), and classification studies, such as that by Nanson and Croke (1992), have contributed to a much better understanding of floodplain processes and sedimentation during the last decades.

Description and classification of ancient floodplain deposits in terms of facies assemblages and architectural elements was attempted by Platt and Keller (1992), and the importance of this environment for coal development has been examined by Flores (1981). Important work on the genesis of paleosols and their stratigraphic context has been carried out by Kraus and Bown (1988) and Wright (1986, 1992). From the point of view of basin analysis, probably the most interesting study of overbank deposits is that of Willis and Behrensmeyer (1994), who used an extremely detailed stratigraphic study of part of the Miocene Siwalik succession in Pakistan as a basis for the building and testing of various models for the development of floodplain stratigraphies.

Only the four main fluvial styles are discussed briefly here. **Meandering rivers** were the subject of one of the first modern facies models, based on the recognition of point bars from their distinctive vertical profile (Bernard and Major 1959; Bernard et al. 1962; Allen 1964). In fact, the point-bar model is an excellent one for the purpose of illustrating the principle of the facies model and the key role played by its characteristic vertical profile (Fig. 4.3). Sediments in meandering rivers may be gravel-, sand- or silt-dominated. The lateral accretion of the point bar is the most distinctive feature of meandering-river deposits, but some bars in braided rivers also exhibit lateral accretion, so a certain amount of caution is needed in interpreting fluvial style from this one characteristic.

The other common fluvial style is that of the **braided river**, characterized by a wide, active valley with several or many component channels of varying sinuosity (Fig. 4.4).

Fig. 4.1 Terminology of the subenvironments of a river

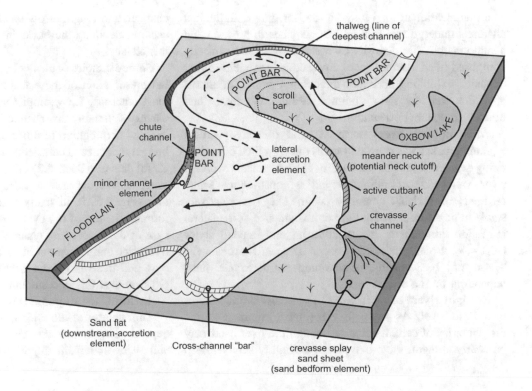

Fig. 4.2 Relationship between grain size and fluvial channel pattern. Architectural elements are those of Miall (1985; see Fig. 4.49). Note that FM has now been replaced by DA: downstream-accretion element (Orton and Reading 1993)

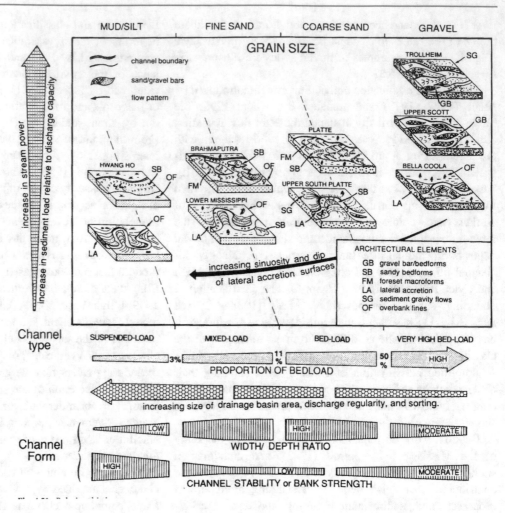

The main sediment load may be sand or gravel. Individual channels undergo rapid migration, switches in position and abandonment. The deposits of braided rivers are typically composed of an assemblage of crossbedded fragments of bar deposits, exhibiting accretion directions oriented in almost any direction, the most common being downstream and oblique to channel orientation.

Anastomosed rivers occur in areas of active vertical aggradation, such as coastal systems during a time of rapidly rising sea level, and in the case of inland systems where a river valley is backfilling behind a constriction. Many channels of variable sinuosity occur. They are relatively stable in position, showing little or no evidence of the lateral migration common to the meandering and braided styles (Fig. 4.5). Broad floodplains occur between the channels. These may be wetlands, peat swamps or desiccated flats, depending on the climate.

Straight rivers occur in areas of very low slope, such as near base level. As such, they are most common as the distributaries of certain deltas. Although the river itself may be nearly straight, there is typically a slightly sinuous deep

channel, and long, narrow bars, called alternate bars, which accumulate along the banks inside the meanders defined by this channel.

Some examples of channel cutbanks are shown in Fig. 2.3, but preservation of these features is, in fact, rather rare. In most outcrops, the geologist will encounter tabular assemblages of variably crossbedded sandstone or conglomerate, and an interpretation of fluvial style must be made from this limited evidence. The orientation of crossbedding may be a useful clue (see Sect. 6.7). Many, but by no means all fluvial successions are characterized by fining-upward vertical profiles (Fig. 3.59c). In many cases, the fine-grained deposits forming the top of the cycle are truncated by erosion at the base of the succeeding channel system. In some sand-laden rivers, the entire channel fill is composed of a similar lithology, and there is no discernible upward fining. On petrophysical logs this is commonly referred to as a "blocky" or "cylindrical" log response (Fig. 3.61).

The nature of floodplain deposits in fluvial systems depends largely on the climate. Figure 4.6 illustrates a variety of floodplain deposits formed in arid to semi-arid

Fig. 4.3 The point-bar model, showing its derivation either from a modern meandering river (top) or from preserved deposits (bottom)

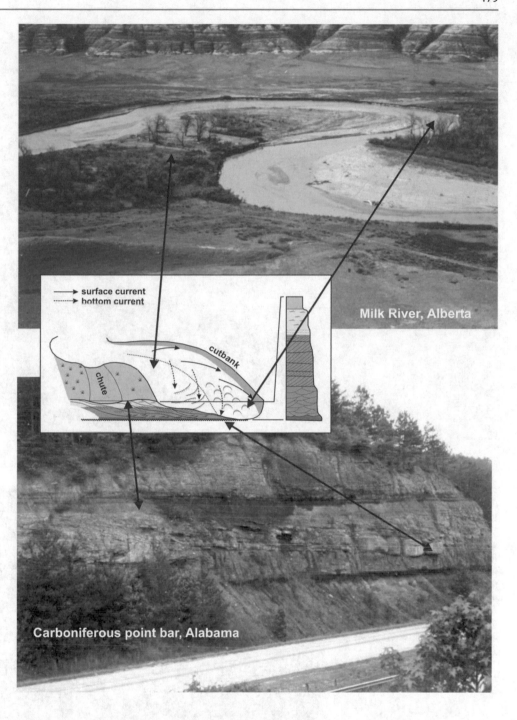

climates. In more humid climates, coals with or without soil horizons may be present.

Caliche (Fig. 4.6a, b) is characteristic of semi-arid climates, where dissolved carbonate accumulates by capillary action near the sediment surface. This example shows large nodules that have coalesced to form an extensive bed of nodular limestone. Such deposits take many thousands of years to develop (Wright 1986).

Fossils in fluvial deposits may include remains of land vertebrates (footprints, bones) and a range of invertebrate trace fossils. Plant remains (logs, leaves, pollen) may be abundant in rocks of Devonian age and younger.

4.2.2 Eolian Environments

The publication of satellite imagery of some of the large and inaccessible deserts of Africa and Asia (McKee 1979) demonstrated a complexity of dune types that has not been recognized in the ancient record. However, it has become

Fig. 4.4 A model for a sandy-braided river, based on the South Saskatchewan River, Canada (Miall 1996)

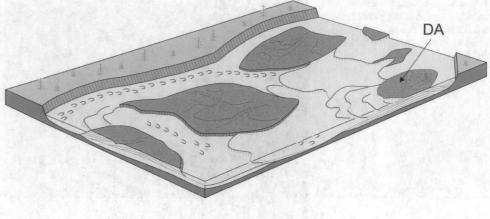

Fig. 4.5 A model for an anastomosed river in a humid environment. Based on the work of Smith and Smith and Smith (1980) in Banff National Park Canada

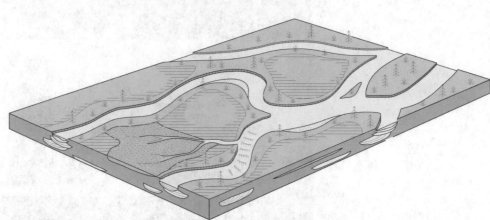

Fig. 4.6 Fluvial floodplain deposits. **a, b** caliche, Triassic, Spain; **c** Interbedded floodplain siltstones and mudstones. Note small channel filled with fine-grained deposits near the base of the section; **d** Thin floodplain muds form the recessive unit at the top of a succession of sheet sandstone formed by flash floods in an ephemeral system; Devonian, Arctic Canada

apparent that many of these dune fields are not currently active, except that minor modifications of the giant bedforms (**draas**) are taking place. The major episodes of sand movement were during the Pleistocene ice age, when climates in desert areas were colder and windier. This curtails the usefulness of modern-process studies of giant dune formation. Fossil dune deposits demonstrate a surprising similarity and relative simplicity of form, as demonstrated by the useful summary papers in McKee (1979). We are only now beginning to understand why.

At the present day, large eolian sand systems tend to be clustered along the Tropic of Cancer and the Tropic of Capricorn (Fig. 4.7), because these belts, centering on 22.5°N and S, are characterized by zones of dry, high-pressure air masses, with steady winds but minimal rainfall. Major sand accumulations are termed sand seas, or **ergs**. These have the superimposed deposits of large dunes at the center, with other facies, such as the deposits of ephemeral rivers, typically occurring closer to the margins (Fig. 4.8). Smaller dune systems are common in coastal areas, where they may form a cap to barrier islands. These occur where winds pick up dry beach sands and blow them back above the high-tide mark. Such dune systems are often prominent features of barrier coastlines, but they have very low preservation potential.

Eolian systems are also present at higher latitudes in continental interiors, such as the Gobi, Taklamakan and Turkestan Deserts of Central Asia. Their presence reflects the tendency toward extreme aridity of high-elevation continental interiors. An ancient analog of the continental interior eolian setting is that of the Rotliegende Sandstone (late

Pennsylvanian to Middle Permian) of the North Sea Basin (an important gas reservoir: Glennie 1972, 1983).

The formation of eolian dunes and the larger stratigraphic features of erg systems have been clarified by the recognition of the importance of the different types of **bounding surface** that may be distinguished within the deposits (Fig. 3.38). Second- and third-order surfaces reflect changing wind patterns within the dunes. First-order surfaces record the migration of the major dunes across the desert surface. As illustrated below, these are commonly surfaces of wind deflation, and are marked by distinctive lithofacies (Fig. 4.9). Most important of all are what are termed **supersurfaces**, which are major stratigraphic surfaces extending for many kilometers through an erg (Kocurek 1988). These may be the product of climate changes or, in the case of coastal ergs, changes in sea level, which affects the level of the water table in the dune deposits (Brookfield 2010; see Fig. 5.24 of this book).

The characteristics of eolian bedforms are described and illustrated in Sect. 3.5.4.7. Here, in Fig. 4.9, we can see some of the other distinctive features of eolian systems. The interbedding of eolian and ephemeral fluvial deposits is common (Fig. 4.9a). Ephemeral interdune ponds may appear after flash floods, and become the site for the accumulation of thin silt and mud lenses as the ponds dry out (Fig. 4.9b). These may be blown away by the wind before they are buried by dune sands, and may not have a very high preservation potential. Temporary saturation may also be the cause of the slumping of dune faces, developing deformed crossbedding (Fig. 4.9c). Damp interdune patches in the lee of active dune faces are seen from the air in Fig. 3.3c.

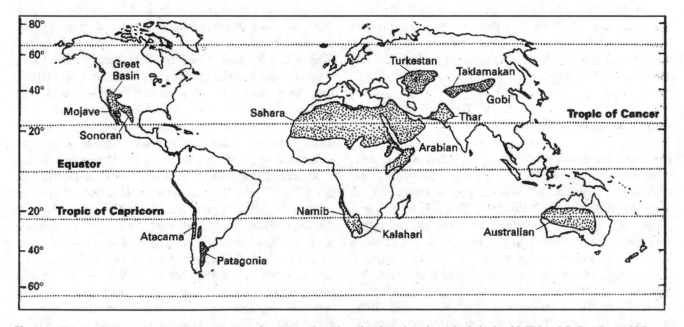

Fig. 4.7 The distribution of major eolian deserts at the present day. Note the clustering along the latitudes 22.5°N and S (Reading, 1996)

Fig. 4.8 Facies model for an erg
(Porter 1986)

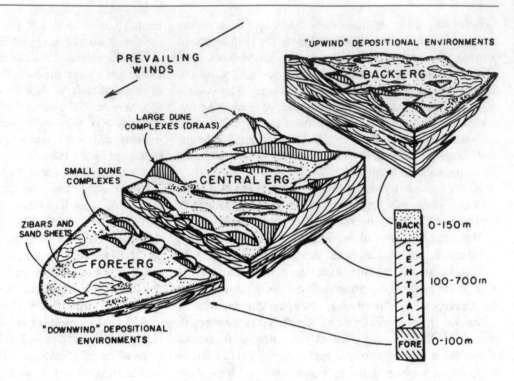

Surfaces of deflation are particularly characteristic of eolian deserts (Fig. 4.9d, e, f), and may, stratigraphically, mark the position of first-order surfaces or supersurfaces. Fig. 4.9d illustrates the rapid movement of barchan dunes across a deflation surface—note the disappearance of modern tire tracks beneath the arms of the dune.

A particularly useful indicator of eolian processes is the **ventifact**, which is a large clast that has been sand blasted by the wind, and periodically turned over (possibly by flash floods), so that it has developed a series of flat faces (Fig. 4.9g). The term **dreikanter** is also used for these objects. Figure 4.9e illustrates a particularly common feature of the desert floor—a deflation surface covered by sand-blasted pebbles. Ancient equivalents of these surfaces are shown in Fig. 4.9h, i.

4.2.3 Lacustrine Environments

Lakes are among the most varied of all depositional environments, even though they occupy a relatively small percentage of the earth's surface at the present day (about 1% according to Talbot and Allen 1996). There are no universal facies models for the lacustrine environment, and recent books and review articles on this subject are not so much syntheses of ideas as catalogs of case studies, each example varying markedly from the next in terms of lithofacies, geochemistry, thickness and extent (Matter and Tucker 1978; Ryder 1980; Picard and High 1981; Katz 1990; Anadón et al. 1991; Talbot and Allen 1996; Renaut and

Gierlowski-Kordesch 2010). The documentation of this variability is perhaps the most important achievement of recent facies studies of lake sediments. This is by no means an academic subject. Vast petroleum reserves occur as oil shales in the Green River Formation of the Uinta Basin and as liquid petroleum in China (e.g., Daqing Oil Field), the Western United States, West Africa and elsewhere (Ryder 1980).

Most major lake deposits owe their origin to tectonic isolation of a sedimentary basin from the sea. The advent of plate tectonics has helped explain the origins of these basins, many of which are surprisingly broad and deep. Typical tectonic settings for lakes include rifts (Lake Baikal, East African Rift System), transform plate margins (Dead Sea, Cenozoic basins of California), remnant ocean basins (Black Sea), some foreland basins (Cenozoic Uinta Basin) and intracratonic depressions (Lake Eyre). Other lakes owe their origins to ice damming or volcanic damming or to glacial scour. Temporary lakes form in various fluvial and coastal settings and may be the site for minor but distinctive deposits.

Another major sediment control is water chemistry. This can vary widely, reflecting variations in the inflow/evaporation balance, nature and quantity of dissolved riverine sediment load, temperature, water level and water body structure (presence of stratification, seasonal overturning, etc.). As Talbot and Allen (1996) noted, lakes are very sensitive to climatic change. They provided a useful review of this topic and suggested a broad classification of lacustrine deposits into those formed in dilute lakes of hydrologically open basins and those formed in saline lakes

Fig. 4.9 Characteristics of eolian environments and deposits. **a** A stream cutting through dune sands; **b** Dried-out interdune pond deposit; **c** Slumped face of dune sandstone; **d** Tire tracks disappearing beneath a barchan; **e, f** surface of deflation. **e** shows a surface covered with ventifacts; **g** close-up of a ventifact; **h, i** Cross-sections of ancient eolian deposits showing surfaces of deflation with ventifacts. Locations: **a** Colorado; **b, d, e, f** Gobi Desert, China; **c** Navajo Sandstone, Arizona; **h** Triassic, Spain; **i** Permian, Scotland

of closed basins. However, lakes may change from one of these states to the other with variations in climate or tectonic setting. Hardie et al. (1978) discussed subenvironments and sub-facies that exist in saline lakes and showed how lacustrine brines evolve during evaporative concentration.

In many ways, lakes represent a microcosm of almost every other environment, usually on a smaller scale than in most marine basins (however, lacustrine carbonates are rather rare). The lake margin may exhibit coastal beach–barrier or delta systems, or there may be coarse fan deltas prograding into the deep basin, with submarine fans forming in the deeper parts of the basin. The lake center may be a quiet-water environment characterized by muds (Fig. 4.10). Storm deposits, including hummocky cross-stratification, have been identified in some ancient lake deposits.

Fig. 4.10 Lacustrine shales of the Eocene Green River Formation, Utah

Because of their sensitivity to climate change and tectonism, and the commonly rapid lateral and vertical facies changes, lacustrine sediments have been much studied by those interested in the processes of allogenic cyclic sedimentation. The processes of **orbital forcing** of climate change (the so-called **Milankovitch mechanisms**) have been studied in lacustrine sediments since their influence was first suspected in the Eocene Green River Shales of Wyoming (Bradley, 1929). Recent studies of such cyclicity (reviewed in Miall 2010) include work on Miocene deposits of northeast Spain (Anadón et al. 1991) and the Triassic Newark Group of the eastern United States (Van Houten 1964; Olsen 1984, 1986, 1990). In the latter, there is considerable variation between humid-climate cycles containing abundant fluvial sandstone deposits, and arid-climate cycles with evaporites. In pro-glacial settings, lakes may contain many of the same facies as those occurring in glaciomarine settings, including sediment-gravity flow deposits from the reworking of glacial tills, and dropstones deposited from floating ice (Fig. 4.11).

Figure 4.12 provides a facies model and hydrological model for arid, intermontane basins, of the type that characterizes the present-day arid southwest of the United States. These basins (especially Death Valley) are commonly cited as analogues of the numerous lacustrine basins of Permian–Triassic age that formed all across what is now eastern North America and Western Europe as Pangea was beginning to break up.

4.2.4 Glacial Environments

Glacigenic deposits were among the last to be studied using facies analysis methods. The analogue method of facies interpretation, in this case the use of Quaternary environmental and facies data, did not at first work well for

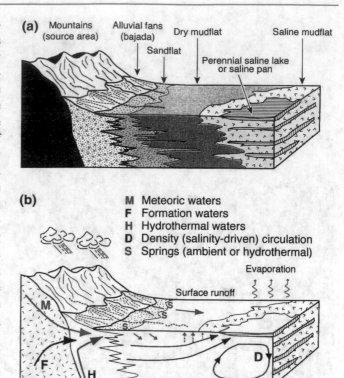

Fig. 4.11 Laminated lacustrine muds and silts formed in a glaciolacustrine setting, and containing dropstones deposited from floating ice. Pleistocene, Ontario

Fig. 4.12 Playa-lake model. **a** General model for closed-basin sedimentation in an arid climate, based on the intermontane basins of the western United States. **b** Hydrology of an intermontane closed basin (Renaut and Gierlowski-Kordesch 2010, Fig. 18, p. 562)

glacigenic deposits. There are several reasons for this. A considerable body of specialized knowledge had built up from the study of Quaternary sediments, but much of this is stratigraphic in nature, rather than sedimentologic, and therefore offers little assistance in the interpretation of the ancient record. A thorough understanding of ancient glacigenic deposits was also hindered by the tendency to describe all coarse-grained, poorly sorted glacigenic deposits as "**till.**" While this term means primarily deposition directly from grounded ice, some authorities have used a much broader definition. For example, Dreimanis (1985) stated that till "is formed and deposited by primary glacial and also by closely associated penecontemporaneous secondary non-glacial processes." It required much sedimentological investigation to document those secondary processes and to demonstrate that in many basins they were dominant. Most ancient glacigenic deposits are formed by the dispersal and redeposition of glacigenic materials by a wide variety of processes in almost every conceivable environment, and it has taken some time for sedimentologists to sort out these various processes, to recognize their importance and to devise criteria for their recognition in the rock record.

Eyles et al. (1983) were among the first to develop systematic facies criteria for this purpose. Most authors now recommend use of the nongenetic term **diamict** (or

diamicton or **mixton**) for the description of poorly sorted gravel-sand-silt-mud mixtures and **diamictite** (or **mixtite**) for the lithified variety. The term till is then restricted for use in an interpretive sense to mean only those deposits that it can be demonstrated were laid down directly by the action of grounded ice and not subsequently reworked (lodgement till). Texturally similar, poorly sorted diamicts also occur in volcaniclastic settings (Eyles and Eyles 2010, p. 78).

The bulk of Quaternary deposits is continental, and tills (in the restricted sense of a primary glacial deposit) are common. However, it is now becoming clear that in the ancient record, subaqueous resedimented glacigenic deposits are much more important, including ice-rafted material and the deposits of sediment-gravity flows. The preservation potential of subaqueous sediments deposited along a continental margin (Fig. 4.13) or in a basin center is much higher than that of subaerial sediments formed on continental interiors, and in subaqueous environments tills are rare. Glacigenic deposits form in almost every clastic environment, and it could be argued that there is no need for a separate class of clastic sedimentary facies labeled "glacigenic." However, there are several unique features about the type of sediment supply, transport processes and facies associations of glacigenic deposits that justify treating them as a separate category.

Several comprehensive syntheses of glacial sedimentary processes and facies are now available, including those by Boulton and Deynoux (1981), Andrews and Matsch (1983; includes an extensive annotated bibliography), Eyles (1984), Eyles and Miall (1984), C.H. Eyles et al. (1985), Ashley et al. (1985), Edwards (1986), Brodzikowski and Van Loon (1987), Eyles and Eyles (1992, 2010) and Miller (1996).

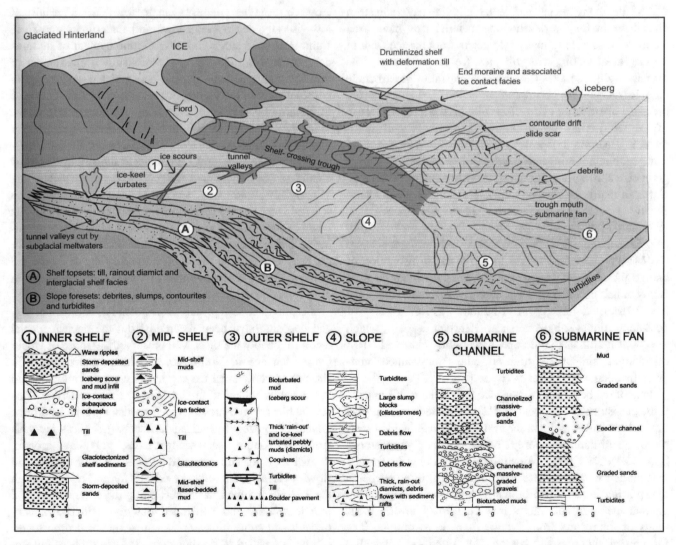

Fig. 4.13 The main environments, processes and deposits of a typical glaciated continental margin. Representative vertical profiles are shown below. Eyles and Eyles (2010, Fig. 3, p. 75) estimate that as much as 90% of the glacigenic deposits that accumulated during the Pleistocene were deposited in environments such as these, and that glaciomarine deposits dominate the record of older glaciations because the terrestrial record is easily eroded

Criteria for the recognition of tills were discussed by Hambrey and Harland (1981a) and Anderson (1983); a comprehensive classification of glacial lithofacies was offered by Eyles et al. (1983) and Earth's glacial record was considered in detail by Frakes (1979), Hambrey and Harland (1981b), Anderson (1983), Chumakov (1985), Frakes et al. (1992) and Eyles (1993, 2008). Interpretation of glacigenic deposits has been facilitated by several major studies of modern subaqueous glacial environments, including the Antarctic shelf and deep sea (Anderson et al. 1983; Wright et al. 1983), the Gulf of Alaska (Molnia, 1983a) and the fiords of southern Alaska (Molnia 1983a; Powell, 1981, 1983, 1984). Detailed facies studies of ancient glacigenic sequences are now too numerous to list. Several are contained in the invaluable research syntheses edited by Molnia (1983b) and Deynoux (1985).

In studying ancient glacial deposits, it must be recognized that while many facies and other criteria may indicate a glacial origin for the detritus, the material may have been redeposited several to many kilometers from the ice front in a **periglacial** setting and still preserve the evidence of its glacial origins. Criteria that indicate glacial origins include particle shape and grain-surface textures, including the presence of striae on clasts. True tills (in the restricted sense of the word) may be distinguished from other diamicts, including ice-rafted deposits and sediment-gravity flows, by such sedimentologic and stratigraphic criteria as internal grain-size sorting (e.g., grading) and sedimentary structures, clast and matrix fabrics, bed thickness, geometry and lateral extent and facies associations. Magnetic characteristics, including the anisotropy of magnetic susceptibility (AMS) and the orientation of natural remanent magnetism (NRM), have become useful tools for the investigation of matrix fabrics, because the results can be related to final depositional processes. For example, lodgement tills may have distorted NRM distributions as a result of post-depositional subglacial shear. Rain-out deposits, which may appear similar in outcrops, have tightly clustered NRM orientations similar to those of other fine-grained, non-glacial, subaqueous deposits (Eyles et al. 1987).

One of the best lines of evidence for glacial environments is the dropstone or lonestone facies—dispersed clasts several millimeters to several meters in diameter, interbedded with sediments of much smaller grain size, commonly mudstones (Fig. 4.11). This very distinctive facies can only be formed in abundance where there are numerous drifting icebergs transporting glacial debris out to sea until dropped by gradual melting. This facies may be formed several thousands of kilometers from the ice front. For example, it is forming at the present day off Newfoundland, from the melting of icebergs that originate at the terminus of valley glaciers in northwest Greenland. Seasonal land-fast ice cannot produce this facies in the same volume. The only

significant nonglacial origin for such a facies is the transport of debris by floating plants, but the volume is likely to be insignificant compared with that forming in polar waters close to floating ice margins.

Striated pavements are a spectacular demonstration of grounded ice, and many examples have been documented in the ancient record, but, as Anderson (1983) warned, they cannot be taken as indicating continental glacial conditions. The Antarctic ice cap is in places grounded at depths of up to 2 km below sea level, and large icebergs are known to form deeply gouged grooves at depths of hundreds of meters on, for example, the continental shelf of Eastern Canada and the floor of Beaufort Sea.

Other traditional glacial indicators must also be used with caution. Rhythmically laminated fine-grained muds and silts formed in glacial settings are routinely described as varves, and interpreted as products of seasonal freeze–thaw cycles. However, similar deposits may form as a result of low-viscosity density interflows and underflows and from slump-generated surge currents, and the number of rhythms does not necessarily bear any relationship to the number of years during which the deposit accumulated. Ashley et al. (1985) described the variations between these different types of rhythm in some detail, with many excellent photographs. Mackiewicz et al. (1984) coined the term **cyclopel** for varve-like laminated interflow and underflow deposits.

Sand-wedge structures may indicate subaerial permafrost conditions, but Eyles and Clark (1985) demonstrated that superficially similar structures are produced by soft-sediment-gravity loading in subaqueous settings. The correct interpretation may be a critical factor in discriminating between terrestrial and subaqueous environments for the beds in which the structures occur, as they demonstrated in the case of a glacigenic unit of Upper Proterozoic age in northwest Scotland.

Despite their low preservation potential, continental glacial deposits have been recorded from the ancient glacial record. Probably the most well preserved are those of Upper Ordovician age in North Africa, which formed around a short-lived continental ice cap in the center of a craton and subsequently remained undisturbed to the present day. Terrestrial tills are well preserved, as are many glacial landforms typical of Quaternary glacial landscapes, such as ice-push structures, kettles, sand-wedge casts, polygonal ground, drumlins, pingos and eskers (Beuf et al. 1971). Marwick and Rowley (1998) reviewed the evidence for glaciation between the Triassic and the Pleistocene, citing the criteria discussed in this section, and concluded that the evidence is very limited, and some indicators, such as scattered dropstones, can be interpreted as detritus carried in the roots of floating trees.

Some ancient glacigenic deposits are interbedded with carbonate sediments, and for many years this has raised interpretive problems because of the assumption that carbonates indicate warm waters and low latitudes. However, it can be demonstrated that some interbedded carbonates are detrital in origin and formed as, for example, continental-slope sediment gravity flow deposits (e.g., Eyles et al. 1985). Anderson (1983) pointed out that the critical factor in carbonate sedimentation is not water temperature but an absence of detritus. Biogenic carbonate production takes place, albeit slowly, in waters of extremely low temperature. For example, the Antarctic continent is ringed with carbonate oozes and shell hash formed in waters of 0 to 8° C.

A variety of glacial environments has now been identified and described. Eyles (1984) and Ashley et al. (1985) synthesized our knowledge of terrestrial glacial landforms and associated terrestrial depositional settings (valley glaciers, outwash plains, glacial lakes). Fiord sedimentation has been described by Powell (1981, 1983, 1984) based on his work in Alaska. Shelf sedimentation has been described by Molnia (1983a) and Anderson et al. (1983). Three main ice-margin models have evolved: that where the ice is grounded on land and the detritus is carried into the sea by outwash rivers, that where the ice is grounded on the shelf below sea level and that where the ice margin floats on tidewater. Distal glacial environments are those characterized mainly by ice-rafted sediments. Glacial marine sedimentation is controlled by a variety of factors, including the rate of berg calving, glacier advance and retreat behavior, the position of debris on the glacier, the position of meltwater streams and oceanographic factors, such as the control of bottom currents on berg drift tracks and sediment plumes, bottom currents, tidal currents and vertical water mixing. Sedimentation rates in front of ice margins can be astonishingly high. Short-term rates up to 9 m/a have been measured in bays near the termini of glaciers in southern Alaska, while shelf sedimentation rates in the Gulf of Alaska typically range from 2 to 10 mm/a, with an average of 4.5 mm/a (Molnia 1983a).

There are two distinct glacial thermal regimes. In temperate environments, including subarctic regions, such as southern Alaska, basal glacier ice slides over a substrate lubricated by meltwater, and abundant meltwater streams move in tunnels through the ice. In frigid polar environments, such as that along most of the Antarctic ice margin, the glacial ice is frozen to the substrate and moves by internal shear and flow. The delivery of detritus to the ice margin is several orders of magnitude slower in such frigid settings, and basin sedimentation rates are considerably less (Eyles et al. 1985, Table I). Most ice margins are probably characterized by both types of thermal regime over time, and sedimentologists have not yet succeeded in developing criteria (other than sedimentation rates) to distinguish the resulting depositional products.

4.2.5 Coastal Wave- and Tide-Dominated Environments

Coastlines, where the land meets the sea, may be either erosional or depositional environments. Erosional coastlines are rarely preserved, although in some cases the architecture of an unconformity may indicate the presence of buried cliffs (e.g., see Fig. 1.15 Just such a feature is visible between units J3.2 and K1.1). Unconformity surfaces are sometimes covered by a deposit of large boulders emplaced by waves at the commencement of a new depositional cycle. A famous example is the Sutton Stone, exposed on coastal cliffs on South Wales (famous because research on this unit has helped to highlight problems in the interpretation of sedimentation rates—see Sect. 8.2 and Fig. 8.5). Coastlines that are tectonically stable or undergoing subsidence are places of net sediment accumulation.

Recent textbooks treat clastic shorelines in different ways. All types of shoreline environment are included in a single chapter on this topic in the third edition of the classic "Reading" textbook (Reading and Collinson 1996), whereas in the later editions of the "Facies models" book the subject matter is subdivided into several chapters. In Walker and James (1992), there are three chapters: "Deltas," "Transgressive barrier island and estuarine systems" and "Tidal depositional systems." In James and Dalrymple (2010), shoreline and shallow-marine systems are treated in two chapters, "Transgressive wave-dominated coasts" and "Wave- and storm-dominated shoreline and shallow-marine systems." This leads to a certain amount of overlap in the case of estuarine environments. An excellent compendium of recent research on tidal sedimentation was compiled by Smith et al. (1991).

This is a highly dynamic environment, where sediment is moved by waves and tides, and episodically affected by storms (Figs. 4.14, 4.15). Rivers may deposit large volumes of sediment at the coast, in the form of deltas, which may or may not be modified by marine processes. However, rising sea levels may cause a river valley to be drowned, forming an estuary. Deltas and estuaries are discussed separately in the next section. Here we describe coastlines characterized by beaches, and beach–barrier–lagoon systems (Fig. 4.14).

The subdivision of coasts into domains on the basis of tide range (**microtidal**, **mesotidal**, and **macrotidal**; Hayes 1976) has been a very useful concept, but Hayes (1979) and Davis and Hayes (1984) showed that it is the relative effectiveness of waves and tides, not their absolute energy levels (wave height, tide range) that is important. Thus, it is possible for a microtidal coast to be tide dominated if wave energy is low enough. Boothroyd et al. (1985) showed how this idea could be used to develop a modified classification

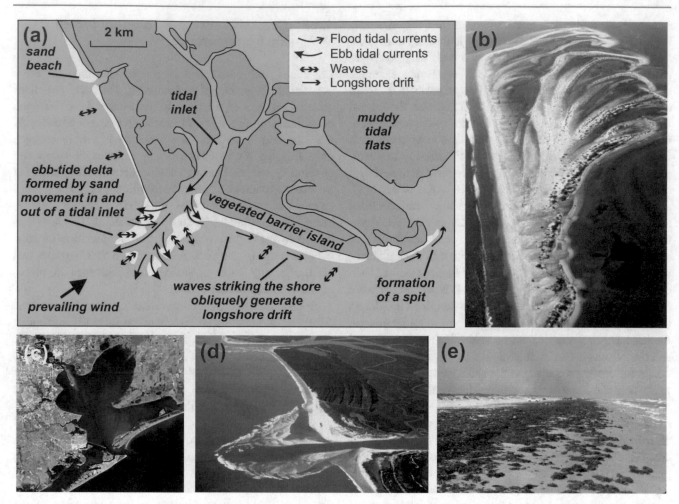

Fig. 4.14 Beach–barrier–lagoon depositional environments. **a** Hypothetical coastline, showing the interaction of waves and tides; **b, d** aerial views of South Carolina coastline; **c** Galveston Island and Galveston Bay, Texas; **e** The beach at South Padre Island, Texas

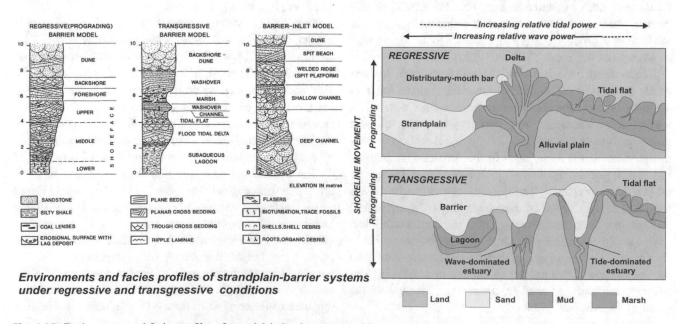

Fig. 4.15 Environments and facies profiles of strandplain-barrier systems under regressive and transgressive conditions. Vertical profiles at left from Reinson (1992); Diagrams at right Adapted from Reading (1996)

of various coastal regions around North America and elsewhere.

Wave- and tide-influenced environments may be recognized in small outcrops or drill core by the distinctive styles of hydrodynamic sedimentary structures (Figs. 3.26–3.31). Wave-formed structures are distinguished by the evidence of oscillating currents, tidal environments by reversing bedforms and the presence of sand-mud couplets (Fig. 3.29).

Typical beach profiles are shown in Fig. 4.15. The profile for a **regressing coastline** is that of a coarsening-upward succession, reflecting the increase in wave energy from lower shoreface to foreshore. The same energy trend is shown in a beach profile in Fig. 3.30. The vertical profile in Fig. 4.15 (see also Fig. 3.58d) was first developed from studies on the Galveston Island barrier, and was one of the first vertical profile models to be established, beginning in the late 1950s (Bernard et al. 1959; see Sect. 1.2.6). Regression by the addition of sediment to the front of a beach under conditions of stable sea level is called **normal regression**, to distinguish this from the regression caused by falling sea level, which has been termed **forced regression** (Plint 1988; see Sect. 5.3.1; Fig. 5.14).

Much research in the area of sequence stratigraphy has focused on shoreline deposits because of their obvious sensitivity to sea-level change, and this has added a new dimension to our knowledge of their long-term response to environmental change. For example, a long-standing controversy regarding the origins of barrier systems and their preservation was largely resolved by Swift et al. (1985) and Niederoda et al. (1985), in a pair of detailed papers dealing with U.S. Atlantic coast barriers, which demonstrated rather conclusively the ability of barriers to migrate landward during transgression, feeding on their own sand to maintain volume. Sequence-stratigraphic reconstructions clearly confirm this process.

There are differences between the geography and sedimentology of coastlines undergoing normal regression versus those undergoing transgression (Fig. 4.15). Transgression across a gently sloping shelf may cause the main wave energy to be dissipated offshore, with the beach deposits isolated from the actual shoreline by a lagoon. Barrier–lagoon systems are mainly the product of transgression, such as the extensive system of such landforms along the southeast Atlantic and Gulf coasts of the United States (Fig. 4.14). Tidal rise and fall cause marine waters to enter and leave the lagoon, and this is accomplished by the maintenance of **tidal inlets** through the barrier system (Fig. 4.14a, d). These are not necessarily associated with river mouths, but may be initiated by storm erosion, which is discussed in greater length below. Tidal inlets may accumulate channel-fill sediment if the inlet is abandoned. Such a deposit is similar to a fluvial channel fill, but may be distinguished by the presence of marine fossils, and crossbedding evidence of reversing currents.

Current patterns on wave- and tide-dominated coastlines may be very complex. The pattern of prevailing winds may be responsible for establishing a dominant angle of attack by waves (Fig. 4.14a). Waves are refracted by friction on the sea bottom as they approach the shore, but still may tend to strike the coast obliquely, driving sediment obliquely up the foreshore. The backwash rolls the grains down the beach slope, and then they are driven obliquely up again by the next wave. In this way, sediment is gradually caused to move along the shore, a process called **longshore drift**. Sediment is driven into the mouths of tidal inlets, which may cause the tidal inflow and outflow to impinge erosively on the opposite bank. In this way, the position of the inlet may gradually shift laterally along the barrier. Sand will also be driven as a growing **spit** into open areas of the sea. The "growth" lines of the spit are commonly visible from the air (e.g., Fig. 4.14b).

Sediment is flushed in and out of the lagoon by the rise and fall of the tides. Where the flow expands into the lagoon (during the flood) or out into the sea (during the ebb) much of this sediment is deposited as **tidal deltas**. An ebb-tidal delta is shown in the diagram of Fig. 4.14a and a South Carolina example in Fig. 4.14d. Current patterns over these deltas are complex, as temporary distributary channels migrate and switch position with the changing weather and seasons, and bedforms migrate and are eroded. Parts of these deltas are commonly dominated by either the ebb or the flood phase of the tides, because of accidents of local topography, the result being that crossbedded tidal delta deposits may be surprisingly unimodal, as in the example shown in Fig. 3.27 (right-hand example).

Lagoons are sites of fine-grained sedimentation. They are typically characterized by broad tidal flats, colonized by various salt-resistant grasses around the margins. The waters may be brackish or, in hot climates, hypersaline. Tidal flood and ebb currents are focused along a network of tidal creeks that function like reversing fluvial channels, complete with meanders and point bars.

Plint (2010) has drawn attention to the importance of mud deposition on continental shelves. Many modern shelves are covered with coarse, relict (**palimpsest**) sediment left by post-glacial sea-level rise, but stratigraphic studies, particularly Plint's own work on the Cretaceous stratigraphy of the Western Interior Basin, have demonstrated that significant thicknesses of mud may be trapped on continental shelves. The long-prevailing idea that mud deposition is characteristic of low-energy, pelagic environments has been demonstrated to be an incomplete interpretation of fine-grained facies. While some mud is deposited in such environments, for example, in lake basins and the deep oceans, mud particles commonly aggregate in saline (marine) water and are redistributed by coastal and geostrophic currents on continental shelves and at their margins, where they may form gently dipping clinoform deposits. For example, huge volumes of mud are currently accumulating on the continental shelf adjacent to the mouth of the Amazon River (check out the story of the escaped convict Papillon for a graphic impression of the coastal mud flats). Application of these new ideas to the interpretation of the preserved record of mud deposits is creating some entirely new approaches to the facies modeling of fine-grained deposits (Plint 2010).

Studies of Cretaceous units in the Western Interior Plains of North America, especially in the Alberta Basin, led to the recognition of a distinctive facies assemblage for which Thomas et al. (1987) coined the term **inclined heterolithic stratification**, and the acronym IHS. This facies consists of sand–mud couplets developed under tidal influence on the point bars of tidal creeks and tidally influenced rivers. Depositional dips reflect the formation of this facies by lateral accretion. Particularly spectacular examples of IHS occur in the Athabasca Oil Sands, near Fort McMurray, Alberta, where tidally influenced point bars up to 25 m thick are exposed (Smith 1987, 1988a, b; Musial et al. 2012). The review by Thomas et al. (1987) provided a useful summation of ideas regarding estuarine channel sedimentation, many of which had been known since the Dutch work of the 1950s, but required reanalysis in the light of modern sedimentological concepts.

Many studies of tidal bedforms have now appeared. Allen's (1980) model of tidal sand waves was an invaluable theoretical study (Fig. 3.27), the results of which have been applied to several ancient deposits (e.g., Blakey, 1984; Teyssen, 1984). Modern estuarine examples of sand waves

have been described by Dalrymple (1984). Harris (1988) reviewed the formation of larger estuarine bedforms, including sand ribbons and sand banks (tidal current ridges of other authors), and confirmed earlier ideas regarding the evolution of the latter by tidally driven racetrack circulation of sand around them. Superb images of tidal sandwaves in a shelf setting were obtained by high-resolution seismic surveying, echo sounding and side-scan sonar surveying of the coast of Normandy, by Berné et al. (1988, 1991). Studies of this type will provide the basis for the next generation of facies modeling of the ancient record, because of the detailed picture of surface bedform patterns and three-dimensional facies architecture which they provide.

As illustrated in Chap. 5, the transition from lowstand to transgression is accompanied in many cases by backstepping or retrogradation of coastal depositional systems, and the partial stripping away of strandplain and barrier systems by the process of **ravinement** (Nummedal and Swift 1987) with the sediment so released forming new beach systems further landward (Fig. 4.16). Wave scour of the sea floor is at its greatest in the shallow offshore, over depths of up to perhaps 10 to 15 m below sea level. At such depths, erosion typically predominates over deposition. Given enough time, and a stable sea level, waves may develop a flat **wave-cut platform** somewhat below the low-tide water level. Under conditions of transgression, the zone of maximum waves scour is gradually translated obliquely upward, as the shoreline is pushed landward. Beach and back-barrier lagoonal deposits are truncated by this process, while shelf sediments begin to cover the surface as water depths increase with the rising sea level. The end result is the development of a major erosion surface that juxtaposes lagoonal and shelf deposits. But it is a time-transgressive surface (as shown by

Fig. 4.16 The process of ravinement. The dashed red line indicates the time correlation between shelf sediments deposited on the ravinement surface and contemporaneous lagoon deposits that are truncated by ravinement erosion as sea level rises (after Nummedal and Swift 1987)

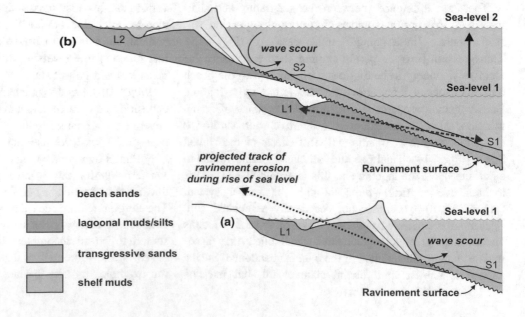

the dashed correlation line in Fig. 4.16). It commonly forms a sequence boundary in the rock record, where erosion cuts down through lowstand and early transgressive deposits.

Episodic storms can dramatically change the configurations of the beach–barrier systems just described. Hurricanes approaching the shore generate wind pressures that can push the sea onshore, raising local water levels by as much as 5 m. Coupled with the high waves generated by the winds, such storm surges can do enormous damage and completely change the landscape. Figure 4.17 is a before-and-after map of a portion of the Matagorda Peninsula barrier system on the Gulf Coast, which was substantially modified by Hurricane Carla in 1964 (Hayes 1967). Moribund channels are opened up to permit a storm surge to pass over the barrier, and considerable volumes of sediment are carried landward, forming **washover fans** on the lagoon side of the barrier. The storm channels may gradually be filled again during subsequent quiet weather, and this can lull property developers into a false sense that these lands are safe to develop. US citizens will be familiar with these processes as a result of the catastrophic hurricanes Katrina (coastal Louisiana 2005) and Sandy (New Jersey, 2012).

Tsunamis are catastrophic events that may radically modify coastal environments, but few examples of tsunami erosion and deposition have been identified in the rock record, possibly because few modern analogues have been sufficiently studied. Reference should be made to Sect. 3.4.3 where the work on this topic by B. R. Pratt is discussed.

4.2.6 Deltas

Deltas typically represent the single largest repository of sediment in many sedimentary basins. Some deltaic complexes are known to have existed for tens of millions of years and have accumulated many kilometers of sediment. Modern work on deltaic sedimentation was summarized by Bhattacharya and Walker (1992), Orton and Reading (1993), Reading and Collinson (1996) and Bhattacharya (2010). A collection of papers on coarse-grained deltas was edited by Colella and Prior (1990).

The most well-known delta is that of the Mississippi (Fig. 4.18). It is one of the world's largest, and has been the subject of detailed study since the 1930s, partly because of its importance to coastal navigation in the Gulf Coast Region of the United States, and partly because it was recognized many years ago that it serves as an excellent modern laboratory for the study of geological processes in deltas (Blum and Roberts 2012). However, it represents only one of three broad classes of deltas (Figs. 4.19 and 4.20). The Mississippi drains into a semi-enclosed sea in which tidal range is low, and wave energy is also typically low (except during the rare hurricanes that affect the coastline). Sediment carried into the

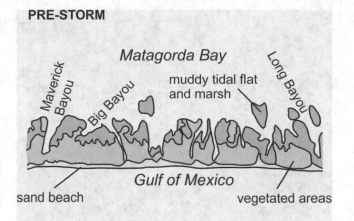

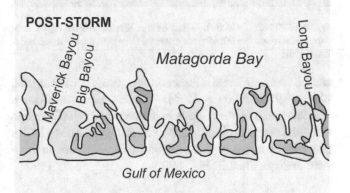

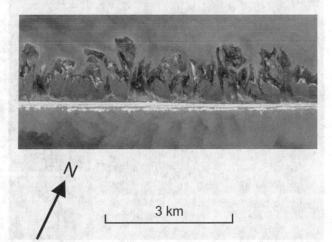

Fig. 4.17 The effects of Hurricane Carla on Matagorda Peninsula, Texas Gulf Coast, 1964 (Hayes 1967). At bottom is a modern aerial image of this barrier system

Gulf by the river has been deposited alongside the distributary channels as long fingers extending out into the sea. In other settings, where marine processes are more important, much deltaic sediment is redistributed by powerful waves and tides, and the resulting architecture of the delta

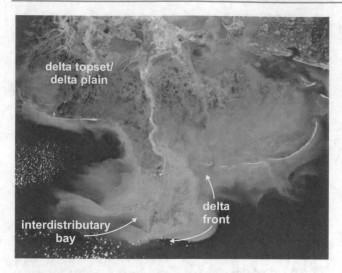

Fig. 4.18 The classic delta—that of the Mississippi River, as seen from space. Note the turbid river water mixing with the clear sea water of the Gulf

can be very different (Fisher et al. 1969; Coleman and Wright, 1975; Galloway 1975). A widely used classification of deltas based on the relative importance of fluvial input and wave and tide reworking was published by Galloway (1975). The various styles of delta shown in Fig. 4.19 are summarized and classified in Fig. 4.20.

New classifications of deltas were offered by Postma (1990) and Orton and Reading (1993). As noted by Orton and Reading (1993), in addition to recognizing the importance of the three major processes governing sediment distribution on deltas, the grain size of the sediment supply is also a critical factor in determining the configuration of the delta and the architecture of the resulting deposits. McPherson et al (1987) were among the first to attempt to incorporate grain size into delta classification, with the recognition of coarse-grained types termed **fan deltas**, and finer grained varieties termed braid deltas and **common deltas**. These ideas were developed further and generalized in the classifications by Postma (1990) and Orton and Reading (1993). Postma (1990) focused on the nature of the fluvial feeder system and the water depth in the receiving basin, which reflects the setting of the delta at the margin of a broad continental shelf or on the shelf margin or slope. This classification departs from Galloways' emphasis on the processes of redistribution of sediment at the delta front by not incorporating data dealing with wave- and tide-generated sediment redistribution. The classification of Orton and Reading (1993) builds on that of Galloway to incorporate data on the grain size of the sediment load. The latter authors also provide a thorough discussion of the importance of other variables such as delta plain slope, and the control of

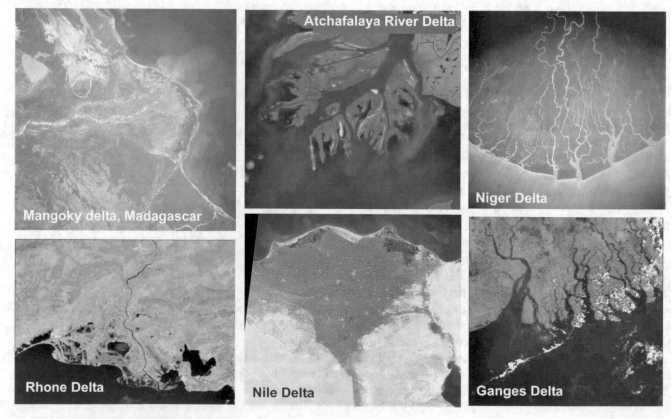

Fig. 4.19 Deltas as seen from space

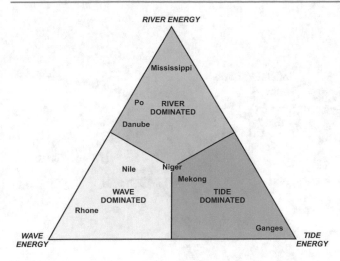

Fig. 4.20 Processes influencing deltas, and a delta classification (Galloway 1975)

grain size and wave height on the tendency of wave energy to be dissipated or reflected at the shoreline, and the effects of these variables on the resulting sedimentary deposit.

Along coastlines where there are strong prevailing winds, there may be a dominant direction of wave attack, and sediment deposited at the mouths of deltaic distributaries is redistributed along the coast, forming arcuate beach deposits. The delta of the Rhone, in the Mediterranean Sea, is commonly cited as a typical example of a wave-influenced delta. The Nile Delta and the Niger Delta also show the smoothly curving stretches of sand-beach coastline that characterizes a wave-influenced delta.

Where tidal influence is strong, deltaic distributaries are shaped by the constant rush of the flood and ebb currents. The volume of water rushing into and out of the delta, in

such situations, requires a greater than usual number of distributaries. Such situations occur where the tidal range is high (macrotidal regimes), such as at the head of wide, but narrowing, bays in the ocean. The Bay of Bengal, with the delta of the Ganges/Brahmaputra River at its head, is a good example (Fig. 4.19). Most deltas experience a combination of influences, by rivers, waves and tides (Fig. 4.20). As indicated in this classification, there are rivers such as the Mississippi, Ganges and Rhone that appear to be dominated by just one of the three main processes, whereas the geography and architecture of others suggest a combination of forces. The Niger, in particular, shows evidence of all three at work. The delta itself has a lobate outline, which is consistent with active seaward progradation by rivers. The delta is fronted by long, narrow beaches, indicating some wave reworking, and the beaches are cut by numerous distributaries that are backed by tidal flats, indicating continual tide-driven flux of water into and out of the delta.

The history of the Mississippi Delta Complex provides a good illustration of how deltas work. The modern delta shown in Fig. 4.18 is only the most recent of a series of deposits formed along this part of the coastline since the last ice age. Based on numerous drill cores, geologists have mapped seven previous deltas comparable in size to the modern lobe, labeled "6" in Fig. 4.21. It appears that each lobe gradually grows, and extends out into the sea, for several thousand years, until it becomes unstable, at which point the main river switches into a different course. The process is illustrated in Fig. 4.22. Delta switching is a process called **avulsion**. It may be triggered during episodes of unusually large runoff, when the flow overtops its banks and finds a shorter, steeper route to the sea.

Fig. 4.21 The seven post-glacial lobes of the Mississippi Delta. Adapted from Penland et al. 1988; after Frazier 1967

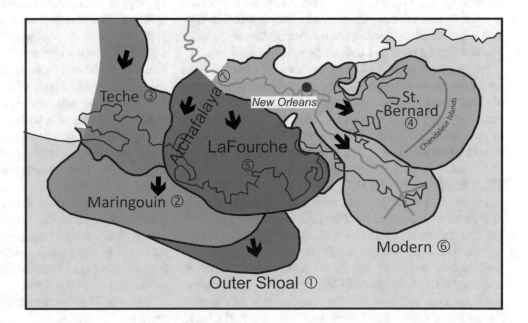

Fig. 4.22 Oblique view of a delta, demonstrating why delta lobes switch in position. (1) As a delta builds out from the coastline the river channel aggrades above sea level (vertical height beneath point A); (2) The path from point A to point B becomes steadily longer and the slope gentler; (3) The path from point A to point C remains shorter and steeper; (4) If the river finds the route to point C by overbank flooding it will result in switching of the river to the shorter course, leading to abandonment of the delta lobe and initiation of a new lobe starting from point C

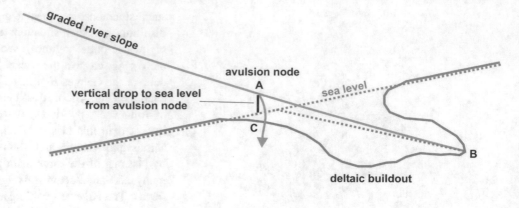

After abandonment, the deposits continue to compact slowly and to subside, allowing the sea to re-advance. The changing depositional geography at the location of each delta lobe is illustrated in Fig. 7.27. The processes are well seen in the case of lobe 4, the St. Bernard lobe, where the lobe has undergone transgression, and much of the sand has been reworked into a series of offshore beach shoals called the Chandaleur Islands. The Atchafalaya lobe is a potential successor to the present, modern lobe. The Atchafalaya River is an old course of the lower Mississippi (located where the word "Atchafalaya" is in Fig. 4.21). When modern river control and navigation works began in the Mississippi River, late in the nineteenth century, the Mississippi was in the process of switching back to this course, which, as can be seen from the map in Fig. 4.21, is a considerably shorter route to the sea than the modern river. Dams and locks were installed to prevent this happening, because of the importance of the flow through New Orleans, a major sea port. The flow that is permitted to divert down the Atchafalaya is nonetheless developing an active delta at its mouth (Fig. 4.19).

The modern delta lobe is an example of what has been called the **birdsfoot** type of delta, narrow, branching fingers of land extending seaward, flanked by bays. The bays themselves gradually fill with sediment deposited in a series of prograding mini-deltas, as shown in Fig. 4.23. Initiation of each of these begins with a **crevasse**, or break in the bank of the main channel. Diversion of sediment and water and filling of the bay is achieved quite rapidly.

Figure 4.24 shows the processes that occur at the mouths of distributaries. Expansion of the flow occurs as it passes out through the mouth. Most of the coarser sediment load is deposited here to form a mouth bar deposit. This may be worked and reworked by fluvial and marine processes, resulting in a well-sorted sand characterized by much internal scouring. Fine sediment stays in suspension a little longer, and is carried further away from the mouth, much of it being

deposited eventually to form the distal portion of the mouth bar. However, the water remains turbid for a considerable distance out from the distributary mouths, as can be seen in the two space photographs (Figs. 4.18 and 4.23). Sediment deposited at the margins of the channel forms a subaqueous extension of the levee system. The incremental additions to the levee and the mouth bar cause the mouth to prograde. The mouth bar therefore develops by a process of continual offlapping, as the mouth is shifted seaward (Fig. 4.24). The mouth bar develops a long, narrow ("**shoestring**") geometry and has an upward-coarsening grain-size trend, reflecting the process whereby coarser sediments, formed on the upper bar, are gradually extended out over the finer sediments that form the lower or more distal bar.

A very useful depositional model for Mississippi-type mouth bars was developed at the Bureau of Economic Geology in Texas, and is shown here as Fig. 4.25. This helps to explain the classic shoestring geometry, a term that has long been used by petroleum geologists in their search for sandstone reservoirs in deltaic systems. Note, also, the various sub-environments illustrated in this diagram, including the interdistributary areas of the delta plain, characterized by fine-grained deposits, possibly peat swamps, that may ultimately become coal deposits.

Up to this point we have been looking primarily at the shape and surface evolution of deltas. Figure 4.26 helps to explain how deltas develop geologically significant thicknesses of sediment. This diagram was drawn following a study of deltaic sedimentation in the Gulf of Mexico, but has a universal applicability. A vertical drill hole placed anywhere along this hypothetical cross-section will penetrate successively more distal and fine-grained deposits the lower it goes. This explains the coarsening-upward nature of most deltaic deposits (Fig. 3.62).

Figure 4.27 illustrates some examples of ancient deltaic deposits.

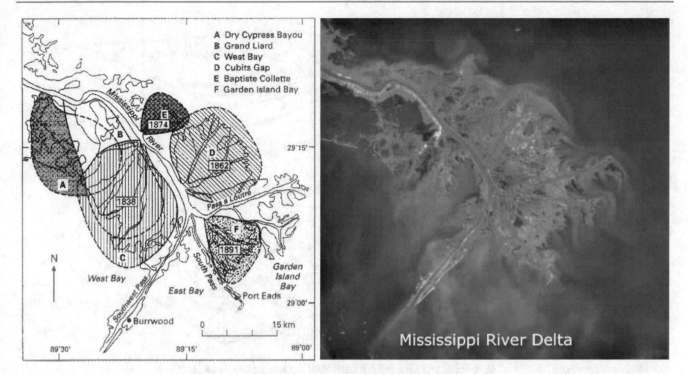

Fig. 4.23 The development of small deltas infilling the bay areas flanking the main distributaries of the modern Mississippi. The amount of sediment in the water is apparent from its turbid appearance in the space photograph. The date of initiation of each of the deltas is shown in the map at the left (diagram from Coleman and Gagliano 1964)

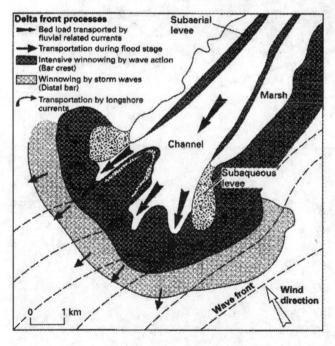

Fig. 4.24 A mouth bar forming at the end of one of the Mississippi Delta distributaries (Coleman and Gagliano 1964)

Coal is commonly associated with deltas, but not all deltaic peat swamps become coal. Much depends on the clastic influx into the swamp. Frequent overbank floods carry silt and clay into the swamp, and the resulting deposit is then a carbonaceous shale rather than coal. Such a deposit is of too poor a quality to be used as coal because of the large amount of non-combustible sediment ("ash") left after combustion. The best coals develop where a peat swamp accumulates plant material at a rate rapid enough that the swamp becomes slightly raised above the level of the surrounding channel system. This elevated position then serves to keep clastic influx out of the swamp, and the peat can then accumulate as a potentially cleaner form of carbon.

Careful subsurface mapping using well data may help to clarify the style of delta that formed a given deposit. The distribution of channel sands, in particular, reveals key information about the transport and reworking of the coarser portion of the sediment load by fluvial and marine processes (Fig. 4.28). The isopachs shown in this figure are net-sand isopachs, which are constructed by careful mapping and correlation of individual deltas in the subsurface. Clearly, maps of this type require good well control. Bhattacharya (2010) provided several examples from the rock record.

Fig. 4.25 The classic
shoestring-sand facies model
(Frazier 1967)

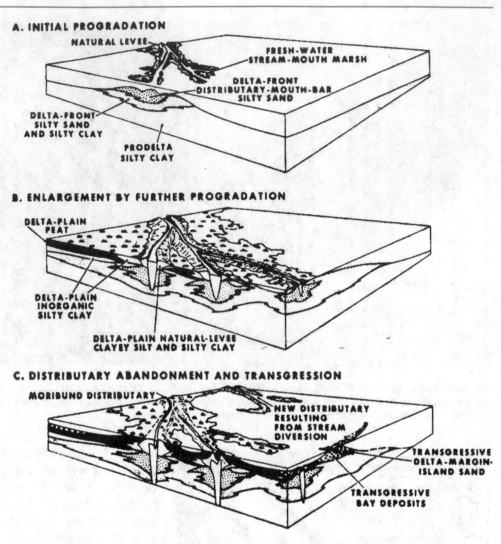

Fig. 4.26 The prograding front
of a delta defines three main
subenvironments: (1) the delta
front, which is the seaward
margin of the delta plain and is
characterized by channel,
overbank and interdistributary
bay environments; (2) the
prodelta, which is the
environment of the outbuilding
distributary. Sloping lines are
successive depositional surfaces,
and are the clinoforms of some
reflection-seismic sections. The
offshore is the deep-marine
environment beyond the influence
of the delta (Scruton 1960).
AAPG© 1960, reprinted by
permission of the AAPG whose
permission is required for further
use

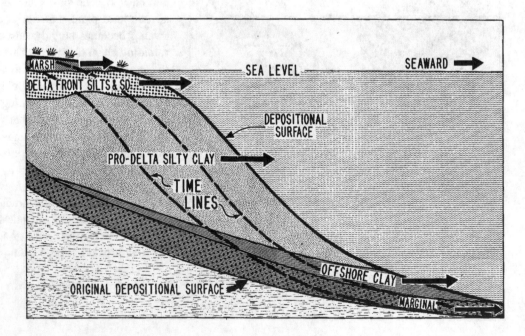

Fig. 4.27 Examples of ancient deltaic deposits. **a** Channel sandstone over coal seam. Jurassic, Alberta; **b** Channel cut into delta-front sandstone, Paleogene, Arctic Canada. Base of channel indicated by dotted red line; **c–f** Carboniferous deposits, Kentucky; **c** Overbank and crevasse-splay deposits; **e, f** coals overlain by crevasse-splay deposits

Fig. 4.28 Isopach patterns of typical deltas, reflecting various combinations of sedimentary processes. Adapted from Coleman and Wright (1975)

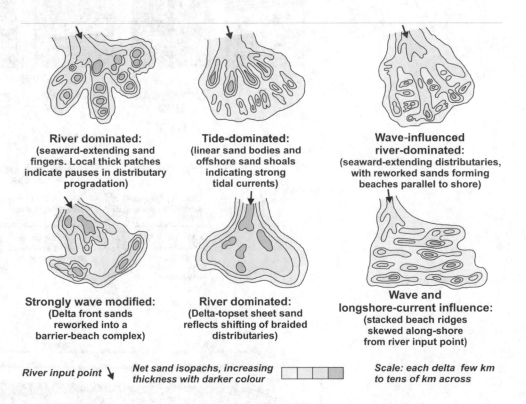

River dominated:
(seaward-extending sand fingers. Local thick patches indicate pauses in distributary progradation)

Tide-dominated:
(linear sand bodies and offshore sand shoals indicating strong tidal currents)

Wave-influenced river-dominated:
(seaward-extending distributaries, with reworked sands forming beaches parallel to shore)

Strongly wave modified:
(Delta front sands reworked into a barrier-beach complex)

River dominated:
(Delta-topset sheet sand reflects shifting of braided distributaries)

Wave and longshore-current influence:
(stacked beach ridges skewed along-shore from river input point)

River input point ↘   *Net sand isopachs, increasing thickness with darker colour*   *Scale: each delta few km to tens of km across*

4.2.7 Estuaries

Estuaries form at the mouths of rivers undergoing flooding or transgression. They are areas of open water and modest sediment input. Facies models and/or collections of papers detailing estuarine deposition have been provided by Dalrymple et al. (1994, 2006) and Boyd (2010).

Sedimentary processes may vary considerably in an estuary. Wave processes, including the formation of a barrier beach deposit, are common at the mouth (Fig. 4.29). At the head, fluvial processes may be important. Tidal influence may be significant throughout the length of an estuary, but are likely to be most important in controlling sedimentary processes near the middle of the estuary, downstream from the main fluvial input and back from the full force of the waves near the mouth. All these various influences may be studied using the indicators of fluvial, wave and tidal processes discussed in Chap. 3. Figure 4.29 provides many of these sedimentological details and stitches them together into two diagrams, a plan view and a cross-section of typical estuaries. The panels showing sedimentary structures and the vertical profiles provide the kinds of practical indicators that field geologists need to assign deposits to the appropriate facies model.

Interest in estuarine sedimentation received a tremendous impetus when it was realized that many producing sandstone

Fig. 4.29 Sedimentary models for estuaries. Top: Reinson (1992); bottom: Dalrymple et al. (1994)

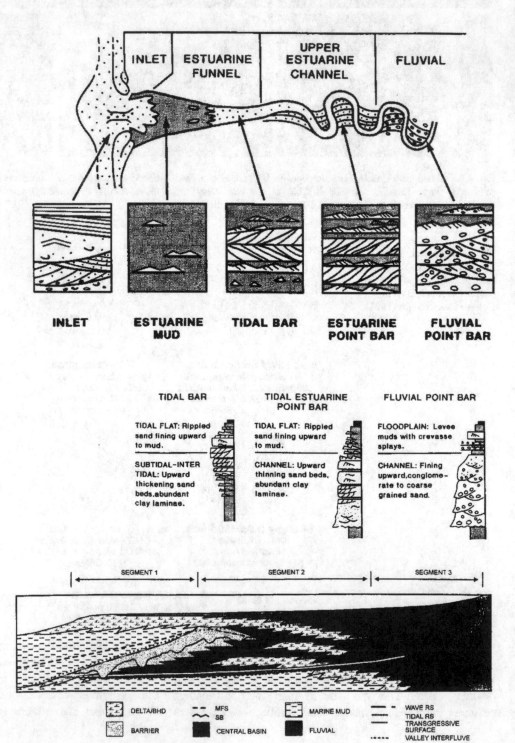

reservoirs in the Western Interior Basin, particularly in the Lower Cretaceous Mannville Group of Alberta, occur in broad linear "channels" and contain evidence of tidal influence during sedimentation. Careful mapping revealed that these channels are deeply incised valleys that form part of regional erosion surfaces, and that they originated as estuaries at the margins of the Western Interior Seaway (e.g., Rosenthal 1988). Incised valleys in the sedimentary record are important indicators of episodes of low relative sea level, and as such may provide important evidence in the mapping of sequences and their component systems tracts (Chap. 5).

4.2.8 Continental Shelf Environment

Most shelves are either tide dominated (e.g., North Sea, Georges Bank) or storm dominated (e.g., U. S. Atlantic shelf, Gulf Coast). A few areas of the continental shelf, such as the Agulhas Bank of southeast Africa, are dominated by strong unidirectional oceanic currents that spill up onto the shelf (e.g., Flemming 1980; Martin and Flemming 1986). In tide-dominated shelves, the principal sedimentary processes are the development of sand waves and sand banks (ridges), whereas in storm-dominated shelves, the main process is geostrophic flow (Fig. 3.33), leading to the development of storm cycles containing HCS (Walker and James 1992; Johnson and Baldwin 1996) and to storm-generated tidal sand ridges. Plint (2010) provided a review of modern work on wave- and storm-dominated shoreline and continental-shelf deposition. Dalrymple (2010) discussed tidally dominated shelf and coastal systems. We now realize that significant volumes of mud are transported across and deposited within shelf settings.

Tidal currents may reach speeds of 1 m/sec, quite adequate to transport sand and mud (the latter in the form of floccules). Sand waves, which may be as much as 15 m high, and large sand ridges, up to 50 m high, occur in areas of particularly vigorous tidal currents, such as the continental shelf around the British Isles (Figs. 4.30, 4.31). In the North Sea, the tides are semi-diurnal, with separate tidal "waves" entering this sea from the English Channel and around Northern Scotland directly from the Atlantic Ocean.

Storms may be a particularly important sedimentary process, scouring the sea floor to depths perhaps as much as 100 m below sea level. The main sedimentary effect of storms is to generate hummocky cross-stratification (HCS) by the combined flow process described in Chap. 3 (see Figs. 3.32, 3.33). Hummocky cross-stratification is the name given to the main, large-scale crossbedded structure formed by storms, but it is commonly only part of a sequence of deposits formed during storm activity. A block diagram model of a **storm cycle** is provided in Fig. 4.32, variations on the storm theme are shown in Fig. 4.33 and

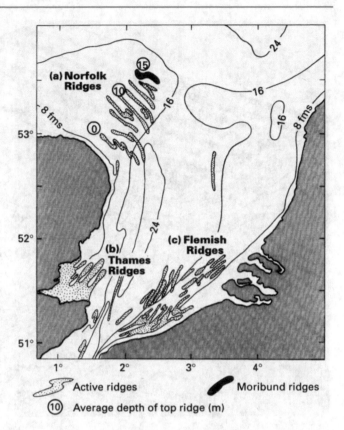

Fig. 4.30 Giant sand ridges, up to 50 km long, in the North Sea, an area characterized by strong semi-diurnal currents (Reading, 1986)

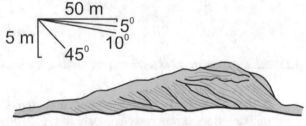

SHELF TIDAL SAND WAVE
(interpreted from seismic-reflection data)

Fig. 4.31 Tidal sand wave on the continental shelf of Brittany, France (diagram drawn from reflection-seismic record). Adapted from Berné et al. (1988)

some examples of HCS in drill core from Alberta are illustrated in Fig. 4.34.

The storm cycle typically begins with a scoured base. This reflects the buildup of the storm to maximum strength, during which gutter casts may be eroded and a basal lag of coarse particles transported. The deposition of HCS reflects the beginning of the waning of the storm, with the rest of the cycle recording the completion of the waning phase. In nearshore environments, it is common for only the bases of the HCS domes to be preserved (Fig. 4.33, column A). The

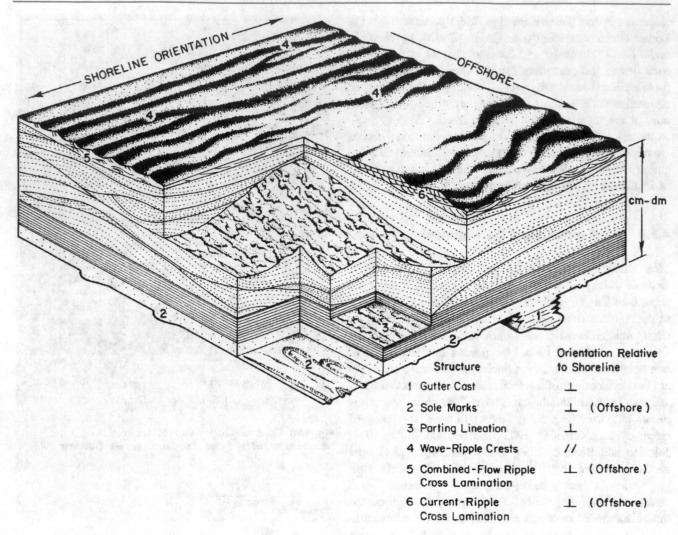

Structure	Orientation Relative to Shoreline
1 Gutter Cast	⊥
2 Sole Marks	⊥ (Offshore)
3 Parting Lineation	⊥
4 Wave-Ripple Crests	//
5 Combined-Flow Ripple Cross Lamination	⊥ (Offshore)
6 Current-Ripple Cross Lamination	⊥ (Offshore)

Fig. 4.32 Block diagram model of a storm cycle. Leckie and Krystinick (1989, Fig. 4, p. 868)

cycle fines upward through plane-laminated and rippled sand, not unlike a Bouma sequence (compare to Fig. 3.48). In deeper water settings of the mid to outer shelf, only the finer grained upper portions of the cycle may be deposited.

4.2.9 Continental Slope and Deep Basin Environment

The ultimate fate of clastic debris is to be shed of the continental shelf and down the continental slope to the deep basin floor of the ocean, to depths of up to 5 km (to even greater depths in the trenches above subduction zones). The most important transport paths are submarine canyons, but detritus may also be swept over the continental shelf by storms and by oceanic currents, forming sediment prisms on the brink of the continental slope that periodically fail by slumping (Fig. 4.35). Such events may or may not evolve

into sediment-gravity flows, as described in Sect. 3.5.5. The most volumetrically significant deposits in deep oceanic basins are accumulations of sediment-gravity flow deposits. The traditional term for these is **submarine fan**, the term implying deposition on radiating patterns of distributaries. However, given the wide variety of depositional patterns that have been identified on modern continental slopes, many workers now refer to these deposits by the more general term **turbidite systems** (Bouma et al. 1985; Arnott, 2010). This is the characteristic mode of sedimentation at the mouths of submarine canyons, which commonly serve as point sources for much of the sediment. Giant submarine fans have developed at the mouths of many of the world's large rivers, as described below. Owing largely to the ability of turbidity currents to travel for thousands of kilometers across the sea floor, these fans include the largest depositional systems on earth, including the Amazon, Indus and Bengal fans (Bouma et al. 1985).

Fig. 4.33 Variations in the composition of storm cycles with depth, across the continental shelf from shoreface to storm wave base. Abbreviatons: WS = wave scour, CS = current scour, B = basal lag, G = grading, P = planar lamination, HCS = hummocky cross-stratification, WX = wave cross-lamination, CX = current cross-lamination, M = bioturbation, A = amalgamation. Brenchley (1985)

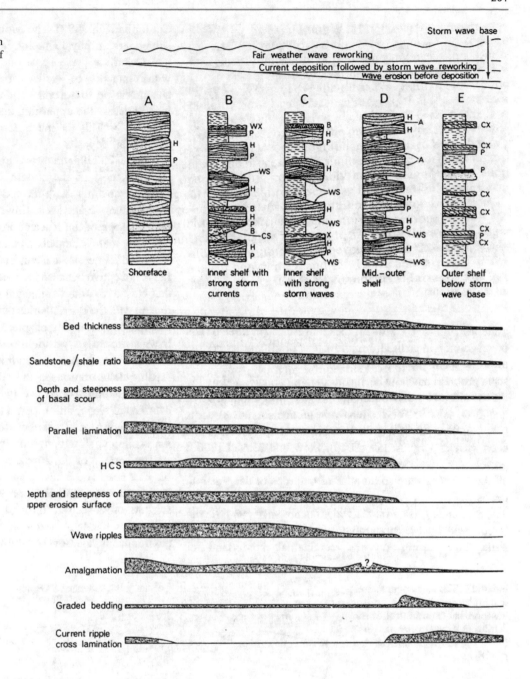

Deposition of large sand sheets also occurs, especially where the deep basin is fed by multiple sediment sources. An example of this would be the floor of the Labrador Sea, between Labrador and Greenland. The continental shelf on both sides of this sea is covered with a thick blanket of glacial debris, which underwent repeated slope failure during and following multiple phases of glaciation and glacioeustatic sea-level change. This resulted in gully erosion and numerous individual sources of sediment, feeding a network of tributaries to a few major channels flowing toward the center and along the axis of the sea, and ultimately out into the Atlantic Ocean (Chough and Hesse, 1976).

Early ideas about the architecture of submarine fans were based almost entirely on ancient fan deposits in California and Italy (Mutti and Ricchi Lucchi, 1972; Walker, 1978). The Navy fan, of California, was one of the first modern fans to be studied in detail, back in the 1970s, but at that time navigational and remote sensing instrumentation was much more primitive than at present, and sedimentary details were based on very limited piston coring and grab sampling. The scale problem has bedeviled the study of submarine fans, because typical outcrops and the range of sizes of features resolvable by marine photography are much too small to encompass the main architectural features of most fan

Fig. 4.34 Examples of hummocky cross-stratification in Cretaceous oil sands, Alberta

systems (Fig. 4.36). Modern fans range through three orders of magnitude in scale, from a few tens of kilometers across, to a few giants several thousands of kilometers across. The scale problem has now been considerably ameliorated by the use of side-scan sonar methods for mapping modern fans, and 3-D seismic for the analysis of ancient fan systems (Sect. 6.3). A modern review of sedimentary processes has been provided by Arnott (2010). Stow and Mayall (2000) presented an overview of fan architecture and models. Nilsen et al. (2007) assembled an atlas of outcrops of deep-marine deposits.

Sedimentary lobes on the Navy fan are small, and the fan is confined by the topography created by extensional and strike-slip faulting on the continental borderland of California (Fig. 4.37). The Monterey fan is more typical of submarine fan physiography, in its radiating wedge of sediment, and the presence of one active distributary channel, with remnants of earlier, inactive channels. The main channel is up to several hundred meters deep, where it is incised across the upper fan, and is up to nearly 4 km wide (Fig. 4.37). This is much larger than most outcrops of ancient fan deposits.

Figure 4.38 illustrates two giant fans. The Mississippi fan occupies most of the eastern Gulf of Mexico, its eastern margin coinciding with the foot of the continental slope of western Florida and its southwest flank lapping up onto the base of slope of the Yucatan Peninsula of Mexico. This fan has one main channel. Shallow-seismic exploration has identified numerous ancient, buried channels. Currently, the fan is not active because its main feeder channel, the canyon that cuts across the continental shelf of Louisiana, does not line up with the distributaries of the Mississippi Delta. As we showed earlier in this chapter, the Mississippi Delta lobes have switched in position several times in the last few thousand years. A few hundred years from now the prograding delta may reach the edge of the continental shelf, and then sediment will once again be shed directly down the continental slope, and a new fan channel may develop dispersing sediment across the Mississippi fan. Until then the fan is simply, slowly, accumulating a thin blanket of mud.

The surface of the Amazon fan is crossed by numerous old channels. These have a surprisingly "fluvial" appearance to them. Figure 4.39 shows a side-scan sonar image and a seismic cross-section through one of the channel-levee complexes of the Amazon. It is now known that in some environments, the flow of a turbidity current may last a long

Fig. 4.35 The range of processes that influence the sorting, transport and deposition of clastic detritus at a continental margin (Stow and Mayall 2000, Fig. 4, p. 128)

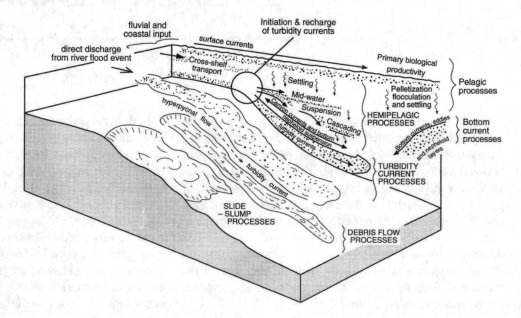

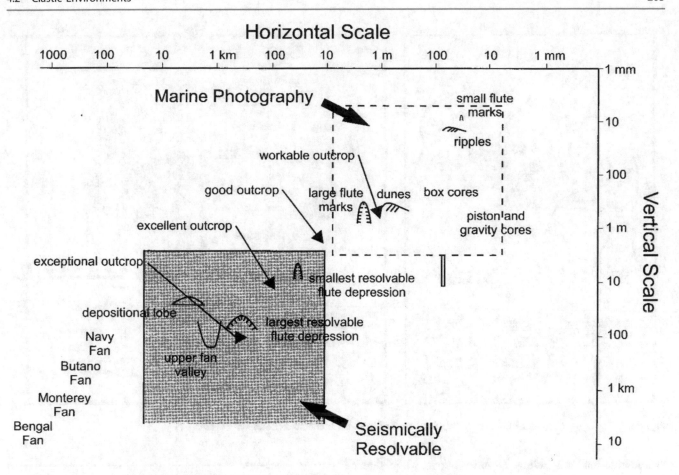

Fig. 4.36 Vertical and horizontal scales of deep-marine systems, modern and ancient, illustrating the scale problem that arises when using available observational methods to reconstruct submarine-fan architecture (Barnes and Normark 1985)

time because of an almost continuous supply of sediment at the upstream source. The hydraulic behavior of the flows, including internal turbulence and the tendency to meander, is therefore similar to that of the traction currents that characterize fluvial flow. In the case of the Amazon, the fan is fed by a very deep canyon, linked directly to the mouth of the Amazon River. The sediment supply was, therefore, not affected by the late Cenozoic fluvioglacial sea-level changes, and has been essentially continuous for millions of years.

The Bengal fan is the world's largest depositional system (Fig. 4.40). This giant depository consists of sediment eroded from the rising Himalayan Mountains, and carried to the sea by the Ganges and Brahmaputra/Jamuna Rivers. It is nearly 3000 km long, and more than 10 km thick. Deposition commenced in the Eocene following the initial collision of India with Asia (Curray 2014).

Whereas the first facies models for submarine fans were based on largely hypothetical vertical profiles (e.g., Walker, 1978), modern marine surveying and 3-D seismic data have indicated that fans have a highly variable composition and architecture. Stow and Mayall (2000) subdivided the larger components of deep-marine systems into a series of

architectural elements, including fan elements, contourite mounds and bounding surfaces (Fig. 4.41). They noted (p. 130)

> These elements are commonly a few hundred metres to several kilometres in width, a few metres to a few hundred metres in relief or thickness and may be irregular in outline, approximately equidimensional or markedly elongate. Each element may occur at a range of scales and within a hierarchy of similar features The sedimentary composition, including facies associations and vertical sequences, that make up any one element, can vary, typically within a limited range.

Stow and Mayall (2000, p. 132) commented that "The natural variability of channels (sensu lato) is enormous: they include steep-sided v-shaped canyons, deeply incised -shaped delta-front troughs, simple straight to slightly sinuous slope gullies, braided to meandering channel complexes, shallow ephemeral high-gradient fan-delta channels, extremely low-gradient ocean basin channels, and so on."

The large-scale architecture of deep-marine systems is highly variable. As Arnott (2010, p. 302) has argued, this variability in shape and composition renders the terms

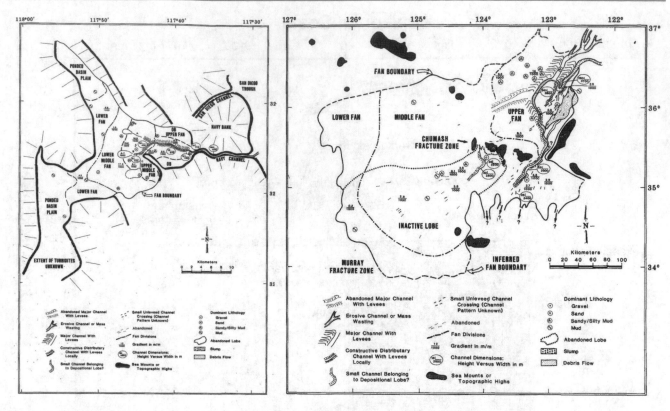

Fig. 4.37 The Navy (left) and Monterey (right) fans, on the California borderland (Bouma et al. 1985)

"submarine fan" misleading. As noted above, his preference is for the term **turbidite system**, following Bouma et al. (1985). A primary control on large-scale architecture is grain size, and the nature of the sediment feeder system is also critical. Classifications of deep-marine systems, based on these two main variables, were developed by Reading and Richards (1994), Richards et al. (1998), Stow and Mayall (2000) and Arnott (2010) (Fig. 4.42). A subdivision of submarine fans into upper, middle and lower fans was suggested by Walker (1978), and is still in use (Fig. 6.34). The upper fan is where the main feeder channel is flanked by levees and is an area of sediment bypass. In some cases, the main channel may be incised into earlier fan deposits within this area. The middle fan is an area of flow expansion and net deposition, typically from a set of radiating distributary channels. Beyond the mouths of these channels is a nearly flat area, where flow is unconfined and deposition is predominantly sheet-like in geometry and this is defined as the lower fan.

Turbidite systems are an important component of many continental-margin stratigraphic sequences. Controversy has surrounded their appropriate place in the development of systems tracts (Sect. 5.3.1), but it is now recognized that their development may occur at any stage in the base-level cycle, depending on the local physiography and the control of sediment supply (e.g., see Fig. 5.15).

4.3 Carbonate Environments

4.3.1 Conditions of Carbonate Sedimentation

The facies analysis of carbonate sediments must, of necessity, take into account some differences in the ways siliciclastic and carbonate sediments accumulate (Fig. 3.12). Carbonate sediments are primarily biochemical in origin. Light intensity, temperature, salinity and nutrient supply are critical controls, and limit the zone of rapid carbonate sedimentation to depths of less than about 20 m (Fig. 4.43a). Warm-water carbonates are generated within the latitudinal range 30°N to 30°S (Jones 2010). Sedimentation follows what Schlager (1992) called the **law of sigmoidal growth** (Fig. 4.43b):

> "Growth is slow at first, then accelerates, often exceeding the rate of change in the forcing function; finally, growth decreases as the system reaches the limits of the newly formed niche. In carbonate sedimentology, this pattern is known as the start-up, catch-up, keep-up stages of growth ..." (Schlager 1992)

The interaction of the controls illustrated in Fig. 4.43 is the reason why most carbonate sedimentation takes place in the so-called **carbonate factory** of the shallow-marine continental shelf. Sedimentation is most rapid in the **photic**

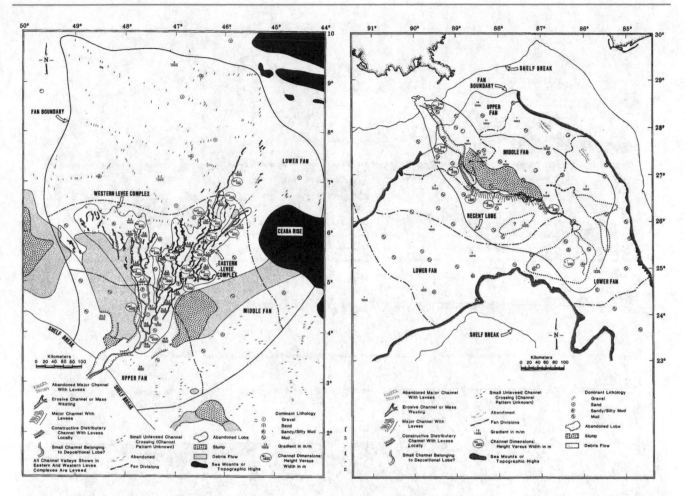

Fig. 4.38 The Amazon (left) and Mississippi (right) submarine fans (Bouma et al. 1985)

zone, which extends to depths of between 80 and 100 m, with maximum productivity at depths of less than 20 m (Fig. 4.44). Most carbonate grains are originally biogenic, but may be modified by contemporary marine processes (e.g., ooliths, intraclasts). Under conditions of warm, clear water, organic productivity is very rapid (Schlager 2005; Jones 2010). In fact, carbonate generation and accumulation under ideal conditions can keep up with almost any geologically imposed rate of subsidence or sea-level change. Near the coast, carbonate sedimentation is inhibited by clastic influxes, and on the continental margin, slower sedimentation under conditions of deeper water gradually exaggerates the differences between shelf and slope as subsidence takes place. This gradually results in the development of a **carbonate platform**, as shown in Fig. 4.44. Commonly this is rimmed by a platform-margin reef, as discussed below. During the Early Paleozoic, much of cratonic North America constituted a vast carbonate platform, to which the term "Great American Bank" has been applied (Sect. 7.9).

Changes in environmental conditions (temperature, nutrient supply, salinity, etc.) may bring carbonate sedimentation to a halt, with the subsequent development of a **hardground**. These are areas of lithified substrate, characterized by submarine cementation, typically with distinctive paleoecological communities, ichnofacies and mineralized crusts (Fig. 2.14). They may extend over substantial distances, and can provide important correlatable surfaces in the rock record (Jones, 2010).

The study of carbonate facies cannot be divorced from considerations of diagenesis. In many environments, carbonate sediments start to undergo cementation and alteration as soon as they are deposited. Rapid cementation of carbonate debris on the shoreface is the most common origin of **beach rock**. The ubiquity of organic processes in carbonate seas introduces a significant complication, and many carbonate deposits are modified penecontemporaneously by submarine cementation and other types of early diagenesis. For example, reef and carbonate-slope buildups commonly have steeply dipping sediment–water interfaces, much

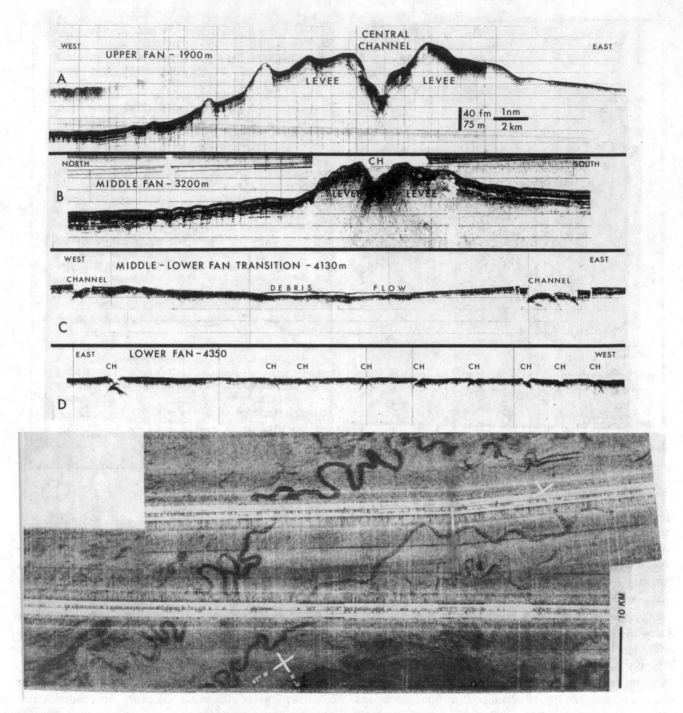

Fig. 4.39 Images of the Amazon fan. *Below*: Side-scan sonar image of several of the old distributaries; *Above*: seismic cross-section through a channel-levee complex (Bouma et al. 1985)

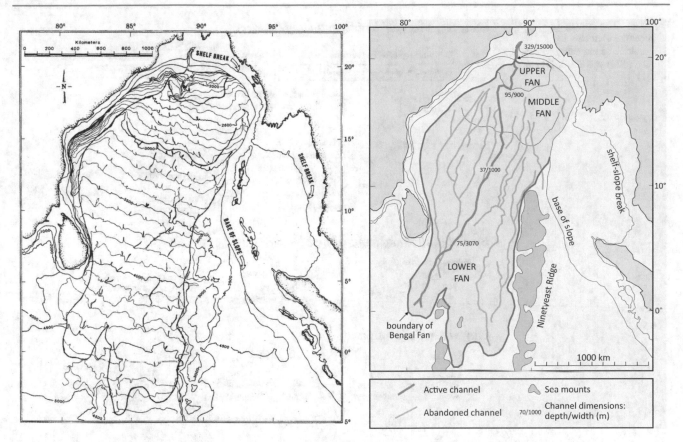

Fig. 4.40   The Bengal fan, between India and Myanmar. Left: bathymetry, right: physiography (Bouma et al. 1985)

steeper than the angle of repose of loose talus, sometimes even vertical or overhanging. This is due to reef growth, organic binding and early cementation. Most research now makes extensive use of special microscopic techniques, such as cathodoluminescence and scanning electron microscopy, plus a range of geochemical techniques, including the analysis of stable isotopes and trace elements, in order to track these diagenetic changes (Dickson, 1985). Cathodoluminescence is widely used to see through diagenetic modifications to primary depositional fabrics (Machel, 1985). Analyses of oxygen and carbon isotopes are used to determine depositional temperature and to distinguish early from late diagenetic cementation (Arthur et al. 1983). In many other respects, carbonate facies studies can be pursued in a similar manner to siliciclastic studies, in that many carbonate deposits are accumulated by clastic processes controlled by waves, tides and gravity.

One of the most important contributions to the study of carbonate facies models was the book by J.L. Wilson (1975) on Phanerozoic carbonates. Wilson demonstrated that carbonate sediments could be described using a standard set of 24 microfacies types (Fig. 3.4). He showed that lithofacies, structures and biota defined nine standard facies assemblages, which could be used to describe practically all

carbonate rocks in the ancient record, although not all nine assemblages are present in every rock unit. Wilson's models are widely used in interpreting the ancient record. More recent syntheses and reviews of carbonate sedimentation have been provided by Scholle et al. (1983), Tucker (1985), Wright and Burchette (1996) and James et al. (2010). Pomar and Hallock (2008) documented the changes in the "carbonate factory" with time, as biogenic carbonate-producing environments changed in response to plate-tectonic movements (the opening and closing of seaways, changes in oceanic currents), climate change and extinctions events (which at critical times, such as at the Permian–Triassic extinction, led to wholesale changes in the assemblage of carbonate-producing organisms).

Sellwood (1986) suggested that five major environmental zones could be recognized in shallow-marine settings: supratidal flats, intertidal flats, the marine shelf, the carbonate sand belt and the reef belt. To these should be added the continental slope and pelagic environments, most of which are deep marine (Jenkyns, 1986). Schlager (2005), following Wilson (1975), broadened the subdivision into nine subenvironments (Fig. 3.13). In many respects, shelf and slope carbonate environments are similar to their siliciclastic counterparts. Thus, the sedimentology of carbonate

Fig. 4.41 The principal
architectural elements in
deep-marine sedimentary systems
(Stow and Mayall 2000, Fig. 7,
p. 131)

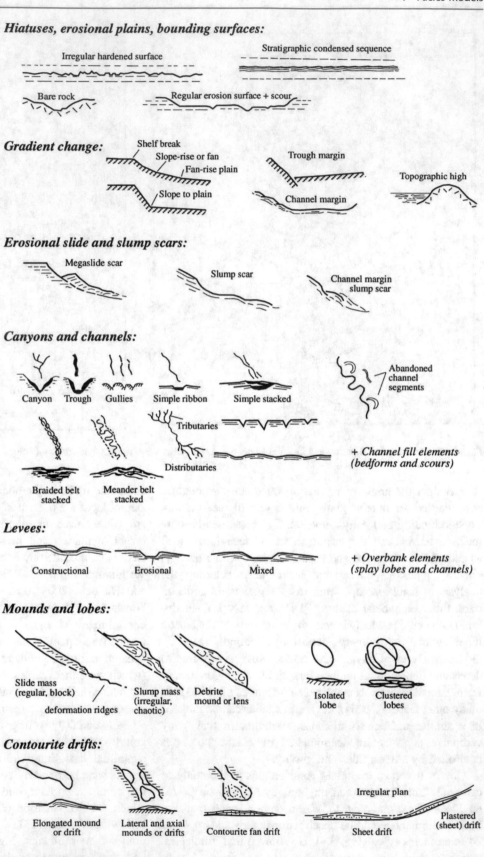

Hiatuses, erosional plains, bounding surfaces:

Irregular hardened surface

Stratigraphic condensed sequence

Bare rock

Regular erosion surface + scour

Gradient change:

Shelf break
Slope-rise or fan
Fan-rise plain
Slope to plain

Trough margin
Channel margin

Topographic high

Erosional slide and slump scars:

Megaslide scar

Slump scar

Channel margin
slump scar

Canyons and channels:

Canyon   Trough   Gullies   Simple ribbon   Simple stacked

Abandoned
channel
segments

Tributaries

Distributaries

+ *Channel fill elements
(bedforms and scours)*

Braided belt
stacked

Meander belt
stacked

Levees:

Constructional   Erosional   Mixed

+ *Overbank elements
(splay lobes and channels)*

Mounds and lobes:

Slide mass
(regular, block)

deformation ridges

Slump mass
(irregular,
chaotic)

Debrite
mound or lens

Isolated
lobe

Clustered
lobes

Contourite drifts:

Elongated mound
or drift

Lateral and axial
mounds or drifts

Contourite fan drift

Irregular plan

Sheet drift

Plastered
(sheet) drift

Sheets and drapes:

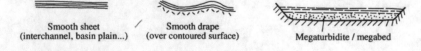

Smooth sheet
(interchannel, basin plain...)

Smooth drape
(over contoured surface)

Megaturbidite / megabed

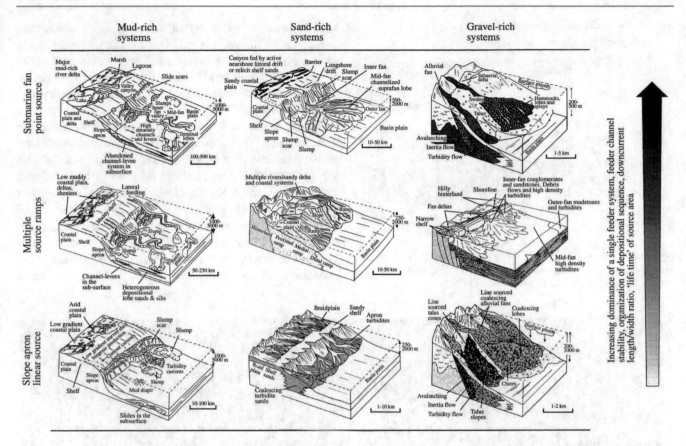

Fig. 4.42 A classification of deep-marine sedimentary systems, based on sediment grain size and the nature of the sediment sources (Stow and Mayall 2000, Fig. 9, p. 133; based on Reading and Richards 1994)

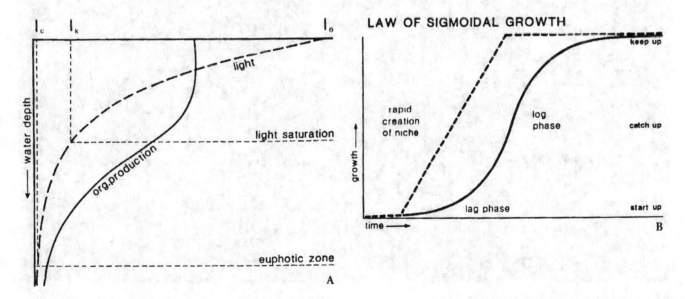

Fig. 4.43 a Growth rates for carbonate sediments, as dependent on light intensity. **b** The law of sigmoidal growth in carbonate sediments. Populations of organisms respond to the opening up of new living space slowly at first. In most situations, sediment production can exceed the rate of increase of new space but is limited by the rate of growth of space (Schlager 1992) AAPG© 1992, reprinted by permission of the AAPG whose permission is required for further use

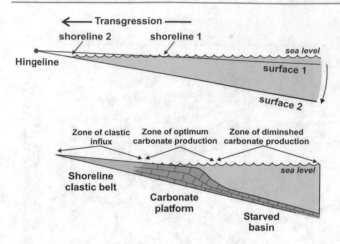

Fig. 4.44 The zone of optimal carbonate production is located between the clastic-contaminated shore zone and the rapidly subsiding continental margin. Adapted from Wilson (1975)

shelves involves the study of storm deposits, some containing HCS (Aigner, 1985; Handford, 1986) and carbonate sand shoals similar to siliciclastic sand waves and banks (Hine, 1977; Cook et al. 1983). Carbonates also occur as eolianites, most of which originated as coastal dune deposits (McKee and Ward, 1983).

By far the most important environments for carbonate sedimentation are warm, clear, shelf (platform) seas. Carbonate generation is most rapid in tropical settings, such as on the Bahamas Banks or Platform (Fig. 4.45), around South Pacific atolls, and along the Great Barrier Reef of Australia (Fig. 4.46). The Great Barrier Reef, at 344,000 km^2, is the largest area of the earth's surface currently undergoing predominantly carbonate sedimentation. However, recent studies have shown that some cooler-water settings, such as the southern continental margin of Australia and the west coast

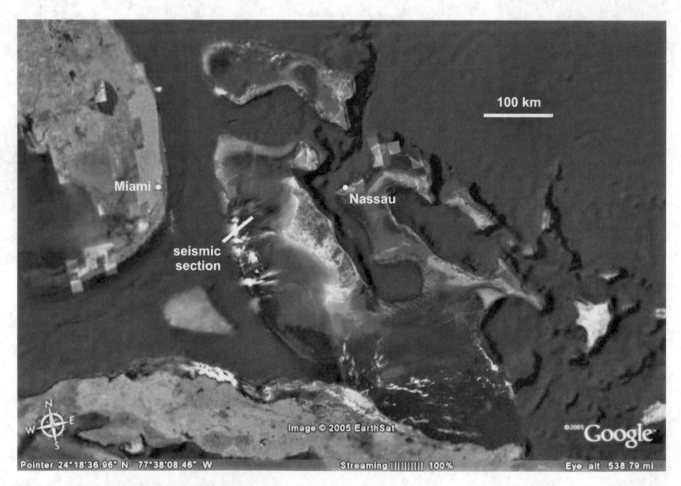

Fig. 4.45 The Bahamas Banks. Areas of shallow water are distinguished by paler colors. The seismic section is Fig. 4.49. From Google Earth

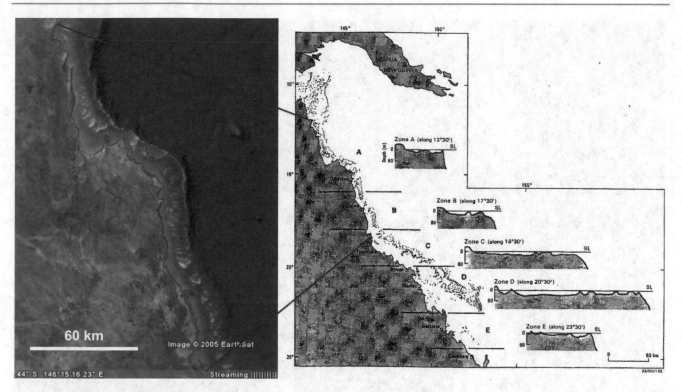

Fig. 4.46 The Great Barrier reef, Queensland, Australia. Diagram at right from Davies et al. (1989)

of Ireland, are also sites of carbonate sediment accumulation, although here the processes of sediment generation and accumulation are proceeding somewhat more slowly (James and Lukasik, 2010). There are no modern analogues for the vast epeiric seas that covered the North American cratonic interior for much of the early Paleozoic (Sect. 7.9).

4.3.2 Platforms and Reefs

An excellent example of the range of environments shown in Fig. 4.44 is illustrated in Fig. 4.47. The Nicaragua continental shelf has a coastal, mud-dominated zone and a marginal barrier reef system. The wide carbonate platform is

Fig. 4.47 The continental shelf of Nicaragua, showing the composition of surface sediments. Note the belt of clastic-dominated sediment along the coast, and the fringing reef at the edge of the shelf (Roberts 1987)

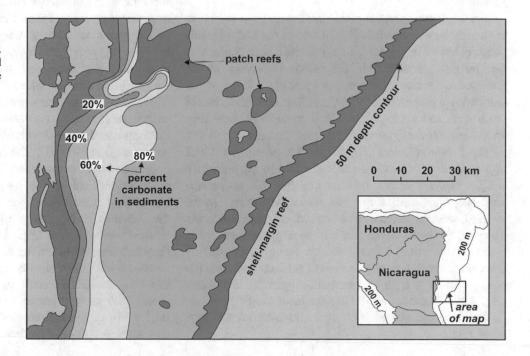

Fig. 4.48 The Florida keys, showing the shallow waters behind the reef, merging with the mangrove swamps of the Everglades. The bar next to the compass diagram is 20 km long

dotted with small reefs or bioherms (small organic buildups). This is a good example of a **rimmed platform**. Many other examples, including a suite of facies models, are provided by Jones (2010).

Satellite images showing two other examples of carbonate environments are shown in Figs. 4.46 and 4.48. The Florida platform has a fringing reef (the Keys) and a back-barrier area that merges gradually into the mangrove swamp of the Everglades. Carbonate production takes place today on the seaward side of the barrier. The Great Barrier Reef is a belt of patch reefs and a fringing reef that extends for more than 1600 km along the Queensland coast of Australia (Fig. 4.46).

The Bahamas Banks (Fig. 4.45) constitutes a broad area of very shallow water (much of it less than 10 m), but with a steeply dipping slopes on the margins. Slopes on the northeast, windward margin are particularly steep. On the leeward, western margin, carbonate sediment production has exceeded the space created for sediment by subsidence, and much of this material has ended up being transported by tides and currents off the edge of the platform, where it has contributed to a zone of westward-prograding clinoforms (Fig. 4.49). These have extended the bank margin by nearly 30 km since sedimentation on the banks commenced in the mid-Cretaceous.

The Nicaragua shelf and the Great Barrier Reef are examples of what are termed **rimmed platforms**, whereas the Bahamas Banks exemplify **isolated platforms**, which are surrounded by deep-water environments on all sides. In the ancient record, rimmed platforms are more commonly located on extensional continental margins. Isolated platforms typically develop over fault blocks. Those forming the basement to the Bahamas developed during the initial break-up of Pangea early in the development of the central Atlantic Ocean.

We touch on the importance of petrology in the facies analysis of carbonate sediments in Sect. 3.5.2. The Bahamas Platform was the site of a classic study of modern carbonate sediments (Purdy 1963). This work demonstrated that carbonate sediment composition is dependent mainly on the water depths and energy levels of the local setting. In high-energy environments, such as the windward margin of the Bahamas Bank, framework-building corals have constructed wave-resistant reefs. On the northern and western (leeward) margins of the bank, reef debris and oolitic accumulations form belts of carbonate sand, while in the interior of the lagoon, micrite muds accumulate, together with allochems such as invertebrate shells and related fecal pellets (Fig. 4.50). A generalized model for the subenvironments that develop around a modern reef is shown in Fig. 4.51,

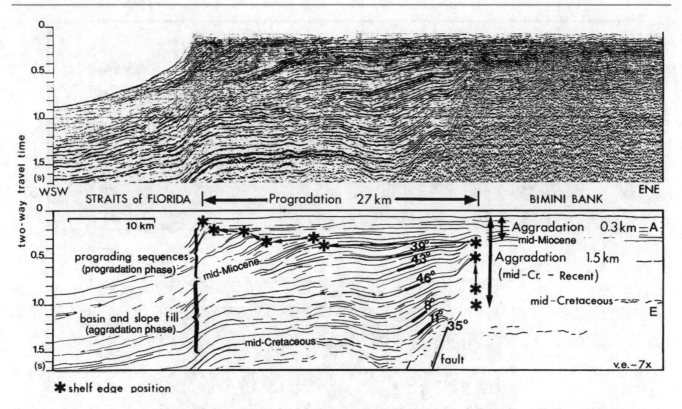

Fig. 4.49 Seismic cross-section through the leeward margin of the Bahamas Bank. The location is shown in Fig. 4.45. Note the extensive lateral progradation of the platform as a series of clinoforms (Eberli and Ginsburg 1989)

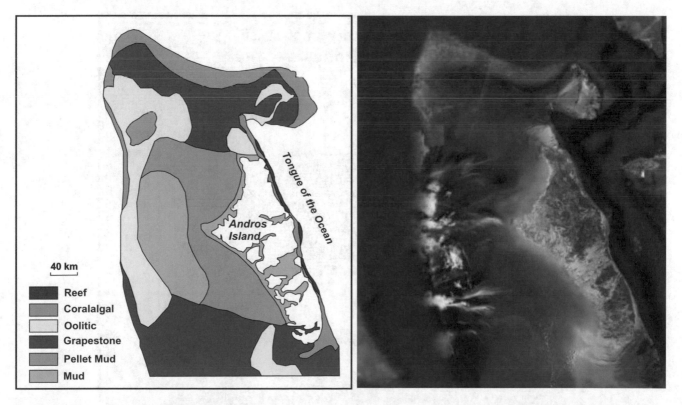

Fig. 4.50 Facies analysis of the surface sediments accumulating over part of the Great Bahamas Bank (after Purdy 1963). The image at right shows the same area as seen on Google Earth

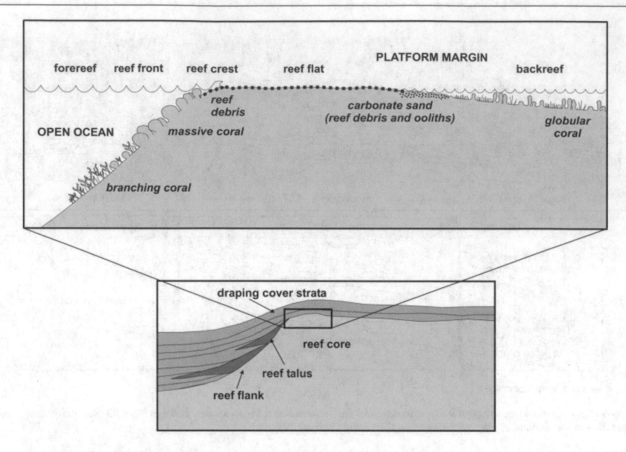

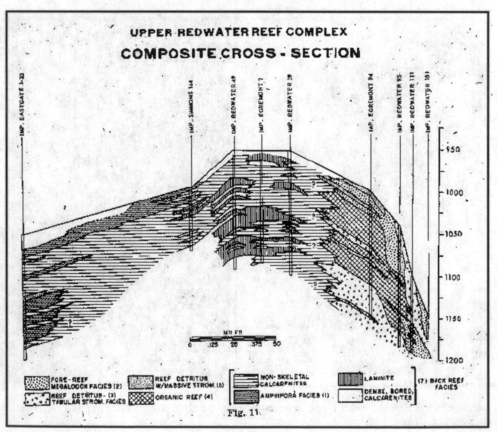

Fig. 4.51 Top: A cross-section through a typical modern reef. In this diagram, the open sea is to the left, and the reef is backed by a lagoon. Adapted from James and Bourque 1992); Bottom: A cross-section constructed from core analysis of a Devonian reef north of Edmonton, Alberta. The forereef slope is to the right, and is characterized by talus deposits. The main reef core does not form the structural high in this example, because of post-depositional structural disturbance (Klovan 1964)

together with a diagram constructed from a detailed core analysis of a Devonian reef in Alberta (Klovan 1964). While corals form the main reef core at the present day, in the geological past other types of organisms fulfilled this function. In the Devonian, for example, stromatoporoids were particularly important.

Examples of Devonian reef limestones and dolomites from Alberta and northern Canada are illustrated in Figs. 4.52 and 4.53. Gyrfalcon Bluff, in Banks Island in the Canadian Arctic, is an example of a small reef that has been "exhumed," that is, modern erosion has caused the original reef to re-appear with approximately its original outline,

Fig. 4.52 Examples of reef deposits. a–e are Devonian, **f** is Ordovician. A, Gyrfalcon Bluff, Banks Island, Arctic Canada, **b**, Crinoidal debris forming the talus slope to a small reef at Princess Royal Islands, Arctic Canada; **c, d, e**, fossiliferous limestones near Norman Wells, NWT; **c** shows Kee Scarp, which exposes the reservoir rock that produces from fractures in the Norman Wells field. **f**, the cavity exposes the rubbly beds of a small bioherm near Niagara Falls, Ontario

Fig. 4.53 The dolomitised, stromatoporoid reef rocks of the Cairns Formation, at Grassi Lakes, near Canmore, Alberta. Blue dotted line in top view outlines stromatoporoid reef bodies. Note the abundant vuggy porosity in the close-up image

following removal of the sediments that covered it following deposition in the mid-Devonian. The Grassi Lakes outcrops, in Alberta, (Fig. 4.53) are classic examples of stromatoporoid reefs. The rocks here have been pervasively dolomitised, and exhibit large-scale vuggy porosity.

Given the importance of these reefs to the petroleum economy of Alberta, it is not surprising that there has been considerable exploration work carried out on these rocks in the subsurface. Figure 4.54 shows two examples of drill cores through dolomitized stromatoporoid reefs rocks. In these examples, much of the vuggy porosity has been filled with calcite cement, reducing their value as reservoir rocks, in contrast to the highly porous example shown in Fig. 4.53.

Reefs constructed by framework-building organisms span all Phanerozoic periods. From the Proterozoic through the Lower Paleozoic, limestones were also constructed by a process of sediment trapping by algae, building fossils called stromatolites. Figure 4.55 illustrates some examples of these. These were not true reefs, in the sense that this word is used today. Stromatolites are still being formed today, for example, in Shark Bay in Western Australia. But the algal mats provide food for some browsing organisms, and they tend, therefore, to grow and survive mainly in environments that are hostile to other forms of life, such as in saline pools.

4.3.3 Tidal Sedimentation

The Paleozoic craton of North America was covered by shallow tropical to subtropical shelf seas for hundreds of millions of years (Pratt and Holden 2008; Pratt 2010). Through most of this time, water circulation by tides and storms was able to maintain an open marine environment. The result was hundreds of meters of platform carbonates constituting units containing facies that, in some cases, can be traced for hundreds of kilometers. There are no modern analogies for depositional systems of this gigantic scale, but modern settings such as the Bahamas Banks can serve to illustrate the major processes and types of sediment that were deposited. Another good analogue is the southern margin of the Persian Gulf. Figure 4.56 is a model of coastal, tidal flat and lagoonal sedimentation generalized from studies of this hot, arid region. Figure 4.57 illustrates some examples of carbonate sediments deposited in environments of this type.

Fig. 4.54 Drill cores through stromotoporoid dolomites from the Middle Devonian Swan Hills formation of northern Alberta. The core in the box at left shows former "vuggy" porosity, now filled with calcite

Fig. 4.55 Algal mats were important rock-forming organisms during the Proterozoic and early Paleozoic. Here are examples of stromatolites from the Proterozoic (Dismal Lakes, NWT, top) and the Silurian (Somerset Island, NWT, bottom)

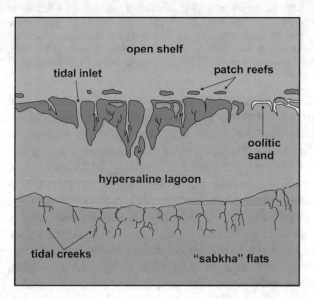

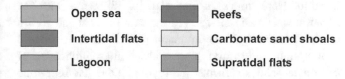

Fig. 4.56 Depositional model for a hot, arid, tidal flat environment, based on the modern Persian Gulf. Adapted from Purser 1973

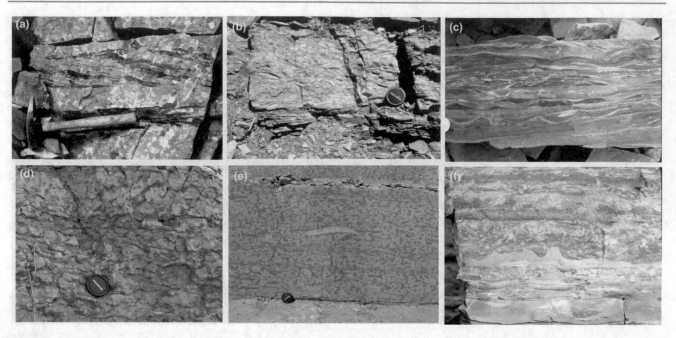

Fig. 4.57 Examples of ancient limestones and dolomites formed in shallow, tidal environments. **a** limestone with crossbedding, in which the bedding is emphasized by chert layers; Lisburne Limestone (Mississippian), North Slope, Yukon; **b** Intraclast dolomite breccia, Ordovician, Somerset Island NWT.; **c**, **f**, rippled dolomites with small scour surfaces, Silurian, Somerset Island, NWT; **d** Mottled limestone, a texture formed by burrowing, Ordovician, Ottawa; **e** Mottled, bioturbated dolomite, Red River Formation, Tyndall Stone, Ordovician, Garson Quarry, near Selkirk, Manitoba

There are some analogies between the Persian Gulf tidal coastline and clastic barrier systems, such as those characterizing the Atlantic continental margins of the southeastern United States. The main point of comparison is the significant sedimentary influence of reversing tidal currents, and their role in forming barrier systems cut by regular tidal channels, through which strong tidal currents reverse direction, typically twice a day. Detrital carbonate material is transported by such currents, and the resulting deposits show abundant evidence of current action, in the form of scour surfaces and cross bedding (Fig. 4.57 a, c, f). Another analogy is the presence of tidal flats, which are exposed to the air during the falling tide, and therefore subject to desiccation. Here, however, lies a major difference with the Atlantic barriers, in the form of the extreme heat and aridity of the Persian Gulf environment, which leads to intense evaporation, and to drying out and incipient lithification of carbonate tidal sediments. Embryonic tidal flat limestone beds formed in such environments are commonly broken up by currents during subsequent high tides, and form fractured or brecciated beds, in which the fragmented pieces may still be almost in their original position, or broken up and dispersed to form an intraclast breccia (Fig. 4.57b). Burrowing (Fig. 4.57e) may lead to pervasive mottling texture, as in some of the Ordovician limestones of the Ottawa area (Fig. 4.57d).

In the supratidal zone, high evaporation rates lead to precipitation of evaporite nodules forming the groundwaters that are continually drawn to the surface by evaporation and capillary action. This environment is termed a **sabkha**, based on the Arabic word for "salt-encrusted desert." Limestones formed here are also prone to early dolomitization, although this is not the primary mode of dolomitization that has affected vast swaths of Paleozoic carbonates across North America. These appear to have been dolomitized by large-scale post-depositional groundwater circulation.

Tidal deposits identified in the rock record are commonly cyclic, consisting of repeated shoaling-up successions (Pratt 2010). The successions commonly begin with subtidal carbonate mudstones or oolitic grainstones, passing up into a range of intertidal facies, including peloidal, bioturbated, stromatolitic or bioclastic wackestones to packstones, and capped by supratidal carbonate muds, commonly containing evidence of desiccation in the form of shrinkage cracks and evaporite nodules or casts (Fig. 4.58). There are several mechanisms that can generate such cycles (Fig. 4.59), including local island growth on the carbonate platform, progradation of the shelf or simple aggradation of an initially subtidal setting. Regional tracing of the cycles may be necessary to determine which of these mechanisms is appropriate in any given case.

Cyclic Sea Level

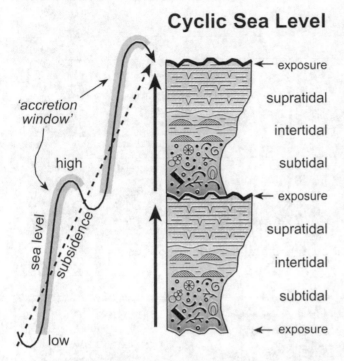

Fig. 4.58 The development of peritidal carbonate cycles by periodic rises in sea level superimposed on a background of uniform subsidence (Pratt 2010, Fig. 19, p. 416)

4.3.4 Carbonate Slopes

Active carbonate platforms where the "carbonate factory" is operating at maximum productivity typically generate carbonate sedimentary material at a rate faster than new space can accommodate it (hence the term **accommodation** for sedimentary space: see Sect. 5.2.1 for a discussion of this term). Much of this sediment eventually is deposited on the continental slope, where it is responsible for the lateral progradation of the platform margin (Figs. 4.49, 4.60). The carbonate factory typically achieves maximum productivity during high stands of sea level, which has given rise to the term **highstand shedding** for this process.

Carbonate slopes vary in the processes and the deposits formed there (Fig. 4.60). Deep-water carbonates comprise a variety of allochthonous, shelf-derived breccias and graded calcarenites, contourite calcarenites and hemipelagic mudstones cut by numerous intraformational truncation (slide) surfaces (Cook and Enos 1977; McIlreath and James 1984; Coniglio and Dix 1992; Playton et al. 2010). On windward slopes, currents may be strong enough to sweep away much or all of the sediment deposited down the slope, resulting in thin sedimentary accumulations, or even disconformity

surfaces. More commonly, carbonate debris accumulates and periodically slides down the slope, triggering sediment-gravity flows. These range from thin-bedded turbidites formed from the finest-grained material, to debris flows, in which boulders meters across may be present. Extreme examples of the slumping of giant pieces of carbonate platform-carbonate rocks are termed **olistoliths**. Sedimentary accumulations are likely to be more extensive on leeward margins, such as the western margin of the Bahamas Platform (Eberli and Ginsburg 1989), as can be seen in the substantial thickness of carbonate clinoform deposits in Fig. 4.49. Principaud et al. (2018) carried a detailed survey of one portion of this margin using sonar and high-resolution seismic surveying methods, and provided many insights into depositional processes.

One of the most well-known examples of carbonate-slope deposition is the Cow Head Breccia of Newfoundland. This was deposited on the extensional continental margin of the Iapetus (proto-Atlantic) Ocean during the Ordovician, and is now well-exposed within Gros Morne National Park in the western part of the province (Fig. 4.61).

In very exceptional circumstances, the transition from platform to slope or from slope to basin may be exposed. Such is the case in a few very large outcrops in the Canadian Arctic Islands. Figure 4.62 illustrates one such example, where carbonate platform and slope deposits of the Upper Paleozoic Nansen Formation pass down slope into the slope deposits of the Hare Fiord Formation. The basin center (Sverdrup Basin) is occupied by deep-water carbonates and by evaporites (Otto Fiord Formation: see next section). Note, in this illustration, the clinoform architecture of the slope deposits something that is rarely seen in outcrops.

4.4 Evaporites

Evaporites are the deposits generated by the evaporation of saline waters. Some vast evaporite deposits, hundreds of meters thick and extending for thousands of kilometers, have been formed at different times during the geological past by the total evaporation of geographically enclosed water bodies. The formation of a new ocean by sea-floor spreading can generate a water body that is partially enclosed early in its history. If this sea is located in a low-latitude setting especially if it straddles the belt of high-pressure dry air at 30°N or 30°S, evaporation can result in the formation of a major evaporite deposit at an early stage in the development of an extensional continental margin. Examples include the Jurassic evaporites of the Gulf of Mexico and the margins of

Fig. 4.59 Three contrasting models for the development of peritidal carbonate cycles (Pratt 2010, Fig. 20, p. 417)

Origin of Peritidal Cycles

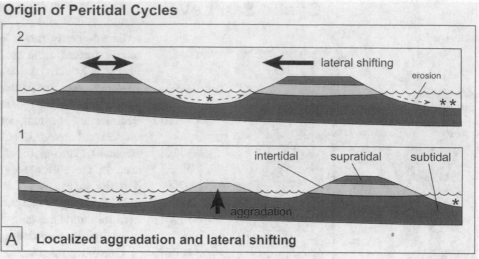

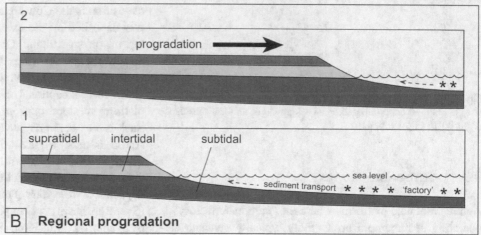

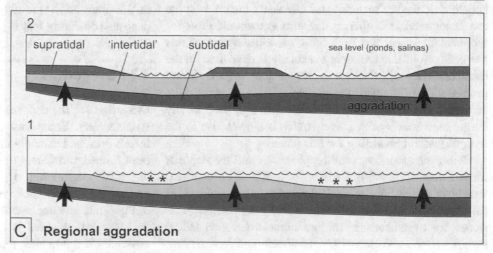

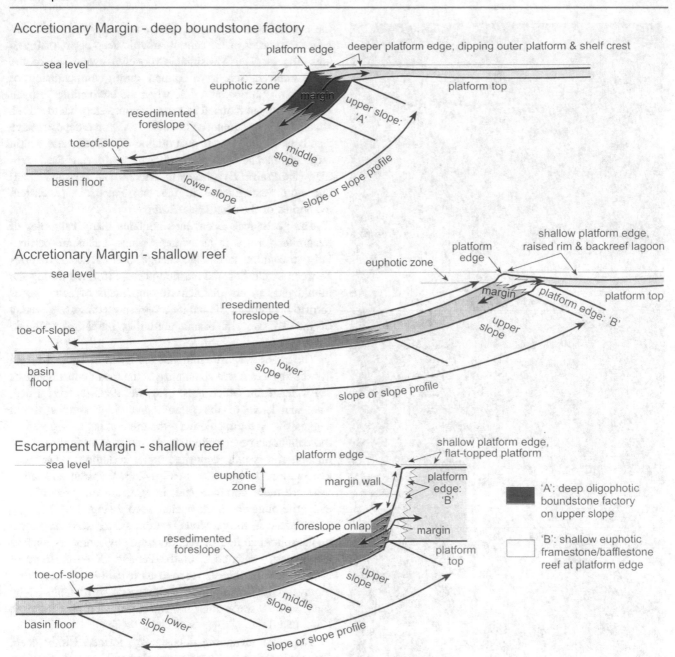

Fig. 4.60 Typical carbonate shelf-to-slope depositional models (Playton et al. 2010, Fig. 7, p. 453)

Fig. 4.61 Examples of carbonate-slope deposits, Cow Head Breccia, Newfoundland. **a** Turbidites, **b** debris flows, **c** a deformed clast within a debris flow

the central Atlantic Ocean, of Nova Scotia, and the Cretaceous evaporites of the central Atlantic deep basin, of Brazil and Angola. The Messinian (Miocene) evaporites of the Mediterranean Sea were formed during the collision of Africa with Europe, at a time when the intervening Tethyan Ocean had been reduced to a nearly completely enclosed sea situated in near-equatorial latitudes. Extensive deposits have also been formed by the evaporation of cratonic seas within large continents, such as North America during the Devonian (the Prairie Evaporites of Alberta and Saskatchewan) and northwest Europe in the late Paleozoic (Zechstein evaporites of the North Sea Basin).

The geological record also contains many examples of evaporites formed in lacustrine settings. Lakes are particularly susceptible to climate change, and cyclic alternations between evaporites and clastic intervals (formed during the humid phases) are distinctive components of some series formed in cratonic, rift and intramontane settings (e.g., many of the Triassic rift basins bordering the North Atlantic Ocean).

The major evaporite minerals are listed in Table 4.1, and 4.2 shows the relative abundance of the major constituents of sea water, from which most evaporite deposits are formed. The abundance of the sodium and chlorine ions would suggest that sodium chloride—the mineral halite—should be the commonest evaporite deposit, but it is a highly soluble mineral and is only deposited under conditions of extreme evaporation, as indicated in Fig. 4.63. Gypsum and anhydrite are more common and, in fact, are the predominant evaporite minerals in the ancient record (Fig. 4.64c).

As noted in the previous section, salt crystals may form on the supratidal flats of very hot, arid environments, such as the modern south coast of the Persian Gulf—the type of environment geologists refer to as a **sabkha**. These commonly dissolve away, and their crystal form replaced by sand, mud or some other mineral acting as a pseudomorph (Fig. 4.64a).

The major evaporite deposits are formed within deep, oceanic basins or on continental shelves, in the basins that may develop behind barriers, such as reefs, that restrict water circulation. Calculations of the amounts of seawater needed to generate the evaporite accumulations in some ancient basins indicate that simple evaporation of a volume of water equivalent to that of the basin is entirely inadequate. Multiple basin volumes of water typically are indicated. This may reflect cyclic drying out and refilling of the basin as a result of climate cyclicity, but it could also simply represent

Fig. 4.62 A dramatic facies transition from platform to basin. The Nansen formation consists of a bioclastic-carbonate debris slope, consisting partly of carbonate turbidites. The Otto Fiord Formation consists of interbedded anhydrite, shales and detrital carbonates. Sverdrup Basin (Carboniferous), Ellesmere Island, Arctic Canada (photo: G. R. Davies)

Table 4.1 Major evaporite minerals

Mineral	Chemical formula
Chlorides	
Halite	NaCl
Sylvite	KCl
Carnallite	$KMgCl_3.6H_2O$
Sulphates	
Anhydrite	$CaSO_4$
Langbeinite	$K_2Mg_2(SO_4)_3$
Polyhalite	$K_2Ca_2Mg(SO_4)_4.2H_2O$
Kieserite	$MgSO_4.H_2O$
Gypsum	$CaSO_4.2H_2O$

Table 4.2 Relative amounts of the major dissolved components of sea water

Cl^-	48.7%
Na^+	41.8
Mg^{2+}	4.7
SO_4^{2-}	2.5
Ca^{2+}	0.9
All others	1.3

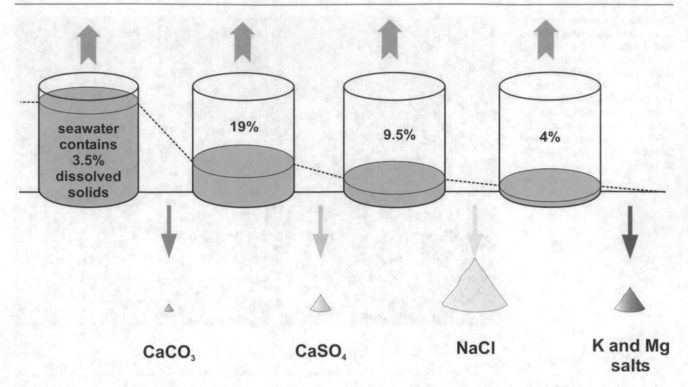

Fig. 4.63 The succession of evaporite salts precipitated from seawater. On evaporation, CaCO₃ is precipitated first. When evaporation has reduced the volume to 19% of the original amount, CaSO₄ begins to precipitate; at 9.5% of the original volume, NaCl starts to precipitate and so on. The volume of the piles represents the relative proportions of the precipitated salts. Adapted from www.geo.utexas.edu/courses

Fig. 4.64 Evaporites in outcrop. **a** Salt pseudomorphs in sandstone, Jurassic, Utah; **b** interbedded gypsum and calcisiltite, Silurian, Somerset Island, Arctic Canada; **c** Interbedded anyhydrite and shale, Carboniferous, Ellesmere Island, Arctic Canada; **d** Cyclic deposits of halite in mine, Devonian, Goderich, Ontario (photos C, D, by N. Eyles)

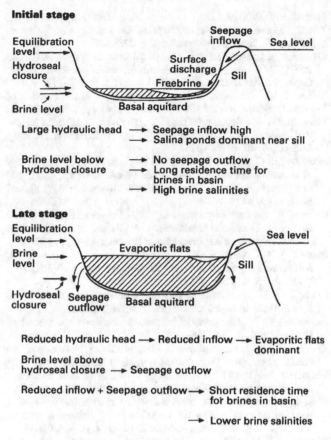

Initial stage
Equilibration level
Hydroseal closure
Brine level
Seepage inflow
Sea level
Surface discharge
Sill
Freebrine
Basal aquitard
Large hydraulic head → Seepage inflow high
→ Salina ponds dominant near sill
Brine level below hydroseal closure → No seepage outflow
→ Long residence time for brines in basin
→ High brine salinities
Late stage
Equilibration level
Brine level
Hydroseal closure
Seepage outflow
Evaporitic flats
Sea level
Sill
Basal aquitard
Reduced hydraulic head → Reduced inflow → Evaporitic flats dominant
Brine level above hydroseal closure → Seepage outflow
Reduced inflow + Seepage outflow → Short residence time for brines in basin
→ Lower brine salinities

Fig. 4.65 Model for the development of an evaporite basin (Kendall 1992)

long-continued seepage of groundwaters into a basin through the basin-margin rocks, with continual evaporation, the groundwaters being replenished by recharge in a distant upland area within a more humid climatic setting.

A model for the development of evaporites in a sedimentary basin is shown in Fig. 4.65. Evaporation may lead to drawdown of the basin waters to well below normal sea level, as is interpreted to have occurred during the Mediterranean Sea during the Miocene (Messinian age). Water depths during evaporite sedimentation may vary from very shallow to below wave base, depending on the salinities and rate of replenishment of the basin waters.

Not all evaporite beds represent simple precipitation and crystallization in place. Once deposited, an evaporite crystal, like any other sediment, is prone to resedimentation as a detrital particle by wind erosion and transportation, by rivers or by marine processes, so long as water concentrations do not drop low enough to permit dissolution of the clastic material. Outcrops of evaporite areas in present-day arid areas, where the surface textures have not been destroyed by rain solution, exhibit a remarkable variety of sedimentary structures and textures. Figure 4.66 illustrates a range of these, as observed in the Messinian evaporites of the Mediterranean Basin. Current transportation is indicated by lamination and ripples, and mass movement downslope is revealed by grain-size cycles and internal structures identical to those of thin-bedded turbidites and debris flows.

Fig. 4.66 Textures and structures in resedimented evaporites, Messinian, Italy (after Schreiber et al. 1976)

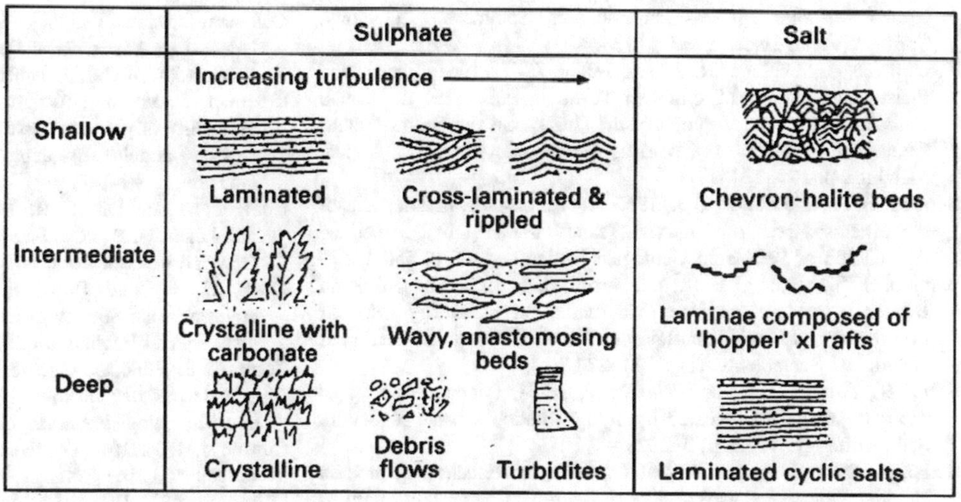

Sulphate
Salt
Increasing turbulence
Shallow
Laminated
Cross-laminated & rippled
Chevron-halite beds
Intermediate
Crystalline with carbonate
Wavy, anastomosing beds
Laminae composed of 'hopper' xl rafts
Deep
Crystalline
Debris flows
Turbidites
Laminated cyclic salts

References

Aigner, T., 1985, Storm depositional systems: Springer-Verlag Inc., New York, Lecture Notes in Earth Sciences 3, 174 p.

Allen, J. R. L., 1964, Studies in fluviatile sedimentation: six cyclothems from the Lower Old Red Sandstone. Anglo-Welsh Basin: Sedimentology, v. 3, p. 163–198.

Allen, J. R. L., 1980, Sand waves: a model of origin and internal structure: Sedimentary Geology, v. 26, p. 281–328.

Anadón, P., Cabrera, L., and Kelts, K., eds., 1991, Lacustrine facies analysis: International Association of Sedimentologists Special Publication 13.

Anderson, J. B., 1983, Ancient glacial-marine deposits: their spatial and temporal distribution, in Molnia, B. F., ed., Glacial-marine sedimentation: Plenum Press, New York, p. 3–92.

Anderson, J. B., Brake, C., Domack, E., Meyers, N., and Wright, R., 1983, Development of a polar glacial-marine sedimentation model from Antarctic Quaternary deposits and glaciological information, in Molnia, B. F., ed., Glacial-marine sedimentation: Plenum Press, New York, p. 233–264.

Andrews, J. T., and Matsch, C. L., 1983, Glacial marine sediments and sedimentation: an annotated bibliography: Geo Abstracts, Norwich, Bibliography II, 227 p.

Arnott, R. W. C., 2010, Deep-marine sediments and sedimentary systems, in James, N. P., and Dalrymple, R. W., eds., Facies Models 4: GEOtext 6, Geological Association of Canada, St. John's, Newfoundland, p. 295–322.

Arthur, M. A., Anderson, T. F., Kaplan, I. R., Veizer, J., and Land, L. S., 1983, Stable isotopes in sedimentary geology: Society of Economic Paleontologists and Mineralogists Short Course 10, 435 p.

Ashley, G. M., Shaw, J., and Smith, N. D., 1985, Glacial sedimentary environments: Society of Economic Paleontologists and Mineralogists Short Course 16, 246 p.

Barnes, N. E., Normark, W. R., 1985, Diagnostic parameters for comparing modern fans and ancient turbidite systems. In: Bouma, A. H., Normark, W. R., and Barnes, N. E., eds., Submarine fans and related turbidite systems: Frontiers in Sedimentary Geology. Springer-Verlag, New York, p. 13–14.

Bernard, H. A., Leblanc, R. J., and Major, C. J., 1962, Recent and Pleistocene geology of southeast Texas, in Rainwater, E. H., and Zingula, R. P., eds., Geology of the Gulf Coast and central Texas: Geological Society of America, Guidebook for 1962 Annual Meeting., p. 175–224.

Bernard, H. A., and Major, C. J., 1963, Recent meander belt deposits of the Brazos River; an alluvial "sand" model (abs): American Association of Petroleum Geologists Bulletin, v. 47, p. 350.

Bernard, H. A., Major, C. F. Jr., and Parrott, B. S., 1959, The Galveston Barrier Island and environs - a model for predicting reservoir occurrence and trend: Transactions of the Gulf Coast Association of Geological Societies, v. 9, p. 221–224.

Berné, S., Auffret, J.-P., and Walker, P., 1988, Internal structure of subtidal sandwaves revealed by high-resolution seismic reflection: Sedimentology, v. 35, p. 5–20.

Berné, S., Durand, J., and Weber, O., 1991, Architecture of modern subtidal dunes (sand waves), Bay of Bourgneuf, France, in Miall, A. D., and Tyler, N., eds., The three-dimensional facies architecture of terrigenous clastic sediments, and its implications for hydrocarbon discovery and recovery: Society of Economic Paleontologists and Mineralogists, Concepts in Sedimentology and Paleontology, v. 3, p. 245–260.

Beuf, S., Biju-Duval, B., De Charpal, O., Rognon, P., Gariel, O., and Bennacef, A., 1971, Les Grès du Paléozoïque Inférieur au Sahara. Sédimentaion et discontinuités, évolution structurale d'un craton: Technip, Paris, 464 p.

Bhattacharya, J., 2010, Deltas, in James, N. P., and Dalrymple, R. W., eds., Facies Models 4: GEOtext 6, Geological Association of Canada, St. John's, Newfoundland, p. 233–264.

Bhattacharya, J., and Walker, R. G., 1992, Deltas, in Walker, R. G., and James, N. P., eds., Facies models: response to sea level change: Geological Association of Canada, Geotext 1, p. 157–177.

Blakey, R. C., 1984: Marine sand-wave complex in the Permian of central Arizona: Journal of Sedimentary Petrology, v. 54, p. 29–51.

Blum, M. D., and Roberts, H. H., 2012, The Mississippi delta region: Past, present, and future: Annual Review of Earth and Planetary Sciences, v. 40, p. 655–683.

Boothroyd, J. C., Friedrich, N. E., and McGinn, S. R., 1985, Geology of microtidal coastal lagoons: Rhode Island, in Oertel, G. F., and Leatherman, S. P. eds., Barrier Islands: Marine Geology, Special Issue, v. 63, p. 35–76.

Boulton, G. S., and Deynoux, M., 1981, Sedimentation in glacial environments and the identification of tills and tillites in ancient sedimentary sequences: Precambrian Research, v. 15, p. 397–422.

Bouma, A. H., Normark, W. R., and Barnes, N. E., eds., 1985, Submarine fans and related turbidite systems: Springer-Verlag Inc., Berlin and New York, 351 p.

Bradley, W. H., 1929, The Varves and Climate of the Green River Epoch: U.S. Geological Survey Professional Paper 158-E, 110 p.

Brenchley, P. J., 1985, Storm influenced sandstone beds: Modern Geology, v. 9, p. 369–396.

Bridge, J. S., 1993, Description and interpretation of fluvial deposits: a critical perspective: Sedimentology, v. 40, p. 801–810.

Brookfield, M. E., 1977, The origin of bounding surfaces in ancient aeolian sandstones: Sedimentology, v. 24, p. 303–332.

Chough, S., and Hesse, R., 1976, Submarine meandering talweg and turbidity currents flowing for 4,000 km in the Northwest Atlantic Mid-Ocean Channel, Labrador Sea: Geology, v. 4, p. 529–533.

Chumakov, N. M., 1985, Glacial events of the past and their geological significance: Palaeogeography, Palaeoclimatology, Palaeoecology, v. 319–346.

Colella, A., and Prior, D. B., eds., 1990, Coarse-grained deltas: International Association of Sedimentologists Special Publication 10, 357 p.

Coleman, J. M., and Gagliano, S. W., 1964, Cyclic sedimentation in the Mississippi River deltaic plain: Transactions of the Gulf Coast Association of Geological Societies, v. 14, p. 67–80.

Coleman, J. M., and Wright, L. D., 1975, Modern river deltas: variability of processes and sand bodies, in Broussard, M. L., ed., Deltas, models for exploration: Houston Geological Society, Houston, p. 99–149.

Coniglio, M., and Dix, G. R., 1992, Carbonate slopes, in Walker, R. G. and James, N. P., eds., Facies models: response to sea-level change: Geological Association of Canada, Geotext 1, p. 349–373.

Cook, H. E., and Enos, P., eds., 1977, Deep-water carbonate environments: Society of Economic Paleontologists and Mineralogists Special Publication 25, 336 p.

Cook, H. E., Hine, A. C., and Mullins, H. T., 1983, Platform margin and deep water carbonates: Society of Economic Paleontologists and Mineralogists Short Course 12, 573 p.

Curray, J. R., 2014, The Bengal depositional system: from rift to orogeny: Marine Geology, v. 352, p. 59–69.

Dalrymple, R. W., 1984, Morphology and internal structure of sandwaves in the Bay of Fundy: Sedimentology, v. 31, p. 365–382.

Dalrymple, R. W., 2010, Tidal depositional systems, in James, N. P., and Dalrymple, R. W., eds., Facies Models 4: GEOtext 6, Geological Association of Canada, St. John's, Newfoundland, p. 201–231.

Dalrymple, R. W., Boyd, R., and Zaitlin, B. A., eds., 1994, Incised-valley systems: origin and sedimentary sequences: SEPM (Society for Sedimentary Geology) Special Publication 51, 391 p.

Dalrymple, R. W., Leckie, D. A., and Tillman, R. W., eds., 2006, Incised valleys in space and time: SEPM Special Publication 85, 348 p.

Davies, P. J., Symonds, P. A., Feary, D. A., and Pigram, C. J., 1989, The evolution of the carbonate platforms of northeast Australia, in Crevello, P. D., Wilson, J. D., Sarg, J. F., and Read, J. F., eds., Controls on carbonate platform and basin development: Society for Sedimentary Geology (SEPM) Special Publication 44, p. 233–258.

Davis, R. A., Jr., and Hayes, M. O., 1984, What is a wave-dominated coast? Marine Geology, v. 60, p. 313–329.

Deynoux, M., ed., 1985, Glacial record: Palaeogeography, Palaeoclimatology, Palaeoecology, v. 51, 451 p. (special issue)

Dickson, J. A. D., 1985, Diagenesis of shallow-marine carbonates, in Brenchley, P. J., and Williams, B. P. J. eds., Sedimentology, recent developments and applied aspects: Blackwell Scientific Publications, Oxford, p. 173–188.

Doyle, J. D., and Sweet, M. L., 1995, Three-dimensional distribution of lithofacies, bounding surfaces, porosity, and permeability in a fluvial sandstone: Gypsy Sandstone of northern Oklahoma: American Association of Petroleum Geologists Bulletin, v. 79, p. 70–96.

Dreimanis, A., 1985, Field criteria for the recognition of till or tillite: Palaeogeography, Palaeoclimatology, Palaeoecology, v. 51, p. 7–14

Eberli, G., and Ginsburg, R. N., 1989, Cenozoic progradation of northwestern Great Bahama Bank, a record of lateral platform growth and sea-level fluctuations, in Crevello, P. D., Wilson, J. L., Sarg, J. F., and Read, J. F., eds., Controls on carbonate platform and basin development: Society of Economic Paleontologists and Mineralogists Special Publication 44, p. 339–351.

Eyles, C. H., and Eyles, N., 2010, Glacial deposits, in James, N. P., and Dalrymple, R. W., eds., Facies Models 4: GEOtext 6, Geological Association of Canada, St. John's, Newfoundland, p. 73–104.

Eyles, N., 1994, Glacial geology: An introduction for engineers and earth scientists: Pergamon Press, Oxford, 409 p.

Eyles, N., 1993, Earth's glacial record and its tectonic setting: Earth-Science Reviews, v. 35, p. 1–248.

Eyles, N., 2008, Glacio-epochs and the supercontinent cycle after ∼3.0 Ga: Tectonic boundary conditions for glaciation: Palaeogeography, Palaeoclimatology, Palaeoecology, v. 258, p. 89–129.

Eyles, N., and Clark, B. M., 1986, Significance of hummocky and swaley cross-stratification in late Pleistocene lacustrine sediments of the Ontario basin, Canada: Geology, v. 14, p. 679–682.

Eyles, N., Day, T. E., and Gavican, A., 1987, Depositional controls on the magnetic characteristics of lodgement tills and other glacial diamict facies: Canadian Journal of Earth Sciences, v. 24, p. 2436–2458.

Eyles, N., Eyles, C. H., and Miall, A. D., 1983, Lithofacies types and vertical profile models; an alternative approach to the description and environmental interpretation of glacial diamict sequences: Sedimentology, v. 30, p. 393–410.

Eyles, C. H., Eyles, N., and Miall, A. D., 1985, Models of glaciomarine sedimentation and their application to the interpretation of ancient glacial sequences: Palaeogeography, Palaeoclimatology, Palaeoecology, v. 51, p. 15–84.

Eyles, N., and Miall, A. D., 1984, Glacial facies, in Walker, R. G., ed., Facies models, 2nd edition: Geoscience Canada Reprint Series 1, p. 15–38.

Farrell, K. M., 1987, Sedimentology and facies architecture of overbank deposits of the Mississippi River, False River region, Louisiana, in Ethridge, F. G., Flores, R. M., and Harvey, M. D., eds., Recent developments in fluvial sedimentology: Society of Economic Paleontologists and Mineralogists Special Publication 39, p. 111–120.

Farrell, K. M., Harris, W. B., Mallinson, D. J., Culver, S. J., Riggs, S. R., Pierson, J., Self-Trail, J. M., and Lautier, J. C., 2012, Standardizing texture and facies codes for a process-based classification of clastic sediment and rock: Journal of Sedimentary Research, v. 82, p. 364–378.

Fisher, W. L., Brown, L. F., Scott, A. J., and McGowen, J. H., 1969, Delta systems in the exploration for oil and gas: University of Texas Bureau of Economic Geology, 78 p.

Flemming, B. W., 1980, Sand transport and bedform patterns on the continental shelf between Durban and Port Elizabeth (southeast Africa continental margin): Sedimentary Geology, v. 26, p. 179–205.

Flores, R. M., 1981, Coal deposition in fluvial paleoenvironments of the Paleocene Tongue River Member of the Fort Union Formation, Powder River area, Powder River basin, Wyoming and Montana, in Ethridge, F. G., and Flores, R. M., eds., Recent and ancient nonmarine depositional environments: models for exploration: Society of Economic Paleontologists and Mineralogists Special Publication 31, p. 161–190.

Frakes, L. A., 1979, Climates throughout geologic time: Elsevier, Amsterdam, 310 p.

Frakes, L. A., Francis, J. E., and Syktus, J. I., 1992, Climate modes of the Phanerozoic: Cambridge University Press, Cambridge, 274 p.

Frazier, D. E., 1967; Recent deltaic deposits of the Mississippi River—their development and chronology: Transactions of the Gulf Coast Association of Geological Societies, v. 17, p. 287–315.

Galloway, W. E., 1975, Process framework for describing the morphologic and stratigraphic evolution of the deltaic depositional systems, in Broussard, M. L., ed., Deltas, models for exploration: Houston Geological Society, Houston, p. 87–98.

Glennie, K. W., 1972, Permian Rotliegendes of northwest Europe interpreted in light of modern desert sedimentation studies: American Association of Petroleum Geologists Bulletin, v. 56, p. 1048–1071.

Glennie, K. W., 1983, Lower Permian Rotliegend desert sedimentation in the North Sea area, in Brookfield, M. E., and Ahlbrandt, T. S., eds., Eolian sediments and processes: Elsevier, Amsterdam, Developments in Sedimentology, v. 38, p. 521–541.

Hambrey, M. J., and Harland, W. B., 1981a, Criteria for the identification of glacigenic deposits, in Hambrey, M. J., and Harland, W. B., eds., Earth's Pre-Pleistocene glacial record: Cambridge University Press, Cambridge, p. 14–27.

Hambrey, M. J., and Harland, W. B., eds., 1981b, Earth's Pre-Pleistocene glacial record: Cambridge University Press, Cambridge, 1004 p.

Handford, C. R., 1986, Facies and bedding sequences in shelf-storm-deposited carbonates - Fayettville Shale and Pitkin Limestone: Journal of Sedimentary Petrology, v. 56, p. 123–137.

Hardie, L. A., Smoot, J. P., and Eugster, H. P., 1978, Saline lakes and their deposits: a sedimentological approach, in Matter, A., and Tucker, M. E., eds., Modern and ancient lake sediments: International Association of Sedimentologists Special Publication 2, p. 7–41.

Harris, P. T., 1988, Large-scale bedforms as indicators of mutually evasive sand transport and the sequential infilling of wide-mouthed estuaries: Sedimentary Geology, v. 273–298.

Hayes, M. O., 1967, Hurricanes as geological agents: Case studies of Hurricanes Carla, 1961, and Cindy, 1963: University of Texas, Bureau of Economic Geology, Austin, Texas, Report of Investigations No. 61.

Hayes, M. O., 1976, Morphology of sand accumulation in estuaries; an introduction to the symposium, in Cronin, L. E., ed., Estuarine Research, v. 2, Geology and Engineering: Academic Press, London, p. 3–22.

Hayes, M. O., 1979, Barrier island morphology as a function of tidal and wave regime, in Leatherman, S. P. ed., Barrier islands—from the Gulf of St. Lawrence to the Gulf of Mexico: Academic Press, New York, p. 1–27.

Hine, A. C., 1977, Lily Bank, Bahamas: history of an active oolite sand shoal: Journal of Sedimentary Petrology, v. 47, p. 1554–1581.

James, N. P., and Bourque, P.-A., 1992, Reefs and mounds, in Walker, R. G. and James, N. P., eds., Facies models: response to sea-level change: Geological Association of Canada, Geotext 1, p. 323–347.

James, N. P., and Dalrymple, R. W., eds., 2010, Facies Modesl 4: GEOtext 6, Geological Association of Canada, St. John's, Newfoundland, 586 p.

James, N. P., and Lukasik, J., 2010, Cool- and cold-water neritic carbonates, in James, N. P., and Dalrymple, R. W., eds., Facies Models 4: GEOtext 6, Geological Association of Canada, St. John's, Newfoundland, p. 371–420.

James, N. P., Kendall, A. C., and Pufahl, P. K., 2010, Introduction to biological and biochemical facies models, in James, N. P., and Dalrymple, R. W., eds., Facies Models 4: GEOtext 6, Geological Association of Canada, St. John's, Newfoundland, p. 323–340.

Jenkyns, H. C., 1986, Pelagic environments. In: Reading, H. G., ed., Sedimentary environments and facies, 2nd edn. Blackwell Scientific Publications, Oxford, p. 343–397.

Johnson, H. D., and Baldwin, C. T., 1996, Shallow clastic seas, in Reading. H. G., ed., Sedimentary environments: processes, facies and stratigraphy: Blackwell Science, Oxford, p. 232–280.

Jones, B., 2010, Warm-water neritic carbonates, in James, N. P., and Dalrymple, R. W., eds., Facies Models 4: GEOtext 6, Geological Association of Canada, St. John's, Newfoundland, p. 341–369.

Katz, B. J., ed., 1990, Lacustrine basin exploration: case studies and modern analogs: American Association of Petroleum Geologists Memoir 50, 340 p.

Kendall, A. C., 1992, Evaporites, in Walker, R. G. and James, N. P., eds., Facies models: response to sea-level change: Geological Association of Canada, Geotext 1, p. 375–409.

Klovan, J. E., 1964, Facies analysis of the Redwater Reef complex, Alberta, Canada. Bull Can Pet Geol., v. 12, p. 1–100.

Kocurek, G. A., 1988, First-order and super bounding surfaces in eolian sequences - bounding surfaces revisited: Sedimentary Geology, v. 56, p. 193–206.

Kraus, M. J., and Bown, T. M., 1988, Pedofacies analysis; a new approach to reconstructing ancient fluvial sequences: Geological Society of America Special Paper 216, p. 143–152.

Leckie, D. A., and Krystinick, L. F., 1989, Is there evidence for geostrophic currents preserved in the sedimentary record of inner to middle shelf deposits? J Sediment Petrol., v. 59, p. 862–870.

Machel, H.-G., 1985, Cathodoluminescence in calcite and dolomite and its chemical interpretation: Geoscience Canada, v. 12, p. 139–147.

Mackiewicz, N. E., Powell, R. D., Carlson, P. R., and Molnia, B. F., 1984, Interlaminated ice-proximal glacimarine sediments in Muir Inlet, Alaska: Marine Geology, v. 57, p. 113–148.

Macquaker, J. H. S., Taylor, K. G., and Gawthorpe, R. L. 2007, High-resolution facies analyses of mudstones: implications for paleoenvironmental and sequence stratigraphic interpretations of offshore ancient mud-dominated successions: Journal of Sedimentary Research, v. 77, p. 324–339.

Martin, A. K., and Flemming, B. W., 1986, The Holocene shelf sediment wedge off the south and east coast of South Africa; in Knight, R. J., and McLean, J. R., eds., Shelf sands and sandstones: Canadian Society of Petroleum Geologists Memoir 11, p. 27–44.

Marwick, P. J., and Rowley, D. B., 1998, The geological evidence for Triassic to Pleistocene glaciation: implications for eustasy, in Pindell, J. L., and Drake, C. L., eds., Paleogeographic evolution and non-glacial eustasy, northern South America: Society for Sedimentary Geology (SEPM) Special Publication 58, p. 17–43.

Matter, A., and Tucker, M., eds., 1978, Modern and Ancient lake sediments: International Association of Sedimentologists Special Publication 2, 290 p.

McIlreath, I. A., and James, N. P., 1984, Carbonate slopes, in Walker, ed., R. G., Facies Models, 2nd edition: Geoscience Canada Reprint Series 1, p. 245–257.

McKee, E. D., ed., 1979, A study of global sand seas: U.S. Geological Survey Professional Paper 1052.

McKee, E. D., and Ward, W. C., 1983, Eolian environment, in Scholle, P. A., Bebout, D. G., and Moore, C. H., eds., Carbonate depositional environments: American Association of Petroleum Geologists Memoir 33, p. 131–170.

McPherson, J. G., Shanmugam, G., and Moiola, R. J., 1987, Fan-deltas and braid deltas: varieties of coarse-grained deltas: Geological Society of America Bulletin, v. 99, p. 331–340.

Miall, A. D., 1977, A review of the braided river depositional environment: Earth Science Reviews, v. 13, p. 1–62.

Miall, A. D., 1978, Lithofacies types and vertical profile models in braided river deposits: a summary, in Miall, A. D., ed. Fluvial Sedimentology: Canadian Society of Petroleum Geologists Memoir 5, p. 597–604.

Miall, A. D., 1985, Architectural-element analysis: A new method of facies analysis applied to fluvial deposits: Earth Science Reviews, v. 22, p. 261–308.

Miall, A. D., 1988a, Reservoir heterogeneities in fluvial sandstones: lessons from outcrop studies: American Association of Petroleum Geologists Bulletin, v. 72, p. 682–697.

Miall, A. D., 1988b, Facies architecture in clastic sedimentary basins, in Kleinspehn, K., and Paola, C., eds., New perspectives in basin analysis: Springer-Verlag Inc., New York, p. 67–81.

Miall, A. D., 1996, The geology of fluvial deposits: sedimentary facies, basin analysis and petroleum geology: Springer-Verlag Inc., Heidelberg, 582 p.

Miall, A. D., 2010, The geology of stratigraphic sequences, second edition: Springer-Verlag, Berlin, 522 p.

Miall, A. D., and Tyler, N., eds., 1991, The three-dimensional facies architecture of terrigenous clastic sediments, and its implications for hydrocarbon discovery and recovery: Society of Economic Paleontologists and Mineralogists Concepts and Models Series, v. 3, 309 p.

Miller, J. N. G. 1996, Glacial sediments, in Reading, H. G., ed., Sedimentary environments: processes, facies and stratigraphy: Blackwell Science, Oxford, p. 454–484.

Molnia, B. F., 1983a, Subarctic glacial-marine sedimentation: a model, in Molnia, B. F., ed., Glacial-marine sedimentation: Plenum Press, New York, p. 95–144.

Molnia, B. F., ed., 1983b, Glacial-marine sedimentation: Plenum Press, New York, 844 p.

Musial, G., Reynaud, J-Y., Gingras, M. K., Féniès, H., Labourdette, R., and Parize, O., 2012, Subsurface and outcrop characterization of large tidally influenced point bars of the Cretaceous McMurray Formation (Alberta, Canada): Sedimentary Geology, v. 279, p. 156–172.

Mutti, E., Ricci-Lucchi, F., 1972, Le turbiditi dell'Appennino settentrionale: introduzione all' analisi di facies: Soc. Geol Ital Mem 11, p. 161–199.

Nanson, G. C., and Croke, J. C., 1992, A genetic classification of floodplains: Geomorphology, v. 4, p. 459–486.

Niederoda, A. W., Swift, D. J. P., Figueiredo, A. G., Jr., and Freeland, G. L., 1985, Barrier island evolution, middle Atlantic shelf, U.S.A. Part II: Evidence from the shelf floor, in Oertel, G. F. and Leatherman, S. P. eds., Barrier islands: Marine Geology Special Issue, v. 63, p. 363–396.

Nilsen, T., Shew, R., Steffens, G., and Studlick, J., eds., 2007, Atlas of deep-water outcrops: American Association of Petroleum Geologists, Studies in Geology 56, 504 p.

Nummedal, D., and Swift, D. J. P., 1987, Transgressive stratigraphy at sequence-bounding unconformities: some principles derived from Holocene and Cretaceous examples, in Nummedal, D., Pilkey, O. H., and Howard, J. D., eds., Sea-level fluctuation and coastal evolution; Society of Economic Paleontologists and Mineralogists Special Publication 41, p. 241–260.

Olsen, P. E., 1984, Periodicity of lake-level cycles in the Late Triassic Lockatong Formation of the Newark Basin (Newark Supergroup, New Jersey and Pennsylvania), in Berger, A., Imbrie, J., Hays, J., Kukla, G., and Saltzman, B., eds., Milankovitch and climate: NATO ASI Series, D. Reidel Publishing Company, Dordrecht, Part 1, p. 129–146.

Olsen, P. E., 1986, A 40-million year lake record of Early Mesozoic orbital climatic forcing: Science, v. 234, p. 842–848.

Olsen, P. E., 1990, Tectonic, climatic, and biotic modulation of lacustrine ecosystems—examples from Newark Supergroup of eastern North America, in Katz, B. J., ed. Lacustrine Basin Exploration: Case Studies and Modern Analogs: American Association of Petroleum Geologists Memoir 50, p. 209–224.

Orton, G. J., and Reading, H. G., 1993, Variability of deltaic processes in terms of sediment supply, with particular emphasis on grain size: Sedimentology, v. 40, p. 475–512.

Penland, S., Boyd, R., and Suter, J. R., 1988, Transgressive depositional systems of the Mississippi delta plain: a model for barrier shoreline and shelf sand development: Journal of Sedimentary Petrology, v. 58, p. 932–949.

Picard, M. D., and High, L. R., Jr., 1981, Physical stratigraphy of ancient lacustrine deposits, in Ethridge, F. G., and Flores, R. M., eds., Recent and ancient nonmarine depositional environments: models for exploration: Society of Economic Paleontologists and Mineralogists Special Publication 31, p. 233–259.

Platt, N. H., and Keller, B., 1992, Distal alluvial deposits in a foreland basin setting — the Lower Freshwater Molasse (Lower Miocene), Switzerland: sedimentology, architecture and palaeosols: Sedimentology, v. 39, p. 545–565.

Playton, T. E., Janson, X., and Kerans, C., 2010, Carbonate slopes, in James, N. P., and Dalrymple, R. W., eds., Facies Models 4: GEOtext 6, Geological Association of Canada, St. John's, Newfoundland, p. 449–476.

Plint, A. G., 1988, Sharp-based shoreface sequences and "offshore bars" in the Cardium Formation of Alberta: their relationship to relative changes in sea level, in Wilgus, C. K., Hastings, B. S., Kendall, C. G. St. C., Posamentier, H. W., Ross, C. A., and Van Wagoner, J. C., eds., Sea-level Changes: an integrated approach: Society of Economic Paleontologists and Mineralogists Special Publication 42, p. 357–370.

Plint, A. G., 2010, Wave- and storm-dominated shoreline and shallow-marine systems, in James, N. P., and Dalrymple, R. W., eds., Facies Models 4: GEOtext 6, Geological Association of Canada, St. John's, Newfoundland, p. 167–199.

Pomar, L., and Hallock, P., 2008, Carbonate factories: a conundrum in sedimentary geology: Earth Science Reviews, v. 87, p. 134–169.

Porter, M. L., 1986, Sedimentary record of erg migration: Geology, v. 14, p. 497–500.

Postma, G., 1990, Depositional architecture and facies of river and fan deltas: a synthesis, in Colella, A., and Prior, D. B., eds., Coarse-grained deltas: International Association of Sedimentologists Special Publication 10, p. 13–27.

Powell, R. D., 1981, A model for sedimentation by tidewater glaciers: Annals of Glaciology, v. 2, p. 129–134.

Powell, R. D., 1983, Glacial marine sedimentation processes and lithofacies of temperate tidewater glaciers, Glacier Bay, Alaska, in Molnia, B. F., ed., Glacial-marine sedimentation: Plenum Press, New York, p. 185–232.

Powell, R. D., 1984, Glaciomarine processes and inductive lithofacies modelling of ice shelf and tidewater glacier sediments based on Quaternary examples: Marine Geology, v. 57, p. 1–52.

Pratt, B. R., 2010, Peritidal carbonates, in James, N. P., and Dalrymple, R. W., eds., Facies Models 4: GEOtext 6, Geological Association of Canada, St. John's, Newfoundland, p. 401–420.

Pratt, B. R., and Holmden, C., eds., 2008, Dynamics of epeiric seas: Geological Association of Canada Special Paper 48.

Principaud, M., Mulder, T., Hanquiez, V., Ducassou, E., Eberli, G. P., Chabaud, L., and Borgomano, J., 2018, Recent morphology and sedimentary processes along the western slope of Great Bahama Bank (Bahamas): Sedimentology, v. 65, p. 2088–2116.

Purdy, E. G., 1963, Recent calcium carbonate facies of the Great Bahama Bank: Journal of Geology, v. 71, p. 334–355, p. 472–497.

Reading, H. G., ed., 1986, Sedimentary environments and facies, second ed.: Blackwell, Oxford, 615 p.

Reading, H. G., ed., 1996, Sedimentary environments: processes, facies and stratigraphy, third edition: Blackwell Science, Oxford, 688 p.

Reading, H. G., and Collinson, J. D., 1996, Clastic coasts, in Reading, H. G., ed., Sedimentary environments: processes, facies and stratigraphy, third edition: Blackwell Science, Oxford, p. 154–231.

Reading, H. G., and Richards, M., 1994, Turbidite systems in deep-water basin margins classified by grain-size and feeder system: American Association of Petroleum Geologists Bulletin, v. 78, p. 792–822.

Reinson, G. E., 1992, Transgressive barrier island and estuarine systems, in Walker, R. G. and James, N. P., eds., Facies models: response to sea-level change: Geological Association of Canada, Geotext 1, p. 179–194.

Renaut, R. W., and Gierlowski-Kordesch, E. H., 2010, Lakes, in James, N. P., and Dalrymple, R. W., eds., Facies Models 4: GEOtext 6, Geological Association of Canada, St. John's, Newfoundland, p. 541–575.

Richards, M., Bowman, M., and Reading, H., 1998, Submarine fan systems I: characterization and stratigraphic prediction: Marine and Petroleum Geology, v. 15, p. 689–717.

Roberts, H. H., 1987, Modern carbonate-siliciclastic transitions: humid and arid tropical examples: Sedimentary Geology, v. 50, p. 25–66.

Rosenthal, L., 1988, Wave dominated shorelines and incised channel trends: Lower Cretaceous Glauconite Formation, west-central Alberta, in James, D. P., and Leckie, D. A., eds., 1988, Sequences, stratigraphy, sedimentology: surface and subsurface: Canadian Society of Petroleum Geologists Memoir 15, p. 207–230.

Ryder, R. T., 1980, Lacustrine sedimentation and hydrocarbon occurrences: a review, in American Association of Petroleum Geologists Fall Education Conference, 103 p.

Schieber, J., Southard, J. B., Kissling, P., Rossman, B., and Ginsburg, R., 2013, Experimental Deposition of Carbonate Mud From Moving Suspensions: Importance of Flocculation and Implications For Modern and Ancient Carbonate Mud Deposition: Journal of Sedimentary Research, v. 83, p. 1025–1031.

Schlager, W., 1992, Sedimentology and sequence stratigraphy of reefs and carbonate platforms: American Association of Petroleum Geologists Continuing Education Course Notes Series 34, 71 p.

Schlager, W., 2005, Carbonate sedimentology and sequence stratigraphy: SEPM Concepts in Sedimentology and Paleontology #8, 200p.

Scholle, P. A., Bebout, D. G., and Moore C. H., eds., 1983, Carbonate depositional environments: American Association of Petroleum Geologists Memoir 33, 708 p.

Schreiber, B. C., Friedman, G. M., Decima, A., and Schreiber, E., 1976, Depositional environments of Upper Miocene (Messinian) evaporite deposits of the Sicilian Basin: Sedimentology, v. 23, p. 729–760.

Scruton, P. C., 1960, Delta building and the delta sequence, in F. P. Shepard, F. B. Phleger, and T. H. van Andel, eds., Recent sediments, northwest Gulf of Mexico, American Association of Petroleum Geologists, p. 82–102.

Sellwood, B. W., 1986, Shallow-marine carbonate environments, in Reading, H. G., ed., Sedimentary environments and facies, 2nd ed.: Blackwell, Oxford, p. 283–342.

Smith, D. G., 1987, Meandering river point bar lithofacies models: modern and ancient examples compared, in Ethridge, F. G., Flores, R. M., and Harvey, M. D., eds., Recent developments in fluvial sedimentology: Society of Economic Paleontologists and Mineralogists Special Publication 39, p. 83–91.

Smith, D. G., 1988a, Modern point bar deposits analogous to the Athabasca Oil Sands, Alberta, Canada, in de Boer, P. L., van Gelder, A., and Nio, S. D., eds., Tide-influenced sedimentary environments and facies: Reidel Publishing Co., Netherlands, p. 417–432.

Smith, D. G., 1988b, Tidal bundles and mud couplets in the McMurray Formation, northeastern Alberta, Canada: Bulletin of Canadian Petroleum Geology, v. 36, p. 216–219.

Smith, D. G., Reinson, G. E., Zaitlin, B. A., and Rahmani, R. A., eds., 1991, Clastic tidal sedimentology: Canadian Society of Petroleum Geologists memoir 16, 387 p.

Smith, D. G., and Smith, N. D., 1980, Sedimentation in anastomosed river systems: examples from alluvial valleys near Banff, Alberta: Journal of Sedimentary Petrology, v. 50, p. 157–164.

Steel, R. J., Milliken. K. L., 2013, Major advances in siliciclastic sedimentary geology, 1960–2012, in Bickford, M. E., ed., The web of geological sciences: Advances, impacts and interactions. Geol Soc Am Spec Paper 500, p. 121–167.

Stow, D. A. V. and Mayall, M., 2000, Deep-water sedimentary systems: new models for the 21st century: Marine and Petroleum Geology, v. 17, p. 125–135.

Swift, D. J. P., Niederoda, A. W., Vincent, C. E., and Hopkins, T. S., 1985, Barrier island evolution, middle Atlantic shelf, U.S.A. Part I: Shoreface Dynamics: Marine Geology, v. 63, p. 331–361.

Talbot, M. R., and Allen, P. A., 1996, Lakes. In: Reading, H. G., ed., Sedimentary environments, processes, facies and stratigraphy, 3rd edn. Blackwell Science, Oxford, p. 83–124.

Teyssen, T. A. L., 1984, Sedimentology of the Minette oolitic ironstones of Luxembourg and Lorraine: a Jurassic subtidal sandwave complex: Sedimentology, v. 31, p. 195–212.

Thomas, R. G., Smith, D. G., Wood, J. M., Visser, J., Calverley-Range, E. A., and Koster, E. H., 1987, Inclined heterolithic stratification -

terminology, description, interpretation and significance: Sedimentary Geology, v. 53, p. 123–179.

Trabucho-Alexandre, J. 2015. More gaps than shale: erosion of mud and its effect on preserved geochemical and palaeobiological signals. in: Smith, D. G., Bailey, R. J., Burgess, P. M., and Fraser, A. J., eds., Strata and Time: Probing the Gaps in Our Understanding: Geological Society, London, Special Publications, 404, p. 251–270.

Tucker, M. E., 1985, Shallow-marine carbonate facies and facies models, in Brenchley P. J., and Williams, B. P. J., eds., Sedimentology: Recent developments and applied aspects: Blackwell Scientific Publications, Oxford, p. 147–169.

Van Houten, F. B., 1964, Cyclic lacustrine sedimentation, Upper Triassic Lockatong Formation, central New Jersey and adjacent Pennsylvania, in Merriam, D. F., ed. Symposium on Cyclic Sedimentation: Geological Survey of Kansas Bulletin 169, p. 495–531.

Walker, R. G., 1978, Deep-water sandstone facies and ancient submarine fans: models for exploration for stratigraphic traps: American Association of Petroleum Geologists Bulletin, v. 62, p. 932–966.

Walker, R. G. and James, N. P., eds., 1992, Facies models: response to sea-level change: Geological Association of Canada, 409 p.

Willis, B. J., and Behrensmeyer, A. K., 1994, Architecture of Miocene overbank deposits in northern Pakistan: Journal of Sedimentary Research, v. B64, p. 60–67.

Wilson, J. L., 1975, Carbonate facies in geologic history: Springer-Verlag, New York, 471 p.

Wright, R., Anderson, J. B., and Fisco, P. P., 1983, Distribution and association of sediment gravity flow deposits and glacial/glacial-marine sediments around the continental margin of Antarctica, in Molnia, B. F., ed., Glacial-marine sedimentation: Plenum Press, New York, p. 265–300.

Wright, V. P., ed. 1986, Paleosols: their recognition and interpretation: Blackwell Scientific Publications, Oxford, 315 p.

Wright, V. P., 1992, Paleopedology, stratigraphic relationships and empirical models, in Martini, I. P., and Chesworth, W., eds., 1992, Weathering, soils and paleosols: Developments in Earth Surface Processes, No. 2, Elsevier, Amsterdam, p. 475–499.

Wright, V. P., and Burchette, T. P., 1996, Shallow-water carbonate environments, in Reading, H. G., ed., Sedimentary environments: processes, facies and stratigraphy, 3rd edition: Blackwell Science, p. 325–394.

Sequence Stratigraphy

Contents

Abstract

This chapter offers a succinct summary of modern concepts and terminology in the area of sequence stratigraphy. An introduction to the relationship between accommodation and sequence architecture includes discussions of the major bounding surfaces and systems tracts of a typical clastic sequence. Brief introductions are provided to sequence models developed for nonmarine, shallow-marine and deep-marine clastic sequences and for carbonate sequences. The chapter closes with a brief discussion of sequence hierarchies and driving mechanisms.

5.1   Introduction

The purpose of this chapter is to present a succinct summary of sequence concepts, focusing on what has become the standard, or most typical, sequence model and making references to exceptions and complexities.

The methods and nomenclature for sequence stratigraphy, including sequence architecture and systems tracts were established initially for use in analyzing and interpreting seismic records by the Exxon group, led by Peter Vail (Vail et al. 1977), and included much new terminology, new concepts and new interpretive methods (Sects. 1.2.10, 1.2.14).

A useful history was provided by Nystuen (1998). This work was revolutionary; it brought about a fundamental change in the methods of stratigraphy. It is important to be familiar with this work, as the terms and concepts came to be used by virtually all stratigraphers. Application of the methods to different types of data (outcrop, well-log, 2-D seismic, 3-D seismic) from around the world permitted a quantum leap in understanding of basin architectures (e.g., Van Wagoner et al. 1990), but also revealed some problems with the Exxon concepts and terminology that have stimulated vigorous debate over the last two decades. The purpose of this chapter is to bring the debate up to date by presenting current concepts, with enough historical background that readers familiar with the debates will be able to comprehend the need for revisions and new approaches.

It is recommended that for an in-depth treatment of sequence stratigraphy the reader should turn to what has (in this author's opinion) become the standard work on sequence stratigraphy, that by Catuneanu (2006). The introduction to Chap. 5 in this book (Catuneanu 2006, p. 165–171) is particularly useful in providing a history of the controversies, the main players and publications, and the concepts and terminologies about which there have been disputes.

The subject has continued to actively evolve since 2006. Catuneanu orchestrated the preparation of several multi-authored publications, the purpose of which was to bring together an international group of specialists to attempt to resolve problems of terminology and definition as part of a necessary approach to the development of mechanisms for the integration of sequence stratigraphy into formal stratigraphic methods (Catuneanu et al. 2009, 2010, 2011). This work is discussed in Sect. 7.7. Other developments touched on in the present chapter include a discussion of stratal geometries and stacking patterns and their interpretation (Catuneanu and Zecchin 2016); the occurrence of sequences over a wide range of time scales and physical scales, and how this is accommodated in description and definition (Catuneanu 2019a, 2019b); a detailed review of the process and products of wave ravinement (Zecchin et al. 2019); a discussion of the controversial term "parasequence" and its relevance in ongoing work (Catuneanu and Zecchin 2020); a discussion of the use of modeling and simulation programs in sequence stratigraphy (Catuneanu 2020a, 2020b); and a discussion of the sequence stratigraphy of deep-water systems (Catuneanu 2020b). Holbrook and Bhattacharya (2012) examined the nature of the subaerial erosion surface, which is commonly employed as the sequence bounding surface, and pointed out some problems and contradictions in our current understanding about this important element of the standard sequence model. Madof et al. (2016) and Burgess (2016) pointed out the problems that can arise when sequence models are based on the premise that basin architecture is orthogonal, with coastal regression, transgression, deltaic

development, clinoform architecture, etc., all arranged perpendicular to a straight shoreline. This is commonly an oversimplification of actual basin configurations.

Sequence-stratigraphic concepts are, to a considerable extent, independent of scale (Schlager 2004; Catuneanu 2006, p. 9). Posamentier and Vail (1988) and Posamentier et al. (1992) gave examples of sequence architecture evolving from base-level changes in small natural systems, and much experimental work has successfully simulated sequence development in small laboratory tanks (e.g., Wood et al. 1993; Paola 2000). Interpretations of the temporal significance of sequences can, therefore, be difficult (Catuneanu 2019a, 2019b). However, as discussed later in this book, some aspects of the driving process, notably tectonism and climate change, generate facies and architecture that are sufficiently distinctive to yield reliable interpretations.

The early work on sequence stratigraphy was inextricably linked to one of Vail's early concepts, that of global sea-level change (eustasy) as the central driving force in the generation of the base-level changes responsible for developing observed sequence architectures. This concept became highly popular in the 1980s, and began to dominate the practical work of stratigraphy. A global sea-level curve and cycle classification based on the global eustasy concept was published by the Exxon group (Haq et al. 1987) and became widely used as an assumed global stratigraphic standard. However, difficulties with the concept began to emerge in the mid-1980s. For example, alternative ideas about the driving force behind base-level change, particularly regional tectonism, began to take hold, and these cast doubt on the universality of any global cycle chart. By the mid-1990s these alternative ideas were entering the mainstream (Miall 1995). A detailed study of what social scientists call the "social construction" of the science behind Exxon's body of work (Miall and Miall 2002) and an examination of the data and research methods that supported it (Miall and Miall 2001) led to the recognition that sequence stratigraphy was characterized by two competing paradigms. Miall and Miall (2001) termed these the global **eustasy paradigm**, that originated by Vail et al. (1977), and the **complexity paradigm**, which asserted that there are multiple causes of the accommodation changes that lead to the generation of sequences. In this book we do not deal in detail with the allogenic driving forces behind sequence development, and we do not discuss the concept of global eustasy. The second edition of my book *The Geology of Stratigraphic Sequences* (Miall 2010) contains an extensive discussion of this issue, and it is assumed here that sequence-generating mechanisms are in accord with the complexity paradigm.

Simmons et al. (2020) reviewed in detail the research that has taken place since the critiques of the global eustasy model appeared in the 1980s. While there is no final agreement on the specific eustatic signatures that exist in any part of the geological record, apart from the Sloss sequences (see Miall 2010), it is clear that detailed study of sequence stratigraphies combined with modern chronostratigraphic methods is closing in on the topic. In particular, it would appear that high-frequency climatic cycles, which may or may not include glacioeustasy, are increasingly being recognized in the Phanerozoic record, and are contributing to the growth of an astrochronologic time scale (Sects. 7.8.7, 8.12, 8.13). A brief discussion of driving mechanisms and a classification of sequences based on these mechanisms is provided in Sect. 7.6.

5.2 Elements of the Model

The practical application of sequence-stratigraphic methods to the mapping of a basin-fill succession, either in outcrop or in the subsurface, involves the following:

- The documentation of a succession of sequences, tracing them by regional correlation, and mapping their major bounding surfaces.
- The study of vertical cyclic facies changes in order to identify the succession of sedimentary environments and in order to subdivide a stratigraphic succession into its component sequences and depositional-systems tracts.
- The mapping of significant internal surfaces, such as maximum flooding surface, basal surface of forced regression, etc.
- The mapping of onlap, offlap and other stratigraphic terminations in order to provide information about the internal architectural development of each sequence.
- Clarification of the relationships, if any, between regional structural geology and the large-scale configuration of sequences, in order to explore the possible influence of tectonism on sequence generation.
- Application of all available methods of dating and correlation in order to determine the aerial extent of the sequences and the time scale they represent, as a necessary basis for interpretations of driving mechanisms.

Regional mapping at a reconnaissance level may be based on seismic-reflection data, well records or outcrops (Chap. 6). This could include the identification of the major systems tracts and bounding surfaces, where seismic resolution is adequate. The use of facies-cycle and other data as described in Chaps. 3 and 4 are an essential component of systematic sequence work (see also Dalrymple 2010b).

Modern sequence stratigraphy began with the work of P. R. Vail and his colleagues, as set out in the groundbreaking memoir edited by Payton (1977). Documentation of the two- and three-dimensional architecture of sequences was one of the most important breakthroughs of the seismic method, as explained in the Sect. 5.2.2. Mitchum et al. (1977a, p. 53) defined a **depositional sequence** as "a stratigraphic unit composed of a relatively conformable succession of genetically related strata and bounded at its top and base by unconformities or their correlative conformities." An unconformity may be traced laterally into the deposits of deep-marine environments, where it may be represented by a **correlative conformity**. As Catuneanu et al. (2011, p. 183) subsequently pointed out, this definition does not cover the genetic sequences defined by Frazier (1974) and Galloway (1989) (sequence definitions are dealt with in Sect. 7.7) and he proposed redefining a sequence as "a succession of strata deposited during a full cycle of change in accommodation or sediment supply" (Catuneanu et al. 2009, p. 19).

An early project to incorporate sequence stratigraphy into formal stratigraphic methods was the proposal for a system of unit names based on the methods of **allostratigraphy**, as noted in Sect. 1.2.13. Major sequences were to be termed alloformations, with allomember sequences nested within them, and so on. Formal definitions and procedures for the employment of allostratigraphy are provided by the North American Commission on Stratigraphic Nomenclature (2005) and are touched on briefly in Sects. 7.6 and 7.7. Allostratigraphy has not proved a popular approach but is regularly used by some workers, notably A. G. Plint, working in the Western Interior Basin of Canada, because it enables classification and labeling without implications as to processes.

Sequences reflect the sedimentary response to **base-level cycles**–the rise and fall in sea level relative to the shoreline, and changes in the sediment supply. Change in sea-level relative to the shoreline may result from **eustasy** (absolute changes in sea-level elevation relative to the center of Earth) or from vertical movements of the basin floor as a result of tectonism. Because of the difficulty in distinguishing between these two different processes, the term **relative sea-level change** is normally used in order to encompass the uncertainty. These basic controls are explained below. In nonmarine settings, upstream controls (tectonism and climate change) are the major determinant of sequence architecture (Sect. 5.3.2).

The cycle of rise and fall of base level generates predictable responses in a sedimentary system, such as the transgressions that occur during rising relative sea level, and the widespread subaerial erosion and delivery of clastic detritus to the continental shelf, slope and deep basin during a fall in relative sea level. The depositional systems that result, and their vertical and lateral relationships, provide the basis for subdividing sequences into systems tracts. These are described and explained in Sect. 5.3.2.

Unconformities provided the basis for the first definitions of sequences, by Blackwelder, Levorsen, Sloss, and Vail

(Sect. 1.2.10). The unconformities that are the key to sequence recognition are those that develop as a result of subaerial exposure. Unconformities may develop below sea level as a result of submarine erosion but are not used as the basis for sequence definition. Where subaerial unconformities are present, as in nonmarine and coastal settings, sequence definition is relatively straightforward. Carrying a correlation into the offshore, including recognition of a correlative conformity, is not necessarily simple; in fact, this has been the cause of considerable debate and controversy, as discussed in Sect. 5.2.2. Several methods of defining sequences evolved from these controversies, and this has inhibited further progress in the establishment of sequence terminology and classification as a formal part of what might be called the official language of stratigraphy. Catuneanu (2006) and Catuneanu et al. (2009, 2010, 2011) have gone a long way toward resolving these problems, and their proposals are summarized in Sect. 7.7.

5.2.1 Accommodation and Supply

A fundamental concept in sequence stratigraphy is that of **accommodation** (Fig. 5.1). Accommodation is defined as "the space available for sedimentation." Jervey (1988) explained the concept this way:

> In order for sediments to accumulate, there must be space available below base level (the level above which erosion will occur). On the continental margin, base level is controlled by sea level and, at first approximation, is equivalent to sea level. ... This space made available for potential sediment accumulation is referred to as accommodation.

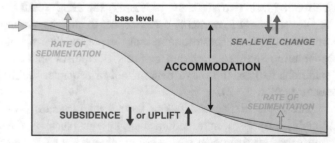

Fig. 5.1 Accommodation, and the major allogenic sedimentary controls. Total accommodation in a basin is that generated by the subsidence of the basin floor (measured by backstripping methods). At any moment in time, the remaining accommodation in the basin represents that not yet filled by sediment, and is measured by water depth (the space between sea level and the sediment–water interface). Changes in accommodation (eustasy+tectonism) almost never correlate with bathymetric changes (eustasy +tectonism+sedimentation). Water depth reflects the balance between simultaneous creation (eustasy + tectonics) and consumption (sedimentation) of accommodation

In marine basins this is equivalent to the space between sea level and the sea floor. In nonmarine basins, a river's graded profile functions as sedimentary base level (Holbrook et al. 2006). Sequences are a record of the balance between accommodation change and sediment supply. As accommodation is filled by sediment, the remaining space is measured by the depth of water from the sea surface to the sediment–water interface at the bottom of the sea. Total accommodation increases when the basin floor subsides or sea level rises faster than the supply of sediment to fill the available space. Barrell (1917) understood this decades before geologists were in a position to appreciate its significance (Sect. 1.2.9; Fig. 5.2). Where supply > accommodation, progradation results. Where supply < accommodation, retrogradation results. These contrasting scenarios were recognized many years ago and are illustrated in Fig. 5.3 with reference to the stacking patterns of deltas on a continental margin. Figure 5.4 illustrates the initial Exxon concept of how "sediment accommodation potential" is created and modified by the integration of a curve of sea-level change with subsidence. A diagram very similar to that of Barrell's was provided by Van Wagoner et al. (1990; see Fig. 5.5) and used to illustrate the deposition of shoaling-upward successions (these have been called **parasequences,** but for reasons explained in Sect. 5.4 the further use of this term is controversial). Zhang et al. (2018) demonstrated by modeling experiments that very similar sequence architectures could be generated by varying either sediment supply or accommodation independently.

The three major controls on basin architecture, subsidence/uplift (tectonism), sea-level change and sediment supply are themselves affected by a range of ultimate causes. Crustal extension, crustal loading and other regional tectonic processes provide the ultimate control on the size and architecture of sedimentary basins (Allen and Allen 2013). Sea-level change is driven by a range of low- and high-frequency processes. Sediment supply is affected by the tectonic elevation of the source area, which controls rates of erosion, and by climate, which affects such factors as rates of erosion, the caliber, volume and type of erosional detrital product, and the rates of subaqueous biogenic carbonate production. These controls and the nature of the resulting sequence record are discussed briefly in Sect. 5.5, and the reader is referred to the book-length treatment of the subject by Miall (2010) for a more complete discussion.

5.2.2 Stratigraphic Architecture

The recognition of unconformities and **stratigraphic terminations** is a key part of the method of **seismic stratigraphy**, as introduced by Vail et al. (1977). This helps to

Time intervals recorded by sedimentation
Columnar section
D
D
D
D
D
A
B
C
A
C
B
FIGURE 5.—Sedimentary Record made by harmonic Oscillations in Baselevel
A-A. Primary curve of rising baselevel.
B-B. Diastrophic oscillations, giving disconformities D-D.
C-C. Minor oscillations, exaggerated and simplified, due largely to climatic rhythms.
Equation of curve C-C: y = sin x —.25 cos 8 x —.05 cos 64 x.

Fig. 5.2 Barrell's (1917) explanation of how oscillatory variations in base-level control the timing of deposition. Sedimentation can only occur when base level is actively rising. These short intervals are indicated by the black bars in the top diagram. The resulting stratigraphic column, shown at the left, is full of disconformities, but appears to be the result of continuous sedimentation

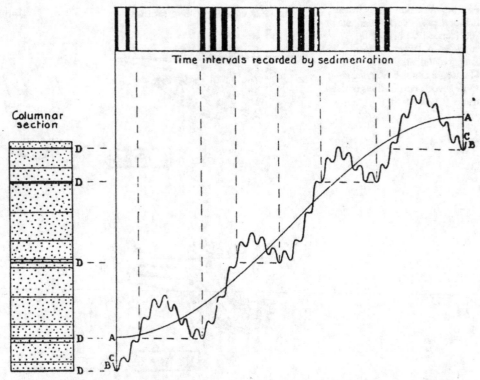

explain why modern sequence stratigraphy developed initially from the study of seismic data, because conventional basin analysis based on outcrop and well data provides little direct information on stratigraphic terminations, whereas these are readily, and sometimes spectacularly, displayed on seismic-reflection cross-sections (e.g., Fig. 1.15).

Unconformities are used to define stratigraphic sequences because they typically define mappable stratigraphic units that may be traced basin-wide, and potentially even have global significance. Because of the processes that generate the unconformity, neither the top of the underlying unit nor the base of the overlying one have a constant age. Erosional downcutting of the surface and subsequent transgression all take a discrete length of time. As seen in Figs. 5.6 and 5.7, transgression generates what is called an **onlap** pattern, which may take a significant proportion of the time represented by a sequence to reach its full extent.

It is generally assumed of an unconformity that the sediments lying above an unconformity are everywhere entirely younger than those lying below the unconformity. However, there are at least three situations in which diachronous unconformities may develop, such that beds below the unconformity are locally younger than certain beds lying above the unconformity. The first case is that where the unconformity is generated by marine erosion caused by deep ocean currents (Christie-Blick et al. 1990, 2007). These can shift in position across the sea floor as a result of changes in topography brought about by tectonism or sedimentation. Christie-Blick et al. (1990) cited the case of the Western Boundary Undercurrent that flows along the continental slope of the Atlantic Ocean off the United States. This current is erosive where it impinges on the continental slope, but deposition of entrained fine clastic material takes place at the margins of the main current, and the growth of this blanket is causing the current to gradually shift up the slope. The result is onlapping of the deposits onto the slope below the current, and erosional truncation of the upslope deposits.

The second type of diachronous unconformity is that which develops at basin margins as a result of syndepositional tectonism. Continuous deformation during sedimentation may lead to migration of a surface of erosion, and subsequent rapid onlap of the erosion surface by alluvial sediments (Riba 1976; Anadón et al. 1986). Typically these unconformities are associated with coarse conglomeratic sediments and die out rapidly into the basin. There is, therefore, little danger of their presence leading to the development of erroneous sequence stratigraphies.

Fig. 5.3 Relationship between rate of deposition (Rd) and rate of subsidence (Rs) in a delta complex. **A** Progradational, **B** aggradational, **C** retrogradational. Maps at right show successive positions of delta fronts (Curtis 1970)

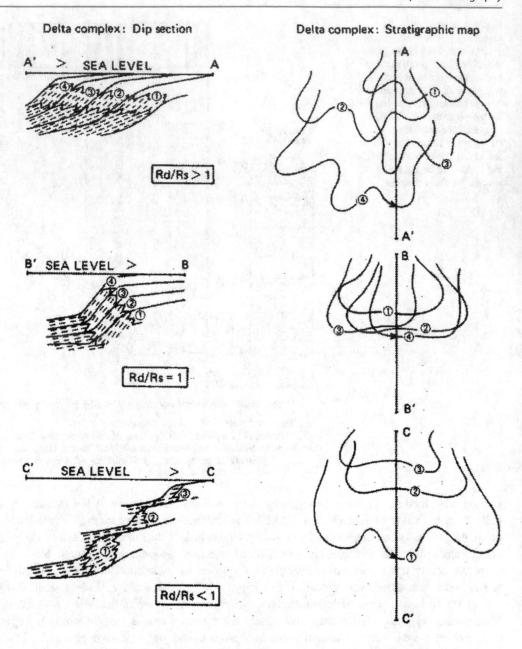

More important than either of these cases is the diachronous nature of the **subaerial erosion surface** that is used to define the sequence boundary in most stratigraphic studies. As Holbrook and Bhattacharya (2012) have demonstrated, the erosion of this surface by fluvial processes during the falling stage of a base-level cycle takes a significant length of time and may include the accumulation of fluvial terraces that are preserved as remnants that survive the subsequent transgression (Fig. 5.8). These authors referred to the resulting surface as a **regional composite scour** (acronym RCS). They argued, in part from flume simulation studies, that the preserved surface never exists as a single topographic surface during its formation, but evolves continuously until final burial. A good example of a

RCS is the surface that forms the basal sequence boundary of the Mesa Rica Sandstone, which is discontinuously exposed long a 400-km dip-oriented transect extending north–south from Colorado to New Mexico. Van Yperen et al. (2021) were able to demonstrate from sedimentological arguments that the basal surface consists of an amalgam of fluvial channel scours, incised valleys and distributary channels in a downdip direction.

Incised alluvial valleys may form a significant component of the subaerial erosion surface, particularly at continental margins (Strong and Paola 2008). Deepening and widening of an alluvial valley initiated during base-level fall may continue during the final stages of the evolution of a subaerial surface, even while a turn-around in the base-level

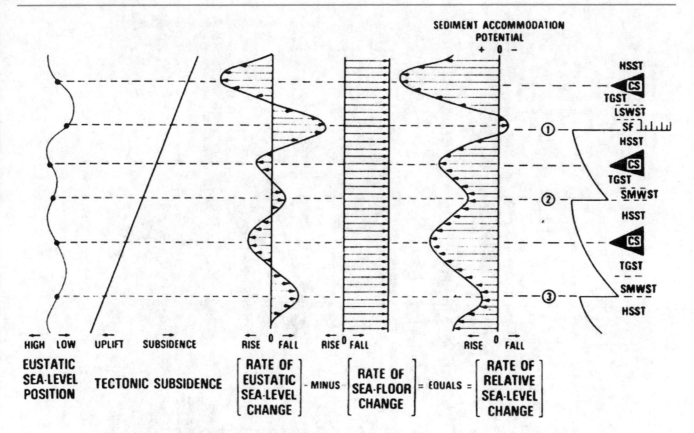

Fig. 5.4 The standard Exxon diagram illustrating the relationship between eustasy and tectonism and the creation of "sediment accommodation potential." Integrating the two curves produces a curve of relative change of sea level, from which the timing of sequence boundaries can be derived (events 1–3 in right-hand column). However, changing the shape of the tectonic subsidence curve will change the shape and position of highs and lows in the relative sea-level curve, a point not acknowledged in the Exxon work. This version of the diagram is from Loutit et al. (1988). CS = condensed section, HSST = highstand systems tract, LSWST = lowstand wedge systems tract, SF = marine fan, SMWST = shelf-margin wedge systems tract, TGST = transgressive systems tract

cycle has begun to transgress and bury the surface during the beginning of a transgression. Channel or overbank deposits that are preserved as terrace remnants, resting on the basal erosion surface, could therefore predate the coastal deposits formed during the final stages of base-level fall, and would therefore be older than the sequence boundary at the coast, although resting on it.

As discussed in Sect. 5.5.3, breaks in sedimentation, called **drowning unconformities**, occur in carbonate sedimentary environments, as a result of environmental change, having nothing to do with changes in relative sea level.

A **correlative conformity** (defined by Mitchum et al. 1977a, 1977b) is the deep-offshore equivalent of a subaerial unconformity. In practice, as noted later in this section, in many cases this surface is conceptual or hypothetical, occurring within a continuous section bearing no indication of the key stratigraphic processes and events taking place contemporaneously in shallow-marine and nonmarine environments. It may be possible to determine an approximate position of the correlative conformity by tracing seismic reflections, but this may be quite inadequate for the purpose

of formal sequence documentation and classification. As pointed out by Christie-Blick et al. (2007, p. 222), given that an unconformity represents a span of time, not an instant, "at some scale, unconformities pass laterally not into correlative conformities but into correlative intervals. Such considerations begin to be important as the resolution of the geological timescale improves at a global scale." The issue of time in stratigraphy is discussed at length in Chap. 8.

Various architectural, or geometric, characteristics record the lateral shift in depositional environments in response to sea-level change and subsidence (Figs. 5.6 and 5.7). **Onlap** typically takes place at the base of the succession, recording the beginning of a cycle of sedimentation. **Offlap** develops when the rate of sedimentation exceeds the rate of accommodation generation. An offlap architecture may predominate in settings of high sediment supply, when sediment is shed off the continental shelf onto the continental slope. **Toplap** represents the abrupt pinch-out of offlapping units at the shelf-slope break. This develops when there is a major difference in sedimentary accumulation between the shelf and slope, for example, when wave, tide or storm processes

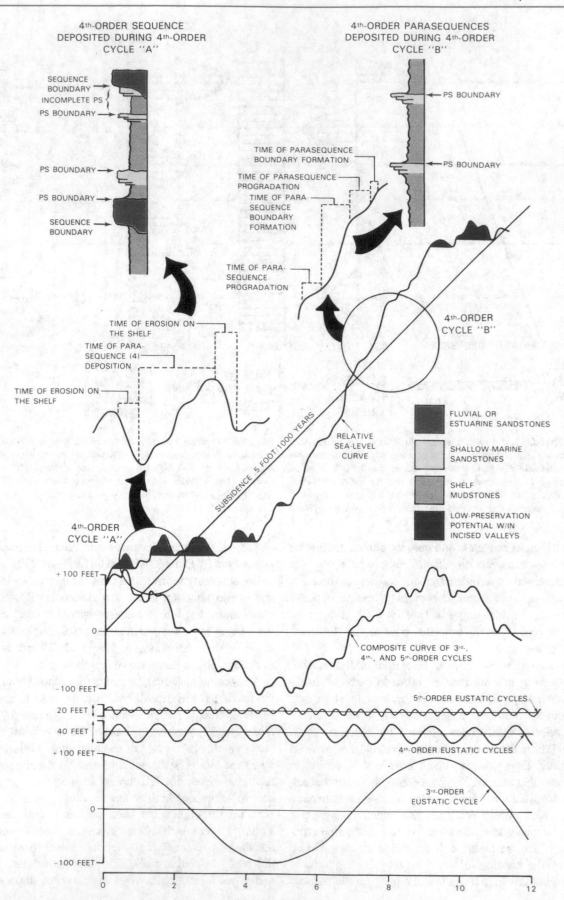

Fig. 5.5 A modern version of Barrell's diagram (Fig. 1.3), showing the relationship between accommodation changes and sedimentation. The "relative sea-level curve" is a composite of three "eustatic" sea-level curves (although this could include other, non-eustatic mechanisms, as discussed by Miall, 2010), integrated with a smooth subsidence curve. The colored areas of the composite curve indicate intervals of time when accommodation is being generated, and parasequences are deposited. Examples of parasequences are indicated by the arrows (Van Wagoner et al. 1990). AAPG © 1990, reprinted by permission of the AAPG whose permission is required for further use

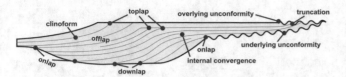

Fig. 5.6 Sequence architecture, showing common characteristics of "seismic-reflection terminations." (redrawn from Vail et al. 1977)

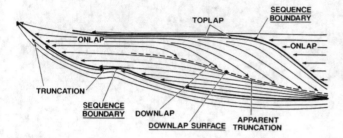

Fig. 5.7 A later diagram of sequence architecture (from Vail 1987), which incorporates the concept of initial onlap followed by progradation across a downlap surface. AAPG © 1987, reprinted by permission of the AAPG whose permission is required for further use

inhibit or prevent accumulation on the shelf. Sediment transported across the shelf is eventually delivered to the slope, a process termed **sediment bypass**. Toplap may represent abrupt thinning rather than truncation, with a thick slope unit passing laterally into a condensed section on the shelf. Discrimination between truncation and condensation may then depend on seismic resolution. **Downlap surfaces** may develop as a result of progradation across a basin floor, and they also develop during a transition from onlap to offlap. They typically develop above flooding surfaces, as basin-margin depositional systems begin to prograde seaward following the time of maximum flooding. The dipping, prograding units are called **clinoforms** and they lap out downward onto the downlap surface as lateral progradation takes place. The word **lapout** is used as a general term for all these types of stratigraphic termination. The term clinoform was borrowed from Rich (1951; see Fig. 1.3) but in his original definition the term was used for the sloping environment situated between the shelf and the deep basin. Rich (1951) termed the deposits that formed in this environment the clinothem, but this term has not survived.

The broad internal characteristics of stratigraphic units may be determined from their **seismic facies**, defined to mean an areally restricted group of seismic reflections whose appearance and characteristics are distinguishable from those of adjacent groups (Sangree and Widmier 1977). Various attributes may be used to define facies: reflection configuration, continuity, amplitude and frequency spectra, internal velocity, internal geometrical relations and external three-dimensional form (Sect. 6.3.3).

Figure 5.9 illustrates the main styles of seismic-facies reflection patterns (Mitchum et al. 1977a), and a further discussion and illustrations of seismic facies are presented in Sect. 6.3.3. Most of these are best seen in sections parallel to depositional dip. Parallel or subparallel reflections indicate uniform rates of deposition; divergent reflections result from differential subsidence rates, such as in a half-graben or across a shelf-margin hinge zone. Clinoform reflections comprise an important class of seismic-facies patterns (Pellegrini et al. 2020). They are particularly common on continental margins, where they commonly represent prograded deltaic or continental-slope outgrowth. Variations in clinoform architecture reflect different combinations of depositional energy, subsidence rates, sediment supply, water depth and sea-level position. **Sigmoid clinoforms** tend to have low depositional dips, typically less than 1°, whereas **oblique clinoforms** may show depositional dips up to 10°. **Parallel-oblique clinoform** patterns show no topsets. This usually implies shallow water depths with wave or current scour and sediment bypass to deeper water, perhaps down a submarine canyon that may be revealed on an adjacent seismic cross-section. Many seismic sequences show complex offlapping stratigraphy, of which the complex sigmoid-oblique clinoform pattern in Fig. 5.9 is a simple example. This diagram illustrates periods of sea-level still stand, with the development of truncated topsets (toplap) alternating with periods of sea-level rise (or more rapid basin subsidence), which allowed the lip of the prograding sequence to build upward as well as outward. Mitchum et al. (1977a) described the **hummocky clinoform** pattern as consisting of "irregular discontinuous subparallel reflection segments forming a practically random hummocky pattern marked by nonsystematic reflection terminations and splits. Relief on the hummocks typically is low, approaching the limits of seismic resolution. The reflection pattern is generally interpreted as strata forming small, interfingering clinoform lobes building into shallow water," such as the upbuilding or offlapping lobes of a delta undergoing distributary switching. Submarine fans may show the same hummocky reflections. **Shingled clinoform** patterns typically reflect offlapping sediment bodies on a continental shelf.

Chaotic reflections may reflect slumped or contorted sediment masses or those with abundant channels or cut-and-fill structures, such as submarine-fan systems. Many carbonate reefs also yield chaotic reflections. Disrupted reflections are usually caused by faults. Lenticular patterns are likely to be most common in sections oriented perpendicular to depositional dip. They represent the depositional lobes of deltas or submarine fans.

A **marine flooding surface** is a surface that separates older from younger strata, across which there is evidence of

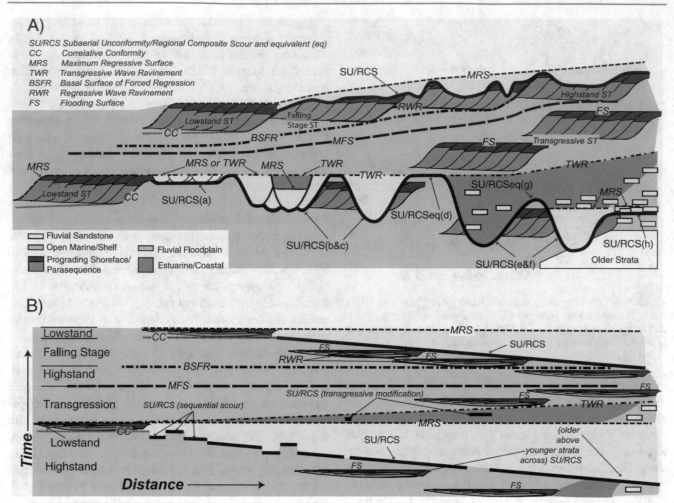

Fig. 5.8 The sequence-stratigraphic surfaces commonly used to define sequences and systems tracts expressed in A) longitudinal cross-section and B) a Wheeler diagram of time vs. distance. The regional composite scour (RCS) of this study is the same as the scoured part of the subaerial unconformity (SU) of other works. The RCS and its subaerial equivalent surface above interfluves together comprise the mapped SU surface that is used as the sequence boundary in depositional sequence stratigraphy (see Catuneanu 2006). The SU/RCS is shown in A) with its many possible variations, labeled here as a–h. SU/RCS variations include: **a** smooth RCS beneath a single-story channel sheet; **b** smooth RCS beneath a multivalley sheet; **c** RCS binding a single simple, complex, or compound valley; **d** RCS incised by and amalgamated with a later transgressive wave ravinement (TWR); **e** RCS binding fully estuarine/marine-influenced strata; **f** RCS cut into strata that predates the transgressive/regressive cycle during which the RCS was formed, **g** SU/RCS-equivalent as interfluve between incised valleys, potentially with well-developed paleosols; and **h** RCS separating fluvial strata. The Wheeler diagram in B) schematically shows the time–distance relationships of the strata in A) as they would be represented assuming the time-depositional concepts reviewed within this paper. The SU/RCS surface is highly diachronous recording its formation over the span of regression and with sequential scouring and some modifying scour during transgression. Likewise, the diachronous nature of scour implies older strata locally are deposited above the surface and younger below at differing locations (from Holbrook and Bhattacharya 2012, Fig. 1)

an abrupt increase in water depth. These surfaces are typically prominent and readily recognizable and mappable in the stratigraphic record. Each of the heavy, arrowed lines within the lower, retrogradational part of the sequence shown in Fig. 5.7 are marine flooding surfaces, as are the heavy lines in Fig. 5.10b. The **maximum flooding surface** records the maximum extent of marine drowning, and separates transgressive units below from regressive units above (the dashed line extending obliquely across the center of the cross-section in Fig. 5.7 is a maximum flooding surface). It commonly is a surface of considerable regional stratigraphic prominence and significance. It may be marked by a widespread shale, or by a condensed section, indicating slow sedimentation at a time of sediment starvation on the continental shelf (Loutit et al. 1988), and may correspond to a downlap surface, as noted above. The prominence of these surfaces led Galloway (1989) to propose that sequences be defined by the maximum flooding surface rather than the subaerial erosion surface. We discuss this, and other alternative concepts, in Sect. 7.7.

Fig. 5.9 Typical
seismic-reflection patterns,
illustrating the concept of seismic
facies (Mitchum et al. 1977b).
AAPG © 1977, reprinted by
permission of the AAPG whose
permission is required for further
use

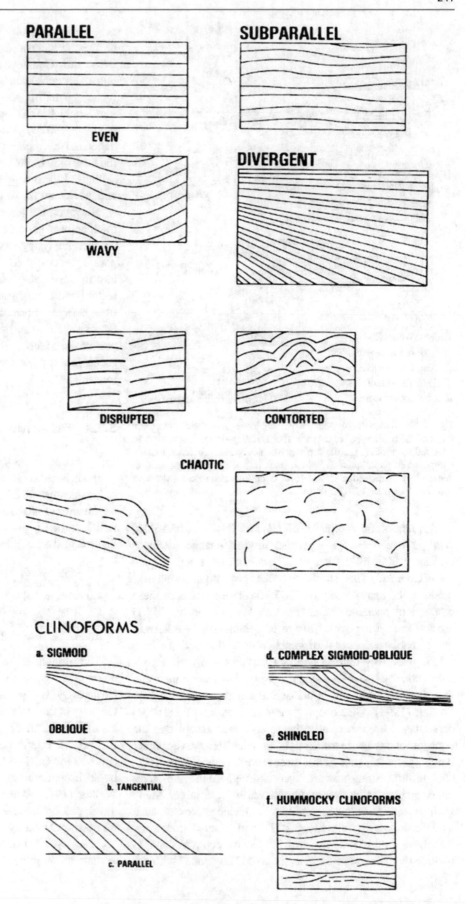

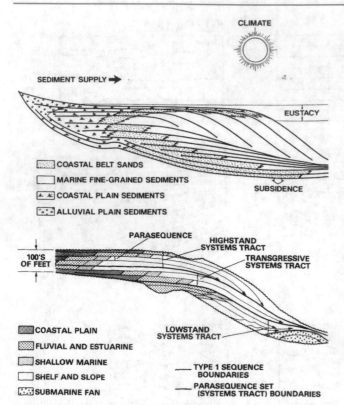

Fig. 5.10 Diagram of sequences, sequence sets, and composite sequences. **a** Parasequences are the shoaling-upward successions, bounded by flooding marine surfaces (the heavy lines). **b** Sequences are composed of parasequences, which stack into lowstand, transgressive, and highstand sequence sets to form composite sequences (Mitchum and Van Wagoner 1991)

Sequences may consist of stacked facies successions, each of which shows a gradual upward change in facies character, indicating a progressive shift in local depositional environments. The small packages of strata contained between the heavy lines in Fig. 5.10a are examples of these component packages of strata. Van Wagoner et al. (1987) erected the term **parasequence** to encompass "a relatively conformable succession of genetically related beds or bedsets bounded by marine flooding surfaces and their correlative surfaces … Parasequences are progradational and therefore the beds within parasequences shoal upward." As Walker (1992) pointed out, "parasequences and facies successions … are essentially the same thing, except that the concept of facies succession is broader." However, other types of facies succession occur within sequences (e.g., channel-fill fining-upward successions), and the term parasequence is therefore unnecessarily restrictive. Many such successions are generated by autogenic processes, such as delta lobe switching (Fig. 7.27), and channel migration, that have nothing to do with sequence controls, and to include them in a term that has the word "sequence" within it

may be misleading. Walker (1992) recommended that the term parasequence not be used. Catuneanu (2006, p. 243–245) and Catuneanu and Zecchin (2020) pointed out numerous problems with the concept of the parasequence, including the imprecise meaning of the term "flooding surface" (which, it is now recognized, may have several different meanings) and the potential confusion with surfaces generated by autogenic processes. He recommended using the term only in the context of progradational units in coastal settings. I have suggested that the term be abandoned altogether. We return to this point in Sect. 5.4.

However, the term parasequence is clearly entrenched in the literature. Colombera and Mountney (2020) carried out a detailed analysis of nearly 1000 units that had been termed parasequences in the published geological record, with the aim of developing useful generalizations about their architecture and deposition. They did not make specific distinctions between those of autogenic and allogenic origin. Li and Schieber (2020) carried out a meticulous outcrop analysis of part of the Mancos Shale in Utah and established a detailed succession of parasequences consisting largely of shelf mudstones.

5.2.3 Depositional Systems and Systems Tracts

The concept of the **depositional-system** and basin-analysis methods based on it were developed largely in the Gulf Coast region as a means of analyzing and interpreting the immense thicknesses of Mesozoic–Cenozoic sediment there that are so rich in oil and gas (Sect. 1.2.3). A depositional system is defined in the Schlumberger online Oilfied Glossary as

> The three-dimensional array of sediments or lithofacies that fills a basin. Depositional systems vary according to the types of sediments available for deposition as well as the depositional processes and environments in which they are deposited. The dominant depositional systems are alluvial, fluvial, deltaic, marine, lacustrine and eolian systems.

The principles of depositional-systems analysis have never been formally stated but have been widely used, particularly by geologists of the Bureau of Economic Geology at the University of Texas (notably W. L. Fisher, L. F. Brown Jr., J. H. McGowen, W. E. Galloway and D. E. Frazier). Useful papers on the topic are those by Fisher and McGowen (1967) and Brown and Fisher (1977). Textbook discussions are given by Miall (1999, Chap. 6) and Walker (1992). The concept of depositional episode was developed by Frazier (1974) to explain the construction of Mississippi Delta by progradation of successive delta lobes. Posamentier et al. (1988, p. 110) defined a **depositional system** as "a three-dimensional assemblage of lithofacies, genetically linked by active (modern) or inferred (ancient) processes and environments." A **systems tract** is defined as

> A linkage of contemporaneous depositional systems ... Each is defined objectively by stratal geometries at bounding surfaces, position within the sequence, and internal parasequence stacking patterns. Each is interpreted to be associated with a specific segment of the eustatic curve (i.e., eustatic lowstand-lowstand wedge; eustatic rise-transgressive; rapid eustatic fall-lowstand fan, and so on), although not defined on the basis of this association (Posamentier et al. 1988).

Elsewhere, Van Wagoner et al. (1987) stated that "when referring to systems tracts, the terms lowstand and highstand are not meant to imply a unique period of time or position on a cycle of eustatic or relative change of sea level. The actual time of initiation of a systems tract is interpreted to be a function of the interaction between eustasy, sediment supply and tectonics." There is clearly an inherent, or built-in contradiction here, that results from the use in a descriptive sense of terminology that has a genetic connotation (e.g., transgressive systems tract implies transgression). We return to this problem below.

Systems tracts are named with reference to their assumed position within the sea-level cycle, and these names incorporate ideas about the expected response of a basin to the changing balance between the major sedimentary controls (accommodation and sediment supply) during a base-level cycle. There are four standard systems tracts. These are the **highstand, falling stage, lowstand** and **transgressive systems tracts.** Standard acronyms for these, which are almost universally used, are HST, FSST, LST and TST. Each is illustrated here by a block diagram model with summary remarks outlining the major sedimentary controls and depositional patterns prevailing at that stage of sequence development (Fig. 5.11). Other terms have been used by different workers, but these four systems tracts and their bounding surfaces provide a useful, easy-to-understand model from which to build interpretive concepts.

The evolution of a typical clastic sequence is shown in Fig. 5.12, by means of twelve timelines. These are shown against the succession of systems tracts in the upper diagram, and keyed to the acronyms for the systems tracts in the lower left. Some typical lithofacies are indicated in the lower diagram, together with the position of some vertical profiles that would be particularly useful as providing information on facies cycles for the interpretation of the sequences from drill core.

The linking of systems tracts to stages of the sea-level cycle is misleading, because it has now been repeatedly demonstrated that the geometric and behavioral features that supposedly characterize each systems tract and its link to the sea-level cycle are not necessarily diagnostic of sea level, but may reflect combinations of several factors. For example, on the east coast of South Island, New Zealand, different

stretches of coast are simultaneously undergoing coastal progradation, reflecting a large sediment supply, and coastal retreat and transgression, because of locally high wave energy (Leckie 1994). Andros Island, in the Bahamas, currently exhibits three different systems tract conditions. Lowstand conditions characterize the eastern (windward) margin of the island, facing the deep-water channel, the Tongue of the Ocean. Transgressive conditions occur along the northwest margin, where tidal flats are undergoing erosion, whereas on the more sheltered leeward margin, along the southwest edge of the Andros coastline, highstand conditions are suggested by tidal-flat progradation (Schlager 2005, Fig. 7.6).

An attempt to remove systems tract terminology from associations with sea-level change was made by Neal and Abreu (2009) and Neal et al. (2016). They substituted the term Aggradation-Progradation-Degradation (APD) for the HST-FSST, Progradation-Aggradation (PA) for the LST, and Retrogradation (R) for the TST, based on the stacking patterns of the systems tracts. All these terms reflect the balance between sediment supply and accommodation and are therefore not fundamentally different from the standard terminology used here.

The use of seismic-stratigraphic data for the construction of regional stratigraphies is discussed in Sect. 6.3.

5.3 Sequence Models in Clastic and Carbonate Settings

In this section, a brief overview of sequence models is provided for the main areas and styles of deposition, marine and nonmarine clastics, and carbonates. A much more complete treatment of this topic forms the core of the book by Catuneanu (2006). In addition, many useful review articles and books provide additional insights into particular areas or themes (e.g., Emery and Myers 1996; Schlager 1992, 1993, 2005; Posamentier and Allen 1999; Hunt and Gawthorpe 2000; Yoshida et al. 2007). What follows corresponds to the model Catuneanu (2006) described as "Depositional Sequence IV" (Fig. 7.26). Variants on this model, and other models, are described briefly in Sect. 7.7.

5.3.1 Marine Clastic Depositional Systems and Systems Tracts

In this section a description is provided of the most typical, or "model" cycle of sequence development, in which it is assumed that the main sediment transport mode is normal to the coastline. Sediment transport and deposition out of this line of section may introduce significant complications, and

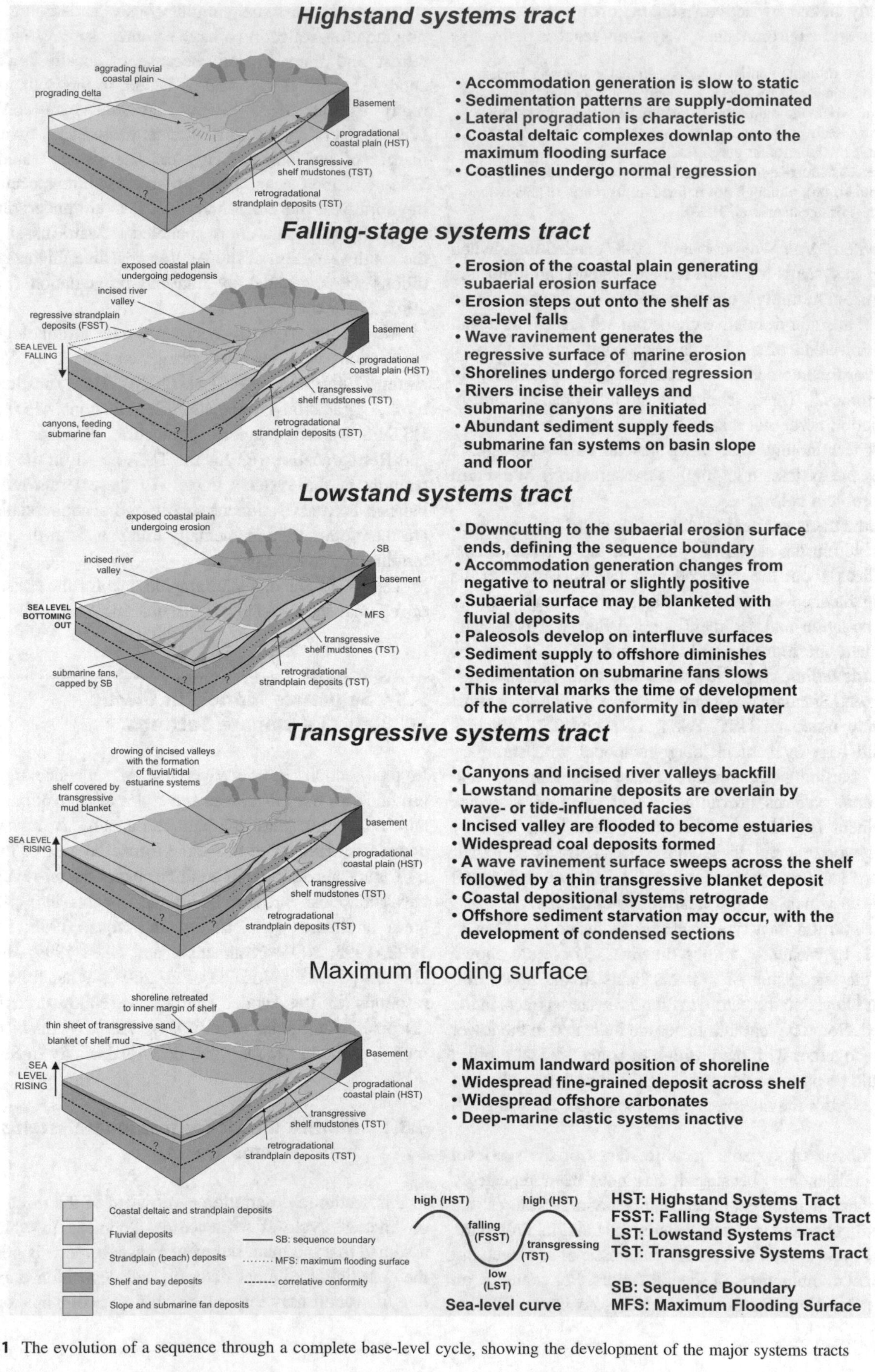

Highstand systems tract
aggrading fluvial coastal plain
prograding delta
Basement
progradational coastal plain (HST)
transgressive shelf mudstones (TST)
retrogradational strandplain deposits (TST)
• Accommodation generation is slow to static
• Sedimentation patterns are supply-dominated
• Lateral progradation is characteristic
• Coastal deltaic complexes downlap onto the maximum flooding surface
• Coastlines undergo normal regression
Falling-stage systems tract
exposed coastal plain undergoing pedogenesis
incised river valley
regressive strandplain deposits (FSST)
SEA LEVEL FALLING
basement
progradational coastal plain (HST)
transgressive shelf mudstones (TST)
retrogradational strandplain deposits (TST)
canyons, feeding submarine fan
• Erosion of the coastal plain generating subaerial erosion surface
• Erosion steps out onto the shelf as sea-level falls
• Wave ravinement generates the regressive surface of marine erosion
• Shorelines undergo forced regression
• Rivers incise their valleys and submarine canyons are initiated
• Abundant sediment supply feeds submarine fan systems on basin slope and floor
Lowstand systems tract
exposed coastal plain undergoing erosion
incised river valley
SB
basement
MFS
SEA LEVEL BOTTOMING OUT
transgressive shelf mudstones (TST)
submarine fans, capped by SB
retrogradational strandplain deposits (TST)
• Downcutting to the subaerial erosion surface ends, defining the sequence boundary
• Accommodation generation changes from negative to neutral or slightly positive
• Subaerial surface may be blanketed with fluvial deposits
• Paleosols develop on interfluve surfaces
• Sediment supply to offshore diminishes
• Sedimentation on submarine fans slows
• This interval marks the time of development of the correlative conformity in deep water
Transgressive systems tract
drowing of incised valleys with the formation of fluvial/tidal estuarine systems
shelf covered by transgressive mud blanket
basement
SEA LEVEL RISING
progradational coastal plain (HST)
transgressive shelf mudstones (TST)
retrogradational strandplain deposits (TST)
• Canyons and incised river valleys backfill
• Lowstand nomarine deposits are overlain by wave- or tide-influenced facies
• Incised valley are flooded to become estuaries
• Widespread coal deposits formed
• A wave ravinement surface sweeps across the shelf followed by a thin transgressive blanket deposit
• Coastal depositional systems retrograde
• Offshore sediment starvation may occur, with the development of condensed sections
Maximum flooding surface
shoreline retreated to inner margin of shelf
thin sheet of transgressive sand
blanket of shelf mud
Basement
SEA LEVEL RISING
progradational coastal plain (HST)
transgressive shelf mudstones (TST)
retrogradational strandplain deposits (TST)
• Maximum landward position of shoreline
• Widespread fine-grained deposit across shelf
• Widespread offshore carbonates
• Deep-marine clastic systems inactive
Coastal deltaic and strandplain deposits
Fluvial deposits
Strandplain (beach) deposits
Shelf and bay deposits
Slope and submarine fan deposits
SB: sequence boundary
MFS: maximum flooding surface
correlative conformity
high (HST)
high (HST)
falling (FSST)
transgressing (TST)
low (LST)
Sea-level curve
HST: Highstand Systems Tract
FSST: Falling Stage Systems Tract
LST: Lowstand Systems Tract
TST: Transgressive Systems Tract
SB: Sequence Boundary
MFS: Maximum Flooding Surface

Fig. 5.11 The evolution of a sequence through a complete base-level cycle, showing the development of the major systems tracts

Fig. 5.12 A typical sequence, with twelve timelines showing the evolution of the four major systems tracts

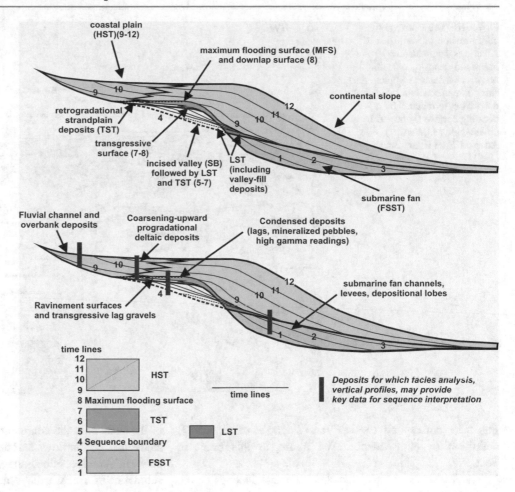

is the reason why sequence analysis should be a completely three-dimensional process (Madof et al. 2016; Burgess 2016). An example of how tectonic and sedimentary processes change dramatically along strike is described briefly in Sect. 6.3.3. (Fig. 6.29).

The acronyms provided in brackets for many of the terms are the standard abbreviations that are now in common use in research literature and in many textbooks.

The highstand systems tract (HST) corresponds to a period when little new accommodation is being added to the depositional environment (timelines 9–12 in Fig. 5.12). As shown in the relative sea-level curve at the lower right corner of Fig. 5.11, base-level rise is in the process of slowing down as it reaches its highest point, immediately prior to commencement of a slow fall. If the sediment supply remains more or less constant, then the rate of deposition (Rd) may be greater than the rate of subsidence (Rs; more correctly, "rate of accommodation increase") at this point (the upper of the three conditions shown in Fig. 5.3). The most characteristic feature of this systems tract is the lateral progradation of coastal sedimentary environments. Major coastal barrier-lagoon and deltaic complexes are the result. **Normal regression** is the term used to describe the seaward

advance of the coastline as a result of the progressive addition of sediment to the front of the beach or the delta systems, developing a broad **topset** environment (the **undathem** of Rich 1951; Fig. 1.3). This is in contrast to the condition of forced regression, which is described below.

Where the terrigenous sediment supply is high, delta systems may largely dominate the resulting sedimentary succession at the coastline, as shown in the accompanying example of the Dunvegan delta, Alberta (Fig. 5.13). The allomember boundaries in this diagram indicate times of relative low sea level, followed by flooding. Each allomember boundary is overlain by a mudstone representing the maximum flooding surface, over which delta complexes prograded. Sedimentary environments characteristically include coastal mangrove swamps and may include significant peat swamps, the sites of future coal development. The numbered subdivisions of each allomember indicate individual deltaic shingles. Subsurface mapping may indicate that shingles of this type shifted laterally as a result of delta switching, in a manner similar to the Mississippi delta (Fig. 4.21) and the Yellow River delta. This points to potential confusion in terminology, because upward-shoaling successions, such as those illustrated in

Fig. 5.13 The Dunvegan alloformation of northwest Alberta is dominated by highstand deposits, reflecting its origins as a deltaic complex. Each shingle represents an individual delta lobe, terminated by a flooding surface (indicated by a downlapping half-arrow). Adapted from Bhattacharya (1991)

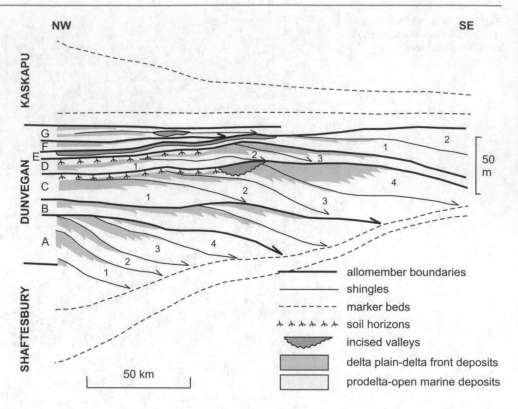

Fig. 5.13 correspond to the Van Wagoner et al. (1987) definition of parasequence. We return to this point in Sect. 7.7.

The thickness of highstand shelf deposits depends on the accommodation generated by marine transgression across the shelf, typically a few tens of meters, up to a maximum of about 200 m. Where the shelf is narrow or the sediment supply is large, deltas may prograde to the shelf-slope break, at which point deltaic sedimentation may extend down slope into the deep basin (Porebski and Steel 2003). High-amplitude clinoforms may result, including significant volumes of sediment gravity flow deposits.

The falling-stage systems tract (FSST). A fall in base level from the highstand position exposes the coastal plain and then the continental shelf, to subaerial erosion (timelines 1–3 in Fig. 5.12). River mouths migrate seaward, and under most conditions, river valleys incise themselves as they continually grade downward to progressively lower sea levels (Fig. 5.11). Deeply incised paleovalleys may result. In the rock record, many of these show evidence of multiple erosional events (Korus et al. 2008; Strong and Paola 2008; Holbrook and Bhattacharya 2012), indicating repeated responses to autogenic threshold triggers (Schumm 1979) or perhaps to minor cycles of base-level change. Significant volumes of sediment are eroded from the coastal plain and the shelf and are fed through the coastal fluvial systems and onto the shelf or down submarine canyons. Eventually, these large sediment volumes may be tipped directly over the edge

of the shelf onto the continental slope, triggering submarine landslides, debris flows and turbidity currents. These have a powerful erosive effect and may initiate development of submarine erosional valleys at the mouths of the major rivers or offshore from major delta distributaries. Many submarine canyons are initiated by this process and remain as major routes for sediment dispersal through successive cycles of base-level change. The FSST is typically a major period of growth of submarine fans.

The falling base level causes basinward retreat of the shoreline, a process termed **forced regression** by Plint (1991) (Fig. 5.14). The occurrence of forced regression, as distinct from normal regression, may be detected by careful mapping of coastal shoreline sandstone complexes. Fall of sea level causes water depths over the shelf to decrease, increasing the erosive power of waves and tides. This typically leads to the development of a surface called the **regressive surface of marine erosion (RSME),** which truncates shelf and distal coastal (e.g., deltaic) deposits that had been formed during the preceding highstand phase (Fig. 5.14b). The first such surface to form, at the commencement of a phase of sea-level fall, is termed the **basal surface of forced regression.** Some specialists used this surface as the basis for sequence definition, as discussed in Sect. 7.7. Given an adequate sediment supply, especially if there are pauses during the fall of sea level (Fig. 5.14c), shoreface sand accumulates above the RSME, forming what have come to be informally termed **sharp-based sandstone**

Fig. 5.14 The process of forced regression, and the development of the regressive surface of marine erosion and "sharp-based sand bodies." Original diagram from Plint (1988)

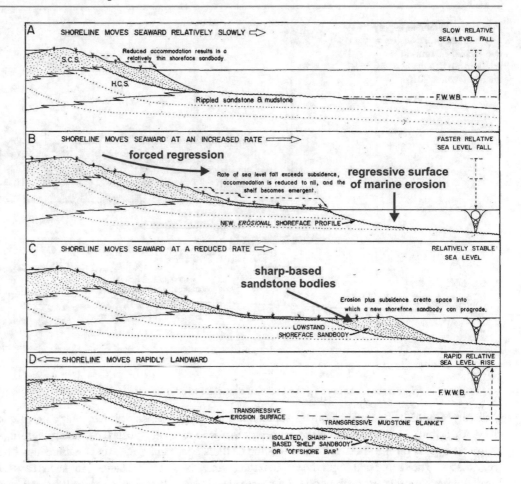

bodies (Plint 1988). These are internally identical to other coastal, regressive sandstone bodies, except that they rest on an erosion surface instead of grading up from the fine-grained shelf sediments, as in the initial coastal sands shown in Fig. 5.14a (which are the product of normal regression). Repeated pulses of sea-level fall punctuated by still stand may develop several offlapping surfaces of marine erosion. Shelf-margin deltas may form where the mouths of major river systems regress to the shelf-slope break during forced regression (Porębski and Steel 2003).

The architecture of the nonmarine–marine interface depends largely on the relative rate of sedimentation versus accommodation generation. As Helland-Hansen and Martinsen (1996) and Helland-Hansen and Hampson (2009) demonstrated, the stacking patterns of shoreline elements reflect the balance between sediment supply and accommodation generation. A shoreline "trajectory" can be reconstructed as successive base-level cycles generate superimposed regressive deposits, with differences depending on whether the shoreline is predominantly depositional, accretionary and stratigraphically rising, or whether it is erosional, non-accretionary and falling.

As noted above, the falling stage is typically the interval during the sea-level cycle when the sediment supply to the continental shelf is at its greatest. However, the timing and magnitude of the sediment flux to the continental slope depends on many factors, including the width and depth of the continental shelf (Fig. 5.15). Most of the early sequence models (e.g., Posamentier et al. 1988; Posamentier and Vail 1988) showed submarine fans resting on a basal sequence boundary, but this configuration now seems unlikely. Given a low-accommodation shelf and adequate sediment supply, submarine fans (or turbidites systems, to use the term that many now prefer) may develop at the base of the continental slope at any time during a sea-level cycle (e.g., Burgess and Hovius 1998; Carvajal and Steel 2006).

On the continental shelf and coastal plain, the sequence boundary is an erosion surface representing the lowest point to which erosion cuts during the falling stage of the base-level cycle. As demonstrated by Holbrook and Bhattacharya (2012), this subaerial surface is a composite feature that continues to develop through the falling-stage cycle and may preserve remnants of fluvial channel and floodplain deposits as terrace remnants resting on the surface (Fig. 5.8). The final preserved surface therefore never existed as a single topographic surface during the falling stage. As sea-level fall slows to its lowest point, sediment delivery from the newly exposed coastal plain and shelf will

Fig. 5.15 Sediment supply and depositional processes on a continental margin, in relationship to a cycle of sea-level change (Arnott 2010)

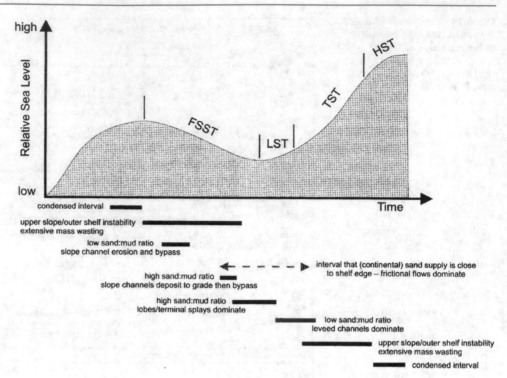

gradually diminish. Sedimentation on submarine fans may slow down, but much depends on the processes that lead to sediment bypass on the shelf and mass wasting of the slope (Fig. 5.15). The sequence boundary, therefore, may be located anywhere within a submarine-fan succession. However, there is unlikely to be an actual mappable break in sedimentation at this level, and it may be difficult to impossible to locate the position in the section corresponding to the turn-around from falling to rising sea level. This horizon is, therefore, what Vail et al. (1977) called a **correlative conformity (CC)**, although his original application of the term was to the fine-grained sediments formed in deep water beyond the submarine-fan wedge, out in the deep basin where it was assumed sedimentation would be continuous throughout a sea-level cycle. As later studies have demonstrated, however, erosion surfaces are common in fine-grained deposits, even in those deposited in deep-marine settings (Bohacs et al. 2002; see additional discussion of this point below).

The sequence boundary (SB) marks the lowest point reached by erosion during the falling stage of the sea-level cycle. On land this is represented by a subaerial erosion surface, which may extend far onto the continental shelf, depending on how far sea-level falls. The sequence boundary cuts into the deposits of the highstand systems tract and is overlain by the deposits of the lowstand or transgressive systems tracts (Fig. 5.8). It is therefore typically a surface where a marked facies change takes place, usually from a relatively lower energy deposit below to a high-energy

deposit above. Mapping of such a surface in outcrop or in the subsurface, using well-logs, is facilitated by this facies change, except where the boundary juxtaposes fluvial on fluvial facies. In such cases, distinguishing the sequence boundary from other large-scale channel scours may be a difficult undertaking.

The *lowstand systems tract (LST):* This systems tract represents the interval of time when sea-level has bottomed out, and depositional trends undergo a shift from seaward-directed (e.g., progradational) to landward-directed (e.g., retrogradational). This corresponds to the time between timelines 4 and 5 in Fig. 5.12. Within most depositional systems there is little that may be confidently assigned to the lowstand systems tract. The initial basal fill of incised river valleys, and some of the fill of submarine canyons are deposited during this phase. Volumetrically the deposits are usually of minor importance, but they may be of a coarser grain size than succeeding transgressive deposits. In parts of the incised valley of the Mississippi River, for example (the valley formed during Pleistocene glacioeustatic sea-level lowstands), the basal fill formed during the initial post-glacial transgression is a coarse braided stream deposit, in contrast to the sandy meandering river deposits that form the bulk of the Mississippi River sediments. The episode of active submarine-fan sedimentation on the continental slope and deep basin may persist through the lowstand phase.

There may be a phase of normal regressive sedimentation at the lowstand coastline. On coastal plains, the lowstand is a time of still stand, or stasis, when little erosion or sedimentation takes place (Tipper 2015). Between the major rivers, on the interfluve uplands, this may therefore be a place where long-established plant growth and soil development takes place. Peat is unlikely to accumulate because of the lack of accommodation, but soils, corresponding in time to the sequence boundary, may be extensive, and the resulting paleosols may therefore be employed for mapping purposes (e.g., McCarthy et al. 1999; Plint et al. 2001). In many nonmarine and shallow-marine environments, where sedimentation rates are negligible during this period of stasis, bedding planes may develop as "true substrates," with rich trace-fossil assemblages, (Davies and Shillito 2018, 2021).

Transgressive systems tract (TST): A rise in base level is typically accompanied by flooding of incised valleys and transgression across the continental shelf (Fig. 5.11; timelines 5–7 in Fig. 5.12). Base-level rise may exceed sediment supply, leading to retrogradation of depositional systems (Rd < Rs in Fig. 5.3), although at the mouths of the largest rivers (such as the modern Amazon) sediment supply may be sufficiently large that deltas or submarine fans may continue to aggrade or prograde.

Flooded river valleys are estuaries; they typically provide ample accommodation for sedimentary accumulation. In estuarine successions, the upward transition from lowstand to transgressive systems tract in estuaries and other coastal river systems is commonly marked by the development of wave- or tide-influenced fluvial facies, such as tidal sand bars containing sigmoidal crossbedding or flaser bedding (Figs. 3.27, 3.28 and 3.29). The sedimentology of this environment has received much attention (Fig. 5.16), because of the potential for the development of stratigraphic sandstone traps, in the form of valley fill ribbon sands. Studies of ancient paleovalley fills have shown that many are complex, indicating repeated cycles of base level change and/or autogenic changes in sediment dispersal (Korus et al. 2008).

On the continental shelf the most distinctive feature of most transgressive systems tracts is the development of a widespread **transgressive surface (TS)**, a flooding surface covered with an equally widespread marine mudstone. A transgressive conglomeratic or sandy lag may blanket the flooding surface. Offshore, rapid transgression may cut the deep-water environment off from its sediment source, leading to slow sedimentation, and the formation of a **condensed section**. This is commonly a distinctive facies, consisting of concentrated shell or fish fragments, amalgamated biozones, and a "hot" (high gamma-ray) response on well-logs, reflecting a concentration of radioactive clays (Loutit et al. 1988; Bohacs et al. 2002). Significant volumes of clastic sediment deposited on the shelf may be reworked during transgression, and mass wasting of the continental slope, with delivery of large volumes of detritus to submarine-fan

complexes at the base of the slope is also common (Fig. 5.15). Posamentier (2002) documented numerous complexes of shelf-sand ridges constituting parts of shelf transgressive systems tracts that were formed by vigorous wave and tide action. Offshore, limestones may be deposited, such as the several Jurassic and Cretaceous limestones and chalks in the Western Interior Seaway (Greenhorn Limestone, Austin Chalk).

In the nearshore setting, wave erosion during transgression is usually the cause of ravinement, with the development of a diachronous **ravinement surface** (Fig. 5.17). The juxtaposition of marine shelf sediments, above, over coastal shoreline or lagoonal sediments below, creates a prominent surface which should not be confused with a sequence boundary. A ravinement surface marks an upward deepening, the opposite of the facies relationships at most sequence boundaries. However, in some cases, ravinement erosion may cut down through lowstand deposits and into the underlying highstand systems tract, and in such cases the ravinement surface becomes the sequence boundary (Nummedal and Swift 1987; Zecchin et al. 2019).

Peat may be deposited on the coastal plain and in deltaic settings at any time during a cycle of base-level change. However, the thickest and most widespread coals are now known to be those formed from peat accumulated during transgression, because of the accommodation provided by rising base level, during a time when clastic influx into the coastal plain is "held back" by the landward-advancing shoreline (Bohacs and Suter 1997; see Fig. 5.18).

The maximum flooding surface (MFS) marks the end of the phase during which the difference between the rate of sea-level rise and the rate of sediment supply is at its greatest (Fig. 5.11; timeline 8 in Fig. 5.12). Sea-level rise continues beyond this point, but as the rate of rise slows, sediment input begins to re-establish progradation at the shoreline, and this defines the transition into the highstand systems tract. The offshore shale commonly formed around the time of the MFS is an excellent mapping marker, because of its widespread nature and distinctive facies. In areas distant from the shoreline, where clastic sediment supply is at a minimum, the MFS is commonly marked by calcareous shale, marl or limestone. In some studies, sequence mapping is more easily accomplished using this surface in preference to the sequence boundary, because of its more predictable facies and its consistent horizontality (Sect. 7.7). This was the preference of Frazier (1974) and Galloway (1989), as shown in Figs. 7.24 and 7.26, but their proposal did not become widely adopted.

The preceding paragraphs constitute a set of useful generalizations. However, there are many exceptions and special cases. For example, consider the ultimate fate of the clastic sediment flux on continental margins during cycles of sea-level change. In the traditional model (Posamentier et al.

Fig. 5.16 Depositional model for estuaries (Reinson 1992; Dalrymple et al. 1994)

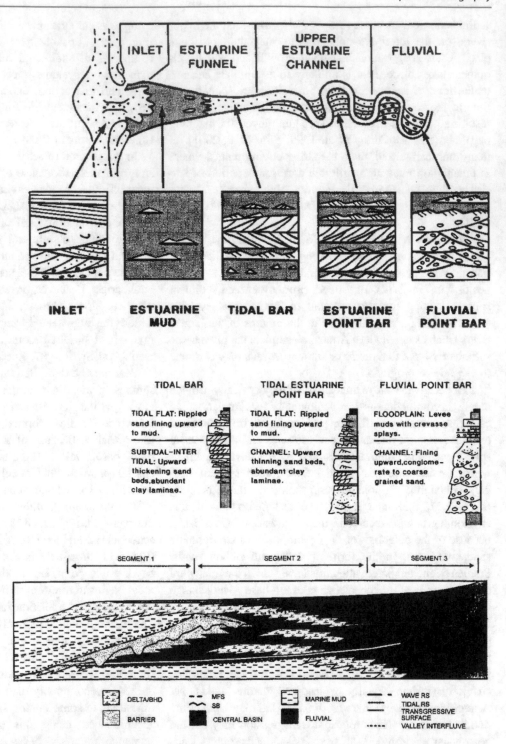

1988), on which this section is largely based, coastal plain complexes, including deltas, typically accumulate during highstand phases, following a period of coastal plain transgression and flooding, and basin slope and plain deposits, including submarine fans, accumulate during the sea-level falling stage and lowstand. However, these generalizations do not necessarily apply to all continental margins. As Carvajal and Steel (2006, p. 665) pointed out,

This model has been challenged using examples from narrow shelf settings (e.g., fans in the California Borderland, Gulf of Corinth, and Mediterranean Sea; see Piper and Normark 2001; Ito and Masuda 1988) or extremely high supply systems (e.g., Bengal Fan; Weber et al. 1997). In these cases slope canyons extending to almost the shoreline may receive sand from littoral drift or shelf currents during rising sea level. In addition, deltas may easily cross narrow shelves and provide sand for deep-water deposits under normal supply conditions during relative sea-level highstand. It has also been postulated that in

Fig. 5.17 The process of transgressive ravinement. The dashed red line indicates the time correlation between shelf sediments deposited on the ravinement surface and contemporaneous lagoon deposits that are truncated by ravinement erosion as sea-level rises. Adapted from Nummedal and Swift (1987)

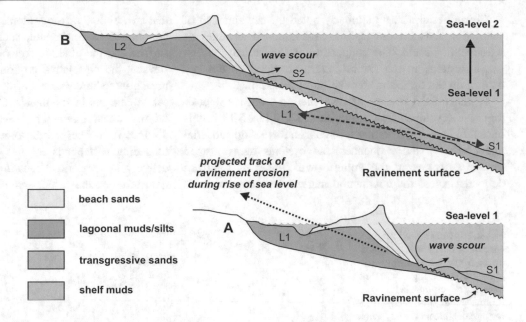

Fig. 5.18 The dependency of the lateral extent and thickness of coal seams on the rate of change of base level (Bohacs and Suter 1997). AAPG © 1997, reprinted by permission of the AAPG whose permission is required for further use

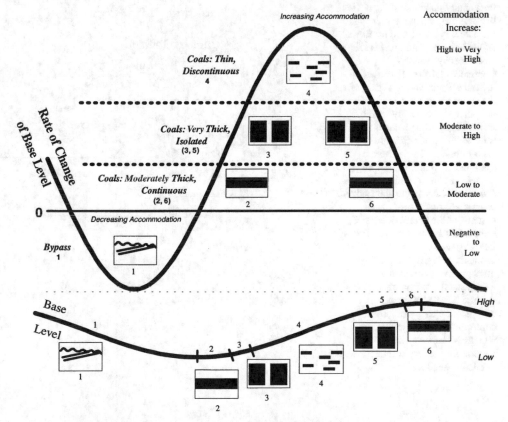

moderately wide (tens of kilometers) to wide shelf (hundreds of kilometers) settings, significant volumes of sand can be bypassed to deep-water areas at highstand through shelf-edge deltas (Burgess and Hovius 1998; Porebski and Steel 2006). Nonetheless, documenting such delivery either in the modern or ancient has been difficult (except for suggestions from studies at the third order time scale, e.g., McMillen and Winn 1991),

biasing researchers to interpret ancient deep-water deposits preferentially following the lowstand model. Thus, focus on this lowstand model has tended to cause us to overlook (1) the dominant role that sediment supply may play in deep-water sediment delivery, and (2) how such supply-dominated shelf margins can generate deep-water fans even during periods of rising relative sea level.

Covault et al. (2007) similarly noted the development of submarine fans on the California borderland at times of sea-level highstand. The connection of canyon and fan dispersal systems to the littoral sediment supply is the key control on the timing of deposition in this setting.

A generalized model for deep-marine sequence deposition was developed by Catuneanu (2020b: Fig. 5.19 in this book). In these environments, of course, there is no subaerial unconformity, and sequences and systems tracts instead reflect variations in the volume and caliber of sediment delivered across the continental margin. Bounding surfaces

that are key to the recognition and definition of sequences in the shallow-marine realm, such as the maximum flooding surface and the offshore equivalent of the subaerial erosion surface, the correlative conformity, typically do not mark breaks in sedimentation but may be indicated, if at all, by subtle changes in lithofacies. Changes in relative sea level, driving transgression and regression, affect the transport distance from river mouth to continental slope. The volume of sediment depends in part on the scale of the fluvial delivery system and the nature of its connection to slope transport systems. For example, at present, sediment

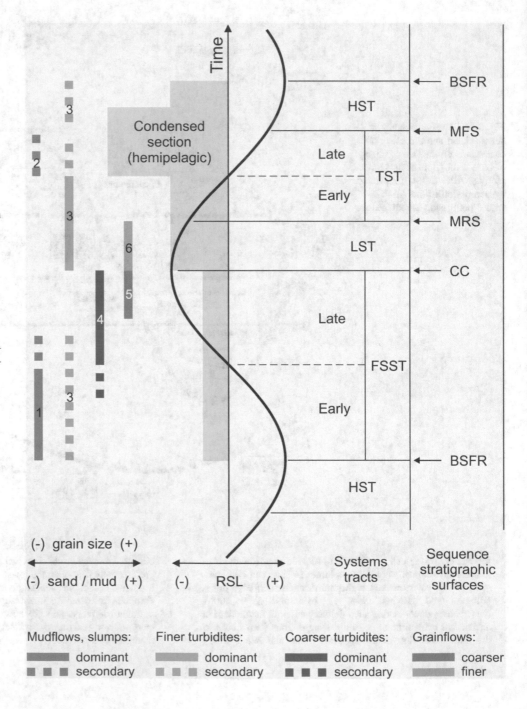

Fig. 5.19 Generalized model describing commonly observed trends of change in deep-water gravity flows during a shoreline transit cycle on the shelf (Catuneanu 2020b). Diagnostic for the identification of systems tracts and bounding surfaces are the trends of change rather than the actual types of gravity flows, as the latter depend on several basin-specific variables. Conducive factors: 1—muds on the outer shelf, and lowering of the storm wave base; 2—muds on the outer shelf, and hydraulic instability; 3—mixed mud and sand at the shelf edge, lower sand-to-mud ratio; 4—mixed sand and mud at the shelf edge, higher sand-to-mud ratio; 5—coastal systems at the shelf edge, coarser sediment; 6—coastal systems at the shelf edge, finer sediment. Abbreviations: FSST—falling-stage systems tract; LST—lowstand systems tract; TST—transgressive systems tract; HST—highstand systems tract; BSFR—basal surface of forced regression; CC—correlative conformity; MRS—maximum regressive surface; MFS—maximum flooding surface; RSL—relative sea level

transported to the Atlantic Ocean by the Amazon is being delivered continuously to the Amazon fan through a deep fluvial channel and submarine canyon, whereas the equally large Mississippi fan is currently virtually dormant because post-glacial sea-level rise has moved the mouths of the Mississippi Delta back from the edge of the continental slope, so that sediment delivery to the fan is negligible (Sect. 4.2.9). In general, sediment supply to turbidite systems is greatest at the end of the period of forced regression and reaches a minimum at the time of maximum flooding of the shelf (Catuneanu 2020b, p. 7), but there are many local variations (Fig. 5.15).

In addition to the physiographic variations noted here, which complicate the relationship between the base-level cycle and systems tract architecture and development, it is quite possible for episodic changes in systems tract development at continental margins to have nothing to do with sea-level change at all. To cite two examples, in the case of the modern Amazon fan, the marked facies variations mapped by the ODP bear no relation to Neogene sea-level changes but reflect autogenic avulsion processes on the upper fan (Christie-Blick et al. 2007). Secondly, White and Lovell (1997) demonstrated that in the North Sea Basin, peaks in submarine-fan sedimentation occurred at times of regional uplift of the crust underlying the British Isles, as a consequence of episodes of magma underplating, resulting in increased sediment delivery to the marine realm.

Given the enormous increase in interest in fine-grained rocks, because of their value as self-sourced gas reservoirs, a special focused study on the sequence stratigraphy of these deposits by Bohacs et al. (2002) is referred to briefly here. Shales and mudstones are typically assumed to be deep-water deposits or to represent continental shelf deposits formed during the flooding phase. However, as this study points out, they may be formed at any stage in the sequence cycle. Some offshore successions are composed entirely of mudstone-dominated sequences. Bohacs et al. (2002, p. 322) noted that:

> sequence boundaries and flooding surfaces continue as recognizable entities well beyond the distal limit of coarse grained lithologies. We also recognize that the correlative "conformity" that is the distal portion of a sequence boundary may exhibit significant erosion, even within completely fine-grained strata and that truly continuous sedimentation at even a fine scale is relatively scarce.

Careful study of some classic black shales, such as the Chattanooga, have shown that they may be subdivided using sequence methods, and that many of the sequence boundaries, some traceable for hundreds of kilometers are clearly erosional.

Recent studies have successfully developed a model for the application of sequence concepts to glaciomarine environments (Fig. 5.20). The falling stage is the stage of glacial freeze-up and subglacial erosion. Most glaciomarine sediments are deposited during the subsequent glacial-retreat phase, which is during the transgressive systems tract, when sea level is rising.

Note, in closing, the caveats at the end of Sect. 5.2.3 regarding the possible confusion between the terminology of systems tracts (highstand, falling stage, etc.) and the actual state of the sea-level cycle which they represent.

5.3.2 Nonmarine Depositional Systems

The early sequence model of Posamentier et al. (1988) and Posamentier and Vail (1988) emphasized the accumulation of fluvial deposits during the late highstand phase of the sea-level cycle, based on the graphical models of Jervey (1988). The model suggested that the longitudinal profile of rivers that are graded to sea level would shift seaward during a fall in base level, and that this would generate accommodation for the accumulation of nonmarine sediments. This idea was examined critically by Miall (1991), who described scenarios where this may and may not occur. The response of fluvial systems to changes in base level was examined from a geomorphic perspective in greater detail by Wescott (1993) and Schumm (1993), and the sequence stratigraphy of nonmarine deposits was critically reviewed in an important paper by Shanley and McCabe (1994). An extensive discussion of the sequence stratigraphy of fluvial deposits was given by Miall (2014a, Chap. 6), and only a brief, updated summary of this material is given here.

Shanley and McCabe (1994) discussed the relative importance of downstream base-level controls versus upstream tectonic controls in the development of fluvial architecture (Fig. 5.21). In general, the importance of base-level change diminishes upstream. In large rivers, such as the Mississippi, the evidence from the Quaternary record indicates that sea-level changes affect aggradation and degradation as far upstream as the region of Natchez, Mississippi, about 220 km upstream from the present mouth. Farther upstream than this, source-area effects, including changes in discharge and sediment supply, resulting from tectonism and climate change, are much more important. In the Colorado River of Texas, base-level influence extends about 90 km upstream, beyond which the river has been affected primarily by the climate changes of Neogene glaciations. Blum (1994, p. 275), based on his detailed work on the Gulf Coast Rivers, stated "At some point upstream rivers become completely independent of higher order relative changes in base level, and are responding to a tectonically controlled long-term average base level of erosion." The response of river systems to climate change is complex. As summarized by Miall (1996, Sect. 12.13.2), cycles of aggradation and degradation in inland areas may be driven

Fig. 5.20 The application of a
sequence model to coastal
glaciomarine sedimentation
(Pedersen 2012, Fig. 1, p. 30)

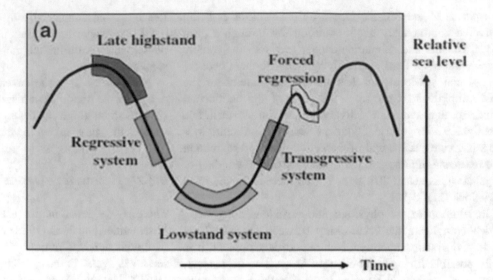

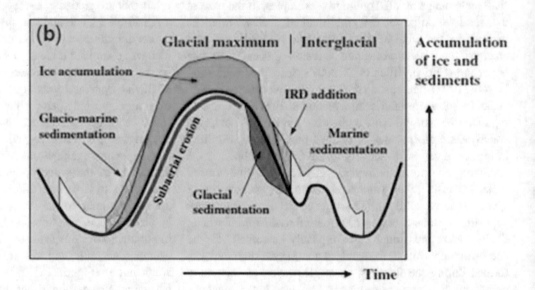

by changes in discharge and sediment load, which are in part climate dependent. These cycles may be completely out of phase with those driven primarily by base-level change.

Holbrook et al. (2006) introduced the useful concepts of **buttresses** and **buffers** to account for longitudinal changes in fluvial facies and architecture upstream from a coastline (Fig. 5.21). A buttress is some fixed point that constitutes the downstream control on a fluvial graded profile. In marine basins this will be marine base level (sea level). In inland basins it will be the lip or edge of a basin through which the trunk river flows out of the basin. The buffer is the zone of space above and below the current graded profile which represents the range of reactions that the profile may exhibit given changes in upstream controls, such as tectonism or climate change, that govern the discharge and sediment load of the river. For example, tectonic uplift may increase the sediment load, causing the river to aggrade toward its upper

buffer limit. A drop in the buttress, for example, as a result of a fall in sea level, may result in incision of the river system, but if the continental shelf newly exposed by the fall in sea level has a similar slope to that of the river profile, there may be little change in the fluvial style of the river. In any of these cases, the response of the river system is to erode or aggrade toward a new dynamically maintained equilibrium profile that balances out the water and sediment flux and the rate of change in accommodation. The zone between the upper and lower limits is the buffer zone and represents the available (potential) **preservation space** for the fluvial system.

Two sequence models developed specifically for fluvial systems were proposed in the early 1990s, those of Wright and Marriott (1993) and Shanley and McCabe (1994). These two papers drew extensively on the concepts relating fluvial architecture to accommodation that had been explored in the

Fig. 5.21 Allogenic controls on fluvial sedimentation. The relative roles of the major depositional controls are based on Shanley and McCabe (1994, Fig. 6); the diagram is intended to suggest how the balance between upstream (tectonic, climatic) and downstream (base level) controls changes from river mouth to source. The buttress and buffer concepts are based on Holbrook et al. (2006), and are discussed in the text

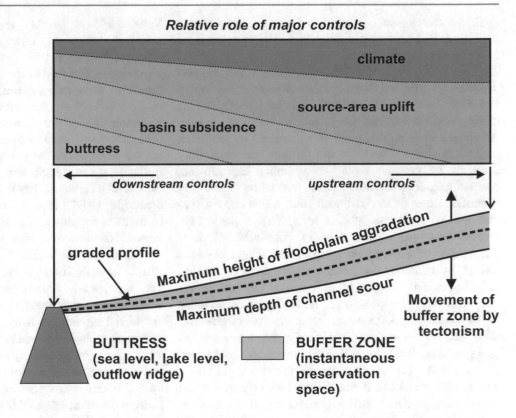

numerical simulation model of Bridge and Leeder (1979). That model, in turn, drew on concepts developed earlier by J. R. L. Allen (Miall 2014a). Both of these models highlight the architecture and stacking pattern of channel sand bodies. It was suggested that the lowstand and highstand systems tracts are characterized by laterally amalgamated sand bodies, whereas the mid-portion of each sequence, including the transgressive systems tract, consists of relatively more isolated sand bodies separated by thicker intervals of floodplain fines. These variations were attributed to changes in the rate of accommodation generation. Amalgamated sand bodies indicate slow rates of accommodation generation. Channels have time to wander freely within the floodplain and erode laterally into each other, whereas intervals of isolated sand bodies develop at times of high rates of accommodation generation, which creates space for floodplain deposits as channels switch in position within the floodplain.

However, these models are based primarily on sedimentary processes studied in modern rivers and the post-glacial record. Because such studies are focused on relatively high-frequency processes and short time scales (typically 10^3–10^5 years), they are of limited value in interpretations of the ancient rock record. Most examples of fluvial systems in the rock record that have been interpreted using these models span time intervals in the 10^6-year range (several examples were described by Miall 2014a, Chap. 6; see also Colombera et al. 2015). Units that have time spans of this duration are characterized by rates of accommodation generation and sedimentation that are at least an order of magnitude slower than those systems on which the Wright and Marriot (1993) and Shanley and McCabe (1994) models were based.

It is suggested that systematic stratigraphic changes in alluvial architecture in the ancient rock record are not the product of changing avulsion rates and changes in fluvial style under the influence of variable rates of accommodation, as in the published models, but reflect regional shifts in facies belts, that themselves are a response to tectonism and to changes in accommodation and other variables (climate, discharge, sediment supply, etc.) at much slower rates, by one to several orders of magnitude, than those that have been assumed (Miall 2014a, Sect. 6.2.4). The published models may be appropriate for high-frequency fluvial cycles generated by orbital forcing, but further work is needed to test this hypothesis.

One of the problems with sequence models for fluvial systems (and no doubt for other environments as well) has been the early assignment of generalized models to inadequate data sets, with results that turn out to be incomplete or misleading. In the first edition of this book, in Chapter 8, I summarized an examination that I had performed on the fluvial Castlegate Sandstone or Utah, as an exercise to demonstrate the possibilities and implications of several alternative sequence models (Miall 2014b). Further work on this unit by Pattison (2020) revealed, in turn, the deficiencies in all of those models, on the basis of the extremely detailed field work that he had just completed. We examine these new ideas in Sect. 8.11.3.

Blum (1994), who studied post-glacial fluvial systems deposited over the last few thousands of years, demonstrated that nowhere within coastal fluvial systems is there a single erosion surface that can be related to lowstand erosion. Additional work on this point was discussed by Holbrook and Bhattacharya (2012). Such surfaces are continually modified by channel scour, even during transgression, because episodes of channel incision may reflect climatically controlled times of low sediment load, which are not synchronous with changes in base level. This is particularly evident landward of the limit of base-level influence. Post-glacial terraces within inland river valleys reveal a history of alternating aggradation and channel incision reflecting climate changes, all of which occurred during the last post-glacial rise in sea level. A major episode of valley incision occurred in Texas not during the time of glacioeustatic sea-level lowstand, but at the beginning of the post-glacial sea-level rise, which commenced at about 15 ka. (Blum 1994). The implications of this have yet to be resolved for inland basins where aggradation occurs because of tectonic subsidence, rather than incision and terrace formation. However, it would seem to suggest that no simple relationship between major bounding surfaces and base-level change should be expected.

Figure 5.22 shows a model of fluvial processes in relationship to glacially controlled changes in climate and vegetation, based on Dutch work. These studies, and those in Texas, deal with periglacial regions, where climate change was pronounced, but the areas were not directly affected by glaciation. Vandenberghe (1993) and Vandenberghe et al. (1994) demonstrated that a major period of incision occurred during the transition from cold to warm phases because runoff increased while sediment yield remained low.

Vegetation was quickly able to stabilize river banks, reducing sediment delivery, while evapotranspiration remained low, so that the runoff was high. Fluvial styles in aggrading valleys tend to change from braided during glacial phases to meandering during interglacials (Vandenberghe et al. 1994). With increasing warmth, and consequently increasing vegetation density, rivers of anastomosed or meandering style tend to develop, the former particularly in coastal areas where the rate of generation of new accommodation space is high during the period of rapidly rising base level (Törnqvist 1993; Törnqvist et al. 1993). Vandenberghe (1993) also demonstrated that valley incision tends to occur during the transition from warm to cold phases. Reduced evapotranspiration consequent upon the cooling temperatures occurs while the vegetation cover is still substantial. Therefore runoff increases, while sediment yield remains low. With reduction in vegetation cover as the cold phase becomes established, sediment deliveries increase and fluvial aggradation is re-established.

It is apparent that fluvial processes inland and those along the coast may be completely out of phase during the climatic and base-level changes accompanying glacial to interglacial cycles. Within a few tens of kilometers of the sea, valley incision occurs at times of base-level lowstand, during cold phases, but the surface may be modified and deepened during the subsequent transgression until it is finally buried. Inland, major erosional bounding surfaces correlate to times of climatic transition, from cold to warm and from warm to cold, that is to say during times of rising and falling sea level, respectively.

The Dakota Group of northeast New Mexico and southeast Colorado provides a good example of a fluvial unit that is internally architecturally complex, that was generated by a combination of upstream tectonic controls and downstream sea-level cycles (Fig. 5.23; Holbrook et al. 2006). At the coastline, progradation and retrogradation creating three

Fig. 5.22 Relationship among temperature, vegetation density, evapotranspiration, precipitation and sedimentary processes in river systems during glacial and interglacial phases, and the relationship to the contemporaneous marine cycle. Based on work in the modern Rhine–Meuse system (Vandenberghe 1993), with the marine cycle added

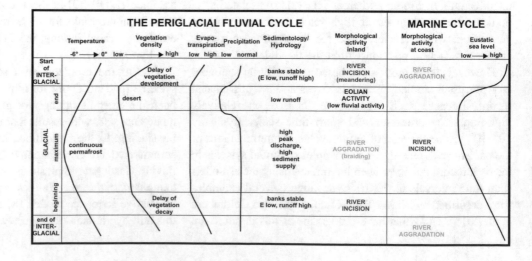

Fig. 5.23 Longitudinal NW–SE section, approximately 250 km in length, through the mid-Cretaceous Dakota Group, from Colorado into the northeast corner of New Mexico. The internal architecture consists of a series of unconformity-bounded sandstone sheets that reflect "deposition during repetitive valley-scale cycles of aggradation and incision." (Holbrook et al. 2006, p. 164)

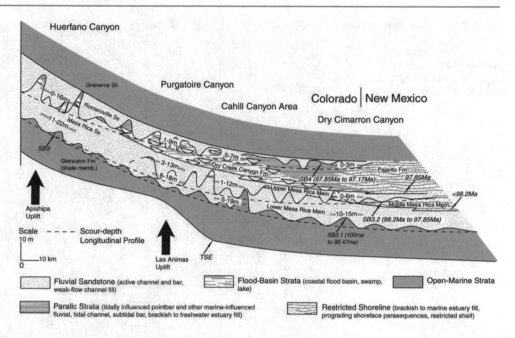

sequences were caused by sea-level cycles on a 10^5-year time scale. Each of these sequences can be traced updip toward the west, where they are composed of repeated cycles of aggradational valley fill successions and mutually incised scour surfaces. These cycles reflect autogenic channel shifting within the limited preservation space available under conditions of modest, tectonically generated accommodation. This space is defined by a lower buffer (in the Holbrook et al. 2006 terminology) set by maximum local channel scour, and an upper buffer set by the ability of the river to aggrade under the prevailing conditions of discharge and sediment load.

Certain eolian environments and deposits can be interpreted using a sequence model. A climate model for erg generation and the formation of the major inter-erg bounding surfaces was first proposed by Talbot (1985). These surfaces were termed **super bounding surfaces** (or **supersurfaces**) by Kocurek (1988). Application of the model, with the discussion of a range of examples, was provided by Clemmensen and Hegner (1991), Kocurek et al. (1991), Brookfield (1992), Kocurek and Havholm (1993), Kocurek (1996) and Brookfield and Silvestro (2010).

Figure 5.24 shows the sequence model developed for the Coconino sandstone of Arizona. A cyclic repetition of facies has been mapped, commencing with the sequence boundary, which corresponds to a supersurface in the sense of Kocurek (1988). Migrating facies succession above this surface indicate a gradual seaward progradation of successively more inland facies, starting with a supratidal Sabkha environment, and continuing through coastal and then large, inland dunes. As noted by Brookfield and Silvestro (2010, p. 159–160), supersurfaces, which may be used to delineate eolian sequences, may reflect climatic or tectonic change, or change in sea level.

A brief note on the importance of orbital forcing as a major influence in the lacustrine environment is provided in Sect. 4.2.3. Cyclic repetition of local depositional environments from coastal fans and fluvial plains to deep lacustrine settings is a common theme.

5.3.3 Carbonate Depositional Systems

Carbonate and clastic depositional systems respond very differently to sea-level change (Sarg 1988; James and Kendall 1992; Schlager 1992, 1993, 2005; James et al. 2010). The differences between carbonates and clastics were not well understood at the time the original Exxon sequences models were developed by Vail et al. (1977) and Vail (1987).

Most carbonate sediments accumulate in shallow water, typically on continental margin **platforms** or isolated **banks**. Large **epeiric platorms** covering much of the North American craton developed during the early Paleozoic during an extended period of high global sea level—a paleogeographic configuration that has not been replicated since. Figure 4.47 illustrates a typical **rimmed carbonate platform**, consisting of a wide, carbonate-dominated shelf with a fringing barrier reef, scattered bioherms or patch reefs in the platform interior and a marginal clastic belt, the width of which depends on the clastic supply delivered to the coast from river mouths, and the strength of the waves and tides redistributing it along the coast. Unrimmed platform margins, lacking a marginal barrier reef, occur on the leeward sides of platforms. Carbonate **ramps** are unrimmed shelves sloping at about 1°.

Figure 5.25 illustrates the differences between the responses of carbonate and clastic systems to sea-level

Fig. 5.24 A sequence model for a coastal eolian succession. Heavy arrows within the cross-sectional point to supersurfaces (Blakey and Middleton 1983)

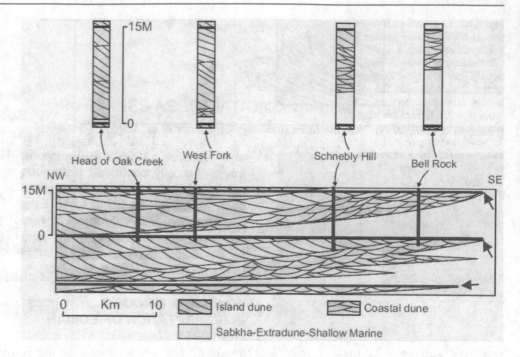

change, which are described briefly here, based on a standard sequence model. At times of sea-level lowstand, terrigenous clastics bypass the continental shelf, leading to exposure, erosion and the development of incised valleys. Submarine canyons are deepened, and sand-rich turbidites systems develop submarine-fan complexes on the slope and the basin floor. Carbonate systems essentially shut down at times of lowstand, because the main "carbonate factory," the continental shelf, is exposed, and commonly undergoes **karstification,** which is the process of subaerial solution, and the formation of caves and below-surface drainages characteristic of limestone terranes. A narrow shelf-edge belt of reefs or sand shoals may occur, while the deep-water basin is starved of sediment or possibly subjected to hyperconcentration, with the development of evaporite deposits. Evaporites may also develop on the continental shelf during episodes of sea-level fall, when reef barriers serve to block marine circulation through to the shelf.

During transgression of a clastic system, incised valleys fill with estuarine deposits and eventually are blanketed with marine shale. There may be a rapid landward translation of facies belts, leaving the continental shelf starved of sediment, so that condensed successions are deposited. By contrast, transgression of a carbonate shelf serves to "turn on" the carbonate factory, with the flooding of the shelf with warm, shallow seas. This may be preceded by a short period of shallow waters in which evaporites may develop. Thick platform carbonate successions are expected to develop; reefs, in particular, being able to grow vertically at extremely rapid rates as sea level rises. At times of maximum transgression, the deepest part of the shelf may pass below the

photic zone, leading to cessation of carbonate sedimentation and development of a condensed section or hardground. The resulting surface is termed a **drowning unconformity** (Schlager 1989; see also next section).

Carbonate and clastic shelves are most alike during times of highstand. The rate of addition of sedimentary accommodation space is low, and lateral progradation will take place if sediment generation exceeds accommodation generation, with the development of clinoform slope architectures (e.g., see Fig. 4.49). Autogenic shoaling-upward cycles are common in both types of environment (e.g., terrigenous deltaic lobes, tidal carbonate cycles). Schlager (1992) stated:

> Prograding [carbonate] margins dominated by offshore sediment transport most closely resemble the classical [siliciclastic] sequence model. They are controlled by loose sediment accumulation and approach the geometry of siliciclastic systems (e.g., leeward margins ...).

Carbonate platforms generate sediment at the highest rate during highstands of sea level, when platforms are flooded and the carbonate factory is at maximum productivity. The sediment volume commonly exceeds available accommodation, and the excess is delivered to the continental slope, where it may provide the sediment for large-scale progradation by carbonate talus and sediment gravity flows. This process, termed **highstand shedding,** is exemplified by the architecture of the Bahamas Platform (Schlager 1992; see Fig. 4.49 of this book). It is the converse of the pattern of siliciclastic sedimentation, within which, as already noted, sediment is fed to the continental slope most rapidly at times of low sea level.

Fig. 5.25 Differences in the response to base-level change between carbonate and clastic systems (adapted from James and Kendall 1992)

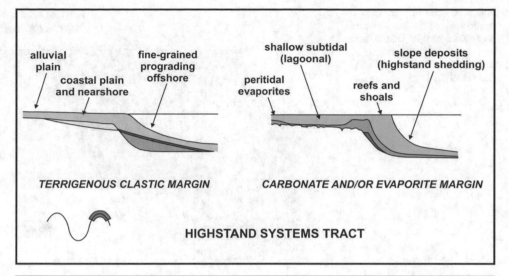

alluvial plain

coastal plain and nearshore

fine-grained prograding offshore

shallow subtidal (lagoonal)

slope deposits (highstand shedding)

peritidal evaporites

reefs and shoals

TERRIGENOUS CLASTIC MARGIN

CARBONATE AND/OR EVAPORITE MARGIN

HIGHSTAND SYSTEMS TRACT

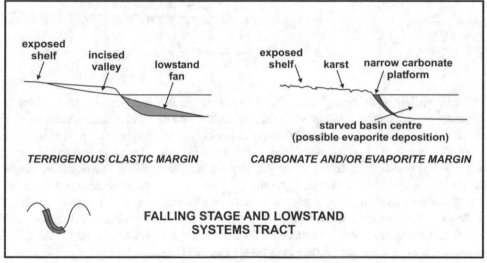

exposed shelf

incised valley

lowstand fan

exposed shelf

karst

narrow carbonate platform

starved basin centre (possible evaporite deposition)

TERRIGENOUS CLASTIC MARGIN

CARBONATE AND/OR EVAPORITE MARGIN

FALLING STAGE AND LOWSTAND SYSTEMS TRACT

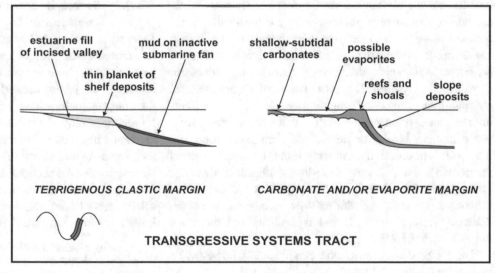

estuarine fill of incised valley

mud on inactive submarine fan

thin blanket of shelf deposits

shallow-subtidal carbonates

possible evaporites

reefs and shoals

slope deposits

TERRIGENOUS CLASTIC MARGIN

CARBONATE AND/OR EVAPORITE MARGIN

TRANSGRESSIVE SYSTEMS TRACT

Fig. 5.26 The stratigraphic architecture of the late Paleozoic continental margin, central Texas, an example of "reciprocal sedimentation" (Galloway and Brown 1973)

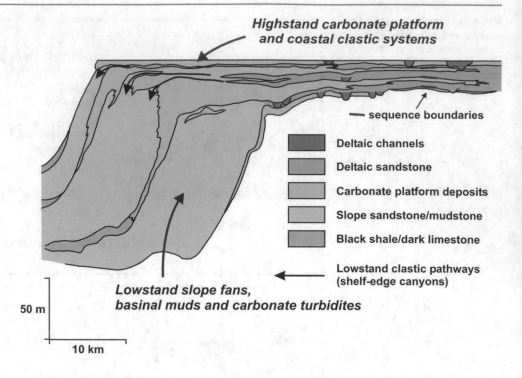

Many ancient shelf deposits are mixed carbonate-clastic successions, containing thin sand banks or deltaic sand sheets interbedded with carbonate platform deposits (Dolan 1989; Southgate et al. 1993). Galloway and Brown (1973) described an example from the Pennsylvanian of northern central Texas, in which a deltaic system prograded onto a stable carbonate shelf (Fig. 5.26). The term **reciprocal sedimentation** has been used for depositional systems in which carbonates and clastics alternate (Wilson 1967). In the example given in Fig. 5.26, deltaic distributary channels are incised into the underlying shelf carbonate deposits. Widespread shelf limestones alternate with the clastic sheets and also occur in some inter-deltaic embayments. Carbonate banks occur on the outer shelf edge, beyond which the sediments thicken dramatically into a clinoform slope-clastics system. This association of carbonates and clastics reflects regular changes in sea level, with the carbonate phase representing high sea level and the clastic phase low sea level. The deltaic and shelf-sand sheets and the slope clinoform deposits represent lowstand systems tracts, while the carbonate deposits are highstand deposits. During episodes of high sea-level, clastics are trapped in nearshore deltas, while during low stands much of the detritus bypasses the shelf and is deposited on the slope (arrows in Fig. 5.26).

The reciprocal-sedimentation model should, however, be used with caution. Dunbar et al. (2000) reported on the results of a detailed sampling program across the Great Barrier Reef, from which they concluded that through the glacioeustatic sea-level cycles of the last 300 ka, the maximum rate of siliciclastic sedimentation on the continental slope occurred during transgression, not during the falling stage. They explained this as the result of very low fluvial slopes across the shelf that became exposed during the sea-level falling stage and lowstand, resulting in sediment accumulation and storage on the shelf during this phase of the sea-level cycle. The sediment stored there was mobilized and transported seaward by vigorous marine processes during the subsequent transgressive phases.

5.3.3.1 Breaks in Sedimentation in Carbonate Environments

Submarine erosion, and other processes, can generate breaks in sedimentation without any change in sea level. This is particularly the case in carbonate sediments, which are very sensitive to environmental change, and may develop drowning unconformities (Schlager 1989, 1992). These may be architecturally similar to lowstand unconformities, and care must be taken to interpret them correctly. They may, in fact, represent an interval of slow sedimentation, represented by thin condensed sections containing many small hiatuses. The stratigraphic details may be indistinguishable on the seismic record from actual unconformable breaks because of limited seismic resolution. Schlager (1992) stated:

Drowning requires that the reef or platform be submerged to subphotic depths by a relative rise that exceeds the growth potential of the carbonate system. The race between sea level and platform growth goes over a short distance, the thickness of the photic zone. Holocene systems indicate that their short-term growth potential is an order of magnitude higher than the rates of long-term subsidence or of third-order sea level cycles ... This implies that drowning events must be caused by unusually rapid pulses of sea level or by environmental change that reduced the growth potential of platforms. With growth reduced, drowning may occur at normal rates of rise.

Schlager (1992) pointed to such environmental changes as the shifts in the El Niño current, which bring about sudden rises in water temperature, beyond the tolerance of many corals. Drowning can also occur when sea-level rise invades flat bank tops, creating shallow lagoons with highly variable temperatures and salinities, plus high suspended-sediment loads due to coastal soil erosion. Oceanic anoxic events, particularly in the Cretaceous, are also known to have caused reef drowning. Schlager (1992) suggested that two Valanginian sequence boundaries in the Haq et al. (1987, 1988) global cycle chart may actually be drowning unconformities that have been misinterpreted as lowstand events. He also noted the erosive effects of submarine currents, and their ability to generate unconformities that may be mapped as sequence boundaries but that have nothing to do with sea-level change. The Cenozoic sequence stratigraphy of the Blake Plateau, off the eastern United States, is dominated by such breaks in sedimentation that do not correlate with the Exxon global cycle chart but have been interpreted as the result of erosion by the meandering Gulf Stream.

5.3.3.2 Platform Carbonates: Catch-Up Versus Keep-Up

The style of carbonate sequence stratigraphy on the platform is mainly a reflection of the balance between sea-level rise and carbonate production, as summarized in reviews by Jones and Desrochers (1992) and Jones (2010). Where sea-level fall exceeds subsidence rates, exposure occurs and karst surfaces may develop (Figs. 5.27 and 5.28). Rapid transgression, on the other hand, leads to the development of condensed sections and may shut down the carbonate factory, resulting in a drowning unconformity. Nutrient poisoning and choking by siliciclastic detritus may also shut down the carbonate factory at times of high sea level, and this can also lead to the development of drowning unconformities, which may be mistaken for sequence boundaries (e.g., Erlich et al. 1993). James and Bourque (1992) argued that poisoning and choking were the processes most likely to cause a shut down of reef sedimentation, because studies have indicated that under ideal conditions vertical reef growth is capable of keeping pace with the most rapid of sea-level rises.

Figure 5.28 illustrates the main variations in platform architecture that develop in response to changes in the controls noted above. Drowning during a rapid rise in relative sea level is typically followed by backstepping. A slightly less rapid transgression may lead to **catch-up** architecture. Here the sea floor remains a site of carbonate production, and as sea-level rise slows, sedimentation is able

to catch up to the new sea level. Vertical aggradation characterizes the first stage of the catch-up, but lateral progradation may occur late in the cycle, when the rate of generation of new accommodation space decreases. Shoaling-upward sequences are the result.

A balance between sea-level rise and sediment-production rates will lead to **keep-up** successions, in which cyclicity is poorly developed. Eventually, typically at the close of a cycle of sea-level rise, carbonate production may exceed the rate of generation of accommodation space. This can lead to lateral progradation, and highstand shedding of carbonate detritus onto the continental slope.

Sanborn et al. (2020) carried out a detailed study of the response of the Great Barrier Reef to the rapid sea-level rise and other environmental changes that took place during the Holocene, following the end of the last ice age. Once the continental shelf was flooded, there was an initial lag of several hundred years during which the "carbonate factory" regenerated, after which catch-up and keep-up of the coral colonies was well able to keep pace with the rapid sea-level rise of 6–7 m/ka that took place between 10 and 7 ka before present.

As with coastal detrital sedimentation, autogenic processes may generate successions that are similar to those that are inferred to have formed in direct response to sea-level change. Pratt et al. (1992) and Pratt (2010) summarized the processes of autogenesis in peritidal environments, where shallowing-upward cycles are characteristic, as a result of short-distance transport and accumulation of carbonate detritus. Lateral progradation and vertical aggradation both may occur. Rapid filling of the available accommodation space may lead to a shut down of the carbonate factory until relative sea-level rises enough to stimulate its reactivation. The alternative to autogenesis is the invocation of a form of rhythmic tectonic movement to generate the required sea-level change, or Milankovitch mechanisms. As Pratt et al. (1992) noted, the scale and periodicity of tectonism required for meter-scale cycles has not yet been demonstrated.

5.4 Sequence Hierarchies

The application of practical sequence methods during the last twenty or more years has generated many examples of sequence successions containing smaller scale sequences nested within them. For example, Fig. 5.29 illustrates a compilation diagram generated for a Cretaceous succession in Oman, for which detailed stratigraphic work has revealed three scales of accommodation cycles within a prograding carbonate shelf succession (from Veeken 2007). In some cases of nested sequences, the smaller scale cycles are clearly the product of autogenic processes, such as channel

Fig. 5.27 The types of sequence that develop, depending on variations in the rate of relative sea-level rise and carbonate sediment production (Jones and Desrochers 1992)

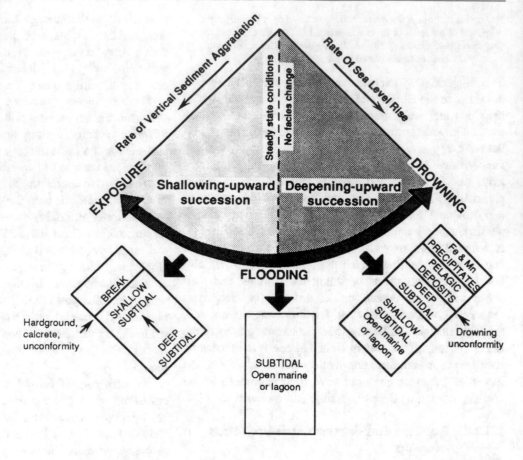

Fig. 5.28 The styles of carbonate platform architecture and their dependence on the balance between the rate of sea-level rise and carbonate productivity (Jones and Desrochers 1992)

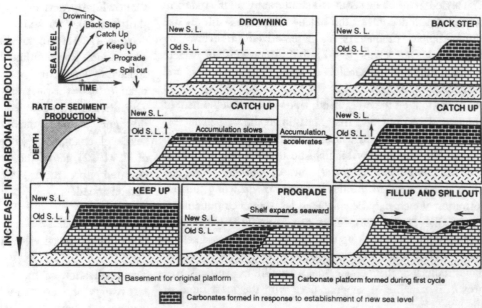

filling or shoreline regression. Tracking the regional extent of such units normally suffices to resolve this basic question of origin. Autogenic products are typically of modest areal extent, reflecting the dimensions of the depositional system within which they occur. These should not be termed sequences. The larger scale cycles in such cases are more

likely to be of regional extent, demonstrating allogenic origin, and such sequences may occur at more than one scale, suggesting more than one allogenic process operating at different time scales.

Another example of a nested sequence hierarchy is shown in Fig. 5.30, in this case a deep-water succession in the Gulf

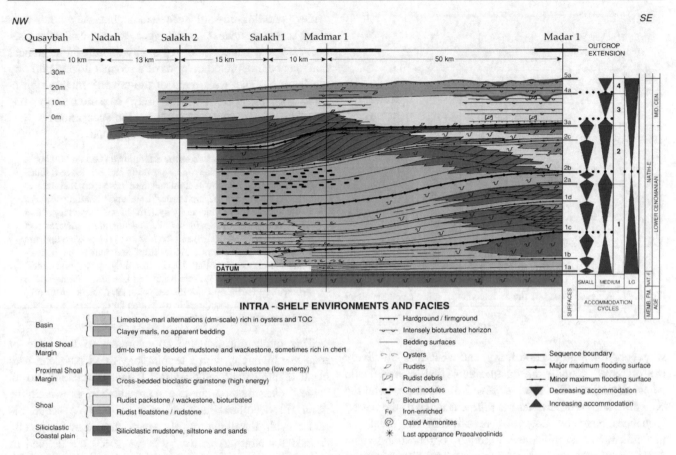

Fig. 5.29 An example of nested sequences. Cretaceous succession in Oman (from Veeken 2007)

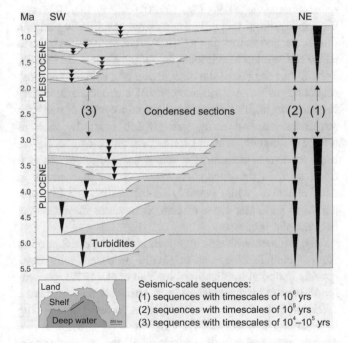

Fig. 5.30 An example of nested sequences constituting a deep-water succession in the deep basin of the Gulf of Mexico (Catuneanu 2019, Fig. 41)

of Mexico. The cyclicity is attributed to changes in sediment supply driven by the regressions and transgressions caused by glacioeustatic sea-level change, and to local changes in sediment dispersal relating to the evolution of individual turbidites systems (Catuneanu 2019a, p. 154).

Figure 5.31 illustrates a general condition, with three scales of cycle or sequence nested within the single major sequence "A." In such cases it has been common practice to term one of the smaller scale cycle sets parasequences, particularly where their facies characteristics indicate they were the product of a shoaling-upward environment and are bounded by flooding surfaces. The "small-scale" cycles in Fig. 5.29 fit this description, as do the numbered shingles in the Dunvegan Formation in Fig. 5.13. The problem is that without detailed mapping it may not be possible to determine the autogenic versus allogenic origin of these cycles, and for this reason this author has long recommended abandoning the term parasequence.

In the original definition of what constitutes a sequence, the phrase "relatively conformable" was used to describe stratal relationships within a sequence, in contrast to the unconformities used to define the boundaries of a sequence (Mitchum et al. 1977b, p. 53). This raises the questions of

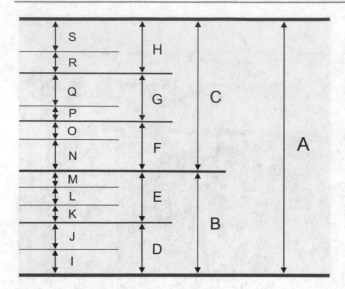

Fig. 5.31 A general case of the architecture of a sequence hierarchy (Catuneanu 2019, Fig. 6)

cycles, shoaling-upward successions that may qualify as small-scale sequences (no longer to be called parasequences), and larger, more regionally significant sequences that may be interpreted in terms of regional allogenic driving mechanisms. The assessment of the unconformable significance of bounding surfaces may constitute only one descriptive item employed to define the succession.

Catuneanu (2019b, p. 369) has suggested:

> Every hierarchy system requires a reference (i.e., an 'anchor'), relative to which smaller or larger units can be defined. Classifications that employ a nomenclature based on hierarchical orders typically use the 'first order' units and bounding surfaces as the anchor for the hierarchy system. In sedimentology, 'first order' designates the smallest units and bounding surfaces that develop at bedform scales, and the hierarchy ranks scale up from bedforms to macroforms, architectural elements, and depositional systems (e.g., Miall 1996). In stratigraphy, 'first order' designates the largest genetic units and bounding surfaces of a sedimentary basin, and the hierarchy ranks scale down to progressively smaller stratigraphic cycles, as far as permitted by the resolution of the data available.

This quote encapsulates an approach to the issue of sequence hierarchy with a terminology that Catuneanu uses in all of his recent publications. It makes reference to an "order" classification, in one representation of which the basin fill becomes the first order. Sloss-type sequences (those with durations of 10^7 years that were originally defined for North America by Sloss 1963) are defined as second order, lesser scale sequences seen on seismic as third order, and those identified on the well-log scale as fourth order (Catuneanu 2019a, Fig. 12; 2019b, Fig. 15). However, the application of this ordering was not specifically tied to a time scale, and Catuneanu's (2019b) approach seems to be to adjust the ranking to suit local flexibility. Catuneanu (2019b, Sect. 8) has suggested the designation of a suite of "anchor" sequences as the main framework of a sequence hierarchy for each basin. The term "stratigraphic cycle" is used for sequences mapped using stratigraphic methods, such as reflection seismic, and the term "sedimentologic cycle" is used for cycles documented using smaller scale observations, such as the documentation of facies successions in core. However, while possibly useful in directing interpretations to particular types of data, these terms do not reflect fundamental processes.

The problem with this approach is that the concept of an "order" classification was originally proposed, with quite specific temporal implications, by Vail et al. (1977), but was found later to be inconsistent in its application (Carter et al. 1991), with the time scales for adjacent "orders" overlapping in practice (Schlager 2004). Vail remained convinced that eustatic sea level was the driver of all the sequence types within his ranked hierarchy, but subsequent work demonstrated that there were no global processes that generated sequences in the 10^6-year time scale (Miall 2010). Vail's use

what constitutes an unconformity, and what does "relatively conformable" actually mean; questions that become acute when situations like that in Fig. 5.31 occur, where the bounding surfaces between the different ranks of the nested sequences may or may not reveal different levels of unconformable significance through chronostratigraphic evidence of missing time, erosion, structural discordance or diagenesis. As discussed elsewhere (Catuneanu 2019a, 2019b; Miall 2016), various terms have been used to suggest different degrees of unconformable separation between the beds above and below an unconformity surface (disconformity, paraconformity, diastem, etc.). However, modern work, particularly that on the Sedimentation Rate Scale (Miall 2015, 2016) has suggested that any distinction between "unconformable," "relatively conformable" and "conformable" is entirely arbitrary. Also, as Catuneanu (2019b) has noted, the degree to which the conformity of a contact can be assessed depends entirely on the resolution of the data being used. Virtually all bedding plane surfaces in sedimentary rocks (even many of those within mudrocks) are small-scale sedimentary breaks, with the breaks varying progressively as a continuum in temporal scale from those representing a few minutes or seconds of scour beneath a bedform to the major angular unconformities that record significant orogenic episodes developed over tens of millions of years. Assessments of the significance of a sedimentary break may very much depend on the method used to document it, with angularity becoming potentially more apparent at a given bounding surface if its mapping can be accomplished with regional stratigraphic correlations or seismic work. Nested successions, such as those shown diagrammatically in Fig. 5.31, may include autogenic facies

Table 5.1 Stratigraphic cycles and their causes

Sequence type	Duration (m.y.)	Other terminology
A. Global supercontinent cycle	200–400	First-order cycle (Vail et al. 1977)
B. Cycles generated by continental-scale mantle thermal processes (dynamic topography), and by plate kinematics, including:	10–100	Second-order cycle (Vail et al. 1977), supercycle (Vail et al. 1977), sequence (Sloss 1963)
1. Eustatic cycles induced by volume changes in global mid-oceanic spreading centers		Sloss cycles
2. Regional cycles of uplift and subsidence induced by dynamic topography		
3. Regional cycles of basement movement induced by extensional downwarp and crustal loading		
C. Regional to local cycles of basement movement caused by regional plate kinematics, including changes in intraplate stress regime, crustal loading	0.01–10	third- to fifth-order cycles (Vail et al. 1977). Third-order cycles also termed: megacyclothem (Heckel 1986), mesothem (Ramsbottom 1979)
D. Global cycles generated by orbital forcing, including glacioeustasy, productivity cycles, etc.	0.01–2	fourth- and fifth-order cycles (Vail et al. 1977), Milankovitch cycles, cyclothem (Wanless and Weller 1932), major and minor cycles (Heckel 1986)

of the term "first order" was to apply to the supercontinent cycle, which has a duration of 10^8 years, and therefore beyond practical use for basin-scale work. An "order" classification does not assist in the separate identification of sequences that may have very similar durations but are generated by very different processes. This is the case with Milankovitch cycles and those generated by high-frequency tectonism (fourth- and fifth-order cycles in the original order definitions: see Table 5.1). Any attempt to reuse or redefine Vail's terms is bound to create confusion with the original classification, and its use has not been recommended by this author for some time (Miall 2010, Sect. 4.2).

Is there a recommended terminology that can be used to describe nested sequences such as those shown in Figs. 5.29 and 5.30? A succession of terms such as sequence-sequence set-composite sequence-megasequence may be useful in some circumstances. A simple system of naming that contains the hierarchy may be employed. For example, in Fig. 5.31 Sequences F, G and H could be renamed Ca, Cb and Cc, indicating that they are lesser sequences nested within sequence C. The lowest rank could then make use of an additional letter. Thus, the subdivisions of Ca (sequences O, N) could be termed Ca1 and Ca2. Ultimately, sequences may be given names based on locations of key sections (stratotypes), just as formations were named under the old traditions of lithostratigraphy. Allostratigraphic terminology (alloformation, allomember, etc.) may be used in order to avoid the formal hierarchical associations of such terms as parasequence, megasequence, etc. Note that in Fig. 5.31

sequences are numbered in depositional order, from bottom to top. It is common practice in petroleum exploration practice to number sequences from the top down, in the order they are penetrated by the drill. This author is not in favor of this practice.

5.5 Driving Mechanisms

There is no space in this book to properly address the issue of sequence driving mechanisms. This important topic was discussed in detail by Miall (2010) in the context of the range of allogenic processes that are now recognized as the global drivers of accommodation change. Table 5.1 provides a classification of sequences according to temporal scale and lists the major sequence-generating mechanisms, and the reader is referred to Miall (2010) for a thorough documentation of these sequence types and their interpretation.

The "first-order cycle" is the cycle of creation and disassembly of supercontinents. Sea levels are characterized by long periods of rise and high levels during the breakup of supercontinents, as occurred during the early Paleozoic, with the breakup of Rodinia, and again during the Cretaceous, during the breakup of Pangea. The period of supercontinent assembly and stasis corresponds to a period of low global sea level, as during the Permian–Triassic, which was the period during which most of the world's continental plates were combined to form Pangea.

Sequences with durations in the 10^7–10^8-year range include the so-called Sloss cycles. These are named after Larry Sloss, who established the concept of six long-term sequence characterizing the Phanerozoic cratonic succession of North America (Sloss 1963). There is some evidence that the causation of the sequences included eustatic sea-level changes, because Sloss (1972) was able to demonstrate crude correlations of the North American sequences with those on the Russian Platform. However, Sloss's documentation of the sequences in the North American interior also included the demonstration of low-angle regional unconformities between the sequences. It now seems likely that the very broad, gentle tectonism that generated this angularity was driven by mantle thermal processes, the so-called process of dynamic topography, superimposed on eustatic sea-level change (Burgess 2019).

Cycles of accommodation change are generated on a regional scale by tectonic processes, including crustal loading during contractional orogeny, by crustal extension, and by changes in intraplate stress related to changes in far-field processes, including sea-floor spreading and subduction. The time scale (episodicity) of tectonic processes ranges from 10^2 years for responses to local structural dislocations, to 10^8 years for regional and continental-scale processes (see summary in Miall 2010).

Lastly, high-frequency cyclothem-like cycles have been documented in many stratigraphic successions spanning the Phanerozoic, and increasingly detailed stratigraphic work is demonstrating that such cycles may be much more common than has traditionally been thought. As with the original late Paleozoic cyclothems, the driving mechanism is thought to be glacioeustasy. It is interesting, therefore, that much of the detailed work that is demonstrating high-frequency (10^4–10^5-year) cyclicity is in rocks of Upper Cretaceous age, a period long thought to have been characterized by a greenhouse climate (Plint 1991; Sageman et al. 2006; Varban and Plint 2008; an example is discussed in Sect. 8.12). Miller et al. (2005) discussed the history of Antarctic glaciation during the last 100 m.y., and demonstrated the possibility of the development of small, short-term ice caps on the continent during the Late Cretaceous that could have been the driver of modest glacioeustatic fluctuations in sea level. The documentation of these cycles is now giving rise to a new astrochronologic time scale with potential accuracy of 10^5 years (Hilgen et al. 2015; see Sects. 7.8.7, 8.13).

5.6 Conclusions

This chapter has attempted to provide an introduction to standard sequence models. Many complications have been omitted, and the reader is recommended to turn to such works as Catuneanu (2006) and Catuneanu et al. (2011) for additional details. Modern developments include the

development of protocols for the integration of sequence stratigraphy into the formal structure of stratigraphy, and in order for this to be accomplished agreement must be reached about how to reconcile the various approaches to sequence mapping and definition that have emerged since the work of Vail et al. (1977). This topic is addressed in Sect.7.7.

References

Allen, P. A., and Allen, J. R., 2013, Basin analysis: Principles and application to petroleum play assessment: Chichester: Wiley-Blackwell, 619 p.

Anadón, P., Cabrera, L., Colombo, F., Marzo, M., and Riba, O, 1986, Syntectonic intraformational unconformities in alluvial fan deposits, eastern Ebro basin margins (NE Spain), in Allen, P. A., and Homewood, P., eds., Foreland basins: International Association of Sedimentologists Special Publication 8, p. 259–271.

Arnott, R. W. C., 2010, Deep-marine sediments and sedimentary systems, in James, N. P., and Dalrymple, R. W., eds., Facies Models 4: GEOtext 6, Geological Association of Canada, St. John's, Newfoundland, p. 295–322.

Barrell, Joseph, 1917, Rhythms and the measurement of geologic time: Geological Society of America Bulletin, v. 28, p. 745–904.

Bhattacharya, J., 1991, Regional to sub-regional facies architecture of river-dominated deltas, Upper Cretaceous Dunvegan Formation, Alberta subsurface, in Miall, A. D., and Tyler, N., eds., The three-dimensional facies architecture of terrigenous clastic sediments and its implications for hydrocarbon discovery and recovery, Society of Economic Paleontologists and Mineralogists, Concepts in Sedimentology and Paleontology, v. 3, p. 189–206.

Blakey, R. C., and Middleton, L. T., 1983, Permian shoreline aeolian complexes in central Arizona; dune changes in response to cyclic sea level changes, in Brookfield, M. E., and Ahlbrandt, T. S., eds., Eolian sediments and processes: Elsevier, Amsterdam, p. 551–581.

Blum, M. D., 1994, Genesis and architecture of incised valley fill sequences: a late Quaternary example from the Colorado River, Gulf Coastal Plain of Texas, in Weimer, P., and Posamentier, H. W., eds., Siliciclastic sequence stratigraphy: Recent developments and applications: American Association of Petroleum Geologists Memoir 58, p. 259–283.

Bohacs, K. M., Neal, J. E., and Grabowski, G. J., Jr., 2002, Sequence stratigraphy in fine-grained rocks: Beyond the correlative conformity: 22nd Annual Gulf Coast Section SEPM Foundation Bob F. Perkins Research Conference, p. 321–347.

Bohacs, K., and Suter, J., 1997, Sequence stratigraphic distribution of coaly rocks: fundamental controls and paralic examples: American Association of Petroleum Geologists Bulletin, v. 81, p. 1612–1639.

Bridge, J. S. and Leeder, M. R., 1979, A simulation model of alluvial stratigraphy: Sedimentology, v. 26, p. 617–644.

Brookfield, M. E., 1992, Eolian systems, in Walker, R. G. and James, N. P., eds., Facies models: response to sea-level change: Geological Association of Canada, Geotext 1, p. 143–156.

Brookfield, M. E., and Silvestro, S., 2010, Eolian systems, in James, N. P., and Dalrymple, R. W., eds., Facies Models 4: GEOtext 6, Geological Association of Canada, St. John's, Newfoundland, p. 139–166.

Brown, L. F., Jr., and Fisher, W. L., 1977, Seismic-stratigraphic interpretation of depositional systems: examples from Brazilian rift and pull-apart basins, in Payton, C. E., ed., Seismic stratigraphy — applications to hydrocarbon exploration: American Association of Petroleum Geologists Memoir 26, p. 213–248.

Burgess, P. M., 2016, The future of the sequence stratigraphy paradigm: dealing with a variable third dimension: Geology, v. 44, p. 335–336.

Burgess, P. M., 2019, Phanerozoic Evolution of the Sedimentary Cover of the North American Craton, in Miall, A. D., ed., The Sedimentary Basins of the United States and Canada, Second edition: Sedimentary basins of the World, v. 5, K. J. Hsü, Series Editor, Elsevier Science, Amsterdam, p. 39–75.

Burgess, P.M., and Hovius, N., 1998, Rates of delta progradation during highstands; consequences for timing of deposition in deep-marine systems: Geological Society, London, Journal, v. 155, p. 217–222.

Carter, R. M., Abbott, S. T., Fulthorpe, C. S., Haywick, D. W., and Henderson, R. A., 1991, Application of global sea-level and sequence-stratigraphic models in southern hemisphere Neogene strata from New Zealand, in Macdonald, D. I. M., ed., 1991, Sedimentation, tectonics and eustasy: sea-level changes at active margins: International Association of Sedimentologists Special Publication 12, p. 41-65.

Carvajal, C. R., and Steel, R. J., 2006, Thick turbidite successions from supply-dominated shelves during sea-level highstand: Geology, v. 34, p. 665–668.

Catuneanu, O., 2006, Principles of sequence stratigraphy: Elsevier, Amsterdam, 375 p.

Catuneanu, O., 2019a, Scale in sequence stratigraphy: Marine and Petroleum Geology, v. 106, p. 128–159.

Catuneanu, O., 2019b, Model-independent sequence stratigraphy, Earth Sciene reviews, v. 188, p. 312–388.

Catuneanu, O., 2020a, Sequence stratigraphy in the context of the 'modeling revolution' Marine and Petroleum Geology, v. 116, #104309, 19 p.

Catuneanu, O., 2020b, Sequence stratigraphy of deep-water systems Marine and Petroleum Geology, v. 114, #104238, 13 p.

Catuneanu, O., and Zecchin, M., 2016, Unique vs. non-unique stratal geometries: relevance to sequence stratigraphy: Marine and Petroleum Geology, v. 78, p.184–195.

Catuneanu, O., and Zecchin, M., 2020, Parasequences: Allostratigaphic misfits in sequence stratigraphy: Earth Science Reviews, v. 208, #103289, 17 p.

Catuneanu, O., Abreu, V., Bhattacharya, J. P., Blum, M. D., Dalrymple, R. W., Eriksson, P. G., Fielding, C. R., Fisher, W. L., Galloway, W. E., Gibling, M. R., Giles, K. A., Holbrook, J. M., Jordan, R., Kendall, C. G. St. C., Macurda, B., Martinsen, O. J., Miall, A. D., Neal, J. E., Nummedal, D., Pomar, L., Posamentier, H. W., Pratt, B. R,. Sarg, J. F., Shanley, K. W., Steel, R. J., Strasser, A., Tucker, M. E., and Winker, C., 2009, Toward the Standardization of Sequence Stratigraphy: Earth Science Reviews, v. 92, p. 1–33.

Catuneanu, O., Bhattacharya, J. P., Blum, M. D., Dalrymple, R. W., Eriksson, P. G., Fielding, C. R., Fisher, W. L., Galloway, W. E., Gianolla, P., Gibling, M. R., Giles, K. A., Holbrook, J. M., Jordan, R., Kendall, C. G. St. C., Macurda, B., Martinsen, O. J., Miall, A. D., Nummedal, D., Posamentier, H. W., Pratt, B. R,. Shanley, K. W., Steel, R. J., Strasser, A., and Tucker, M. E., 2010, Sequence stratigraphy: common ground after three decades of development: First Break, v. 28, p. 21–34.

Catuneanu, O., Galloway, W.E., Kendall, C.G.St.C., Miall, A.D., Posamentier, H.W., Strasser A., and Tucker M.E., 2011, Sequence Stratigraphy: Methodology and Nomenclature: Report to ISSC: Newsletters on Stratigraphy, v. 4 (3), p. 173–245.

Christie-Blick, N., Mountain, G. S., and Miller, K. G., 1990, Seismic stratigraphy: record of sea-level change, in Revelle, R., ed., Sea-level change: National Research Council, Studies in Geophysics, Washington, National Academy Press, p. 116-140.

Christie-Blick, N., Pekar, S. F., and Madof, A. S., 2007, Is there a role for sequence stratigraphy in chronostratigraphy? Stratigraphy, v. 4, p. 217–229.

Clemmensen, L. B., and Hegner, J., 1991, Eolian sequence and erg dynamics: the Permian Corrie Sandstone, Scotland: Journal of Sedimentary Petrology, v. 61, p. 768–774.

Colombera L, Mountney NP (2020) Accommodation and sediment-supply controls on clastic parasequences: a meta-analysis. Sedimentology 67:1667–1709

Colombera, L., and Mountney, N. P., 2020, Accommodation and sediment-supply controls on clastic parasequences: a meta-analysis: Sedimentology, v. 67, p. 1667–1709.

Covault, J. A., Normark, W. R., Romans, B. W., and Graham, S. A., 2007, Highstand fans in the California borderland: the overlooked deep-water depositional systems: Geology, v. 35, p. 783–786.

Curtis, D. M., 1970, Miocene deltaic sedimentation, Louisiana Gulf Coast, in Morgan, J. P., ed., Deltaic sedimentation modern and ancient: Society of Economic Paleontologists and Mineralogists Special Publication 15, p. 293–308.

Dalrymple, R. W., Boyd, R., and Zaitlin, B. A., eds., 1994, Incised-valley systems: origin and sedimentary sequences: SEPM (Society for Sedimentary Geology) Special Publication 51, 391 p.

Dalrymple, R. W., 2010b, Introduction to siliciclastic facies models, in James, N. P., and Dalrymple, R. W., eds., Facies Models 4: GEOtext 6, Geological Association of Canada, St. John's, Newfoundland, p. 59–72.

Davies, N. S., and Shillito, A. P., 2018, Incomplete but intricately detailed: The inevitable preservation of true substrates in a time-deficient stratigraphic record: Geology, v. 46, p. 679-682.

Davies, N. S., and Shillito, A. P., 2021, True substrates: the exceptional resolution and unexceptional preservation of deep-time snapshots on bedding surfaces: Sedimentology, in press.

Dolan, J. F., 1989, Eustatic and tectonic controls on deposition of hybrid siliciclastic/carbonate basinal cycles: American Association of Petroleum Geologists Bulletin, v. 73, p. 1233–1246.

Dunbar, G. B., Dickens, G. R. and Carter, R. M., 2000, Sediment flux across the Great Barrier reef to the Queensland Trough over the last 300 ky: Sedimentary Geology, v. 133, p. 49-92.

Emery, D., and Myers, K. J., 1996, Sequence stratigraphy: Blackwell, Oxford, 297 p.

Erlich, R. N., Longo, A. P., Jr., and Hyare, S., 1993, Response of carbonate platform margins to drowning: evidence of environmental collapse: in Loucks, R. G., and Sarg, J. F., Jr., eds., Carbonate sequence stratigraphy – recent developments and applications: American Association of Petroleum Geologists, memoir 57, p. 241–266.

Fisher, W. L., and McGowen, J. H., 1967, Depositional systems in the Wilcox Group of Texas and their relationship to occurrence of oil and gas: Transactions of the Gulf Coast Association of Geological Societies, v. 17, p. 105–125.

Frazier, D. E., 1974, Depositional episodes: their relationship to the Quaternary stratigraphic framework in the northwestern portion of the Gulf Basin: Bureau of Economic Geology, University of Texas, Geological Circular 74-1, 26 p.

Galloway, W. E., 1989, Genetic stratigraphic sequences in basin analysis I: Architecture and genesis of flooding-surface bounded depositional units: American Association of Petroleum Geologists Bulletin, v. 73, p. 125–142.

Galloway, W. E., and Brown, L. F., Jr., 1973, Depositional systems and shelf-slope relations on cratonic basin margin, uppermost Pennsylvanian of north-central Texas: American Association of Petroleum Geologists Bulletin, v. 57, p. 1185–1218.

Haq, B. U., Hardenbol, J., and Vail, P. R., 1987, Chronology of fluctuating sea levels since the Triassic (250 million years ago to present): Science, v. 235, p. 1156–1167.

Haq, B. U., Hardenbol, J., and Vail, P. R., 1988, Mesozoic and Cenozoic chronostratigraphy and cycles of sea-level change, in Wilgus, C. K., Hastings, B. S., Kendall, C. G. St. C., Posamentier,

H. W., Ross, C. A., and Van Wagoner, J. C., eds., Sea-level Changes: an integrated approach: Society of Economic Paleontologists and Mineralogists Special Publication 42, p. 71–108.

Heckel, P. H., 1986, Sea-level curve for Pennsylvanian eustatic marine transgressive-regressive depositional cycles along midcontinent outcrop belt, North America: Geology, v. 14, p. 330–334.

Helland-Hansen, W., and Hampson, G. J., 2009, Trajector analysis: concepts and applications: Basin Research, v. 21, p. 454–483.

Helland-Hansen, W., and Martinsen, O., 1996, Shoreline trajectories and sequences: description of variable depositional-dip scenarios: Journal of Sedimentary Research, v. 66, p. 670–688.

Hilgen, F. J., Hinnov, L. A., Aziz, H. A., Abels, H. A., Batenburg, S., Bosmans, J. H. C., de Boer, B., Hüsings, S. K., Kuiper, K. F., and Lourens, L. J., 2015, Stratigraphic continuity and fragmentary sedimentation: the success of cyclostratigraphy as part of integrated stratigraphy in Smith, D. G., Bailey, R., J., Burgess, P., and Fraser, A., eds., Strata and time: Geological Society, London, Special Publication 404, p. 157–197.

Holbrook, J. M., and Bhattacharya, J. P., 2012, Reappraisal of the sequence boundary in time and space: Case and considerations for an SU (subaerial unconformity) that is not a sediment bypass surface, a time barrier, or an unconformity: Earth Science Reviews, v 113, p. 271–302.

Holbrook, J., Scott, R. W., and Oboh-Ikuenobe, F. E., 2006, Base-level buffers and buttresses: a model for upstream versus downstream control on fluvial geometry and architecture within sequences: Journal of Sedimentary Research, v. 76, p. 162–174.

Hunt, D., and Gawthorpe, R. L., eds., 2000, Sedimentary response to forced regressions: Geological Society of London Special Publication 172, 383 p.

Ito, M., and Masuda, F., 1988, Late Cenozoic deep-sea to fan-delta sedimentation in an arc-arc collision zone, central Honshu, Japan; sedimentary response to varying plate-tectonic regime, in Nemec, W., and Steel, R.J., eds., Fan deltas; sedimentology and tectonic settings: Glasgow, UK, Blackie and Son, p. 400–418.

James, N. P., and Bourque, P.-A., 1992, Reefs and mounds, in Walker, R. G. and James, N. P., eds., Facies models: response to sea-level change: Geological Association of Canada, Geotext 1, p. 323–347.

James, N. P., and Kendall, A. C., 1992, Introduction to carbonate and evaporite facies models, in Walker, R. G., and James, N. P., eds., Facies models: response to sea level change: Geological Association of Canada, Geotext 1, p. 265–275.

James, N. P., Kendall, A. C., and Pufahl, P. K., 2010, Introduction to biological and biochemical facies models, in James, N. P., and Dalrymple, R. W., eds., Facies Models 4: GEOtext 6, Geological Association of Canada, St. John's, Newfoundland, p. 323–340.

Jervey, M. T., 1988, Quantitative geological modeling of siliciclastic rock sequences and their seismic expression, in Wilgus, C. K., Hastings, B. S., Kendall, C. G. St. C., Posamentier, H. W., Ross, C. A., and Van Wagoner, J. C., eds., Sea level Changes - an integrated approach: Society of Economic Paleontologists and Mineralogists Special Publication 42, p. 47–69.

Jones, B., 2010, Warm-water neritic carbonates, in James, N. P., and Dalrymple, R. W., eds., Facies Models 4: GEOtext 6, Geological Association of Canada, St. John's, Newfoundland, p. 341–369.

Jones, B., and Desrochers, A., 1992, Shallow platform carbonates, in Walker, R. G., and James, N. P., eds., Facies models: response to sea level change: Geological Association of Canada, Geotext 1, p. 277–301.

Kocurek, G. A., 1988, First-order and super bounding surfaces in eolian sequences - bounding surfaces revisited: Sedimentary Geology, v. 56, p. 193–206.

Kocurek, G. A., 1996, Desert aeolian systems, in Reading, H. G., ed., Sedimentary environments, processes, facies, stratigraphy, third edition: Blackwell Science, Oxford, p. 125–153.

Kocurek, G., and Havholm, K. G., 1993, Eolian sequence stratigraphy —a conceptual framework, in Weimer, P., and Posamentier, H. W., eds., Siliciclastic sequence stratigraphy: American Association of Petroleum Geologists Memoir 58, p. 393–409.

Kocurek, G. A., Havholm, K. G., Deynoux, M., and Blakey, R. C., 1991, Amalgamated accumulations resulting from climatic and eustatic changes: Akchar Erg, Mauritania: Sedimentology, v. 38, p. 751–772.

Korus, J. T., Kvale, E. P., Eriksson, K. A., and Joeckel, R. M., 2008, Compound paleovalleys fills in the Lower Pennyslvanian New River Formation, West Virginia, USA: Sedimentary Geology, v. 208, p. 15–26.

Leckie, D. A., 1994, Canterbury Plains, New Zealand—implications for sequence stratigraphic models: American Association of Petroleum Geologists Bulletin, v. 78, p. 1240–1256.

Li, Z., and Schieber, J., 2020, Application of sequence stratigraphic concepts to the Upper Cretaceous Tunuk Shale Member of the Mancos Shale formation, south-central Utah: parasequence styles in shelfal mudstone strata: Sedimentology, v. 67, p. 118–151.

Loutit, T. S., Hardenbol, J., Vail, P. R., and Baum, G. R., 1988, Condensed sections: the key to age dating and correlation of continental margin sequences, in Wilgus, C. K., Hastings, B. S., Kendall, C. G. St. C., Posamentier, H. W., Ross, C. A., and Van Wagoner, J. C., eds., Sea-level Changes: an integrated approach: Society of Economic Paleontologists and Mineralogists Special Publication 42, p. 183–213.

Madof, A. S., Harris, A. D., and Connell, S. D., 2016, nearshore along-strike variability: IS the concept of the systems tract unhinged? Geology, v. 44, p. 315–318.

McCarthy, P. J., Faccini, U. F., and Plint, A. G., 1999, Evolution of an ancient coastal plain: palaeosols, interfluves and alluvial architecture in a sequence stratigraphic framework, Cenomanian Dunvegan Formation, NE British Columbia, Canada: Sedimentology, v. 46, p. 861–891.

McMillen, K.M., and Winn, R.D., Jr., 1991, Seismic facies of shelf, slope, and submarine fan environments of the Lewis Shale, Upper Cretaceous, Wyoming, in Weimer, P., and Link, M.H., eds., Seismic facies and sedimentary processes of submarine fans and turbidite systems: New York, Springer-Verlag, p. 273–287.

Miall, A. D., 1991, Stratigraphic sequences and their chronostratigraphic correlation: Journal of Sedimentary Petrology, v. 61, p. 497–505.

Miall, A. D., 1995, Whither stratigraphy? Sedimentary Geology, v. 100, p. 5–20.

Miall, A. D., 1996, The geology of fluvial deposits: sedimentary facies, basin analysis and petroleum geology: Springer-Verlag Inc., Heidelberg, 582 p.

Miall, A. D., 1999, Principles of sedimentary basin analysis, Third edition: New York, N. Y.: Springer-Verlag Inc., 616 p.

Miall, A. D., 2010, The geology of stratigraphic sequences, second edition: Springer-Verlag, Berlin, 522 p.

Miall, A. D., 2014a, Fluvial depositional systems: Springer-Verlag, Berlin 316 p.

Miall, A. D., 2014b, The emptiness of the stratigraphic record: A preliminary evaluation of missing time in the Mesaverde Group, Book Cliffs, Utah: Journal of Sedimentary Research, v. 84, p. 457–469.

Miall, A. D., 2015, Updating uniformitarianism: stratigraphy as just a set of "frozen accidents", in Smith, D. G., Bailey, R., J., Burgess, P., and Fraser, A., eds., Strata and time: Geological Society, London, Special Publication 404, p. 11–36.

Miall, A. D., 2016, The valuation of unconformities: Earth Science Reviews, v. 163, p. 22–71.

Miall, A. D., and Miall, C. E., 2001, Sequence stratigraphy as a scientific enterprise: the evolution and persistence of conflicting paradigms: Earth Science Reviews, v. 54, #4, p. 321–348.

Miall, C. E., and Miall, A. D., 2002, The Exxon factor: the roles of academic and corporate science in the emergence and legitimation of a new global model of sequence stratigraphy: Sociological Quarterly, v. 43, p. 307–334.

Miller, K. G., Wright, J. D., and Browning, J. V., 2005, Visions of ice sheets in a greenhouse world: Marine Geology, v. 217, p. 215–231.

Mitchum, R. M., Jr., Vail, P. R., and Thompson, S. III, 1977a, Seismic stratigraphy and global changes of sea level, Part 2, The depositional sequence as a basic unit for stratigraphic analysis, in Payton, C. E., ed., Seismic stratigraphy—applications to hydrocarbon exploration: American Association of Petroleum Geologists Memoir 26, p. 53–62.

Mitchum, R. M., Jr., Vail, P. R., and Sangree, J. B., 1977b, Seismic stratigraphy and global changes of sea level, Part six: Stratigraphic interpretation of seismic reflection patterns in depositional sequences, in Payton, C. E., ed., Seismic stratigraphy—applications to hydrocarbon exploration; American Association of Petroleum Geologists Memoir 26, p. 117–133.

Mitchum, R. M., Jr., and Van Wagoner, J. C., 1991, High-frequency sequences and their stacking patterns: sequence-stratigraphic evidence of high-frequency eustatic cycles: Sedimentary Geology, v. 70, 131–160.

Neal, J., and Abreu, V., 2009, Sequence stratigraphy hierarchy and the accommodation succession method: Geology, v. 37, p. 779–782.

Neal, J. E., Abreu, V., Bohacs, K. M., Feldman, H. R., and Pederson, K. H., 2016, Accommodation succession ($\delta A/\delta S$) sequence stratigraphy: observational method, utility and insights into sequence boundary formation: Journal of the Geological Society, London, v. 173, p. 803–816.

North American Commission on Stratigraphic Nomenclature (NACSN), 2005, North American Stratigraphic Code: American Association of Petroleum Geologists Bulletin, v. 89, p. 1547–1591.

Nummedal, D., and Swift, D. J. P., 1987, Transgressive stratigraphy at sequence-bounding unconformities: some principles derived from Holocene and Cretaceous examples, in Nummedal, D., Pilkey, O. H., and Howard, J. D., eds., Sea-level fluctuation and coastal evolution; Society of Economic Paleontologists and Mineralogists Special Publication 41, p. 241–260.

Nystuen, J. P., 1998, History and development of sequence stratigraphy, in Gradstein, F. M., Sandvik, K. O. and Milton, N. J., eds., Sequence stratigraphy — concepts and applications: Norwegian Petroleum Society Special Publication 8, p. 31–116.

Paola, C., 2000, Quantitative models of sedimentary basin filling: Sedimentology, v. 47 (supplement 1) p. 121–178.

Pattison, S. A. J., 2020, No evidence for an unconformity at the base of the lower Castlegate Sandstone in the Campanian Book Cliffs, Utah-Colorado, United States: Implications for sequence models: American Association of Petroleum Geologists Bulletin, v. 104, p. 595–628.

Payton, C. E., ed., 1977, Seismic stratigraphy—applications to hydrocarbon exploration: American Association of Petroleum Geologists Memoir 26, 516 p.

Pedersen, S. A. S., 2012, Glaciodynamic sequence stratigraphy, in Huuse, M., Redfern, J., LeHeron, D. P., Dixon, R. J., Moscariello, A., and Craig, J., eds., Glaciogenic reservoirs and hydrocarbon systems, Geological Society, London, Special Publication 368, p. 29–51.

Pellegrini, C., Patruno, S., Helland-Hansen, W., Steel, W. J., and Trincardi, F., 2020, Clinforms and clinothems: Fundamental elements of basin infill: Basin Research, v. 32, p. 187–205.

Piper, D.J.W., and Normark, W.R., 2001, Sandy fans; from Amazon to Hueneme and beyond: American Association of Petroleum Geologists Bulletin, v. 85, p. 1407–1438.

Plint, A. G., 1988, Sharp-based shoreface sequences and "offshore bars" in the Cardium Formation of Alberta: their relationship to relative changes in sea level, in Wilgus, C. K., Hastings, B. S., Kendall, C. G. St. C., Posamentier, H. W., Ross, C. A., and Van Wagoner, J. C., eds., Sea-level Changes: an integrated approach: Society of Economic Paleontologists and Mineralogists Special Publication 42, p. 357–370.

Plint, A. G., 1991, High-frequency relative sea-level oscillations in Upper Cretaceous shelf clastics of the Alberta foreland basin: possible evidence for a glacio-eustatic control? in Macdonald, D. I. M., ed., Sedimentation, tectonics and eustasy: sea-level changes at active margins: International Association of Sedimentologists Special Publication 12, p. 409–428.

Plint, A. G., McCarthy, P. J., and Faccini, U. F., 2001, Nonmarine sequence stratigraphy: updip expression of sequence boundaries and systems tracts in a high-resolution framework: Cenomanian Dunvegan Formation, Alberta foreland basin, Canada: American Association of Petroleum Geologists Bulletin, v. 85, p. 1967–2001.

Porebski, S. J., and Steel, R. J., 2003, Shelf-margin deltas: their stratigraphic significance and relation to deepwater sands: Earth-Science Reviews, v. 62 p. 283–326.

Porebski, S.J., and Steel, R.J., 2006, Deltas and sea-level change: Journal of Sedimentary Research, v. 76, p. 390–403.

Posamentier, H. W., and Allen, G. P., 1999, Siliciclastic sequence stratigraphy—concepts and applications: Society for Sedimentary Geology (SEPM), Concepts in sedimentology and paleontology 7, 210 p.

Posamentier, H. W., and Vail, P. R., 1988, Eustatic controls on clastic deposition II—sequence and systems tract models, in Wilgus, C. K., Hastings, B. S., Kendall, C. G. St. C., Posamentier, H. W., Ross, C. A., and Van Wagoner, J. C., eds., Sea level Changes - an integrated approach: Society of Economic Paleontologists and Mineralogists Special Publication 42, p. 125–154.

Posamentier, H. W., Jervey, M. T., and Vail, P. R., 1988, Eustatic controls on clastic deposition I—Conceptual framework, in Wilgus, C. K., Hastings, B. S., Kendall, C. G. St. C., Posamentier, H. W., Ross, C. A., and Van Wagoner, J. C., eds., Sea level Changes - an integrated approach: Society of Economic Paleontologists and Mineralogists Special Publication 42, p. 109–124.

Posamentier, H. W., Allan, G. P., and James, D. P., 1992, High-resolution sequence stratigraphy – the East Coulee Delta, Alberta: Journal of Sedimentary Petrology, v. 62, p. 310–317.

Posamentier, H. W., 2002, Ancient shelf ridges—a potentially significant component of the transgressive systems tract: case study from offshore northwest Java: American Association of Petroleum Geologists Bulletin, v. 86, p. 75–106.

Pratt, B. R., 2010, Peritidal carbonates, in James, N. P., and Dalrymple, R. W., eds., Facies Models 4: GEOtext 6, Geological Association of Canada, St. John's, Newfoundland, p. 401–420.

Pratt, B. R., James, N. P., and Cowan, C. A., 1992, Peritidal carbonates, in Walker, R. G. and James, N. P., eds., Facies models: response to sea-level change: Geological Association of Canada, Geotext 1, p. 303–322.

Ramsbottom, W. H. C., 1979, Rates of transgression and regression in the Carboniferous of NW Europe: Journal of the Geological Society, London, v. 136, p. 147–153.

Reinson, G. E., 1992, Transgressive barrier island and estuarine systems, in Walker, R. G. and James, N. P., eds., Facies models: response to sea-level change: Geological Association of Canada, Geotext 1, p. 179–194.

Riba, O., 1976, Syntectonic unconformities of the Alto Cardener, Spanish Pyrenees, a genetic interpretation: Sedimentary Geology, v. 15, p. 213–233.

Rich, J. L., 1951, Three critical environments of deposition and criteria for recognition of rocks deposited in each of them: Geological Society of America Bulletin, v. 62, p. 1–20.

Sageman, B. B., Myers, S. R., and Arthur, M. A., 2006, Orbital time scale and new C-isotope record for Cenomanian-Turonian boundary stratotype: Geology, v. 34, p. 125–128.

Sanborn, K. L., Webster, J. M., Webb, G. E., Braga, J. C., Humblet, M., Nothdurft, L., Patterson, M. A., Dechnik, B., Warner, S., Graham, T., Murphy, R. J., Yokoyama, Y., Obrochta, S. P. Zhao, J., and Salas-Saavedra, M., 2020: A new model of Holocene reef initiation and growth in response to sea-level rise on the Southern Great Barrier Reef: Sedimentary Geology, v. 397, #105556, 18 p.

Sangree, J. B., and Widmier, J. M., 1977, Seismic stratigraphy and global changes of sea level, part 9: Seismic interpretation of clastic depositional facies, in Payton, C. E., ed., Seismic stratigraphy — applications to hydrocarbon exploration: American Association of Petroleum Geologists Memoir 26, p. 165–184.

Sarg, J. F., 1988, Carbonate sequence stratigraphy, in Wilgus, C. K., Hastings, B. S., Kendall, C. G. St. C., Posamentier, H. W., Ross, C. A., and Van Wagoner, J. C., eds., Sea level Changes - an integrated approach: Society of Economic Paleontologists and Mineralogists Special Publication 42, p. 155–181.

Schlager, W., 1989, Drowning unconformities on carbonate platforms, in Crevello, P. D., Wilson, J. L., Sarg, J. F., and Read, J. F., eds., Controls on carbonate platforms and basin development: Society of Economic Paleontologists and Mineralogists Special Publication 44, p. 15–25.

Schlager, W., 1992, Sedimentology and sequence stratigraphy of reefs and carbonate platforms: American Association of Petroleum Geologists Continuing Education Course Notes Series 34, 71 p.

Schlager, W., 1993, Accommodation and supply—a dual control on stratigraphic sequences: Sedimentary Geology, v. 86, p. 111–136.

Schlager, W., 2004, Fractal nature of stratigraphic sequences, Geology, v. 32, p. 185–188.

Schlager, W., 2005, Carbonate sedimentology and sequence stratigraphy: SEPM Concepts in Sedimentology and Paleontology #8, 200p.

Schumm, S. A., 1979, Geomorphic thresholds: the concept and its applications: Transactions of the Institute of British Geographers, v. 4, p. 485–515.

Schumm, S. A., 1993, River response to baselevel change: implications for sequence stratigraphy: Journal of Geology, v. 101, p. 279–294.

Shanley, K. W., and McCabe, P. J., 1994, Perspectives on the sequence stratigraphy of continental strata: American Association of Petroleum Geologists Bulletin, v. 78, p. 544–568.

Simmons, M. D., Mller, K. G., Ray, D. C., Davies, A., van Buchem, F. S. P., and Gréselle, B., 2020, Phanerozoic eustasy, in Gradstein, F. M., Ogg, J. G., Schmitz, M. D., and Ogg, G. M., eds., Geologic Time Scale 2020, Elsevier, Amsterdam, p. 357–400.

Sloss, L. L., 1963, Sequences in the cratonic interior of North America: Geological Society of America Bulletin, v. 74, p. 93–113.

Sloss, L. L., 1972, Synchrony of Phanerozoic sedimentary-tectonic events of the North American craton and the Russian platform: 24th International Geological Congress, Montreal, Section 6, p. 24–32.

Southgate, P. N., Kennard, J. M., Jackson, M. J., O'Brien, P. E., and Sexton, M. J., 1993, Reciprocal lowstand clastic and highstand carbonate sedimentation, subsurface Devonian reef complex, Canning Basin, Western Australia, in Loucks, R. G., and Sarg, J. F., eds., Carbonate sequence stratigraphy: American Association of Petroleum Geologists Memoir 57, p. 157–179.

Strong, N., and Paola, C., 2008, Valleys that never were: time surfaces versus stratigraphic surfaces: Journal of Sedimentary Research, v. 78, p. 579–593.

Talbot, M. R., 1985, Major bounding surfaces in aeolian sandstones—a climatic model: Sedimentology, v. 32, p. 257–265.

Tipper, J. C., 2015, The importance of doing nothing: stasis in sedimentation systems and its stratigraphic effects: in Smith, D. G., Bailey, R., J., Burgess, P., and Fraser, A., eds., Strata and time: Probing the Gaps in Our Understanding: Geological Society, London, Special Publication 404, p. 105–122.

Törnqvist, T. E., 1993, Holocene alternation of meandering and anastomosing fluvial systems in the Rhine-Meuse delta (central Netherlands) controlled by sea-level rise and subsoil erodibility: Journal of Sedimentary Petrology, v. 63, p. 683–693.

Törnqvist, T. E., van Ree, M. H. M., and Faessen, E. L. J. H., 1993, Longitudinal facies architectural changes of a Middle Holocene anastomosing distributary system (Rhine-Meuse delta, central Netherlands): Sedimentary Geology, v. 85, p. 203–219.

Vail, P. R., 1987, Seismic stratigraphy interpretation using sequence stratigraphy, Part 1: seismic stratigraphy interpretation procedure, in Bally, A. W., ed., Atlas of seismic stratigraphy: American Association of Petroleum Geologists Studies in Geology 27, v. 1, p. 1–10.

Vail, P. R., Mitchum, R. M., Jr., Todd, R. G., Widmier, J. M., Thompson, S., III, Sangree, J. B., Bubb, J. N., and Hatlelid, W. G., 1977, Seismic stratigraphy and global changes of sea-level, in Payton, C. E., ed., Seismic stratigraphy - applications to hydrocarbon exploration: American Association of Petroleum Geologists Memoir 26, p. 49–212.

Vandenberghe, J., 1993, Changing fluvial processes under changing periglacial conditions: Z. Geomorph, N.F., v. 88, p. 17–28.

Vandenberghe, J., Kasse, C., Bohnke, S., and Kozarski, S., 1994, Climate-related river activity at the Weichselian-Holocene transition: a comparative study of the Warta and Maas rivers: Terra Nova, v. 6, p. 476–485.

Van Wagoner, J. C., Mitchum, R. M., Jr., Posamentier, H. W., and Vail, P. R., 1987, Seismic stratigraphy interpretation using sequence stratigraphy, Part 2: key definitions of sequence stratigraphy, in Bally, A. W., ed., Atlas of seismic stratigraphy: American Association of Petroleum Geologists Studies in Geology 27, v. 1, p. 11–14.

Van Wagoner, J. C., Mitchum, R. M., Campion, K. M. and Rahmanian, V. D. 1990, Siliciclastic sequence stratigraphy in well logs, cores, and outcrops: American Association of Petroleum Geologists Methods in Exploration Series 7, 55 p.

Van Yperen, A. E., Holbrook, J. M., Poyatos-Moré, M., Myers, C., and Midtkandal, I., 2021, Low-accommodation and backwater effects on sequence stratigraphic surfaces and depositional architecture of fluvio-deltaic settings (Cretaceous Mesa Rica Sandstone, Dakota Group, USA): Basin Research, v. 33, p. 513–543.

Varban, B. L. and Plint, A. G. 2008, Sequence stacking patterns in the Western Canada foredeep: influence of tectonics, sediment loading and eustasy on deposition of the Upper Cretaceous Kaskapau and Cardium Formations: Sedimentology, v. 55, p. 395–421.

Veeken, 2007, Seismic stratigraphy, basin analysis and reservoir characterization: Elsevier, Amsterdam, Seismic Exploration, v. 37, 509 p.

Walker, R. G., 1992, Facies, facies models and modern stratigraphic concepts, in Walker, R. G. and James, N. P., eds., Facies models: response to sea-level change: Geological Association of Canada, p. 1–14.

Wanless, H. R., and Weller, J. M., 1932, Correlation and extent of Pennsylvanian cyclothems: Geological Society of America Bulletin, v. 43, p. 1003–1016.

Weber, M.E., Wiedicke, M.H., Kudrass, H.R., Huebscher, C., and Erlenkeuser, H., 1997, Active growth of the Bengal Fan during sea-level rise and highstand: Geology, v. 25, p. 315–318.

Wescott, W. A., 1993, Geomorphic thresholds and complex response of fluvial systems—some implications for sequence stratigraphy: American Association of Petroleum Geologists Bulletin, v. 77, p. 1208-1218.

White, N., and Lovell, B., 1997, Measuring the pulse of a plume with the sedimentary record: Nature, v. 387, p. 888-891.

Wilson, J. L., 1967, Cyclic and reciprocal sedimentation in Virgilian strata of southern New Mexico: Geological Society of America Bulletin, v. 78, p. 805-818.

Wood, L. J., Ethridge, F. G., and Schumm, S. A., 1993, The effects of rate of base-level fluctuations on coastal plain, shelf and slope depositional systems: an experimental approach, in Posamentier, H. W., Summerhayes, C. P., Haq, B. U., and Allen, G. P., eds.,

Sequence stratigraphy and facies associations: International Association of Sedimentologists Special Publication 18, p. 43-53.

Wright, V. P., and Marriott, S. B., 1993, The sequence stratigraphy of fluvial depositional systems: the role of floodplain sediment storage: Sedimentary Geology, v. 86, p. 203–210.

Yoshida, S., Steel, R. S, and Dalrymple, R. W., 2007, Changes in depositional processes – an ingredient in a new generation of

sequence stratigraphic models: Journal of Sedimentary Research, v. 77, p. 447–460.

Zecchin, M., Catuneanu, O., and Caffau, M., 2019, Wave-ravinement surfaces: classification and key characteristics: Earth Science Reviews, v. 188, p. 210–239.

Zhang, J., Burgess, P. M., Granjeon, D., and Steel, R., 2018, Can sediment supply variations create sequences? Insights from stratigraphic forward modelling: Basin Research, v. 31, p. 274-289.

Basin Mapping Methods

Contents

Abstract

Geological mapping methods comprise a range of techniques for extending geological observations beyond individual outcrops or subsurface data points in order to provide a regional synthesis of stratigraphic variability of lithology, facies, thickness or any other significant stratigraphic variable. The mapping of petrophysical and reflection-seismic data is among the most important modern techniques for basin analysis. A paleogeographic analysis is facilitated by the mapping of detrital clastic petrofacies, particularly detrital zircons, and by the use of paleocurrent data, where available.

A. D. Miall, *Stratigraphy: A Modern Synthesis*, Springer Textbooks in Earth Sciences, Geography and Environment,
https://doi.org/10.1007/978-3-030-87536-7_6

6.1 Introduction

In the introduction to this chapter in the third edition of *"Principles of sedimentary basin analysis,"* I said (Miall 1999, p. 249):

> The art of basin mapping is one of reconstructing paleogeography and fill geometry from very limited evidence. Being skilled at data synthesis, interpretation, and extrapolation is of prime importance. Maps drawn only from outcrop data are rarely sufficiently accurate or useful for subsurface predictions, because they do not contain an adequate, three-dimensional distribution of data points.

This statement now qualifies as a historically interesting comment, because the maturation of sequence stratigraphy and developments in reflection-seismic methods have caused a complete change in the practice of basin mapping. In 2000, seismic stratigraphy had assumed considerable importance, but reliance on scattered exploration wells was still essential. Sequence methods have advanced substantially since then, and now comprise a comprehensive set of predictive tools for the tracking of stratigraphic units in the surface and subsurface. The basic methods, including an introduction to the topic of seismic facies, are described in this chapter. Modern, sophisticated, digital methods for the processing and display of seismic data have made the work of the subsurface geologist immeasurably easier. These new methods include an increasingly common use of three-dimensional seismic data volumes, from which cross-sections in any orientation and stratigraphically horizontal seismic maps can be readily constructed, enabling the three-dimensional visualization of complex stratigraphic objects. Such objects reflect the depositional, structural and diagenetic processes by which stratigraphy is constructed, and this new ability for three-dimensional visualization has led to the growth of an entirely new tool, that of **seismic geomorphology**—the study of ancient landscapes and depositional systems based on 3-D seismic data. Petrophysical and core data are essential in providing the "ground truthing" of seismic interpretations, such as confirming complex correlations and providing necessary lithofacies and biostratigraphic detail, but in many cases they no longer provide the primary data set from which the subsurface geophysicists and geologists do their work.

Seismic data are used to generate and extend a stratigraphic synthesis across a broad project area. Older methods, such as facies and isopach mapping, and paleocurrent analysis (mainly useful for surface work) do not provide this level of detail and are useful mainly in order to explore regional basinal trends, including the direction of paleoslope, differential subsidence patterns and regional facies trends. These methods are included in this chapter for completeness because not all mapping projects have access to seismic data.

Paradoxically, the growth in the sophistication of subsurface methods has revealed the importance of outcrop work. Modern seismic data and the increasing sophistication of sequence stratigraphic analysis have raised many questions regarding the fine detail of stratigraphic reconstructions. Questions about the local response to issues such as base-level change, changes in sediment supply or the significance of autogenic influences on sequence architecture may point to the need for very detailed stratigraphic mapping, which often can be only answered if a good outcrop exists in exactly the right locations. Several recent studies (which we return to in Chap. 8) have demonstrated that large-scale outcrop mapping, that is, the literal old-fashioned walking out of bed contacts for hundreds of meters, or even kilometers, can still provide essential insights into sequence models and allogenic processes.

6.2 Stratigraphic Mapping with Petrophysical Logs

6.2.1 Log Shape and Electrofacies

Well-log data are used for stratigraphic correlation and lithologic interpretation. They may also be used directly for facies mapping purposes. As noted briefly in Sect. 2.4, various combinations of logs can be employed in crossplots to yield lithologic information. These relationships can be converted into computer algorithms, where log data are digitized and stored in data banks, and a powerful automated mapping technique becomes available. Digital log data can also be displayed and manipulated using interactive computer graphics routines, a facility that permits easy comparisons for stratigraphic correlation purposes, plus checking for accuracy, and for the normalizing of logs made at different times using different hole conditions, and many other purposes. Well-log service companies have devoted considerable energy to devising and marketing automated processing and display techniques for use in basin analysis and petroleum development, but all such techniques suffer from the limitation that petrophysical responses are very location-specific. The techniques cannot be used without much initial careful calibration to local petrographic and groundwater conditions.

An early example of the use of petrophysical log data for mapping purposes was given by McCrossan (1961). He constructed a resistivity map for the lower and middle Ireton Formation of central Alberta that can be used to predict the proximity to reefs (Fig. 6.1). The Ireton is an Upper Devonian basinal shale that drapes against the reefs. Calcareous content and thus resistivity increase close to the reefs because of the presence of reef talus. The resistivity data, therefore, provide a useful analog paleogeographic map.

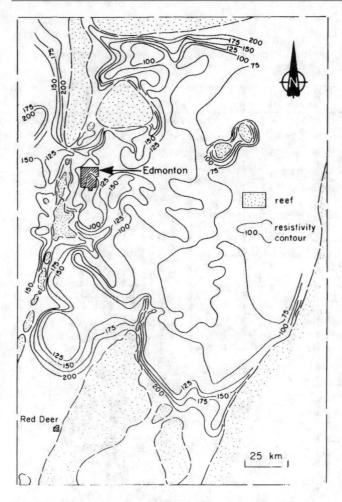

Fig. 6.1 Resistivity map of the Ireton Formation (Devonian), central Alberta. High values indicate an increase in carbonate content adjacent to reef bodies (McCrossan 1961)

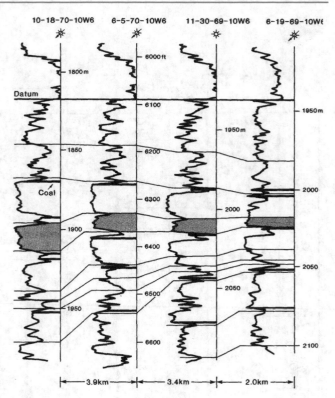

Fig. 6.2 A suite of gamma ray logs, showing how log shape may be used to carry out subsurface lithostratigraphic correlation (Cant 1992)

The recognition of characteristic log "shapes" and distinctive vertical profile character in siliciclastic deposits (**electrofacies**) may be a useful tool for correlation purposes (Figs. 6.2 and 6.3). Regardless of their lithology, facies or environment, lithologic units typically have a recognizable profile on petrophysical logs, which enables them to be traced and correlated from well to well. The ease and reliability of this mapping technique depend on how rapidly lithologic units change in thickness and facies laterally. Where there is considerable lateral variability, as in channelized deposits such as fluvial systems and submarine fans, it may be difficult or impossible to correlate reliably between drill holes. Much depends on hole spacing. Figure 6.2 is an example of a lithologic succession that shows little change from hole to hole, over distances of several kilometers, which permits every major unit to be defined and correlated on the stratigraphic cross-section. In contrast, in Fig. 6.3, with greater well spacing, several units thin, change lithology or thin out altogether across the section, which makes

correlation more difficult. The use of log shape analysis to interpret vertical facies successions and characterize certain sedimentary environments is described in Sect. 3.5.10.

The application of sequence-stratigraphic techniques, including the identification of systems tracts, may considerably improve the practice of log correlation because of their predictive value. The various elements of many standard sequence models commonly have distinctive lithologies and geometries, and typically occur in the same vertical succession, which provides useful clues for interpretation and correlation (Chap. 5; see Fig. 5.12). For example, incised valleys that commonly cut down into highstand deposits may consist of distinctive coarse channel-fill deposits incised into layered, fine-grained shelf deposits; sequence boundaries in clastic successions are commonly marked by widespread coarse-grained deposits; maximum flooding surfaces may be recognized by widespread "hot" (radioactive) shales (usually detected by the use of high readings on the gamma-ray log) or by condensed deposits, and so on.

Log shape analysis may be a useful aid for interpreting depositional environment (Figs. 3.61, 3.62 and 3.63). The analysis depends on the distinctiveness of the vertical profile through a sandstone body, and our ability to interpret this profile in terms of depositional processes and environment. The well-known bell-shaped gamma ray or spontaneous potential log response yielded by a typical fluvial fining

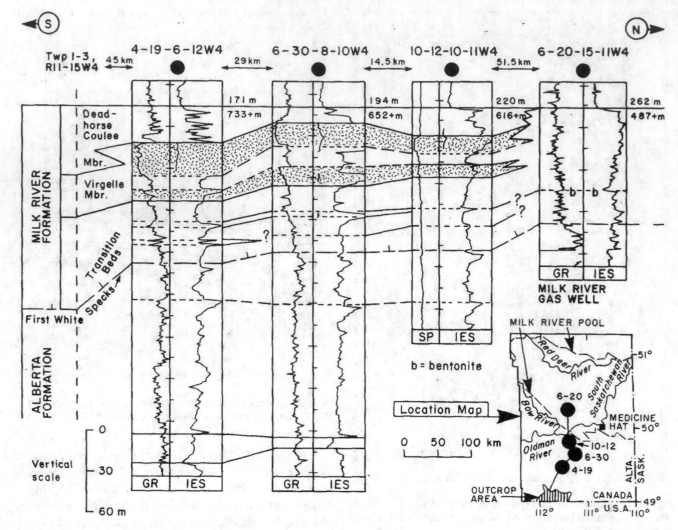

Fig. 6.3 Use of gamma ray (GR), spontaneous potential (SP) and induction electric survey (IES) logs to correlate subsurface sections. This diagram shows that where lithostratigraphic units retain similar thickness and lithology, they can be readily correlated using geophysical logs. Cretaceous, Alberta (Myhr and Meijer-Drees 1976)

upward cycle is a classic example (Visher 1965). The study of an interpreted subsurface point-bar sandstone by Berg (1968) is an early example of the application of these ideas to a practical exploration problem.

Despite the dramatic increase in the volume and quality of reflection-seismic data, the correlation of petrophysical logs is still the best means to construct a sequence stratigraphy for any stratigraphic volume that has an adequate network of exploration holes. For example, the huge data base of well-logs stored by the Alberta government in Calgary has enabled Plint and his students and colleagues to extend parts of the high-resolution sequence framework of the Cretaceous succession over the entire province of Alberta and to extend it southward into Montana, where it can be tied into American work on the same part of the stratigraphic record. Shank and Plint (2013) constructed a

network of petrophysical log-correlation cross-sections of the Cardium Alloformation (Turonian-Coniacian) across the entire southern half of the province, based on a subsurface database of 1200 wells, and 25 outcrop sections in the foothills belt to the west. Figure 6.4 illustrates a portion of one of the sections, which shows how the outcrop geology is tied to the subsurface allostratigraphic framework. Walaszczyk et al. (2014) tied this framework into the biostratigraphic and sequence framework of Montana, from whence it can be related to the cyclostratigraphic research underway across the Rocky Mountains states, using the so-called HiRES techniques first developed by Kauffman (1986, 1988; see also Cramer et al. 2015). As discussed in Sect. 8.13, Locklair and Sageman (2008) and Sageman et al. (2014) are developing a tightly constrained astrochronological time scale for equivalent rocks in the latter area.

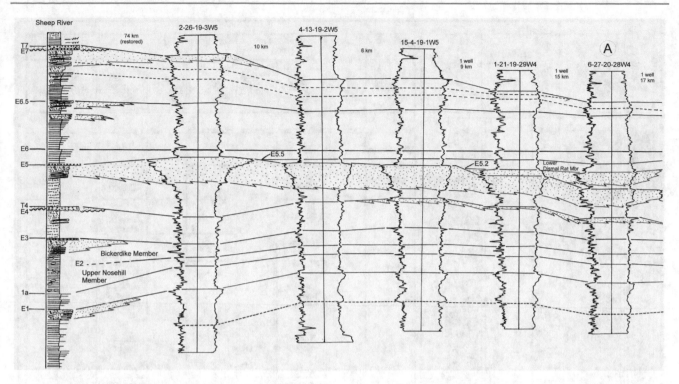

Fig. 6.4 A portion of one of the regional stratigraphic cross-sections through the Cardium Formation constructed by Shank and Plint (2013). The major allostratigraphic bounding surfaces are the E/T surfaces (E-erosion, T = transgression) first mapped by Plint et al. (1986)

6.2.2   Examples of Stratigraphic Reconstructions

Three modern examples of the use of petrophysical log correlation are discussed in this section. In the first, core and log data are used to provide a facies classification of the shallow-marine Bakken Formation in Saskatchewan where log correlation is made considerably easier by the recognition of key sequence surfaces and systems tracts. In the second example, a new model for the Marcellus Shale of New York shows how the correct choice of datum and the application of a new depositional model not only clarifies the stratigraphy but provides useful information for the understanding of the distribution of organic-rich shales, the source of the shale gas in this unit. In the last example, log shape analysis is used to map the complex pattern of facies that constitute a tertiary fluvial system in Texas.

The Bakken Formation is a marine siliciclastic unit of Upper Devonian and Lower Carboniferous age that extends across the Williston Basin (mainly North and South Dakota, Saskatchewan, southeastern Alberta and Montana). In recent years, fracking operations in this unit in North Dakota have turned that state into one of the leading US producers of oil, gas and gas liquids. In this section, we refer to a core and petrophysical log study of the unit carried out across southern Saskatchewan (Angulo and Buatois 2012). The formation has been subdivided into three members: lower and upper organic-rich black shale members and a middle

member of dolomitic siltstone and sandstone, which constitutes the main hydrocarbon reservoir. Further subdivisions of the middle member are shown in Fig. 6.5. The three members may be traced on petrophysical logs throughout the Williston Basin, with the upper and lower members characterized by unusually high gamma ray readings, reflecting the concentration of radioactive elements in the organic shales. Eleven lithofacies types have been recognized by Angulo and Buatois (2012), falling into two broad facies associations: open marine and brackish-water marginal marine. As shown in Fig. 6.5, a sequence analysis demonstrates that the lower member and the lower part of the middle member comprises a regressive (transgressive to highstand) systems tract, with a sequence boundary occurring at the base of a brackish-marine shoreface sandstone in the middle of the middle member, and the upper part of the Bakken Formation interpreted as a transgressive systems tract. Figure 6.6 shows how core and petrophysical-log data may be used to correlate these units across Saskatchewan. The distinctive deflections and shapes of the gamma ray, sonic and resistivity logs through each of these members are apparent from this diagram.

The second example described here is a re-analysis of the basin setting and configuration of the stratigraphic section that contains the important gas-bearing unit, the Marcellus Shale (Smith and Leone 2014). Well-logs were first calibrated against calculated values for total organic carbon and

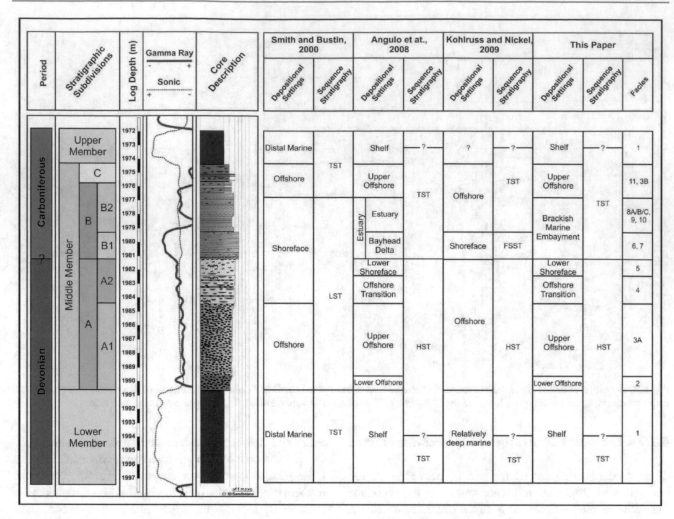

Fig. 6.5 Core and log characteristics and stratigraphic subdivision of the Bakken Formation of southern Saskatchewan (Angulo and Buatois 2012). AAPG © 2012, reprinted by permission of the AAPG whose permission is required for further use

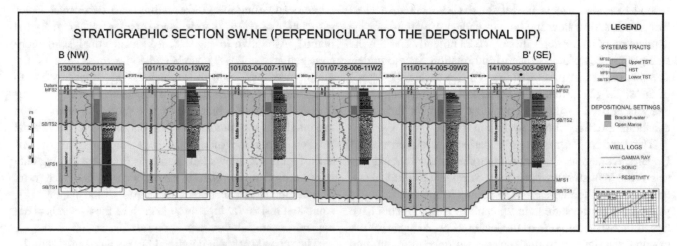

Fig. 6.6 Stratigraphic cross-section through the Bakken Formation, Saskatchewan (Angulo and Buatois 2012). AAPG © 2012, reprinted by permission of the AAPG whose permission is required for further use

Fig. 6.7 Calibration of petrophysical logs against compositional data derived from core analyses, Marcellus Shale, New York (Smith and Leone 2014). AAPG © 2014, reprinted by permission of the AAPG whose permission is required for further use

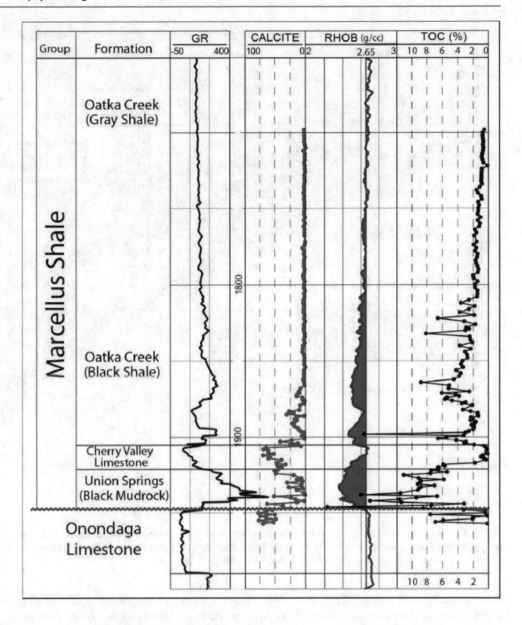

calcium carbonate obtained from core analyses (Fig. 6.7). Earlier basin models had suggested that the Marcellus Shale was a distal component of the Catskill clastic wedge that downlapped westward into an anoxic basin up to 250 m deep constituting the Appalachian foreland basin. However, the stratigraphic setting of nearby outcrops of similar organic-carbon-rich rocks suggests that they were deposited in shallow water over the basin forebulge in water depths probably less than 50 m. Instead of basing stratigraphic interpretations on the concept of a basal downlap surface, Smith and Leone (2014) used markers higher in the section as datum and constructed a different stratigraphic model (Fig. 6.8). In this model organic shales are thin but become more organic-rich westward as they interbed with progressively thicker carbonate units. This configuration led to a different depositional model. Instead of a deep anoxic basin

in which the organic shales were deposited, the authors suggested the occurrence of seasonal anoxia in a shallow water basin, with recycling of nutrients delivered from the Catskill "delta" complex. The richest organic shales are those further to the west, the most distal from the clay influx from the delta. They onlap multiple unconformities westward onto a tectonically positive forebulge. In an eastward direction they are interbedded with turbidites derived from the Appalachian orogen. For other examples of petrophysical (and seismic) mapping of fine-grained rocks, the reader is referred to Bohacs et al. (2002).

The third example discussed briefly here demonstrates how petrophysical log correlation and the analysis of log shape was used to document the stratigraphy of a coastal clastic system and identify and map the array of lithofacies constituting the deltaic–barrier–lagoon system within which

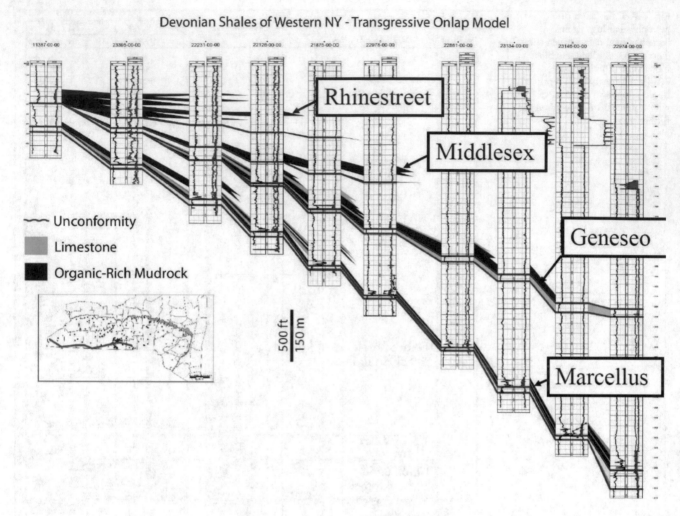

Fig. 6.8 A cross-section through the Marcellus and other distal parts of the Catskill complex in western New York State showing the westward change in facies and thicknesses toward the onlap of the basin forebulge (Smith and Leone 2014). AAPG © 2014, reprinted by permission of the AAPG whose permission is required for further use

these deposits were formed (Zeng et al. 1996). Figure 6.9 is a dip-oriented cross-section through two progradational shelf–shoreface–coastal successions separated by a surface of transgression. Fourteen sandstone units have been identified and correlated across the study area. These show considerable lateral persistence but display significant differences in log shape across the area. Facies analysis of each sandy depositional unit was derived from SP log character (and sometimes from resistivity log shape) and sandstone geometry. Some sidewall cores provided information on grain size and mineral identification. No conventional cores were available for this study. SP log patterns for 11 sandy units in the middle of the modeled section distinguished three major log facies: blocky, upward fining, and mixed (Fig. 6.10). These patterns were used to group sandy units developed in similar depositional environments. The distribution of these patterns within three of the sandstone units is shown in the maps in Fig. 6.10, and sandstone thickness,

measured from SP log response, is shown in Fig. 6.11. The thicknesses and relative arrangements of the various facies, with supplementary information on lithology from sidewall cores, were then used to make paleoenvironmental interpretations for each of the sandy units identified in the cross-section. Three of the resulting maps are shown in Fig. 6.12.

6.2.3 Problems and Solutions

There are several difficulties inherent in the use of petrophysical logs for stratigraphic work, which stem from the fact that drill holes necessarily represent a very limited sample of the rocks under study (the same problems arise with the use of outcrop sections for correlation except in areas of exceptional exposure, where correlations may be "walked out" on the ground). Even with the very close well

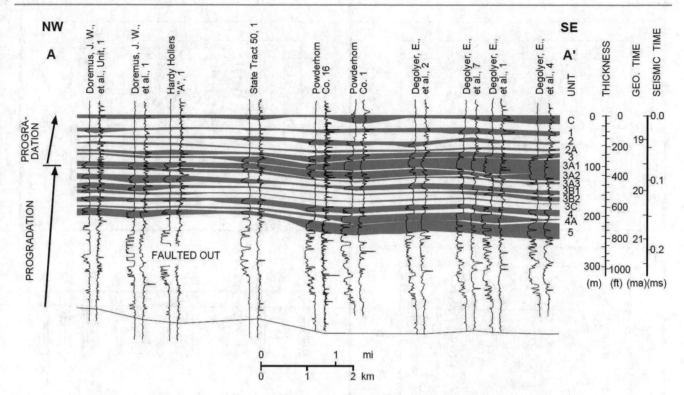

Fig. 6.9 Dip-oriented stratigraphic cross-section through a Miocene coastal deltaic system in the Powderhorn field, Texas, showing the definition of a series of sand-dominated lithostratigraphic units, grouped into two progradational successions (Zeng et al. 1996, Fig. 4, p. 21). AAPG © 1996, reprinted by permission of the AAPG whose permission is required for further use

spacing that characterizes enhanced recovery projects (a few hundred meters), it is commonly the case that the lateral dimensions of lithofacies units may be less than the distance between wells, which means that correlations may be unreliable, and attempts to characterize and model lithofacies architecture for production purposes are subject to wide margins of error. Martin (1993) provided some graphic examples of this problem in the case of fluvial reservoir architecture. An example is illustrated in Fig. 6.13, which shows three different approaches to the correlation of a suite of fluvial sand bodies in the subsurface where well spacing averages 1.54 km. This example is discussed by Miall (2014, Sect. 2.3.2), as part of a general discussion of the difficulty of carrying out facies modeling in the subsurface. In other environments where lateral variability is not so marked, stratigraphic correlations may be more reliable. The example of a complex facies mosaic is illustrated in Figs. 6.9, 6.10, 6.11 and 6.12, which shows where a degree of certainty may be assumed from the reconstructions showing that the extent of most of the facies units is considerably greater than well spacings, which range up to 5 km. In the case of some open-marine facies, such as transgressive sand bodies or turbidites, correlations of individual units over even greater distances may be possible.

Log correlation is usually guided by the identification of key surfaces and units that display characteristic log responses and maybe aided by biostratigraphic, chemostratigraphic or other tools, some of which are discussed later in this chapter. The development of sequence stratigraphy has made an enormous difference to the work of subsurface stratigraphic correlation. Sequence models have a predictive power that can help to clarify what might otherwise be very confusing vertical and lateral lithofacies relationships. Key surfaces, such as subaerial erosion surfaces, maximum flooding surfaces or transgressive sands, may be traceable for many kilometers and may serve to vertically align well sections in an arrangement that reveals the larger stratigraphic architecture. Much may depend on the choice of the datum on which to hang correlations.

In stratigraphic work the choice of datum may be key to the successful resolution of stratigraphic complexity. This is a surface selected because of its assumed original horizontality. Lining up sections by aligning a key surface at a horizontal datum removes structural disturbance from a section, and may move facies volumes and their interpreted depositional environments into relationships that make their interpretation much simpler. Bhattacharya (2011) provided an extensive discussion on this point and provided many examples of how confusing facies and stratigraphic reconstructions can be clarified by the appropriate choice of datum and the application of the appropriate sequence model. Several examples follow. He noted that in marine deposits, stratigraphic

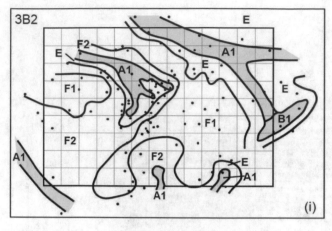

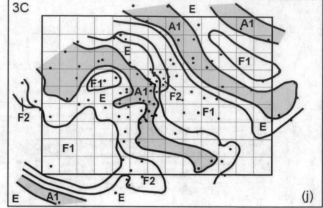

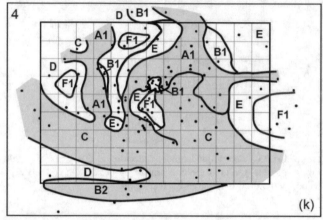

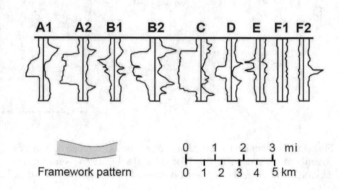

Fig. 6.10 SP log patterns of three of the sandy depositional units identified in Fig. 6.9. Log shapes at each location are identified at lower right: A1 = fining upward, A2 = fining upward with shoulder, B1 = coarsening upward, B2 = coarsening upward and blocky, C = blocky, D = fingerlike, E = serrate, F1 = straight with low resistivity, F2 = straight with high resistivity (Zeng et al. 1996, part of Fig. 5, p. 23). AAPG © 1996, reprinted by permission of the AAPG whose permission is required for further use

packages exhibiting significant depositional dip are extremely common. These are termed **clinoforms**, a term coined by Rich (1951; see Fig. 1.3). They range in a scale from the dipping units at the front of deltas, typically as little as a few meters in amplitude where the delta progrades into shallow water, to the shelf clinoforms developed where a continental margin is building into an ocean, with vertical amplitudes up to several kilometers. Reconstructing clinoforms from outcrop or petrophysical data is extremely difficult because the depositional dip may be less than 1°. In fact, it was not until the advent of seismic stratigraphy that earth scientists fully appreciated the ubiquitous nature of the clinoform architecture in the stratigraphic record. Figure 6.14 illustrates the importance of correlating correctly. The reconstruction shown in Fig. 6.14a is based on traditional lithostratigraphic principles, using a prominent facies change as a marker and datum. However, Fig. 6.14b is the correct interpretation, in which units of gradually changing lithofacies with a significant depositional dip prograded from left to right.

What may be the first reconstruction of clinoform architecture to be based on petrophysical data is illustrated in Fig. 6.15. Earlier mapping of the reefal Woodbend Group (Upper Devonian) in central Alberta by McCrossan (1961) had suggested the presence of depositional dips in the Ireton Formation, a mixed shale-limestone unit which was recognized as constituting the talus deposits flanking prominent reefs carbonate units. However, it was the later work of Oliver and Cowper (1963, 1965) that specifically identified the three depositional environments of Rich (1951)—the undaform, clinoform and fondoform. Their reconstruction (Fig. 6.15) is based on the correlation of picks in resistivity logs. Thin, resistive beds are carbonate talus units deposited on the flanks of the reef. In this study, the top of the Ireton Formation was selected as the datum, but Bhattacharya (2011) argued that in some cases selecting an underlying surface as the datum is necessary for the correct reconstruction to be recognized. He made this point by presenting a redrawing of the seismic reflections within a modern delta

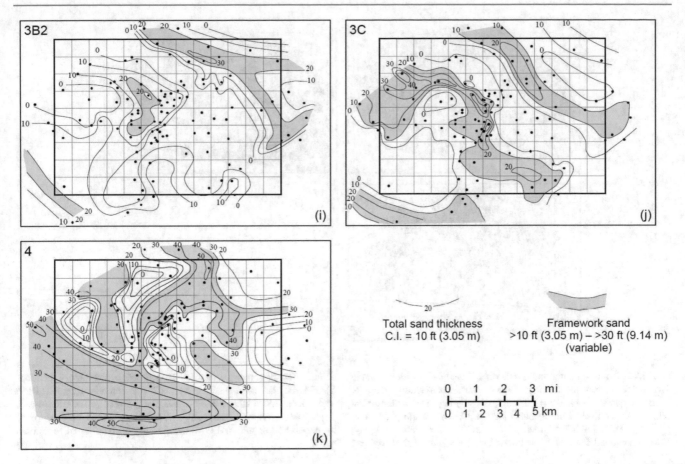

Fig. 6.11 Total sandstone thickness of sandy depositional units (Zeng et al. 1996, part of Fig. 6, p. 26). AAPG © 1996, reprinted by permission of the AAPG whose permission is required for further use

(Fig. 6.16). The clinoform architecture of the delta is obvious from the original cross-section, in which every surface dips seaward, showing offlapping clinoform wedges and a dipping modern depositional surface (Fig. 6.16a). But if one were to imagine that each reflection was a pick on a petrophysical log and that the picks were correlated based on a flattening of the top surface, the architecture becomes severely distorted, with some beds actually dipping landward (Fig. 6.16b).

The evolving interpretation of the Wapiabi Formation in southern Alberta demonstrates the difficulty in making the appropriate reconstruction when several different types of criteria could be used to guide correlations. This example (Fig. 6.17) actually makes use of outcrop sections in an area where walking out correlations was not possible, but the problems are identical to those encountered in the subsurface. In the first reconstruction (Fig. 6.17a) the top of the Chungo Member is used as datum. The Chungo Member constitutes a coarsening-upward and shallowing-upward succession in which diachronous and interfingering facies changes are indicated, but a clinoform architecture is not

suggested or implied. In the second attempt (Fig. 6.17b) the top of the shoreface sandstones within the Chungo Member is used as datum, but this reconstruction provides no easily interpreted facies architectures. The third configuration (Fig. 6.17c) returns to the use of the top of the Chungo Member as the datum, and this time the clinoform configurations are specifically addressed in the suggested unit correlations. Given that the progradation of the Chungo shoreface must represent a significant elapsed time, this implies a substantial time break above the fluvial Chungo preceding the Nomad transgression. Bhattacharya (2011, p. 130) reported that ancillary chronostratigraphic data support this third reconstruction.

The reader is referred to Sect. 8.9 and to Bhattacharya's (2011) paper for further discussion of these issues, including the problems inherent in the mapping, correlation and chronostratigraphic interpretation of coastal fluvial and shallow-marine systems and incised valleys, and the use of supplementary correlation criteria, including coal seams and (particularly useful) bentonite beds. In Sect. 8.10 we discuss the use of Wheeler diagrams to illustrate correlations in

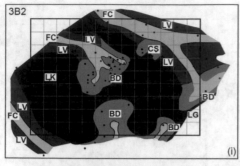

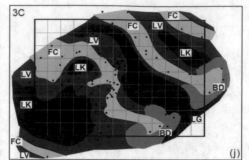

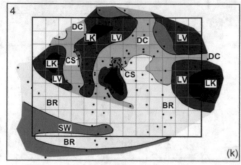

Fig. 6.12 Environmental interpretation of sandy depositional units. Symbols for various panels are as follows. Barrier bar/lagoon system (**e, f**): BC = barrier core, IF = inlet fill, IN = inlet, FTD = flood-tidal delta, ETD = ebb-tidal delta, LG = lagoon. Coastal stream plain (**a–c, g, i, j**): FC = fluvial channel, LV = levee, CS = crevasse splay, AC = abandoned channel, FP = flood plain, LK = lake, BD = bayhead delta. Wave-dominated delta (**d, h, k**): DC = distributary channel, LV = levee, CS = crevasse splay, AC = abandoned channel, DP = delta plain, LK = lake, SPL = strand plain, SW = swale, BR = beach ridge (Zeng et al. 1996, part of Fig. 7, p. 27). AAPG © 1996, reprinted by permission of the AAPG whose permission is required for further use

terms of elapsed time—a technique that, in throwing light on chronostratigraphic relations, may raise important questions about sedimentation rates, timing and the correlation of lithofacies units through areas of stratigraphic uncertainty.

6.3 Seismic Stratigraphy

The seismic-reflection method began with the work of two French brothers early in the twentieth century, who carried out some of the early experiments in the transmission of sound waves into the subsurface, and the recording of energy returned by reflection from rock layers underground. However, some of the more important work in this field was made by German scientists, who had discovered, during the First World War, that trigonometric calculations made on the surface sound waves yielded by artillery fire could be used to locate enemy gun positions (AAPG 1999, p. 42). It is widely agreed that the discovery of the Spindletop field in Texas, in a structural trap above a salt dome, was the first successful application of scientific methods, including gravity and seismic techniques, and scientific concepts, such as the *anticlinal theory* (North 1985, p. 19; AAPG 1999). Galey (1985, p. 440) cited a publication of the Spindletop 50th

Anniversary Commission which suggested that Spindletop was "where oil became an industry" (see Fig. 1.1).

Until the 1970s, the seismic-reflection method was primarily about the mapping of subsurface structure. Seismic stratigraphy, the study of stratigraphic units defined on the basis of their seismic characteristics (Cross and Lessenger 1988; Hart 2013), was pioneered by several companies in the 1970s, but first became widely known and used following the landmark publication of Exxon's work in the field (Vail et al. 1977).

An introduction to the use of large-scale facies architecture (and its representation by seismic facies) to map depositional systems and systems tracts is provided in Sect. 5.2. Problems of sequence definition are discussed in Sect. 7.7. In this section the practical methods of sequence correlation and mapping are described and illustrated. This section does not deal with data processing, which is a very specialized field beyond the scope of the stratigraphic practitioner (see Veeken 2007, for an extensive introduction). The examples selected for discussion here make use of seismic data that has already received the necessary processing to highlight the relevant stratigraphic detail. If necessary, reference may be made back to the original articles from which the examples used here have been derived.

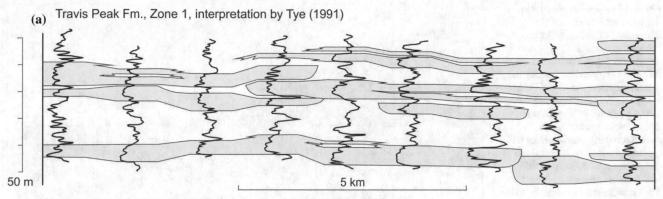

(a) Travis Peak Fm., Zone 1, interpretation by Tye (1991)

50 m   5 km

Wells are shown regularly spaced. Actual spacing ranges from 0.8 to 2.2 km, averaging 1.54 km

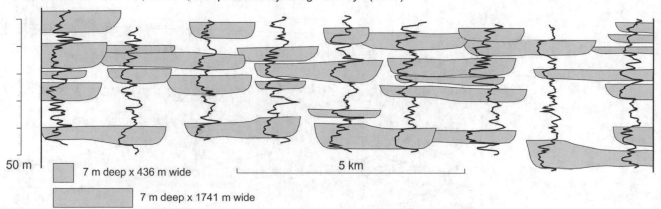

(b) Travis Peak Fm., Zone 1, interpretation by Bridge and Tye (2000)

50 m   5 km

7 m deep x 436 m wide

7 m deep x 1741 m wide

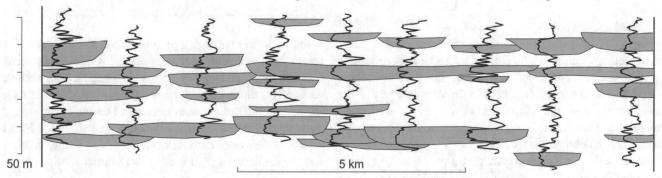

(c) . Travis Peak Fm., Zone 1, model based on high sand and minimal relief on channel bounding surfaces

50 m   5 km

Fig. 6.13 Three interpretations of the braided-fluvial deposits of the Travis Peak Formation, Zone 1 (Early Cretaceous, East Texas). **a** An initial interpretation, by Tye (1991), based on a detailed core and isopach mapping study. Arbitrary equal well spacing is used in this and the subsequent diagrams; **b** A reinterpretation by Bridge and Tye (2000), based on assumptions of narrower channel belts. Rectangular boxes at the base of this panel indicate the range of channel belt sizes predicted from estimated bankfull depth, using the equations of Bridge and Mackey (1993). Their own model, shown here, does not make use of this range of values; **c** An alternative model developed by the present writer (Miall 2006), based on two basic guidelines for interpreting petrophysical logs: **a** channels normally have flat bases, and **b** the main sand bodies are indicated only by blocky-shaped, low-value gamma ray signatures (from Miall 2014, Fig. 2.16)

6.3.1   The Nature of the Seismic Record

A seismic reflection is generated at the interface between two materials with different acoustic properties. The latter is described as **acoustic impedance** and is given by the product of density and velocity. The larger the difference in impedance between two lithologies, the stronger is the reflection. Where a soft lithology lies on a harder one, the deflection

Fig. 6.14 An example of how the choice of datum for the correlation of petrophysical logs can obscure or clarify depositional relationships. **a** In this reconstruction the assumption is made that the top of the marine sandstone constitutes a timeline. **b** The presence of significant lateral facies change and a depositional dip characterizes this reconstruction. The succession constitutes a clinoform set within which facies change laterally from nonmarine to shoreface to offshore (Bhattacharya 2011, Fig. 5, p. 127)

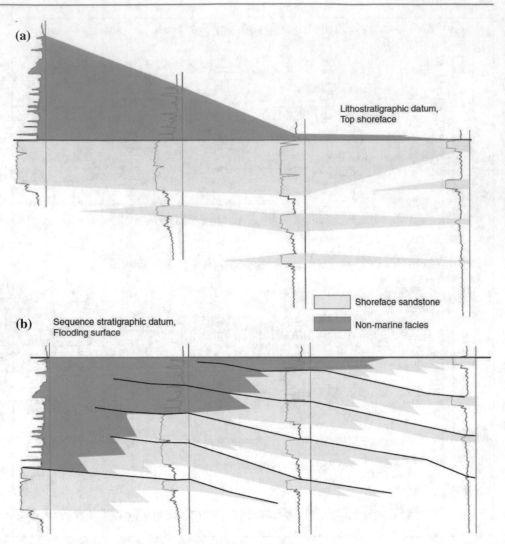

polarity is positive and is normally represented by a waveform deflection to the right, colored to render it more visible. Where the reverse situation applies, the polarity of the deflection is to the left (left blank). This is the origin of the familiar variable-area **wiggle traces**. Colors are now commonly used to exaggerate amplitude and other differences, and these may emphasize major facies variations.

A seismic wave trace recorded at a geophone represents the modification of a source pulse by reflection from various layers below the surface. The nature of each wavelet depends on the nature of the impedance contrasts at each bed contact. An example of a synthetic seismogram illustrating the production of a seismic signal is shown in Fig. 6.18. Such seismograms are time sections; that is, their long axis is proportional to time, not depth. Complex calculations must be performed to convert time to depth. This is because seismic velocities generally increase downward as rock density increases and porosity diminishes because of mechanical compaction, pressure solution and cementation. Depth scales are rarely provided for published seismic sections because

variation in the stratigraphic succession across the section means that velocities, and consequently, the depth scale, differ from point to point across the section. The seismic section constituting Fig. 1.15 is an exception. Note that the increments marked for each kilometer become closer spaced downward. An empirical relationship, known as the **Faust equation**, has been demonstrated between seismic velocity and the depth and age of rock units (Sharma 1986):

$$V = 46.5 \, (ZT)^{1/6} \, \text{m/s},$$

where V is seismic velocity, Z is the depth of burial and T is geological age in years.

It is important to remember that the seismic technique cannot resolve small-scale stratigraphic anomalies or very thin beds and that the resolution decreases with depth, as seismic velocity increases. A general rule is that two reflecting horizons must be a minimum of about one-fourth of a wavelength apart if they are to be resolvable in a typical seismic section (Sheriff 1985). Thus, at shallow depths in loosely consolidated sandstones and mudstones, a typical

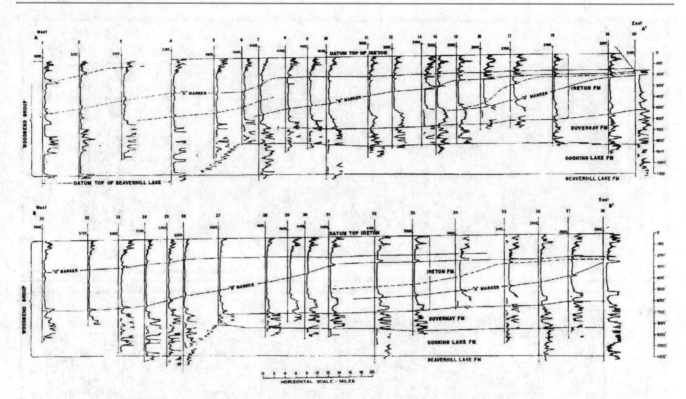

Fig. 6.15 This is one of the first reconstructions of clinoform stratigraphy based on petrophysical log correlation (Oliver and Cowper 1965, Fig. 4, p. 1414). AAPG © 1965, reprinted by permission of the AAPG whose permission is required for further use

Fig. 6.16 **a** Seismic reflection through a modern shelf-edge delta (Rhone delta, Mediterranean Sea), showing a clear clinoform architecture. **b** Flattening the top surface of the delta, a common practice in the subsurface correlation of deposits of this type, results in a severe distortion of the internal geometry of the beds (Bhattacharya 2011, Fig. 10, p. 134)

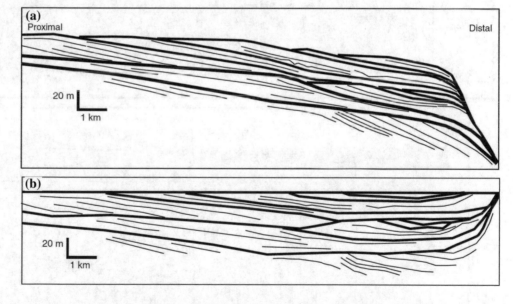

seismic velocity would be on the order of 1800 m/s, and the typical frequency would be around 60 Hz. One-quarter wavelength is therefore 7.5 m. Resolution is much poorer in older and deeper formations. Consider, for example, a Paleozoic carbonate section, with a seismic velocity of 4500 m/s and a frequency of 15 Hz. The thinnest unit resolvable in this situation would be 75 m. This lower limit is much larger (thicker) than many of the important facies

details that can be seen in outcrops and high-resolution petrophysical logs, as shown dramatically in Fig. 7.1. High-resolution techniques are available for studying sediments at shallow depths and are becoming widely used in detailed studies of stratigraphic architecture in modern environments (Bouma et al. 1987), but they are not yet available for the study of deep-basin structure and stratigraphy.

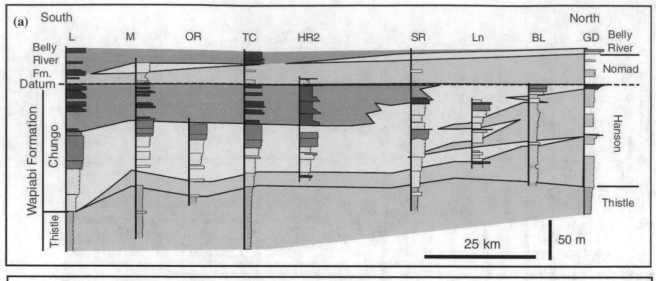

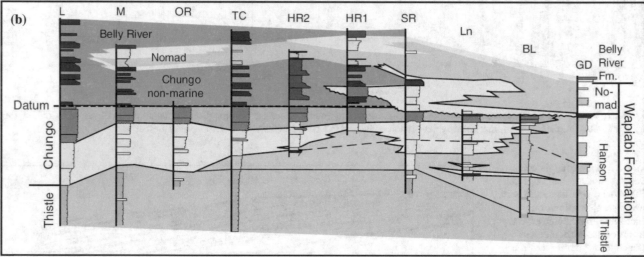

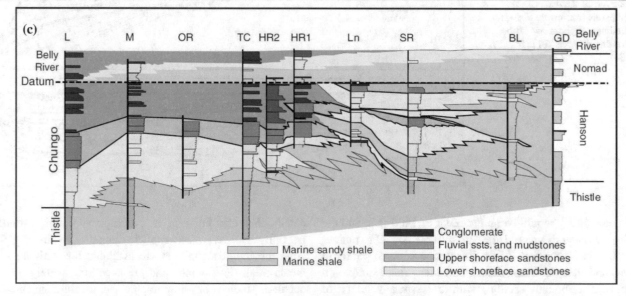

Fig. 6.17 Three different stratigraphic reconstructions of the Upper Cretaceous Wapiabi Formation in Alberta. These outcrop sections are tens of kilometers apart. **a** From Rosenthal et al. (1984; **b** from Rosenthal and Walker (1987); **c** From Leckie et al. (1994). See text for discussion and explanation

Fig. 6.18 Synthetic reflection seismogram generated from a simple geological section. At each interface, the input wavelet is modified by the reflection effects (Sharma 1986)

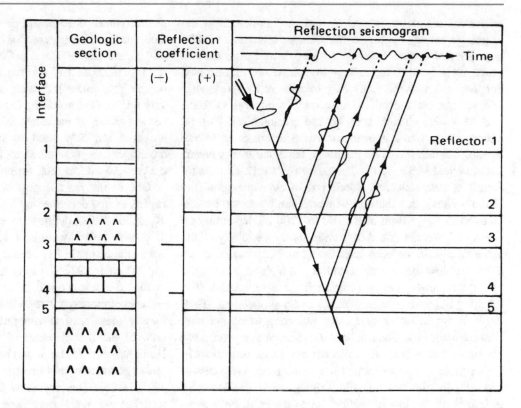

Published seismic sections show a series of vertical wave traces, each representing the recording made at a shot point. These are usually located about 25–50 m apart. Each trace is obtained by combining (**stacking**) the records of many adjacent geophones. The vertical scale of the section is **two-way travel time**, not depth. For each time horizon, depth varies according to velocity, as noted previously, so that, unless corrected, the structure revealed on a section may be spurious. For example, a mass of high-velocity rock, such as a dense carbonate reef or channel fill, may overlie what looks like a gentle anticline. This is a case of **velocity pull-up**, where the more rapid acoustic transmission through the dense material returns reflections from the underlying layers sooner than occurs on either side. Computer processing can produce a depth-corrected section if accurate velocity information is available, typically by making use of sonic petrophysical logs. This is discussed briefly below.

It is important to remember that it is the contacts that produce reflections, not rock bodies as such. For example, tight, cemented sandstone and dense, nonporous carbonate have higher impedances than most mudrocks so that a contact of one of these with an overlying mudstone will yield a strong positive reflection. Conversely, a gas-filled sandstone has a very low impedance. Where it is overlain by a mudstone there will be a strong negative deflection. This appears as a blank area in the seismic record and is the basis of the **bright-spot technique** for locating gas reservoirs. However, there are many conditions where interbedded units

may have contrasting lithology but very similar impedances so that they yield little or no seismic reflections. Examples would include a contact between a tight sandstone and a porous carbonate and a very porous oil- or water-saturated sandstone and a mudstone.

A very important assumption, first emphasized by Vail et al. (1977), and reemphasized by Cross and Lessenger (1988), and Hart (2013) is that in most cases seismic reflections from conformable bedding surfaces are chronostratigraphic (isochronous) in nature. Facies contacts are diachronous, but they are too gradual to yield reflections. Facies changes may be deduced from subtle lateral changes in reflection characteristics within a single bed. The assumption that continuous seismic reflections represent "time lines" is of profound importance in the study of basin structure and stratigraphy and the analysis of stratigraphic sequences (Chap. 5).

Imagine an angular unconformity with lateral facies changes in the overlying unit and a gently dipping, truncated sequence of different units below. The seismic reflection from the unconformity could vary from strongly positive to strongly negative, with some areas completely invisible. This would be a difficult reflection to follow accurately on seismic records.

Seismic records are also complicated by **multiples** and **diffractions**. Multiples are reflections from a shallow layer that are reflected downward again from a near-surface level such as the base of the weathering zone. This internal reflection is then redirected toward the surface from the original point and appears in the seismic record as a parallel layer several tenths of a second lower down in the section. Multiples may be strong enough to obscure deep reflections, but they are now relatively easy to remove by processing. Diffractions are reflections from steeply dipping surfaces, such as faults, channel margins and erosional relief on an unconformity. Their appearance in the record often occurs because seismic waves are not linear, laser-like energy beams but spherical wave fronts. The pulse from each shot point, therefore, sees reflection events over a circular region (the **Fresnel zone**), the diameter of which can be shown by calculation to range from about 100 to 1200 m, depending on depth and wavelength. A reflection on the periphery of the wave front will be seen later than one at the same depth directly below the shot point and, if at a high angle from the horizontal, may produce complex internal reflections. The resulting diffraction patterns may be very pronounced. They are a useful indicator of a steep reflecting event but may confuse other reflections and can be removed by a processing routine called **migration**, which repositions the reflections in their correct spatial position. The circular reflection area also accounts for the fact that diffractions can be produced by reflectors off the line of section, for example, if the seismic line skims past a reef body or a buried fault line scarp.

The seismic record can be considerably improved, and the interpretation is facilitated by the use of two special techniques: the construction of **synthetic seismograms** and the use of **vertical seismic profiling** (VSP) (Hardage 1985). Synthetic seismograms are generated by the conversion of sonic and density data from petrophysical logs into a series of reflection coefficients (see Sect. 2.4 for a discussion of petrophysical logs). Vertical seismic profiling is the recording and analysis of seismic signals received from a geophone-lowered downhole. Signals generated at the surface are recorded as they descend and as they return to the surface. The geophone is moved in increments, the length of each being set at less than half the shortest wavelength of the signals being analyzed. A new recording is made at each position. A comparison between synthetic, VSP and field-recorded data is shown in Fig. 6.19. Both synthetic and VSP data are extremely valuable for the purposes of calibrating the seismic record. For example, they enable a detailed record of seismic velocities to be generated that can be used to produce depth-corrected cross-sections. However, the VSP technique is far superior for various reasons. It can, for example, be used to extend the calculations below the total depth (TD) of the well, which synthetic seismograms cannot (Fig. 6.19). The synthetic method relies on well-log data, which only reflect conditions immediately adjacent to the hole. In the case of poor hole conditions or seismic reflectors having a very small horizontal extent (less than one Fresnel zone), the synthetic method may not provide a

record typical of the area. For example, in Fig. 6.19 there is a discontinuous calcareous sandstone (denoted by the letters SD) at a depth of about 3070 ft. It is visible as a strong reflector at about 2.15 s in the synthetic seismogram, but not on the VSP trace, indicating that the bed does not extend very far from the borehole. Other possible occurrences of the bed are shown by the letters SD on the field record. VSP data are now routinely used to improve subsurface imaging around critical boreholes, to carry out precise time-depth conversions and to remove multiples.

One of the most important features of **seismic stratigraphy**, as first described by Vail et al. (1977), is the concept **of seismic facies,** defined to mean an areally restricted group of seismic reflections whose appearance and characteristics are distinguishable from those of adjacent groups (Sangree and Widmier 1977). This is introduced in Sect. 5.2.2, and discussed in some detail in Sect. 6.3.3.

A revolutionary advance in seismic methods is the use of **three-dimensional seismic data**. According to Davies et al. (2004b) the first commercial 3-D survey was run in the North Sea in 1975, but it was not until the mid-1980s that the power of the new tool became more widely known (Brown 1985, 1986). Developments in computer processing and visualization were necessary before this could happen. A tight grid of seismic lines is run across an area at line spacings of a few tens of meters (25 m is typical). The correlation of key reflectors from line to line permits the plotting of horizontal seismic sections at selected depths. Correlation of key reflectors can be carried out by computers using automatic tracking techniques, and this provides a facility for plotting seismic maps of selected stratigraphic horizons. Where the structure is simple or can be removed by suitable processing routines, the resulting plots display areal amplitude variations for a single stratigraphic horizon, which may be highlighted by the use of false colors. These variations depend on lithology so that the sections may reveal detailed subsurface facies variations that can readily be interpreted in terms of depositional environment and paleogeography. Channels, sand bars and other depositional features may be clearly delineated. For example, Zeng et al. (1996) explored the potential for individual depositional elements to be revealed by 3-D seismic by developing synthetic models based on petrophysical data (see Figs. 6.9, 6.10, 6.11 and 6.12 and discussion in Sect. 6.2.2).

One of the most important aspects of 3-D data is the three-dimensional visualization it provides to depositional systems. This is a hugely important development. Traditionally, 2-D cross-sections have been used to illustrate proximal–distal variability in stratigraphy and interpreted depositional systems. Typically, thicknesses and facies undergo significant changes from basin-margin to basin-center, and dip-oriented 2-D cross-sections have been the preferred way to illustrate this. However, sediment-dispersal is a three-dimensional process—long-shore drift, geostrophic currents, diverging distributaries on deltas and submarine fans, are all examples of where significant sediment transport may take place out of the plane of such 2-D sections. For example, Moscardelli et al. (2012) discuss this important point with respect to a suite of shelf-margin deltas in Trinidad (a paper to which we refer below). In addition, orthogonal, linear subsidence or rotational subsidence on an axis parallel to the basin margin may not be the norm for any given basin, which is an important attribute that is difficult to capture with 2-D sections. Interpretations of the timing of important stages in sequence development, and quantitative modeling of a basin, such as source-to-sink calculations, may be in significant error if these three-dimensional characteristics of the basin cannot be fully taken into account.

Processing and interpretation of 3-D data require the use of sophisticated computer workstations and interactive graphics. Color display and plotting are essential. Savit and Changsheng (1982) provided an early example of the application of these ideas; the latest edition of the classic memoir by Brown (2011) brings them up to date. Hart (2013) offered a very useful review of modern seismic-stratigraphic methods and potential. Many examples of the interpretation of depositional features, such as channels and bars, from horizontal seismic sections have now been published (Posamentier 2000). Henry Posamentier has led the introduction of an entire new subdiscipline, termed **seismic geomorphology**, which uses horizontal seismic sections to explore and interpret entire depositional systems (Posamentier 2000; Posamentier and Kolla 2003; Davies et al. 2004a; 2007; Sect. 6.3.4). Chronostratigraphic interpretations of 3-D seismic volumes may be constructed using techniques of "computational seismic chronostratigraphy" (Stark et al. 2013), as discussed in Sect. 8.10.2.

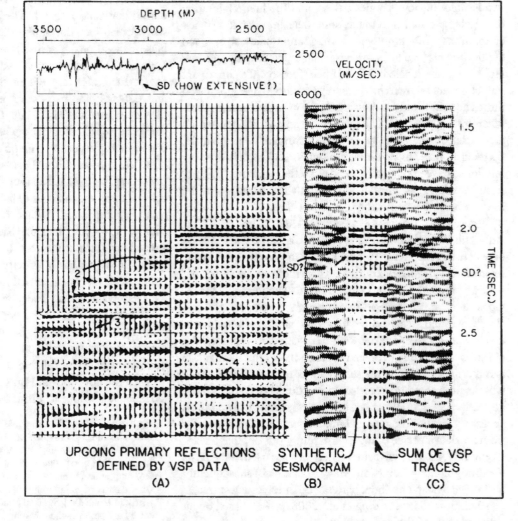

Fig. 6.19 Comparison of VSP data with a synthetic seismogram and that produced by normal surface field recording. The surface section is split into two at the location of a well down which the synthetic record and the VSP traces were generated. At left is shown the VSP traces and above is the sonic log of the well. The total well depth is at 2.41 s. Notes: 1, calcareous sandstone (see text); 2, curvature of VSP reflections implies structural dip; 3, termination of reflection implies presence of a fault near the borehole; 4, VSP data from below TD (Hardage 1985)

6.3.2 Constructing Regional Stratigraphies

The use of conventional 2-D seismic-reflection data to provide structural information in frontier basins has been standard practice since shortly after the end of the First World War. Since 1968, the year the *Glomar Challenger*, the first ship of the Deep Sea Drilling Project set sail, marine seismic has been a standard, essential component of the global effort to map the geology of the continental margins and deep oceans. Seismic data are calibrated against well and core records. In the case of onshore and offshore frontier exploration, the seismic data are correlated to the first suites of exploration drill holes. Seismic stratigraphy revolutionized the practice of petroleum exploration and development when it was introduced in the late 1970s, a development credited mainly to Peter Vail and his colleagues at Exxon Corporation (Vail et al. 1977) and to Bert Bally at Shell (see, in particular, Bally 1987).

What follows are five examples of what has become standard subsurface methodology. Reconnaissance 2-D seismic is used first, in the exploration of the structure and stratigraphy of new frontier basins, and the first exploration drill holes are tied into the seismic data in order to provide calibrations of lithologies, biostratigraphic ages and depth-to-sonic velocity relationships. Prominent seismic reflections are labeled and codified (commonly, important reflectors are shown in distinctive colors) and are traced throughout the study area using the loop-correlation method, whereby horizons are followed as far as possible around an area using intersecting survey lines, ultimately bringing the correlation back to its starting point to ensure that the correlations have not inadvertently shifted into a different stratigraphic level. Further ideas about the workflow to be used in basin analysis are provided in Sect. 1.5.

Figure 6.20 illustrates a seismic line from the West Antarctic Rift System, a series of marine rift basins located between the Antarctic Peninsula and the Ross Sea. Earlier seismic and drilling activity had provided the basis for the stratigraphic subdivisions shown in this section, which provides a clear picture of the subsurface stratigraphy and structure of the area. Seismic facies and structural relations in this and other sections may be interpreted in terms of the phases of development of the rift basins: early rift, main phase and thermal-relaxation phase. The illustrated section follows what is now standard practice, to use different colors for the key reflectors. Initially, this adheres to the empirical approach necessary for building a stratigraphic reconstruction directly from the data and free of regional assumptions or other biases. Eventually, these color-coded reflections are used as the basis for stratigraphic subdivision. Many of them will subsequently be defined as sequence boundaries or other important surfaces.

As described by Fielding et al. (2008, p. 13), based on an interpretation of the section shown here as Fig. 6.20:

In the area of Roberts Ridge, the Early Rift interval forms an eastward-thickening wedge that is abruptly truncated against the first major basement-offsetting fault. Within the Early Rift section, the principal seismic facies are 1. irregular to chaotic, discontinuous, moderate to high amplitude reflections in the updip area, passing downdip into 2. mainly concordant, parallel to somewhat irregular, semi-continuous reflections of moderate to high amplitude and moderate frequency, with local development of clinoform sets, in turn passing into 3. chaotic to reflection-free facies in the immediate hanging wall region of the half-graben-bounding fault.

Wherever possible, the calibration between petrophysical logs or cores and seismic data should be clearly shown. Core and log data help to clarify the interpretation of seismic facies, and both together may be used in the definition of the sequence stratigraphy and regional tectonic reconstructions. Figure 6.21 is an example from a very complex area, the Sigsbee Knolls area, on the continental slope of Texas, in the northwest Gulf of Mexico. Vast amounts of clastic debris have been shed into the Gulf since the Jurassic, with many major depositional systems having prograded southward, extending the continental margin by tens of kilometers during that time interval. The Brazos-Trinity slope system, which developed during the Pleistocene, is one of the last of these depositional phases. It represents the distal end of a nonmarine-shallow-marine and shelf-edge delta system, the source for which lay within the Brazos-Trinity watershed, that extends northwestward across Texas and into New Mexico (Pirmez et al. 2012). The Sigsbee Knolls is an area along the continental slope where evaporites of Jurassic age rise in multiple complex diapiric bodies to form topographic highs on the sea floor, between which clastic sediment is ponded in multiple intra-slope basins. The sediment influx is controlled by the shifting deltaic depositional systems to the north, by the modifications to the basins by salt tectonics, and by sea-level changes. Steep, local intra-basin slopes created by salt activity are thought to be capable of triggering local sediment gravity flows. Considerable exploration work has been carried out in this broad area because the deep-water sands deposited by these depositional systems have proved to be very hydrocarbon rich. Local stratigraphic variability is therefore considerable, and explorationists must rely on, and reconcile, multiple data sets in order to generate reliable regional reconstructions.

The reflectors seen in Fig. 6.21 may be traced throughout the area. According to Pirmez et al. (2012, p. 115):

Each key horizon is mappable throughout the area and has distinguishing geometric or seismic facies characteristics. Horizon 10 separates laterally continuous, moderate- to high-amplitude reflectors below, from a laterally continuous low-amplitude to transparent unit above. Horizon 20 is defined on the basis of onlapping reflections above it, observed particularly in Basin IV [Fig. 6.21] and Basin …. Horizon 20 can be traced to Basin II, where it shows conformable reflections above and below. Horizon 30 is a laterally continuous reflector defined

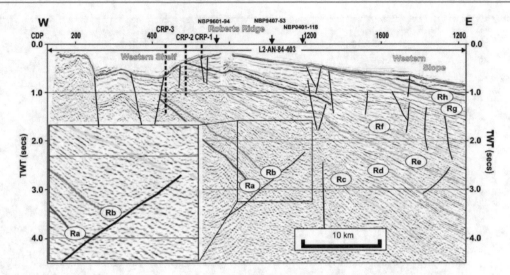

Fig. 6.20 Seismic stratigraphy of the Antarctic margin. This section shows rift architecture close to the western edge of the Victoria Land Basin (Western Shelf). Inset shows close-up view of the principal seismic facies of the Early Rift (navy blue—Ra to yellow—Rb reflectors) and Main Rift (yellow—Rb to purple-grey—Rc) successions. Thermal relaxation phase. Purple reflectors and Rf-Rg reflectors.

In the Early Rift section, mainly concordant, moderately continuous, high amplitude events with some low-angle clinoforms pass downdip into a chaotic seismic fabric adjacent to the half-graben-bounding fault. In the Main Rift section, mainly concordant events of similar character to the underlying section dominate, and are continuous across the top of earlier extensional topography (Fielding et al. 2008, Fig. 2, p. 13)

by onlapping reflections above it in all basins; locally it is truncated by an acoustically transparent unit [Fig. 6.21]. Horizons 40, 50, and 60 are laterally continuous horizons that mark a vertical change in seismic facies. Horizon 40 is an erosional surface locally near the exit of Basin II … Horizon 61 [Fig. 6.21] is an erosional surface with onlapping reflections above. It is a very weak reflection and is mappable only in Basin IV. Horizon 70 is well defined in Basin IV, where it forms a baselap surface at the base of a fan wedge sourced from the eastern channel [Fig. 6.21].

This example of basin analysis is discussed further in Sect. 8.10.1, where some of the details of the advanced chronostratigraphic dating and correlation techniques are briefly summarized.

The third example discussed briefly here is a regional study of the Cretaceous Chalk in the Norwegian portion of the North Sea's Central Graben (from Gennaro et al. 2013). As the authors noted (p. 236), "In the last decade, the classical paradigm of chalk as a monotonous and flat-lying deposit has been significantly revised and increasing evidence from seismic data has revealed the role that bottom currents have on chalk deposition. These currents sculpted the chalk sea-floor relief creating important topographic features such as channels, drifts and valleys." Sediment gravity flows and contourites composed of chalk have been mapped, and there is increasing evidence for syndepositional tectonic activity. The stratigraphic subdivisions and petrophysical and seismic character of the chalk are shown in Fig. 6.22. These are based on seismic-facies characteristics and the mapping of reflection terminations. Examples of the regional stratigraphic architecture are illustrated in Fig. 6.23,

and an enlarged portion of this section, showing facies characteristics in more detail, is shown in Fig. 6.24.

Detailed mapping of the thicknesses and facies variations of the sequences identified in this study have been interpreted in terms of three tectonic phases: pre-inversion, syn-inversion and post-inversion. Although the details of this study are not relevant for our present purposes, it is of interest to note how the seismic data illustrates the onlap of the top Narve and top Thud reflectors onto anticlinal highs in Fig. 6.23. These features of the seismic architecture are interpreted as the result of syndepositional uplift during the inversion phase, following which erosion of the resulting highs provided a source for allochthonous chalk during the post-inversion phase.

One of the more important improvements in the seismic-reflection technique has been the development of the ability to image the complex structures associated with diapiric salt, including the edges and underlying boundaries of salt bodies and the stratigraphy beneath. A description of the processing techniques that have permitted this is beyond the scope of this book, but includes the use of what is termed "pre-stack migration." Traditionally, seismic traces are stacked to form a single trace, and then undergo migration to restore distortion caused by varying seismic velocities. With the increased processing speed and power of modern computer methods, it is now possible to migrate each trace before combining, or "stacking" them, which preserves considerably more detail. An example is shown in Fig. 6.25. It is this breakthrough in processing that has permitted

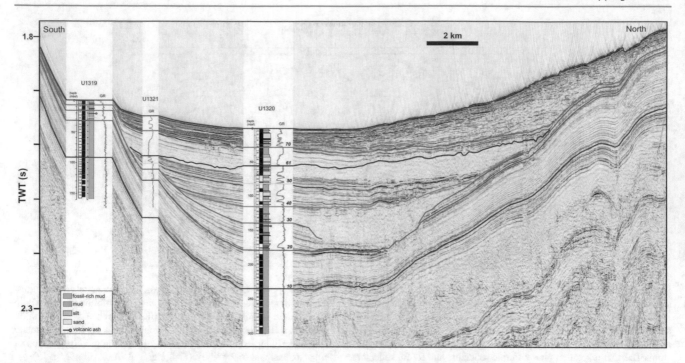

Fig. 6.21 Seismic section through the Brazos-Trinity depositional system (Pleistocene) in an intra-slope basin of the Texas margin continental shelf. High-resolution 2D seismic data are correlated to relevant well-logs, showing key surfaces, and positions of cores. Logs are plotted in depth scaled using an average velocity; note the good relationship between the key-mapped horizons and major lithologic changes. Core columns contain, from left to right: (with respective core number), core recovery (black = recovered, white/blue = no core, red = core interval with gas expansion), and lithologic column (with legend key to colors) (Pirmez et al. 2012, Fig. 4, p. 16)

deep-water exploration of pre-salt structure and stratigraphy in such basins as the Gulf of Mexico and the deep offshore Atlantic Ocean of Brazil, Angola, etc.

The final example in this section touches on one of the most frequently encountered seismic facies, that of the complex clinoform set. The example illustrated (Fig. 6.26) is from the Barents shelf margin, located between northern Norway and Svalbard. An active phase of subsidence commenced in the mid-Eocene as rifting accompanied the opening of the Norwegian-Greenland Sea. High relief, a narrow shelf and an abundant sediment supply led to the development of thick prograding clinoform sets within which slumping and subsidence remobilized large volumes of sand to form injectites. A single well with two short cores has been drilled within the study area, although not along the line of the illustrated section. The major stratigraphic subdivisions that may be recognized in this section consist of prograding clinoform units on the continental slope, consisting of resedimented shelf deltaic and other marine sediment. These units pass down into what is labeled as "depositional units" that have been interpreted as submarine fan complexes. Analysis of the architecture of these clinoforms provides considerable information about the history of subsidence and sea level, a subject touched on in the next section.

6.3.3 Seismic Facies

The early work on seismic stratigraphy, in the 1970s, provided many useful broad generalizations about stratigraphic architecture that focused on large-scale geometric features, particularly the nature of reflection terminations (Sect. 5.2.2). These were illustrated in what is by now the classic diagram shown in Fig. 5.8, and constitute the larger scale elements of what is termed **seismic facies**, a term defined to mean "areally restricted group[s] of seismic reflections whose appearance and characteristics are distinguishable from those of adjacent groups" (Sangree and Widmier 1977). Reflection amplitude and continuity are additional elements of seismic facies that have become increasingly important as processing and visualization power have improved. The concept of seismic facies is most usefully applied where the primary data consists of 2-D cross-sections, for which stratigraphic and sedimentological interpretations may not be immediately obvious. But the ground truthing of seismic facies in terms of lithofacies may be difficult in the absence of well data (particularly cores) in key locations because few seismic facies have unique interpretations (clinoform facies are the simplest to interpret directly). This problem is much less acute where 3-D data are available, because of the essential attribute of the

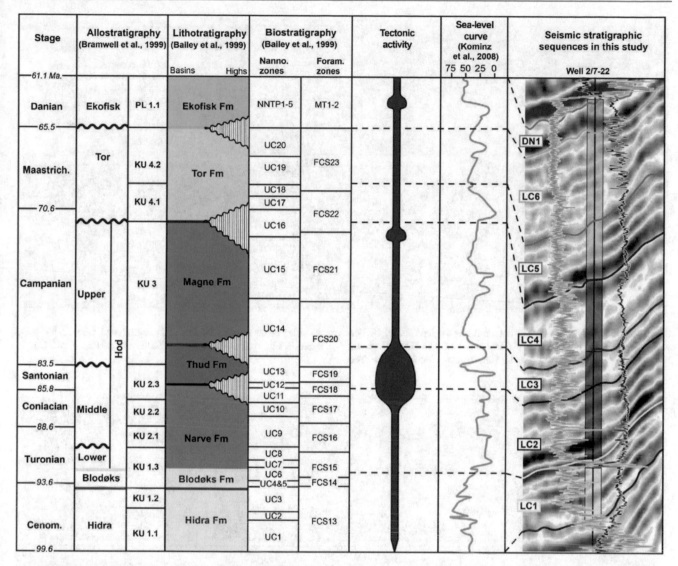

Fig. 6.22 Allostratigraphic and lithostratigraphic subdivision of the Chalk Group in the southern Norwegian sector of the North Sea and the corresponding seismostratigraphic units used in this study. The graph on the tectonic activity indicates the relative magnitude of the tectonic inversion. The tectonic activity is considered to be continuous, but with culminations. Green well curve, gamma ray log increasing from left to right; black well curve, sonic log increasing from right to left (Gennaro et al. 2013, Fig. 2, p. 238)

3-D volume that entire depositional systems may be visualized.

The other essential aspect of seismic-stratigraphic data, particularly 3-D data, is the scale of features that can now be imaged in their entirety. Complete depositional systems may be sliced and viewed in any vertical orientation and at different horizontal levels, over horizontal dimensions of tens of kilometers. This is orders of magnitude greater than even the best outcrop sections, which means that we can now "see" things that have never been seen before. This is particularly the case with deep-marine depositional systems, such as submarine fans and contourite drifts, for which outcrop data has mostly been very scrappy. The first depositional models for these features (Mutti and Ricci-Lucchi 1972; Walker 1978) were based on limited outcrop data and, as has become apparent, provide very little real guidance for the interpretation of the large submarine channel and fan systems that form important reservoirs in many deep-water base-of-slope frontier basins, in part because of the enormous variability we now know exists in the size, shape and composition of submarine fans. This point can readily be appreciated by comparing these early, simple models, with the realities of a wide array of modern fans surveyed using side-scan sonar technology, summary details of which were assembled by Bouma et al. (1985). Figure 4.42 provides a current classification of turbidite systems, and Figs. 4.37, 4.38 and 4.40 illustrate a range of modern examples.

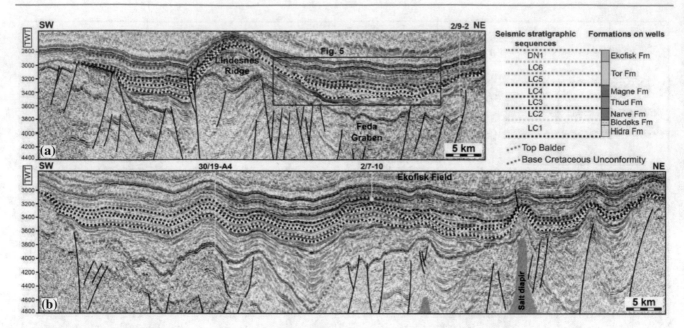

Fig. 6.23 Seismic sections of the southern Norwegian Central Graben with interpreted intra-chalk seismic sequences LC1 to DN1. Regional structural elements and tectonic lineaments are also indicated. The rectangular areas labeled as Fig. 5 is reproduced here as the lower diagram in Fig. 6.23 (Gennaro et al. 2013, part of Fig. 4, p. 242)

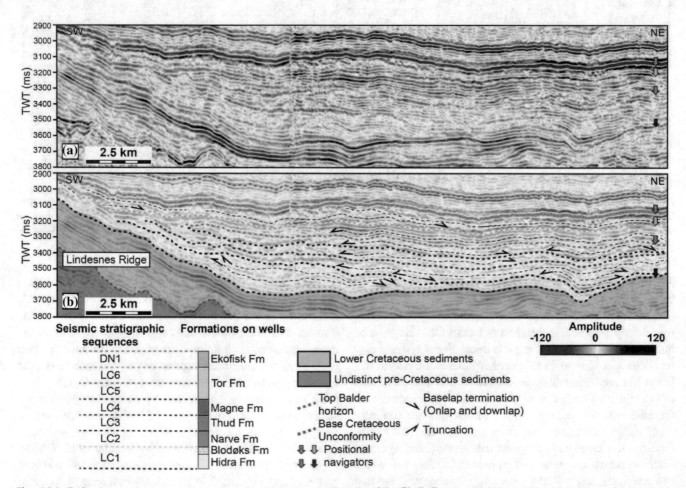

Fig. 6.24 Reflection configuration and terminations within the seismic units of the Chalk Group. **a** Uninterpreted and **b** interpreted 3-D seismic profile (Gennaro et al. 2013, Fig. 5, p. 244)

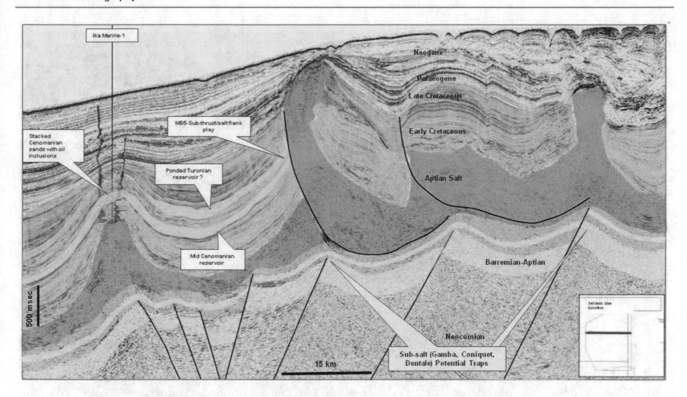

Fig. 6.25   An interpreted seismic line from offshore Gabon, showing the extremely complex structure and stratigraphy associated with diapiric salt (from Presalt.com; a seismic line from joint exploration by Ophir and Petrobras)

Davies et al. (2004b) discussed the practicalities and pitfalls of the use of 3-D seismic data. The knowledge and experience of both geophysics and geology are required, in order to fully understand what can be done with processing, and the limitations inherent in the seismic method (geophysics) and the realities of stratigraphy and sedimentological interpretation (geology). A common error is an attempt at the direct interpretation of clastic lithofacies from reflection strength and continuity. There are many criteria that govern this feature of seismic facies (fluid content and diagenetic modification being obvious ones), and all of these need to be taken into account.

Facies models and sequence models provide guidance for the interpretation of the real world (Chaps. 4 and 5). It would be an ideal world if seismic facies could be categorized in the same way. Many researchers have provided categorizations of the seismic facies in their project areas, but as several authors have noted (e.g., Futalan et al. 2012), calibration between seismic attributes and lithofacies is to a considerable extent basin-specific, and so generalizations must be accorded a considerable latitude for inherent variability. Two seismic-facies classification tables are shown in Figs. 6.27 and 6.28.

The first illustration (Fig. 6.27) shows a facies classification for a mixed carbonate-clastic continental-margin assemblage in the Philippines. The lithologic attributes of most of these facies vary from clastic to carbonate or mixed,

as indicated by well calibration, but the geometrical attributes are distinctive. For example, facies D represents turbidites, facies E clearly represents clinoforms and facies F images channels.

Sayago et al. (2012, p. 1844) stated: "It is known that 40% or more of recoverable hydrocarbons in carbonate deposits are trapped at stratigraphic unconformities, and in most cases, they are of karst origin." Their study is of a largely carbonate succession of Pennsylvanian–Permian age on the Barents shelf in the Norwegian Sea. The analysis commenced with a multiattribute classification study of the seismic volume. Eighteen seismic attributes covering the whole range of seismic information were computed (e.g., time, amplitude, frequency and attenuation). Comparisons and crossplots between attributes were used to define the seismic facies. This classification was then applied to samples of the seismic survey volume. An initial "unsupervised" classification represents an automated analysis of the seismic data; a "supervised" classification includes some operator adjustments to better correlate the seismic data to information from the core. The "training data" used in this classification is shown in Fig. 6.28. An example of a 2-D seismic cross-section analyzed in this way is shown in Fig. 6.29a. Four seismic attributes have been calculated for the classification (instantaneous bandwidth, gradient magnitude, envelope and dominant frequency), as shown in diagram B. Diagram C shows an "unsupervised classification" based

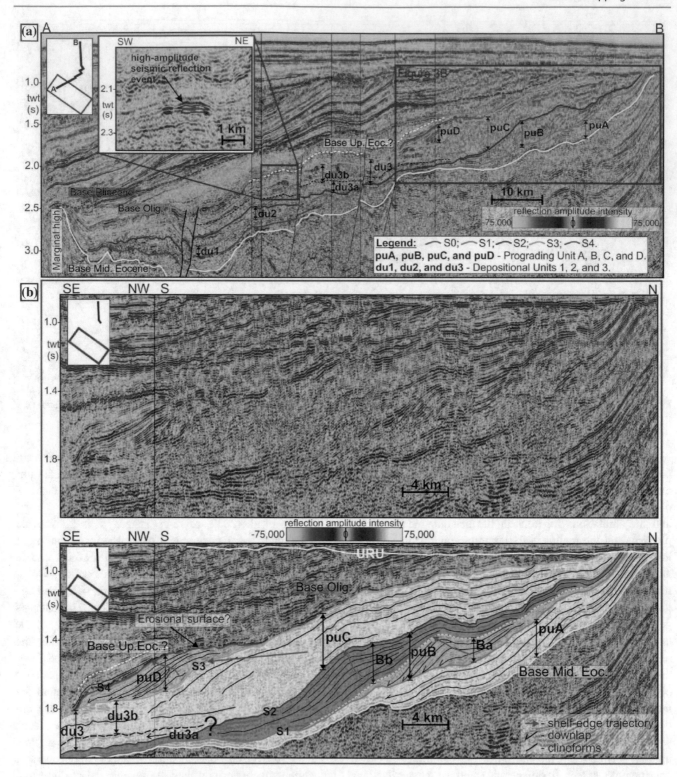

Fig. 6.26 Clinoform sets of Eocene–Oligocene age on the margin of the Barents shelf. **a** In this section, three deep-water depositional units (1–3) pass up dip into prograding units, four of which (units A–D) have been mapped. **b** Uninterpreted and interpreted sections showing prograding units A–D in the northern part of the study area (the section outlined by the blue rectangle in the upper-right portion of diagram a) (Safronova et al. 2014, Fig. 3, p. 520). AAPG © 2014, reprinted by permission of the AAPG whose permission is required for further use

entirely on numerical manipulations of the digital seismic data. Seismic facies 1 (SF1) is present not only in the flat-topped carbonate interval but also in the vicinity of the main seismic reflectors occupying the outer slopes of the structural high, that is, top basement, top Gipsdalen and top Bjarmeland (white arrows). Strongly chaotic patterns and complex alternation of seismic facies 2 (SF2), seismic facies 3 (SF3), seismic facies 4 (SF4) and seismic facies 5 (SF5) are observed. "Supervised classification" results are shown in Diagram D. Seismic facies 1 is mostly concentrated in the topmost area (crest) of the Loppa High where it correlates with the parallel high-amplitude training points (the volumes selected by the user to fine tune the classification). It is interpreted as consisting of paleokarst breccia units and its distribution in the supervised classification is associated with a major regional unconformity. Seismic facies 2 is restricted to the slopes of the structural high, highlighting the main seismic boundaries.

An instructive clastic example of seismic facies and sequence interpretation is that of Moscardelli et al. (2012) who examined the development of shelf-edge deltas on the Atlantic shelf margin of Trinidad, in an area where the continental-margin tectonic environment changes from transpressive to extensional within a space of 100 km. Figure 6.30 illustrates a composite northwest–southeast-oriented 3-D seismic line along strike, showing the shelf-edge region in eastern offshore Trinidad. Diagram A) shows the seismic section; diagram B) is a line drawing showing the seismic facies (SF) distribution and sequence-stratigraphic interpretation of this section. The seismic character and distribution of the main sequence-stratigraphic units significantly vary from south to north; these variations are thought to be associated with underlying structural controls and changes in sedimentation rates across the margin. Contrasts in tectonic settings between the northern and southern parts of the project areas described in this study are marked. Steep thrust faults are important in the north; extensional growth faults are common in the south. These faults have guided the development of submarine canyons and therefore the downslope distribution of detritus. Tidal and along-shore currents are important in sediment dispersal, and gravity sliding and slumping have had a significant effect on the ultimate architecture of depositional units. This is clearly an example where the traditional 2-D approach to stratigraphic and sedimentological interpretation would not be successful. A seismic-facies classification developed for this area is shown in Fig. 6.31.

Much fine detail may be added to sequence-stratigraphic interpretations by the use of high-resolution three-dimensional seismic data. Paumard et al. (2020) established the fine-scale variations on the seismic architecture of rapidly deposited clinoform sets in a Lower Cretaceous unit offshore Australia and used these data to establish quantitative and statistical relationships between the shelf-margin architecture, paleoshoreline processes and deep-water system types (i.e., quantitative 3D seismic stratigraphy). The results confirm that low values of rate of accommodation/rate of sediment supply ($\delta A/\delta S$) conditions on the shelf are associated with sediment bypass, whereas high $\delta A/\delta S$ conditions are linked to increasing sediment storage on the shelf.

In this study and that of Safronova et al. (2014), analysis of the internal composition and architecture of clinoform units provides essential clues for the analysis of depositional and structural history. For example, in Fig. 6.26, the vertical trajectory of the shelf-slope break in dip-section is highlighted in the bottom diagram, providing insights into the changes in sediment supply and changes in relative sea level through the accumulation of this unit.

Zeng et al. (2013) pointed out that seismic resolution may be a problem in the identification and mapping of clinoform sets, particularly those developed by deltaic progradation, which typically develop in shallow water and are much smaller than shelf-margin clinoforms.

It seems that the type of seismic clinoform configuration may also be related to data frequency. An oblique clinoform seismic configuration in higher frequency data … tends to become a shingled configuration in the lower frequency data … As a result, shingled facies observed in seismic data are not necessarily truly representative of geologic clinoform architecture. The merging of seismic responses of the thinner, low-angle downdip portion of clinoforms with that from underlying flat host rocks in low-frequency data appears to distort the seismic facies (Zeng et al. 2013, p. SA45).

A process termed spectral balancing may be used to enhance the high-frequency end of the data spectrum, and this can reveal a subtle clinoform signature (Fig. 6.32).

The last example in this section is derived from the work of Burgess et al. (2013) who provided criteria for the identification of isolated carbonate buildups in the stratigraphic record. As they pointed out, these may be difficult to distinguish from volcanoes, erosional remnants and tilted fault blocks. In tropical and subtropical areas, such buildups may be common features of the continental shelf, but in many areas, such as continental margins of Southeast Asia, characterized by rapid tectonism and a high sediment supply, local conditions may change over geologically rapid time scales, resulting in highly complex stratigraphic relationships. Aids to the identification of carbonate buildups (which can provide excellent petroleum reservoirs) have therefore been found useful. Figure 6.33 provides some examples of the distinctive seismic-facies characteristics of carbonate buildups developed by Burgess et al. (2013) from a large seismic data bank assembled from offshore southeastern Asia. Key criteria include geometry (the vertical nature of

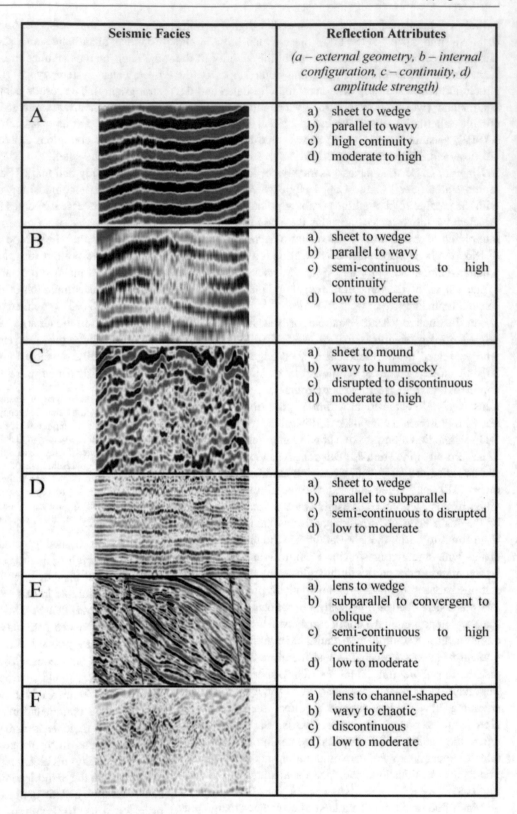

Seismic Facies	Reflection Attributes
	(a – external geometry, b – internal configuration, c – continuity, d) amplitude strength)
A	a) sheet to wedge b) parallel to wavy c) high continuity d) moderate to high
B	a) sheet to wedge b) parallel to wavy c) semi-continuous to high continuity d) low to moderate
C	a) sheet to mound b) wavy to hummocky c) disrupted to discontinuous d) moderate to high
D	a) sheet to wedge b) parallel to subparallel c) semi-continuous to disrupted d) low to moderate
E	a) lens to wedge b) subparallel to convergent to oblique c) semi-continuous to high continuity d) low to moderate
F	a) lens to channel-shaped b) wavy to chaotic c) discontinuous d) low to moderate

Fig. 6.28 A seismic-facies classification for a carbonate succession, including paleokarst breccias. See text for explanation (Sayago et al. 2012, Table 4, p. 1860). AAPG © 2012, reprinted by permission of the AAPG whose permission is required for further use

Seismic Facies	Reflection geometry	Amplitude characteristic	Spatial distribution	Example (Vertical bars represent 100 ms)
SF1	Parallel continuous	High amplitude	Occurs mainly at the crest of the structural high and at the top of basement	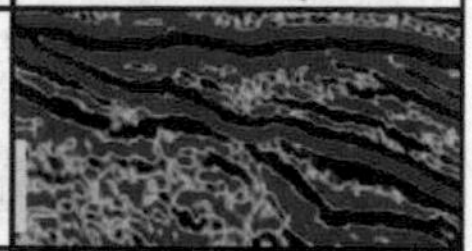
SF2	Parallel continuous	Medium to low amplitude	Occurs mostly towards the flanks of the Loppa High	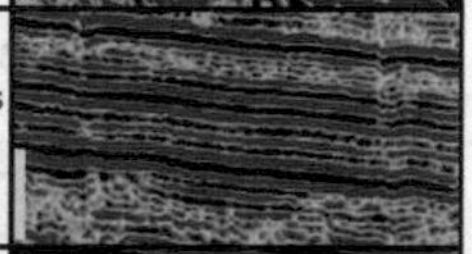
SF3	Parallel discontinuous	Medium to low amplitude	In overlying Triassic clastics and some areas of the carbonate intervals	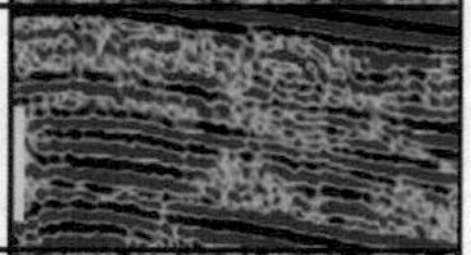
SF4	Chaotic	Low amplitude	Present at the core of the buildups and in the basement	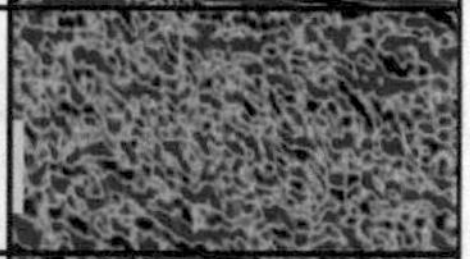
SF5	Semiparallel dipping discontinuous	Medium amplitude	Occurs mostly in the slopes of buildups	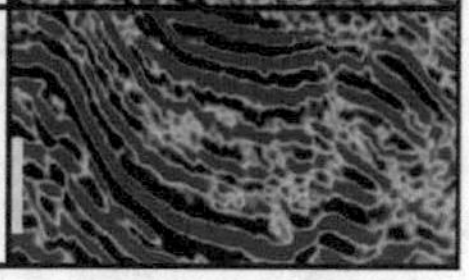

the buildup, commonly with an overlying structural drape), seismic terminations (e.g., onlap of buildup margin by draping clastic units) and contrasts in reflection amplitude and continuity between the buildup and flanking strata.

6.3.4 Seismic Geomorphology

The development of seismic geomorphology represents a huge step forward in methods for petroleum exploration and development. High-quality 3-D data are required, with processing by skilled geophysicists, but a deep understanding of sequence stratigraphy and sedimentology is essential for in-depth interpretations of these data. Figures 5.11 and 5.12 illustrate how the analysis of depositional systems paleo-geographically and in vertical drill hole profiles may be integrated with the architecture of a sequence. Where such a combination of data and skillsets is available it may render all earlier sedimentological approaches to subsurface analysis obsolete (except for teaching purposes). The simplistic facies models used through the 1960s to 1990s have been replaced by actual visualizations of entire depositional systems. However, the understanding of depositional processes that evolved during the facies models era forms an essential background to seismic analysis. For example, the earlier ideas about inner, mid and outer fans that were introduced in the 1970s (Mutti and Ricci-Lucchi 1972; Walker 1978) were still found to be useful when Bouma et al. (1985) compiled numerous studies of modern fans using side-scan sonar, and

some of the defining elements of these three subdivisions do appear on seismic images of some submarine fans, a category of a depositional system that we now know, through such studies, to be vastly more complex and varied than was imagined a few decades ago. For example, Fig. 6.34 illustrates three cross-sections through a fan complex on the Borneo continental margin, in which the three-fold subdivision that had been proposed for fans appears to be applicable—an inner fan characterized by a prominent channel-levee complex (commonly a single main channel fed from a canyon point source), a mid fan with a much more subdued distributary channel-level system and an outer fan with a smooth, unchannelized, convex-up cross-section representing the accumulation of sheet-like turbidites.

The most spectacular 3-D interpretations of depositional systems are those of channelized environments, particularly fluvial and submarine fan settings, with their complexes of levees, crevasse splays and overbank areas (e.g., Posamentier and Kolla 2003; see Figs. 6.35, 6.36, 6.37 and 6.38 of this book). For fluvial systems there is, of course, a virtually limitless suite of modern analogs for comparison with the ancient examples. Comparisons of horizontal seismic sections with selected images from Google Earth or some comparable source of aerial imagery make for very powerful and effective interpretations. Many such studies have now been published, many including quantitative analyses of channel and meander characteristics. The same is the case for deltaic and shoreface environments and platform carbonate environments.

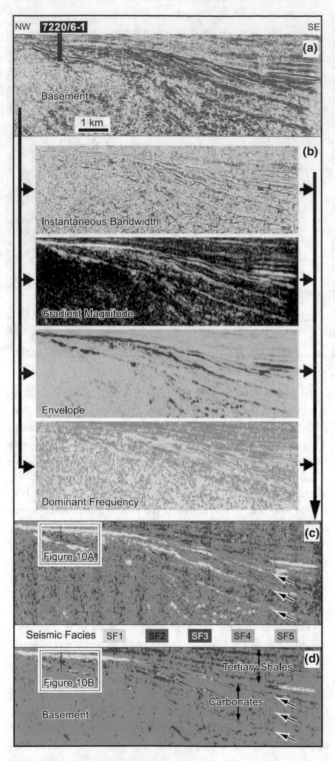

Fig. 6.29 Shelf-margin seismic section showing how the seismic-facies classification is derived. **a** The uninterpreted seismic section; **b** four seismic attributes calculated for the classification; **c** unsupervised classification of the original seismic data; **d** supervised classification (Sayago et al. 2012, Fig. 9, p. 1858). AAPG © 2012, reprinted by permission of the AAPG whose permission is required for further use

For submarine fans, imagery of modern fans is limited to what can be obtained from side-scan sonar data, and ancient-to-modern comparisons are less abundant. Descriptions and illustrations of some of the world's best outcrops of ancient deep-marine systems were collected in a large-format volume by Nilsen et al. (2007).

A few seismic-geomorphic studies of carbonate environments have been published, but few make use of horizontal sections to image a carbonate system. It would appear that such sections are of limited usefulness, compared to traditional 2-D seismic.

The other major use of seismic geomorphology has been to image erosion surfaces, including unconformities and such features as incised valleys and paleokarst surfaces (Figs. 6.39, 6.40, 6.41 and 6.42). A range of examples of modern applications is contained in the volumes edited by Davies et al. (2004a, 2007).

Turning first to fluvial examples, Ethridge and Schumm (2007) provided a discussion of modern concepts in fluvial geomorphology, focusing on what could be deduced from the imagery of a random segment of a fluvial system. The use of modern aerial photographs or satellite imagery makes such interpretations relatively straightforward. Figure 6.35 is one of the first illustrations of a horizontal seismic section through a fluvial system to be published. It shows a meandering system in the shallow subsurface of the Gulf of Thailand. False-color imagery is used to highlight the amplitude differences between the meandering channels and the overbank areas. Miall (2002) used successive time slices from a Pleistocene system to explore the architecture of an incised-valley system and its badland-style tributaries.

A recent example of the use of horizontal seismic sections to explore an ancient fluvial system was provided by Hubbard et al. (2011) from the Alberta Oil Sands (Fig. 6.36). The strata in this study area are between 300 and 400 m below the surface, and seismic resolution is less than 5 m. A modern analog of the meandering fluvial style was suggested, in the form of an aerial photograph of the Sittang Estuary, Myanmar. The addition of a map of gamma ray readings from wells that penetrate this section provides an essential cross-calibration between seismic and gamma ray data that can then be used for the interpretation of lithofacies and reservoir properties in other areas. Hubbard et al. (2011) provided a detailed analysis of the lithofacies and architectural elements of this section, in terms of point bars, counterpoint bars and other features.

The use of perspective views of depositional systems can considerably enhance the visual effectiveness of 3-D seismic imagery. Figure 6.37 displays a turbidite channel in the Gulf of Mexico. The relevant portion of a vertical 2-D section that intersects the horizontal section provides the necessary

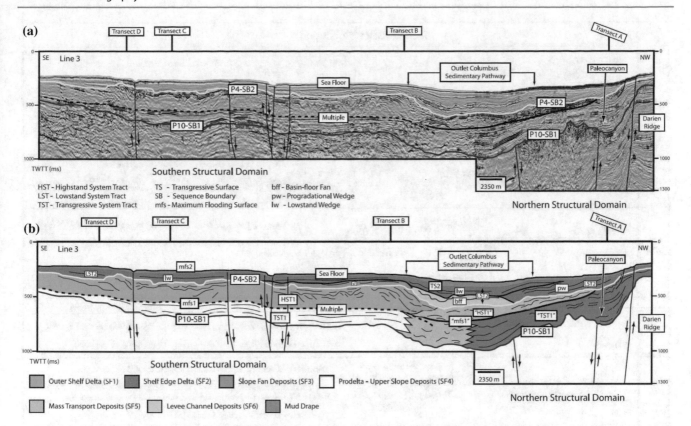

Fig. 6.30 Strike-oriented seismic section (**a**) and its interpretation (**b**), along the Atlantic shelf margin of Trinidad. Terms in quotation marks indicate that the indicated units and surfaces do not present the typical seismic characteristics defined by the sequence-stratigraphic approach; however, these intervals are time equivalent to units to the south that fit the traditional sequence-stratigraphic definition for key surfaces and system tracts. TWT = two-way traveltime; LST = lowstand systems tract; HST = highstand systems tract; TST = transgressive systems tract; mfs = maximum flooding surface; lw = lowstand wedge; bff = basin-floor fan (Moscardelli et al. 2012, Fig. 3, p. 1489). AAPG © 2012, reprinted by permission of the AAPG whose permission is required for further use

information on the vertical seismic characteristics of the succession. False-color variations, which display seismic amplitude variations, may be correlated locally to lithologic variability across the area, from channel-fill to overbank.

Another example of a submarine fan is shown in Fig. 6.38. This is a small fan, subdivided into upper, mid and lower fan segments, according to Walker's (1978) model. The stratal slice image shown in Fig. 6.38b is not a structurally horizontal seismic section, but follows the synclinal dip of the strata, as shown by the dashed line in the vertical section. This image was derived from the top of a seismically chaotic section, which is interpreted by Zeng et al. (1998) as representing the lithologically heterogeneous assemblage characteristic of the fan setting.

The remaining examples in this section deal with the imaging of important surfaces in the subsurface. Unconformities may contain significant information about contemporary erosional patterns, and their topography may have useful implications for petroleum trapping configurations. A particularly interesting unconformity is the surface that marks the top of the Upper Miocene Messinian stage in the Mediterranean basin. Since the first deep-sea drilling project in the Mediterranean Sea in 1970 geologists have known about a prominent, widespread unconformity that spans most of the basin margins, and a thick evaporite section occupying the basin center. During the Miocene the Mediterranean basin became isolated from the Atlantic Ocean, and evaporated to dryness several times (a drop in sea level of up to 2 km), leaving a deeply eroded basin margin cut by fluvial canyons (Ryan and Cita 1978). Figure 6.39 illustrates a seismic section across the basin margin of the Ebro delta in northeastern Spain. The erosional relief on what is interpreted as the Messinian unconformity is clear in this section, as is the truncation of the units below and the progradation of the Pliocene section that draped and buried the unconformity. A 3-D seismic survey of this area yielded the perspective map view shown in Fig. 6.40 The dendritic drainage pattern revealed by this map is very similar to that of "badland" topography in desert areas, such as that in southern Spain and parts of the southwest United States. About 400 m of erosional relief is visible in the seismic data from this area (Martinez et al. 2004).

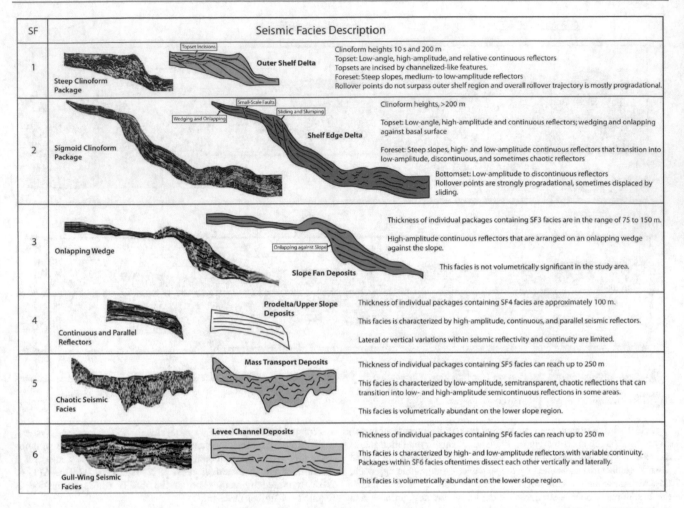

Fig. 6.31 A seismic-facies classification of the shelf-margin depositional systems, Atlantic margin of Trinidad (Moscardelli et al. 2012, Fig. 6, p. 1493). AAPG © 2012, reprinted by permission of the AAPG whose permission is required for further use

As noted in the previous section, paleokarst deposits may be very important as petroleum traps, and 3-D seismic data has proved very successful in identifying and mapping these deposits (Sayago et al. 2012; Zeng et al. 2011; Ahlborn et al. 2014). Figure 6.41 is a 2-D seismic section through a succession of Lower Paleozoic carbonate deposits cut by several unconformities that are underlain by what is interpreted as paleokarst deposits. As each surface was developing by surface exposure and erosion, fluctuating water tables led to the development of subterranean drainage systems, with sinkholes, tunnels and caves, many of which underwent significant collapse, resulting in disruption of the stratigraphic layering. Surface erosion, in some cases, ultimately led to the development of fluvial gullies and channels, with isolated karts towers left behind. An analogy with the karst topography of parts of the Guilin Province in southern China was suggested by Zeng et al. (2011, p. 2076). Extraction of stratal slice data from seismic volumes through these types of succession may provide detailed maps of the surface karst topography. Figure 6.42 provides an oblique view of such a

surface and contains graphic imagery of the surface erosional and collapse features of a paleokarst surface of Ordovician age.

6.4 Directional Drilling and Geosteering

Historically, most drill holes have been drilled vertically or as near vertically as possible, but for many reasons, directional drilling may have advantages, both geologically and for practical reasons. For example, access to potential reservoirs beneath environmentally sensitive lands, or underwater, may be accessed by directional drilling, and the technology now allows multiple holes to be drilled in every direction from a single well pad, which reduces the surface impact of the operation. Drilling multiple slanted holes from a single offshore drilling platform reduces the costs considerably. Geologically, the advantage of directional drilling is that it creates the possibility to access specific subsurface targets by the use of real-time downhole steering. In the

a)

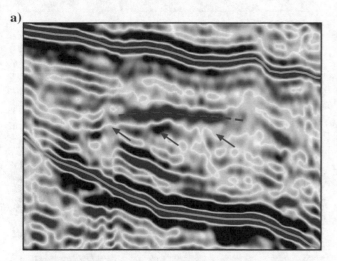

b)

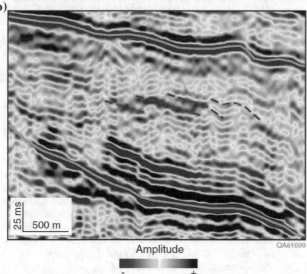

25 ms | 500 m

Amplitude

QAe1699

- +

Fig. 6.32 Reducing ambiguity in interpreting clinoform prograding sequences by spectral balancing. **a** Original stacked and migrated seismic section in the Abo Kingdom carbonate field of west Texas, with a flat (dashed line) event and some toplapped events (arrows) underneath. **b** The same section after spectral balancing processing. The flat event in the original data has been broken up into clinoforms (dashed lines) having slopes similar to those of surrounding events. The toplaps disappear (Zeng et al. 2013, Fig. 18, p. SA48). AAPG © 2013, reprinted by permission of the AAPG whose permission is required for further use

1980s, the development of the technology permitted wells to be completed with horizontal segments that could be steered to penetrate a reservoir along its length, generating substantially greater exposure of a hydrocarbon-bearing unit to the producing well. Horizontal segments of up to 4 km are now possible (up to 7 km in exceptional circumstances). Drilling motors are mounted at the end of the drill pipe, driven by the drilling mud, and subsurface navigation is achieved by the use of gyroscopes, with steering achieved on

the basis of navigational information sent electronically from the drill head in real time. Horizontal drilling has become widely used since the 1990s in the production of shale gas and tight oil, and is also an integral part of the technique of in situ production of oil from the Alberta Oil Sands.

Geological mapping techniques may be used as an aid in subsurface navigation. This is termed **geosteering**, and may be carried out using detailed mapping information obtained before drilling commences, or may be done in real time, for example by using petrophysical log data recorded as the well is being drilled or by using well cuttings that may be monitored for lithological content, microfossils or palynomorphs that are known to characterize the target unit and the overlying and underlying formations. The objective is to ensure that the well continues to penetrate the desired stratigraphic level as drilling proceeds, although allowance must be made for the delay caused by the travel time of the cuttings from the drill face. Three examples are briefly discussed here. In each case the operator stressed that considerable care needs to be taken to construct a very detailed and accurate three-dimensional model of the subsurface target volume before drilling commences because the real-time data needs to be matched against the model as drilling proceeds.

Taylor (2010) described the use of real-time petrophysical monitoring during the drilling of a horizontal hole. The project began with the establishment of the petrophysical characteristics of the rocks to be drilled. From this information a model is developed of the petrophysical values to be predicted along the projected path of the hole, based on the three-dimensional reconstruction of the stratigraphic architecture (Fig. 6.43). Note how, in this diagram, the planned well path first passes nearly vertically through the stratigraphy, and is then designed to drill out straight legs at slightly different angles in order to follow the mapped structure, with the predicted petrophysical response along the course of the hole shown in the two graphs at the top. Figure 6.44 shows the results being obtained as drilling is in progress. Deviations from the modeled petrophysical responses may be fed back into the structural model of the target volume in order to permit corrections to be made, if necessary, to the path followed by the drill. Some earlier examples of the use of real-time "logging while drilling" were described by Bristow (2002).

Southcott (2014) described briefly how a 3-D seismic data volume was subjected to rigorous calibration and time-depth conversion with the use of relevant well petrophysical and core data, and then used to plot the desired path of a projected lateral well.

In Sect. 6.6.3, a brief reference is made to a study of "forensic chemostratigraphy" that was used by Hildred et al. (2014) to characterize the units that had been penetrated along a horizontal well through the Bakken petroleum system.

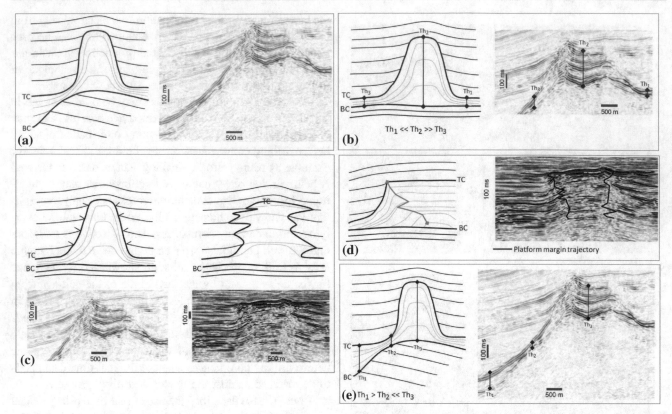

Fig. 6.33 Seismic facies of isolated carbonate buildups. **a** An antecedent topographic high beneath an isolated carbonate buildup. **b** Significant localized thickening within an isolated carbonate buildup. **c** Onlap of overburden onto the margins of an isolated carbonate buildup (left), contrasted with a situation where depositional relief on the margins of the isolated carbonate buildup was lower because of contemporaneous infill of the adjacent basin (right). In this case, carbonate material from the platform top was transported away from the platform margin to produce depositional wings that interfinger with the basin-fill strata. **d** Platform margin trajectories with phases of progradation, aggradation and retrogradation, which can be indicative of an isolated carbonate buildup. **e** The thin–thick–thin pattern commonly developed on isolated carbonate buildups that are shedding material from the platform top through a bypass zone to be redeposited in deeper water adjacent to the platform. TC = top carbonate; BC = base carbonate. (diagrams assembled from several illustrations in Burgess et al. 2013). AAPG © 2013, reprinted by permission of the AAPG whose permission is required for further use

6.5 Older Methods: Isopleth Contouring

In the absence of seismic data, the explorationist and the basin analyst must maximize the data available from well or outcrop data. This was the situation after the Second World War before the seismic technique had evolved to the point that stratigraphic information could be extracted from the data. As noted in Sect. 1.2.3, the need for a "big picture" of a sedimentary basin was the drive that underlay the work of W. C. Krumbein and his colleagues at Northwestern University (from which modern ideas about sequence stratigraphy emerged: see Sects. 1.2.9, 1.2.10). Krumbein (1948) devised a suite of quantitative lithologic indices and ratios that could be calculated from the lithologic succession at each sample point. Contouring of the resulting data created "isopleths," which formed maps that could be then interpreted in terms of paleogeography. Some of these approaches have become part of the standard methodology

of basin analysis. Net sand maps or sand–shale ratio maps are good examples. For example, Fig. 4.28 shows how the distribution of sand thicknesses may be used to define patterns that may be interpreted in terms of the sedimentary controls acting on deltas. Figure 4.47 shows how the ratio of clastic to carbonate lithologies defines the transition from a coastal clastic belt to a broad carbonate platform. The gamma radiation map of Fig. 6.36c shows how petrophysical data may be used for a similar purpose, in this case the definition of net sand thickness, a parameter that could (in this case) be interpreted in terms of the patterns of channels and bars within a meandering fluvial system. Figure 6.45 illustrates the use of the net sand mapping technique to develop highly detailed facies interpretations, where adequate drill data are available. Figure 6.46 illustrates a modern example of the use of this technique. Seismic data were used to restore Middle Miocene topography in this basin in Trinidad. Net sand values reveal two major northeast to southwest sand trends and their relationship to growing

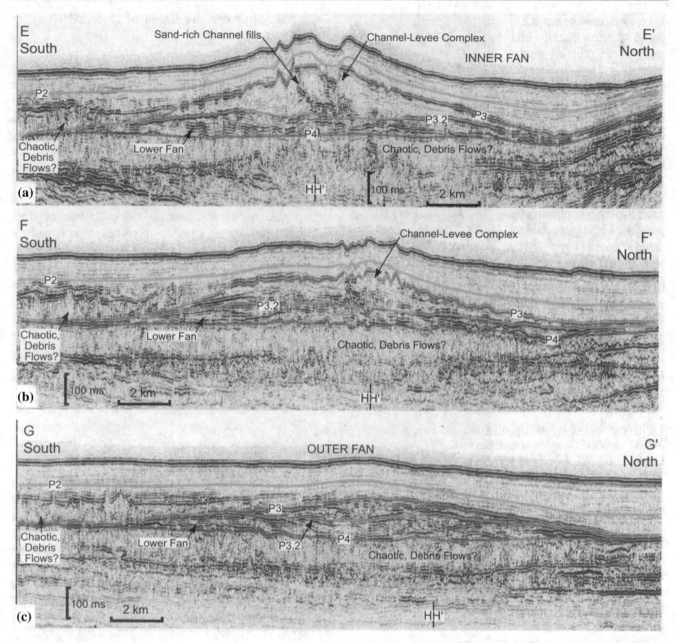

Fig. 6.34 Three strike sections across the submarine fan formed on the basin floor, Borneo (Saller et al. 2004). AAPG © 2004, reprinted by permission of the AAPG whose permission is required for further use

structures. "Zones of absence of Middle Miocene Herrera sands coincided with paleo-highs on the restored Middle Miocene topography" (Moonan 2011).

Krumbein (1948) developed a clastic ratio parameter, calculated from the sum thickness of clastic lithologies divided by the sum thickness of carbonate and evaporite lithologies in a vertical section. Figure 6.47 is an example of the use of this ratio. He stated that, of this map (p. 1913):

> … the upper map shows generalized isopachs, clastic ratio lines, and sand-shale ratio lines. The three sets of lines render the map somewhat cumbersome, and in practice the clastic ratio may be

used alone with the isopachs, if the section shows a balance between clastics and nonclastics. … The lower map … is a combination of the clastic and sand-shale ratios in the upper map, and it shows the statistical lithologic differentiations of the map area. The map is based on combinations of the clastic ratio and the sandshale ratio in terms of certain limiting values.

The regional pattern revealed by this map is of a broad clastic belt in the east (Montana-Colorado) bordering the Canadian Shield, passing gradually westward into a carbonate-dominated area near the continental margin, running north–south through Idaho. As an aid to constructing

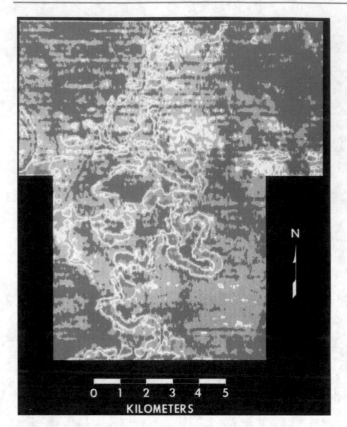

Fig. 6.35 A meandering fluvial channel in shallow sediments in the Gulf of Thailand. From the first (1986) edition of Brown's AAPG memoir (Brown 2011 is the latest edition). AAPG © 2011. Reprinted by permission of the AAPG whose permission is required for further use

6.6 Mapping on the Basis of Detrital Composition

Traditional methods of correlating petrophysical logs are discussed in Sect. 6.2. A wide range of additional techniques has evolved over the years for the examination of subsurface petrophysical, core and sample data. Compilations of such studies have been published by Hailwood and Kidd (1993) and Ratcliffe and Zaitlin (2010). These methods have several purposes: not just to aid in correlation, but to provide additional information, such as provenance studies, which can help in reconstructions of paleogeography, regional stratigraphic trends and tectonic history. Some of these techniques provide information on numerical ages and are discussed in Chaps. 7 and 8. Others are useful for local or regional correlation or provenance studies but do not necessarily provide such age information. A list of some of the latter is as follows:

Characterization by clastic petrofacies, including
  Heavy mineral stratigraphy
  Geochemical fingerprinting of sandstones and mudstones (chemostratigraphy)
  Magnetostratigraphic susceptibility
  Detrital zircon provenance studies

6.6.1 Clastic Petrofacies

The grain size, grain shape and composition of the clastic detritus present in a stratigraphic unit depends initially on the nature of the source, or **provenance**, of the detritus, but the detritus may undergo many modification processes during its derivation, transport, deposition and burial so that the characteristics of the original source become obscured. By careful study, the effects of some of these modification processes can be identified, providing much valuable information to a basin analysis.

Detrital composition depends on these main factors:

1. Geological composition of the source area
2. Climate and relief of the source area and the area through which the detritus is transported
3. Detrital dispersal and mixing patterns brought about by the transport processes
4. Chemical and mechanical abrasion, winnowing and breakdown that occur during transport and at the depositional site
5. Diagenetic changes during burial.

regional paleogeography this is a useful map, but clearly many data points would be required to plot individual depositional systems on a more local, 10^3–10^4 m scale.

Extensive use of these various isopleth maps was made in the early regional stratigraphic and sequence work of Sloss et al. (1949), but these techniques have largely fallen into disuse in recent decades, because of the more powerful methods of facies mapping that emerged in the 1960s and 1970s, and the seismic-stratigraphic and sequence-stratigraphic techniques that evolved more recently. An exception would be the net-sand map, which is still used to illustrate sand reservoirs in stratigraphic traps (e.g., Fig. 6.46). It has become apparent that the most useful isopleth maps for lithofacies mapping are those which are based on individual depositional units, such as parasequences or allomembers. Thicker units would typically combine the facies information from more than one base-level cycle, during which autogenic processes may have significantly changed the local paleogeography.

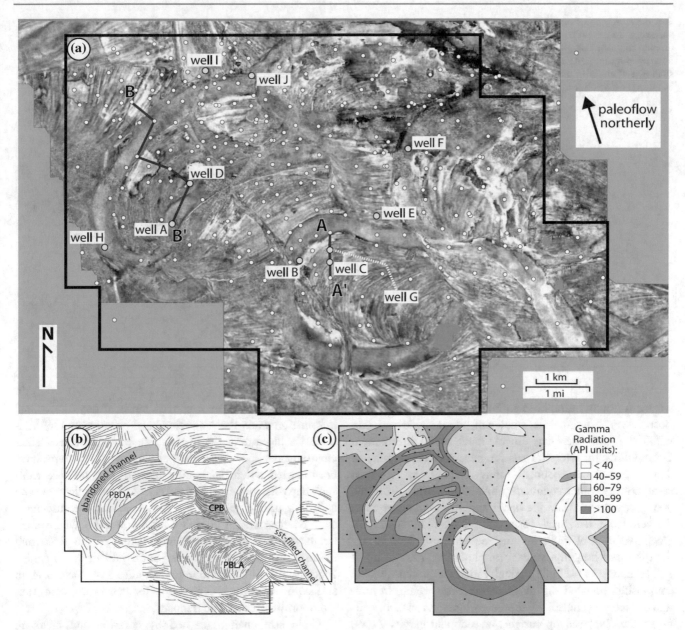

Fig. 6.36 Seismic exploration of the Alberta Oil Sands. Mannville Group, south of Fort McMurray, Alberta. **a** Time slice taken at a horizon, 8 ms (approx. 8 m) below a flooding surface at the top of the oil-bearing unit; **b** Interpretation in terms of scroll bars, point bars and channels; **c** Gamma ray values at the selected horizon. From Hubbard et al. (2011, Fig. 2, p. 1126). AAPG © 2011. Reprinted by permission of the AAPG whose permission is required for further use

Despite the many changes that the detritus undergoes during transport and deposition, the nature of the original source area can usually be identified. Certain compositional ranges are characteristic of particular types of tectonic source environment. This is the subject of **petrofacies analysis** and is not discussed here (see Miall 1999, Sect. 9.4). Without going into the details of plate-tectonic setting, information regarding the geology of the local source area may be of considerable value in a basin analysis, because of the information it provides regarding transport directions, drainage areas, location of sources of economically valuable

minerals (such as those occurring in placer deposits), uplift and **unroofing history** of the source area, and so on. This is the main subject for discussion in this section.

Initially, of course, the petrographic composition of primary detritus is similar to that of the rocks exposed in the sediment source area, but changes begin to occur even as the rocks weather and become part of the subsoil. After only a short transport distance, the petrography of the detritus may not necessarily bear a simple relationship to that of the source area because of the destruction of unstable grains in transit, particularly in humid environments where chemical

Fig. 6.37 A map of amplitude variations draped on a gently dipping stratal surface, illustrating a meandering turbidite channel in shallow sediments in the Gulf of Mexico (Posamentier et al. 2007, Fig. 2, p. 3)

weathering may be rapid (e.g., Franzinelli and Potter 1983; Basu 1985). Nevertheless, studies of petrographic variation within the basin itself may provide much useful information. Downcutting by rivers reveals progressively older rocks so that the derived sediments may display an **inverted stratigraphy**, older sediments containing detritus from the young cover rocks and the upper strata containing material derived from the local basement (Graham et al. 1986). Documentation of this unroofing sequence may help clarify the timing of orogeny, volcanism, plutonism, rates of uplift etc. (Cerveny et al. 1988). Areal variations in petrographic composition depend on transport directions which, in turn, reflect tectonic control of the paleoslope. As clastic sediments are dispersed by currents or sediment gravity flows, they are subjected to size sorting, abrasion, breakage and rounding. They may retain a distinctive detrital composition, although easily weathered minerals, such as ferromagnesian grains, will gradually be destroyed. As a result of these processes, a particular lithosome may become fingerprinted with a distinctive size range and composition, and down-current changes in the various petrographic parameters may be used as paleocurrent indicators. Interpretation of the source area for marine rocks may be a more complex procedure. For example, submarine fans may contain deltaic detritus plus shelf carbonate or siliciclastic material swept down to the continental rise via submarine canyons.

Petrographic data are the simplest to use where the sediments have been deposited by unidirectional currents, as in fluvial, deltaic and submarine-fan environments. Size and sorting parameters do not convey much, if any, directional information when the sediments have been formed by multidirectional currents, as in most shallow-marine and shelf environments. Few petrographic studies of this type have been carried out in submarine fans, in part because most such deposits contain abundant paleocurrent indicators such as sole markings (Sect. 2.2.2.7), and in part because most exposed fans are in complex structural belts where areal trends are difficult to reconstruct. Most of the remaining discussion in this section relates to fluvial and fluvio-deltaic deposits, for which petrographic studies may have uses in surface or subsurface exploration for petroleum, coal, uranium, or other economic deposits.

Grain size changes caused by abrasion and breakage during downstream transport are rapid in gravel detritus, but the rate of change with the distance of transport drops off asymptotically, becoming slow in coarse sand and negligible in fine sand to silt. Much experimental work has been carried out in an attempt to quantify the rate of change so that grain size in the field could be interpreted directly in terms of distance of transport. However, each detrital lithology responds differently to current transport, and rivers vary markedly in competency at any one point, so that research of this type is unlikely to meet with any success. Nevertheless, mapping of relative grain size changes in conglomeratic and coarse sandy sediments has considerable use as a paleocurrent indicator.

What measurements should be made? For grain size studies in conglomeratic rocks, the simplest and the most

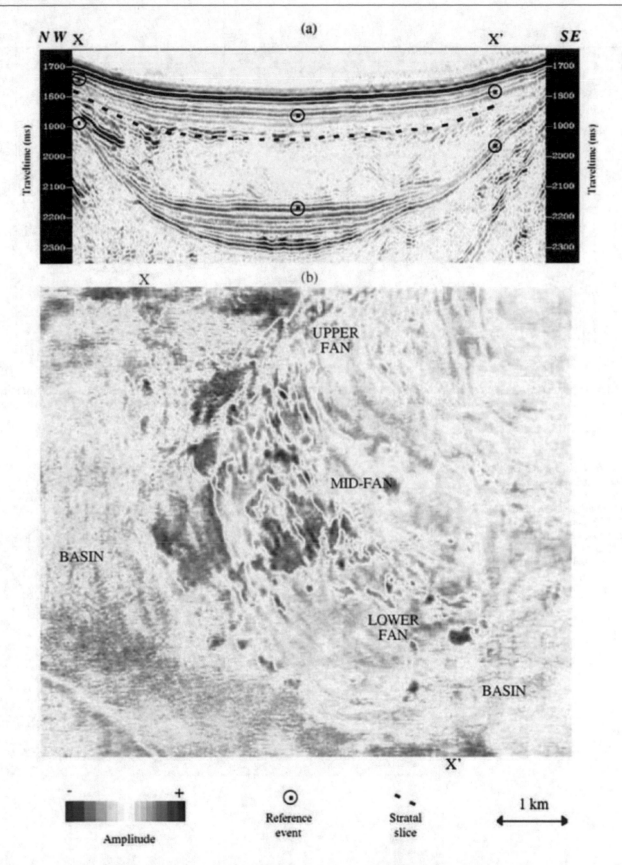

Fig. 6.38 A submarine fan shown on a stratal slice taken from the Pleistocene, deep-water offshore Louisiana. **a** A vertical profile with a dominant frequency of 40 Hz; **b** A stratal slice from the 1400 to 1800 m (1750–1950 ms) interval, along the dashed line in (**a**) (Zeng et al. 1998, Fig. 6, p. 521)

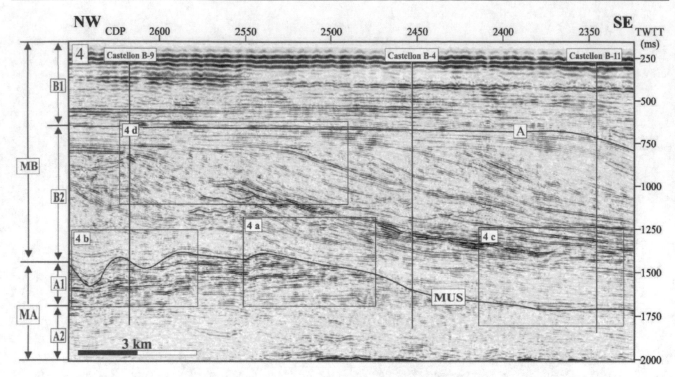

Fig. 6.39 Dip-oriented seismic cross-section across the Mediterranean continental margin of northeastern Spain, showing the deeply incised Messinian unconformity (MUS). Megasequence A (MA) consists of a shallow-upward deep-marine to the deltaic succession of Miocene age. Megasquence B (MB) comprises a Prograding slope wedge of Pliocene age. The labeled rectangles refer to enlarged portions of this section that are not shown here (Martinez et al. 2004, Fig. 4, p. 93)

Fig. 6.40 A perspective view of the Messinian unconformity on the Mediterranean continental margin of northeastern Spain. The 2-D section shown in the previous figure crosses this area from center-left to lower right (Martinez et al. 2004, Fig. 8, p. 96)

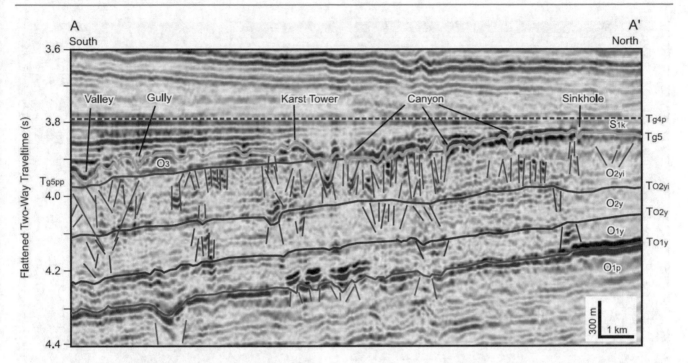

Fig. 6.41 Vertical seismic section through an Ordovician–Silurian carbonate succession in the Tarim Basin, western China, flattened on the Lower Silurian (Tg4p reflection) in order to show the original topography at the top of the Ordovician unconformity (Tg5). At this and older paleokarst surfaces, collapsed caves and sinkhole are shown by the short colored vertical lines, which mark disrupted or discontinuous reflections (Zeng et al. 2011, Fig. 10, p. 2074). AAPG © 2011, reprinted by permission of the AAPG whose permission is required for further use

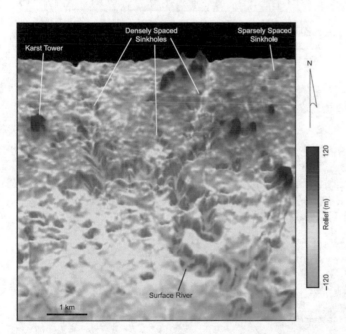

Fig. 6.42 A perspective view of the Ordovician unconformity extracted from 3-D seismic data, showing the morphology of the eroded, paleokarst surface (Zeng et al. 2011, Fig. 14, p. 2077). AAPG © 2011, reprinted by permission of the AAPG whose permission is required for further use

useful mapping parameter is maximum clast size, which is usually given as the average of the intermediate clast diameter of the 10 largest clasts present at a sample point. For sand-sized grains, mean or maximum grain size may be used. Both can be measured in an outcrop or a thin section by visual estimation against a grain size chart (Fig. 2.5). This permits measurements to the nearest 0.5Φ, which is quite accurate enough for our purposes (sieve analyses and other laboratory studies are time-consuming, and usually add little).

The detrital composition of conglomerates can be studied by counting and identifying 100 or more clasts at each sample point. Sandstone composition should be studied in a thin section using a moveable microscope stage and a point counter. It is recommended that at least 300 grains be identified and counted per sample point. The stage should be set to advance at each count by an amount equal to or slightly greater than the mean grain diameter. The number of points counted can be adjusted according to the compositional variability of the sample. Pure quartz arenites do not need such intensive study, but immature, lithic arenites with many varieties of rock fragments may need more detailed examination. Care should be taken to recognize and

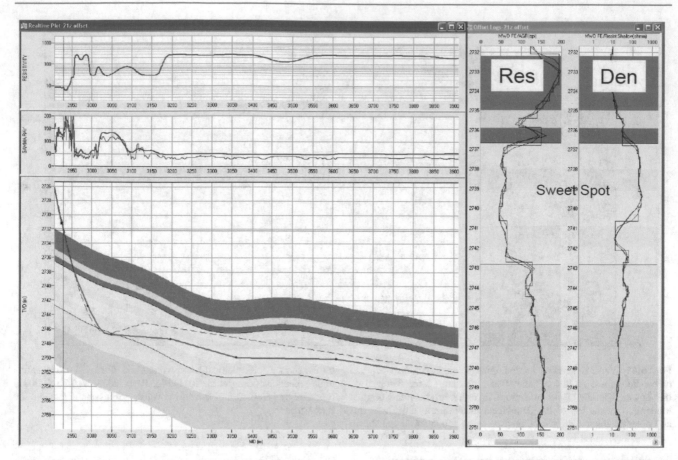

Fig. 6.43 At right: density and resistivity logs through a target interval; at lower left, a two-dimensional cross-section of the modeled structure and the planned path of the directed wellbore; at top left, the expected log response along the well (Taylor 2010). AAPG © 2010, reprinted by permission of the AAPG whose permission is required for further use

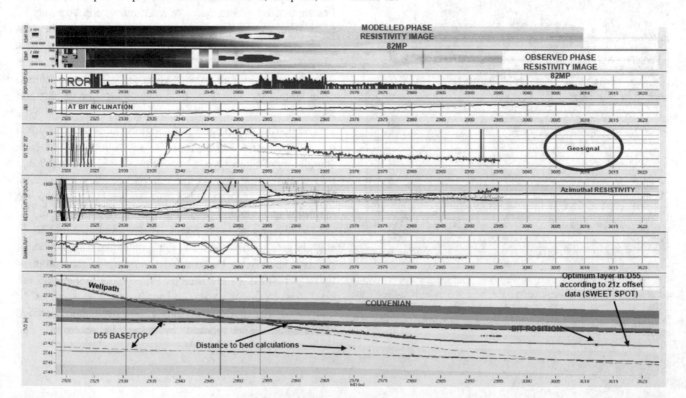

Fig. 6.44 Data being obtained in real time as the directional well planned in Fig. 6.42 was being drilled (Taylor 2010). AAPG © 2010, reprinted by permission of the AAPG whose permission is required for further use

Fig. 6.45 Definition of a crevasse splay and a channel sand body by close-spaced drilling in Cenozoic fluvial deposits of the Gulf Coast (Galloway 1981)

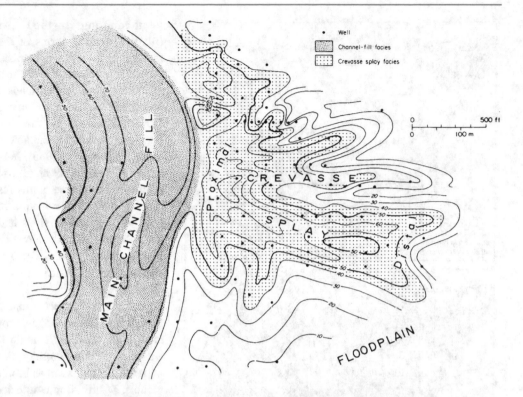

Fig. 6.46 Net sand map, derived from well data in an oil field, southwest Trinidad (Moonan 2011). AAPG © 2011, reprinted by permission of the AAPG whose permission is required for further use

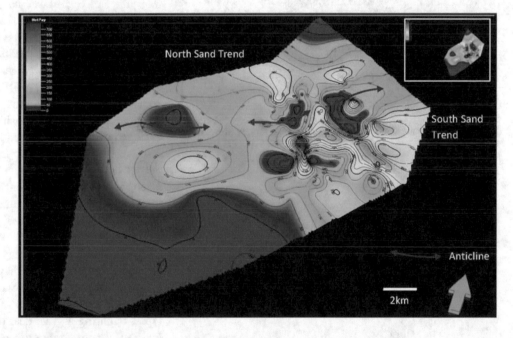

distinguish **extrabasinal** and **intrabasinal** detritus, particularly in the case of carbonate and volcanic fragments. Zuffa (1985b, 1987) discussed criteria for distinguishing these classes of particles. Mineral types have different densities and so they vary in their hydrodynamic behavior. They also have different fragmentation patterns. The detrital composition of a clastic rock may therefore vary markedly with grain size (Fig. 6.48) and so, for mapping purposes, it is advisable to control measurements according to a standard grain size

range, for example, 1ϕ. class. These methods will usually result in measurement repeatability of two or three percentage points or better, and this, again, is accurate enough for our purposes. Analytical methods are discussed in detail by Dickinson (1970), Pettijohn et al. (1973) and Zuffa (1985a, b, 1987). It is important to standardize laboratory procedures in order to facilitate comparisons between data sets. Zuffa (1985b, 1987) and Ingersoll et al. (1984) recommended the use of the so-called **Gazzi-Dickinson**

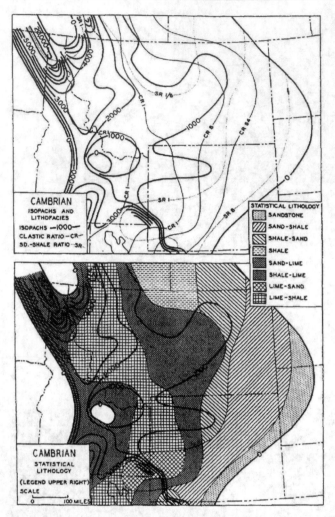

Fig. 6.47 A clastic ratio map for Cambrian strata in the western United States (Krumbein 1948, Fig. 3, p. 1914). AAPG © 1948, reprinted by permission of the AAPG whose permission is required for further use

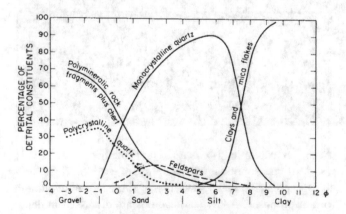

Fig. 6.48 Relationship between grain size and composition of the detrital fraction in siliciclastic sediments (From Blatt H, Middleton GV, Murray RC (1980) Origin of sedimentary rocks, 2nd edn. p 321. Reprinted by permission of Prentice-Hall Inc., Englewood Cliffs, New Jersey)

point-counting method, which counts coarse grains (>0.0625 mm) in terms of their component mineral grains rather than as rock fragments. The method also requires that the observer attempt to reconstruct the original composition of altered grains. This leads to a better clustering of data points and increased accuracy in provenance determinations. However, there is no universal agreement that the Gazzi–Dickinson method is suitable for all purposes. For example, Suttner and others (1985), in two separate discussions of the paper by Ingersoll et al. (1984), pointed out that the method of classifying coarse grains results in a loss of certain types of data, such as information regarding lithic grains. Careful study of these discussions and the replies by Ingersoll and his colleagues is recommended for anyone contemplating detailed petrographic work. Howard (1993) discussed the statistics of clast counting, sources of error and field procedures.

Much use has been made of heavy minerals (density greater than about 2.8) in provenance studies, even though most sands typically contain less than 1% by weight. They are extracted from loose or disaggregated sand by separation in dense liquids, such as bromoform, and examined in grain mounts. Distribution is affected by hydraulic sorting during transport and by intrastratal solution following deposition (Morton 1985a). Considerable skill is needed to identify grain types correctly, and the technique is, in general, more difficult to use than thin-section analysis of the light fraction. We do not discuss this research theme further (see Morton and Hallsworth 1994, 1999), except for the specialized zircon studies described below.

Grain size and compositional measurements may be averaged over a vertical section at each mapping location or restricted to individual correlatable stratigraphic units. Plotting of both areal and vertical changes may provide useful paleogeographic information, and it is up to the basin analyst to exploit the data in the most effective way. Petrographic data may be analyzed manually, graphically or statistically to define petrographic assemblages or provinces, typifying a particular source area geology (Suttner 1974). Alternatively, particular mineral species may be treated separately.

Some workers have concentrated on identifying and mapping varieties of quartz, feldspar, zircon or other grain types, requiring a great deal of highly specialized research, but possibly providing provenance information unobtainable in any other way (Allen 1972; Heim 1974; Schnitzer 1977; Helmold 1985). Folk's (1968) work on quartz, chert, chalcedony and their varieties is particularly noteworthy. Sophisticated analytical techniques are increasingly being used to define mineral varieties based on trace-element composition (Matter and Ramseyer 1985; Morton 1985a, b). For example, Morton (1985b) employed an electron microprobe to discriminate three populations of detrital garnets in a stratigraphic unit and related them to probable sources. Heller and Frost (1988) explored a huge new area: the study of stable and radiogenic isotope distributions in detrital grains and their use in fingerprinting source terranes.

Resistant grains, such as quartz and chert, plus the heavy minerals zircon, tourmaline, rutile, and garnet, are commonly recycled through several episodes of erosional derivation, sedimentation, lithification, uplift and reerosion. Usually, they show abundant evidence of this in the form of a high degree of rounding and the presence of authigenic overgrowths. It is usually difficult or impossible to determine the source of such grains because, in most cases, they could have come from several different stratigraphic units in the source area. Variations in color, crystal habit or inclusions may help, but require more detailed research (Zuffa 1985b, 1987). Zircon studies have recently assumed a new importance as provenance indicators, as discussed in Sect. 6.6.2.

Extensive treatments of sandstone petrology and its application in sedimentological studies were given by Pettijohn et al. (1973) and Blatt et al. (1980). Zuffa (1985a) edited a useful compilation of research and review papers, several of which are quoted above. In the remainder of this section, some examples of field studies are discussed to illustrate the kind of contribution petrography can make to regional basin analysis.

Two examples of the analysis of clastic detritus are provided here in order to illustrate the main uses of the method. Miall (1979) analyzed the petrography of the Isachen Formation, a Lower Cretaceous braided river deposit in Banks Basin, Arctic Canada. The bulk of the formation consists of texturally and mineralogically mature medium to coarse sandstone, showing abundant evidence of a recycled origin. Samples were obtained from outcrops divided into two broad groups, from the north and south parts of the island. Paleocurrent evidence indicates northward flowing trunk rivers with tributaries entering from the east (the west side of the basin is covered by younger strata but may also have been a source area). Initial attempts to recognize petrographic provinces in the sandstone were unsuccessful. Ternary plots show that the samples from the north and south end of the basin contain essentially the same mineral composition. Cluster analysis revealed a weak trend for feldspar to be more common in the north, and sedimentary rock fragments are slightly more abundant in the south. This was then emphasized by re-plotting the thin-section data in terms of three minor components, which shows that chert is also slightly more abundant in the north. It was found that a few rare components, such as detrital quartz-feldspar intergrowths, quartz with mica, tourmaline or zircon inclusions, and zoned quartz, are all present almost exclusively in the

north end of the basin. Analyses of these data in terms of potential sediment sources resulted in the map shown in Fig. 6.49. The bulk of the quartz sand and the sedimentary rock fragments were derived from mature Proterozoic sediments, chert was fed into the north end of the basin from various Lower Paleozoic sources and the feldspar plus rare quartz and feldspar components appear to have been derived from a small Archean granodiorite pluton. A Middle to Upper Devonian sandstone unit that outcrops over wide areas adjacent to the north end of the basin appears to have contributed little detritus. It is too fine-grained and has the wrong texture and mineralogy.

Rahmani and Lerbekmo (1975) carried out a detailed heavy mineral analysis of the Cretaceous–Paleocene molasse sandstones derived from the Canadian Rocky Mountains. Factor analysis was used to determine mineral associations, and these resulted in the definition of a series of petrographic provinces (Fig. 6.50). It was found, unexpectedly, that these provinces were distributed in belts subparallel to structural strike. It was concluded that the dominant fluvial systems flowed longitudinally down the depositional basin from source areas to the west and northwest, not transversely, as would initially have been expected. The mineral provinces could be distinguished because of variations in source area geology.

In southern Alberta the Lower Cretaceous stratigraphy is characterized by a suite of thin marine–nonmarine sequences featuring multiple cycles of incision, which have created a very complex stratigraphy in which it may be quite difficult to correctly correlate individual units. Petrographic and chemostratigraphic techniques have been employed to supplement conventional petrophysical log correlation work. We report here on a petrographic study by Zaitlin et al. (2002). Figure 6.51 was compiled from thin section studies of core samples. Zaitlin et al. (2002, p. 38) described this compilation as follows:

> Summary lithostratigraphic comparison of the Basal Quartz as proposed in this study. A ternary diagram with quartz, chert, and clay-rich grains at the apices is effective in partitioning the petrographic data into distinctive populations. A second set of ternary diagrams, with intergranular, intragranular, and microporosity pore types at the apices is used to illustrate porosity fabric and reservoir quality. Key diagnostic grains and alteration features are used to exhibit the two cycles of mineralogical, textural, and reservoir quality cyclicity.

The stratigraphy is defined as two cycles, as shown in Fig. 6.51, but in practice the cycles consist of a complex of sheet sandstones with mutually erosive incised valleys. The petrographic work helps to differentiate these stratigraphic fragments. An example of the resulting reconstructions is shown in Fig. 6.52.

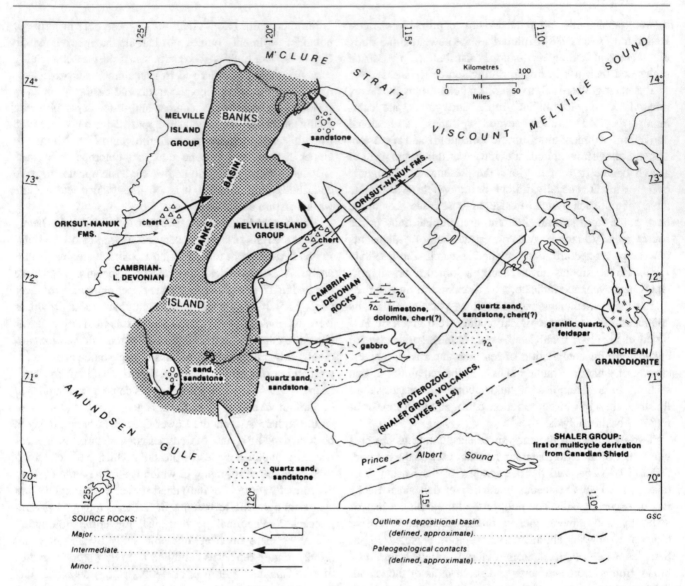

Fig. 6.49 The depositional basin (shaded area) of the Lower Cretaceous Isachsen Formation of the Banks Island area, Arctic Canada showing interpreted source rocks of the Isachsen sandstones (Miall 1979)

6.6.2 Provenance Studies Using Detrital Zircons

The heavy mineral zircon has been found to be a particularly valuable subject for provenance studies. It is highly resistant and survives multiple recycling events, which means that it can be used to identify the ultimate source of the detritus in which it is found. Zircon grains are now readily dated with the uranium/lead method, and the advent of sensitive high-resolution ion microprobe (SHRIMP) and laser-ablation–inductively coupled plasma–mass spectrometry (LA-ICP-MS) means that large populations of detrital zircon grains can be analyzed quickly (Thomas 2011). With

care, it is possible to use zircon assemblages to map sediment sources as an aid to paleogeographic reconstruction. It is, of course, necessary to take into account the issue of recycling, which can be done by examining all aspects of a sandstone's detrital petrography. For example, in the case of the Banks Island study discussed briefly above (Fig. 6.49), it is highly relevant that the sand grains of the unit under discussion (the Isachsen Formation) are much coarser and more mature than those of the Devonian Melville Island Group, which is the immediately underlying unit around much of the basin margin, indicating that this unit could not have provided the bulk of the sand in the Isachsen

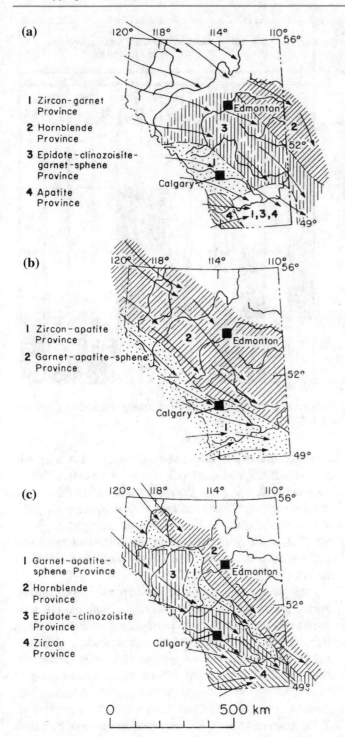

(a)

1 Zircon-garnet
   Province

2 Hornblende
   Province

3 Epidote-clinozoisite-
   garnet-sphene
   Province

4 Apatite
   Province

(b)

1 Zircon-apatite
   Province

2 Garnet-apatite-sphene
   Province

(c)

1 Garnet-apatite-
   sphene Province

2 Hornblende
   Province

3 Epidote-clinozoisite
   Province

4 Zircon
   Province

0   500 km

Fig. 6.50 Heavy mineral provinces, as determined by factor analysis, in three formations within the upper molasse wedge of Alberta. Arrows show interpreted dispersal directions. **a** Belly River Formation (Upper Cretaceous); **b** Edmonton Formation (Cretaceous-Tertiary); **c** Paskapoo Formation (Tertiary) (Rahmani and Lerbekmo 1975)

Formation. Zotto et al. (2020) carried out a detailed study of zircon recycling within the North American stratigraphic record. Moecher et al. (2019) demonstrated that it is simpler to identify recycling in the detrital mineral monazite and recommended its wider use in provenance studies. Given these types of caveats, zircon studies in North America have generated some surprising results, as discussed below.

Zircon studies on the North American continent began with a detailed study of the distribution and ages of zircons in potential sources terranes in the Cordilleran province along the western margin of the continent (Gehrels et al. 1995). Given the long-continued history of terrane accumulation that has built the continent, this information provides a quantitative basis for assigning zircons to ultimate sources based on their ages. In most sandstones, analysis demonstrates that there are several zircon populations exhibiting distinct age ranges, and it is usually a simple matter to relate these age populations to the age distributions of the ultimate sources. Dickinson and Gehrels (2003) were among the first to realize the importance of distant sediment sources, such as the Grenville and Appalachian orogens, in the sourcing of much of the sand now constituting the Late Paleozoic and Mesozoic clastic successions of the Colorado Plateau area of the southwestern United States.

Grenvillian ages dominate Neoproterozoic to Paleozoic detrital zircon populations across eastern Laurentia and persist through the present. The persistence of this dominance is inferred to result from recycling of zircon grains ultimately sourced from exceptionally Zr-rich and zircon-fertile Grenvillian granitoids (Zotto et al. 2020). Recycling events include dispersal of post-Grenvillian sediment during deposition of Neoproterozoic to Cambrian strata (formation of the "Great Unconformity") (cycle 1); subsequent erosion of metamorphosed Neoproterozoic to Cambrian strata generating detritus for Appalachian Pennsylvanian arenites (cycle 2) and modern erosion of those arenites (cycle 3). Pancontinental river systems facilitated dispersal of sediment of ultimate Grenvillian age during or after each cycle (Zotto et al. 2020).

Figure 6.53 illustrates the distribution of zircon ages in five samples from the Oil Sands of northern Alberta. This shows the importance of the Grenville as a sediment source, a feature of most zircon distributions from North American sources, and which indicates dramatic changes in sediment sources and the paleogeography of the major river systems through the Phanerozoic.

Raines et al. (2013) investigated the zircon geochronology of Upper Jurassic and lowermost Cretaceous strata in the

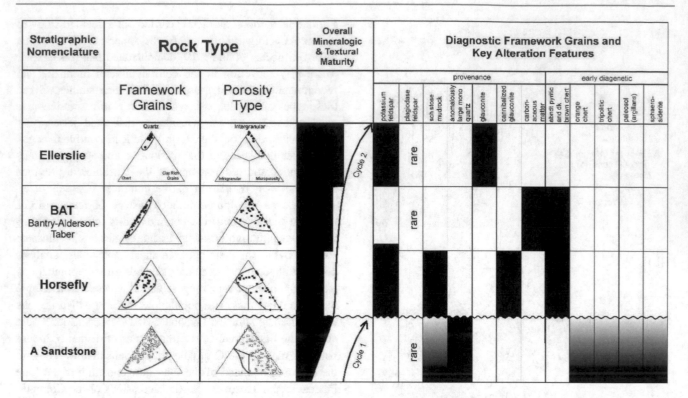

Fig. 6.51 Characterization of the petrography of units within the Lower Cretaceous "Basal Quartz" in southern Alberta. These data are used to identify and correlate these units in areas of complex stratigraphy (Zaitlin et al. 2002, Fig. 5, p. 38)

Alberta foreland basin. Conventional petrographic and paleocurrent studies had long been interpreted in terms of sources to the south and west, but considerable additional detail was obtained by these new methods. The zircon data indicate sediment in the early foreland basin was delivered via two principal sedimentary systems: a south-to-north axial river system and transverse fluvial systems that emanated from the adjacent Cordillera (Fig. 6.54). Of particular interest are derived zircons from contemporary (Jurassic) strata preserved in the Michigan basin, which confirm the importance of an Appalachian source for some of the detritus. Speculating from this single data point, it is suggested that a network of tributaries fed sediment into the axial system flowing through the Alberta basin from a wide area to the southeast (Fig. 6.54).

Combining these studies and others, of sands in the southern United States, Blum and Pecha (2014) generated two speculative maps that showed the likely complete reorganization of the interior drainage of the North American continent that evolved with the elevation of the Cordilleran orogen in the mid-Mesozoic (Fig. 6.55). Prior to this time, large-scale drainage systems appeared to have delivered abundant detritus from the Appalachian orogen to the west of the continent (Dickinson and Gehrels 2003). Large-scale river systems developed oriented in the reverse direction once significant sediment sources arose in the paths of these rivers, with significant transport eastward. Drainage into Hudson Bay, as outlined in the Paleocene map, was earlier suggested by McMillan (1973) and Duk-Rodkin and Hughes (1994).

There is not the space to develop this topic further. Detrital zircon studies are now having a profound effect on reconstructions of large-scale paleogeography through time. Here are a few other noteworthy recent studies: Craddock et al. (2021) compiled zircon age spectra for Eocene deposits along the Gulf Coast, and refined the reconstructions of large-scale dispersal trends within North America that occurred in response to Cordilleran orogeny. Garzanti et al. (2018) examined the detrital zircon composition of Nile river sediments and reconstructed the shifting source terranes for this giant river across central and northern Africa. Morón et al. (2019) employed zircon data to reconstruct transcontinental drainage patterns across Gondwana before early Mesozoic breakup.

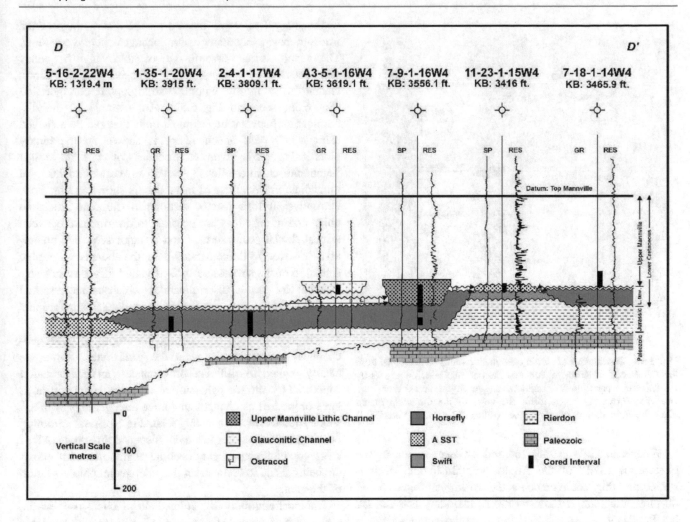

Fig. 6.52 An example of the complex Lower Cretaceous stratigraphy of southern Alberta. Individual lithosomes have been identified using the petrographic data shown in Fig. 6.50 (Zaitlin et al. 2002, Fig. 16, p. 50)

6.6.3 Chemostratigraphy

The development of high-precision analytical methods in the last few decades has facilitated the development of a range of new techniques for supplementing lithostratigraphic correlations in regional studies, and several detailed studies have been reported.

Pearce et al. (1999) provided a general discussion of chemostratigraphic methods, based on the few published sources available at that time, and also drawing on unpublished, proprietary studies carried out in petroleum industry laboratories. They made a number of useful general points. For example, studies of sandstones need to take into account grain size effects. Absolute elemental concentrations decrease in coarser-grain size fractions owing to dilution by quartz. Also, it is difficult to develop homogenous samples in coarser sandstone fractions, and so finer sands are preferred. Diagenesis may modify compositions, with Ca, Na, K, Pb, Rb, Sr and certain REEs liable to reduce in concentration during the dissolution of feldspars. Therefore, chemostratigraphic work makes preferential use of the more immobile elements Al, Ti, Zr, Nb and Cr. Wright et al. (2010) pointed out that sandstone and mudstones have completely different whole-rock geochemistrys. Sandstones, which typically contain a significant proportion of quartz, are silica-dominated, whereas mudrocks are rich in the cations that constitute the lattice in clay minerals.

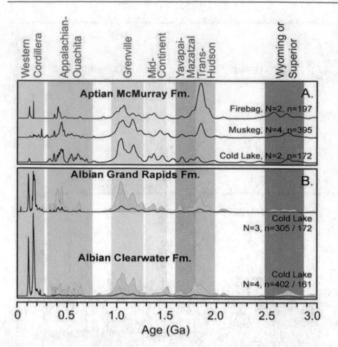

Fig. 6.53 Distribution of zircon ages in five sandstone samples from the Oil Sands of northern Alberta. Zircons of Grenville age (1.0–1.3 Ga) are particularly common, suggesting sources from the Grenville Province, which forms the edge of the Canadian Shield along the entire eastern margin of the continent (Blum and Pecha 2014)

Wray and Gale (1993) focused on marl bands in the Cretaceous Chalk of the Anglo-Paris Basin and demonstrated that they could be characterized (fingerprinted) based on their trace-element compositions, including Ba, Cu, Li, Na, Sc, Ti, V, Y, Zn and Zr. They used this characterization to trace individual bands from the deeper parts of the basin into "poorly correlated, condensed successions" deposited over structurally positive areas. These correlations yielded insights into syndepositional fault activity and differential subsidence across the project area.

Wright et al. (2010) presented a detailed chemostratigraphic study of the Mannville and Colorado Group (Aptian-Albian) strata of southern Alberta, covering much the same stratigraphic internal in much the same area as the petrographic study by Zaitlin et al. (2002) reported in the previous section. In both cases, the objective was to develop practical tools for correlation in areas of complex stratigraphy. During the Early Cretaceous this area was undergoing repeated cycles of small-scale relative changes in sea level, but because this part of the Western Canada

Sedimentary Basin was located on the distal fringe of the foreland basin, accommodation changes were very slow. The result is an assemblage of thin, high-frequency sequences with multiple, mutually erosive incised valleys, leading to a highly complex stratigraphic configuration (Fig. 6.56; see also Fig. 6.52). An example of the discrimination between depositional units that can be achieved using whole-rock geochemistry is shown by the binary plots in Fig. 6.57. Figure 6.58 consists of what are termed "synthetic chemical logs" based on core samples. An immediate practical use of these data is shown in Fig. 6.59. Petrophysical logs cannot distinguish different sandstone units, and in core it is not possible to discriminate between a local fluvial scour surface and a major sequence boundary. However, as demonstrated here, the chemostratigraphic signature of the successive incised valley fills is sufficiently different to permit discrimination, which is an essential starting point for constructing regional stratigraphic syntheses.

Wright et al. (2010) discussed the origins of these geochemical differences between the sandstones. Some are clearly related to different provenances, an interpretation supported by surface paleocurrent data, but some appear to have originated as changes in source area geology within a single source area, presumably related to erosional unroofing or to upstream changes in fluvial dispersal directions. All of these potential causes may become of importance if efforts are being made to reconstruct the paleogeographic evolution of the area.

Chemostratigraphy has immediate practical uses in the correlation of deposits that contain no other readily available tool for regional mapping. Franzel et al. (2020) described an example of this, the use of chemostratigraphic profiling to construct a detailed stratigraphy of a fluvial unit in Spain, as a precursor to sedimentological and paleogeographical interpretations.

Experimental use of "forensic chemostratigraphy" has been used to map the path of a horizontal well. Hildred et al. (2014) used chemostratigraphic analysis of cuttings from vertical drill holes, to characterize four units, including a producing dolomite in the Bakken petroleum system of southwestern Alberta, and the formations that bracket it. They then took a series of cuttings that had been generated along a horizontal well through this system, and were able to show how the hole had passed up and down through this stratigraphy, which contains several faulted offsets.

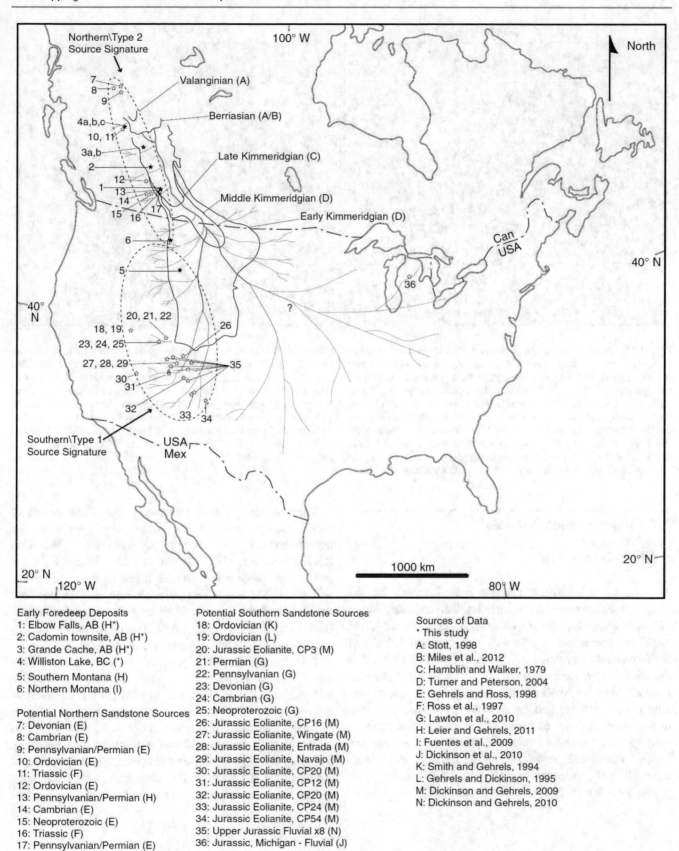

Early Foredeep Deposits
1: Elbow Falls, AB (H*)
2: Cadomin townsite, AB (H*)
3: Grande Cache, AB (H*)
4: Williston Lake, BC (*)
5: Southern Montana (H)
6: Northern Montana (I)

Potential Northern Sandstone Sources
7: Devonian (E)
8: Cambrian (E)
9: Pennsylvanian/Permian (E)
10: Ordovician (E)
11: Triassic (F)
12: Ordovician (E)
13: Pennsylvanian/Permian (H)
14: Cambrian (E)
15: Neoproterozoic (E)
16: Triassic (F)
17: Pennsylvanian/Permian (E)

Potential Southern Sandstone Sources
18: Ordovician (K)
19: Ordovician (L)
20: Jurassic Eolianite, CP3 (M)
21: Permian (G)
22: Pennsylvanian (G)
23: Devonian (G)
24: Cambrian (G)
25: Neoproterozoic (G)
26: Jurassic Eolianite, CP16 (M)
27: Jurassic Eolianite, Wingate (M)
28: Jurassic Eolianite, Entrada (M)
29: Jurassic Eolianite, Navajo (M)
30: Jurassic Eolianite, CP20 (M)
31: Jurassic Eolianite, CP12 (M)
32: Jurassic Eolianite, CP20 (M)
33: Jurassic Eolianite, CP24 (M)
34: Jurassic Eolianite, CP54 (M)
35: Upper Jurassic Fluvial x8 (N)
36: Jurassic, Michigan - Fluvial (J)

Sources of Data
* This study
A: Stott, 1998
B: Miles et al., 2012
C: Hamblin and Walker, 1979
D: Turner and Peterson, 2004
E: Gehrels and Ross, 1998
F: Ross et al., 1997
G: Lawton et al., 2010
H: Leier and Gehrels, 2011
I: Fuentes et al., 2009
J: Dickinson et al., 2010
K: Smith and Gehrels, 1994
L: Gehrels and Dickinson, 1995
M: Dickinson and Gehrels, 2009
N: Dickinson and Gehrels, 2010

Fig. 6.54 Proposed paleogeography showing the sediment sources and interpreted (schematic) rivers systems for the Western Interior Basin during the Late Jurassic and earliest Cretaceous (Raines et al. 2013, Fig. 11, p. 752)

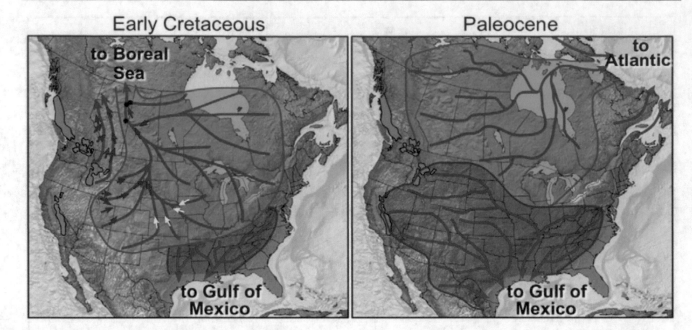

Fig. 6.55 Early Cretaceous to Paleocene continental-scale drainage reorganization. The major Early Cretaceous (Aptian, ca. 125–113 Ma) drainage system was sourced in the Appalachian Mountains, and routed through the Assiniboia paleovalley to the Boreal Sea (green). Tributaries joined from the southwestern United States and eastern Canada. Only a small part of North America south of the Appalachian-Ouachita cordillera (orange) was routed to the Gulf of Mexico. A similar map for the early Albian (ca. 113–108 Ma) would include tributaries from arc-dominated terrains in the northwestern United States and southwestern Canada. Red arrows indicate paleocurrents for basal Cretaceous fluvial deposits, whereas yellow arrows indicate orientations of bedrock paleovalleys that define the sub-Cretaceous unconformity. By the Paleocene, fluvial systems of the western United States were routed directly to the Gulf of Mexico, or to the Mississippi embayment, where they joined fluvial systems from the Appalachian Mountains: much of western Canada may have drained to the Atlantic. Locations of fluvial axes are schematic. Pinkish areas indicate locations of Cordilleran arc batholiths (after DeCelles 2004) (Blum and Pecha 2014, Fig. 4, p. 609)

6.7 Paleocurrent Analysis

6.7.1 Introduction

The art of paleocurrent analysis was invented by the quintessential nineteenth century English amateur, Henry Clifton Sorby, who published his first paper on the subject in 1852. He understood the formation of "ripple drifted" and "drift bedded" structures (climbing ripples and large-scale crossbedding, respectively) and, in seven years, claimed to have made about 20,000 observations, a total that has probably never been equaled. Most of Sorby's work remained unpublished. He was far ahead of his time, as paleocurrent work did not become a routine analytical procedure until the 1950s. His work remained unappreciated and largely unknown until exhumed by Pettijohn (1962) and reinterpreted by Allen (1963b) (see also Miall 1996, Chap. 2).

The technique could not become a routine component of basin analysis until proper attention was paid to the description and classification of sedimentary structures (McKee and Weir 1953; Pettijohn 1962; Allen 1963a, b), and a start was made on the investigation of bedform hydraulics. The major breakthrough here was the development of the flow-regime concept by Simons and Richardson (1961) and Simons et al. (1965). These two topics are discussed at some length in Sects. 1.2.5 and 3.5.4.

Paleocurrent analysis is primarily an outcrop study. Oriented core can be used, but is rarely available. Subsurface information may also be obtained with the use of borehole imaging tools. The dipmeter, which makes use of four resistivity logs run at 90° around a borehole, may be used to reconstruct the attitude of surfaces intersecting the hole, including bedding planes, and fractures. The orientation of crossbedding is, theoretically, obtainable from such data, and service companies have promoted this use for the tool

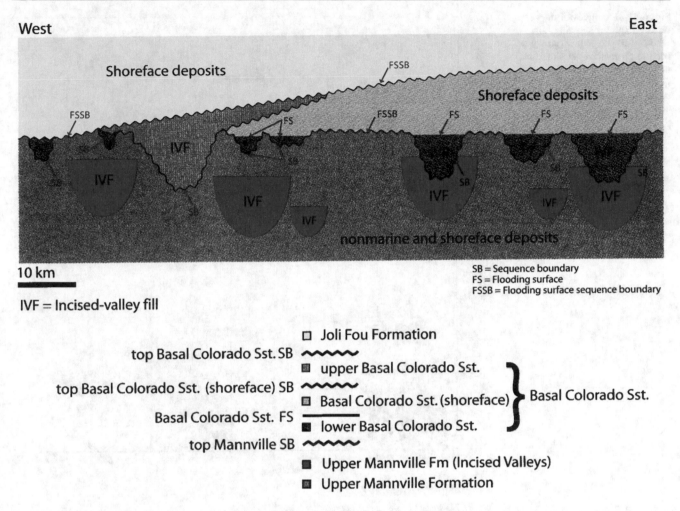

West

East

Shoreface deposits

FSSB

Shoreface deposits

FSSB

FS

IVF

FS

FSSB

FS

FS

FS

IVF

SB

IVF

IVF

SB

IVF

SB

IVF

SB

IVF

SB

IVF

nonmarine and shoreface deposits

10 km

IVF = Incised-valley fill

SB = Sequence boundary
FS = Flooding surface
FSSB = Flooding surface sequence boundary

☐ Joli Fou Formation

top Basal Colorado Sst. SB 〜〜〜
▨ upper Basal Colorado Sst.

top Basal Colorado Sst. (shoreface) SB 〜〜〜
▨ Basal Colorado Sst. (shoreface) } Basal Colorado Sst.

Basal Colorado Sst. FS ——
▨ lower Basal Colorado Sst.

top Mannville SB 〜〜〜

▨ Upper Mannville Fm (Incised Valleys)

▨ Upper Mannville Formation

Fig. 6.56 The stratigraphy of the Upper Mannville-Colorado Group (Albian) of south-central Alberta, showing the complexity of the numerous incised valley systems that developed in this low-accommodation setting (Wright et al. 2010, Fig. 2, p. 96)

since the 1970s (see also Miall 1999, Chap. 5), but thin-bed resolution may be a problem, and the tool has not often been used. Modern digital imaging tools, including Schlumberger's Formation MicroScanner (FMS) and other, more recently developed tools, provide a full scan of the hole, and may be more practical (Pöppelreiter et al. 2010). In petroleum exploration wells, unoriented core may be oriented using a dipmeter, but correlation between the two is commonly difficult (Davison and Haszeldine 1984; Nelson et al. 1987). Figure 6.60 illustrates a modern borehole image of planar-tabular crossbedding and the dips calculated from it, indicating consistent westerly to northwesterly flow directions. Borehole imaging has not been widely used, mainly because of cost considerations, but as Fox and Vickerman (2015) argued, this may be a false economy. For coal, metals and other types of mining, where there may be large surface

or underground outcrops, paleocurrent analysis may, in contrast, be extremely useful (e.g., Minter 1978; Long 2006).

Paleocurrent analysis can provide information on five main aspects of basin development:

1. The direction of local or regional paleoslope, which reflects tectonic subsidence patterns
2. The direction of sediment supply
3. The geometry and trend of lithologic units
4. The depositional environment
5. The architecture of bars and bedforms at the outcrop scale

Examples of these applications are discussed in the ensuing paragraphs.

Fig. 6.57 Examples of the binary diagrams showing how whole-rock geochemistry may be used to characterize stratigraphic units in the Mannville-Colorado Group (Aptian-Albian) of south-central Alberta (Wright et al. 2010, Fig. 3, p. 97)

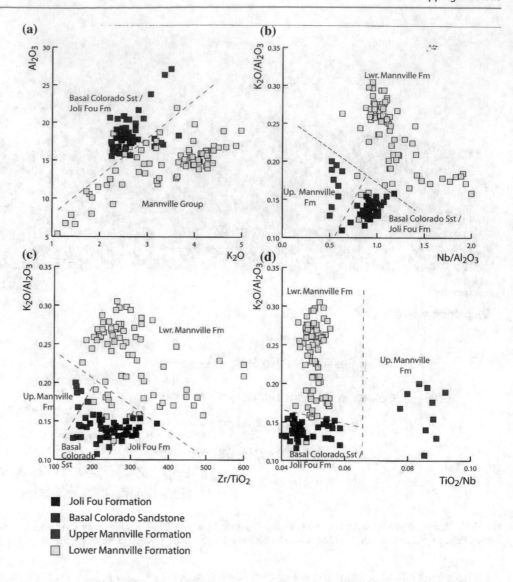

Joli Fou Formation
Basal Colorado Sandstone
Upper Mannville Formation
Lower Mannville Formation

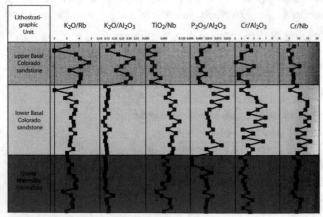

■ = Core samples, sandstone lithologies
◨ = upper Basal Colorado sandstone, incised-valley-fill sandstones
◻ = lower Basal Colorado sandstone, incised-valley-fill sandstones
■ = upper Mannville Formation, incised-valley-fill sandstones

Fig. 6.58 "Synthetic chemical logs" developed from the sampling of cores through the Upper Mannville Formation and Colorado Group of south-central Alberta (Wright et al. 2010, Fig. 6, p. 100)

6.7.2 Types of Paleocurrent Indicators

Sedimentary structures and fabrics used in facies and paleocurrent studies are described and illustrated in Sect. 2.2.2.6. The following notes explain briefly how they yield current directions:

1. For ripple marks and crossbedding (Figs. 2.8 and 2.9), the inclination of foreset directions is generally downcurrent, because of the grain avalanching mechanism. Needless to say, it is not always this simple. Smith (1972) demonstrated that planar-crossbed sets in rivers commonly advance obliquely to the flow direction. Trough crossbeds do not yield accurate flow directions unless the analyst can observe the orientation of the trough axis (see below). Hunter (1981) discussed the flow of air around large eolian dunes and suggested that these, too, commonly advance obliquely to the wind

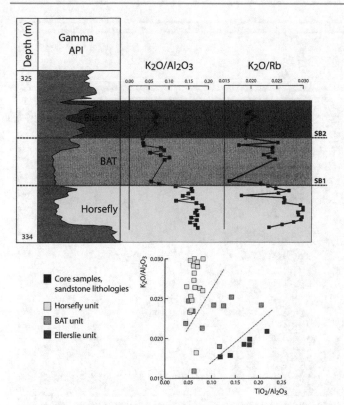

Fig. 6.59 Whole-rock geochemical data are used here to differentiate incised valley fill systems within a single well-bore through the Lower Manville Formation, south-central Alberta (Wright et al. 2010, Fig. 9, p. 105)

direction. The use of statistical procedures and appropriate facies data usually circumvents these problems (this is also discussed below).

2. Channels and scours (Fig. 2.13) occur in many environments and may indicate the orientation of major erosive currents, such as those generating river or tidal channels and delta or submarine fan distributaries. However, the larger channels, which are those most likely to be of regional significance, are usually too large to be preserved in outcrops. They may be readily apparent on seismic records, particularly 3-D, as discussed in this chapter (which obviates the need for paleocurrent studies!).

3. Parting lineation or primary current lineation, the product of plane-bed flow conditions (Fig. 2.9d), is only visible on bedding-plane exposures. It usually yields directional readings of low variance because it forms during high-energy flow in river, delta or tidal channels, when bars or other obstructions to flow are under water and flow sinuosity is low. The structure indicates orientation but not direction of flow, because of the ambiguity between two equally possible readings at 180° to each other. Usually, this can be resolved with reference to other structures nearby or to regional facies trends.

4. Clast transport by traction or in sediment gravity flows commonly produces a measurable gravel fabric. Imbrication in traction current deposits occurs where platy clasts are stacked up in a shingled pattern, with their flattest surface dipping upstream and resting on the next clast downstream (Fig. 2.12). Because gravel is only moved under high-energy flow conditions, it tends to show low directional variance, like parting lineation. Rust (1975) found that in gravel rivers variance decreased with increasing clast size, suggesting reduced flow sinuosity with increased stream power. Interpretation of clast fabrics in subaqueous, poorly sorted conglomerates (diamictites) can help distinguish depositional mechanisms, such as debris flow, ice rain-out and glacial lodgment, as well as provide paleocurrent information (Eyles et al. 1987).

5. Sole markings (Fig. 2.15) are typically associated with the deposits of turbidity currents and fluidized or liquified flow, where vortex erosion may occur at the base of a flow, and "tools" can be swept down a bedding surface. They also occur less commonly in other clastic environments. Their greatest use, however, is in the investigation of submarine-canyon and fan deposits. They are best seen on the undersides of bedding surfaces, where sandstone has formed a cast of the erosional feature in the underlying bed. Tool markings yield information on orientation but not direction, like parting lineation. Flute marks are longitudinally asymmetric, with their deepest end lying upstream (Fig. 2.16).

6. Oriented plants, bones, shells etc. do not respond systematically to the aligning effects of currents unless they are elongated. There may be ambiguity as to whether they are oriented transverse or perpendicular to current patterns, and there are other difficulties, for example, the tendency of fossils such as high-spired gastropods to roll in an arc. Fossils are usually only useful for local, specialized paleocurrent studies.

7. Slump structures generated on depositional slopes contain overfolds that may be aligned parallel to strike. They are therefore a potential paleoslope indicator, particularly on the relatively steep slopes of prodeltas and actively prograding continental margins. Friction at the margin of the slump may cause it to rotate out of strike alignment so that a statistical approach to measurement is desirable. Potter and Pettijohn (1977, Chap. 6), Rupke (1978) and Woodcock (1979) reviewed their use as paleoslope indicators.

Several other paleocurrent indicators are reviewed by Potter and Pettijohn (1977), including some, such as sand-grain orientation and magnetic anisotropy, requiring detailed laboratory analysis of oriented samples. They are

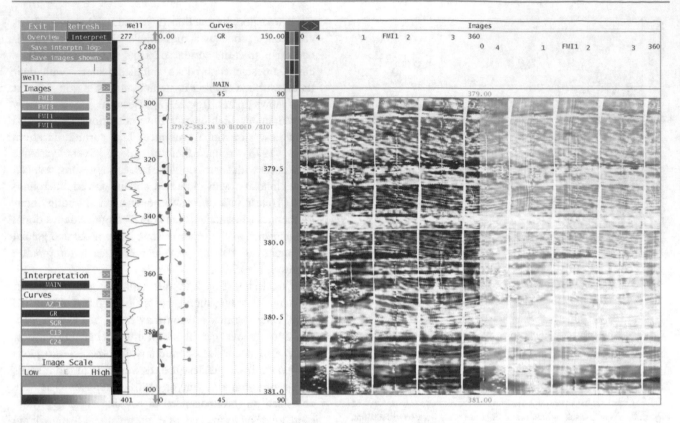

Fig. 6.60 A borehole image log, showing planar-tabular crossbedding and the dip readings obtained from image interpretation. The red tadpoles indicate stratification and the green tadpoles are from crossbed foresets. Two separate representations of the same scan are shown, which were developed using different false-color processing and visualization of the resistivity data. This image is of bitumen-saturated sandstone in the McMurray Formation of northern Alberta (Fox and Vickerman 2015, reproduced with permission)

mainly of academic interest, where they may help solve a particular local problem, although laboratory techniques have improved in recent years to the point where studies of magnetic fabric could become routine (Hrouda and Stránik 1985; Eyles et al. 1987). The seven types of paleocurrent indicators listed above can be quickly and routinely measured in the field and have been widely used in basin studies.

6.7.3 Data Collection and Processing

Paleocurrent data should be carefully documented in the field. In the past, paleocurrent trends have sometimes been reported on the basis of a geologist's mental estimates of the range of indicated directions, but even if this is correct, it may result in a significant loss of useful information. For every paleocurrent observation recorded in field notes, the following information should be included:

1. Location and (if relevant) precise position in a stratigraphic section
2. Structure type

3. Indicated current direction
4. Scale of structure (thickness of crossbed, depth and width of channel, mean or maximum clast size in imbricate gravels) and
5. Local structural dip.

Current directions should be measured to the nearest 5° with a magnetic or sun-compass and corrected to true north wherever necessary. In the case of parting lineation and tool markings, the correct orientation of two possibilities at 180° can usually be identified by referring to other types of current structures nearby. Measurement accuracy greater than ±5° is difficult to achieve and is, in any case, rarely important. Difficulties in field measurement commonly arise because of incomplete exposure of crossbed sets. For example, two-dimensional exposures in flat outcrop faces cannot provide precise orientation information. DeCelles et al. (1983) and Bradley (1983) discussed the problem of incomplete or partial exposure of trough crossbed sets and ripple cross-lamination, and offered some solutions involving field measurement techniques and statistical treatment of the data.

Indicated current directions may need to be corrected for structural dip and fold plunge, otherwise significant directional distortions may result (Ramsey 1961). This should be carried out as soon as possible, preferably after every day's work, in order that possible errors can be detected while there is still time to rectify them in the field. For linear structures such as parting lineation or sole markings, structural dips as high as 30° can be safely ignored, as they result in errors of less than 4°. However, the foreset dip orientation of crossbedding is significantly affected by structural dip, and should be corrected wherever the structural dip exceeds 10°. This subject is discussed further by Potter and Pettijohn (1977, Chap. 10), who illustrated a correction technique using a stereonet. Ramsey (1961) provided graphical solutions. Parks (1970) presented computer routines for the necessary trigonometric calculations. Wells (1988) provided a practical guide to field procedures, and recommended the use of a pocket calculator to perform structural corrections and data synthesis in the field.

Sooner or later the question will arise as to how many readings should be made? There is no single or simple answer to this problem because it depends on how many measurable current indicators are available for observation and what the objectives of the study are. Olson and Potter (1954) discussed the use of grid-sampling procedures and random selection of structures to measure, followed by calculation of reliability estimators to determine how many readings were necessary in order to be sure of determining correct directional trends. In this study, they were concerned only with determining regional paleoslope. We now understand a great deal about air and water flow patterns in different depositional environments, and a case could be made for measuring and recording every visible sedimentary structure. Such detailed data can be immensely useful in amplifying environmental interpretations and clarifying local problems, as discussed Sect. 6.7.5. Some selection may have to be made in areas of particularly good exposure. A practical compromise is to record every available structure along measured stratigraphic sections and to fill in the gaps between sections with spot (gridded or random) samples. This procedure permits the elaboration of both local and regional paleocurrent trends. When constructing lateral profiles of large outcrops, it is essential to document these with abundant paleocurrent determinations. Such data provide the third dimension to a two-dimensional outcrop and yield essential information regarding the shape and orientation of architectural elements (Sect. 3.5.11).

If the trend itself is important, 25 readings per sample station is commonly regarded as the minimum necessary for statistically significant small samples. However, the same or fewer readings plotted in map or section form can yield a great deal of environmental detail, whether or not their mean direction turns out to be statistically significant. Several

hundred or a few thousand readings may be necessary for a thorough analysis of a complete basin.

A variety of statistical data-reduction and data-display techniques is available for paleocurrent work. The commonest approach is to group data into subsets according to stratigraphic or areal distribution criteria, display them visually in current-rose diagrams, and calculate their mean and standard deviation (or variance). The method of grouping the data into subsets has an important bearing on the interpretations to be made from them, as discussed in the next section. A current rose diagram is simply a histogram converted to a circular distribution. The compass is divided into 20, 30, 40 or 45° segments, and the rose is drawn with the segment radius proportional to number of readings or percent of total readings. A visually more correct procedure

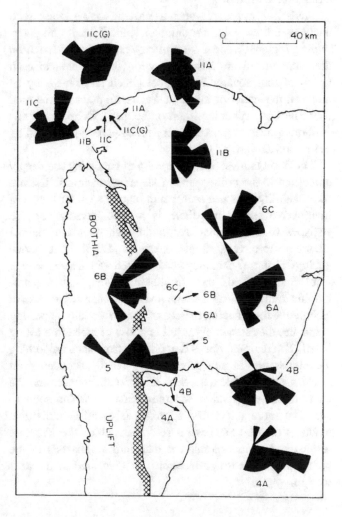

Fig. 6.61 Typical regional paleocurrent map, showing current rose diagrams and vector mean arrows for each of 10 outcrop areas, with station numbers. The map represents 165 field readings. Devonian fluvial sandstones, Somerset Island, Arctic Canada Most readings are consistent with an interpretation of deposition taking place on ephemeral sandy alluvial fans radiating from multiple point sources along the front of the Boothia Uplift. Station 11C(G) represents giant crossbedding, interpreted as eolian dune stratification (Miall and Gibling 1978)

is to draw the radius proportional to the square root of the percent number of readings so that the segment area is proportional to percent. Examples are given in Fig. 6.61. Wells (1988) warned that the choice of origin for rose diagrams and the method of subdivision of the segments can affect the visual appearance of the resulting diagram and the orientation of the modes. The calculation of means, which is done from raw data, is not affected, but Wells' warning indicates the need for caution in the visual interpretation of rose diagrams.

Potter and Pettijohn (1977) and Curray (1956) discussed arithmetic and vector methods for calculating mean, variance, vector strength and statistical tests for randomness. Miall (1976) and Cant and Walker (1976) demonstrated the display and interpretation of data collected from vertical stratigraphic sections.

Statistical procedures, such as moving averages and trend analysis, are available for smoothing local detail and determining regional trends. An arbitrary grid may be drawn on the map; the mean current direction for the data in each group of four squares is then calculated and shown by an arrow at the center of this area. Each data point is thus used four times (except at the edges of the map). Examples of this method applied to paleocurrent data have been published by Potter and Pettijohn (1977).

The basin analyst should be wary of becoming too deeply enmeshed in the refinements of statistical methods. The use of probability tests is a useful curb on one's wilder flights of interpretive fancy, but there is a vast literature on the statistics of the circular distribution that seems to detract attention from very simple questions: Do the data make geological sense? Can they be correlated with trends derived from other methods, such as lithofacies mapping or petrographic data? Moving average and trend analyses are useful techniques for reducing masses of data to visually appealing maps, but they inevitably result in a loss of much interesting detail. The regional trends that emerge from such smoothing techniques are, in any case, usually readily deduced from modern stratigraphic data, such as grain size trends, and the patterns of progradation and retrogradation within sequences. The more interesting approach is to relate individual readings to the architectural setting in which the structure resides and to incorporate the data into an analysis of the bedform and lithosome hierarchy, as discussed in the next section.

6.7.4 The Bedform Hierarchy

A geologist collects paleocurrent measurements over individual bedding planes, through a local sequence in a quarry, along a lengthy dipping section in a river cut, or from several such sections throughout a map area. These readings differ from each other by varying amounts at different levels of this sampling hierarchy, but what do these differences mean? Statistically this is a classic example of a problem in analysis of variance (Olson and Potter 1954; Kelling 1969). But what do the differences mean geologically? The larger the sampling area or the thicker the sampled section, the greater will be the number of depositional events that contributed to the data set, and normally it will mean an increase in directional variance. This generalization can be systematized by the concept of the **bedform hierarchy,** as first noted by Allen (1966, 1967). Flow fields and the sedimentary structures arising from them can be defined by bounding surfaces that constitute a six-fold hierarchy, ranging from individual ripples to entire depositional systems. Allen (1966) and Miall (1974) applied this idea to river systems (Figs. 3.67 and 6.62). In Table 3.3 an attempt is made to expand it to other environments. Jackson (1975) proposed that bedforms be grouped into three dynamic types:

1. **Microforms** are structures controlled by turbulent eddies in the fluid boundary layer. The time scale of variation is on the order of hours to seconds (deposits of SRS 1–3).
2. **Mesoforms** are structures that form in response to what Jackson (1975) termed **dynamic events**, such as hurricanes, seasonal floods or eolian sand storms, when disproportionately large volumes of sediment are moved in geologically instantaneous time periods. The system may remain virtually unchanged between dynamic events (SRS 4–5).
3. **Macroforms** represent the long-term accumulation of sediment in response to major tectonic, geomorphic and climatic controls, such as major bars and channels (SRS 6–7).

Data constituting the sedimentation rate scales (SRS) (see Sect. 8.4) indicate that the time scales represented by these three architectural categories range over ten orders of magnitude, from seconds to hundreds of years, with sedimentation rates varying over three orders of magnitude. It is important, also, to note that the bounding surfaces of bedforms, sequences and basin fills constitute a range of sedimentary breaks ranging from the almost instantaneous scour created by a migrating bedform, to the angular unconformity created by orogenesis (Sect. 7.6).

It is important for mapping purposes to attempt to relate the sampling scale to this hierarchy so that the sources of directional variance can be interpreted intelligently. For example, a small quarry might be located entirely within the deposit of a single point bar or tidal delta, whereas a township (6 miles or 9.6 km square) could encompass an entire delta lobe or barrier-inlet system. When interpreting

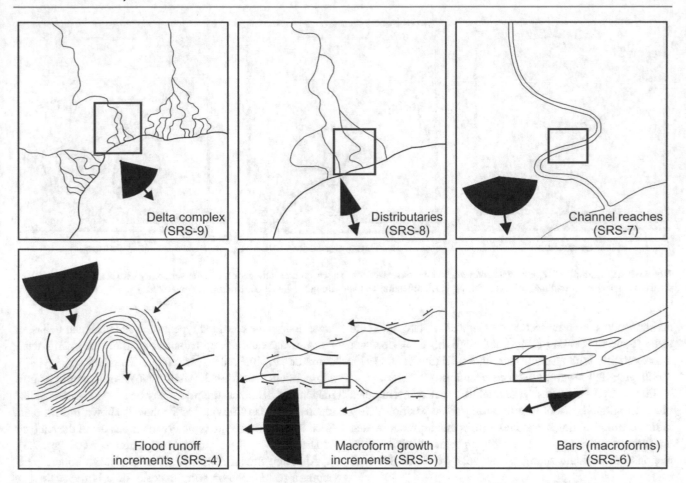

Fig. 6.62 Paleocurrent data plotted according to geographic scale, encompassing increasingly large areas moving anticlockwise from the bottom left. Each diagram corresponds to the area shown in the red square in the next diagram in an anticlockwise direction (adapted from Miall 1974)

microforms and mesoforms (the usual source of paleocurrent data) over large areas, it is important to remember that directional variance within each rank is summed to that of the higher ranks as far as is locally appropriate. Miall (1974) demonstrated the implications of this for fluvial–deltaic systems, such as those shown in Fig. 6.62, by compiling data for the various ranks summed to that of the single river or meander belt. A further discussion of the hierarchical nature of sedimentary facies is presented in Sect. 3.5.11

6.7.5 Environment and Paleoslope Interpretations

A recommended first level of analysis is to plot current-rose diagrams and mean vectors for each outcrop or local outcrop group, if necessary separating the readings according to major facies variations. An example is illustrated in Fig. 6.61. If contemporaneous basin shape and orientation can be deduced from these data or from other mapping techniques, the paleocurrent data can be used interactively

with lithofacies and biofacies criteria to interpret the depositional environment and to outline major depositional systems, such as deltas or submarine fan complexes. The geologist should examine the relationships between mean current directions in different lithofacies and in different outcrops. The number of modes in the rose diagrams (modality) and their orientation with respect to assumed shoreline or lithofacies contours are also important information. Another useful approach is to plot individual readings or small groups of readings collected in measured sections at the correct position in graphic section logs (e.g., Fig. 6.66). Some examples of how to use these data are discussed below. Potter and Pettijohn (1977) provided extensive examples and an annotated bibliography. Advanced methods for use in the analysis of fluvial systems are described by Miall (1996, Chap. 9).

Useful introductions to paleocurrent models have been published by Pettijohn (1962) and Selley (1968), but our increased understanding of clastic (siliciclastic, plus clastic-carbonate and evaporite) depositional systems requires a fresh look at the problem, because some of the

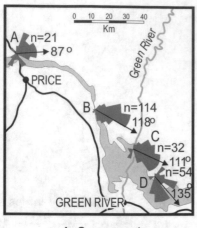

A. Sequence 1

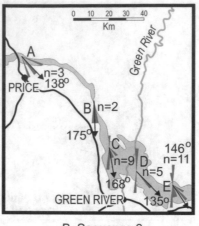

B. Sequence 2

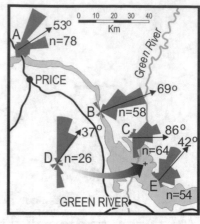

C. Sequence 3

Fig. 6.63 An example of regional paleocurrent distributions. Each map shows outcrop summaries, and the variability between the maps indicates significant changes in regional paleocurrent dispersal, reflecting tectonic tilting (Willis 2000; Miall and Arush 2001)

earlier models are now seen to be simplistic. Many ideas are to be found in Reading (1996), although this book does not deal explicitly with paleocurrent models. Harms et al. (1975, 1982) provided several useful case studies.

Figure 6.63 provides examples of paleocurrent data plotted regionally according to stratigraphic position. They indicate broad regional changes in dispersal directions that are interpreted in terms of changes in regional dispersal in response to tectonic tilting of the margin of the foreland basin.

Local paleocurrent distributions are commonly categorized as unimodal, bimodal, trimodal and polymodal. Each reflects a particular style of current dispersion. For example, the rose diagrams in Fig. 6.61 are unimodal to weakly bimodal, and vector mean directors are all oriented in easterly directions, roughly normal to the basin margin, with the exception of station 11C(G). The data were derived mainly from trough and planar crossbedding and parting lineation, and the environment of deposition is interpreted as braided fluvial (Miall and Gibling 1978). In some areas, for example, stations 6A, B and C, there is a suggestion of a fanning out of the current systems, indicating possible deposition on large, sandy alluvial fans. The anomalous data set of station 11C(G) was derived from giant crossbed sets ranging from 1 to 6 m in thickness. They are interpreted as eolian dunes.

In high-sinuosity rivers, point bars dip at high angles to channel trends, whereas minor structures (mesoforms and microforms) migrating down the point bars are oriented subparallel to channel direction and therefore parallel to the strike of point-bar (epsilon) crossbedding (Fig. 6.64). This paleocurrent pattern is a useful diagnostic criterion for recognizing point bars and distinguishing them from other

large, low-angle crossbed types, such as Gilbertian deltas, or the downstream-accretion surfaces that are particularly common in braided rivers (e.g., Puigdefábregas 1973; Mossop and Flach 1983; Miall 1996). An example of how meticulous field measurements may be used to make detailed reconstructions of fluvial architecture is shown in Fig. 6.65. Such measurements are typically made as part of the analysis of large two- or three-dimensional outcrops (e.g., Fig. 2.3).

Although fluvial deposits typically yield unimodal paleocurrent patterns on an outcrop scale, on a larger scale they may show much more complex patterns, for example centripetal patterns indicating internal drainage (e.g., Friend and Moody-Stuart 1972) or bimodal patterns with the two modes at 90° and occurring in different lithofacies assemblages. Rust (1981) reported two examples of the latter case. In each case, one mode, occurring in coarse conglomerates, was interpreted as the product of alluvial fans prograding transversely out from the basin margin. The other mode, in interbedded sandstones, represented a trunk-river system draining longitudinally.

Coastal regions where rivers debouch into an area affected by waves and tides can give rise to very complex paleocurrent patterns (Fig. 4.14). Bimodal, trimodal or polymodal distributions may result, for example, in the tide-swept offshore bar described by Klein (1970). However, time-velocity asymmetry of tidal currents can result in local segregation of currents so that they are locally unimodally ebb- or flood-dominated (Fig. 3.27). This could cause confusion at the outcrop level, because the tidal deltas showing such paleocurrent patterns may consist of lithofacies that are very similar to some fluvial deposits (although recent work has demonstrated the existence of a range of very subtle

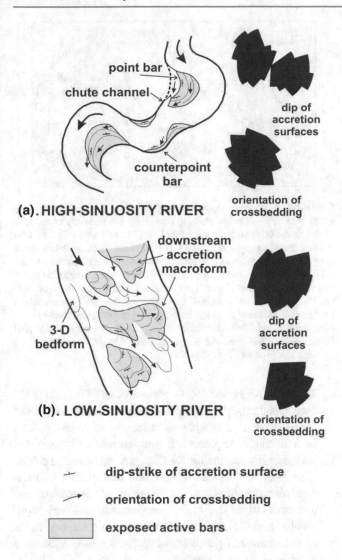

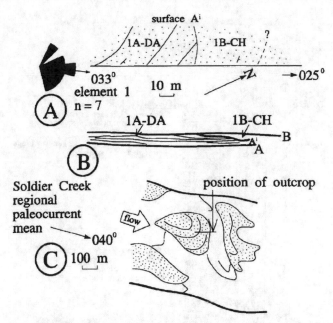

Fig. 6.65 An example of the type of architectural interpretation that may be made from the analysis of a vertical outcrop face, using paleocurrent data. A sketch of the relevant portion of the outcrop is shown in diagram B. Diagram A, above, shows a reconstruction of a plan view of the top bounding surface (surface B), with the suggested orientations of accretion surface, based on the measurements of dip-and-strike made at the outcrop face (dip-and-strike symbols). Diagram C shows the suggested interpretation of the outcrop, incorporating all orientation measurements of crossbedding (summarized in the rose diagram at top left). This is one of several worked examples provided by Miall (1994)

Fig. 6.64 Hypothetical examples of fluvial styles, indicating the range of orientations of dipping accretion surfaces and the variance of crossbed orientations. **a** High-sinuosity river, such as a typical meandering river. Note that although there is a regional overlap of crossbed and accretion-surface orientations, locally the two dips are oriented nearly perpendicular to each other, and the rose diagrams reflect this divergence. A counterpoint bar is shown. Such bars can be distinguished from the more common point bars by the fact that the accretion surfaces show curvature in plan view that is concave in a downdip direction, contrasting with the convex curvature of point-bar surfaces. Commonly such bars are characterized by finer grain sizes than other macroforms. **b** Low sinuosity river—a typical sandy-braided river. Such rivers may show relatively high local channel sinuosity, in which case macroforms accreting downstream may also be oriented at a high angle to the regional trend. In the example shown here, mean directions are skewed to the southeast, diverging from the overall south-southeast channel orientation because of a major channel and bar complex (near the center of the reach) oriented in an eastward direction (Miall 1994)

facies criteria that can be used to recognize a tidal signature: see Sect. 4.5.4). In wave-dominated environments, current reversals generate distinctive internal ripple-lamination patterns and herringbone crossbedding, as illustrated in Figs. 2.9c and 3.27. Paleocurrent analyses of these bimodal crossbeds may yield much useful information on the direction of wave attack and hence on shoreline orientation. The term paleoslope means little in this environment because there are a diversity of local slopes and because waves and tides are only marginally influenced by the presence and orientation of bottom slopes.

Several examples of sandy shoreline deposits and their paleocurrent patterns are given by Harms et al. (1975, 1982). One of the most instructive is the section through the Cretaceous Gallup Sandstone, New Mexico, illustrated in Fig. 6.66. Individual trough and ripple orientations are shown at the left (north toward top). The shoreline is known from regional mapping to be oriented northwest-southeast. The top 2–3 m of section consists of low-angle, cross-stratified, beach-accretion sets dipping at a few degrees toward the northeast (offshore). This structure is typical of intertidal wave-swash zones (Fig. 4.14) Facies 2 (about 3–14 m in depth) consists of fine- to medium-grained sandstone with trough and some planar cross-stratification. Foreset-dip and trough-axis orientations are mainly toward the northwest, southwest and south to southeast, with the

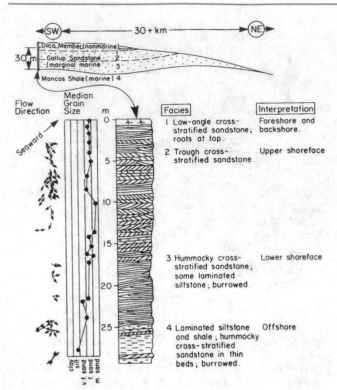

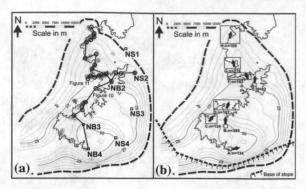

Fig. 6.67 **a** Isopach thickness map of Fan 3 constrained from integration of the core and outcrop data. Note the high rate of thinning to the south compared to the rest of the isopach map. Circles with crosses are outcrop logs, squares are borehole sites. **b** Paleocurrent arrows illustrate the general paleocurrent readings from Fan 3 across the entire study area from outcrop and Formation MicroImager measurements from boreholes. Overall the trend is to the north; however, note the strong eastward trend in the south of the study area where stratigraphic thinning rates are highest. The east-directed readings are interpreted as turbidity-current directions prior to deflection northwards of an ENE–WSW trending slope, possibly the basin-margin slope (Hodgson et al. 2006, Fig. 9, p. 28)

Fig. 6.66 Lithofacies profile, crossbed data and interpreted depositional environments of the Gallup Sandstone (Cretaceous), New Mexico (Harms et al. 1975)

latter predominating, suggesting onshore-directed, wave-generated currents with local reversals and longshore currents directed toward the southeast. The same longshore current pattern is apparent in the subtidal lower-shoreface deposits, which are dominated by storm-generated hummocky cross-stratification. Wave ripples are present in the lowest facies, again showing predominantly onshore-directed currents.

Submarine-fan and other deep-marine deposits show three main paleocurrent patterns:

1. Individual submarine fans prograde out from the continental slope and therefore show radial paleocurrent patterns with vector mean directions oriented perpendicular to the regional basin strike (e.g., Stow et al. 1996; see Fig. 6.67). This pattern is typical of many continental margins, particularly divergent margins.
2. In narrow oceans and many arc-related basins, deep-water sedimentation takes place in a trough oriented parallel to tectonic strike. Sediment gravity flows, particularly low-viscosity turbidity currents, emerge from submarine fans and turn 90°, to flow longitudinally downslope, possibly for hundreds of kilometers. Current directions may be reversed by tilting of the basin. Many examples of this pattern have been published (e.g.,

Trettin 1970; Hesse 1974). Pickering and Hiscott (1985) used detailed grain-orientation measurements as a basis for reinterpreting anomalous paleocurrent measurements in a turbidite sequence. It was shown that supposed backset cross-lamination, attributed to upcurrent antidune migration, was in fact the product of turbidity currents that reversed direction as a result of deflection and reflection from the basin margins and internal topographic highs. Here is a case where subtle use was made of paleocurrent data to improve the sophistication of a basin-analysis case study.

3. Contour currents or boundary undercurrents flow parallel to continental margins and generate paleocurrent patterns oriented parallel to basin margins. Stow and Lovell (1979) and Stow and Holbrook (1984) discussed the use of facies criteria and paleocurrent data in distinguishing these deposits.

References

Ahlborn, M., Stemmerik, L., and Kalstø, T.-K., 2014, 3D seismic analysis of karstified interbedded carbonates and evaporites, Lower Permian Gipsdalen Group, Loppa High, southwestern Barents Sea: Marine and Petroleum Geology, v. 56, p. 16-33.

Allen, J. R. L., 1963a, The classification of cross-stratified units, with notes on their origin: Sedimentology, v. 2, p. 93–114.

Allen, J. R. L., 1963b, Henry Clifton Sorby and the sedimentary structures of sands and sandstones in relation to flow conditions: Geologie en Mijnbouw, v. 42, p. 223–228.

Allen, J. R. L., 1966, On bed forms and paleocurrents: Sedimentology, v. 6, p. 153–190.

Allen, J. R. L., 1967, Notes on some fundamentals of paleocurrent analysis, with reference to preservation potential and sources of variance: Sedimentology, v. 9, p. 75–88.

Allen, P., 1972, Wealden detrital tourmaline: implications for north-western Europe: Journal of the Geological Society of London, v. 128, p. 273–294.

American Association of Petroleum Geologists, 1999, A century: AAPG Explorer, Special Issue: Tulsa, Oklahoma, American Association of Petroleum Geologists.

Angulo, S., and Buatois, L. A., 2012, Integrating depositional models, ichnology, and sequence stratigraphy in reservoir characterization: The middle member of the Devonian-Carboniferous Bakken Formation of subsurface southeastern Saskatchewan revisited: American Association of Petroleum Geologists Bulletin, v. 96, p. 1017–1043.

Bally, A. W., ed., 1987, Atlas of seismic stratigraphy: American Association of Petroleum Geologists Studies in Geology 27, in 3 vols.

Basu, A., 1985, Influence of climate and relief on composition of sands released at source areas, in Zuffa, G. G., ed., Provenance of arenites: D. Reidel Publishing Company, Dordrecht, p. 1-18.

Berg, R. R., 1968, point-bar origin of Fall River Sandstone reservoirs, northeastern Wyoming: American Association of Petroleum Geologists Bulletin, v. 52, p. 2116–2122.

Bhattacharya, J. P., 2011, Practical problems in the application of the sequence stratigraphic method and key surfaces: integrating observations from ancient fluvial-deltaic wedges with Quaternary and modelling studies: Sedimentology, v. 58, p. 120–169.

Blatt, H., Middleton, G. V., and Murray, R., 1980, Origin of sedimentary rocks; 2nd edition: Prentice-Hall Inc., Englewood Cliffs, New Jersey, 782 p.

Blum, M., and Pecha, M., 2014, Mid-Cretaceous to Paleocene North American drainage reorganization from detrital zircons: Geology, v. 42, p. 607–610.

Bohacs, K. M., Neal, J. E., and Grabowski, G. J., Jr., 2002, Sequence stratigraphy in fine-grained rocks: Beyond the correlative conformity: 22nd Annual Gulf Coast Section SEPM Foundation Bob F. Perkins Research Conference, p. 321–347.

Bouma, A. H., Normark, W. R., and Barnes, N. E., eds., 1985, Submarine fans and related turbidite systems: Springer-Verlag Inc., Berlin and New York, 351 p.

Bouma, A. H., Stelting, C. E., and Feeley, M. H., 1987, High-resolution seismic reflection profiles, in Bally, A. W., ed., Atlas of seismic stratigraphy: American Association of Petroleum Geologists Studies in Geology 27, v. 1, p. 72–94.

Bradley, D. C., 1983, Paleocurrent directions from two-dimensional exposures of cross laminae in the Devonian flysch of Maine: Journal of Geology, v. 95, p. 271–279.

Bridge, J. S., and Mackey, S. D., 1993, A theoretical study of fluvial sandstone body dimensions, in Flint, S. S., and Bryant, I. D., eds., The geological modelling of hydrocarbon reservoirs and outcrop analogues: International Association of Sedimentologists Special Publication 15, p. 213–236.

Bridge, J. S., and Tye, R. S., 2000, interpreting the dimensions of ancient fluvial channel bars, channels, and channel belts from wire-line logs and cores: American Association of Petroleum Geologists Bulletin, v. 84, p. 1205–1228.

Bristow, J. F., 2002, Real-time formation evaluation for optimal decision making while drilling: examples from the southern North Sea, in Lovell, M., and Parkinson, N., eds., Geological applications of well logs: American Association of Petroleum Geologists Methods in Exploration 13, p. 1–13.

Brown, A. R., 1985, The role of horizontal seismic sections in stratigraphic interpretation, in Berg, O. R., and Woolverton, D. G., eds., Seismic stratigraphy II: American Association of Petroleum Geologists Memoir 39, p. 37–47.

Brown, A. R., 1986, Interpretation of three-dimensional seismic data: American Association of Petroleum Geologists Memoir 42, 194 p.

Brown, A. R., 2011, Interpretation of three-dimensional seismic data, seventh edition, American Association of Petroleum Geologists Memoir 42, 646 p.

Burgess, P. M., Winefield, P., Minzoni, M., and Elders, C., 2013, Methods for identification of isolated carbonate buildups from seismic reflection data: American Association of Petroleum Geologists, v. 97, p. 1071–1098.

Cant, D. J., 1992, Subsurface facies analysis, in Walker, R. G., and James, N. P., eds., Facies models: response to sea level change: Geological Association of Canada, St. John's, Newfoundland, p. 27–45.

Cant, D. J., and Walker, R. G., 1976, Development of a braided-fluvial facies model for the Devonian Battery Point Sandstone, Quebec; Canadian Journal of Earth Sciences, v. 13, p. 102-119.

Cerveny, P. F., Naeser, N. D., Zeitler, P. K., Naeser, C. W., and Johnson, N. M., 1988, History of uplift and relief of the Himalaya during the past 18 million years: evidence from fission-track ages of detrital zircons from sandstones of the Siwalik Group, in Kleinspehn, K. L., and Paola, C., eds., New perspectives in basin analysis: Springer-Verlag Inc., Berlin and New York, p. 43–61.

Craddock, W. H., Coleman, J. L., and Kylander-Clark, A. R. C., 2021, Detrital zircon age spectra of middle and upper Eocene outcrop belts, U.S. Gulf Coast region: Basin Research, v. 33, p. 250-269.

Cramer, B. D., Vandenbroucke, T. R. A., and Ludvigson, G. A., 2015, High-resolution event stratigraphy (HiRES) and the quantification of stratigraphic uncertainty: Silurian examples of the quest for precision in stratigraphy: Earth Science Reviews, v. 141, p. 136–153.

Cross, T. A., and Lessenger, M. A., 1988, Seismic stratigraphy: Annual Review of Earth and Planetary Sciences, v. 16, p. 319–354.

Curray, J. R., 1956, The analysis of two-dimensional orientation data: Journal of Geology, v. 64, p. 117–131.

Davies, R. J., Cartwright, J. A., Stewart, S. A., Lappin, M., and Underhill, J. R., eds., 2004a, 3D Seismic technology: Application to the exploration of sedimentary Basins: Geological Society, London, Memoir 29, 355 p.

Davies, R. J., Stewart, S. A., Cartwright, J. A., Lappin, M., Johnston, R., Fraser, S. I., and Brown, A. R., 2004b, 3D seismic technology: are we realising its full potential? In Davies, R. J., Cartwright, J. A., Stewart, S. A., Lappin, M., and Underhill, J. R., eds., 3D Seismic technology: Application to the exploration of sedimentary Basins: Geological Society, London, Memoir 29, p. 1–9.

Davies, R. J., Posamentier, H. W., Wood, L. J., and Cartwright, J. A., eds., 2007, Seismic geomorphology: applications to hydrocarbon exploration and production: Geological Society, London, Special Publication 277, 274 p.

Davison, I., and Haszeldine, R. S., 1984, Orienting conventional cores for geological purposes: a review of methods: Journal of Petroleum Geology, v. 7, p. 461–466.

DeCelles, P. G., 2004, Late Jurassic to Eocene evolution of the Cordilleran thrust belt and foreland basin system, western USA: American Journal of Science, v. 304, p. 105–168.

DeCelles, P., Langford, R. P., and Schwartz, R. K., 1983, Two new methods of paleocurrent determination from trough cross-stratification: Journal of Sedimentary Petrology, v. 53, p. 629–642.

Dickinson, W. R., 1970, Interpreting detrital modes of graywacke and arkose: Journal of Sedimentary Petrology, v. 40, p. 695–707.

Dickinson, W. R., and Gehrels, G. E., 2003, U-Pb ages of detrital zircons from Permian and Jurassic eolian sandstones from the Colorado Plateau, USA: paleogeographic implications: Sedimentary Geology, v. 163, p. 29–66.

Duk-Rodkin, A., Hughes, O. L., 1994. Tertiary–Quaternary drainage of the pre-glacial Mackenzie Basin. Quaternary International v. 22/ 23, p. 221-241.

Ethridge, F. G., and Schumm, S. A., 2007, Fluvial seismic geomorphology: a view from the surface, in Davies, R. J., Posamentier, H. W., Wood, L. J., and Cartwright, J. A., eds., Seismic geomorphology: applications to hydrocarbon exploration and production: Geological Society, London, Special Publication 277, p. 205–222.

Eyles, N., Day, T. E., and Gavican, A., 1987, Depositional controls on the magnetic characteristics of lodgement tills and other glacial diamict facies: Canadian Journal of Earth Sciences, v. 24, p. 2436–2458.

Fielding, C. R., Whittaker, J., Henrys, S. A., Wilson, T. J., and Naish, T. R., 2008, Seismic facies and stratigraphy of the Cenozoic succession in McMurdo Sound, Antarctica: implications for tectonic, climatic and glacial history, Palaeogeography, Palaeoclimatology, Palaeoecology, v. 260, p. 8-29.

Folk, R. L., 1968. Petrology of Sedimentary Rocks. Hemphill's, Austin, 170 p.

Fox, A., and Vickerman, K., 2015, The value of borehole image logs: Reservoir, Canadian Society of Petroleum Geologists, January, p. 30–34.

Franzel, M., Jones, S. J., Meadows, N., Allen, M. B., McCaffrey, K., and Morgan, T., 2020, Basin-scale fluvial correlation and response to the Tethyan marine transgression: an example from the Triassic of central Spain: Basin Research, v. 33, p. 1–25.

Franzinelli, E., and Potter, P. E., 1983, Petrology, chemistry and texture of modern river sands, Amazon river system: Journal of Geology, v. 91, p. 23-40.

Friend, P.F., and Moody-Stuart, M., 1972, Sedimentation of the Wood Bay Formation (Devonian) of Spitsbergen: regional analysis of a late orogenic basin: Norsk Polarinstitutt Skrifter Nr. 157.

Futalan, K., Mitchell, A., Amos, K., and Backe, G., 2012, Seismic facies analysis and structural intepretation of the Sandakan sub-basin, Sulu Sea, Philippines: American Association of Petroleum Geologists, Search and Discovery Article 30254.

Galey, J. T. 1985, The anticlinal theory of oil and gas accumulation: its role in the inception of the natural gas and modern oil industries in North America, in Drake, E. I., and Jordan, W. M., eds., Geologists and ideas: A history of North American Geology: Geological Society of America Centennial Special Volume 1, p. 423–442.

Galloway, W. E., 1981, Depositional architecture of Cenozoic Gulf Coastal Plain fluvial systems, in Ethridge, F. G., and Flores, R. M., eds., Recent and ancient nonmarine depositional environments: models for exploration: Society of Economic Paleontologists and Mineralogists Special Publication 31, p. 127–156.

Garzanti, E., Vermeesch, P., Rittner, M., and Simmons, M., 2018, The zircon story of the Nile: time-structure maps of source rocks and discontinuous propagation of detrital signals: Basin Research, v. 30, p. 1098–1117.

Gehrels, G. E., Dickinson, W. R., Ross, G. M., Stewart, J. H., Howell, D. G., 1995. Detrital zircon reference for Cambrian to Triassic miogeoclinal strata of western North America. Geology v. 23, p. 831-834.

Gennaro, M., Wonham, J. P., Gawthorpe, R. and Saelen, G., 2013, Seismic stratigraphy of the Chalk Group in the Norwegian Central Graben, North Sea: Marine and Petroleum Geology, v. 45, p. 236-266.

Graham, S. A., Tolson, R. B., DeCelles, P. G., Ingersoll, R. V., Bargar, E., Caldwell, M., Cavazza, W., Edwards, D. P., Follo, M. F., Handschy, J. F., Lemke, L., Moxon, I., Rice, R., Smith, G. A., and

White, J., 1986, Provenance modeling as a technique for analyzing source terrane evolution and controls on foreland sedimentation, in Allen, P. A., and Homewood, P., eds., Foreland basins: International Association of Sedimentologists Special Publication 8, p. 425–436.

Hailwood, E. A., and Kidd, R. B., eds., 1993, High resolution stratigraphy: Geological Society, London, Special Publication 70, 357 p.

Hardage, B. A., 1985, Vertical seismic profiling—measurement that transfers geology to geophysics, in Berg, O. R., and Woolverton, D. G., eds., Seismic stratigraphy II: American Association of Petroleum Geologists Memoir 39, p. 13–34.

Harms, J. C., Southard, J. B., Spearing, D. R., and Walker, R. G., 1975, Depositional environments as interpreted from primary sedimentary structures and stratification sequences: Society of Economic Paleontologists and Mineralogists Short Course 2, 161 p.

Harms, J. C., Southard, J. B., and Walker, R. G., 1982, Structures and sequences in clastic rocks: Society of Economic Paleontologists and Mineralogists Short Course 9, Calgary.

Hart, B. S., 2013, Whither seismic stratigraphy: Interpretation: A journal of subsurface characterization: Society of Exploration Geophysicists and American Association of Petroleum Geologists, v. 1, p. SA3-SA20.

Heim, D., 1974, Uber die feldspate in Germanischen Buntsandstein, ihre Korngrossenabhangigkeit, verbreitung undpaleogeographische bedeutung: Geol. Rundschau, v. 63, p. 943-970.

Heller, P. L., and Frost, C. D., 1988, Isotopic provenance of clastic deposits: applications of geochemistry to sedimentary provenance studies, in Kleinspehn, K. L., and Paola, C., eds., New perspectives in basin analysis: Springer-Verlag, New York, p. 27–42.

Helmold, K. P., 1985, Provenance of feldspathic sandstones—the effect of diagenesis on provenance interpretations: a review: in Zuffa, G. G., ed., Provenance of arenites: D. Reidel Publishing Company, Dordrecht, p. 139–163.

Hesse, R., 1974, Long-distance continuity of turbidites: possible evidence for an Early Cretaceous trench-abyssal plain in the East Alps: Geological Society of America Bulletin, v. 85, p. 859–870.

Hildred, G. V., Martinez-Kulikowski, N., and Zaitlin, B. A., 2014, forensic chemostratigraphy: A tool to determine lateral wellbore placement: Search and Discovery Article 41359.

Hodgson, D. M., Flint, S. S., Hodgetts, D., Drinkwater, N. J., Johannessen, E. P., and Luthi, S. M., 2006, Stratigraphic evolution of fine-grained submarine fan systems, Tanqua depocenter, Karoo Basin, South Africa: Journal of Sedimentary Research, v. 76, p. 20-40.

Howard, J. L., 1993, The statistics of counting clasts in rudites: a review, with examples from the upper Paleogene of southern California, USA: Sedimentology, v. 40, p. 157–174.

Hrouda, F., and Stránik, Z., 1985: The magnetic fabric of the Zdanice thrust sheet of the Flysch belt of the West Carpathians: sedimentological and tectonic implications: Sedimentary Geology, v. 45, p. 125–145.

Hubbard, S. M., Smith, D. G., Nielsen, H., Leckie, D. A., Fustic, M., Spencer, R. J., and Bloom, L., 2011, Seismic geomorphology and sedimentology of a tidally influenced river deposit, Lower Cretaceous Athabasca oil sands, Alberta, Canada: American Association of Petroleum Geologists Bulletin, v. 95, p. 1123-1145.

Hunter, R. E., 1981, Stratification styles in eolian sandstones: some Pennsylvanian to Jurassic examples from the Western Interior U.S. A., in Ethridge, F. G., and Flores, R., eds., Recent and ancient nonmarine depositional environments: models for exploration: Society of Economic Paleontologists and Mineralogists Special Publication 31, p. 315–329.

Ingersoll, R. V., Bullard, T. F., Ford, R. L., Grimm, J. P., Pickle, J. D., and Sares, S. W., 1984, The effect of grain size on detrital modes: a test of the Gazzi-Dickinson point-counting method: Journal of Sedimentary Petrology, v. 54, p. 103–116.

Jackson, R. G., II, 1975, Hierarchical attributes and a unifying model of bed forms composed of cohesionless material and produced by shearing flow: Geological Society of America Bulletin, v. 86, p. 1523–1533.

Kauffman, E. G., 1986, High-resolution event stratigraphy: regional and global bio-events, in Walliser, O. H. ed., Global bioevents: Lecture Notes on Earth History: Springer-Verlag, Berlin, p. 279–335.

Kauffman, E. G., 1988, Concepts and methods of high-resolution event stratigraphy: Annual Reviews of Earth and Planetary Sciences, v. 16, p. 605–654.

Kelling, G., 1969, The environmental significance of cross-stratification parameters in an Upper Carboniferous fluvial basin: Journal of Sedimentary Petrology, v. 39, p. 857–875.

Klein, G. deV., 1970, Depositional and dispersal dynamics of intertidal sand bars: Journal of Sedimentary Petrology, v. 40, p. 1095–1127.

Krumbein, W. C., 1948, Lithofacies maps and regional sedimentary-stratigraphic analysis: American Association of Petroleum Geologists, v. 32, p. 1909–1923.

Leckie, D. A., Bhattacharya, J., Gilboy, C. F. and Norris, B., 1994, Colorado\Alberta Group strata of the Western Canada Sedimentary Basin, in: Mossop, G. D., and Shetsen, I., eds., Geological Atlas of the Western Canada Sedimentary Basin: Canadian Society of Petroleum Geologists and Alberta Research Council, Calgary. p. 335–352.

Locklair, R. E., and Sageman, B. B., 2008, Cyclostratigraphy of the Upper Cretaceous Niobrara Formation, Western Interior, U.S.A.: A Coniacian-Santonian timescale: Earth and Planetary Science Letters, v. 269, p. 539–552.

Long, D. G. F., 2006, Architecture of pre-vegetation sandy-braided perennial and ephemeral river deposits in the Paleoproterozoic Athabasca Group, northern Saskatchewan, Canada as indicators of Precambrian fluvial style: Sedimentary Geology, v. 190, p. 71–95.

Martin, J. H., 1993, A review of braided fluvial hydrocarbon reservoirs: the petroleum engineer's perspective, in: Best, J. L., and Bristow, C. S., eds., Braided rivers: Geological Society, London, Special Publication 75, p. 333–367.

Martinez, J. F., Cartwright, J. A., Burgess, P. M., and Bravo, J. V., 2004, 3D seismic interpretation of the Messinian unconformity in the Valencia Basin, Spain, in Davies, R. J., Cartwright, J. A., Stewart, S. A., Lappin, M., and Underhill, J. R., eds., 3D Seismic technology: Application to the exploration of sedimentary Basins: Geological Society, London, Memoir 29, p. 91–100.

Matter, A., and Ramseyer, K., 1985, Cathodoluminescence microscopy as a tool for provenance studies of sandstones, in Zuffa, G. G., ed., Provenance of arenites: D. Reidel Publishing Company, Dordrecht, p. 191-211.

McCrossan, R. G., 1961, Resistivity mapping and petrophysical study of Upper Devonian inter-reef calcareous shales of central Alberta, Canada: American Association of Petroleum Geologists Bulletin, v. 45, p. 441–470.

McKee, E. D., and Weir, G. W., 1953, Terminology for stratification in sediments: Geological Society of America Bulletin, v. 64, p. 381–389.

McMillan, N.J., 1973, Shelves of Labrador Sea and Baffin Bay, Canada, in: McCrossan, R. G., ed., The Future Petroleum Provinces of Canada; Their Geology and Potential, Canadian Society of Petroleum Memoir 1, p. 473–517.

Miall, A. D., 1974, Paleocurrent analysis of alluvial sediments: a discussion of directional variance and vector magnitude: Journal of Sedimentary Petrology, v. 44, p. 1174–1185.

Miall, A. D., 1976, Palaeocurrent and palaeohydrologic analysis of some vertical profiles through a Cretaceous braided stream deposit: Sedimentology, v. 23, p. 459–483.

Miall, A. D., 1979, Mesozoic and Tertiary geology of Banks Island, Arctic Canada: the history of an unstable craton margin: Geological Survey of Canada Memoir 387.

Miall, A. D., 1994, Reconstructing fluvial macroform architecture from two-dimensional outcrops: examples from the Castlegate Sandstone, Book Cliffs, Utah: Journal of Sedimentary Research, v. B64, p. 146-158.

Miall, A. D., 1996, The geology of fluvial deposits: sedimentary facies, basin analysis and petroleum geology: Springer-Verlag Inc., Heidelberg, 582 p.

Miall, A. D., 1999, Principles of sedimentary basin analysis, Third edition: New York, N. Y.: Springer-Verlag Inc., 616 p.

Miall, A. D., 2002, Architecture and sequence stratigraphy of Pleistocene fluvial systems in the Malay Basin, based on seismic time-slice analysis: American Association of Petroleum Geologists Bulletin, v. 86, p. 1201–1216.

Miall, A. D., 2006, Reconstructing the architecture and sequence stratigraphy of the preserved fluvial record as a tool for reservoir development: a reality check; American Association of Petroleum Geologists Bulletin v. 90, p. 989-1002.

Miall, A. D., 2014, Fluvial depositional systems: Springer-Verlag, Berlin 316 p.

Miall, A. D., and Arush, M., 2001, The Castlegate Sandstone of the Book Cliffs, Utah: sequence stratigraphy, paleogeography, and tectonic controls: Journal of Sedimentary Research, v. 71, p. 536–547.

Miall, A. D., and Gibling, M. R., 1978, The Siluro-Devonian clastic wedge of Somerset Island, Arctic Canada, and some regional paleogeographic implications: Sedimentary Geology, v. 21, p. 85-127.

Minter, W. E. L., 1978, A sedimentological synthesis of placer gold, uranium and pyrite concentrations in Proterozoic Witwatersrand sediments, in Miall, A. D., ed., Fluvial Sedimentology: Canadian Society of Petroleum Geologists Memoir 5, p. 801-829.

Moecher, D. P., Kelly, E. A., Hietpas, J., and Samson, S. D., 2019, Proof of recycling in clastic sedimentary systems from textural analysis and geochronology of detrital monazite: implications for detrital mineral provenance analysis: Geological Society of America Bulletin, v. 131, p. 1115–1132.

Moonan, X. R., 2011, 4-D understanding of the evolution of the Penal/Barrackpore Anticline, southern sub-basin, Trinidad: American Association of Petroleum Geologists, Search and Discovery Article 90135.

Morón, S., Cawood, P. A., Haines, P. W., Gallagher, S. J., Zahirovic, S., Lewis, C. J., and Moresi, L., 2019, Long-lived transcontinental sediment transport pathways of East Gondwana: Geology, v. 47, p. 513–516.

Morton, A. C., 1985a, Heavy minerals in provenance studies, in Zuffa, G. G., ed., Provenance of arenites: D. Reidel Publishing Company, Dordrecht, p. 249-277.

Morton, A. C., 1985b, A new approach to provenance studies: electron microprobe analysis of detrital garnets from Middle Jurassic sandstones of the northern North Sea: Sedimentology, v. 32, p., 553–566.

Morton, A. C., and Hallsworth, C. R., 1994, Identifying provenance-specific features of detrital heavy mineral assemblages in sandstones: Sedimentary Geology, v. 90, p. 241–256.

Morton, A. C., and Hallsworth, C. R., 1999, Processes controlling the composition of heavy mineral assemblages in sandstones: Sedimentary Geology, v. 124, p. 3–29.

Moscardelli, L., Wood, L. J., and Dunlap, D. B., 2012, Shelf-edge deltas along structurally complex margins" A case study from eastern offshore Trinidad: American Association of Petroleum Geologists, v. 96, p. 1483–1522.

Mossop, G. D., and Flach, P. D., 1983, Deep channel sedimentation in the Lower Cretaceous McMurray Formation, Athabasca Oil Sands, Alberta: Sedimentology, v. 30, p. 493-509.

Mutti, E., and Ricci-Lucchi, F., 1972, Le turbiditi dell'Appennino settentrionale: introduzione all' analisi di facies: Soc. Geol. Ital. Mem. 11, p. 161-199.

Myhr, D. W., and Meijer-Drees, N. C., 1976, Geology of the southeastern Alberta Milk River gas pool; in Lerand, M. M., ed., The sedimentology of selected clastic oil and gas reservoirs in Alberta: Canadian Society of Petroleum Geologists, p. 96–117.

Nelson, R. A., Lenox, L. C., and Ward, B. J., Jr., 1987, Oriented core: its use, error, and uncertainty: American Association of Petroleum Geologists Bulletin, v. 71, p. 357–367.

Nilsen, T., Shew, R., Steffens, G., and Studlick, J., eds., 2007, Atlas of deep-water outcrops: American Association of Petroleum Geologists, Studies in Geology 56, 504 p.

North, F. K., 1985, Petroleum geology: Boston: Allen & Unwin, 607 p.

Oliver, T. A., and Cowper, N. W., 1963, Depositional environments of the Ireton Formation, central Alberta: Bulletin of Canadian petroleum Geology, v. 11, p. 183–202.

Oliver, T. A., and Cowper, N. W., 1965, Depositional environments of the Ireton Formation, central Alberta: Bulletin of the American Association of Petroleum Geologists, v. 49, p. 1410–1425.

Olson, J. S., and Potter, P. E., 1954, Variance components of crossbedding direction in some basal Pennsylvanian sandstones of the eastern Interior Basin: statistical methods: Journal of Geology, v. 62, p. 26–49.

Parks, J. M., 1970, Computerized trigonometric solution for rotation of structurally tilted sedimentary directional features: Geological Society of America Bulletin, v. 81, p. 537–540.

Paumard, V., Bourget, J., Payenberg, T., George, A. D., Ainsworth, R. B., Lang, S., and Posamentier, H.W., 2020, Controls on deep-water sand delivery beyond he shelf edge: accommodation, sediment supply, and deltaic process regime: Journal of Sedimentary Research, v. 90, p. 104–130.

Pearce, T. J., Besly, N. M., Wray, D. S, and Wright, D. K., 1999, Chemostratigraphy: a method to improve interwell correlation in barren sequences — a case study using onshore Duckmantian/Stephanian sequences (West Midlands, U.K.): Sedimentary Geology, v. 124, p. 197–220.

Pettijohn, F. J., 1962, Paleocurrents and paleogeography: American Association of Petroleum Geologists Bulletin, v. 46, p. 1468–1493.

Pettijohn, F. J., Potter, P. E., and Siever, R., 1973, Sand and sandstone: Springer-Verlag, New York, 618 p.

Pickering, K. T., and Hiscott, R. N., 1985, Contained (reflected) turbidity currents from the Middle Ordovician Cloridorme Formation, Quebec, Canada: an alternative to the antidune hypothesis: Sedimentology, v. 32, p. 373-394.

Pirmez, C., Prather, B. E., Mallarino, G., O'Hayer, W. W., Droxler, A. W, and Winker, C. D., 2012, Chronostratigraphy of the Brazos-Trinity depositional system, western Gulf of Mexico: implications for deepwater depositional models, in Applications of the Principles of seismic geomorphology to continental slope and base-of-slope systems: case studies from seafloor and near-seafloor analogues: Society for Sedimentary Geology Special Publication 99, p. 111–143.

Plint, A. G., Walker, R. G., and Bergman, K. M., 1986, Cardium Formation 6. Stratigraphic framework of the Cardium in subsurface: Bulletin of Canadian Petroleum Geology, v. 34, p. 213–225.

Pöppelreiter, M., Garcia-Carballido, C., and Kraaijveld, M. A., eds. 2010, Dipmeter and borehole image log technology, American Association of Petroleum Geologists, Memoir 92.

Posamentier, H. W., 2000, Seismic stratigraphy into the next millennium: a focus on 3-D seismic data: American Association of Petroleum Geologists, Annual Conference, New Orleans, p. 16–19.

Posamentier, H. W., Davies, R. J., Cartwright, J. A., and Wood, L., 2007, Seismic geomorphology – an overview, in Davies, R. J., Posamentier, H. W., Wood, L. J., and Cartwright, J. A., eds., Seismic geomorphology: applications to hydrocarbon exploration and production: Geological Society, London, Special Publication 277, p. 1–14.

Posamentier, H. W., and Kolla, V., 2003, Seismic geomorphology and stratigraphy of depositional elements in deep-water settings: Journal of Sedimentary Research, v. 73, p. 367–388.

Potter, P. E., and Pettijohn, F. J., 1977, Paleocurrents and basin analysis, 2nd edition: Academic Press, San Diego, California, 296 p.

Puigdefábregas, C., 1973, Miocene point-bar deposits in the Ebro Basin, northern Spain: Sedimentology, v. 20, p. 133–144.

Rahmani, R. A., and Lerbekmo, J. F., 1975, Heavy mineral analysis of Upper Cretaceous and Paleocene sandstones in Alberta and adjacent areas of Saskatchewan, in Caldwell, W. G. E. ed., The Cretaceous system in the Western Interior of North America: Geological Association of Canada Special Paper 13, p. 607–632.

Raines, M. K., Hubbard, S. M., Kukulski, R. B., Leier, A. L., and Gehrels, G. E., 2013, Sediment dispersal in an evolving foreland: detrital zircon geochronology from Upper Jurassic and lowermost Cretaceous strata, Alberta Basin, Canada, Geological Society of America Bulletin, v. 125, p. 741-755.

Ramsey, J. G., 1961, The effects of folding upon the orientation of sedimentation structures: Journal of Geology, v. 69, p. 84–100.

Ratcliffe, K. T., and Zaitlin, B. A., eds., 2010, Application of modern stratigraphic techniques: theory and case histories: Society for Sedimentary Geology (SEPM) Special Publication 94, 241 p.

Reading, H. G., ed., 1996, Sedimentary environments: processes, facies and stratigraphy, third edition: Blackwell Science, Oxford, 688 p.

Rich, J. L., 1951, Three critical environments of deposition and criteria for recognition of rocks deposited in each of them: Geological Society of America Bulletin, v. 62, p. 1–20.

Rosenthal, L. R. P., Leckie, D. A. and Nadon, G., 1984, Depositional cycles and facies relationships within the Upper Cretaceous Wapiabi and Belly River formations of west-central Alberta: Canadian Society of Petroleum Geologists, Field Trip Guide Book, 54.

Rosenthal, L. R. P. and Walker, R. G., 1987, Lateral and vertical facies sequences in the Upper Cretaceous Chungo Member, Wapiabi Formation, southern Alberta: Canadian Journal of Earth Sciences, v. 24, p. 771–783.

Rupke, N. A., 1978, Deep clastic seas, in Reading, H. G., ed., Sedimentary environments and facies: Blackwell, Oxford, p. 372–415.

Rust, B. R., 1975, Fabric and structure in glaciofluvial gravels, in Jopling, A. V., and McDonald, B. C., eds., Glaciofluvial and glaciolacustrine sedimentation: Society of Economic Paleontologists and Mineralogists Special Publication 23, p. 238–248.

Rust, B. R., 1981, Alluvial deposits and tectonic style: Devonian and Carboniferous successions in eastern Gaspe, in Miall, A. D., ed., Sedimentation and tectonics in alluvial basins: Geological Association of Canada Special Paper 23, p. 49-76.

Ryan, W. B. F., and Cita, M. B., 1978, The nature and distribution of Messinian erosional surfaces — indicators of a several-kilometres-deep Mediterranean in the Miocene: Marine geology, v. 27, p. 193–230.

Safronova, P. A., Henriksen, S, Andreassen, K., Laberg, J. S., and Vorren, T. O., 2014, Evolution of shelf-margin clinoforms and deep-water fans during the middle Eocene in the Sørvestsnaget Basin, southwest Barents Sea: American Association of Petroleum Geologists Bulletin, v. 98, p. 515–544.

Sageman, B. B., Singer, B. S., Meyers, S. R., Siewert, S. E., Walaszczyk, I., Condon, D. J., Jicha, B. R., Obradovich, J. D., and Sawyer, D. A., 2014, Integrating $^{40}Ar/^{39}Ar$e,U-Pb and astronomical clocks in the Cretaceous Niobrara Formation, Western Interior Basin, USA: Geological Society of America Bulletin, v. 126, p. 956-973.

Saller, A. H., Noah, J. T., Ruzuar, A. P., and Schneider, R., 2004, Linked lowstand delta to basin-floor fan deposition, offshore Indonesia: an analog for deep-water reservoir systems: American Association of Petroleum Geologists Bulletin, v. 88, p. 21–46.

Sangree, J. B., and Widmier, J. M., 1977, Seismic stratigraphy and global changes of sea level, part 9: Seismic interpretation of clastic depositional facies, in Payton, C. E., ed., Seismic stratigraphy — applications to hydrocarbon exploration: American Association of Petroleum Geologists Memoir 26, p. 165–184.

Savit, C. H., and Changsheng, W. V., 1982, Geophysical characterization of lithology—application to subtle traps; in Halbouty, M. T., ed., The deliberate search for the subtle trap: American Association of Petroleum Geologists Memoir 32, p. 11-30.

Sayago, J., Di Lucia, M., Mutti, M., Cotti, A., Sitta, A., Broberg, K., Przybylo, A., Buonaguro, R. and Zimina, O., 2012, Characterization of a deeply buried paleojarst terrain in the Loppa High using core data and multiattribute seismic facies classification: American Association of Petroleum Geologists, v. 96, p. 1843–1866.

Schnitzer, W. A., 1977, Die quarzkornfarben-methoden und ihre Bedeutung fur die stratigraphische und palaogeographische erforschung psammitischer sedimente: Erlanger Geol. Abh., Heft 103, 28 p.

Selley, R. C., 1968, A classification of paleocurrent models: Journal of Geology, v. 76, p. 99–110.

Shank, J. A., and Plint, A. G., 2013, Allostratigraphy of the Upper Cretaceous Cardium Formation in subsurface and outcrop in southern Alberta, and correlation to equivalent strata in northwestern Montana: Bulletin of Canadian Petroleum Geology, v. 61, p. 1-40.

Sharma, P. V., 1986, Geophysical methods in geology: 2nd. Edition: Elsevier Scientific Publications, Amsterdam, 442 p.

Sheriff, R. E., 1985, Aspects of seismic resolution, in Berg, O. R., and Woolverton, D. G., eds., Seismic stratigraphy II: American Association of Petroleum Geologists Memoir 39, p. 1–10.

Simons, D. B., and Richardson, E. V., 1961, Forms of bed roughness in alluvial channels: American Society of Civil Engineers Proceedings, v. 87, No. HY3, p. 87-105.

Simons, D. B., Richardson, E. V., and Nordin, C. F., 1965, Sedimentary structures generated by flow in alluvial channels, in Middleton, G. V., ed., Primary sedimentary structures and their hydrodynamic interpretation: Society of Economic Paleontologists and Mineralogists Special Publication 12, p. 34-52.

Sloss, L. L., Krumbein, W. C., and Dapples, E. C., 1949, Integrated facies analysis; in Longwell, C. R., ed., Sedimentary facies in geologic history: Geological Society of America Memoir 39, p. 91-124.

Smith, N. D., 1972, Some sedimentological aspects of planar cross-stratification in a sandy braided river: Journal of Sedimentary Petrology, v. 42, p. 624–634.

Smith, T., and Leone, J., 2014, Shallow onlap model for Ordovician and Devonian organic-rich shales, New York State: American Association of Petroleum Geologists Search and Discovery Article #50911.

Southcott, A., 2014, 3D seismic proves its value in Bakken geosteering: Search and Discovery article 41435.

Stark, T. J., Zeng, H., and Jackson, A., 2013, An introduction to this special section: Chronostratigraphy: The Leading Edge, v. 32, p. 132–138.

Stow, D. A. V., and Holbrook, J. A., 1984, North Atlantic contourites: an overview, in Stow, D. A. V., and Piper, D. J. W., eds., Fine-grained sediments: deep-water processes and facies: Geological Society of London Special Publication 15, p. 245–256.

Stow, D. A. V., and Lovell, J. P. B., 1979, Contourites: their recognition in modern and ancient sediments: Earth Science Reviews, v. 14, p. 251–291.

Stow, D. A. V., Reading, H. G., and Collinson, J. D., 1996, Deep seas, in Reading, H. G., Sedimentary environments: processes, facies and stratigraphy, 3rd edition: Blackwell Science, Oxford, p. 395–453.

Suttner, L. J., 1974, Sedimentary petrographic provinces: an evaluation, in Ross, C. A., ed., Paleogeographic provinces and provinciality: Society of Economic Paleontologists and Mineralogists Special Publication 21, p. 75-84.

Suttner, L. J., Basu, A., Decker, J., Helmhold, K. P., Ingersoll, R. V., et al., 1985: The effect of grain size on detrital modes: a test of the Gazzi-Dickinson point-counting method: discussions and replies: Journal of Sedimentary Petrology, v. 55, p. 616–621.

Taylor, M. S. G., 2010, Visualization and the use of real time data while geosteering – onshore Algeria: Search and Discovery Article 40592.

Thomas, W. A., 2011, detrital-zircon geochronology and sedimentary provenance: Lithosphere, Geological Society of America, v. 3, p. 304-308.

Trettin, H. P., 1970, Ordovician-Silurian flysch sedimentation in the axial trough of the Franklinian Geosyncline, northeastern Ellesmere Island, Arctic Canada, in Lajoie, J., ed., Flysch sedimentology in North America: Geological Association of Canada Special Paper 7, p. 13-35.

Tye, R. S., 1991, Fluvial-sandstone reservoirs of the Travis Peak Formation, East Texas Basin, in Miall, A. D., and Tyler, N., eds., The three-dimensional facies architecture of terrigenous clastic sediments, and its implications for hydrocarbon discovery and recovery: Society of Economic Paleontologists and Mineralogists Concepts and Models Series, v. 3, p. 172–188.

Vail, P. R., Mitchum, R. M., Jr., Todd, R. G., Widmier, J. M., Thompson, S., III, Sangree, J. B., Bubb, J. N., and Hatlelid, W. G., 1977, Seismic stratigraphy and global changes of sea-level, in Payton, C. E., ed., Seismic stratigraphy - applications to hydrocarbon exploration: American Association of Petroleum Geologists Memoir 26, p. 49-212.

Veeken, 2007, Seismic stratigraphy, basin analysis and reservoir characterization: Elsevier, Amsterdam, Seismic Exploration, v. 37, 509 p.

Visher, G. S., 1965, Use of vertical profile in environmental reconstruction; American Association of Petroleum Geologists Bulletin: v. 49, p. 41-61.

Walaszczyk, I., Shank, J. A., Plint, A. G., and Cobban, W. A., 2014, Interregional correlation of disconformities in Upper Cretaceous strata, Western Interior Seaway: Biostratigraphic and sequence-stratigraphic evidence for eustatic change: Geological Society of America Bulletin, v. 126, p. 307-316.

Walker, R. G., 1978, Deep-water sandstone facies and ancient submarine fans: models for exploration for stratigraphic traps: American Association of Petroleum Geologists Bulletin, v. 62, p. 932–966.

Wells, N. A., 1988, Working with paleocurrents: Journal of Geological Education, v. 35, p. 39–43.

Willis, A. J., 2000, Tectonic control of nested sequence architecture in the Sego Sandstone, Neslen Formation, and Upper Castlegate

Sandstone (Upper Cretaceous), Sevier Foreland Basin, Utah, U.S. A.: Sedimentary Geology, v. 136, p. 277–318.

Woodcock, N. H., 1979, The use of slump structures as palaeoslope orientation estimators: Sedimentology, v. 26, p. 83–99.

Wray, D. S., and Gale, A. S., 1993, Geochemical correlation of marl bands in Turonian chalks of the Anglo-Paris basin, in Hailwood, E. A., and Kidd, R. B., eds., High resolution stratigraphy: Geological Society, London, Special Publication 70, p. 211–226.

Wright, A. M., Ratcliffe, K. T., Zaitlin, B. A., and Wray, D. S., 2010, The application of chemostratigraphic techniques to distinguish compound incised valleys in low-accommodation incised-valley systems in a foreland-basin setting: an example from the Lower Cretaceous Mannville Group and Basal Colorado Sandstone (Colorado Group), Western Canadian Sedimentary Basin, in Ratclife, K. T., and Zaitlin, B. A., eds., Application of modern stratigraphic techniques, Society for Sedimentary geology, Special Publication 94, p. 93–107.

Zaitlin, B. A., Warren, M. J., Potocki, D., Rosenthal, L., and Boyd, R., 2002, depositional styles in a low accommodation foreland basin setting: an example from the Basal Quartz (Lower Cretaceous), southern Alberta: Bulletin of Canadian Petroleum Geology, v. 50, p. 31–72.

Zeng, H., Backus, M., Barrow, K. T., and Tyler, N., 1996, Facies mapping from three-dimensional seismic data: potential and guidelines from a Tertiary sandstone-shale sequence model, Powderhorn field, Calhoun County, Texas: American Association of Petroleum Geologists Bulletin, v. 80, p. 16-46.

Zeng, H., Henry, S. C., and Riola, J. P., 1998, Stratal slicing, Part II: Real 3-D seismic data: Geophysics, v. 63, p. 514–522.

Zeng, H., Loucks, R., Janson, X., Wang, G., Xia, Y., Yuan, B., and Xu, L., 2011, Three-dimensional seismic geomorphology and analysis of the Ordovician paleokarst drainage system in the central Tabei Uplift, northern Tarim Basin, western China: American Association of Petroleum Geologists, v. 95, p. 2061–2083.

Zeng, H., Zhu, X., and Zhu, R., 2013, new insights into seismic stratigraphy of shallow-water progradational sequences: subseismic clinoforms: Interpretation: A journal of subsurface characterization: Society of Exploration Geophysicists and American Association of Petroleum Geologists, v. 1, p. SA35-SA51.

Zotto, S. C., Moecher, D. P., Niemi, N. A., Thigpen, J. R., and Samson, S. D., 2020, persistence of Grenvillian dominance in Laurentian detrital zircon age systematics explained by sedimentary recycling: evidence from detrital zircon double dating and detrital monazite textures and geochronology: Geology, v. 48, p. 792–797.

Zuffa, G. G., ed., 1985a, Provenance of arenites: D. Reidel Publishing Company, Dordrecht, 408 p.

Zuffa, G. G., 1985b, Optical analyses of arenites: influence of methodology on compositional results; in Zuffa, G. G., ed., Provenance of arenites: D. Reidel Publishing Company, Dordrecht, p. 165-189.

Zuffa, G. G., 1987, Unravelling hinterland and offshore paleogeography from deep-water arenites, in Leggett, J. K., and Zuffa, G. G., eds., Marine clastic sedimentology, concepts and case studies: Graham and Trotman, London, p. 39–61.

Contents

7.1 Introduction

Stratigraphy is defined as the study of layered rocks. In the context of sedimentary geology in general, and of this book in particular, stratigraphy is the discipline that pulls everything together. In Chaps. 2–5 of this book we deal with increasingly large and complex sedimentological concepts, and in Chap. 6 we discuss mapping methods, which are essentially the methods for extending our interpretations beyond our immediate data points by interpolation and extrapolation. In this chapter we add the elements of chronostratigraphic dating and correlation and demonstrate the interdisciplinary nature of modern stratigraphic methods.

A. D. Miall, *Stratigraphy: A Modern Synthesis*, Springer Textbooks in Earth Sciences, Geography and Environment,
https://doi.org/10.1007/978-3-030-87536-7_7

Interpretations that focus on local to regional facies studies or sequence stratigraphy rely on the principles of stratigraphy to create and confirm local and regional correlations. Rock units and the depositional environments in which they formed may be related to each other in a way that enables additional correlations to be made to, for example, regional and global climatic and tectonic events. Stratigraphic methods are required to construct such correlations, and a dependence on these methods increases with the scale of a project, from the local to the regional to the continental to the global.

Formal stratigraphic practices, including the definition of formations and stages, had their origins in nineteenth-century field geology, beginning with William Smith (Conkin and Conkin 1984; Miall 2004), and have evolved into a set of carefully defined procedures for naming and correlating the various kinds of stratigraphic unit (Salvador 1994; NACSN 2005). They have also, for many years, been applied successfully to subsurface well data. These methods are mainly based on detailed lithostratigraphic and biostratigraphic information, the analysis of which is discussed later in this chapter, along with other important aids to correlation, including radioisotopic dating and magnetic reversal stratigraphy. Lithostratigraphic classification of the sedimentary record remains the basic descriptive process for stratigraphic documentation, simply because of the large body of historical documentation based on lithostratigraphy; but the ultimate goal is now to develop stratigraphic frameworks based on sequence stratigraphy (Catuneanu 2006; Catuneanu et al. 2011; see Sect. 7.7).

A broader, more regional approach to correlation generally is taken by those dealing with reflection seismic data (Veeken 2007). Commercial seismic work in frontier regions and the deep reflection profiles produced by groups such as the Consortium for Continental Reflection Profiling (COCORP) in the United States, Canada's Lithoprobe Project, and many other international projects, provide sweeping regional cross-sections, within which correlations at the detailed level may be far from clear. This work may be supplemented by detailed three-dimensional reconstructions created using 3-D seismic methods applied to prospective rock volumes (Brown 2011).

With outcrop and well data, the problem may be to establish the regional framework from a mass of local detail, whereas with seismic data it is the detail that may be hard to see (depending on depth, on which seismic resolution depends). The ideal combination is, of course, a basin with a network of key exploration holes tied to regional seismic lines, plus local 3-D seismic volumes. Most advanced petroleum exploration projects now achieve this level of detail. Examples are provided in Chap. 6.

These differences in data type and scale have led to two different approaches to stratigraphic correlation. In industry

exploration work in frontier regions, particularly the great offshore basins, a rather informal, pragmatic approach may be taken to such topics as biostratigraphic zonation and the naming of formations, at least in the early stages of basin development. The broad picture can be derived from seismic cross-sections, and the details gradually resolve themselves as more well data become available. Application of various basin mapping methods (Chap. 6) and use of the genetic, depositional systems approaches and sequence stratigraphy concepts (Chap. 5) are of particular value here. Some examples of detailed mapping and correlation and the kinds of research questions these projects raise are discussed in Chap. 8.

In the absence of seismic data, it is necessary to construct the forest from the individual trees (the second approach). The stratigraphic framework in most well-explored (mature) basins was built up this way using lithostratigraphic methods, and in the past the work has usually been accompanied by considerable controversy, as local specialists have argued about the relationships between the successions in different parts of a basin or between outcrop and subsurface units. Sequence concepts, because of their predictive power, can now make this task easier, although the work is usually not without its problems.

Whether a basin analysis exercise starts from seismic sections or outcrop work, it is desirable, eventually, to document the fine detail of the stratigraphy by establishing a sequence-stratigraphic framework and, ideally, to tie this to the global time scale (formal sequence-stratigraphic methods have yet to be finalized; see Sect. 7.7). Every local biostratigraphic, radioisotopic and magnetostratigraphic study can potentially contribute to the long-term effort to perfect a global chronostratigraphic (time) scale, which goes by the formal name of the **Geologic Time Scale (GTS)**. This last step is beyond the needs of most exploration companies and is an area of research commonly taken over by state geological surveys and individuals in academic institutions, although the data base and expertise built up in industry may form an essential component. This is one area in which the government core and data repositories can prove their usefulness.

Several commissions and subcommissions and numerous working groups of the *International Union of Geological Sciences* (IUGS), mostly under the auspices of the *International Commission on Stratigraphy* (ICS), have been carrying this work for many years, like many national groups. The *North American Commission on Stratigraphic Nomenclature* has been particularly influential. Some of the results are reported in this chapter, and the interested reader may wish to examine the IUGS journal *Episodes*, which reports on the activities of these various groups and announces important publications. Other important publications include *Newsletters on Stratigraphy* (published by E. Schweizerbart, founded in 1965), and the journal *Stratigraphy* (published by

Micropaleontology Press, founded by William Berggren and John A. Van Couvering in 2004). They both publish original research articles and reviews and may also include background information on the ongoing work of ICS and the *International Subcommission on Stratigraphic Classification* (ISSC). The North American Commission published an updated, comprehensive guide to stratigraphic procedure in 2005. Recent synthesis publications describing the geological time scale have brought this topic to a high level of sophistication (Gradstein et al. 2004a, 2012, 2020). We discuss this later in this chapter (Sect. 7.8). The website https://stratigraphy.org is the official website of the ICS.

This chapter is intended to provide an introduction to practical working methods. Chronostratigraphic (including biostratigraphic) research must form an integral part of any ongoing basin analysis project unless it is strictly local in scope. The work is usually performed by specialists. Correlation methods based on lithostratigraphy and sequence stratigraphy are also compared and contrasted here, although the procedures for erecting formal, named units (included here for consistency and completeness) can be left to advanced stages of the analysis (and do not yet include procedures relating to sequence stratigraphy). Such naming is best carried out by individuals with sedimentological training so that the depositional systems and sequence concepts described in Chaps. 3–5 can be incorporated into the work. Research trends, emerging problems and some new developments are discussed in Chap. 8.

As discussed in detail in Chap. 8, the sedimentary record is "more gap than record." For 100 years, since Barrell (1917) published his ground-breaking work on the rates of sedimentary processes, the fragmentary nature of the record has largely been ignored by practicing stratigraphers and sedimentologists. Typically, in many stratigraphic sections, only about 10% of elapsed time is represented by the preserved rock at a 10^6-year time scale. All the developments in dating and correlation methods and the emergence of the powerful new descriptive-interpretive methods of sequence stratigraphy have all been accomplished despite this fact. Stratigraphy continues to "work" as a practical method of basin mapping and resource exploration. How can this be? It is because there is a limited suite of natural processes that deposit and preserve sediment, and these predominate in the development of the stratigraphic record, including both the accumulated sedimentary successions and the hiatuses that separate them (Miall 2015). These are widespread and are now well known. For example, modern work on sequence stratigraphy has identified a limited range of allogenic processes, all characterized by a specific range of time scales, that are now known to generate sequences (Miall 1995; see Sect. 7.6). These typically extend through a sedimentary basin and may correlate to other successions regionally, or even globally. The orders-of-magnitude range of time scales

over which these processes operate was the basis for the original hierarchical "order" classification of sequences (Vail et al. 1977). The range of natural processes that build the sedimentary record, from the burst-and-sweep turbulence of traction current to the accumulation of a basin-fill over millions of years, constitutes a crudely fractal distribution of rates and time scales, but these processes are genetically unrelated (Miall 2015). This is important because it means that the distribution of relevant time scales is not mathematically precise, and that, therefore, quantification of processes, for example on the basis of fractal theory, may not be particularly useful, illuminating or, indeed, relevant. Also, as noted elsewhere, interpretations of sedimentary processes that imply continuity, such as sediment-transport mass-balance calculations and the reconstruction of shoreline trajectories through transgressions and regressions, need to take into account the interruptions in the record represented by hiatuses, the frequency and duration of which have consistently been ignored.

A formal approach to the issue of unconformities is discussed in Sect. 7.6. We address rate and time scale issues in more detail in Chap. 8.

7.2 Types of Stratigraphic Unit

Rocks may be described in terms of any of their physical, chemical, organic or other properties, including lithology, fossil content, geochemistry, petrology, mineralogy, electrical resistivity, seismic velocity, density (gravity), magnetic polarity or age. Theoretically, any of these properties may be used for description and correlation, and most are so used for various purposes. In practice, lithology is the most important criterion; fossil content is also crucial for rocks of Phanerozoic age. Magnetic polarity has become useful as a correlation tool, particularly for the younger Mesozoic and the Cenozoic, and radioisotopic ages are used to assign numerical ("absolute") ages to biostratigraphic, magnetic and other chronostratigraphic units. A standard oxygen isotope scale is available for the late Cenozoic (Pliocene to present), and increasing use is being made of other isotopic signatures, particularly carbon and strontium. Later sections of this chapter deal with all these techniques in more detail. Other geophysical properties are used in the early reconnaissance stages of exploration of a sedimentary basin. Not all these properties will necessarily give rise to the same correlations of a given rock body; for example, it is commonly difficult to relate geophysical properties precisely to lithology. Therefore, no single type of stratigraphic unit can be used to define all the variability present in nature.

Reflection-seismic data have been widely used in the exploration of sedimentary basins since the 1970s, making use of the concepts of seismic stratigraphy (Sect. 6.3), and with particular application to the interpretive field of sequence stratigraphy (Chap. 5). Sheriff (1976) pointed out that for reflecting events to be distinguished on seismic data they must each represent a clear velocity contrast and must be at least the equivalent of a quarter wavelength apart. At shallow depths, velocities are in the range of 1.5–2.5 km/s and reflections are of relatively high frequency, about 5–100 Hz, so that a quarter wavelength is on the order of 5–12 m. At greater depths, typical reflection wavelengths increase considerably. Therefore, stratigraphic resolution is fairly coarse (Fig. 7.1). Early work in seismic sequence stratigraphy had portrayed sequence boundaries as distinctive reflection surfaces corre-latable over wide distances (Mitchum et al. 1977), but improved acquisition and processing techniques have demonstrated that the broad, through-going correlations on which this early work was based may in many cases be suspect. Cartwright et al. (1993) demonstrated that many of the critical features of sequence architecture, including the sequence boundaries, break down upon detailed examination into complex reflection patterns representing local facies variability, and major through-going surfaces may not be present or easy to trace. Seismic data are essential for studying large-scale stratigraphic features such as depositional systems and regional (or global) sequences, but may be of less use in the development of the refined stratigraphic subdivisions that are the subject of this chapter. The reader interested in the application of seismic methods to basin analysis may wish to consult standard textbooks in this area, such as Catuneanu (2006) or Veeken (2007).

Two basic categories of stratigraphic information are essential for the complete documentation of the stratigraphic record: (1) descriptive lithic units and (2) geochronologic information, dealing with correlation and age of the strata (Harland 1993).

The most important types of stratigraphic units are:

Lithostratigraphic units: These are strictly empirical, based on observable lithologic features including composition and grain size, and possibly also including certain basic sedi-mentological information, such as types of sedimentary structures and cyclic successions.

Biostratigraphic units: These are based on fossil content. Life forms evolve with time, permitting subdivision into **biozones** on the basis of changes in the fauna or flora. The first and last appearance of particular species or variants may also serve as useful time markers. When used on their own, biostratigraphic units provide relative ages. Numerical ("absolute") ages are derived by cross-correlation with chronostratigraphic data, as noted below.

Unconformity-bounded units: These are units bounded above and below by unconformities. They may consist of any kinds of rocks, igneous, metamorphic or sedimentary. Unconformity-bounded sedimentary successions may be formalized using the empirical, descriptive classification procedures of **allostratigraphy** (see NACSN 2005), but increasingly geologists now employ the interpretive proce-dures and models of **sequence stratigraphy** as the main basis for subdivision and mapping of the basin fill.

Chronostratigraphic units: These comprise an interpretive stratigraphy, in contrast to lithostratigraphic and biostrati-graphic units, which are strictly descriptive. Chronostratig-raphy concerns itself with correlation and the age of the strata in years ("absolute" age), which may be determined by a variety of means, of which the most important are fossil content, radioisotopic dating, magnetic polarity (for the post Middle Jurassic), and the isotopic record of oxygen, carbon and strontium (for different parts of the Phanerozoic record). The principal chronostratigraphic units, which form the main

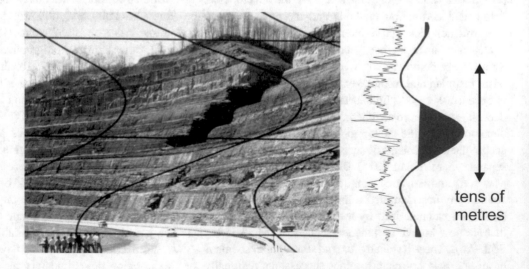

Fig. 7.1 Scale of a typical seismic wave form as compared to an outcrop (left), and compared to a wireline log (right). The frequencies characteristically used in petroleum exploration seismic work (10–60 Hz) have long wavelengths. Seismic resolution is therefore limited to large-scale stratigraphic features (based on an idea from A. E. Pallister and A. E. Wren)

tens of metres

foundation of the **Global Time Scale** (or **Geologic Time Scale**) are **stages**. Increasing use is being made of the cyclostratigraphy of selected stratigraphic sections to build a highly accurate **astrochronologic time scale**.

Both lithostratigraphic and biostratigraphic units may be local in extent. Lithologic character depends on the depositional environment, sediment supply, climate, rate of subsidence etc., all of which can vary over short distances. Lithostratigraphic units are diachronous to a greater or lesser degree, that is, they represent a different time range in different places, reflecting gradual shifts in the environment, for example, during transgression or regression. The limits of a stratigraphic unit are either its erosional truncation at the surface or beneath an unconformity or a facies change into a contemporaneous unit of different lithology. A special type of lithostratigraphic unit is that formed by short-lived **stratigraphic event**s, which are those that have widespread depositional effects within very short time spans (Ager 1981). Examples of such events are volcanic ash falls, the deposits of violent storms and tsunamis, and certain regional or global sea-level changes. Such event deposits may prove very useful as local correlation tools, as discussed in Sect. 7.8.4.

Biostratigraphic units are based on fauna or flora, the distribution of which is ecologically controlled. Also, contemporaneous faunas located in ecological niches that are similar but geographically isolated may show subtly different evolutionary patterns, making comparisons or correlations between the areas difficult (this is the subject of **biogeography**). All life forms evolve with time so that faunas and floras show both spatial and temporal limits on their distribution.

The unconformities that demarcate unconformity-bounded units are caused by subaerial erosion during times of low stands of sea level, by erosion following tectonic uplift, submarine erosion or sudden environmental change (Sect. 7.6). These events are typically widespread. Sea-level change may be caused by tectonic elevation of the basement or by eustatic changes in sea level. In either case, the unconformity surfaces define units of considerable lateral extent. Rapid changes in the depositional environment may generate what Schlager (1989) termed **drowning unconformities**. Some unconformities may be of global significance, although this is typically difficult to demonstrate (Miall 2010, Chap. 14) Unconformities provide an excellent basis for the regional subdivision of basin fill, and their interpretation may throw considerable light on regional tectonic evolution. **Sequence stratigraphy** has become the method of choice for subsurface mapping by the petroleum industry precisely because of its practical utility in focusing on and documenting these unconformity-bounded

successions (Chap. 5). Unconformities and unconformity-bounded units are discussed in Sect. 7.6.

Chronostratigraphy attempts to resolve the difficulties in regional and global correlation by establishing a global, time-based reference frame. A standard **Geological Time Scale (GTS)** has gradually evolved since the discovery of radioisotopic dating early in the twentieth century. However, the accuracy of chronostratigraphic correlation is only as good as that of the time-diagnostic criteria on which it is based. Imprecision and error remain (Sects. 7.8 and 8.10).

The evolution of these four types of units has had a long and complex history (Hancock 1977; Conkin and Conkin 1984; Miall 2004) and there has been controversy about definitions. Hedberg (1976), Hancock (1977) and Harland (1978, 1993) discussed some of the early practical and philosophical problems. The nineteenth-century geological practice did not distinguish lithology from age, causing severe correlation problems, wherever a facies change or a diachronous boundary occurred. More recently, there has been controversy over whether the rocks (lithologic units) or interpreted age range should form the primary basis of chronostratigraphy (e.g., Zalasiewicz et al. 2004). The discussions are likely to seem somewhat academic and theoretical to the average basin analyst and will not be discussed at length here. A historical summary of the methods that gradually evolved for the construction of the geological time scale is provided by Gradstein and Ogg (in Gradstein et al. 2020, Chap. 2).

7.3 The Six Steps Involved in Dating and Correlation

Six main "steps" are involved in the dating and correlation of stratigraphic events (Miall 1994). Figure 7.2 summarizes these steps and provides generalized estimates of the magnitude of the uncertainty associated with each aspect of the correlation and dating of the stratigraphic record. Some of these errors may be cumulative, as discussed in the subsequent sections. The assignment of ages and correlations with global frameworks is an iterative process that, in some areas, has been underway for many years. There is much feedback and cross-checking from one step to another. What follows should be viewed, therefore, as an attempt to break down the practical business of dating and correlation into more readily understandable pieces, all of which may be employed at one time or another in the unravelling of regional and global stratigraphies. The main steps are as follows:

1. Identification of the units or stratigraphic events to be correlated, and development of regional correlation frameworks, including the mapping of hiatuses,

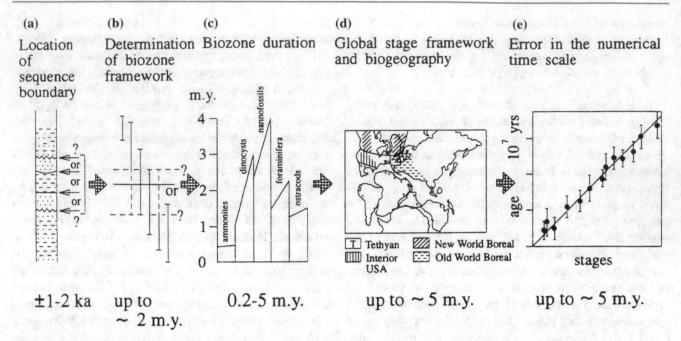

(a) Location of sequence boundary

(b) Determination of biozone framework

(c) Biozone duration

(d) Global stage framework and biogeography

(e) Error in the numerical time scale

±1-2 ka   up to ~ 2 m.y.   0.2-5 m.y.   up to ~ 5 m.y.   up to ~ 5 m.y.

Fig. 7.2 Steps in the correlation and dating of stratigraphic events. e = typical range of error associated with each step. **a** In the case of the sequence framework, the location of sequence boundaries (step 1) may not be a simple matter but depends on the interpretation of the rock record using sequence principles. **b** Assignment of the boundary event to the biozone framework (step 2). An incomplete record of preserved taxa (almost always the case) may lead to ambiguity in the placement of biozone boundaries. **c** The precision of biozone correlation depends on biozone duration (step 3). Shown here is a simplification of Cox's (1990) summary of the duration of zones in Jurassic sediments of the North Sea Basin. **d** The building of a global stage framework (step 4) is fundamental to the development of a global time scale (step 5). However, global correlation is hampered by faunal provincialism. Shown here is a simplification of the faunal provinces of Cretaceous ammonites, shown on a mid-Cretaceous plate-tectonic reconstruction. Based on Kennedy and Cobban (1977) and Kauffman (1984). **e.** The assignment of numerical ages to stage boundaries and other stratigraphic events (step 6) contains an inherent experimental error and also the error involved in the original correlation of the datable horizon(s) to the stratigraphic event in question. Diagrams of this type are a standard feature of any discussion of the global time scale (e.g., Haq et al. 1988; Harland et al. 1990). The establishment of a global biostratigraphically-based sequence framework involves the accumulation of uncertainty from step (**a**) through (**d**). Potential error may be reduced by the application of radioisotopic, magnetostratigraphic or chemostratigraphic techniques which, nonetheless, contain their own inherent uncertainties (step 6)

unconformities and other key surfaces. Local correlations may be based on lithostratigraphy, but sequence stratigraphic concepts and methods are now practically universal. Correlations may be guided or constrained by supplementary data, such as biostratigraphic zonation. Determining the position of events such as sequence boundaries may or may not be a straightforward procedure, and requiring the application of facies mapping and sequence mapping techniques (Chaps. 3, 4, 5).

2. Determining the extent and chronostratigraphic significance of unconformities. Unconformities, including sequence boundaries, represent finite time spans which vary in duration from place to place. In any given location this time span could encompass the time span represented by several different sedimentary breaks at other locations. Resolving such problems may require that some of the other steps be completed, particularly step 3.

3. Determination of the biostratigraphic framework. One or more fossil groups is used to assign the selected event to a biozone framework, and zones are defined and correlated from section to section. Error and uncertainty may be introduced because of the incompleteness of the fossil record. Graphic correlation or other quantitative techniques may be employed (Sect. 7.5).

4. Assessment of relative biostratigraphic precision. The length of time represented by biozones depends on such factors as faunal diversity and rates of evolution. Durations of biozones vary considerably through geological time and between different fossil groups. Steps 3 and 4 may be aided by the availability of numerical ages obtained from the radioisotopic dating of igneous material, such as lava flows and ash beds. Increasing use is now being made of chemostratigraphic data to cross-reference with and calibrate biostratigraphic data.

5. Correlation of biozones with the global stage framework (Sect. 7.8.3). Much of the existing stage framework was initially, with notable exceptions, built from the study of macrofossils in European-type sections, although

microfossils have become increasingly important for subsurface work and global studies (McGowran 2005). Correlation with this framework raises questions of environmental limitations on biozone extent, our ability to inter-relate zonal schemes built from different fossil groups, and problems of global faunal and floral provinciality and diachroneity.

6. Assignment of numerical ("absolute") ages (Sect. 7.8.2). The use of radioisotopic and magnetostratigraphic dating methods, plus the increasing use of chemostratigraphy (oxygen, carbon and strontium isotope concentrations) permits the assignment of numerical ages in years to the biostratigraphic framework (in addition to the possible direct dating of stratigraphic units, as noted in point 4, above). Such techniques also constitute methods of correlation in their own right, especially where fossils are sparse. The geological time scale (GTS) has become an instrument of considerable geological importance and practical utility in recent years, contributing to the emergence of what Miall (2013) termed "Sophisticated Stratigraphy" (Sect. 7.8; see Fig. 7.33).

7.4 Lithostratigraphy

Until the 1980s it was standard practice to describe and map stratigraphic successions on the basis of lithostratigraphic principles (Fig. 7.3). In the field, particularly in arid regions where the rocks are well exposed, it is still the historically established formations that are the basis for field location and identification. Such is the case, for example, in the Grand Canyon and Canyonlands areas of the United States, and the Front Ranges of the Rocky Mountains in Alberta. Some lithostratigraphic names have long been part of

geological language and are unlikely to be deposed for a long time (e.g., in the UK: Carboniferous Limestone, Old Red Sandstone; in the United States Austin Chalk, Mancos Shale; in Canada: Leduc Formation, Rundle Group). It is necessary, therefore, to be able to read older publications and maps and understand what type of information they convey.

Among the problems with lithostratigraphy as a method of description is that the defined units carry no meaning regarding the origins or age of the units. Formations are commonly diachronous, and many stratigraphic names were established many decades ago, long before the advent of modern facies and sequence analysis. Older literature may therefore be replete with the names of local, poorly defined units, with a given body of rocks defined and named differently in different parts of a basin. Procedures are available (e.g., see NACSN 2005) for the revision and redefinition of units as new information becomes available from surface mapping or subsurface exploration.

7.4.1 Types of Lithostratigraphic Units and Their Definition

A hierarchy of units has been developed based on the **formation**, which is the primary lithostratigraphic unit (NACSN 2005).

Group

Formation

Member

Tongue or lentil

Bed

The formation. An important convention has long since been established that all sedimentary rocks should be subdivided (when sufficient data have been collected) into formations. No other types of lithostratigraphic subdivisions

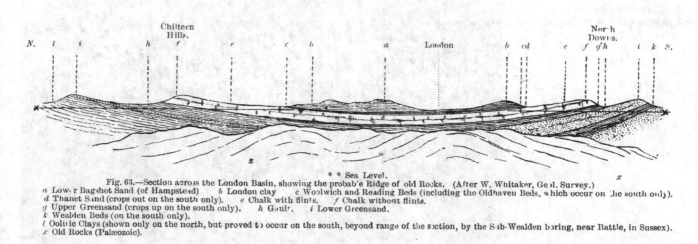

Fig. 63.—Section across the London Basin, showing the probable Ridge of old Rocks. (After W. Whitaker, Geol. Survey.)
a Lower Bagshot Sand (of Hampstead)   *b* London clay   *c* Woolwich and Reading Beds (including the Oldhaven Beds, which occur on the south only).
d Thanet Sand (crops out on the south only).   *e* Chalk with flints.   *f* Chalk without flints.
g Upper Greensand (crops up on the south only).   *h* Gault.   *i* Lower Greensand.
k Wealden Beds (on the south only).
l Oolitic Clays (shown only on the north, but proved to occur on the south, beyond range of the section, by the Sub-Wealden boring, near Battle, in Sussex).
x Old Rocks (Palæozoic).

Fig. 7.3 A cross-section of the London Basin, England, showing the development of descriptive terminology for stratigraphic units. From "Geology of the counties of England and Wales", by Jerome Harrison, 1882 need to be used, although convenience of description may require them.

What is a formation? There are no fixed definitions that deal with the scale or variability of what should constitute a formation, although the procedures for establishing limits (contacts) and names are well established (e.g., see NACSN 2005). Figure 7.4 provides a good example of the way in which stratigraphic successions are subdivided on the basis of lithology. The lithologies, colors and weathering characteristics of the rocks suggest a fourfold subdivision of the exposure. Comparison with other exposures nearby and the presence of distinctive fossils permit three of the subdivisions to be assigned to previously existing formations, while the fourth (oldest) unit is different from the local succession, and has yet to be given a name. This outcrop is large enough that the angular unconformity between two of the units (the Nansen and the Barrow formations) can clearly be seen.

The degree of lithologic variability required to distinguish a separate formation tends to reflect the level of information available to the stratigrapher. Formations may be only a few meters or several thousands of meters in thickness; they may be traceable for only a few kilometers or for thousands of kilometers. Formations in frontier basins usually are completely different in physical magnitude from those in populated, well-explored basins, such as much of western Europe and the United States. As an exploration in frontier basins proceeds, some of the larger formations first defined on a reconnaissance basis may subsequently be subdivided into smaller units and the ranking of the names changed. NACSN (2005) provides the procedures for making these kinds of revisions.

The most important criteria for establishing a formation are its usefulness in subdividing stratigraphic cross-sections and its "mappability." For reconnaissance mapping, a thin unit that cannot accurately be depicted at a scale of, for example, 1:250,000, may be of little use, although the definition and mapping of thin but widespread marker units may be of considerable utility. For more detailed work, mappability at a scale of 1:50,000 or even 1:10,000 may be a more useful criterion. Problems of consistency may arise when detailed work is conducted around a mine site within what is otherwise a poorly explored frontier basin.

Formations should not contain major unconformities, although minor disconformities may be acceptable (indeed, as we now recognize, they are all but unavoidable). The contacts of the formation should be established at obvious lithologic changes. These may be sharp or gradational. An unconformity is a logical choice for a formation contact. Where lithologies change gradually, either vertically or laterally, it may be difficult to choose a logical place to draw the contact. For example, a mudstone may pass up into a sandstone through a transitional succession with sandstone beds becoming thicker and more abundant upward. The mudstone–sandstone formation contact could be drawn at the oldest thick, coarse sandstone (with thickness and coarseness carefully spelled out), at the level where sandstone and mudstone each constitute 50% of the section, or at the youngest extensive mudstone bed. The choice is arbitrary, and it is immaterial which method is selected as long as the same method is used as consistently as possible throughout the extent of the formation.

Fig. 7.4 An example of a lithostratigraphic subdivision of a rock succession. Stratigraphic units exposed in the mountains of northern Ellesmere Island, Arctic Canada (photo: A. F. Embry)

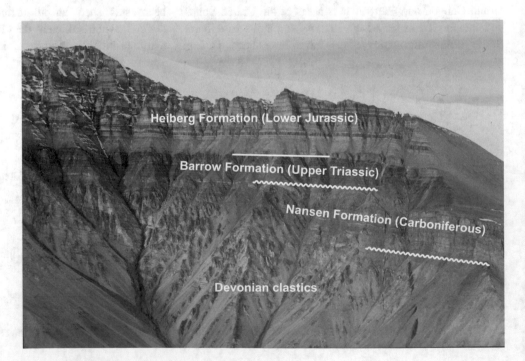

Other problems of definition arise where there are lateral lithologic changes, requiring a definition of a new formation. A simple diachronous contact is not a problem, but where the two units intertongue with one another, it may be virtually impossible to draw a simple formation contact. One solution is to give each tongue the same name as the parent formation. A section passing through the transition region may then show the two formations succeeding each other several times. The only problem this causes is if formation contact and thickness data are stored in a data bank and used in automated contouring programs. Without additional input from the operator, a computer program might not be able to handle this type of data. Other alternatives are to define the whole transitional rock volume in terms of one of the parent units, to separate the transitional lithologies as a separate lithostratigraphic entity or to give separate tongues their own bed, tongue or member names. Published stratigraphic codes (e.g., Salvador 1994; NACSN 2005) provide procedures and practical solutions but do not specify any rigid rules for the resolution of such problems. The main criteria should be practicality, convenience and consistency.

The sometimes arbitrary nomenclatural issues raised by lithostratigraphic methods may be clarified or avoided by the use of modern sequence methods, but it should not be forgotten that sequence stratigraphy is an interpretive approach to the rocks, and the inductive, empirical nature of lithostratigraphy will likely remain as an essential underpinning of basinal stratigraphic frameworks for some time to come.

A range of other terms is used to group or subdivide stratigraphic successions on the basis of lithostratigraphy (Fig. 7.5). Vertical and lateral contacts between units may be defined on the basis of clear lithologic change, but are commonly somewhat arbitrary, and as noted above, such subdivisions contain no useful information about the depositional relationships of the strata. We address these issues later, in the discussion on sequence stratigraphy (Sect. 7.6), where it is demonstrated how this genetic approach to

stratigraphy can lead to much more meaningful reconstructions and interpretations.

The group. All other stratigraphic units are based on the formation. A group consists of two or more formations related lithologically. In the past, named groups have been established for thick and varied successions without first defining the constituent formations. This is not recommended practice. In contrast, formations defined during reconnaissance exploration may be subdivided into constituent formations and the original name retained and elevated to group status if detailed mapping subsequently provides appropriate data. Groups should not contain major unconformities.

The component formations of a group may not be the same everywhere. Lateral facies changes requiring the definition of different formations can occur within a single group. In contrast, a component formation may extend laterally from one group to another. Groups are normally defined for regions of complex stratigraphy. Toward the basin margin or basin center, the component formations may lose their individuality, in which case the group may be "demoted" to a formation, while still retaining the same name.

The terms supergroup and subgroup are occasionally used to provide an additional hierarchy of subdivisions. Usually, there are historical reasons for this; some of the higher ranking names may have started out as member, formation or group names, with reclassification and promotion being required as additional work demonstrated the need for further subdivision.

The member. This is the next ranking unit below the formation. Not all formations need to be divided into members, and formal names need to be used for only a few, one or none of the constituent members, depending on the convenience or the level of information available.

There are no standards for the thickness or extent of members, and commonly it is difficult to decide whether to define a given lithostratigraphic unit as a member or a

Fig. 7.5 A correlation table for a hypothetical basin fill, drawn to illustrate the various ways in which lithostratigraphic terminology may be adapted to best capture existing stratigraphic variability

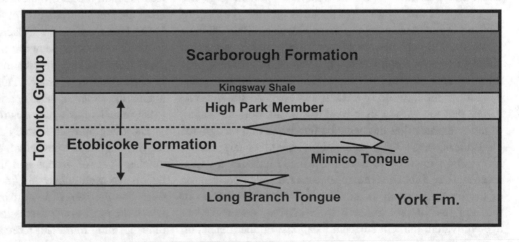

formation. However, the recommended practice is that all parts of a succession be subdivided into formations, and so this is the best level at which to start. A member cannot be defined without its parent formation.

For mapping and other purposes, it is commonly convenient to establish informal units, such as the lower sandstone member, which do not require formal names.

Tongue or lentil. These are similar to members. Because of their geometric connotations, the terms are useful for parts of formations where they interfinger with each other. Formal names may be established for one, several or all such units, depending on convenience and practicality.

Bed. This is the smallest formal, named unit in the hierarchy of lithostratigraphic units. Normally, only a few parts of a stratigraphic succession will be subdivided into named beds. Coal seams in mine areas, prominent volcanic tuff horizons and other marker beds are typical examples. Certain stratabound ore-bearing beds, such as placer units, may also be named.

7.4.2 The Names of Lithostratigraphic Units

When establishing a named unit, it is standard practice to give it a geographical name, chosen to suggest the location or areal extent of the unit. This may be a river, lake, bay, headland, hill, mountain, town, village, etc. Permanent names are preferable. Subsurface work in frontier basins, particularly in offshore areas, may rapidly use up all the available names, in which case the name of the well chosen as the type section may be used. Failing this, names may have to be invented.

In most cases, the geographical name will be followed by the rank designation, for example, Wilcox Group, Pocono Formation. For beds, this is commonly not done, particularly in the case of coal seams, for which a complex mine terminology may have evolved. Many older stratigraphic units use a lithologic term instead of a rank term, for example, Gault Clay, Dakota Sandstone and Austin Chalk (e.g., see Fig. 7.3), but this is not recommended because the rank of the unit is not clear from the name alone (NACSN 2005).

Workers should beware of using a geographic name that has already been employed in a different context or renaming units without justification. Geological survey organizations commonly retain a file of current and obsolete stratigraphic names that the worker may wish to consult. Formal naming of units requires that the name be published in a recognized publication, such as a national or international journal. Information required to establish a name includes a designated type section or **stratotype**, with a detailed description of the succession and information about the distribution of the unit and its relationship to overlying, underlying or age-equivalent units in adjacent locations. Further details on

the establishment of stratotypes are provided in Sect. 7.8: Chronostratigraphy.

Lithostratigraphic units may be changed in rank as the level of knowledge improves. For example, the Cornwallis Group in the Canadian Arctic Islands started as the Cornwallis Formation and was raised to group rank when it was realized that it contained three mappable units of formation rank. Conversely, the Eureka Sound Group was named for a thick and varied clastic succession, but it was never subdivided into named constituent formations by the original author and was reduced to formation status. The unit has now been subdivided and has been formally redefined as a group once again.

When a unit is raised in rank, the original name should not be used for any of the subdivisions but is best retained for the higher ranking unit or abandoned altogether.

7.5 Biostratigraphy

Biostratigraphy is the study of the relative arrangement of strata based on their fossil content. Descriptive or empirical biostratigraphy is used in erecting zones for local or regional stratigraphic correlation and forms the basis for a global system of chronostratigraphic subdivision. Gradstein (in Gradstein et al. 2020, Chap. 3) provided a succinct description of the major fossil groups used in biostratigraphic studies.

Fossil content varies through a stratigraphic succession for two main reasons: evolutionary changes and ecological differences, such as changes in climate or depositional environment. Biostratigraphy should be based only on evolutionary changes, but it is always difficult to distinguish these from changes that take place in a biostratigraphic assemblage as a result of ecological modifications, and this problem is a cause of continuing controversy for many fossil groups.

Biostratigraphy obviously can only be studied, and a classification erected where fossils are present. This rules out all of the Precambrian, except for the concluding subdivision of the Proterozoic—the Ediacaran (Knoll et al. 2006). Even in the Phanerozoic, there are many rock units for which the fossil record is very sparse, and biostratigraphic subdivision is correspondingly crude. This is particularly the case in nonmarine strata or those (particularly carbonates) in which fossil remains have been destroyed by diagenesis.

Biostratigraphy is a study for specialists. Refined work requires intimate knowledge of the phylogeny of a large number of fossil groups and their regional or global distribution. To accumulate this knowledge may take half a lifetime, and the subject is an excellent example of science in which the practitioner seems to spend inordinate amounts of time "learning more and more about less and less." Some of the leading authorities in a particular fossil group may be able to discuss the cutting edge of their research with only half a dozen other colleagues around the world. This gives them considerable value if one happens to find their kind of fossil, but it may somewhat restrict their scientific scope. Geologists engaged in basin analysis of the Phanerozoic are very rarely such specialists. Biostratigraphers are therefore employed by many organizations to provide these specialized service skills, or they function independently as consultants. They may be engaged much of the time in pursuing paleontological research but are able to provide biostratigraphic diagnoses for selected fossil types over a specified age range.

Professional biostratigraphic work may take a great deal of field and laboratory time. Sections that a sedimentologist may dismiss as sparsely fossiliferous may yield hundreds or even thousands of specimens to the careful collector. Laboratory extraction of microfossils or palynomorphs may yield similar numbers. It is this kind of work that is necessary for modern, refined biostratigraphic studies. Much of the submitted material, particularly that from frontier exploration wells, may itself provide the basis for new biostratigraphic zoning schemes.

Basin analysts should understand what they are getting when they submit their own material to a specialist for identification. Commonly, they are interested in two items of information: (1) age of the enclosing rock, the information that can be used for correlation purposes, and (2) information regarding the ecological environment of the fossils, which can aid in the interpretation of depositional environments. Age is a chronostratigraphic interpretation based on taxonomic descriptions, but commonly there are problems of fossil identification or interpretation, particularly where the fossil record is sparse, or the material is from a new, poorly studied area. It is particularly important that fossil collections (or the rock containing microfossils or palynomorphs for laboratory extraction) be located as precisely as possible in the outcrop section or well samples from which they were taken. The types of increased stratigraphic precision that are now required for advanced sequence-stratigraphic analysis require this (this is further discussed in Chap. 8). The purpose of this section, therefore, is to discuss some of the problems of the biostratigraphic record and to describe the methods that biostratigraphers use in plying their trade.

7.5.1 The Nature of the Biostratigraphic Record

Biofacies and Biogeography: The geographical distribution of taxa reflects the restriction of ranges due to ecological variations and the geographical isolation of populations. Two topics are included under this heading, the facies control of faunas and floras and the problem of faunal and floral provincialism.

Some taxa are adapted to a benthonic (bottom-dwelling) mode of life and others to a nektonic (swimming) or planktonic (floating) habit. In principal, nektonic or planktonic forms should be preferred for biostratigraphic purposes because of the likelihood of being more widely distributed and therefore more broadly useful. Benthonic forms tend to be more facies dependent because of their need for certain water conditions or sediment types for feeding and dwelling behavior. However, in practice, benthonic forms are widely used by professional biostratigraphers. Even such static forms as corals, burrowing mollusks, and anchored brachiopods have been found to be invaluable for zoning the deposits of the continental shelves. Many benthonic taxa have a planktonic larval stage that ensures wide distribution via marine circulation. Conversely, many planktonic forms, such as the graptolites, are too fragile to survive in agitated, shallow-water environments and are therefore just as facies-bound as their benthonic contemporaries. In practice, virtually every taxonomic group has some biostratigraphic utility, although considerable problems may arise in attempts to determine the relationships between the various facies-bound faunas, unless environmental fluctuations cause lithofacies of different types, with their accompanying faunas or floras, to become interbedded (Fig. 7.6). Where well exposed, such mixed successions are of great value in establishing a global chronostratigraphic framework. Figure 7.6 illustrates such a scenario schematically, where marine foraminiferal zones can be correlated to nonmarine palynomorphs because of the interfingering of these facies zones. The zig-zag "shazam" interfingering configuration used to illustrate facies relationships in this diagram is an overly simplistic representation of the progradation and retrogradation that occurs as a result of changes in relative sea level but helps to make the point that the vertical range of a given assemblage may be **diachronous**, changing in age laterally as a result of shifts through time of facies belts.

Classic examples of facies-bound faunas widely used by biostratigraphers are the **shelly** and **graptolitic** faunas of the lower Paleozoic. The shelly fauna actually includes two more or less distinct subfaunas, one in the inner, shallower shelf dominated by brachiopods and the other on the outer shelf, characterized by trilobites. The term appears to have arisen with Russian work that defined a "small shelly fauna," consisting of small exoskeletons of many different types (Matthews and Missarzhevsky 1975). The graptolitic fauna is confined mainly to low-energy deposits of the continental slope, rise and abyssal plain (Berry 1977). In Newfoundland, on what was the ancient eastern continental margin of North America during the Cambrian and Ordovician, these two facies interfinger. Carbonate turbidites and debris flows,

Fig. 7.6 The interbedding of three biofacies resulting from shifting of environments. Typically, such interbedding is caused by a sea-level change (McGowran 2005, Fig. 1.4, p. 8)

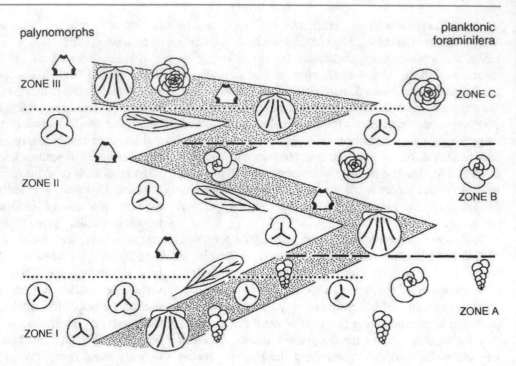

Fig. 7.7 Interbedding of a shelly fauna in limestone sediment-gravity flows and a graptolitic fauna in interbedded shales, Cambrian–Ordovician continental-margin deposits, Green Point, Newfoundland. The beds here are overturned. This location has been designated as a stratotype for the Cambrian–Ordovician boundary

derived from the collapse of the continental margin, are interbedded with graptolitic shales at the base of the continental slope (Fig. 7.7). At Green Point, on the west coast, this assemblage straddles the Cambrian–Ordovician boundary, and the location has been established as the stratotype for this important chronostratigraphic boundary (the GSSP for the base of the Tremadocian Series: Cooper et al. 2001; see Sect. 7.8.1 for a discussion of the GSSP).

A good example of facies control of what might appear at first sight to be a recurrent, biostratigraphically controlled fauna is provided by the brachiopod communities of the Upper Ordovician to Middle Silurian of the Welsh Borderlands. Ziegler et al. (1968) showed that at many localities there is a sequence of assemblages containing, in upward stratigraphic order, *Lingula, Eocoelia, Pentamerus, Stricklandia* and *Clorinda*, followed by a graptolitic fauna. Careful correlation of these sections using graptolites showed that the brachiopod sequence is markedly diachronous. This is shown in Fig. 7.8, in which the graptolite zones are shown by horizontal correlation lines labeled A1 to C7. One might ask, why are the graptolites trusted more than the brachiopods for the purposes of chronostratigraphic correlation? The answer is that brachiopods are benthonic organisms known to be prone to facies control, whereas graptolites are tried and tested biostratigraphic indicators. The correlations shown in Fig. 7.8 are supported by additional work on two of the brachiopod genera, *Eocoelia* and *Stricklandia*. When examined in detail, evolutionary trends can be detected within the populations of these genera as they are traced from west to east across the line of the section shown in Fig. 7.8 (and other sections not shown). The overall interpretation of these faunal data is that the sections reveal a gradual eastward marine transgression and deepening of the water such that successive brachiopod communities represent ecological adjustments to increased depths (*Lingula* inhabited brackish waters, and in part, the subsequent succession represents changing shell thicknesses in response to wave and tidal energy). There is no

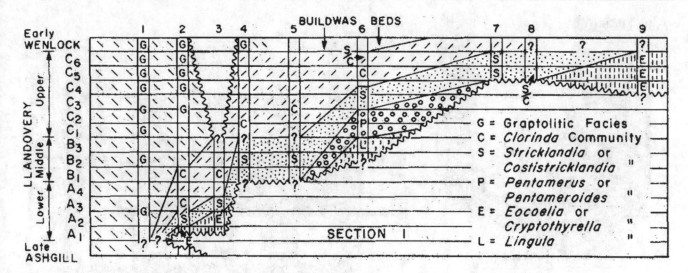

Fig. 7.8 An example of facies-bound faunas. The succession of brachiopod communities in each of these sections is the same (L, E, P, S, C), but the use of graptolite biostratigraphy (biozones A1 to C7) shows that they are markedly diachronous and therefore facies-controlled. Silurian, Welsh Borderland (Ziegler et al. 1968; McKerrow 1971)

obvious relationship between brachiopod assemblage and sediment type in this case.

Biogeography and evolution: William Smith recognized the significance of the succession of faunas in the stratigraphic record long before Darwin's theory of evolution was established. The point is that it is not necessary to understand evolutionary relationships of fossil groups in order to make use of them as tools in the inductive establishment of stratigraphic order and correlation. However, it can certainly help to explain the nature of taxonomic change and the distribution of distinct groups through time and space.

Molecular biology shows that the underlying process of evolution is genetic drift, the gradual accumulation of random incremental change in gene variants. Natural selection favors some mutations over others, which generates steady change. Isolation of populations will also tend to increase genetic divergence, eventually to the point that populations will be unable to interbreed and then constitute distinct species (see the review of current ideas by Kelley et al. 2013).

Three styles of evolution were described by Eldridge and Gould (1977), and are illustrated in Fig. 7.9. The first, termed **phyletic gradualism** or **transformational evolution** (McGowran 2005, p. 382), refers to long-term evolutionary change, typically in response to geographical, climatic or other environmental pressures (Fig. 7.9a). Certain varieties of a species may be favored by these changes so that there is a gradual adjustment in the stock until a distinctive new species appears.

Kauffman (1977) described two examples of phyletic gradualism in Cretaceous pelecypods. Figure 7.10 illustrates

a series of histograms of height–width ratios of the *Inoceramus pictus* lineage, derived from populations collected at about 40 cm intervals (lower graph), and the number of growth ridges in the first 25 mm of shells of *Mytiloides labiatus* (upper graph). The data permit subdivision of the population into species (S), subspecies (SS) and morphological zones (MZ), as indicated in the adjacent columns.

Another example of gradual evolution is provided by the foraminifera *Globigerina* and *Orbulina* (Fig. 7.11). The species listed in the top are an evolutionary series, three of which are illustrated (1: *Globigerina quadrilobatus*; 2: *G. bisphericus*; and 3: *Orbulina universa*). The gradual variation between these types has permitted the erection of seven zones, as indicated by the horizontal lines at the left.

Two types of environmental adaptation can occur, that which is accompanied by permanent genetic change and that which can, to some extent (never precisely), reverse itself to recreate the same variety or race of a species more than once, whenever the same environmental conditions are repeated (homeomorphy). Clearly, the first type is the only one of use to biostratigraphers, but the literature is replete with ambiguous biostratigraphic determinations that may be falsely based on diachronous environmental change. For example, this has been a serious problem with the ammonites (Kennedy and Cobban 1977), one of the best biostratigraphic indicators.

Taxa that evolve by phyletic gradualism have the most potential for refined biostratigraphic zonation, but they require specialist study to recognize the very subtle changes between the varieties. This type of work is beyond the abilities of the generalist basin analyst.

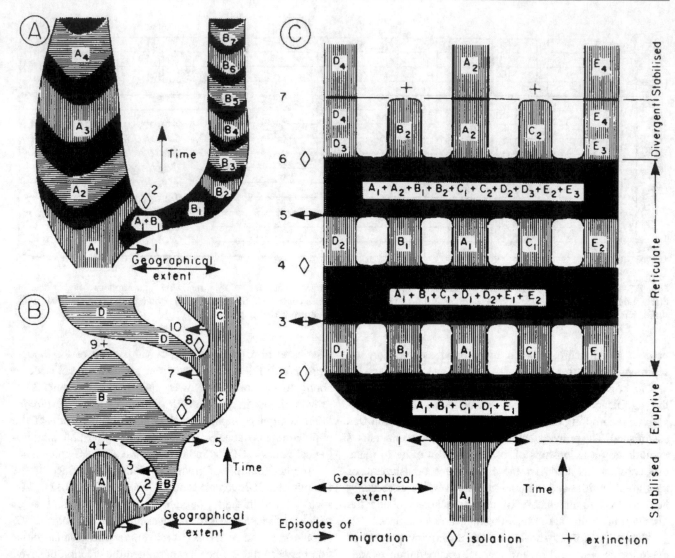

Fig. 7.9 Three models of evolution: A phyletic gradualism; B punctuated equilibrium; and C. reticulate speciation. A, B, etc., refer to successive varieties or species; 1, 2, etc., refer to the chronology of

events. In B, the same area at the left of the diagram is successively occupied by three species A, B, and C, which evolve elsewhere and migrate in (Sylvester-Bradley, 1977)

The second style of evolution was named **punctuated equilibrium** by Eldridge and Gould (1972, 1977). The concept was adapted from an earlier term, **allopatric speciation**, and is based on the premise that in a successful, widely distributed taxon the population is genetically conservative. Evolution is thought to occur only where extreme variants are selected by environmental pressures on the fringes of the species range. Rather than a gradual adaptation to an ecological niche or a broadening of a species range by extending slowly into subtly different niches, as in phyletic gradualism, the hypothesis of punctuated equilibrium proposes the spasmodic occurrence of bursts of relatively rapid evolutionary change. Extreme variants of a species can only evolve into a new species if they become isolated by changes in the environment, climate or geography, as through the

rifting and drifting apart of continental plates. Also, the catastrophic extinction of organisms by bolide impacts or other catastrophes empties out many ecological niches and permits rapid adaptive radiation and the explosive development of many new taxa in the period immediately following these extinctions (Fig. 7.9b).

Sylvester-Bradley (1977) proposed a third style of evolution, which he termed **reticulate speciation** (Fig. 7.9c). This combines, on a small scale, the mechanisms of both the other two evolutionary styles. Gene transfer may take place by several processes, including symbiosis, lateral transfer and hybridization. Sylvester-Bradley offered the modern common vole as an example of a taxon that has evolved in this way. The vole is distributed virtually globally and comprises numerous races reflecting adaptation to local

Fig. 7.10 Evolutionary trends in two pelecypod species, Cretaceous of Western Interior. Symbol X in the lithologic column indicates bentonite beds used for radioisotopic age dating. Suggested systematics and zonal subdivisions are shown in the numbered columns: MZ, morphological biozone; SS, subspecies; S, species (Kauffman 1977)

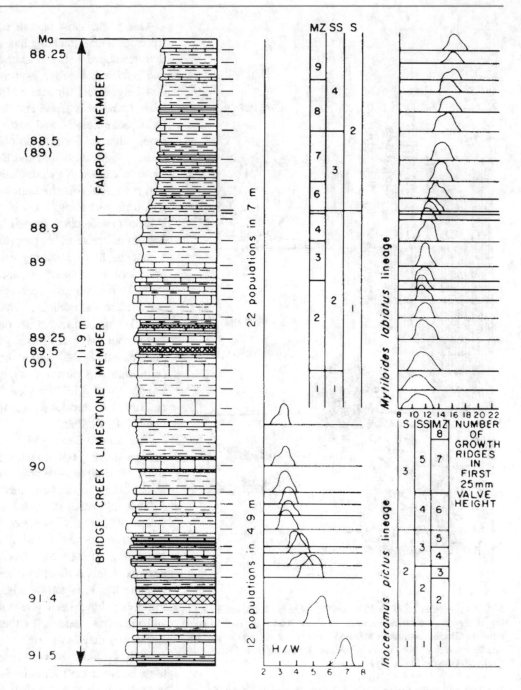

variations in climate, vegetation, altitude, isolation on islands etc. These varieties have evolved in response to rapid global changes following the Pleistocene ice age, and they demonstrate the rapidity with which geographical and ecological changes may bring about evolution. Grant and Grant (2008) similarly observed rapid reticulate speciation in Darwin's finches on the Galapagos Islands. The apparent stability of the species of many taxonomic groups for several million years or more at times during the geological past contrasts with the rapid adaptability of the modern vole and the finch. To what extent reticulate speciation will be

recognized for fossil groups remains to be determined. However, to recognize this style of evolution would seem to require an immense bank of detailed descriptive data, and therefore it is very much a subject of study for specialists.

A review of modern concepts in evolution as applied to the fossil record is provided by Kelley et al. (2013).

In the geological record, many distinct populations have been recognized, based on geographical distributions, and are defined as **faunal provinces**. These are much discussed by biostratigraphers. To the nonspecialist such concepts as the Malvinokaffric Province or the Tethyan or Boreal

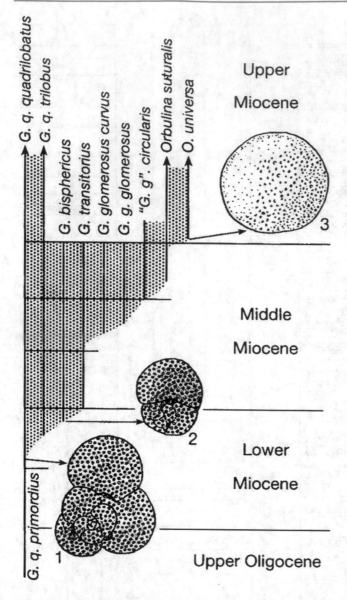

Fig. 7.11 Evolution of the foraminifera *Globigerina*, culminating in the different genus *Orbulina*, in southern Australia. The stippling pattern indicates continuous variation between the various morphotypes and their ranges. Seven zones are defined by the horizontal lines at left (from McGowran 2005, Fig. 1.4, p. 95)

Realms are sometimes difficult to understand. The definition of what constitutes a given faunal province requires a great deal of specialist knowledge, but even the specialists had difficulty before the advent of plate tectonics in comprehending why many of these provinces existed. Until the 1970s there was much discussion of the appearance and disappearance of strange, narrow "land bridges" to explain the merging and divergence of provincial variations. But in fact, faunal provincialism provides some of the most convincing geological lines of evidence for plate tectonics (Tarling 1982).

One of the most famous of these is the example of the trilobite faunas flanking Iapetus, the proto-Atlantic Ocean that developed between Laurentia (ancient North America) and Baltica and Africa in the earliest Paleozoic. Specialists noted significant differences in Cambrian trilobites between those found in England and those in Scotland, and between those from western and eastern occurrences in Newfoundland, whereas Scottish trilobites are similar to those in western Newfoundland, and English trilobites are similar to those in eastern Newfoundland (Fig. 7.12). Given the current geographic distributions of the fossils, these similarities and differences make no biogeographic sense. However, Wilson (1966) cited these distributions as one of several lines of evidence in his proposal—now universally accepted—that the line demarking the two distinct trilobite faunal provinces constitutes an ancient continental suture that formed when a former ocean closed as a result of subduction and continental collision. That ocean, now called Iapetus, occupied approximately the position of the present Atlantic Ocean, but many significant continental fragments changed margins when the present ocean developed; that is, the continental rift occurred along a somewhat different line. Landing et al. (2013) reviewed in detail the faunal provincialism that developed in the Cambrian following the breakup of Rodinia.

The ammonites provide an excellent example of the various biogeographic styles that can occur in organisms, some offering considerable advantages to the biostratigrapher, others a severe hindrance. Many ammonites underwent a planktonic larval stage that may have lasted from hours to weeks. Where this occurred, it would have been of some importance to the distribution of the species. Not all ammonites showed this. Distribution patterns and varying degrees of facies independence show that some ammonites were benthonic in adult life habitat, some were nektonic and some may have been planktonic. Their facies distribution and provincial tendencies thus varied considerably. Some ammonites may have drifted long distances after death. The modern *Nautilus*, the only living relative of the ammonites, has a buoyant shell after death, and observations in modern oceans suggest that the shell may drift for hundreds, if not thousands of kilometers. Geologically this could be of great importance, but Kennedy and Cobban (1977) suggested that many ammonites in fact became rapidly waterlogged after death and did not float appreciable distances.

Kennedy and Cobban (1977) summarized much of the data for Cretaceous ammonite distribution and concluded that there were five types of faunal provinces. Some genera have a virtually worldwide, or **pandemic**, distribution. Pandemic taxa would seem to offer the best possibilities for global correlation. They are relatively facies independent,

Fig. 7.12 The Cambrian trilobite and graptolite faunas of the Iapetus margins. Based on Wilson (1966)

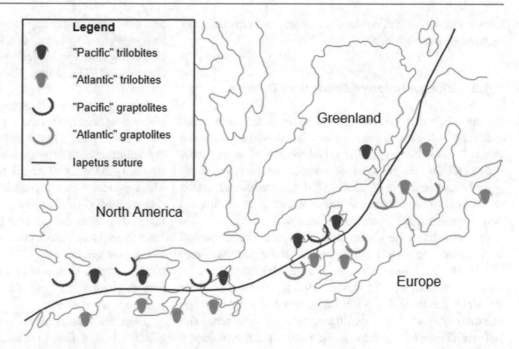

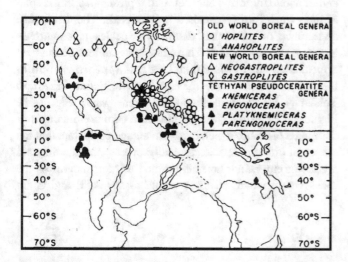

Fig. 7.13 The biogeography of selected Cretaceous ammonites, plotted on a Cretaceous plate-tectonic reconstruction of the continents. Climatic tolerances underlie the different geographic spread of Boreal versus Tethyan forms (an example of latitudinally restricted distributions), while the gradually widening Atlantic Ocean caused the gradual isolation and separate evolutionary development of ammonite faunas on different sides of the ocean (longitudinally restricted distributions). These faunas occasionally mixed in the Western Interior Seaway of North America because of shifts in climate and changes in sea level that caused local faunal migrations along a north–south axis (Kennedy and Cobban 1977)

but it turns out that many are long-ranging forms and thus of limited biostratigraphic usefulness.

Some ammonites have **latitudinally restricted** distributions, reflecting their preference for waters of a certain temperature or salinity and their tolerance of seasonal fluctuations. Some examples are shown in Fig. 7.13. They

define two provinces, the northern, colder water Boreal province (open symbols) and the more tropical Tethyan province (closed symbols). In many parts of Europe and North America, faunal fluctuations through stratigraphic successions between Tethyan and Boreal (and other) faunas have been cited as evidence of the existence of connecting seaways and transgressions across otherwise barren areas.

Longitudinal restrictions on distribution, such as the presence of land masses or large ocean basins, are a cause of further provincialism. These are added to latitudinal restrictions in the generation of the third type of faunal province: **endemic** distributions. Note that in Fig. 7.13 the Tethyan genera show no longitudinal restriction, whereas the four Boreal genera are typical endemic taxa, restricted to either Eurasia or North America (these are examples chosen to illustrate a point and should not be taken to define a universal difference between the Boreal and Tethyan provinces). Endemic ammonites have been shown to have evolved rapidly and thus are of prime biostratigraphic importance, although their provincialism has hindered intercontinental correlation.

Disjunct distributions are those of scattered but nevertheless widely distributed taxa. The distributions are not thought to represent inadequate data or severe facies control but probably reflect very low population densities.

As noted, some ammonite taxa may drift in oceanic currents after death. In extreme cases, where an endemic form is involved, such drifted or **necrotic** distributions may prove invaluable for long-distance correlation.

In general, the provinciality of taxa increases the difficulty by which they may be related to the global time scale, because provincialism reduces the variety of forms that may be used to establish relative ages.

7.5.2 Biochronology: Zones and Datums

By the mid-nineteenth century, the work of the early stratigraphers, following Smith, had clearly established the value of fossil assemblages for the establishment of stratigraphic order and for the purposes of comparison between stratigraphic sections. The similarity of the succession of faunas or floras between sections in different basins, even different continents—termed **homotaxis**—was well established. The main elements of the Phanerozoic geological time scale had been defined, including all the names of the periods (Berry 1987), and the concepts of the biozone and the stage were already developed (Hancock 1977). However, there remained the issue of age. In the absence of a clear understanding of how faunas and floras changed with time, and without the tools to establish numerical age, it remained a legitimate question whether homotaxis could be equated with **synchrony**. Darwin's *The Origin of Species* was published in 1859, yet in 1862 T. H. Huxley stated "for anything that geology or palaeontology are able to show to the contrary, a Devonian fauna or flora in the British Isles may have been contemporaneous with Silurian life in North America, and with a Carboniferous fauna and flora in Africa."

Eventually, this philosophical dilemma was resolved by developments in the understanding of the processes of evolution, coupled with the establishment of even more detailed systems of zonation and correlation, which left little room for doubt regarding the reality of the principle of relative age and time correlation based on fossil content. Some of the steps in the evolution of thought are described by Hancock (1977), Miall (2004) and McGowran (2005, pp. 54–65).

It is now universally accepted that formally established biozones represent specific intervals of time, subject to two important caveats: (1) the changes in taxonomy that lead to the definition of discrete datum planes of change, or zones of similarity in the fossil record, are not globally instantaneous. The appearance of a successful evolutionary step requires a discrete period of time for it to spread throughout its full range. This time period may be short, in geological terms, but it is not instantaneous and may be measurable. We return to this problem in Sect. 7.5.3. (2) The geographic range of a biostratigraphic datum or zone is limited by ecological factors. A biozone may not, therefore, have exactly the same age range everywhere.

There are various methods available for making the most efficient use of fossil occurrences. Particularly distinctive and/or abundant forms may serve to represent a specific span of time. Such forms are called **index fossils**. The first or last appearance of particular, distinctive species is commonly employed as biomarkers. These horizons are termed the **first appearance datum (FAD)** and the **last appearance datum (LAD)**. Suites of fossils may be used to define **biozones**. This may be done in several different ways (Fig. 7.14). Figure 7.15 illustrates one of the more common methods of defining a biozone, which takes advantage of the fact that the ranges of different species typically overlap. In this diagram a **concurrent range zone** is defined as that interval of the rocks within which all three of the fossils A, B and C are

Fig. 7.14 Types of biozone, as defined in the International Stratigraphic Guide (Salvador 1994. Diagram from Pearson (1998), Fig. 5.2, p. 126)

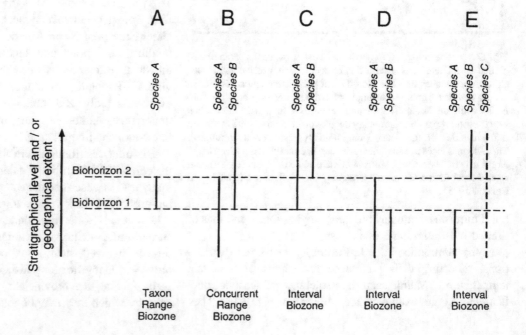

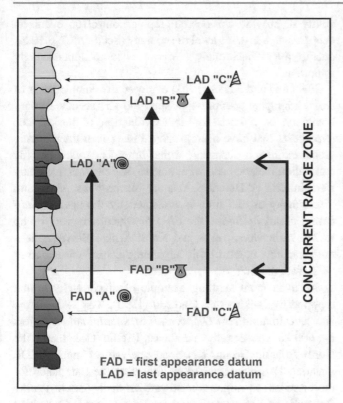

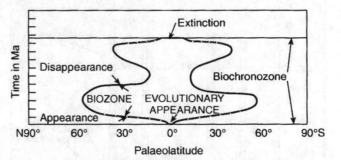

Fig. 7.16 The diachroneity of a biozone. A taxon first appears near the equator and takes a discrete amount of time (up to 2 m.y.) to migrate into higher latitudes north and south of the equator. The range does not then remain constant because of ecological factors, in this case, climate change, which leads to contraction and then renewed expansion of the range, until the taxon becomes extinct. The biochronozone (or biochron) of this taxon (the time span it represents) is global in extent but, in practice, the time span is that which is indicated by the actual presence of the taxon, which varies from place to place and, in this illustration, is absent altogether in high latitudes (McGowran 2005, Fig. 2.11, p. 38; after Loutit et al. 1988)

Fig. 7.15 The use of first and last appearance datums to define a concurrent range biozone. The area colored in grey is the only part of the section where all three species, A, B and C are expected to be present

present, and falls between the FAD of species B and the LAD of species A.

It is important to be aware that the time range indicated by a biozone is not necessarily the same everywhere. Figure 7.16 illustrates a hypothetical example where the diachronous spread of a fossil taxon and subsequent variations in its range owing to climatic factors has led to significant variability in the local range of the biozone. Diachroneity is discussed further in the next section. There are sophisticated methods for managing these issues and, as discussed in Sect. 7.8, biostratigraphy is still the major foundation of the geological time scale for the Phanerozoic.

In addition to ecological factors, there are considerations of preservation and potential sampling bias. Diagenetic destruction of fossils is common, and sampling bias might simply reflect bad luck in the choice of sampling site, or a bias induced by poor collection practices (Fig. 7.17). For example, whereas a field geologist undertaking a reconnaissance mapping exercise might be satisfied with a cursory examination of an outcrop for fossil content, a professional biostratigrapher is likely to carry out more thorough investigations. For example, soft sands and clays may be run through a sieve or water-washed on site to isolate macrofossils, or large lithologic samples may be collected at

routine stratigraphic intervals for water or acid treatment to extract microfossils or palynomorphs. In the subsurface, drilling disturbance may constitute a major problem. The drilling process penetrates layers from the top down, so the stratigraphically last appearance of a fossil (the LAD) will be the first encounter with a given taxon, and the level of this horizon is one that can be trusted. However, the tendency for holes to cave can lead to fossils (and rock cuttings) being fed into the mud stream after the drill bit has passed on down their point of origin (Fig. 2.28). For this reason, the FAD of a fossil taxon in the subsurface needs to be treated with caution.

As Sadler et al. (2014, p. 4) stated: "Signals and noise mingle among these contradictions. Records of real ecological patch dynamics, biogeographical habitat shifts and evolutionary turnover are confounded by incomplete preservation and collection."

Figure 7.18 is an example of palynological zonation of two wells through a Cretaceous succession in Delaware (from Doyle 1977). The wells are represented by their gamma ray logs with sample collection depths given in feet. The ranges of the principal angiosperm pollen types are shown by vertical bars and are shown dashed where identification is uncertain. Concurrent range zones are delimited by dashed lines perpendicular to the depth scale and are numbered I to IV next to the Series, Stage and Formation designations. It was found that the zones could be most easily defined on the basis of the first (oldest) appearance of a taxon, partly because extinct species tended to be reworked, and partly because taxa were found to die out slowly at the upper limit of their range. Work of this type required the counting and documentation of several

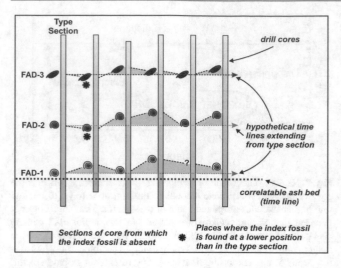

Fig. 7.17 First and last appearance data define ideal timelines in the rocks. However, in this example, many first occurrences are higher (younger) than the hypothetical timelines would predict and a few are lower (older). The reasons for this are discussed in the text

hundreds of individual pollen grains in each sample. Several or many complete sections through the succession of interest may be required before the data are adequate for the definition of the biozones. Range charts, such as that illustrated here, must be prepared for each and carefully compared.

Another way to define biozones focuses on the gradual change in the anatomy of a particular evolutionary lineage. This is called a **lineage zone**. McGowran (2005) used an example from the late Cenozoic mammalian evolution to illustrate this concept (Fig. 7.19). The evolution of dental morphology provided much of the information on which this zonal scheme was based.

7.5.3 Diachroneity of the Biostratigraphic Record

Another common item of conventional wisdom is that evolutionary changes in faunal assemblages are dispersed so rapidly that, on geological time scales, they can be essentially regarded as instantaneous. This argument is used, in particular, to justify the interpretation of FADs as time-stratigraphic events (setting aside the problems of preservation discussed above). However, this is not always the case. Some examples of detailed work have demonstrated considerable diachroneity in important pelagic fossil groups. Landing et al. (2013) provided an overview of the problem, focusing in particular on the Cambrian. Cramer et al. (2015) pointed out that many "events" that are assumed to be instantaneous on a geological time scale, may, in fact, be diachronous on a finer time scale. With our increasing

ability to provide chronostratigraphic control on events in deep time to a $\pm 0.1\%$ level of accuracy (Sects. 7.8.2, 8.10.3), quantifying diachroneity may become increasingly important.

MacLeod and Keller (1991) explored the completeness of the stratigraphic sections that span the Cretaceous-Tertiary boundary, as a basis for an examination of the various hypotheses that have been proposed to explain the dramatic global extinction occurring at that time. They used graphic correlation methods and were able to demonstrate that many foraminiferal FADs and LADs are diachronous. Maximum diachroneity at this time is indicated by the species *Subbotina pseudobulloides,* the FAD of which may vary by up to 250 ka between Texas and North Africa. However, it is not clear how much of this apparent diachroneity is due to preservational factors.

An even more startling example of diachroneity is that reported by Jenkins and Gamson (1993). The FAD of the Neogene foraminifera *Globorotalia truncatulinoides* differs by 600 ka between the southeast Pacific Ocean and the North Atlantic Ocean, based on analysis of much DSDP material. This is interpreted as indicating the time taken for the organism to migrate northward from the South Pacific following its first evolutionary appearance there. As Jenkins and Gamson (1993) concluded:

> The implications are that some of the well documented evolutionary lineages in the Cenozoic may show similar patterns of evolution being limited to discrete ocean water masses followed by later migration into other oceans ... If this is true, then some of these so-called 'datum planes' are diachronous.

Cody et al. (2008) provided another example of diachroneity. They reported on the distribution of diatoms in 32 Neogene cores from the Southern Ocean, which allowed an estimate to be made of the differences between the levels of the observed local first occurrence and last occurrence events and the projected levels of the global FADs and LADs for the same species. Around 50% of local event levels do not accurately record the timing of the global event: a few are off by 4 Ma or more from the total global FAD and LAD, due mostly to a small set of individually incomplete local ranges.

These conclusions are of considerable importance, because the results were derived from excellent data, and can, therefore, be regarded as highly reliable. They relate to some of the most universally preferred fossil groups for Mesozoic–Cenozoic biostratigraphic purposes, diatoms and foraminifera. It would appear to suggest a limit of up to about one-half million years on the precision that can be expected of any biostratigraphic event. Figure 7.16 illustrates the general problem.

The cases reported here may or may not be a fair representation of the magnitude of diachroneity in general,

Fig. 7.18 Use of palynological concurrent range biozones to correlate two subsurface wells (as shown by gamma ray logs). Correlation brackets (double-headed arrows) terminate just above and below samples in the other well that bracket the age of the indicated sample (Doyle 1977)

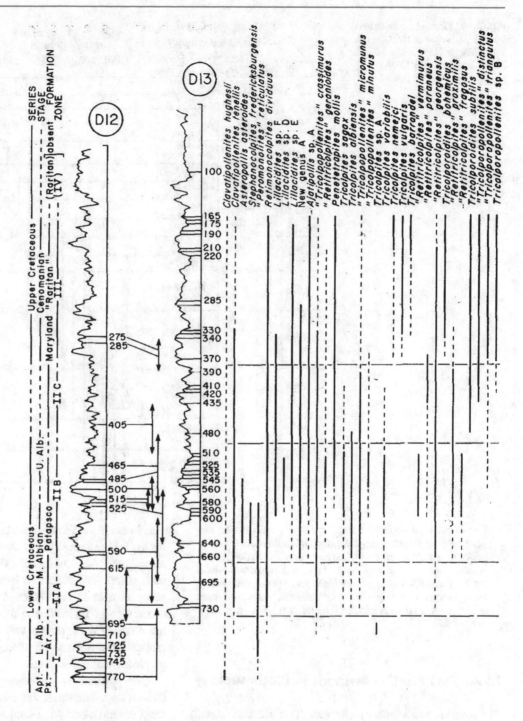

although this is recognized as a general problem (Smith et al. 2015). After a great deal of study, experienced biostratigraphers commonly determine that some species are more reliable or consistent in their occurrence than others. Such forms may be termed **index fossils**, and receive a prominence reflecting their usefulness in stratigraphic studies. Studies may indicate that some groups are more reliable than others as biostratigraphic indicators. For example, Ziegler et al. (1968) demonstrated that brachiopod successions in the

Welsh Paleozoic record were facies controlled and markedly diachronous, based on the use of the zonal scheme provided by graptolites as the primary indicator of relative time (Fig. 7.8). Armentrout (1981) used diatom zones to demonstrate that molluscan stages are time transgressive in the Cenozoic rocks of the northwest United States. Wignall (1991) demonstrated the diachroneity of Jurassic ostracod zones. Landing et al. (2013, p. 136) offered this general caution.

Fig. 7.19 Lineage zones defined
by rodent evolution (McGowran
2005, Fig. 4.31, p. 150, based on
Fejfar and Heinrich 1989)

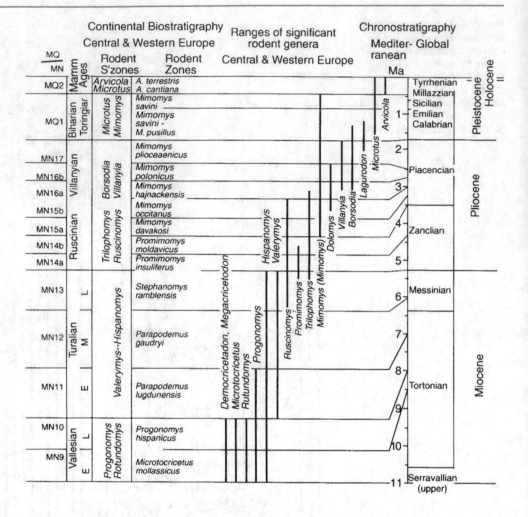

Use of the local FADs of a fossil for correlation between sections without rigorous supplementary information will lead to errors in correlation or poorly defined chronostratigraphic units because significant time intervals likely will separate the local FADs. Each FAD must mirror biological phenomena ranging from evolutionary origination, to dispersal, successful local colonisation, and appearance of facies that allow a species' fossilisation

7.5.4 Quantitative Methods in Biochronology

The graphic correlation technique. This method was first described in a landmark book by Shaw (1964). Useful explanations of the technique were given by Miller (1977) and Edwards (1984, 1985). Mann and Lane (1995) edited a research collection devoted to the application of this topic to practical problems in basin analysis. Gradstein et al. (2004b) discussed the use of the method in the construction of the geological time scale. The example used herein has been borrowed from Miller (1977).

As with conventional biostratigraphy, the graphic method relies on the careful field or laboratory recording of

occurrence data. However, only two items of data are noted for each taxon, the first (oldest) and last (youngest) occurrence (the FAD and LAD). These define a local range for each taxon. The objective is to define the local ranges for many taxa in at least three complete sections through the succession of interest. The more sections that are used, the more nearly these ranges will correspond to the total (true) ranges of the taxa. To compare the sections, a simple graphical method is used.

One particularly complete and well-sampled section is chosen as a **standard reference section**. Eventually, data from several other good sections are amalgamated with it to produce a **composite standard reference section**. A particularly thorough paleontologic study should be carried out on the standard reference section, as this enables later sections, for example, those produced by exploration drilling, to be correlated with it rapidly and accurately.

The graphic technique, which will be now described, is used both to amalgamate data for the production of the composite standard and for correlating the standard with new sections. Figure 7.20 shows a two-dimensional graph in which the thicknesses of two sections X and Y have been

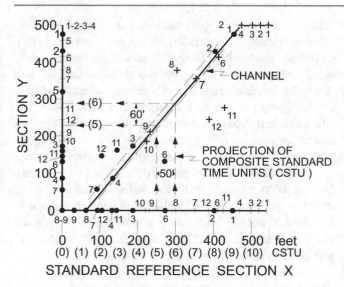

Fig. 7.20 This and the next three figures illustrate Shaw's (1964) graphic correlation method as discussed by Miller (1977). This plot shows the distribution of first occurrences (open circles) and last occurrences (crosses) in two sections and the positioning of the line of correlation. The channel is the zone on either side of the line of correlation encompassing observation error

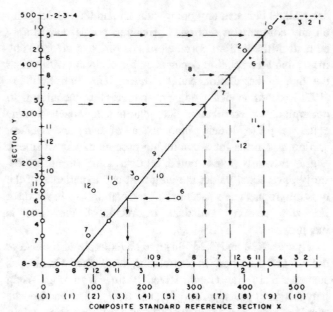

Fig. 7.21 The method used to compile a composite standard reference section. Data from new sections may be used to extend the range of occurrence of taxa that do not show their full range in the standard section (lowest occurrence of species 6, highest occurrence of species 8) and may also be used to transfer data on to the standard section, such as the range of species 5, which does not occur in the latter (Miller 1977)

marked of on the corresponding axes. The first occurrence of each taxon is marked by a circle on each section and the last occurrence by a cross. If the fossil taxon occurs in both sections, points can be drawn within the graph corresponding to the first and last occurrences by tracing lines perpendicular to the X and Y axes until they intersect. For example, the plot for the top of fossil 7 is the coincidence of points X = 350 and Y = 355.

If all the taxa occur over their total range in both sections and if sedimentation rates are constant (but not necessarily the same) in both sections, the points on the graph fall on a straight line, called the line of correlation. In most cases, however, there will be a scatter of points. The X section is chosen as the standard reference section, and ranges will presumably be more complete there. The line of correlation is then drawn so that it falls below most of the first occurrence points and above most of the last occurrence points. The first occurrence points to the left of the line indicate the late first appearance of the taxon in section Y. Those to the right of the line indicate the late first appearance in section X. If X is the composite standard, it can be corrected by using the occurrence in section Y to determine where the taxon should have first appeared in the standard. The procedure is shown in Fig. 7.21. Arrows from the first occurrence of fossil 6 show that in section X the corrected first appearance should be at 165 ft. The same arguments apply to the points for last appearances. Corrections of the kind carried out for fossil 6 (and also the last occurrence of fossil 8) in Fig. 7.21 enable refinements to be made to the reference section. Combining several sections in this manner is

the method by which the composite standard is produced. Data points can also be introduced for fossils that do not occur in the reference section. In Fig. 7.21, arrows from the first and last occurrences of fossil 5 in section Y show that it should have occurred between 320 and 437 ft in section X.

If the average, long-term rate of sedimentation changes in one or other of the sections, the line of correlation will bend. If there is a hiatus (or a fault) in the new, untested sections (sections Y), the line will show a horizontal terrace. Obviously, the standard reference section should be chosen so as to avoid these problems as far as possible. Harper and Crowley (1985) pointed out that sedimentation rates are in fact never constant and that stratigraphic sections are full of gaps of varying lengths (we discuss this problem in Chap. 8). For this reason, they questioned the value of the graphic correlation method. However, Edwards (1985) responded that when due regard is paid to the scale of intraformational stratigraphic gaps versus the (usually) much coarser scale of biostratigraphic correlation, the presence of gaps is not of critical importance. Longer gaps of the scale that can be detected in biostratigraphic data (e.g., missing biozones) will give rise to obvious bends in the line of correlation, as noted previously.

The advantage of the graphic method is that once a reliable composite standard reference section has been drawn up it enables chronostratigraphic correlation to be determined between any point within it and the correct point on any comparison section. Correlation points may simply be read of the line off correlation. The range of error arising from such a correlation depends on the accuracy with which the line of correlation can be drawn. Hay and Southam (1978) recommended using linear regression techniques to determine the correlation line, but this approach assigns equal weight to all data points instead of using one standard section as a basis for a continuing process of improvement. But as Edwards (1984) noted, all data points do not necessarily have equal value; the judgment and experience of the biostratigraphers are essential in evaluating the input data. For this reason, statistical treatment of the data is inappropriate.

Figures 7.22 and 7.23 illustrate an example of the use of the graphic method in correlating an Upper Cretaceous succession in the Green River Basin, Wyoming, using palynological data (from Miller 1977). The composite standard reference section has been converted from thickness into **composite standard time units**, by dividing it up arbitrarily into units of equal thickness. As long as the rate of sedimentation in the reference section is constant, these time units will be of constant duration, although we cannot determine by this method alone what their duration is in years. Figure 7.22 shows the method for determining the position of selected time lines on each test section, and in Fig. 7.23 the time units are used as the basis for drawing correlation lines between four such sections. Note the unconformity in each illustration and the variation in sedimentation rates in Fig. 7.23.

The value of the graphic method for correlating sections with highly variable lithofacies and no marker beds is obvious, and it is perhaps surprising that the method is not more widely used. An important difference between this method and conventional zoning schemes is that zoning methods provide little more than an ordinal level of correlation (biozones, as expressed in the rock record, have a finite thickness which commonly cannot be further subdivided), whereas the graphic method provides interval data (the ability to make graduated subdivisions of relative time). Given appropriate ties to the global time frame the composite standard time units can be correlated to absolute ages in years and used to make precise interpolations of the age of any given horizon (such as a sequence boundary) between fossil occurrences and tie points (Gradstein et al. 2004b). The precision of these estimates is limited solely by the accuracy and precision obtainable during the correlation to the global standard. MacLeod and Keller (1991) provided excellent examples of this procedure, and their results suggest an obtainable precision of less than ±100 ka.

Constrained optimization (CONOP): Gradstein et al. (2004b) pointed out several disadvantages of the basic graphic correlation method. It relies on only a few sections, placing particular importance on a single section that becomes the basis for the composite standard. A superior, automated correlation method, called constrained optimization, has been used in the construction of several parts of the Phanerozoic time scale. This method automates graphic correlation so that multiple sections are compared and correlated simultaneously. In this way, gaps and changes in sedimentation rate in the initial standard section do not influence the outcome. The method is described by Kemple et al. (1995) and Sadler (1999) and critically evaluated by Smith et al. (2015).

Several of the time scales for the Phanerozoic periods described in detail in Gradstein et al. (2004a) were developed using these techniques. For example, the Ordovician scale in Gradstein et al. (2004b) made use of 669 graptolite taxa in 119 sections. In the recent study of Ordovician–Silurian graptolite biostratigraphy by Sadler et al. (2009),

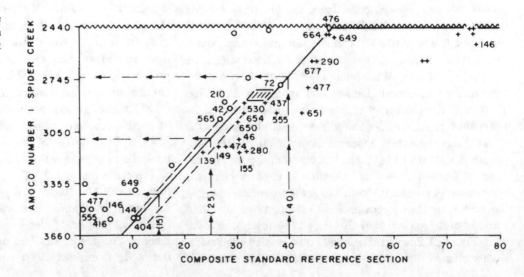

Fig. 7.22 An example of the use of the graphic method, showing a plot of data from one well against the composite standard. *Break* and *shaded areas* on the line of correlation are interpreted as an unconformity. The reference section has been divided into arbitrary thickness units (composite standard time units) (Miller 1977)

Fig. 7.23 Correlation of four
wells in an area of marked lateral
facies change using the composite
standard time units from Fig. 7.22
(Miller 1977)

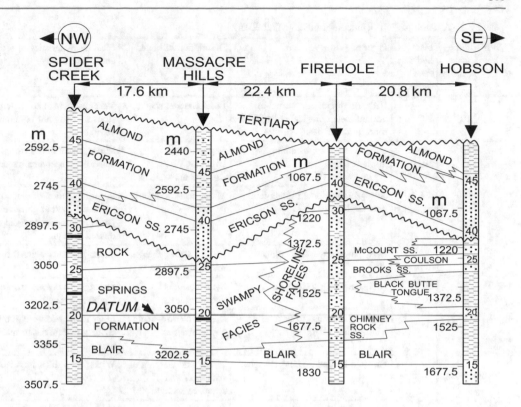

they noted (p. 887) that "The Graptolite zones vary widely in
duration from as short as 0.1 m.y. to nearly 5.0 m.y. The
mean duration of zones or zonal groupings calibrated here is
1.44 m.y. in the Ordovician and 0.91 m.y. in the Silurian."
Current developments in the biostratigraphic basis of
chronostratigraphy are discussed further in Sects. 8.9 and
8.10.

Calibration against the chemostratigraphic record. The
global development of chemostratigraphic methods
(Sect. 7.8.2) has provided an entirely new method for the
quantitative calibration of biostratigraphic data. It is now a
common practice among stratigraphers developing detailed
time scales for specific basins and those working on the
geological time scale to carry out chemostratigraphic sampling through the sections from which biostratigraphic data
have been obtained. The increasing reliability of the oxygen,
carbon and strontium isotopic global time scales has provided the basis for the numerical dating of biozones, which
can be then used on their own, if necessary, as independent
indicators of numerical age. For example, Cramer et al.
(2010) employed $\delta^{13}C$ isotope chemostratigraphy to calibrate conodont and graptolite biozones across the Llandovery–Wenlock (early-middle Silurian) boundary in
several sections in the Baltic area, with cross-referencing to
sections in Britain and the Niagara area of New York state.
They determined that the seven conodont and four graptolite

zones investigated spanned about one million years and that
the biozones each represented less than 500,000 years.

7.6 Unconformities and Unconformity-Bounded Units

The stratigraphic record is replete with sedimentary breaks
ranging from the localized scours swept out by migrating
bedforms, to the major angular unconformities that record
significant orogenic events. As discussed in Chap. 5,
unconformities have served as the primary basis for the
subdivision of stratigraphic successions into sequences. In
many basins there are hierarchies of sequences nested within
each other and separated by sedimentary breaks of varying
time significance (e.g., Figs. 5.30 and 5.31). Various terms
have been erected in attempts to reflect the significance of
different sedimentary breaks, in terms of the sedimentologic
or structural discordance represented by the break (e.g.,
diastem, paraconformity etc.), but none of these have been
rigorously defined, and their use is not recommended.
Likewise, the expression "relatively conformable" has been
used to define the nature of the successions that characterize
the sequence contained between the unconformities, but as
discussed in Sect. 5.4, this term, also, has not been defined
and, as we now know from several decades worth of

Table 7.1 A classification of unconformities (Miall 2016)

SRS	Time scale (yrs)	Inst Sed Rate (m/ka)	Process	Description of break	Field characteristics of sedimentary break and/or of beds above and below
1–4	10^{-6}–10^{-1}	10^4–10^6	Bedform migration; diurnal to normal meteorological changes in runoff; tidal cycles	Local channel scours	Nesting of channels, macroforms and bedforms within a structure of minor bounding surfaces (ranks 1–5 of Miall, 1996)
5–7	10^0–10^3	10^0–10^3	Subtle tectonism, including in-plane stress	Migration and switching of depositional systems	Minor cut-and-fill erosion, early cementation, "unconformity paleosols" in nonmarine settings
			Autogenic seasonal to long-term geomorphic processes	Minor erosion, mature paleosols	Superimposition of delta and shelf-margin clinoform lobes separated by transgressive ravinement surfaces, rare preservation of falling-stage incised distributary channels, incised valleys;
			Rare extreme weather events	Marked facies change, minor regional erosion	Facies blanket, regional marker horizon
7-9	10^4–10^5	10^{-2}–10^0	High-frequency tectonism	Syndepositional unconformities	Strong but very localized angularity, coarse clastic wedges ("growth strata")
			Regional response to flexural loading/unloading	Basin-wide low-angle unconformities	Low- to very-low-angle clinoform sets. Evidence of fluvial or marine erosion, transgressive lag deposits at breaks
			Far-field intraplate stress changes	Tilting and warping of sequences and sequence sets	Widespread shifts in paleocurrent patterns, shoreline trends
			Orbital forcing	Continental (potentially global)—scale, nonangular break	Cyclothemic facies changes, potentially deep erosion of unconformity surface, coastal and shelf-margin clinoform onlap-offlap cycles
9–12	10^6–10^7	10^{-3}–10^{-1}	Orogenic tectonism	Regional angular unconformity	May be associated with deep erosional relief, clastic wedges
			Dynamic unconformities associated with basin formation	Onlap and offlap caused by basin subsidence	Onlap of extensional margins during flexural subsidence. Onlap/offlap during motion of foreland-basin forebulge
			Dynamic topography	Sub-continental unconformity	Low-angularity (units above and below have similar dip). Commonly little field evidence of major time break
			Global eustasy	Global unconformity	Similar to above
			Long-term environmental change	Regional disconformities	Eolian supersurfaces, drowning unconformities (carbonates)

SRS = Sedimentation rate scale (from Miall 2015)

research on sequence stratigraphy, is misleading to the extent that all stratigraphic successions contain sedimentary breaks of varying temporal magnitude and physical extent (Table 7.1).

Table 7.1 provides a classification of sedimentary breaks based on their temporal significance and the mechanisms that generated them (from Miall 2016). They are conveniently classified into four broad classes on the basis of their temporal significance. Firstly, there are the minor breaks formed by bedform turbulence, representing minutes to seconds during normal bedform migration, and days to weeks as channels migrate laterally. These need careful examination, because subtle indicators, such as evidence of organic activity may indicate the passage of significant time (Davies and Shillito 2018). The second group represents breaks of up to a few hundred years duration caused by seasonal to long-term geomorphic processes, such as the migration and switching of depositional systems. The third group comprises the products of high-frequency tectonism and those caused by orbitally forced climate change, including glacioeustatic fluctuations in base level. These represent 10^4–10^5 years. Those caused by tectonism may be angular; those generated by glacioeustasy will be structurally conformable. Lastly, there are the angular unconformities formed by orogenesis, basin development and dynamic topography, and those formed by long-term eustasy, with time significance of 10^6–10^7 years.

Unconformities, typically appear in the rock record as apparently simple time planes, such as a stratal surface marked by contrasting lithologies having a significant difference in weathering characteristics. However, a great deal of information may be gained by a regional study of such surfaces (Miall 2016). The following are examples of "group 1" unconformities in Miall's (2016) classification, that is, those caused by regional tectonism on a 10^6–10^7-year time scale (Table 7.1). The continent-wide unconformity that defines the contact between the Canadian Shield and the Phanerozoic sedimentary record across North America is an ancient paleo-landscape surface, with remnants of fluvial drainage valleys and shorelines, remnant sea stacks and boulder beaches of late Precambrian age. Over much of western Australia, there is evidence of a regional Cretaceous surface. Erosional valleys with preserved regolith grade down into valleys occupied by lavas of Eocene age; Precambrian granite domes emerge as hills from beneath a remnant cover of Jurassic rocks that onlap their lower slopes (Twidale 1997). In the Rocky Mountain states a series of regional unconformities in the Triassic and Jurassic succession was recognized and named by Pipiringos and O'Sullivan (1978). These contain a great deal of information about the tectonic and paleogeographic evolution of the Mesozoic Western Interior Seaway. Zuchuat et al. (2019) carried out a detailed study of one of these surfaces, the J-3 Unconformity, separating the Middle Jurassic Entrada Sandstone from the Upper Jurassic Curtis Formation (and laterally equivalent units) in east-central Utah (USA). This is a laterally variable surface, generated by either erosion-related processes, such as eolian deflation, and water-induced erosion, or by deformational processes. The J-3 Unconformity is a composite surface formed by numerous processes that interacted and overlapped spatially and temporally.

Unconformities have commonly served as convenient boundaries for various types of stratigraphic units, particularly those based on lithostratigraphy. In many areas, regional unconformities have long been used to define natural subdivisions of the stratigraphic record (Blackwelder 1909). As originally defined by Sloss et al. (1949), sequences were defined as operational units separated by "marked discontinuities in the stratal record of the craton which may be traced and correlated for great distances on the objective bases of lithologic and faunal 'breaks' ...." The use of unconformities as boundaries is now avoided in the definition of chronostratigraphic units, for reasons explained in Sect. 7.7. However, it is increasingly, being recognized that the stratigraphic record is subdivided into unconformity-bounded units of regional and possibly even global extent, caused by widespread changes in sea level or by regional tectonic or climatic events (Vail et al. 1977; Emery and Myers 1996; Miall 1995, 2010). As noted by Miall (2015), sequence stratigraphy "works" because there is

a limited number of processes that operate at the 10^4–10^8-year time scale to create the sequence record and the unconformities that serve to subdivide it into convenient packages for description and interpretation. The methods of sequence stratigraphy have now become virtually universal (Catuneanu 2006).

Unconformity-bounded units include other types of units within them, such as biostratigraphic and lithostratigraphic units. They are not the same as these and are not necessarily equivalent to chronostratigraphic units, because the ages of the bounding unconformities may change from place to place. However, unconformity-bounded units have a certain chronostratigraphic significance because, with certain unusual exceptions, all the rocks below an unconformity are older than all of those above, and time lines do not cross unconformity surfaces. Some exceptions to these rules include (1) where a disconformity surface is caused by submarine erosion by a deep oceanic current that changes position with time, as a result of changing the configuration of the ocean basin (Christie-Blick et al. 1990), but this is not a problem that is likely to be encountered very frequently; (2) Ravinement surfaces are time-transgressive (Nummedal and Swift 1987); (3) Large-scale-incised valleys, caused by base-level fall at a coastline, may be significantly time-transgressive (Strong and Paola 2008); (4) The subaerial erosion surfaces that form the basis for most sequence subdivisions are typically highly complex and diachronous surfaces, as demonstrated by Holbrook and Bhattacharya (2012; see Fig. 5.8).

Proposals for the formal definition of unconformity-bounded units were given by the International Subcommission on Stratigraphic Classification (1987; see also Salvador 1994) and are also contained in the North American Commission on Stratigraphic Nomenclature (1983, 2005). The term **sequence**, as originally used by Sloss et al. (1949), was not recommended by these authorities, because this term had come to be used in a number of slightly but significantly different ways. In common geological parlance, the term *sequence* has commonly been used as a synonym for *succession*, a practice that needs to be discouraged because of the value now associated with the sequence method. The ISSC (1987) recommended the formal term **synthem**, a proposal followed by Salvador (1994), whereas the North American Stratigraphic Commission proposed the definition of **allostratigraphic units**, including alloformation, allogroup and allomember (NACSN 1983). The term synthem has not been accepted by the stratigraphic community. **Allostratigraphy**, as a formal system of definition and naming, has had modest success. The intent was that major unconformity-bounded units would be termed alloformations, with minor units nested within them, such as parasequences, labeled allomembers (NACSN 2005). The intent of this approach was to be purely descriptive, with no genetic connotations built into the terminology or methods.

Several groups of workers have made use of allostratigraphic methodology. For example, Autin (1992) subdivided the terraces and associated sediments in a Holocene fluvial floodplain succession into alloformations. R. G. Walker and his coworkers employed allostratigraphic terminology in their study of the sequence stratigraphy of part of the Alberta Basin, Canada. Their first definition of unconformity-bounded units is described in Plint et al. (1986), where the defining concepts were referred to as **event stratigraphy**, following the developments of ideas in this area by Einsele and Seilacher (1982). With the increasing realization that sequences and their bounding surfaces may be markedly diachronous, we no longer refer to sequences and sequence boundaries as "events." Walker (1990) discussed some of the practical problems in making use of sequence and allostratigraphic concepts. Explicit use of allostratigraphic terms appears in later papers by this group (e.g., Plint 1990). A text on facies analysis that built extensively on the work of this group recommends the use of allostratigraphic methods and terminology as a general approach to the study of stratigraphic sequences (Walker 1992). Martinsen et al. (1993) compared lithostratigraphic, allostratigraphic and sequence concepts as applied to a stratigraphic succession in Wyoming. As they were able to demonstrate, each method has its local advantages and disadvantages. Sequence stratigraphy is characterized by powerful, genetic concepts and interpretive methods, which provide it with a major advantage where appropriate. However, some workers, notably A. G. Plint, have found that the basic mapping of unconformities, and flooding surfaces is practical and efficient, and he continues to use allostratigraphic terminology in his work (see, for example, Shank and Plint 2013, and Fig. 6.4, in which correlations are focused on the mapping of surfaces of erosion and transgression; what they term E/T surfaces).

7.7   The Development of Formal Definitions for Sequence Stratigraphy

Allostratigraphy represented one of the first attempts to incorporate unconformity-bounded units into the formal framework of stratigraphy (see Sect. 1.2.9 for earlier ideas on this topic). Acceptance of its successor, sequence stratigraphy, by the "official" international community—the *International Commission on Stratigraphy*—has nearly been achieved. Controversies about how to define sequences have hindered the development of formal procedures.

One of the commonest complaints about sequence stratigraphy was that it is "model-driven." Catuneanu (2006, pp. 6–9) summarized the various approaches that have been taken to defining sequences and argued the case that the differences between the various models are not important, so long as sequences are described properly with reference to a selected standard model, with correct and appropriate recognition of systems tracts and bounding surfaces. His diagram comparing the various approaches is reproduced here as Fig. 7.24, and the suite of important surfaces that are used in sequence and systems-tract definition is shown in Fig. 7.25. The major difference between the sequence models is where different workers have chosen to place the sequence boundary. It should be noted that in each of the sequence definitions shown in Fig. 7.26, a similar set of systems tracts is shown in much the same relationship to each other. Exceptions include the T-R sequence, which makes use of a simplified definition of systems tracts, and such differences as that between the "late highstand" of depositional sequence III and the "falling stage" of depositional sequence IV.

There has been extensive discussion between the original proponents of the modern sequence models (P. R. Vail, H. Posamentier, J. Van Wagoner and their colleagues at Exxon) and others, regarding sequence definitions, centered on such characteristics as the facies shifts that take place within sequences and their significance with regard to the base-level cycle. This discussion has led to a number of different ways of defining sequences (Figs. 7.24, 7.25 and 7.26). In a masterly synthesis of the controversies, Catuneanu (2006) showed how to resolve these differences, and in three major collaborative publications (Catuneanu et al. 2009, 2010, 2011) discussed proposals for formal definitions that could be accepted by all workers (Fig. 7.26). As he demonstrated, "all approaches are correct under the specific circumstances for which they were proposed" (Catuneanu et al. 2011, pp. 232–233). Additional details, including descriptions and illustrations of the key bounding surfaces that develop on continental shelves, are described by Zecchin and Catuneanu (2013).

In the original Exxon model (Vail et al. 1977) the sequence boundary (commonly abbreviated as SB on diagrams) was drawn at the subaerial unconformity surface, following the precedent set by Sloss (1963), an approach that readily permits the sequence framework to be incorporated into an allostratigraphic terminology, at least for coastal deposits, where the subaerial erosion surface is readily mapped. Offshore may be a different story. The first sequence model (Vail et al. 1977) did not recognize the falling-stage systems tract. The highstand of one sequence was followed directly by the lowstand of the next sequence, with the sequence boundary falling

Fig. 7.24 The evolution of sequence definitions. T-R = transgressive–regressive. From Catuneanu et al. (2011)

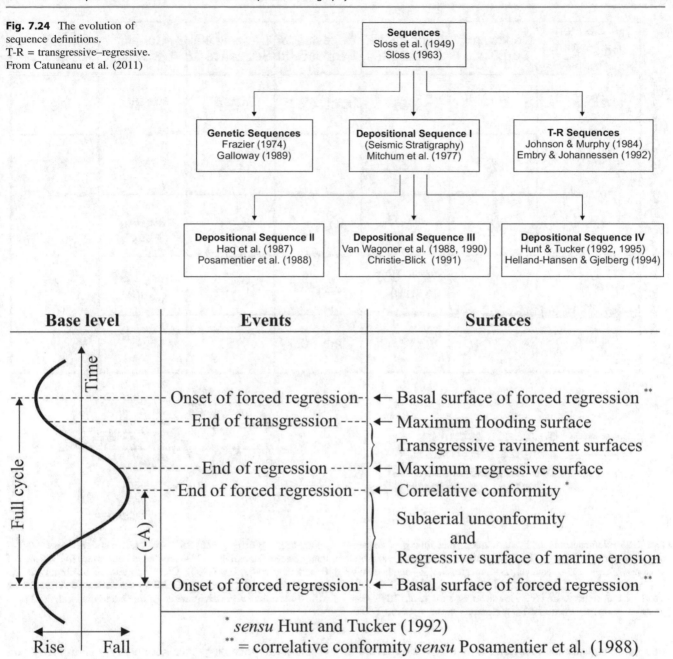

Fig. 7.25 Stratigraphic surfaces used in the definition of sequences and systems tracts, and their timing, relative to the cycle of base-level change (Catuneanu 2006, Fig. 4.7)

between the two systems tracts. Plint (1988) was the first to recognize the importance of the process of **forced regression** that generated by a fall in base level, and that included the formation of "sharp-based sandstone bodies" and the **regressive surface of marine erosion** (Fig. 5.14). The forced regressive deposits could be assigned either to a "late highstand" or an "early lowstand." Based on assumptions about the changing rate of sea-level fall, the sequence boundary—the coastal equivalent of the sub-aerial erosion surface—was initially placed at the **basal**

surface of forced regression (assumed commencement of forced regression). This placement of the sequence boundary is the basis for what Catuneanu (2006) refers to as "depositional sequence II" (Fig. 7.26). The problem with this definition is addressed below. In addition, the early Exxon work defined several different types of sequence-bounding unconformity. Vail and Todd (1981) recognized three types, but later work (e.g., Van Wagoner et al. 1987) simplified this into two, termed type-1 and type-2 unconformities, based on assumptions about the

Events and stages (Sequence model)	Depositional Sequence I	Depositional Sequence II	Depositional Sequence III	Depositional Sequence IV	Genetic Sequence	T-R Sequence
HNR		HST	early HST	HST	HST	RST
end of T					——MFS—	
T	Sequence	TST	TST	TST	TST	TST
end of R						—MRS—
LNR		late LST (wedge)	LST	LST	late LST (wedge)	
end of RSL fall		- - - - - -	- - - - -	CC**	- - - - - -	
FR		early LST (fan)	late HST	FSST	early LST (fan)	RST
onset of RSL fall	——CC*—	CC* - - - - -				
HNR		HST	early HST	HST	HST	

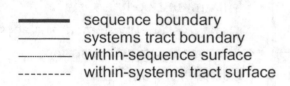

———— sequence boundary
———— systems tract boundary
·········· within-sequence surface
- - - - - - within-systems tract surface

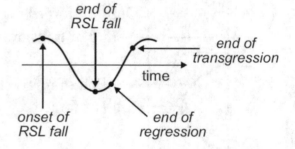

end of RSL fall

end of transgression

time

onset of RSL fall

end of regression

Fig. 7.26 Nomenclature of systems tracts, and timing of sequence boundaries for the various sequence stratigraphic approaches. Abbreviations: RSL—relative sea level; T—transgression; R—regression; FR—forced regression; LNR—lowstand normal regression; HNR—highstand normal regression; LST—lowstand systems tract; TST—transgressive systems tract; HST—highstand systems tract; FSST—falling-stage systems tract; RST—regressive systems tract; T-R—transgressive–regressive; CC*—correlative conformity in the sense of Posamentier and Allen (1999); CC**—correlative conformity in the sense of Hunt and Tucker (1992); MFS—maximum flooding surface; MRS—maximum regressive surface (from Catuneanu et al. 2011)

rate of change of sea level and how this was reflected in the sequence architecture. In the rock record, they would be differentiated on the basis of the extent of subaerial erosion and the amount of seaward shift of facies belts. However, Catuneanu (2006, p. 167) pointed out the long-standing confusions associated with these definitions and recommended that they be abandoned. These "types" are not discussed further in this book. Schlager (2005, p. 121) recommended the separate recognition of the third type of sequence boundary, the **drowning unconformity**, which forms "when sea level rises faster than the system can aggrade, such that a transgressive systems tract directly overlies the preceding highstand tract often with a significant marine hiatus. … Marine erosion frequently accentuates this sequence boundary, particularly on drowned carbonate platforms."

Hunt and Tucker (1992) were among the first (since Barrell!) to point out that during sea-level fall, subaerial erosion continues until the time of sea-level lowstand, with the continuing transfer of sediment through clastic delivery systems to the shelf, slope and basin, and with continuing downcutting of the **subaerial erosion surface** throughout this phase. The age of the subaerial erosion surface, therefore, spans the time up to the end of the phase of sea-level fall, a time substantially later than the time of initiation of forced regression. The use of the basal surface of forced regression as a sequence boundary, as in "depositional sequence II" is, therefore, not an ideal surface at which to define the sequence boundary, although, as Catuneanu (2006) reports, it is commonly a prominent surface on seismic-reflection lines. In fact, as Embry (1995) pointed out (see Fig. 7.29), there is no through-going surface associated with forced regression that can be used to extend the subaerial erosion surface offshore for the purpose of defining a sequence boundary. He argued that "from my experience I have found that the most suitable stratigraphic surface for the conformable expression of a sequence boundary is the transgressive surface" (Embry, 1995, p. 4). This meets his criterion—one with which all stratigraphers would agree—that "one of the main purposes of sequence definition [is] a coherent genetic unit without significant internal breaks" (Embry 1995, p. 2). His preferred definition of sequences, the T-R sequence, places the sequence boundary at the TR surface, at the end of the phase of regression and the time of initial transgression (Figs. 7.28, 7.29). There is, of course, a delay in time between the end of downcutting of the subaerial erosion surface during the falling stage, and the flooding of the same surface during transgression. The results of the two processes may coincide in the rocks, which is why this surface may provide a good stratigraphic marker, but it is important to remember that the surface is not a time marker and represents a time gap, with the gap decreasing in duration basinward.

Highlighting the timing of development of the subaerial erosion surface by Hunt and Tucker (1992) also served to highlight the inconsistency of assigning the main succession of submarine fan deposits on the basin floor to the lowstand systems tract, as shown by the Vail et al. (1977) and Posamentier et al. (1988) models, in which these deposits are shown resting on the sequence boundary. Notwithstanding the discussion of the Hunt and Tucker (1992) paper by Kolla et al. (1995), who defended the original Exxon models, this is an inconsistency that required a redefinition of the standard sequence boundary. It is now recognized that in the deep offshore, within submarine-fan deposits formed from sediment delivered to the basin floor during a falling stage (which is often the most active interval of sediment delivery to the continental margin; see Fig. 5.15), there may be no sharp definition of the end of the falling stage nor of the turn-around and subsequent beginning of the next cycle of sea-level rise and, therefore, no distinct surface at which to draw the sequence boundary. The boundary here would be a **correlative conformity**, and may be very difficult to define in practice.

An alternative sequence model, termed the **genetic stratigraphic sequence** (Figs. 7.26, 7.28), was defined by Galloway (1989), building on the work of Frazier (1974). Although Galloway stressed supposed philosophical differences between his model and the Exxon model, in practice, the difference between them is simply one of where to define the sequence boundaries. The Exxon model places emphasis on subaerial unconformities, but Galloway (1989) pointed out that under some circumstances, unconformities may be poorly defined or absent and, in any case, are not always easy to recognize and map. For example, in all but the largest outcrops a fluvial channel scour surface may look exactly like a regional subaerial erosion surface. Galloway's (1989) preference was to draw the sequence boundaries at the **maximum flooding surface**, which corresponds to the highstand downlap surfaces. He claimed that these surfaces are more prominent in the stratigraphic record, and therefore more readily mappable.

Galloway's proposal has not met with general acceptance. For example, Walker (1992) disputed one of Galloway's main contentions, that "because shelf deposits are derived from reworked transgressed or contemporary retrogradational deposits, their distribution commonly reflects the paleogeography of the precursor depositional episode." Galloway (1989) went on to state that "these deposits are best included in and mapped as a facies element of the underlying genetic stratigraphic sequence." However, as Walker (1992) pointed out, most sedimentological parameters, including depth of water, waves, tides, basin geometry, salinity, rates of sediment supply and grain size, change when an unconformity or a maximum flooding surface is crossed. From the point of view of genetic linkage, therefore, the only sedimentologically related packages lie (1) between a subaerial unconformity and a maximum flooding surface, (2) between a maximum flooding surface and the next younger unconformity or (3) between a subaerial erosion surface and the overlying unconformity (an incised valley fill)(Walker 1992, p. 11).

However, some workers have found Galloway's use of the maximum flooding surface much more convenient for sequence mapping, for practical reasons. For example, it may yield a prominent gamma ray spike in wireline logs (Underhill and Partington 1993), or it may correspond to widespread and distinctive goniatite bands (Martinsen 1993), or it may provide a more readily traceable marker, in contrast to the surface at the base of the lowstand systems tract, which may have irregular topography and may be hard to distinguish from other channel-scour surfaces (Gibling and Bird 1994). In nonmarine sections it may be hard to find the paleosol on interfluves that correlates with the sequence-bounding channel-scour surface (Martinesen 1993). In some studies (e.g., Plint et al. 1986; Bhattacharya 1993) it has been found that ravinement erosion during transgression has removed the transgressive systems tract so that the marine flooding surface coincides with the sequence boundary.

Catuneanu et al. (2011, p. 183) pointed out that however the sequences are defined, there may be unconformities contained within them. For example, Galloway's (1989) genetic stratigraphic sequences are explicitly defined with the sequence boundary at the maximum flooding surface, not the subaerial erosion surface. Therefore, sequences can no longer be defined as "unconformity-bounded." Given the need to encompass all types of sequence model, Catuneanu et al. (2009, p. 19) proposed redefining a sequence as "a succession of strata deposited during a full cycle of change in accommodation or sediment supply." As discussed in Sect. 5.4, sequences commonly occur in nested hierarchies, with thinner sequences of shorter duration nested within larger sequences that represent longer time spans. The boundaries between all of these sequences are unconformities of varying time significance (Figs. 5.29–5.31).

As described in Chaps. 3 and 4, many stratigraphic successions contain small-scale cycles nested within a sequence. Van Wagoner et al. (1988, 1990) erected the term **parasequence** to encompass the shoaling-upward successions, capped by flooding surfaces, that are common in coastal clastic successions. The term was proposed originally as part of a hierarchy of terms, the bed, bedset, parasequence, parasequence set and sequence. Prograding delta lobes, regressing clastic shorelines and peritidal carbonate cycles are examples of parasequences that are particularly common in the geological record. Catuneanu et al. (2009, p. 19) noted that the term has also been used for cyclic deposits in some fluvial and deep-marine deposits, where the concept of "flooding surface" is irrelevant. A particular source of confusion comes from the incorporation of the word "sequence" within the term parasequence. The nomenclature problem was not improved by the usage employed by Mitchum and Van Wagoner (1991), who equated parasequences with "4th-order paracycles." Sequences are allogenic products of regional controls, whereas parasequences may be a product of autogenic processes, such as delta lobe switching (Fig. 7.27). This has been demonstrated to be the case in the example of the Dunvegan delta illustrated in Fig. 5.11 (Bhattacharya 1991). The shingles and their bounding flooding surfaces are therefore local in distribution, and their development has little, if anything, to do with the allogenic mechanisms that generate sequences. However, to apply to these successions a term that contains the word "sequence" in it is inevitably to introduce the implication that they are allogenic in origin and constitute regionally correlatable units. The correct interpretation clearly depends on good mapping to determine the extent and correlatability of each shingle, and it would seem advisable not to use a term in a descriptive sense that carries genetic implications. Yet Mitchum and Van Wagoner (1991) illustrated parasequences/paracycles with a strike-oriented

cross-section 400 km long, which claims to show correlations between several separate depositional systems. This, and the implication that parasequences are the same thing as minor sequences, added to the confusion. Furthermore, some high-frequency sequences formed by Milankovitch processes are comparable in thickness and outcrop appearance. It would be this author's preference that the term parasequence be abandoned entirely, but Catuneanu et al. (2009, 2011) (papers to which this author contributed!) recommended that the term may continue to be used if it is restricted to its original definition. An additional discussion of the term parasequence is provided in Sect. 5.4.

The selection of which sequence definition or model to use in any given stratigraphic setting is a matter of choice. The original Sloss/Vail **depositional sequence** model (Sloss 1963; Vail et al. 1977) and the **T-R model** of Embry and Johannessen (1992) use the subaerial erosion surface as the sequence boundary (Figs. 7.28, 7.29). The sequence then represents a full cycle of increasing and decreasing accommodation preserved between two erosion surfaces. The Galloway (1989) **genetic stratigraphic sequence** model uses the maximum flooding surface (MFS) as the sequence boundary (Figs. 7.28, 7.29). In many stratigraphic settings, recognizing and correlating a maximum flooding surface (MFS) is much more easily accomplished than mapping the subaerial erosion surface. The MFS is commonly represented by a marine shale or a condensed section which, because it is deposited at a time of maximum transgression, is typically widespread and forms a distinctive marker bed between packages of coarser clastic or carbonate/evaporite facies. Subaerial erosion surfaces may be characterized by significant erosional relief, which may make them difficult to trace within suites of wireline logs. Furthermore, where this surface occurs within successions of nonmarine strata it can be very difficult to distinguish the sequence boundary from local fluvial channel scour surfaces (see Fig. 5.8). However, these differences in mappability do not need to be reflected in the choice of sequence model.

The TR sequence model uses the subaerial erosion surface as the sequence boundary for the nonmarine and nearshore portions of a sequence but differs from the depositional sequence where sequences are traced into the offshore.

In a nonmarine to coastal to marine section, the maximum seaward extent of the subaerial erosion surface depends on the amplitude of relative sea-level change. Traced far enough seaward there will be a point beyond which water depths are such that exposure and erosion do not take place during base-level fall. Correlating sequence boundaries and systems tracts into the offshore may be difficult. The erosion surface generated by forced regression (Fig. 5.14; RSE in Fig. 7.29) defines the base of the falling-stage systems tract but, as demonstrated by Plint (1988), this surface is not at a

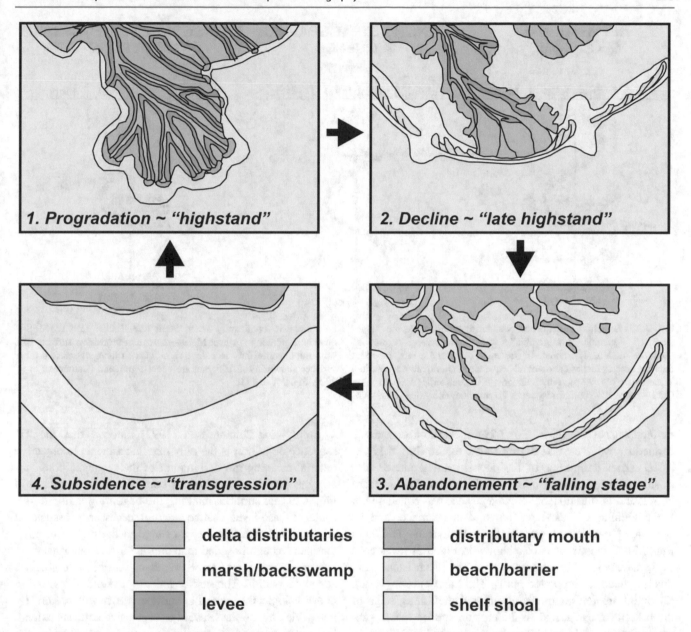

Fig. 7.27 The development and abandonment of delta lobes in a river-dominated, Mississippi-type delta (e.g., see Fig. 4.21), based on a detailed analysis of the Mississippi delta system by Boyd and Penland (1988). In stage 1, progradation develops an upward-shoaling deltaic succession. Abandonment, followed by subsidence (resulting from compaction) cause the upper layers of the succession to be reworked (stage 2), resulting in the development of an extensive barrier island system (stage 3). Finally, the deposit undergoes transgression and is covered by marine shale (stage 4). Repetition of this succession of events when a new delta lobe progrades back over the older deposit results in shoaling upward successions bounded by transgressive flooding surfaces, that is, parasequences. In this case, however, they are clearly of autogenic origin. Systems-tract designations for each of the four stages are indicated in parenthesis

stratigraphically consistent level, but may recur at stratigraphically higher positions in the offshore direction as base level continues to fall. The offshore limit of the RSE records the end of the phase of base-level fall, and this point is used to define the sequence boundary in **depositional sequences III** and **IV** (Fig. 7.29). Regressive sedimentation is likely to continue until the rate of rising accommodation during the subsequent cycle of base-level rise equals the rate of sediment supply. At this time, regression ends, transgression begins, and this point is used by Embry (1995) and Embry and Johannessen (1992) to define the sequence boundary for **TR cycles**. The transgressive surface which then develops marks the beginning of the next cycle of base-level rise (TS in Fig. 7.29). In many shelf settings, the TS is marked by lag deposits or a condensed section and is readily recognizable in cores and on wireline logs. It passes landward into the

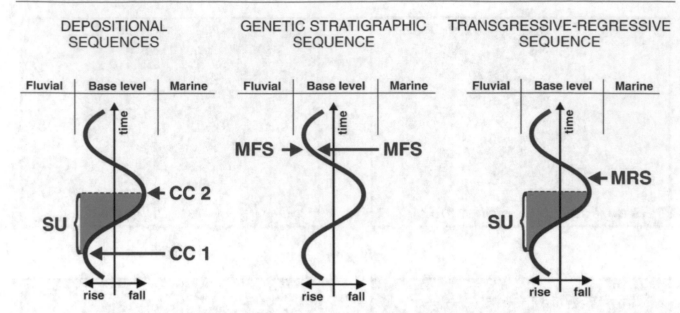

DEPOSITIONAL SEQUENCES **GENETIC STRATIGRAPHIC SEQUENCE** **TRANSGRESSIVE-REGRESSIVE SEQUENCE**

Fig. 7.28 Selection of sequence boundaries according to the "depositional," "genetic stratigraphic" and "transgressive–regressive" sequence models. The choice of sequence boundary is less important than the correct identification of all sequence stratigraphic surfaces in a succession (Fig. 7.10). Abbreviations: SU—subaerial unconformity; CC 1—correlative conformity *sensu* Posamentier and Allen (1999); CC 2—correlative conformity *sensu* Hunt and Tucker (1992); MFS—maximum flooding surface; MRS—maximum regressive surface. The subaerial unconformity is a stage-significant surface, whereas all other surfaces shown in this diagram are event-significant (Catuneanu et al. 2009, Fig. 24, p. 17)

ravinement surface (R in Fig. 7.29), a diachronous surface formed by wave erosion during transgression (Fig. 5.17).

In earlier definitions of the depositional sequence (**depositional sequences I** and **II** in Fig. 7.26), the seaward correlation of the subaerial erosion surface was equated with the beginning of base-level fall and this surface, and its seaward extension as correlative conformity (CC1 in Fig. 7.28; this surface is also shown in Fig. 7.29) was used to demarcate the sequence boundary. The inconsistencies in this approach were pointed out by Hunt and Tucker (1992). Subaerial erosion continues throughout the falling stage of the base-level cycle, and so the erosion surface that is subsequently preserved by being transgressed and buried during the succeeding base-level rise will correspond in age to the end of the falling stage, not the beginning. The first regressive surface marking the beginning of the falling stage (e.g., as in Fig. 5.14) may be followed by others, further seaward and stratigraphically younger, as the falling stage proceeds. Another of the inconsistencies of the early sequence definitions was the assignment of offshore submarine fan deposits primarily to the "lowstand wedge" formed (according to this earlier interpretation) during the beginning of the next cycle of base-level rise. Given that much of the sediment supply for submarine fans come from subaerial erosion during falling base level (Fig. 5.15), this does not seem plausible.

Embry and Johannessen (1992) claimed that the TR sequence definition is the only one that adheres to objective criteria, meaning the designation of the sequence boundary at a mappable surface. This is correct in that, as shown in Fig. 7.29, the surfaces marking the beginning and end of a cycle of base-level change are not necessarily associated with any recognizable facies change in the rocks, meaning that their recognition and mapping in surface outcrops or in subsurface cores, logs or seismic data, requires interpolation and extrapolation. The surface marking the end of regression is a lithologically mappable surface, but it will be directly mappable only so far seaward as regressive deposits extend. Beyond that point, a correlation may be possible by the tracing of reflections in seismic data, but recognition of the surface of maximum regression in cores or logs through deep marine deposits is not likely to be possible, so this particular feature of the TR definition is not significantly better than that of the depositional sequence.

A generalized sequence model is shown in Fig. 7.30 (see also Fig. 5.11). In this model the base-of-slope submarine fan and other deep-water deposits are assigned to the falling stage systems tract, contrary to the first sequence definitions (depositional sequences I and II). The sequence boundary (timeline 4) includes an incised valley, which is filled by the lowstand systems tract and during transgression (timelines 5–7). Commonly, incised valleys are filled toward the end of

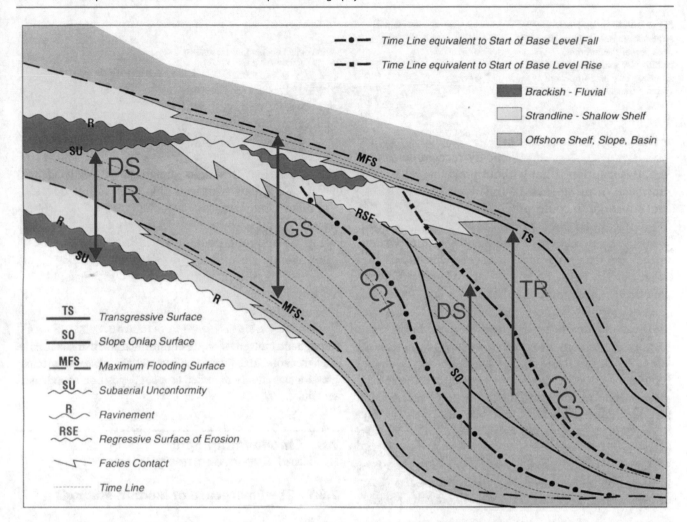

Fig. 7.29 Schematic cross-section through an ideal continental-margin sequence, showing the relationships between the major surfaces. Red arrows and letters denote three alternative ways to define sequences. In coastal and nearshore deposits, the sequence boundary for depositional sequences (DS) and TR sequences (TR) is defined by the subaerial erosion surface (SU). Correlating this surface offshore may be difficult in the absence of high-quality seismic-reflection data. Genetic stratigraphic sequences (GS) are defined by the maximum flooding surfaces. Depositional sequences III and IV use the correlative conformity corresponding to the end of base-level fall as the sequence boundary (CC** of Fig. 7.26, CC2 of Fig. 7.28 and this figure). This may be indicated by a change from erosional to aggradational deposition in the forced-regressive deposits. The time of initiation of base-level fall corresponds to correlative conformity CC* of Fig. 7.26 and CC1 of Fig. 7.28 and this figure, and defines the contact between early and late highstand in depositional sequence III. The top of a DS is indicated here by the arrow labeled DS that terminates at the surface marking the beginning of base-level rise. TR sequences are defined offshore by the end of regression. In this diagram the wedge of regressive deposits is truncated by the transgressive surface, which here marks the top of a TR sequence. In depositional sequence terminology, this latest wedge of regressive deposits is classified as the lowstand systems tract (adapted from Embry 1995, Fig. 1)

the transgressive phase, with deposition extending beyond the margins of the valley. Elsewhere on the continental margin transgressive deposits are characteristically thin and may consist of condensed deposits or a coarse lag. Hardgrounds and extensive bioturbation are common.

A significant flaw of the TR model is that the surface of maximum regression may not correlate with the subaerial erosion surface. As Embry and Johannessen (1992) noted, regression may continue in time beyond the end of the cycle of base-level fall, which corresponds to the age of the subaerial erosion surface. The deposits that form after this time are assigned to the lowstand systems tract and may comprise a substantial thickness of regressive deposits, accumulated on the shelf, slope and deep basin during the beginning of the next cycle of transgression. Two seismic lines are shown here which contain this feature (Figs. 7.31, 7.32). The lowstand normal regressive deposits (LNR) in Fig. 7.31 are the deposits that form after the end of base-level fall—

Fig. 7.30 A generalized sequence model. Twelve time lines reveal the internal architecture and evolution of the sequence and its component systems tracts

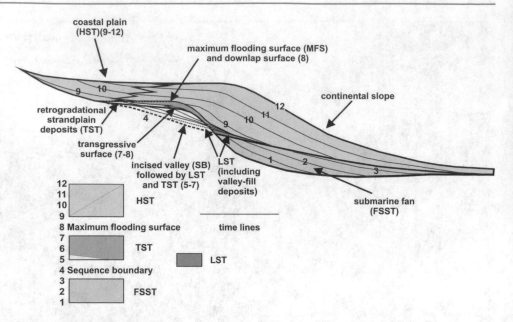

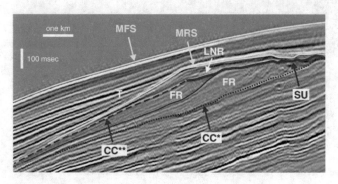

Fig. 7.31 Seismic line in the Gulf of Mexico showing different genetic types of deposits (forced regressive, normal regressive, transgressive) and stratigraphic surfaces that may serve as sequence boundaries according to different sequence stratigraphic models (modified from Posamentier and Kolla, 2003). Abbreviations: FR—forced regressive; LNR—lowstand normal regressive; T—transgressive; SU—subaerial unconformity; CC*—correlative conformity sensu Posamentier and Allen, 1999 (= basal surface of forced regression); CC**—correlative conformity sensu Hunt and Tucker 1992; MRS—maximum regressive surface; MFS—maximum flooding surface. The line displays the typical stacking patterns and stratal terminations associated with forced regression (offlap, downlap, toplap, truncation), normal regression (downlap, topset), and transgression (onlap) (Catuneanu et al. 2009, Fig. 7, p. 6)

indicated by the correlative conformity (CC**) extending basinward from the subaerial erosion surface—and are capped by the **maximum regressive surface (MRS)**. A particularly thick succession of shelf and slope deposits is assigned to the LNR in Fig. 7.32. This section illustrates the migration through time of the shelf margin and the shoreline. In this example a high sediment supply was maintained

throughout transgression, resulting in a thick transgressive shelf section. This is commonly not the case, with transgression commonly recorded as a lag deposit or a condensed section.

7.8 Chronostratigraphy and Geochronometry

7.8.1 The Emergence of Modern Methods

Geochronometry is the study of the continuum of geologic time. Geologic events and rock units may be fixed within this time frame by a variety of methods, of which radioisotopic dating is the most direct.

Chronostratigraphy is the study of the standard stratigraphic scale, comprising the familiar eras, periods and ages (e.g., Paleozoic, Triassic, Campanian) (Table 7.2).

For the Phanerozoic, biostratigraphy is the main basis of the chronostratigraphic method, but radioisotopic methods

Table 7.2 The conventional hierarchy of formal chronostratigraphic and geochronologic terms

Chronostratigraphic	Geochronologic
Eonothem	Eon
Erathem	Era
System	period
Series	Epoch
Stage	Age
Substage	Subage

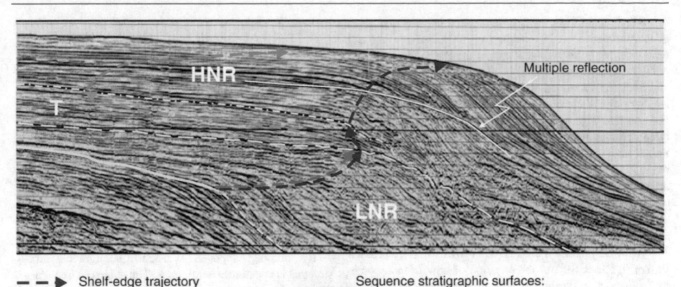

--- ▶ Shelf-edge trajectory

Shoreline trajectory:
--- ▶ lowstand normal regression
--- ▶ transgression
--- ▶ highstand normal regression

Sequence stratigraphic surfaces:

——— subaerial unconformity
- - - - correlative conformity (*sensu* Hunt and Tucker, 1992)
— — maximum regressive surface
—·—·— maximum flooding surface

Fig. 7.32 Dip-oriented regional seismic profile from the Pelotas Basin, southern Brazil (modified from Abreu 1998), showing large-scale (high-rank) lowstand normal regressive (LNR), transgressive (T) and highstand normal regressive (HNR) systems tracts. Lower rank sequences are nested within these higher rank systems tracts. The transgressive systems tract thickens landward, which reflects the direction of shift of the depocenter. Individual backstepping parasequences are difficult to observe within the transgressive systems tract due to the limitation imposed by vertical seismic resolution. The shoreline trajectory and the shelf-edge trajectory may coincide during lowstand normal regression but are separate during transgression and highstand normal regression. The change in depositional trends from dominantly progradational to dominantly aggradational is typical for lowstand normal regressions. Conversely, the change in depositional trends from dominantly aggradational to dominantly progradational is typical for highstand normal regressions. Horizontal scale: approximately 50 km. Vertical scale: 2 s two-way travel time (Catuneanu et al. 2009, Fig. 19, p. 14)

and magnetic reversal stratigraphy are essential for providing numerical ("absolute") ages, and chemostratigraphy is becoming increasingly important. Radioisotopic dating is the only chronostratigraphic tool available for the Precambrian, and is essential for providing a calibration scale of biostratigraphic subdivisions in the Phanerozoic (Harland et al. 1964, 1990; Berggren et al. 1995). Magnetic reversals are now of great importance for studying Upper Cretaceous and Cenozoic strata but are difficult to use in older rocks owing to the difficulty of obtaining complete reversal sequences and the problems of post-depositional modification (Kennett 1980).

Early speculations about the age of the Earth, and of rates of geological processes, began to be situated within a modern quantitative framework with the development of radioisotopic dating by Ernest Rutherford in 1905. The English geologist Arthur Holmes made the development of a radioisotopically calibrated geologic time scale a central part of his life's work, and it could be argued that the first modern era in the quantification of geologic time ended with his death in 1965. Two major events occurred at about this time. The second edition of his great book "*Principles of Physical Geology*" appeared (Holmes 1965), containing, among other things, a lengthy treatment of radioisotopic dating (plus what were then his very advanced ideas about continental drift), and the Geological Society of London published the first comprehensive compilation of data in support of a modern geological times scale (Harland and Francis 1964, 1971).

It is important to distinguish between the two quite different, but interrelated concepts of time, geochronometry, the measurement of time, in standard units such as the year and the second, and chronostratigraphy, the compilation of standard rock units. A distinction between "time" and "rocks" is essential, given the incompleteness of the stratigraphic record, and the need to continually revise, expand and update the means by which we relate the chronostratigraphic to the geochronometric scale. Harland (1978) and Harland et al. (1990, Chaps. 1–3) provided detailed explanations of the theory and terminology surrounding these terms. Because of the very fragmentary nature of the sedimentary record, there is still no chronostratigraphic standard for the Precambrian, which is subdivided mainly on a geochronometric basis, except for the Ediacaran System (635–542 Ma), the youngest part of the Precambrian, which is the only exception (Knoll et al. 2006).

The subdivision of the stratigraphic record into even smaller units based on detailed studies of their fossil content had reached a remarkably sophisticated level by the early twentieth century, based on the specialized study of some unique units in which the fossils are abundant and contain readily measurable indicators of rapid evolution. The ammonites of the Jurassic in southern Britain figure prominently in this history, as described by Callomon (1995, 2001). It can be demonstrated that local subdivisions of relative stratal time representing time spans in the range of 10^5 years are possible based on such work (this is discussed further in Chap. 8). But the question of how to incorporate this information systematically in the construction of a chronostratigraphic scale that could be used worldwide continued to be controversial and problematic until the 1970s (this early history is summarized by Miall 2004, p. 6–11). There are three obvious reasons: the record is everywhere locally incomplete, fossils are facies bound, and with very rare exceptions it is not possible to assign specific numerical ages to fossil horizons without interpolations and extrapolations that might incorporate the other two problems. Some boundaries, including many established early in the history of the science, had been very poorly defined. For example, the Silurian–Devonian boundary, first recognized in Britain, occurs at a major angular unconformity, or within a marine to nonmarine transition, within which biostratigraphic correlation was problematic, raising (but only much later, when time began to be quantified) the question of how to define and categorize the time undocumented at the break.

Solutions to these problems began to emerge in the 1960s and might be said to constitute the beginning of the second phase of the modern era of modern stratigraphy. The key development was the evolution of the **Global Stratigraphic Sections and Points (GSSP)** concept. The ideas appear to be primarily British in origin (e.g., Ager 1964; Bassett 1985; Cowie 1986; Holland 1986). They encompass two important concepts: the idea of designating key marker boundaries at specific type locations within continuous sections, and the idea that multiple criteria—biostratigraphy and radioisotopic dating, and more recently magnetostratigraphy and chemostratigraphy, wherever applicable, should be used to nail down the precise age of the boundary "by all available means," to quote Torrens (2002, p. 256). McLaren (1970, p. 802) explained the desirability of defining boundaries within continuous sections in this way:

There is another approach to boundaries, however, which maintains that they should be defined wherever possible in an area where "nothing happened." The International Subcommission on Stratigraphic Classification, of which Hollis Hedberg is Chairman, has recommended in its Circular No. 25 of July, 1969, that "Boundary-stratotypes should always be chosen within sequences of continuous sedimentation. The boundary of a chronostratigraphic unit should never be placed at an unconformity. Abrupt and drastic changes in lithology or fossil content should be looked at with suspicion as possibly indicating gaps in the sequence which would impair the value of the boundary as a chronostratigraphic marker and should be used only if there is adequate evidence of essential continuity of deposition. The marker for a boundary-stratotype may often best be placed *within* a certain bed to minimize the possibility that it may fall at a time gap." This marker is becoming known as "the Golden Spike."

By "nothing happened" it meant a stratigraphic succession that is apparently continuous. The choice of boundary is then purely arbitrary and depends simply on our ability to select a horizon that can be the most efficiently and most completely documented and defined (just as there is nothing about time itself that distinguishes between, say, February and March, but to define a boundary between them is useful for purposes of communication and record). This is the epitome of an empirical approach to stratigraphy. Choosing to place a boundary where "nothing happened" is to deliberately avoid having to deal with some "event" that would require interpretation (see Miall 2004 for a discussion of the importance of this point). This recommendation was accepted in the first International Stratigraphic Guide (Hedberg 1976, pp. 84–85). The concept also includes the proviso that only the base of a unit is so defined at the chosen location, not the top of the underlying unit, lest future work determines that at the type section the boundary is marked by a hiatus—hence the term **topless stage**. The "missing time" so identified is then assigned to the underlying unit, permitting continuous revision without the need for new boundary definitions.

The concept of the Global Stratigraphic Section and Point (GSSP), informally called the **Golden Spike** concept, was rapidly accepted, and has led to an explosion of specialized work under the auspices of the *International Commission on Stratigraphy* (ICS), a division of the *International Union of Geological Sciences* (IUGS), to correlate key sections worldwide in order to develop internationally recognized markers for epochs and stages that could then become part of the standard chronostratigraphic scale (e.g., Remane 2000a). This work is regularly reported in the IUGS journal *Episodes* and, nowadays, on the website https://stratigraphy.org/ maintained by the ICS. The criteria for the selection and ratification of a GSSP are as follows (https://stratigraphy.org/gssps/):

- A GSSP has to define the lower boundary of a geologic stage.
- The lower boundary has to be defined using a primary marker (usually the first appearance datum of a fossil species).
- There should also be secondary markers (other fossils, chemical, geomagnetic reversal).
- The horizon in which the marker appears should have minerals that can be radioisotopically dated.
- The marker has to have a regional and global correlation in outcrops of the same age
- The marker should be independent of facies
- The outcrop has to have an adequate thickness
- Sedimentation has to be continuous without any changes in facies
- The outcrop should be unaffected by tectonic and sedimentary movements, and metamorphism
- The outcrop has to be accessible to research and free to access
- This includes that the outcrop has to be located where it can be visited quickly (International airport and good roads) and has to be kept in good condition (ideally a national reserve), in accessible terrain, extensive enough to allow repeated sampling and open to researchers of all nationalities.

A discussion of selected GSSPs is presented in Sect. 7.8.3. As of 2021, 102 Phanerozoic boundaries had been established by the definition of a GSSP. Of these, only 72 had been agreed upon and ratified by the Commission. The first to be ratified, in 1972, was that for the base of the Lochkovian Stage, which defines the base of the Devonian. The GSSP was defined at Klonk, in the Czech Republic (Fig. 7.42). All the GSSPs for the Silurian and Devonian have been ratified, and those for the Silurian were completed in 1984. Only 4 of the 12 Cretaceous GSSPs have been ratified, a list that does not include that for the base of the Berriasian, which defines the base of the Cretaceous. This is surprising, considering the wealth of biostratigraphic and chemostratigraphic data now available for the Cretaceous and younger intervals in the Phanerozoic.

The importance of using "all available means" to identify, calibrate and date GSSPs cannot be over-emphasized. As noted in Sect. 7.5.3, the use of single criteria, such as first appearance data (FAD) and last appearance data (LAD) can lead to errors associated with biostratigraphic diachroneity. Quantitative methods, particularly graphic correlation and its advanced version, constrained optimization, make use of multiple criteria. As Smith et al. (2015) noted, despite the best efforts of stratigraphers, it has emerged that some GSSPs have been defined at sedimentary breaks, and others still over-rely on few chronostratigraphic criteria, such as the

range of a single key taxon. As of 2015, some 74 of currently defined GSSPs actually only make use of biostratigraphic criteria. Smith et al. (2015) also noted that most GSSPs have been defined at outcrop sections of shallow-marine rocks (where the probability of multiple hiatuses and much missing section is greatest: see Chap. 8), with little use being made of subsurface drill-core sections, and none have been established in deep-marine sediments, where the preserved record is likely to contain fewer hiatuses.

It is instructive to review the development of the modern geologic time scale from 1964 to the present. Among the first to appear were the scales for the Jurassic and Cretaceous developed by Van Hinte (1976a, b). Useful discussions of stratigraphic concepts were compiled by Cohee et al. (1978). Numerous attempts to synthesize existing data for the Phanerozoic have been made, notably by British (Harland et al. 1982, 1990), French (Odin 1982) and American (Berggren et al. 1995) groups. An important review of modern Chinese work was provided by Ogg (2019). Each group drew on its own data base and made different interpolations and extrapolations, with the result that there are significant differences in the assigned ages of many of the important chronostratigraphic boundaries.

A quantum leap forward was achieved by the ICS with the publication in 2004 of its updated geologic time scale (GTS2004: Gradstein et al. 2004a). Gradstein et al. (2012) subsequently published their own updated version (although this is not an official product of the International Stratigraphic Commission), and a further major revision appeared in 2020 (Gradstein et al. 2020). The 2004 version incorporated numerous new data points, documented with the use of quantitative biostratigraphy, much-improved radioisotopic dating methods, chemostratigraphy and (for the Neogene) cyclostratigraphy (Fig. 7.33). The new scale (now updated to GTS2020) presents us with unprecedented opportunities for the comparison and calibration of detailed local and regional studies of rates and processes. All Cambrian to Paleogene ages are given to the nearest 100,000 years, although for much of the Cambrian–Devonian scale, potential errors of > 1 m.y. remain. Post-Paleogene ages are given to within 0.01 m.y. This scale, like all before it, incorporates numerous revisions of assigned ages. Almost all major chronostratigraphic boundaries in the Mesozoic and Paleozoic have been revised by several million years relative to earlier scales, such as that of Berggren et al. (1995), reflecting new data or changing interpretations of earlier data. There is no sign, yet, that the time scale has finally stabilized, although the incremental changes from one scale to the next do appear to be getting smaller. We return to this point later.

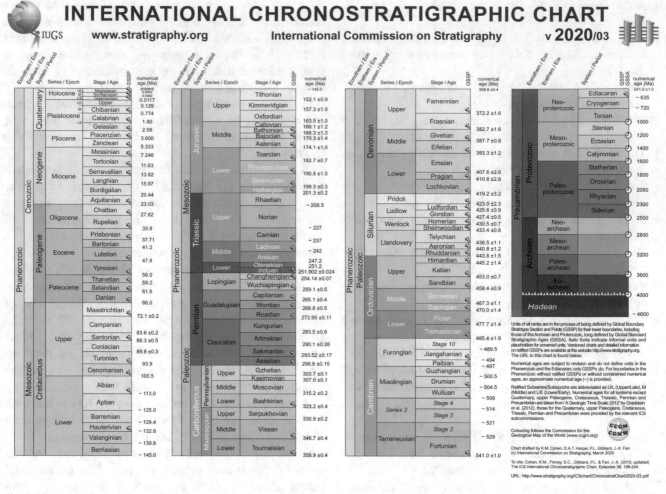

Fig. 7.33 The geological time scale prepared by the International Commission on Stratigraphy. Internationally agreed GSSPs are marked with a golden-spike icon. This figure shows the version current as of March 2021 (https://stratigraphy.org)

The time scale now undergoes continuous revision, under the auspices of the ICS. For example, a 2012 version was published in time for the 34th international Geological Congress at Brisbane (Cohen et al. 2012). (The latest version available at the time of going to press is provided in Fig. 7.33.) An example of the detail involved in the construction of the scale is shown in Fig. 7.34.

7.8.2 Determining the Numerical ("Absolute") Age of a Stratigraphic Horizon

Direct dating of sedimentary rocks by radioisotopic dating can only be carried out on a few potassium-bearing minerals, such as glauconite, because potassium is the only constituent of authigenic sedimentary minerals that contains naturally occurring radiogenic isotopes. Potassium–argon and argon–argon methods are the most common ones employed. However, the ages determined by the use of this method may

relate to diagenetic age rather than depositional age, and there are problems associated with the loss of the daughter product, argon.

More commonly, radioisotopic ages are determined for interbedded volcanic horizons, especially ash beds or bentonites, and the stratigraphic age of the rocks of interest are then determined by interpolation, as shown in Fig. 7.35. In order to make use of this method, two major assumptions have to be invoked, firstly, that sedimentation was continuous and, secondly, that sedimentation was at a constant rate. In most cases, neither of these assumptions can be assumed to be fulfilled. This does not mean that the technique is useless, only that it must be employed with care. Ideally, many radioisotopic determinations should be made on successive volcanic units, and then interpolations and extrapolations can be fine-tuned by methods of averaging and by making local corrections. We come back to this issue in Chap. 8, where we examine further some deeper questions regarding the interpretation of time as preserved in the rock record.

Fig. 7.34 Part of the GTS2020 scale for the Late Cretaceous, showing the magnetic polarity chrons and selected biozone schemes. Arrows in the microfossil columns indicate FADs (pointing upward) and LADs (pointing downward) (Gale et al., in Gradstein et al. 2020, Fig. 27.9, p. 1044)

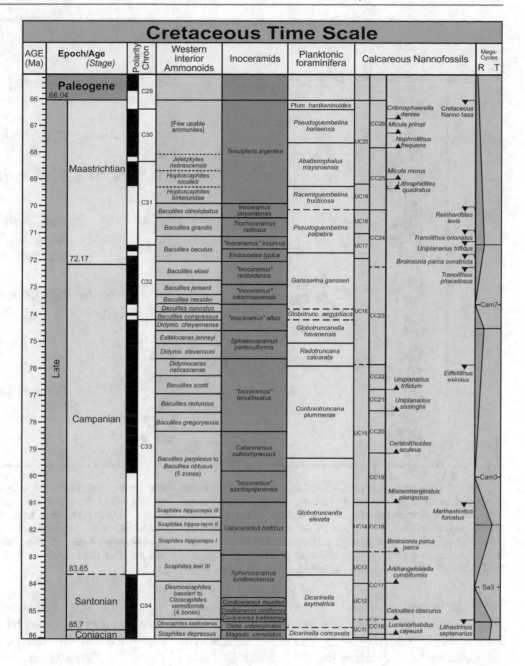

Modern radioisotopic dating methods. Not surprisingly, the methods of radioisotopic dating and the data base have undergone orders-of-magnitude transformations over the last half-century (Mattinson 2013). Holmes (1960) provided a list of 63 ages, mostly obtained by the K/Ar method. Of these, 19 were used to construct a linear time scale, adjusting the space allotted to each system according to cumulative stratal thickness, it being assumed at the time that this constituted an indication of the duration of the period. By the time of the 1964 Geological Society of London scale, the data base consisted of some 337 measurements. The

compilation of Harland et al. (1990, Table 4.2) lists some 750 entries. Figure 7.36 illustrates the increased accuracy and precision of the dating of the Permian–Triassic boundary over the 22-year period from 1991 to 2011. The current ICS date for this boundary is 252.17 ± 0.06 Ma.

Among the improvements in methods has been the establishment of the chronogram method by Harland et al. (1982, pp. 3–4) for plotting and visualization of measurement error, a standardization of decay constants, much enhanced analytical methods and a better appreciation of the systematics of the various isotopic decay paths. The K/Ar

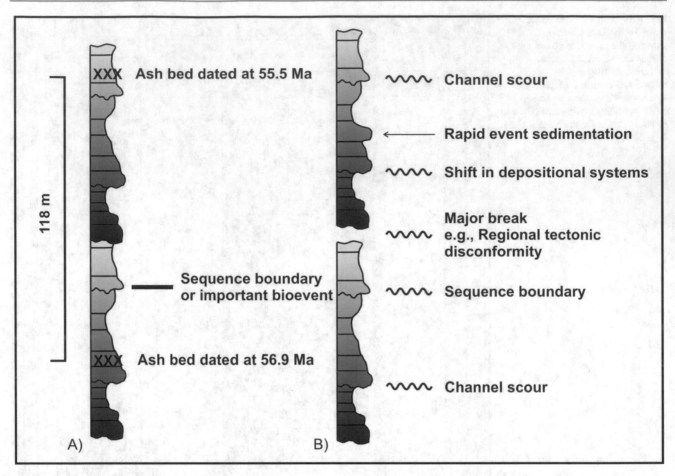

Fig. 7.35 The determination of "absolute" ages of stratigraphic events by using the method of "bracketing" the event by a pair of radioisotopic ages; The principle of the method (left) and the reality in practice (right). A) On the left, 118 m of beds are shown accumulating in 56.9–55.5 m.y., = 1.4 m.y. The average sedimentation rate is therefore 118/1.4 m/m.y., = 84 m/m.y. The sequence boundary of interest is 30.3 m above the lower ash bed. Assuming a constant sedimentation rate and no hiatuses, the 30.3 m of beds accumulated in 30.3/84 = 0.36 m.y. Therefore the age of this boundary is 56.9–0.36 = 56.54 Ma. However, as shown in B), at right, real sections are full of sedimentary breaks, which represent missing time, and variations in sedimentation rate, all of which render the concept of the average sedimentation rate suspect

and Rb/Sr methods have been found to be less reliable than those obtained by the $^{40}Ar/^{39}Ar$ and U–Pb methods (Villeneuve 2004; Mattinson 2013). The use of ages derived from glauconites, popular in the 1980s (Odin 1982), has decreased, because of the increasing realization of the unreliability of the method. It has been demonstrated repeatedly that daughter products, Sr or Ar, may be lost from the mineral grains, yielding ages that are too young.

Radioisotopic age determinations are characterized by a normal experimental error. The current practice may achieve a precision of ± 0.1% or better (Mattinson 2013, p. 310; Smith et al. 2015; Cramer et al. 2015; Schmitz et al., in Gradstein et al. 2020, Chap. 6), e.g., ± 100,000 years at 100 Ma. However, the attainment of such accuracy and precision in the stratigraphic record depends on the availability of appropriate datable material at the rights places in the rock record, and the accuracy of the existing time scale varies from stage to stage

because of this. Refinements in decay constants and inter-laboratory calibration become even more important. As Sageman et al. (2014) demonstrated in their discussion of U–Pb and $^{40}Ar/^{39}Ar$ data sets from an Upper Cretaceous interval in the Western Interior Basin, dating accuracy and precision are now such that differences between the two methods may relate to internal differences in the processes that set the final isotopic ratios—differences in Ar closure temperature between different minerals, and the length of the cooling phases of igneous bodies from which such minerals as zircon are ultimately derived. There may be differences of several hundred thousand years between the results from different methods, which then need to be reconciled by calibration against other data sets, including biochronology, chemostratigraphy and magnetostratigraphy. An important multi-authored review of the $^{40}Ar/^{39}Ar$ dating method was provided by Schaen et al. (2021).

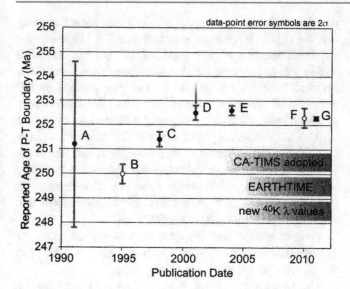

Fig. 7.36 The improvements in accuracy and precision of the dating of the Permian–Triassic boundary from 1991 to 2011. Solid circles are dates based on U–Pb zircon analysis; open circles are dates based on Ar–Ar analysis of sanidine (Mattinson 2013, Fig. 11, p. 315)

Even given a very precisely dated GSSP for a key chronostratigraphic boundary, it is then quite another matter to exploit this information to date and calibrate new sections elsewhere. This exercise requires a biostratigraphic, radioisotopic or chemostratigraphic record capable of yielding correlations of equivalent accuracy and precision.

Magnetostratigraphy: The first major development, the erection of a satisfactory geomagnetic polarity scale for the last 4.5 Ma, was published by Cox (1969), based primarily on the sampling of successions of lava flows. The standard scale now in use was developed primarily from the study of cores through undisturbed deep-sea sediments. The first use of such cores predated the *Deep Sea Drilling Project* (DSDP), and the cores were very short (see review by Kennett, 1980). Harrison and Funnell (1964) were the first to combine biostratigraphy and chronostratigraphy in an attempt to correlate a reversal event. Later, Opdyke et al. (1966) studied some longer cores, up to 12 m in length, and were able to correlate reversal events with the land-based lava sequence. Radiolarian zone boundaries closely parallel the reversal correlations (Fig. 7.37). The DSDP started in 1968 and began to have an important effect on chronostratigraphy. However, difficulties were encountered in establishing magnetic stratigraphy directly from the cores because of drilling disturbance and bioturbation. The practice developed of calibrating polarity and biostratigraphic data by correlating deep marine with exposed on-land sections and dating the latter radioisotopically.

Since the 1960s, magnetostratigraphy has become almost equal in importance to biostratigraphy, chemostratigraphy and radioisotopic dating in establishing a time scale for the last 160 Ma of Earth's history. The establishment of a reliable reversal sequence requires close sampling of an assumed, undisturbed, continuous section, and relies on the ability to recognize a kind of "bar-code" pattern of normal and reversed intervals. However, undisturbed sections for the pre-Late Jurassic are rare—there is no undisturbed sea

Fig. 7.37 Correlation of magnetic stratigraphy and radiolarian zones in seven cores from the Antarctic Ocean (Opdyke et al. 1966)

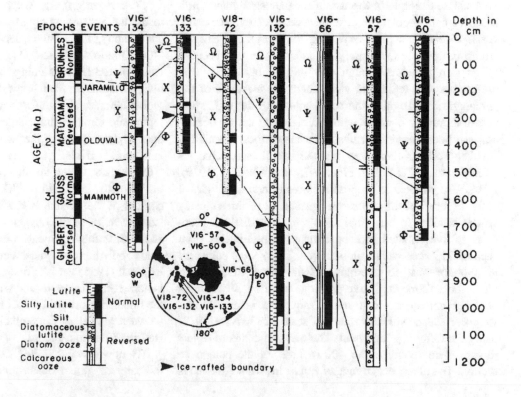

floor of greater age—and it is therefore much more difficult to standardize the scale for these older rocks.

The technique is most useful for time spans characterized by frequent reversals. During the Cenozoic, reversals occurred two or three times each million years, providing a very distinctive pattern that has enabled the establishment of 29 reversal intervals, named **chrons** for that era (Ogg and Smith 2004). The availability of age information on a 10^4–10^5-year time scale has provided powerful new tools for exploring the rates of sedimentary and stratigraphic processes in the geological record. Reversal sequences for older parts of the Mesozoic and the Paleozoic have been assembled from partial sections (Langereis et al. 2010), but are not yet reliable enough to become part of the standard geological time scale. Figure 7.34 shows part of the GTS2020 time scale for the Cretaceous, including the numbered polarity chrons.

Magnetostratigraphy may also be used to correlate local sections with each other without regard to the global scale. However, because reversal events are not unique, it is not possible to correlate them by matching sequences from different stratigraphic sections unless they contain particularly distinctive long or short polarity intervals. Some supplementary criteria may be required to assist in matching, such as marker beds or biostratigraphic zonation. Picard (1964) and Irving (1966) were among the first Western workers to use paleomagnetic correlation for sediments on the continents. The technique has become widely used for nonmarine sediments because of the scarcity of other means of precise correlation.

A single example of the use of magnetostratigraphy in a practical field problem is described here briefly. For many years, a team has been exploring the nonmarine Siwalik Group (Oligocene-Quaternary) of Pakistan, in part because of its rich vertebrate fauna and in part because of the information the sediments yield about the tectonics of the Himalayas, from which the sediments were derived. Magnetostratigraphy, coupled with radioisotopic dating of several ash beds, provided a useful means of local correlation between sections that show marked lateral facies changes (Keller et al. 1977; Barndt et al. 1978; Johnson et al. 1979). It was also possible to propose a correlation with the global scale. Figure 7.38 illustrates the correlation of three closely spaced sections in the Pabbi Hills area, near Jhelum. From three to five oriented rock specimens were collected from each of 113 sites within the sections. These were subjected to laboratory tests to determine the stability of the field and the absence of magnetic overprinting, and the pole positions obtained corrected for structural dip. The reversal zones were correlated with the standard scale of Opdyke (1972) using the following argument. The oldest remains of *Equus* (horse) were found at the 400 m level in the composite section. The oldest occurrence of *Equus* in North America is

dated 3.5 Ma and in Asia 2.5 Ma. It is considered unlikely that the Pabbi Hills fossils are older than 3.5 Ma. Note that the fossil locality is in a short normal polarity sequence within a long reversed interval. Only two such dominantly reversed intervals are present in the magnetostratigraphic scale, the Matuyama and the Gilbert zones. The Gilbert zone extended from 5.1 to 3.3 Ma, and the Matuyama from 2.41 to 0.70 Ma (revised ages of Opdyke 1972; modified from Cox 1969). The evidence of *Equus* suggests that this is the Matuyama zone. The two short normal events then correlate with the Olduvai and Jaramillo subzones. The Gilsa event (subzone) of Fig. 7.38 is not universally recognized and does not appear on Opdyke's (1972) chart.

Four composite sections have been correlated using the presence of two tuff horizons (Visser and Johnson 1978) and the polarity zones, as shown in Fig. 7.38.

The importance of this work is that it permits precise local and global correlation of vertebrate localities, permits accurate calculations of sedimentation rates and provides accurate control for studying sedimentological characteristics, basin architecture and tectonic events. Some of these aspects have been explored in later papers by this research group and other workers in the area. For example, Johnson et al. (1985) were able to determine the various time scales represented by fluvial cycles and to explore the implications for rates of channel wandering and the nature of tectonic and climatic controls on sedimentation. Behrensmeyer (1987) developed detailed two-dimensional reconstructions of the stratigraphic architecture as a basis for an examination of the taphonomy of the vertebrate remains.

Chemostratigraphy: **Oxygen isotope stratigraphy** has made an enormous contribution to the development of stratigraphy. The method depends on measurements of the $^{16}O/^{18}O$ ratio. Because ^{16}O is the lighter of the two isotopes, water molecules containing this light oxygen are preferentially evaporated from seawater. During times of ice-free global climate, they are recycled to the oceans and the isotopic ratio remains in a stable balance corresponding to the natural proportions of the two isotopes in the hydrosphere. However, glacial ice is composed of condensed ^{16}O-enriched water so that continental ice buildups are preferentially enriched in ^{16}O, with the result that the $\delta^{18}O$ content of the oceans is increased. Sediments preserve the isotopic ratios of the oxygen that existed in the hydrosphere at the time the sediments were formed. Measurements are made on the carbonate comprising foraminiferal tests. Emiliani (1955) was the first to demonstrate that the oxygen isotope record is cyclic, and was among the first to argue that the fluctuations should reflect the high-frequency oscillations between glacial and interglacial stages that have characterized the Cenozoic record.

We now know that the $^{16}O/^{18}O$ ratio is a highly sensitive indicator of global ocean temperatures and ice cover, and

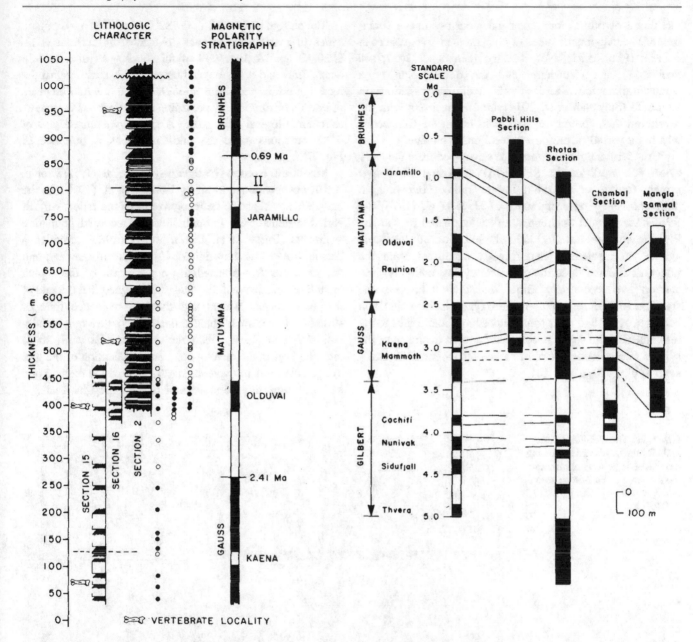

Fig. 7.38 The practical use of magnetostratigraphy to correlate sections of nonmarine strata in Pakistan. Three partial sections were compiled at the left to provide a single composite section for the Pabbi Hills region. Key vertebrate locations are shown, and field samples showing normal and reversed magnetic polarities are indicated by the vertically arranged suites of black and white circles. Correlation of the Pabbi Hills section with the standard scale and with other nearby stratigraphic sections is shown at right. Note the presence of missing reversals in some of the sections, indicating either incomplete sampling or reversals that are actually missing because of local erosion (redrawn from Johnson et al. 1979)

can therefore be used as an analog recorder of ice volumes (Hays et al. 1976; Shackleton and Opdyke 1976; Matthews 1984, 1988). Matthews (1984) suggested a calibration value of $\delta^{18}O$ variation of about 0.011‰ per meter of sea-level change. Miller et al. (2005) compiled the available data for the Late Cretaceous to present and used this compilation to develop a scenario for the growth and variation in the extent of the ice cover of Antarctica (see Sect. 7.9 and Fig. 7.55).

The appearance of large ice caps in the early Oligocene, and the beginning of the northern hemisphere glaciation at about 5 Ma are clearly indicated by large stepwise increases in $\delta^{18}O$, but also of interest are the fluctuations in the isotopic data that would suggest the development of temporary, small ice caps in the Antarctic in the Late Cretaceous. This provides important independent evidence to support the growing body of sequence stratigraphic work that suggests the likelihood of orbital forcing of glacioeustasy in the Cretaceous. The mapping of suites of high-frequency sequences in Alberta (Plint 1991), New Mexico (Lin et al. 2021; see Sect. 8.12) and elsewhere, and the development of a chemostratigraphic record of orbital climate variations in Colorado (Sageman et al. 2014) are all consistent with the revelation that, contrary to long-held ideas, the Cretaceous was not a period of uniform "greenhouse" climate.

A modern oxygen isotope scale developed from the early systematic work of the SPECMAP (SPEctral MAPping) project (Imbrie et al. 1984; Imbrie 1985). Following, in particular, the pioneering work of Hays et al. (1976) the SPECMAP project established a detailed record for the late Pleistocene (the last 780 ka) which calibrated the scale against an insolation energy index calculated from the integration of the three major orbital cycles, obliquity, precession and eccentricity. Since that time it has become standard procedure to record the oxygen isotope signal in deep-sea cores, and numerous studies have contributed to the refinement of the scale. A complete modern treatment of the subject is provided by Grossman and Joachimski (in Gradstein et al. 2020, Chap. 10).

The current scale used in GTS2020 includes a systematic series of marine cycles back to 5.5 Ma (Gradstein et al. 2020, Chap. 10; Fig. 7.39 of this book). As discussed in Sects. 7.8.7 and 8.11, the calibration of this scale against the record of orbital variations through the last few million years of Earth's history is now providing the main basis for a new, astrochronological time scale. A more generalized plot of $\delta^{18}O$ variations since the mid-Cretaceous is provided in Fig. 7.55.

Strontium isotope stratigraphy is a relatively new topic. It was not mentioned at all by Harland et al. (1990). In the late 1970s, it began to be recognized that the ratio $^{87}Sr/^{86}Sr$ varied systematically through time, as recorded in marine sediments (Burke et al. 1982). Veizer (1989) provided a detailed review of the origins of strontium in seawater and the processes that affected the preservation of the signal, including the effects of diagenesis. He argued that the rate of mixing of ocean waters and the long-residence time of strontium in seawater could potentially yield a reliable signal. McArthur (1994, 1998; McArthur and Howarth 2004) took the lead in the development and application of the tool to age determination and standardization of the method and its integration into the process of refining the time scale.

Fig. 7.39 The GTS2020 time scale for the last 5.3 Ma, showing the magnetic polarity chrons and the marine oxygen isotope record (Gradstein et al. 2020, Fig. 10.1, p. 284)

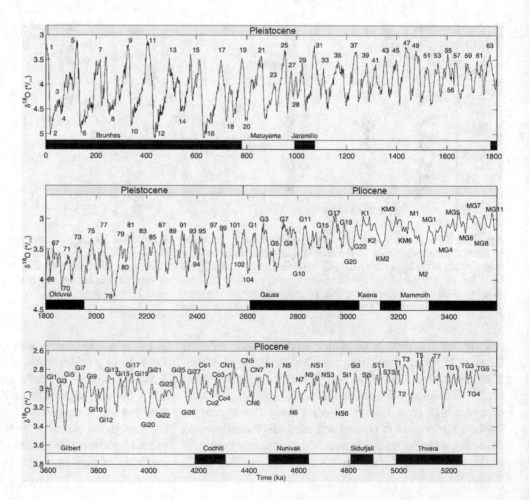

Fig. 7.40 Stable isotope profiles for the Bohemian Cretaceous Basin, showing the main named carbon excursion events (Jarvis et al. 2015, Fig. 7)

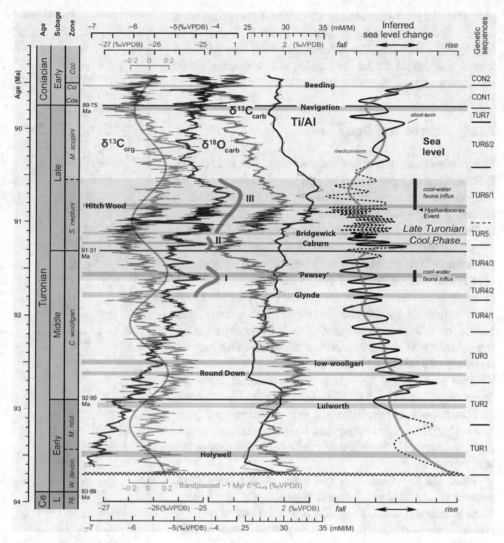

The $^{87}Sr/^{86}Sr$ ratio varies with age in the carbonate of various marine shells, mainly because of fluctuations in the rates and types of continental weathering, and the measurement of this value enables a sample to be situated on a graph showing the variation in composition with time. The graph for the Phanerozoic, as currently in use (McArthur and Howarth 2004, Fig. 7.1) is remarkably similar to that compiled by Veizer (1989), which indicates that the technique rapidly reached maturity. However, El Meknassi et al. (2018) warned that water mixing from continental or submarine groundwater sources may introduce inaccuracies in the dating method.

The values of the ratio are not unique with respect to time, because it has risen and fallen in a crudely cyclic manner within the range 0.7070–0.7090 during the last 450 m.y. so that some intermediate values correspond to several different ages on the curve. It is, therefore, necessary to know, within a few million years, which part of the graph to use for reading off the age against the calculated ratio.

Another chemostratigraphic tool of increasing importance is the **carbon isotope record**, $\delta^{13}C$. Research on this topic began with the work of Scholle and Arthur (1980) on certain Cretaceous intervals. The $\delta^{13}C$ record was calibrated against a rich biostratigraphic record in order to ensure chronostratigraphic precision. It was found that many carbon excursions could be correlated globally, and this was interpreted as a result of "oceanographic changes driving biotic turnover in the marine fossil record" (Jarvis et al. 2006, p. 565). Jarvis et al. (2006) expanded the work to define 39 **carbon excursion events** spanning the Cenomanian to Santonian interval, from which they developed a global reference curve, and it was speculated that the events were generated by Milankovitch-band orbital forcing of climate change. Additional research on the Turonian interval, with an analysis of pCO_2 changes and climate variability, was reported by Jarvis et al. (2015). An example of this work is shown here (Fig. 7.40) to indicate the potential for the tool to identify and correlate carbon and oxygen isotope values globally. Note a large number of named carbon excursion events. Cramer and Jarvis (2020) provided an overview of the topic for the GTS2020 volume, in which they provided scatter plots for all the data assembled to date (nearly 60,000 measurements) for the entire Phanerozoic.

7.8.3 Stages and Boundaries

The evolution of the **stage** concept was a confused one. D'Orbigny and Oppel in the mid-nineteenth century were the first to use the term with basically its present meaning (Hancock 1977; McGowran 2005).

Stages are effectively convenient groupings of biozones. Stage boundaries may be drawn at the top or base of a particularly well or widely developed biozone or a prominent faunal change. Many of our modern stage names were rather loosely defined, perhaps on only a handful of taxa when originally established, but have subsequently been refined by a more detailed study using many different life forms. Biozones are now usually established using only the members of a single phylum (except, perhaps, Oppel zones), whereas many stages have now been defined in many different ways. For example, Devonian stages are based mainly on brachiopods, corals, trilobites, fish, conodonts and palynomorphs; Cretaceous stages are based mainly on ammonites, pelecypods, brachiopods, foraminifera, nannofossils and palynomorphs.

What has happened over the years is that stage terms became so useful that geologists began to define the same stage in different ways. They added descriptions of suites of different kinds of fossils, which helped to define parallel successions of biozones and to transcend biogeographic problems, and then when radioisotopic and magnetostratigraphic data became available, this information was added in as well. Chemostratigraphic data are built-in wherever they are available. Gradually the rock-term stage became a more broadly based term referring globally to a particular interval of time. The modern use of stage concepts is well illustrated in current treatments of the geologic time scale, e.g., in Gradstein et al. (2004a, 2012, 2020). The stage is now regarded as "the basic working unit of chronostratigraphy (Hedberg 1976; Salvador 1994; Smith et al. 2015; Gradstein et al. 2020; Chap. 2). A total of 102 stages is now used to subdivide the Phanerozoic. Their average duration is 5.3 m.y.

In any given sedimentary basin, the recognition of a stage and its component chronozones depends on the nature of the fossil and other evidence that can be compiled for analysis. Surface studies may be able to benefit from the very detailed studies of macrofossils that have been carried out over the last 150 years. For example, paleontologists have found that ammonites evolved so rapidly during the Mesozoic that they have been able to erect scores of biozones. Modern dating

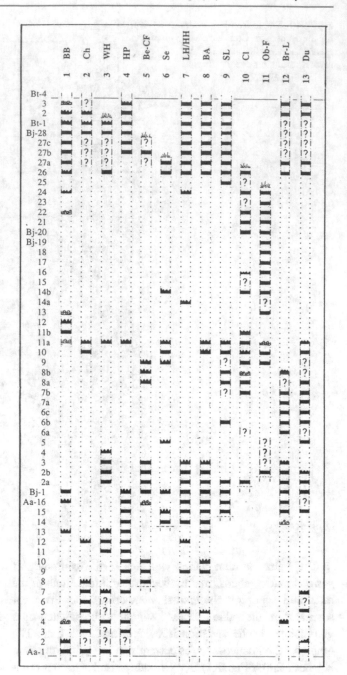

Fig. 7.41 The 56 ammonite faunal horizons recognized in 13 sections of the Inferior Oolite of Dorset and Somerset, England. The sections average 5 m in thickness are spaced out over a total distance of about 80 km, and span about 5 Ma of the Middle Jurassic; therefore each faunal horizon represents an average of about 90,000 years. Aa = Aalenian, Bj = Bajocian, Bt = Bathonian stages (Callomon 1995, Fig. 5, p. 143)

methods have shown that for some parts of the Mesozoic these zones may represent time intervals as small as 90,000 years, which permits an astonishing precision in dating where the record is complete enough (Callomon 1995). An extreme example is shown in Fig. 7.41 (we return to a discussion of this particular example in Chap. 8). For parts of the early Paleozoic, trilobites and graptolites offer similar biostratigraphic precision (see Sect. 7.5.4), with graptolites now providing some of the most tightly defined and numerically dated zonal schemes in the Phanerozoic (Sadler et al. 2009).

Subsurface studies must rely on microfossil or palynological evidence (McGowran 2005). Foraminifera evolved rapidly during the Cenozoic, but for most of the geologic time microfossils, though commonly occurring in great abundance, seemed to have evolved more slowly, and so individual biozones necessarily represent longer intervals of time.

Most fossil groups tend to be facies bound, and the key to successful stage correlation is to locate sections where more than one useful fossil group is present or where repeated facies variations cause ecologically incompatible taxa to occur close together by interdigitation. This is largely a matter of chance. Much depends on lucky exposure of the right section, and the literature is replete with obscure geographic localities that have attained a specialized kind of fame because of the excellence of the biostratigraphic work carried out there. Examples are such small Welsh towns as Llanvirn and Llandeilo (Ordovician), the Eifel district, Belgium (Devonian), and the Barrandian area of the Czech Republic (Silurian-Devonian boundary). Geologists worldwide knew about Maastricht long before the average European politician had heard about this minor city. It seems like a happy accident of onomatopoeia that the hammering in of the first golden spike to locate a GSSP (the base Lochkovian 1972) would be at a place called Klonk! (Fig. 7.42). Most stages are named after such places. Many of the classic stages that were erected in the nineteenth or early twentieth century in Europe and North America have been replaced during the international work to develop the GTS as it was determined that the defining faunas were endemic or too facies-bound. Plate-tectonic movements cause faunal provincialism to vary in many taxonomic groups simultaneously, and so the perfection of this correlation varies from place to place and time to time. For example, Berry (1977) reported that the correlation of early and middle Ordovician graptolitic faunas between Europe and North America has been fraught with controversy because of faunal provincialism. At that time, the proto-Atlantic Ocean (Iapetus) was at its widest. We earlier referred to the comparable problem of trilobite correlation across Iapetus (Fig. 7.12). The end result of this extended effort is that most stages can now be recognized globally, with varying degrees of confidence.

As an example of modern work in establishing stages by means of detailed faunal work on several animal groups, a brief discussion is presented here of the Pridolian and Lochkovian stages, as defined in the Barrandian area of the Czech Republic. Chlupác (1972) published a detailed description of the faunas and revised the earlier biostratigraphic subdivisions of these rocks. His faunal list for the two stages includes over 300 species, of which graptolites, conodonts and trilobites constitute the most important biostratigraphic indicators. Other groups providing subsidiary control include eurypterids, phyllocarids, ostracodes, echinoids, cephalopods, gastropods, pelecypods and brachiopods. Figure 7.42 illustrates one of several short but critical sections measured through the Pridolian-Lochkovian boundary near Klonk. Limestone beds are numbered 1 to 53. This section constitutes the very first internationally agreed GSSP, having been ratified by the International Union of Geological Sciences in 1972. It constitutes the stratotype for the base of the Devonian epoch. A vital characteristic of the Barrandian area is that here the section straddling the Silurian–Devonian boundary is basinal-marine in origin, and therefore preserves a more continuous marine section relative to the marginal-marine, nonmarine or unconformable contact in the British type areas of the Silurian-Devonian contact.

The Pridolian–Lochkovian succession in this area consists mainly of thinly interbedded, grayish-black, calcareous mudstone and grayish-black to dark gray, fine-grained (micritic), skeletal, platy, weathering limestone, with subordinate beds of pale limestone and coarser, detrital limestone (calcarenite to calcirudite). Graptolites are abundant in the mudstones; trilobites and conodonts occur sparsely in the limestones, particularly in the paler, less muddy units, and become more abundant north of Klonk, where the rocks undergo a facies change into a predominantly pure carbonate succession. This interbedding of different facies with their contrasting faunas is one of the most important features of these central European sections from the point of view of biostratigraphic stratotype definition.

Two graptolite biozones have been recognized in the Pridolian and form the main basis for the definition of the unit. They are the *Monograptus ultimus* zone below and the *M. transgrediens* zone above. The latter does not reach the top of the Pridolian but is followed by a graptolite interregnum containing only sparse, nondiagnostic forms (Fig. 7.42). The base of the Lochkovian is defined by the sudden widespread appearance and abundance of *Monograptus uniformis* in bed 20. Other species of *Monograptus* and *Linograptus* appear in the upper part of the lower Lochkovian, while in the upper Lochkovian, *M. hercynicus* is typical. Some of these species occur in the pure carbonate facies, permitting close correlation with the shelly fauna. It is interesting to note that in spite of the effort biostratigraphers have made to formalize their biozone types, Chlupác's work is typical of many in that no attempt is made to state what kind of biozone is in use. The older of these graptolite biozones appear to be single-taxon range biozones, with concurrent-range biozones for the upper part of the lower Lochkovian and the upper Lochkovian.

Fig. 7.42 The principle of the Golden Spike. The boundary between the Silurian and the Devonian was defined here in 1968 at a bed a certain distance below the first occurrence (FAD) of the graptolite *Monograptus uniformis*. This definition also serves to define the boundary between the Pridolian and the Lochkovian stages. Adapted from Chlupác (1972). This became the first internationally recognized GSSP

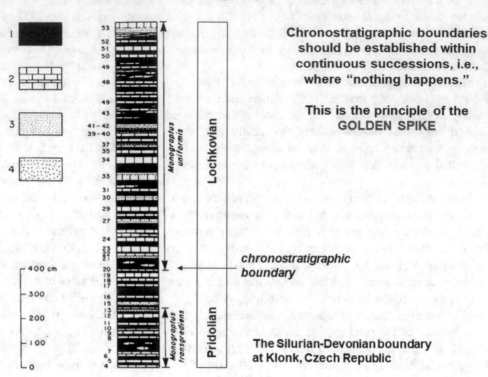

Chronostratigraphic boundaries

Chronostratigraphic boundaries should be established within continuous successions, i.e., where "nothing happens."

This is the principle of the GOLDEN SPIKE

chronostratigraphic boundary

The Silurian-Devonian boundary at Klonk, Czech Republic

The trilobite *Warburgella (Podolites) rugulosa rugosa* is of primary importance in delineating the lower boundary of the Lochkovian. In the Klonk section, it appears in limestone bed 21, immediately above the first appearance of *M. uniformis* in the upper part of bed 20 (Fig. 7.42).

The conodont *Icriodus woschmidti* defines a range biozone corresponding approximately with the lower part of the Lochkovian, although in the Barrandian area, it ranges down through the graptolite interregnum into the top of the *M. transgrediens* biozone. Conodonts are not common in the somewhat argillaceous facies at Klonk.

This discussion could be extended considerably into a consideration of other faunal groups and some of the subsidiary species that define concurrent range zones. It is to be hoped, however, that by this time the reader can discern the main threads of a procedure that has now been followed innumerable times by many different workers.

Barnes et al. (1976) discussed the correlation between graptolites, conodonts, trilobites and brachiopods for the Ordovician rocks of Canada. This is an interesting case in that the benthonic fauna was used to establish a North American stage nomenclature in the craton (the Richmond, Maysville, etc.), whereas graptolites occur mainly in deeper water deposits and were correlated with the classic British stages (Ashgill, Caradoc, etc.), in spite of the difficulties of

faunal provincialism across the Iapetus Ocean referred to earlier. To be of any regional use, these biostratigraphic schemes had to be integrated with each other, and this depended on finding locations, as at Klonk, where the different biofacies are interbedded. In order to cover the entire Ordovician system, it was necessary to study partial sections in the Canadian Rocky Mountains, Nevada, Texas, Newfoundland, the St. Lawrence Lowlands, and parts of the Canadian Arctic Islands. One of the key sections was the Lower to Middle Ordovician Cow Head Group of Newfoundland. At Green Point, where the beds are overturned, the stratotype for the Cambrian–Ordovician boundary has been established (Fig. 7.7; Cooper et al. 2001). Here, trilobites and conodonts occur in limestone boulders slumped from the shelf into the deep water basin, where graptolite-bearing shales buried them. Integration of the biozone schemes had to allow for the fact that the boulders were probably slightly younger than the enclosing shales. Elsewhere, transgressions and regressions caused the intermingling of faunas from different facies and different faunal provinces.

Another example of GSSP is that for the base of the Turonian at Pueblo, Colorado, the description of which was provided by Kennedy et al. (2005). The succession here consists of alternating limestones and shales (Figs. 7.43,

Fig. 7.43 The type section of the base of the Turonian. The lithological succession is that of the Bridge Creek Member of the Greenhorn Limestone on the north side of the Pueblo Reservoir State Recreation area, Colorado. The GSSP for the base of the Turonian is placed at the base of bed 86 (Kennedy et al. 2005, Fig. 8, p. 101)

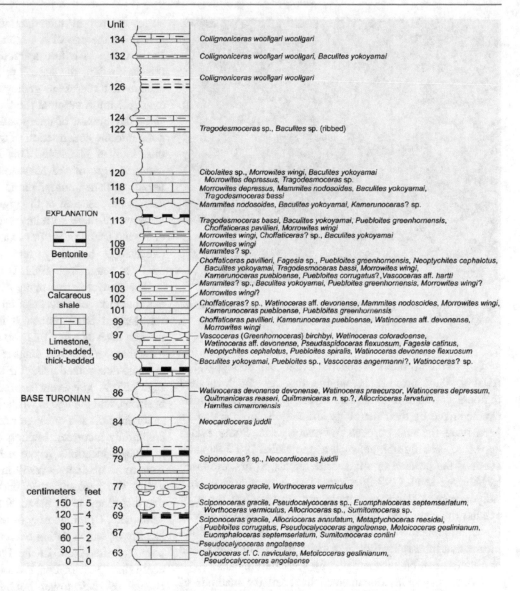

7.44). The limestones are fossiliferous biomicrites. Much of the succession is bioturbated. The alternation of the two major lithologies is attributed to orbitally forced climatic cyclicity. The base of the Turonian has been placed at the base of bed 86, which corresponds to the first occurrence of the ammonite *Watinoceras devonense* (Fig. 7.43). There are many secondary biostratigraphic indicators of the boundary, including bivalves, dinoflagellates and other ammonites. Figure 7.43 shows the ranges of the inoceramids in this section, and the inoceramid and ammonite zones. The boundary is also indicated by a carbon-isotope excursion which can be correlated worldwide, and corresponds to an Oceanic Anoxic Event. The numerical age of the boundary has been derived from radioisotopic dating of several bentonite beds that be traced throughout much of the Western Interior Basin. The bentonites at the Pueblo location are too weathered to be dated, but Kennedy et al. (2005) provide

dates determined from six other samples collected from correlative units in other parts of the basin.

7.8.4 Event Stratigraphy

The term **event stratigraphy** has been attributed to Ager (1973), although, as noted by Torrens (2002, p. 258), geologists have been aware of the importance of sudden events for some time. Typical geological "events" include sudden sedimentary events, such as storms and sediment gravity flows; volcanic events, generating widespread ash beds; earthquakes; biologic events such as first- and last-appearances of taxa, and mass extinction events; and chemostratigraphic events, such as "carbon excursions." Comprehensive treatments of this topic have been provided by Einsele and Seilacher (1982) and Kauffman (1988).

Fig. 7.44 The GSSP for the base of the Turonian. See Fig. 7.42 for stratigraphic location (photo by Brad Sageman)

Where the products of specific events can be recognized reliably, they provide invaluable markers for local and regional correlation. Bentonite layers, because they can be characterized by their petrology and radioisotopic age, have been used for this purpose for many years. Some events appear to be truly unique, such as the global K-T boundary event as the product of a meteorite impact, as first proposed by Alvarez et al. (1980); a hypothesis now almost universally accepted. However, there are many other types of events that are repeatable (e.g., major sediment-gravity flows; liquefaction events attributable to earthquakes), which means that their reliability as markers may be limited unless correlation can be substantiated by secondary means.

Kauffman (1988) documented in detail the methods of what he termed **high-resolution event stratigraphy,** in which he made use of bentonites, biomarkers and other events to construct regional stratigraphic correlations using the methods of graphic correlation, with an estimated accuracy of ± 100 ka. In Sect. 7.8.6 we discuss a debate regarding the use of event stratigraphy in the construction of the standard suite of GSSPs. We return to this topic in Sect. 8.12 as part of a brief introduction to modern research to develop the astrochronological time scale.

7.8.5 Absolute Ages: Their Accuracy and Precision

Time scales like the one in Fig. 7.33 look finished. It would appear that if fossils can be assigned to one or other of the stages shown in this diagram, it should be possible to determine the age of a section almost anywhere within the Phanerozoic to within a fraction of a million years. Often this is, in fact, the case, but the precision of the scale is deceptive. It represents geologists' best guesses, given all the available information at the time of compilation. It does not show the amount of interpolation and extrapolation that has gone into the construction of the scale. To get an idea about this, look at Fig. 7.45. This is a table that compares the assigned ages of the Mesozoic stages between eleven different published compilations. Each of these syntheses represents the research of the highest international standard.

Why are there so many variations in assigned age from one scale to the next? For example, the age of the Jurassic–Cretaceous boundary has moved up and down by 15.3 Ma years since 1982. How can this be?

The answer is that many of the assigned ages depend on extrapolation or interpolation from a limited number of well-known fixed points. It is rare for an important biomarker, such as a key FAD, to occur stratigraphically next to a datable ash bed or a magnetic reversal event. Many event ages are determined using the kinds of calculations shown in Fig. 7.35. A key event is bracketed by two or more dated horizons, and the age is worked out by assuming continuous sedimentation at a constant rate. Both these assumptions are commonly incorrect. Making the same kind of calculations at several locations where the same kinds of assumptions have to be made may result in a range of ages for the same event. The geologist then makes a best guess, picking the one that seems to be the most reliable, or averaging them in some way. Opinions may differ about which particular field locality or which dating exercise offers the most reliable tie point for the scale. In the Introduction to GTS2020 it is noted that "30% of Phanerozoic stage boundary ages have a change of their lower boundary by more than 0.5 Myr [relative to GTS2012], and in some cases much more" (Gradstein et al. 2020, p. 5). For example, some of the stage boundaries in the Early Cretaceous and Late Jurassic have shifted by more than 2 m.y. since the 2004 version of the GTS, and are characterized by potential errors of more than 0.5 m.y. Gradstein et al. (2020, p. 11) stated that "the new philosophy, which was started with GTS2004 and GTS2012, is to select analytically precise radioisotopic dates with high stratigraphic resolution. More than 330 radioisotopic dates were thus selected for their reliability and stratigraphic importance to calibrate the geologic record in linear time." Further, "Ages and durations of Cenozoic stages derived from orbital tuning are considered to be accurate to within a precession cycle (~20 ka) assuming that all cycles are correctly identified, and that the theoretical astronomical tuning for progressively older deposits is precise."

Fig. 7.45 Comparison of Mesozoic time scales (Gradstein et al. 2020, Fig. 1.4)

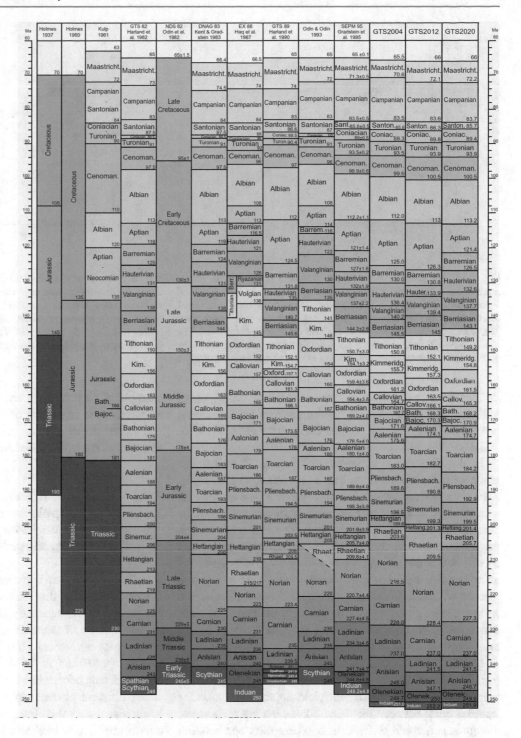

In the case of the Jurassic–Cretaceous boundary, Gale et al. (in Gradstein et al. 2020, p. 1024) stated:

> The Cretaceous is the only Phanerozoic system that does not yet have an accepted global boundary definition, despite over a dozen international conferences and working group meetings dedicated to the issue since the 1970s .... Difficulties in assigning a global Jurassic/Cretaceous boundary are the product of historical usage, the lack of any major faunal change between the latest Jurassic and earliest Cretaceous, a pronounced provincialism of marine fauna and flora and a concentration of previous studies on often endemic ammonites. Another problem is the occurrence of widespread hiatuses or condensations in many European and Russian epicontinental successions caused by the long-term "Purbeckian regression."

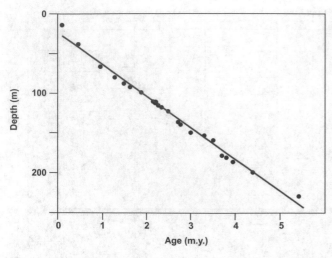 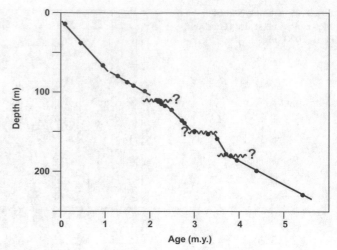

Fig. 7.46 The problem of calibration. The data points in both graphs represent the same set of real data from a well in the Gulf Coast of Florida (from Roof et al. 1991). Each point is a biomarker that has been dated against the available time scale. The question then arises, does this data represent continuous sedimentation at a constant rate? The straight line of correlation in the left-hand figure suggests that this is the case, making allowance for small errors in the dating of the fossils. But it is possible to look at the data a different way, as seen in the right-hand figure. Sedimentation rate may vary, and there may be several minor unconformities. Careful examination of the section is necessary to determine the nature of facies changes, the presence of breaks in sedimentation, and so on, in order to test the validity of the interpretations suggested in the right-hand diagram

This is a particularly severe example of the types of problem that have affected all the projects to develop an international agreement on the designation of GSSPs for stage boundaries.

In the absence of sufficient datable field locations, all kinds of assumptions have had to be made. For example, assumptions that the species of a given group of animals evolve at a constant rate so that their zones can all be assumed to be of the same duration. The first work to date the reversal events in the magnetostratigraphic scale, was carried out on the magnetized oceanic crust, and made use of an assumption that sea-floor spreading rates were constant so that the width of the reversal zone could be transformed into a value for its duration. This particular assumption has since been proved to be completely wrong—sea-floor spreading rates vary almost continuously.

Figure 7.46 illustrates the problem, using some real data from a well of the Florida coast. Which of the two interpretations is correct? Both will yield approximate ages for samples lying between the dated points, but the differences, though small, could be important. Where the slope between two points in this diagram is low, this means that the thickness of strata between the points is untypically small, given the indicated age span. A simple explanation would be that there is a local disconformity. An unusually steep slope could mean the occurrence of a rare event, such as the passage of a turbidity current, which locally thickens the stratal record. Careful reexamination of the original rock record could help to resolve this problem. This is a good example of why the **field context** of the rocks is important, as noted at the beginning of this book (Sect. 1.1).

Further developments in chronostratigraphic accuracy and precision are discussed in Sects. 8.10 and 8.11.

7.8.6 The Current State of the Global Stratigraphic Sections and Points (GSSP) Concept and Standardization of the Chronostratigraphic Scale

Despite the apparent inductive simplicity of the approach described here to the refinement of the time scale, the completion of the necessary suite of GSSPs has been slow, in part because of the inability of some working groups to arrive at an agreement (Vai 2001). In addition, two contrasting approaches to the definition of chronostratigraphic units and unit boundaries have now evolved, each emphasizing different characteristics of the rock record and the accumulated data that describes it (Smith et al. 2015). Castradori (2002) provided an excellent summary of what has become a lively controversy within the International Commission on Stratigraphy. The first approach, which Castradori described as the **historical and conceptual approach**, emphasizes the historical continuity of the erection and definition of units and their boundaries, the data base for which has continued to grow since the nineteenth century by a process of inductive accretion. Aubry et al. (1999, 2000) expanded upon and defended this approach. As noted by

Smith et al. (2015), in some cases, precedence and historical continuity have had to be set aside in favor of choosing new stratotypes that provide a greater data base for correlation purposes.

The alternative method, which Castradori terms the **hyper-pragmatic approach** (a very misleading label, in this writer's opinion), focuses on the search for and recognition of significant "events" as providing the most suitable basis for rock-time markers, from which correlation and unit definition can then proceed. The followers of this methodology (see the response by Remane 2000b, to the discussion by Aubry et al. 2000) suggest that in some instances historical definitions of units and their boundaries should be modified or set aside in favor of globally recognizable event markers, such as a prominent biomarker, a magnetic reversal event, an isotopic excursion, or, eventually, events based on cyclostratigraphy. This approach explicitly sets aside McLaren's (1970) recommendation (cited above) that boundaries be defined in places where "nothing happened," although it is in accord with suggestions in the first stratigraphic guide that "natural breaks" in the stratigraphy could be used or boundaries defined "at or near markers favorable for long-distance time-correlation" (Hedberg 1976, pp. 71, 84). The virtue of this method is that where appropriately applied it may make field recognition of the boundary easier. The potential disadvantage is that it places prime emphasis on a single criterion for definition, and relies on assumptions about the superior time-significance of the selected boundary event. The deductive flavor of the hypothesis is therefore added to the methodology. In this sense the method is not strictly empirical (as discussed below, assumptions about global synchroneity of stratigraphic events may in some cases be misguided. See Miall and Miall 2001). Very few "events" are likely to be global in scope, which means that where they are absent, boundary determination has to revert to the historical, inductive approach.

Smith et al. (2015) noted that even where the traditional approach has been used to define a GSSP, there has tended to be an overreliance on single biostratigraphic criteria for the definition, which may limit their usefulness and flexibility. To the key defining criterion of (for example) the first or last appearance datum of a chosen taxon could be added the stratigraphic distance above or below the FAD or LAD of other taxa. They also noted the importance of supplementary criteria.

The hyper-pragmatic approach builds assumptions into what has otherwise been an inductive methodology free of all but the most basic of hypotheses about the time-significance of the rock record. The strength of the historical and conceptual approach is that it emphasizes multiple criteria, and makes use of long-established practices for reconciling different data bases, and for carrying correlations into areas where any given criterion may not be recognizable. For this reason, this writer is not in favor of the proposal by Zalasiewicz et al. (2004) to eliminate the distinction between time-rock units (chronostratigraphy) and the measurement of geologic time (geochronology). Their proposal hinges on the supposed supremacy of the global stratotype boundary points. History has repeatedly demonstrated the difficulties that have arisen from the reliance on single criteria for stratigraphic definitions, and the incompleteness of the rock record, which is why "time" and the "rocks" are so rarely synonymous in practice (see also Aubry 2007, on this point; and Heckert and Lucas 2004, for other comments on the Zalasiewicz et al. proposal). Some of the current controversies surrounding the placement of GSSPs in the Cenozoic are discussed by Berggren (2007) and Walsh (2004). The latter paper also contains a lengthy discussion regarding the controversies surrounding the definitions and usages of the key terms, including stage, boundary stratotypes, GSSP etc., most of which are beyond the concerns of the practicing stratigrapher.

A different debate has arisen since the power of astrochronological calibration of the time scale became evident, a topic we take up in the next section, and again in Sect. 8.13. Astrochronology is based on the assertion that, in certain, carefully selected sections, a complete record of orbital forcing is preserved by cyclic variations in facies and thicknesses in the sedimentary record, and that by counting the cycles and correlating them to numerical ages derived from radioisotopic dating of some other means, a precise time scale with accuracy and precision in the 10^4-year range may be established.

As noted above, the traditional (at least since the 1970s) method of defining the time scale has been by the erection of GSSPs and topless stages, the purpose of the latter feature being to automatically allow for the presence of hidden hiatuses in the succession at the stratotypes. However, the relevance of this issue has been called into question when the method of selecting and dating a GSSP involves the necessary assumption of stratigraphic completeness, either in the actual stratotype or in the composite section upon which the stratotype is based. Such is the case with astrochronology. As noted by Hilgen et al. (2006, p. 117):

> … all late Neogene GSSPs are by now defined in land-based deep marine sections. All these sections have an integrated high-resolution stratigraphy, uniting detailed cyclo-, magneto- and biostratigraphies and have been astronomically tuned. Moreover, they cover the entire interval of the stage in a demonstrable continuous succession. As such, the sections perfectly embody the concept of a stage and may serve as unit stratotype for that stage in addition to accommodating its GSSP.

These authors argued for the extension of the concept to the remainder of the Cenozoic and also to the Mesozoic, as the astrochronological data base becomes more complete (Hilgen et al. 2015, 2020; see Sect. 8.11). They also pointed out that if the GTS is based on complete sections then the distinction between "rock" and "time" becomes unnecessary. However, the determination of "completeness" is always difficult, and as we now know, all sections contain some sedimentary breaks, although in deep-marine sections they may be of minor importance. I remain convinced that we need to be careful about abandoning the long-held cautions that support this dual terminology. The preservation of sections that are complete at the 10^4–10^5-year time scale is likely to be very unusual, requiring especially undisturbed basinal conditions, and the identification of the sections that are complete enough to be used in the establishment of an astrochronological time scale should be considered a rare event (see next section). The issue of the incompleteness of the stratigraphic record is discussed further in Chap. 8.

7.8.7 Cyclostratigraphy and Astrochronology

Cyclostratigraphy: the subdiscipline of stratigraphy that deals with the identification, characterization, correlation and interpretation of cyclic variations in the stratigraphic record.

Astrochronology: The dating of sedimentary units by calibration of the cyclostratigraphic record with astronomically tuned time scales.

Tuning: Adjusting the frequencies, including harmonics, of a complex record preserved in natural succession to best-fit a predicted astronomical signal.

Croll (1864) and Gilbert (1895) were the first to realize that variations in the Earth's orbital behavior may affect the amount and distribution of solar radiation received at the Earth's surface, by latitude and by season, and could be the cause of major climate variations. Several classic studies were undertaken to search for orbital frequencies in the rock record, and theoretical work on the distribution of insolation was carried out by the Serbian mathematician Milankovitch (1930, 1941), who showed how orbital oscillations could affect the distribution of solar radiation over the Earth's

surface (the mathematical work of Milankovitch was so advanced and important for its time that an image of Milankovitch is now used on one of the Serbian currency bills). However, it was not for some years that the necessary data from the sedimentary record was obtained to support his model. Emiliani (1955) was the first to discover periodicities in the Pleistocene marine isotopic record, and the work by Hays et al. (1976) is regarded by many (e.g., de Boer and Smith 1994) as the definitive study that marked the beginning of a more widespread acceptance of orbital forcing, the so-called **Milankovitch processes**, as a major cause of stratigraphic cyclicity on a 10^4–10^5-year frequency—what is now termed the **Milankovitch band**. The model is now firmly established, particularly since accurate chronostratigraphic dating of marine sediments has led to the documentation of the record of faunal variations and temperature changes in numerous upper Cenozoic sections (an early summary was provided by Miall 2010, Sects. 7.2, 11.3. Later reports were given by Gradstein et al. 2004a; Hilgen et al. 2015). These show remarkably close agreement with the predictions made from astronomical observations. Many high-frequency sequence records are now interpreted in terms of the orbital-forcing model (summaries and reviews in Miall 2010, Chap. 11; Hilgen et al. 2015), and there is increasing evidence from detailed stratigraphic studies of orbital forcing and glacioeustasy in the Cretaceous record. We return to this work in Sect. 8.11.

There are several separate components of orbital variation (Fig. 7.47). The present orbital behavior of the Earth includes the following cyclic changes (Schwarzacher 1993).

1. Variations in orbital **eccentricity** (the shape of the Earth's orbit around the sun). Several "wobbles," which have periods of 2035.4, 412.8, 128.2, 99.5, 94.9 and 54 ka. The major periods are those at around 405 and 100 ka.

2. Changes of up to 3° in the **obliquity** of the ecliptic, with a major period of 41 ka, and minor periods of 53.6 and 39.7 ka.

3. **Precession** of the equinoxes. The Earth's orbit rotates like a spinning top, with a major period of 23.7 ka. This affects the timing of the perihelion (the position of the closest approach of the Earth to the sun on an elliptical orbit), which changes with a period of 19 ka.

Imbrie (1985, p. 423) explained the effects of these variables as follows:

Variations in obliquity alter the income side of the radiation budget in two fundamental ways: they modulate the intensity of the seasonal cycle, and they alter the annually integrated pole-to-equator insolation gradient on which the intensity of the atmospheric and oceanic circulations largely depend. (Low values of obliquity correspond to lower seasonality and steeper

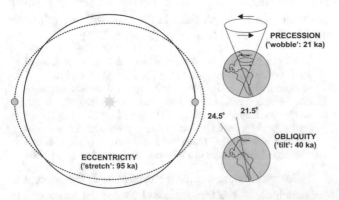

Fig. 7.47 Perturbations in the orbital behavior of the earth, showing the causes of Milankovitch cyclicity. Adapted from Imbrie and Imbrie (1979)

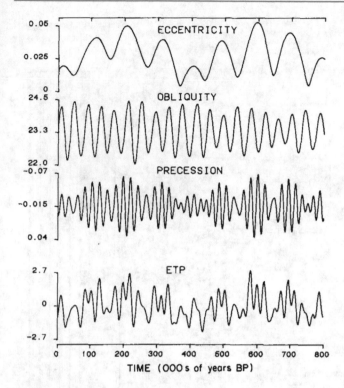

Fig. 7.48 The three major orbital-forcing parameters, showing their combined effect in the eccentricity-tilt-precession (ETP) curve at the bottom. Absolute eccentricity values are shown. Obliquity is measured in degrees. Precession is shown by a precession index. The ETP scale is in standard deviation units (Imbrie 1985)

insolation gradients.) Variations in precession, on the other hand, alter the structure of the seasonal cycle by moving the perihelion point along the orbit. The effect of this motion is to change the earth-sun distance at every season, and thereby change the intensity of incoming radiation at every season.

Each of these components is capable of causing significant climatic fluctuations given an adequate degree of global sensitivity to climate forcing. For example, when obliquity is low (rotation axis nearly normal to the ecliptic), more energy is delivered to the equator and less to the poles, giving rise to a steeper latitudinal temperature gradient and lower seasonality. Variations in precession alter the structure of the seasonal cycle, by moving the perihelion point along the orbit. This changes the Earth–Sun distance at every season, thus changing the intensity of insolation at each season. "For a given latitude and season typical departures from modern values are on the order of ∼5%" (Imbrie, 1985, p. 423). Because the forcing effects have different periods they go in and out of phase (Fig. 7.48). One of the major contributions of Milankovitch was to demonstrate these phase relationships on the basis of laborious time-series calculations. These can now, of course, be readily carried out by computer. The success of modern stratigraphic work has been to demonstrate the existence of curves of change in temperature, redox state, carbonate content, organic productivity and other variables in the Cenozoic record that can be correlated directly with the curves of Fig. 7.48. For this purpose, sophisticated time-series spectral analysis is performed on various measured parameters, such as oxygen-isotope content or cycle thickness. This approach has led to the development of a special type of quantitative analysis termed **cyclostratigraphy** (House 1985).

The Earth became highly sensitized to orbital variations during the cool climates of the Late Cenozoic, possibly as a result of a northern hemisphere cooling of air masses by the uplift of the Tibetan plateau (Ruddiman 1997), and it is now generally accepted, following the work of Hays et al. (1976) that the fluctuations in glaciation that characterized the Neogene were driven by orbital forcing, a process that gave us the oxygen isotope time scale (Emiliani 1955; Imbrie and Imbrie 1979; Imbrie 1985; Imbrie et al. 1984). The last ice age was ended by a phase of increasing solar insolation, which peaked about 10,000 years ago. A period of climatic warming, called the Holocene Optimum, existed from about 9000 to 6000 years ago. Since then there has been a very slow, long-term cooling trend (Fig. 7.56; possible anthropogenic influences are not discussed here).

To develop a time scale from the orbital record in sedimentary successions requires several important assumptions, given that, unlike, for example, bioevents or datable ash beds, cyclostratigraphy consists of a succession of identical, or near-identical, cyclic fluctuations in some primary depositional characteristic, such as carbon or calcium carbonate content, redox state, bed thickness or, more generally, facies. These assumptions include the following:

1. The section is continuous, or
2. (alternate): Discontinuities in the section can be recognized and accounted for in the subsequent analysis.
3. Sedimentation rate was constant, or event beds (such as turbidites) can all be recognized and discounted.
4. Orbital frequencies may be reconstructed for the distant past based on astronomical calculations of planetary motions.
5. Thickness can be converted to time using a simple sedimentation rate transformation.
6. The variabilities in stratigraphic preservation (facies changes, hiatuses) can be effectively managed by pattern-matching techniques.
7. Orbital frequencies can be reconstructed from the rock record of the distant geological past, based on independent age-bracketing of the section.

In general, these assumptions are more likely to be met in deep-marine and lacustrine settings, where it may be

Fig. 7.49 Composite stratigraphic columns for the Nukumaru and Castlecliff coastal sections, Wanganui Basin, New Zealand, showing lithostratigraphy, sequence stratigraphy, and correlations with the oxygen isotope timescale (Naish et al. 2005). This particularly complete Pliocene to Pleistocene section serves as a regional standard for geological time (Naish et al. 2005)

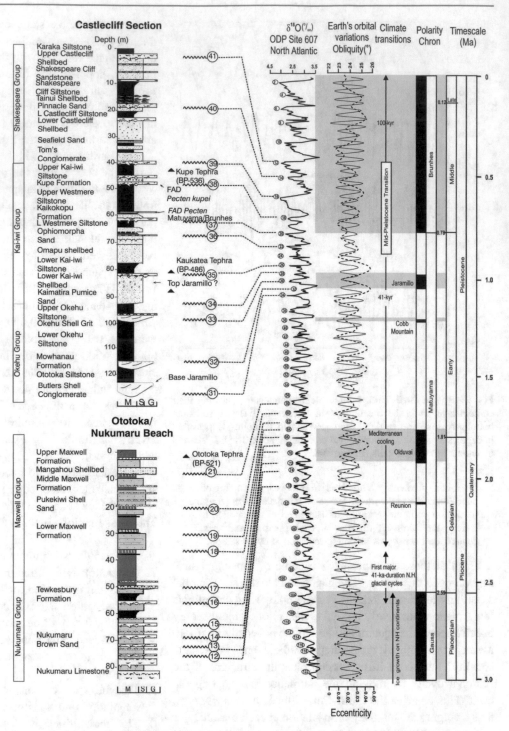

expected that the allogenic forcing of sedimentary processes by orbital mechanisms might be expected to overwhelm local autogenic influences. However, one of the more convincing studies of a high-frequency sequence stratigraphic record being used as the basis for a cyclostratigraphic time scale is that of the shallow-marine Plio-Pleistocene cycles of the Wanganui Basin in North Island, New Zealand

(Fig. 7.49). Also, Hilgen et al. (2015) cite several studies of fluvial systems in which it would appear that orbital forcing is the primary determinant of stratigraphic cyclicity.

The first major modern study of cyclostratigraphy (Hays et al. 1976) demonstrated that the last 800 ka of Earth time was dominated by a 100-ka cyclicity, as indicated by the cyclic climatic fluctuations from glacial to interglacial

Fig. 7.50 The Punta di Maiata section on Sicily. Punta di Maiata is the middle partial section of the Rossello Composite and part of the Zanclean unit stratotypes, which defines the base of the Pliocene (Van Couvering et al. 2000; Hilgen et al. 2006). Larger-scale eccentricity-related cycles are clearly visible in the weathering profile of the cape. Small-scale quadripartite cycles are precession-related; precession-obliquity interference patterns are present in particular in the older 400-kyr carbonate maximum indicated in blue. All cycles have been tuned in detail and the section has excellent magnetostratigraphy, calcareous plankton biostratigraphy and stable isotope stratigraphy (Hilgen et al. 2006)

stages. These are recorded in fine detail by the oxygen-isotope record, particularly as this is measured in ODP cores (Fig. 7.39). In the longer term, extending back into the Paleogene and earlier, the 405-ka eccentricity cycle seems to be the most stable and the most likely to provide the basis for astrochronology.

There were several early focused attempts to examine the use of the orbital "pacemaker" as the basis for a high-precision time scale (Herbert et al. 1995; House and Gale, 1995; Shackleton et al. 1999), and several important regional studies were carried out that began to make a substantial contribution to the growth of this field of research. The stratigraphic and sedimentologic basis for this research was summarized by Miall (2010, Chap. 11). A range of indicators may be used to examine for orbital cyclicity, beyond the physical "cyclic" appearance of the rocks themselves. These include bed thickness, oxygen isotope ratios, weight-percent calcium carbonate or organic carbon, grayscale pixel data (from core scans), magnetic susceptibility, and resistivity data from a microimaging scanner.

Pioneering studies to establish an astrochronological time scale have been led by Fritz Hilgen. A reliable cyclostratigraphic (astrochronologic) time scale was first established for the youngest Cenozoic strata, back to about 5 Ma (Hilgen 1991; Berggren et al. 1995; Hilgen et al. 2006; see Figs. 7.50, 7.51). Over the succeeding decade, astronomically calibrated sections were used to extend the astrochronological time scale back to 14.84 Ma, the base of the Serravallian stage, in the mid-Miocene (https://stratigraphy.org), and research is proceeding to extend the time scale not only to the base of the Cenozoic, but through at least the Mesozoic (Hinnov and Ogg 2007; Hilgen et al. 2006, 2015). Figure 7.52 illustrates the correlation of Upper Miocene sections in the Mediterranean basin to the orbital scale. Westphal et al. (2008) offered a sharply critical review

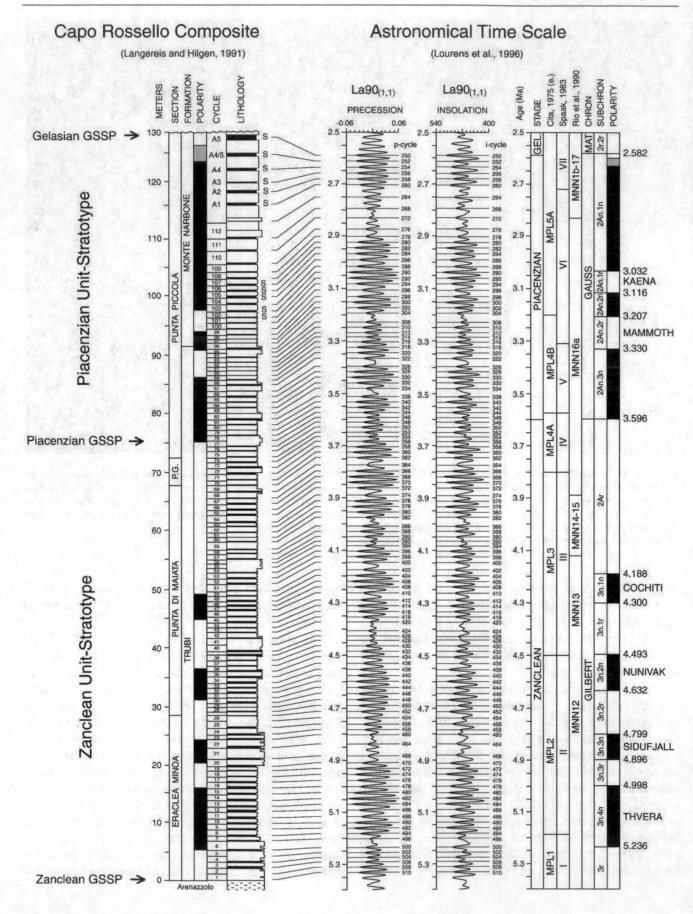

◀ **Fig. 7.51** The Rossello composite section (RCS, Sicily, Italy,) the unit stratotypes for series spanning the base of the Pliocene, incorporating the orbital tuning of the basic precession-controlled sedimentary cycles and the resulting astronomical time scale with accurate and precise astronomical ages for sedimentary cycles, calcareous plankton events and magnetic reversal boundaries. The Zanclean and Piacenzian GSSPs are formally defined in the RCS while the level that time-stratigraphically correlates with the Gelasian GSSP is found in the topmost part of the section. The well-tuned RCS lies at the base of the early–middle Pliocene part of the astrochronological time scale and the Global Standard Chronostratigraphic Scale and as such could serve as unit stratotype for both the Zanclean and Piacenzian Stage (Hilgen et al. 2006)

Fig. 7.52 Astronomical tuning of sapropels and associated grey marls in land-based deep marine sections in the Mediterranean for the interval between 10 and 7 Ma (upper Miocene). Colors in the lithological columns indicate sapropels (black), associated grey marls (grey) and homogeneous marls (yellow). Colors in the magnetostratigraphic columns indicate normal polarities (black), reversed polarities (white) and uncertain polarities (grey). Sapropels and associated grey marls have been numbered per section and lumped into large-scale groups (roman numerals) and small-scale groups. The initial age model is based on magnetobiostratigraphy. Phase relations between sapropel cycles and the orbital parameters/insolation used for the tuning are based on the comparison of the sapropel chronology for the last 0.5 myr with astronomical target curves (Hilgen et al. 2015, Fig. 2)

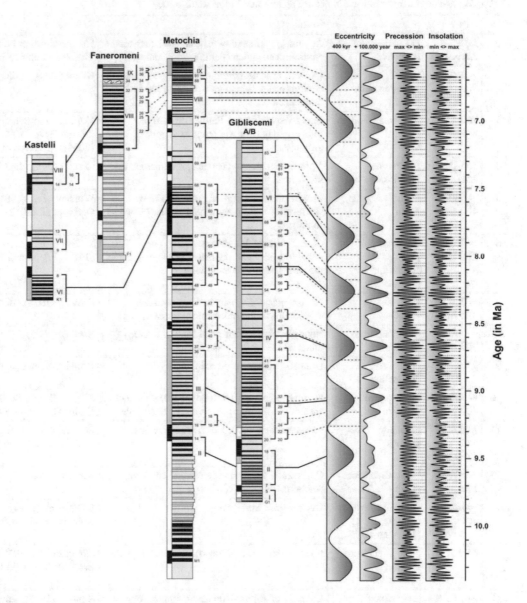

of the field data base on which part of the astrochronological time scale is based, pointing to problems of diagenesis and differential compaction of contrasting lithologies, that render direct one-for-one correlations between the critical field sections problematic. A careful comparison of the correlations between two critical field sections shows that the correlations of the astronomical cycles are not always supported by the correlations of bioevents. In some cases, there are different numbers of cycles between the occurrences of key bioevents. Multiple cross-checks are required to evaluate and, if possible, correct for such discrepancies.

Astronomical studies suggest that because of the long-term chaotic nature of the Earth's response to the gravitational influence of the planets, it is not possible to extend the present orbital frequencies back beyond about 60 Ma, with the exception of the eccentricity harmonic of

405 ka, which appears to have been stable through the Phanerozoic (Laskar et al. 2004). A modern treatment of the astronomical solutions is provided by Laskar (2020).

For the older part of the geological record (particularly the Mesozoic and Paleozoic), several studies have now established convincing "floating" scales for specific stratigraphic intervals; that is, scales that exhibit reliable orbital frequencies, once tuned, but that cannot be precisely correlated to the numerical time scale because of residual

Table 7.3 Major changes in Earth's history as revealed in the stratigraphy

#	Age	Process	Result	References
1	Late Archean	Generation of cratons	Appearance of shallow-marine and nonmarine environments, microbial life, stromatolites	Hawkesworth et al. (2017), Beall et al. (2018), Eriksson et al. (1998, 2013)
2	~2.4–2.0 Ga	Great Oxygenation Event	Increase in atmospheric oxygen led to oxidized environments, appearance of much red Fe^{3+} mineralization	Cloud (1968), Eriksson et al. (1998, 2013)
3	780–630 Ma	Snowball Earth?	Hypothesis of frozen Earth. Disputed on sedimentological and chronostratigraphic grounds	Hoffman and Schrag (2002), Allen and Etienne (2008), Eyles (1993, 2008), Eyles and Januszczak (2004), Le Heron et al. (2019)
4	555–500 Ma	Cambrian Explosion	Apparent sudden appearance of numerous new life forms.	Cloud (1948), Smith and Harper (2013)
5	~500–400 Ma	Great cratonic seas	Continental-scale cratonic seas caused by high sea levels. Widespread shallow-water carbonate sediments	Worsley et al. (1984), Pratt and Holmden (2008), Derby et al. (2012)
6	~400–250 Ma	Creation of Pangea (Caledonian orogenies)	Global changes in geology. Great unconformities and mismatch of biogeographic provinces. Old Red Sandstone.	Wilson (1966), Scotese (2001), Miall and Blakey (2019), Friend and Williams (2000)
7	~360–300 Ma	Gondwana Glaciation	Marine glacial deposits in Gondwana; cyclothems in the Northern hemisphere	Crowell (1978), Heckel (1986), Eyles (2008)
8	~350–300 Ma	Carboniferous System	Near-global extent of great forests. Source of much of the world's coal	Stanley (2005)
9	252 Ma	End Permian extinction	Dramatic change in global faunas	Esmeray-Senlet (in Gradstein et al. 2020, Chap. 3L)
10	~300–200 Ma	Breakup of Pangea	Nonmarine rift basins throughout Europe and eastern North America. New Red Sandstone	Ziegler (1988), Scotese (2001), Stanley (2005), Withjack et al. (1998)
11	Jurassic and Cretaceous	Oceanic anoxic events	Preservation of mudrocks rich in organic carbon	Schlanger and Jenkyns (1976), Schlanger et al. (1987), Cramer and Jarvis (in Gradstein et al. 2020)
12	~100–65 Ma	Late Cretaceous high sea levels	Continental flooding. Widespread chalk in southern US and NW Europe	Stanley (2005)
13	~100–65 Ma	Orbital forcing of climate, glacioeustasy	Cyclothemic stratigraphy	Elder et al. (1994), Plint and Kreitner (2007), Sageman et al. (2006), Shank and Plint (2013)
14	65 Ma	End Cretaceous impact and extinction	Iridium clay, tsunami deposits, shocked quartz	Alvarez et al. (1980), Hildebrand et al. (1991), Esmeray-Senlet (in Gradstein et al. 2020, Chap. 3L)
15	~40-5 Ma	Alpine and Himalayan orogeny	Creation of Alpine-Himalayan ranges, Tibetan Plateau, global cooling	Kennett (1977), Raymo and Ruddiman (1992), Ruddiman (2008), Summerhayes (2015)
16	2.5-0 012 Ma	Northern hemisphere Cenozoic glaciation	Multiple glacial episodes	Ruddiman (2008), Summerhayes (2015)
17	0.12-0.00 Ma	The Holocene	Variable post-glacial climates	Bradley (1999), Anderson et al. (2013), Plimer (2009), Carter (2010)
18	Post-WW2	The Anthropocene	Increasing dominance of human influence on global processes	

imprecisions in numerical dating methods (Hilgen et al. 2015).

Gradstein and Ogg (in Gradstein et al. 2020, p. 24) stated:

> Now, recent developments in integrated high-resolution stratigraphy and astronomical tuning of continuous deep marine successions combine potential unit stratotypes and boundary stratotypes for global stages as basic building blocks of the standard Global Chronostratigraphic Scale (GCS) (Hilgen et al. 2006, 2020). For the late Neogene with its outstanding orbitally tuned stratal record, some of the Global Stratotype Section and Point (GSSP) sections may also serve as unit stratotypes.

As this quote indicates, the pressure to recognized "unit stratotypes" and remove the distinction between "rock" and "time" is building. However, recent work on the generally fragmentary nature of stratigraphic preservation, which we detail in Chap. 8, means that this debate continues. A discussion of some current research in the area of astrochronology is provided in Sect. 8.11.

7.9 Stratigraphy Reflects Changing Earth Environments

The Earth's record of deep time is contained in its stratigraphy. The changing thermal structure of the planet, which led to the initiation of plate tectonics at about 3 Ga, the development of cratonic crust, the growth of continents by accretion and their migration, collisions and separations to form and disaggregate supercontinents, and Earth's ever-changing climate are all recorded in the deep structural geology and sedimentary cover of Earth. As sea-floor spreading moved continents through different climatic regimes, and continents collided, forming mountain belts that were uplifted, eroded and shed huge volumes of clastic debris, all was recorded more or less faithfully in the stratigraphic record. The evolution of plants led to dramatic changes in Earth's atmosphere and this, in turn, helped to fuel the appearance and diversification of animal life. At times, it seems, whole continents were characterized by particular types of sedimentary records and their contained fossils, reflecting the particular paleogeography and regional paleoclimate of the period. Some of the unique environmental conditions displayed by rocks from the distant past are discussed in Sect. 3.4.3. In this section we follow, very briefly, some of these changes that helped to create some particularly distinctive regional stratigraphies. Eighteen specific events and periods are discussed in this section, as summarized in Table 7.3

The evidence from the Earth's crust and modeling of mantle thermal behavior suggest that the distinctive form of mantle and crustal evolution that we group under the heading of plate tectonics began around 3 Ga (Hawkesworth et al. 2017; Beall et al. 2018). This was when the first cratons

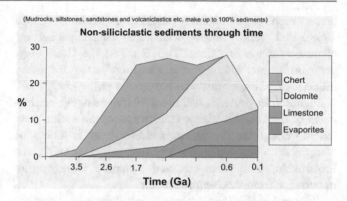

Fig. 7.53 The relative abundance of non-siliciclastic sediments through time, showing dolomites to be more abundant than limestones during much of the Proterozoic when microbial ecosystems dominated the biosphere (Eriksson et al. 2013, Fig. 7)

began to form; that is, the skins of stable sialic crust that rest isostatically on the mantle, just above sea level. We can recognize their appearance from the first evidence of shallow water, even nonmarine rocks in their sedimentary cover, where this has not been deformed out of recognition by subsequent tectonism or metamorphism. Eriksson and Mazumder (2020) provided an overview of modern work on Archean Earth processes based on recent research around the globe on preserved Archean structural belts that preserve protocontinental fragments. For example, the 3.55–3.2 Ga Barberton greenstone belt of southern Africa is characterized by "carbonaceous cherts with filamentous, spheroidal, and lenticular microstructures; traces of hydrothermal biofilms; pseudocolumnar stromatolites; large spheroidal microfossils; and apparently photosynthetic microbial mats" preserved in sediments of intertidal, supratidal and fluvial origin (Eriksson and Mazmunder 2020, p. 2).

Eriksson et al. (1998) provided an extensive overview of Precambrian sedimentation systems. The atmospheric composition changed significantly through the Precambrian. The Archean atmosphere consisted primarily of carbon dioxide, nitrogen and methane, with minor amounts of water, hydrogen, carbon monoxide and reduced sulfur gases. CO_2 content was much higher than at any time during the Phanerozoic, likely several thousand parts per million. Atmospheric oxygen began to be generated by photosynthesis when cyanobacteria evolved, somewhere around 3 Ga, but for hundreds of millions of years the quantities were modest, and the atmosphere continued to be dominated by the volcanic outgassing of CO_2, and gases of nitrogen and sulfur (Eriksson et al. 1998, 2013). Banded iron formation, generated in a reducing atmosphere, was an important sedimentary product during this phase (Archean to Paleoproterozoic; ~3700–2450 Ga). Photosynthesis increased in importance through the Paleoproterozoic, and about 2.4 billion years ago the atmosphere became essentially an oxidizing environment, with the increased formation and preservation of sedimentary ferric iron minerals (Cloud 1968). Red, yellow and brown-colored sandstones became common for the first time. This is called the Great Oxygenation Event, although the transition lasted for many millions of years, so this was not strictly an "event."

Figure 7.53 provides estimates of the changing composition of the non-siliciclastic constituents of the sedimentary record through the Proterozoic and Phanerozoic. As can be seen, cherts and dolomite were abundant through the Proterozoic, with limestone and evaporite increasing in importance from the Neoproterozoic.

Regional glaciations occurred several times through the Precambrian, as evidenced by deposition of thick glacio-marine sediment-gravity-flow and ice-rafted deposits. Hoffman and Schrag (2002) focused on a series of widespread glaciations that took place during the Neoproterozoic, and developed the hypothesis of "Snowball Earth," the central idea of which is that the Earth descended into a deep freeze that also partly froze the oceans and essentially shut down hydrological systems and biological evolution. This idea has been challenged by several groups of workers. Allen and Etienne (2008) focused on sedimentological evidence for the continuation of active marine sedimentation through the supposed snowball period, indicating the continuing presence of open oceans and the aqueous transport and deposition of siliciclastic sediments. Many supposed glacial deposits, on close sedimentological examination, prove to have little or no connection to glaciation (e.g., Kennedy et al. 2019). Eyles (1993, 2008) and Eyles and Januszczak (2004) argued that the Earth needed to be conditioned to widespread

glaciation by regional uplift, which cooled the atmosphere. The evidence from the location and timing of major glacial episodes indicates that the uplift that occurs during continental separation and extension is often associated with regional glacial episodes. Chronostratigraphic data indicate that glaciation during the Neoproterozoic, while widespread, consisted of regional episodes occurring over an extended period of tens of millions of years, with no single intervals when the Earth could be said to be largely frozen (Le Heron et al. 2020).

At the beginning of the Phanerozoic, the atmosphere still contained an order of magnitude more CO_2 than at the present day, at nearly 7000 ppm (Fig. 7.54). The CO_2 level gradually dropped through the Paleozoic, largely because of the rise of land plants, which increased the extraction of this gas from the air for use as a nutrient source by photosynthesis (Summerhayes, 2015, p. 160). It has been demonstrated that, over the Phanerozoic, there has been no clear relationship between CO_2 levels and global temperature (Berner and Kothavala 2001; Davis 2017). The stratigraphic record contains qualitative paleoclimate indicators, such as certain plants and animals, which have an expected latitudinal range; for example, corals, which grow best between 30° north and south. Such sediments as glacial deposits, eolian sands and evaporites also have paleoclimatic significance if they are regionally widespread. When regional paleoclimates reconstructed using this evidence are related to their paleolatitude, based on reconstructions of the plate-tectonic history, the familiar global climate belts appear (Summerhayes 2015, Chap. 6). Global climates have

Fig. 7.54 Changes in atmospheric carbon dioxide composition of the atmosphere through the Phanerozoic. Compiled by Rhode (2019), https://commons.wikimedia.org/wiki/File:Phanerozoic_Carbon_Dioxide.png. GEOCARBIII is from Berner and Kothavala (2001)

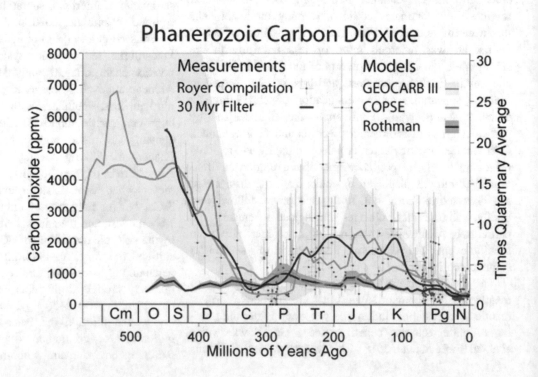

experienced extremes of heat and cold, such as the glacial to interglacial fluctuations of the late Cenozoic. The terms "greenhouse climate" and "icehouse climate" were proposed for long-term Phanerozoic global climate states by Fischer (1984), and he related these to cycles of "high" and "low" CO_2, a model generally supported by Summerhayes (2015, Chap. 10). However, in detail, these models become hard to support, in part because modern data on the ancient atmospheric composition (Fig. 7.54) indicate a more complex history than was available to Fischer, and in part, modern stratigraphic data likewise reveal substantial short- to medium-term variability in climate through the Phanerozoic. We discuss some of these variations below.

One of the most distinctive features of the Phanerozoic sedimentary record that distinguishes it from the Precambrian is the abundance of fossil invertebrates. This led to the coining of the term "Cambrian Explosion" for the supposed burst of evolution that led to the faunal abundance (Cloud 1948). However, a detailed examination of the fossil record and the date of the first appearance of the various phyla and classes in the rock record reveals that the so-called explosion took tens of millions of years. For example, Smith and Harper (2013) summarized the enormous growth in diversity between the appearance of the Ediacaran fauna (645 Ma) and the completion of the first biodiversity expansion at the end of the Ordovician (443 Ma), a period of some 192 m.y. The first mollusks, phyoliths, brachiopods, archeocyatha and trilobites appeared over a span of some 15 million years in the Early Cambrian; hardly an explosion. Nonetheless, some explanations seem necessary for the eventual abundance of life forms. Smith and Harper (2013, p. 1356) suggested that a range of

> interacting processes generated an evolutionary cascade that led to the rapid rise in diversity. The initiating event is likely to have been the early Cambrian sea-level rise that led to inundation of continental margins and interiors and the rapid input of erosional by-products. This sea-level rise would also have generated a very large increase in habitable area lying between the base of wave turbulence and the depth to which light penetrates, providing a further driver for large increases in diversity.

Clearly, the processes of evolution had to have reached a point where spontaneous mutations led to advantages for organisms that were already poised for diversification. One of these advantages was the development of biomineralization, which provided both defensive and predatory hard tissue nearly simultaneously and, as a byproduct, greater preservability for exoskeletons in the fossil record.

Sea levels underwent a major long-term eustatic rise commencing at the end of the Neoproterozoic and continuing into the Ordovician, remaining at high levels until the Carboniferous (Vail et al. 1977; Fischer 1981). It is widely accepted that this was the result of the breakup of the supercontinent Rodinia, with accelerated sea-floor spreading that led to the production of large areas of young oceanic crust at isostatically shallow levels, and the resultant displacement of ocean waters in the form of a long-continued global transgression (Worsley et al. 1984). The early Paleozoic was characterized by some of the most extensive cratonic epeiric seas that the Earth has experienced. As described by Pratt and Holmden (2008) and Derby et al. (2012), among the stratigraphic products was thick and areally extensive shallow-marine carbonate deposits. In the United States the term "Great American Bank" has been used to describe this stratigraphy (Derby et al. 2012).

The construction of Pangea began with regional orogenies, such as the early Taconic orogeny in eastern North America in the late Cambrian. A long series of regional to continental orogenic episodes, collectively referred to as the Caledonian Orogeny, marked the suturing of continental fragments during the Cambrian to Carboniferous period. The major episode of continental accretion was the suturing of Laurentia with Baltica, the Scandian phase, which lasted from mid-Silurian to early Devonian (Scotese 2001; Miall and Blakey 2019). The closure of large oceans, such as the Iapetus and Rheic oceans brought long-separated and distinctive biogeographic provinces together, creating proximity between mismatched faunas and floras. One of the first of these mismatches to be recognized and mapped was that of early Paleozoic trilobite and graptolite provinces, which constituted part of the evidence used by Wilson (1966) to postulate the closing and reopening of the Atlantic Ocean, as discussed in Sect. 7.5.1 (Fig. 7.12). Stratigraphically, among the most distinctive products of this era was the deposition of the largely nonmarine Devonian Old Red Sandstone in a wide variety of syn- to post-orogenic tectonic settings within or adjacent to the Caledonian orogen, including Maritime and Arctic Canada, Svalbard, Greenland, the British Isles and Norway (Friend and Williams 2000). One of the most well known of these occurrences is at Hutton's famous unconformity at Siccar Point in southeast Scotland, where the sandstone rests on deep-marine Silurian arenites, close to the Laurentian-Baltic suture.

During the Paleozoic, the Gondwana continent, comprising Africa, South America, India, Australia and Antarctica, drifted across the south pole (Crowell 1978; Scotese 2001). Widespread continental glaciation occurred on these continental areas, commencing in Africa and South America in the Late Devonian, extending to southern Africa in the Mississippian and to India, Australia and Antarctica in the Pennsylvanian and Permian. Thick glaciomarine deposits are widespread in these continents (Eyles 2008). The near-polar locations of the continents constituted an important precondition for glaciation, but as Eyles (2008) emphasized, tectonism, leading to broad areas of regional uplift was also important in generating long-term cooling.

Stratigraphic research in Britain, Germany and North America led to the recognition of the cyclic nature of Carboniferous stratigraphy, and Wanless and Weller (1932) coined the term "cyclothem" for these deposits in the mid-continental United States (see Sect. 1.2.2). Shepard and Wanless (1935) attributed the cyclicity to glacioeustasy driven by the growth and decay of Gondwana ice caps, based on comparisons to the high-frequency glacial-interglacial climatic fluctuations documented for the late Cenozoic glaciation. This interpretation has been confirmed and extended by subsequent research. Heckel (1986) developed a sea-level curve for the midcontinent deposits based on the cyclic repetition and areal extent of key lithofacies, such as conodont-bearing shales, open-marine limestones, prograding fluvial-deltaic deposits and unconformities, many marked by paleovalleys indicating short-term base-level fall. The base-level fluctuations reconstructed for this curve suggest high-frequency sea-level cycles with periodicities comparable to modern orbital frequencies (Heckel 1986).

Coal of Carboniferous age occurs throughout northern Europe, Asia, and midwestern and eastern North America (Stanley 2005). In fact, the term "Carboniferous" comes from England, in reference to the rich deposits of coal that occur there. The evolution of land plants led to the first appearance of woody tissue and bark, which facilitated the rapid development and spread of widespread forests. Most of the coal is of Pennsylvanian age, the term now used for the Upper Carboniferous. It was this coal that was the basis for the Industrial Revolution, first in Britain toward the end of the eighteenth century, and then in continental Europe and North America.

There were five great biological extinctions during the Phanerozoic: (1) Late Ordovician, (2) late Devonian, (3) end-Permian, (4) end-Triassic, and (5) end-Cretaceous (Esmeray-Senlet, in Gradstein et al. 2020). Of these, the end-Permian extinction is considered the most severe, with 80% of marine genera and 75% of terrestrial genera extinguished. Among the major groups that were terminated at this time were eurypterid arthropods, trilobites, acanthodians and blastoid echinoderms. Other groups that were rendered nearly extinct, but which underwent a recovery during the Triassic include the ammonoids, brachiopods, corals, bryozoans, anthozoans, crinoids, gastropods, foraminifera and radiolaria. Over two-thirds of terrestrial labyrinthodont amphibians, sauropsid reptiles and therapsid proto-mammals also became extinct. Several kill mechanisms have been proposed for the end-Permian mass extinction, including carbon cycle disruption, ocean anoxia, ocean acidification, global warming, acid rain, ozone destruction and toxic metal poisoning. Many of these kill mechanisms were linked to the eruption and emplacement of the Siberian Trap Large Igneous Province, the largest volume of preserved continental basaltic magmatism generated in the Phanerozoic (Esmeray-Senlet, in Gradstein et al. 2020, p. 129). None of the proposed extinction mechanisms has widespread support at this time.

The breakup of Pangea commenced with a diachronous Late Permian–Triassic rift along the northern margin of Gondwana (Arabia–India–Australia), with the separation of a Cimmerian continent (Turkey–Iran–Tibet) and the opening of the Neotethyan ocean (Scotese 2001). Rifting extended to the site of the future north Atlantic Ocean in the Triassic, with the development of an enormous series of faulted rift basins extending from Florida to New England and Atlantic Canada, Greenland, North Africa and most of western Europe (Ziegler 1988). These were mostly located in low-latitude settings, and are characterized by nonmarine redbed deposits (fluvial, lacustrine and eolian deposits, some with evaporites). Some old stratigraphic terms, including the New Red Sandstone (Britain), and the Keuper and Buntsandstein (Germany), became familiar names for these deposits. Many of these basins subsided as flexural subsidence accompanied the opening of the Atlantic Ocean, and are now deeply buried beneath the resulting extensional-margin sedimentary wedge. See, for example, the transects across the Atlantic margins of North America prepared by Withjack et al. (1998).

As noted by Schlanger et al. (1987, p. 372): "One of the prime results of the Deep Sea Drilling Project during the 1970s was the discovery that the major ocean basins, during Cretaceous time, were the sites of deposition of sediments anomalously rich in organic carbon in comparison to the average organic-carbon content of Phanerozoic sediments." Schlanger and Jenkyns (1976) were among the first to note the "anomalous stratigraphic concentration of carbonaceous sediments, loosely described as 'black shales', and [they] came to the general conclusion that whatever the mechanism involved, the Cretaceous oceans, from roughly Hauterivian through Santonian time, were locally or regionally oxygen deficient" (Schlanger et al. 1987, p. 372). They termed these periods Oceanic Anoxic Events. At first, two broad time envelopes were recognized within which these events occurred, late Barremian through Albian and late Cenomanian through early Turonian time. Modern work (Cramer and Jarvis, in Gradstein et al. 2020) now recognizes a single Jurassic event (Toarcian) and five Cretaceous events. These "events" are marked by high $\delta^{13}C$ values in the ocean waters, caused by the enhanced preservation of ^{12}C in biomass deposited on the ocean floor. The explanation for these events is interpreted as the spread of oxygen-deficient waters in the world ocean, due to global high temperatures and low latitudinal temperature gradients which, coupled with the decreased solubility of oxygen in these warm waters,

Fig. 7.55 The long-term trend in $\delta^{18}O$ values from the mid-Cretaceous to the present, as measured in deep-sea cores. The computed relationship between $\delta^{18}O$, ocean temperature and continental ice volume provides the basis for the estimates of changing long-term Antarctic ice cover. Not shown are the high-frequency glacial to interglacial fluctuations generated by orbital forcing (from Miller et al. 2005)

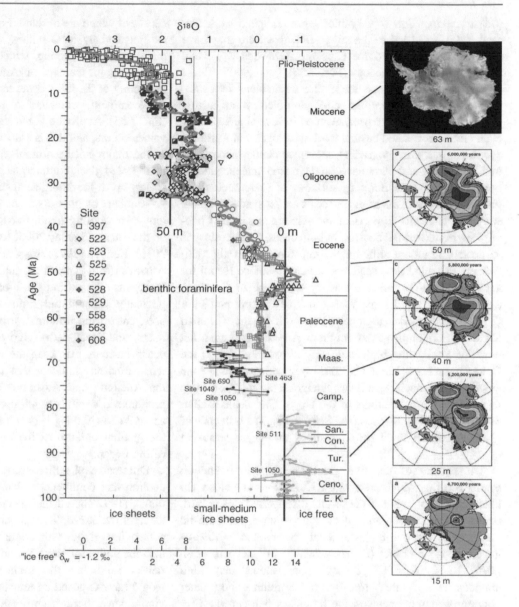

decreased the rate of reoxygenation of bottom waters (Fischer and Arthur 1977). A link with the great Cretaceous transgressions has also been suggested (Schlanger and Jenkyns 1976). It is now realized that these event beds are among the most important source beds for the world's petroleum.

The Cretaceous transgressions constituted the second long Phanerozoic period of high global sea level, and are attributed to the active sea-floor spreading that led to the breakup of Pangea (Worsley et al. 1984). Once again, as during the early Paleozoic, the world's cratons were flooded, leading to the development of widespread epeiric seas. It has also long been thought that much of the Cretaceous period was characterized by a greenhouse climate, that is, one that is significantly warmer than today, and globally more equable (Stanley 2005). Much of western Europe was

covered by epeiric seas warmed by equatorial currents flowing westward from the Tethys Ocean that lay between Africa and Eurasia. One of the most distinctive sedimentary products of this period is the Upper Cretaceous Chalk, famous for forming the white cliffs of "Albion" (from the Latin word for white) along the south coast of England (Stanley 2005). This unit underlies most of southeast England and also occurs through much of western Europe, including France, Germany and Denmark. Its formation is attributed to the flourishing of microscopic organisms, notably the coccoliths, in warm-temperate marine waters. A comparable unit is the Austin Chalk of southern Texas, also of the Late Cretaceous age.

The long-term persistence of greenhouse climates began to be questioned when oxygen isotope measurements suggested cool periods during the Cenomanian and the

Maastrichtian. High $\delta^{18}O$ values were compiled by Miller et al. (2005) and led to the suggestion that there may have been small, short-lived ice caps on Antarctica at several times during the Cretaceous (Fig. 7.55).

Stratigraphic evidence for cyclic successions of cyclothem type, with durations and periodicities suggesting orbital frequencies have been mapped in several areas. Elder et al. (1994) correlated basinal marl-shale cycles of Turonian age in Kansas with prograding shoreface clastic successions in Utah. Plint and Kreitner (2007) traced thin sequences bounded by marine flooding surfaces of Cenomanian age without changes in thickness or facies across syndepositional structural elements in parts of the Alberta Basin, and likewise suggested orbital control. In both cases, glacioeustasy is suspected. In a quite different setting, Sageman et al (2006) defined an orbital time scale and a new C-isotope record for a Cenomanian–Turonian boundary stratotype in Colorado (see Sect. 7.8.3 and Figs. 7.43 and 7.44). Shank and Plint (2013) constructed a regional framework for the Turonian Cardium Formation across southern Alberta and began the work of linking the high-frequency allostratigraphy to the Colorado sections. Lin et al. (2021) were able to estimate the amplitude of glacial sea-level change at up to 50 m from backstripped cross-sections of the Upper Cretaceous Gallup Sandstone in New Mexico (see Sect. 8.12). The influence of glacioeustasy in the Late Cretaceous no longer seems in doubt.

The second of the five great Phanerozoic extinction events was that which brought the Cretaceous to a close. On land, land dinosaurs and pterosaurs disappeared; many bird, lizard, snake, insect and plant groups underwent drastic changes. In the oceans, $\sim 75\%$ of species and $\sim 40\%$ of genera became extinct (Esmeray-Senlet, in Gradstein et al. 2020, Chap. 3L). The discovery of anomalously high abundances of the rare earth iridium and other platinum-group elements at the boundary between the Cretaceous and the Cenozoic led to the hypothesis that a large bolide had impacted the Earth, creating a major environmental catastrophe, including tsunamis, wildfires and a lengthy "nuclear winter" (Alvarez et al. 1980). The discovery of a major crater located offshore from the coast of Yucatan, Mexico suggested a possible site for the impact (Hildebrand et al. 1991), and subsequent studies have confirmed the likelihood of this interpretation. Other interpretations, including environmental perturbations caused by large volcanic eruptions, have not withstood detailed examination. The boundary event was initially referred to as the K-T boundary, after the abbreviations for the Cretaceous and Tertiary, but with the abandonment of the term Tertiary for reasons of chronostratigraphic consistency, the boundary is now abbreviated as the K-Pg event (Pg = Paleogene).

The Earth underwent a short but pronounced warming episode during the late Paleogene and early Eocene, the causes of which are obscure. Following this, at about 50 Ma, as indicated by $\delta^{18}O$ values, the Earth commenced a dramatic long-term cooling, which initially led to the development of major ice caps on Antarctica, and then, around the beginning of the Oligocene, to ice coverage of almost all of the continent, comparable to the ice coverage today (Fig. 7.55). At around 5 Ma a further sharp increase in $\delta^{18}O$ values occurs, and this is associated with the commencement of the major northern hemisphere glaciation, characterized by repeated glacial to interglacial fluctuations on orbital time scales, up to the interglacial of the present day (Fig. 7.39). A number of processes have been suggested for this long-term cooling trend, but there is no general agreement on the causes (Stanley 2005; Ruddiman 2008; Summerhayes 2015). Kennett (1977) suggested that the separation of South America from Antarctica, and then the further separation from Australia during the breakup of Pangea, would have gradually isolated the Antarctic continent over the south pole, cut off the flow of warm ocean waters from low latitudes, and created the vigorous circum-Antarctic air and ocean currents that keep the continent cold. Plate tectonic reconstructions indicate that complete separation between the Antarctic and Australia occurred around 30 Ma. The pronounced cooling trend suggested in Fig. 7.55 began at about 34 Ma, the lag between these dates perhaps indicating the gradual buildup of the circum-Antarctic currents once rifting began.

Different explanations have been offered for the further cooling that commenced at about 5 Ma. Raymo and Ruddiman (1992) based their proposal on the supposed importance of the atmospheric concentration of carbon dioxide in controlling global temperature variations. They suggested that the uplift of the Tibetan plateau and the Himalayan ranges following the India-Asia plate-tectonic collision would have exposed an unusual area of the Earth's crust to vigorous weathering, a process that takes carbon dioxide out of the atmosphere and therefore, according to this hypothesis, could have led to major cooling. The emergence of these elevated areas in the midst of the global air circulation patterns would, by itself, have tended to cool the atmosphere, so it is possible that multiple processes were at work.

As recently as the 1960s mainstream interpretations still followed the hypothesis of Penck and Bruckner (1909), that there were four major glaciations in Europe, named the Gunz, Mindel, Riss and Wurm phases. However, since the development of oxygen isotope stratigraphy, and particularly since the recovery of continuous Late Cenozoic sections from deep-sea cores, it has been realized that there have been multiple glaciations. The current $\delta^{18}O$ chronostratigraphic chart contains 104 Quaternary cycles and 105 for the Pleistocene (Fig. 7.39). The sparse fourfold chronology initially developed in Europe in part reflects the very fragmentary nature of the record of continental glaciation, a

Fig. 7.56 Top: Holocene temperature variations over central Greenland, based on the GISP2 ice core. Bottom: atmospheric CO_2 measurements, from the EPICA Dome C ice core. Diagram from http://www. climate4you.com/ (downloaded April 9, 2021). Ice core data ends in the mid-nineteenth century. Dashed red line in the temperature graph shows the beginning of the modern warm period

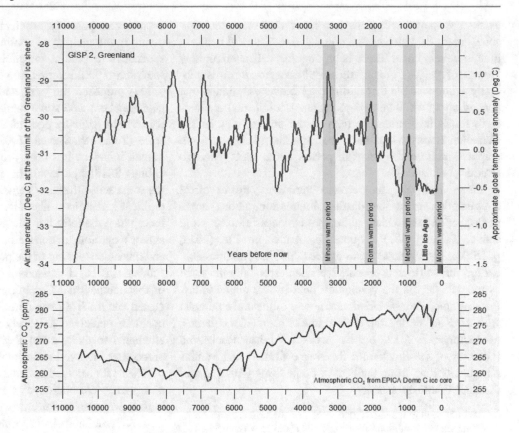

process that is predominantly erosional, and in part reflects the lack of accurate dating methods until the advent of the oxygen isotope method coupled with radioisotopic dating.

Until about 1 Ma the $\delta^{18}O$ values fluctuated on an approximately 41 ka cycle, corresponding to the obliquity period. At around 1 Ma the amplitude of the cyclicity became much more pronounced and the periodicity shifted to a 100-ka cycle, modulated by 400-ka periodicity, corresponding to the eccentricity cycle and its prominent harmonic. The reasons for this are still not well understood (Ruddiman 2008; Summerhayes 2015) and are, in any case, beyond the scope of this book.

The Holocene epoch should be of particular interest to Earth scientists, because this is the era, since the end of the last ice age, about 12,000 years ago, when human society underwent its rapid evolution to the present complex industrial era in which we live. In particular, this era offers the opportunity to compare past natural climate variations with the climate of the present day, under long-term geological conditions, such as plate-tectonic setting, the elevations of mountains and continents, ocean currents, weather systems, etc., that have barely changed in 12,000 years. H. H. Lamb was the pioneer in this field, and his many books and articles helped to establish such terms as the Medieval Warm Period and the Little Ice Age by the 1970s (e.g., Lamb 1972). Yet the approach of the Earth science

community to this question since Lamb has been curiously ambiguous. Some paleoclimatologists have published very detailed treatises on the Holocene (e.g., Bradley 1999; Anderson et al. 2013), yet the crucial question: what can we learn from Holocene climate change to help us to understand present-day variability, is typically not directly addressed.

Figure 7.56, which is from the website of Ole Humlum (http://www.climate4you.com), is an exception, and a very similar graph was published by Lewis and Maslin (2015, Fig. 2). It compares two post-glacial data sets derived from ice core data, temperature and atmospheric CO_2 composition. The temperature data are from Greenland and the CO_2 are from Antarctica. It is not to be expected that data sets from opposite sides of the Earth should be directly correlated because atmospheric and oceanic processes are commonly hemispheric and out of phase; yet the long-term (>10,000 year) trends that this illustration reveals are in direct opposition to those which the anthropogenic global warming model would predict. Atmospheric CO_2 shows a slow trend of modest decrease from 10,500 to 7,500 years ago, followed by a slow but steady increase. The Greenland temperature trend shows the opposite: a dramatic post-glacial increase of 5° from 11,000 to 8,000 years ago, followed by a slow decrease to the very recent. The long-term temperature trend illustrated in Fig. 7.56 is attributed to orbitally forced changes in solar insolation.

There are many oscillations into short warm and cool episodes, many of which can be correlated to historical trends in human development There is an ongoing debate about how "regional" versus "global" these episodes are, but it is generally acknowledged that during the Holocene Optimum that peaked about 8000 years ago regional continental ice caps were much less extensive than at the present day. In the Canadian Rocky Mountains, ice caps and glaciers did not reappear until the "Neoglacial" period, about 3000 years ago (Rutter et al. 2006).

These Holocene temperature variations are in direct contradiction to a long-lasting impression among some geologists and the wider public, that Holocene climates were stable. Raikes (1967) is quoted by Anderson et al. (2013, p. 155) as stating that "From at least 7000 BC, and possibly earlier, the worldwide climate has been essentially the same as today." This is the perception held by the public today. For example, Thomas Friedman, a very influential columnist with the New York Times, based one of his regular columns, on October 7 2015, on the work of Johan Rockström (Director of the Stockholm Resilience Center) and Mattias Klum. Quoting from their new book "*Big World, Small Planet*," Friedman stated:

> It's only been in the last 10,000 years that we have enjoyed the stable climate conditions allowing civilizations to develop based on agriculture that could support towns and cities. This period, known as the Holocene, was an 'almost miraculously stable and warm interglacial equilibrium, which is the only state of the planet we know for sure can support the modern world as we know it.' It finally gave us "a stable equilibrium of forests, savannahs, coral reefs, grasslands, fish, mammals, bacteria, air quality, ice cover, temperature, fresh water availability and productive soils. 'It is our Eden,' Rockström added, and now 'we are threatening to push earth out of this sweet spot.'

Anderson et al. (2013, p. 155) strongly disagreed. They state that, based on all the paleoclimatological evidence, which is described in detail in their book, "clearly … the concept of a stable Holocene environment is quite untenable." It is my contention that were Holocene climates fully discussed and understood by the general public, the current obsession with the urgency of climate change would never have developed, because there is nothing about present-day climates, including temperature extremes and severe weather events, that has not been preceded at some time during the Holocene. It is likely that the current global temperature regime is no warmer than that which prevailed during the Medieval warm period, at around 1000 AD. For a more balanced view of Holocene climates, see Plimer (2009) and Carter (2010).

The concept of the "Anthropocene" has received a great deal of attention in recent years, defining a period in Earth history when human influence on surface processes became global in scope. Such issues as pollution, deforestation, loss of habitat and biodiversity, climate change, extreme weather events, and possibly accelerated rise in global sea level owing to the melting of continental ice caps, have led some specialists to predict impending catastrophe. Mungall and McLaren (1990) referred to a "planet under stress." Crutzen (2002) proposed the term **Anthropocene**, to define a new geological era, and some Earth scientists have proposed that the term be formally defined and added to the geologic time scale (Zalasiewicz et al. 2008). There is no question that human influence has been profound (e.g., Ruddiman 2013; Gibling 2018; Koster 2020), and the term is a useful one, in the same sense that the "Renaissance" identifies a period of political and artistic liberation in European history. But to incorporate the term into the geological time scale is, in this author's opinion, an unnecessary step that would be a misuse and misunderstanding of the purpose of the GTS. Finney and Edwards (2016) discussed the proposal from the perspective of the International Commission on Stratigraphy, and suggested that the ICS should not be involved in an activity that could be perceived as purely political. As this chapter has attempted to make clear, the purpose of the GTS and the procedures involved in its construction, such as the selection of GSSPs for selected boundaries, is to facilitate the accurate and precise correlation of the sedimentary record of geological processes and events worldwide. There is no need for a formally defined Anthropocene to facilitate the procedure of using geological methods to correlate processes that are largely within historical memory.

References

Abreu, V., 1998, Evolution of the conjugate volcanic passive margins: Pelotas Basin (Brazil) and offshore Namibia (Africa): Implications for global sea-level changes. Unpublished Ph.D. thesis, Rice University, Houston.

Ager, D. V., 1964, The British Mesozoic Committee: Nature, v. 203, p. 1059.

Ager, D. V., 1973, The nature of the stratigraphical record: New York, John Wiley, 114 p.

Ager, D. V., 1981, The nature of the stratigraphical record (second edition): John Wiley, New York, 122 p.

Allen, P. A., and Etienne, J. L., 2008, Sedimentary challenge to Snowball Earth, Nature Geoscience, v. 1, p. 817–825

Alvarez, L. W., Alvarez, W., Asaro, F., and Michel, H. V. 1980, Extraterrestrial cause for the Cretaceous-Tertiary extinction: Science, v. 208, p. 1095–1108

Anderson, D. E., Goudie, A. S., and Parker, A. G., 2013, Global environments through the Quaternary, Second edition: Oxford University Press, Oxford, 406 p.

Armentrout, J. M., 1981, Correlation and ages of Cenozoic chronostratigraphic units in Oregon and Washington: Geological Society of America Special Paper 184, p. 137–148.

Aubry, M.-P., 2007, chronostratigraphic terminology: Building on Principles: Stratigraphy, v. 4, p. 117–125.

Aubry, M.-P., Van Couvering, J., Berggren, W. A., and Steininger, F., 1999, Problems in chronostratigraphy: stages, series, unit and boundary stratotype section and point and tarnished golden spikes: Earth-Science Reviews, v. 46, p. 99–148.

Aubry, M.-P., Van Couvering, J., Berggren, W. A., and Steininger, F., 2000, Should the gold spike glitter: Episodes, v. 23, p. 203–210.

Autin, W. J., 1992, Use of alloformations for definition of Holocene meander belts in the middle Amite River, southeastern Louisiana: Geological Society of America Bulletin, v. 104, p. 233–241.

Barndt, J., Johnson, N. M., Johnson, G. D., Opdyke, N. D., Lindsay, E. H., Pilbeam, D., And Tahirkheli, R. A. H., 1978, The magnetic polarity stratigraphy and age of the Siwalik Group near Dhok Pathan Village, Potwar Plateau, Pakistan: Earth Planetary Science Letters, v. 41, p. 355–364.

Barnes, C. R., Jackson, D. E., and Norford, B. S., 1976, Correlation between Canadian Ordovician zonations based on graptolites, conodonts and benthic macrofossils from key successions, in Bassett, M. G., ed., The Ordovician System: proceedings of a Palaeontological Association symposium, Birmingham, September 1974: University of Wales and National Museum of Wales, Cardiff, p. 209–225.

Barrell, Joseph, 1917, Rhythms and the measurement of geologic time: Geological Society of America Bulletin, v. 28, p. 745–904.

Bassett, M. G., 1985, Towards a "common language" in stratigraphy: Episodes, v. 8, p. 87–92.

Beall, A. P., Moresi, L., and Cooper, C. M., 2018, Formation of cratonic lithosphere during the initiation of plate tectonics: Geology, v. 46, p. 487–490.

Behrensmeyer, A. K., 1987, Miocene fluvial facies and vertebrate taphonomy in northern Pakistan, in Ethridge, F. G., Flores, R. M., and Harvey, M. D., eds., Recent developments in fluvial sedimentology: Society of Economic Paleontologists and Mineralogists Special Publication 39, p. 169–176.

Berggren, W. A., 2007, Status of the hierarchical subdivision of higher order marine Cenozoic chronostratigraphic units: Stratigraphy, v. 4, p. 99–108.

Berggren, W. A., Kent, D. V., Aubry, M.-P., and Hardenbol, J., eds., 1995, Geochronology, time scales and global stratigraphic correlation: Society for Sedimentary Geology Special Publication 54, 386 p.

Berner, R. A., and Kothavala, Z., 2001, GEOCARB III: A revised model of atmospheric CO_2 over Phanerozoic time: American Journal of Science, v. 301, p. 182–204.

Berry, W. B. N., 1977, Graptolite biostratigraphy: a wedding of classical principles and current concepts; in Kauffman, E. G., and Hazel, J. E., eds., Concepts and methods of biostratigraphy: Dowden, Hutchinson and Ross Inc., Stroudsburg, Pennsylvania, p. 321–338.

Berry, W. B. N., 1987, Growth of prehistoric time scale based on organic evolution, revised edition: Oxford: Blackwell Science, 202 p.

Bhattacharya, J., 1991, Regional to sub-regional facies architecture of river-dominated deltas, Upper Cretaceous Dunvegan Formation, Alberta subsurface, in Miall, A. D., and Tyler, N., eds., The three-dimensional facies architecture of terrigenous clastic sediments and its implications for hydrocarbon discovery and recovery, Society of Economic Paleontologists and Mineralogists, Concepts in Sedimentology and Paleontology, v. 3, p. 189–206.

Bhattacharya, J. P., 1993, The expression and interpretation of marine flooding surfaces and erosional surfaces in core; examples from the Upper Cretaceous Dunvegan Formation, Alberta foreland basin, Canada, in Posamentier, H. W., Summerhayes, C. P., Haq, B. U., and Allen, G. P., eds., Sequence stratigraphy and facies associations: International Association of Sedimentologists Special Publication 18, p. 125–160.

Blackwelder, E., 1909, The valuation of unconformities: Journal of Geology, v. 17, p. 289–299.

Bradley, R. S., 1999, Paleoclimatology: Reconstructing the climates of the Quaternary, Second edition: International Geophysics Series, Amsterdam, v. 68, 614 p.

Brown, A. R., 2011, Interpretation of three-dimensional seismic data, seventh edition, American Association of Petroleum Geologists Memoir 42, 646 p.

Burke, W. H., Denson, R. E., Hetherington, E. A., Koepnick, R. B., Nelson, H. F., and Otto, J. B., 1982, Variations of seawater $^{87}Sr/^{86}Sr$ through Phanerozoic time: Geology, v. 10, p. 516–519.

Callomon, J. H., 1995, Time from fossils: S. S. Buckman and Jurassic high-resolution geochronology, in Le Bas, M. J., ed., Milestones in Geology: Geological Society of London Memoir 16, p. 127–150.

Callomon, J. H., 2001, Fossils as geological clocks, in C. L. E. Lewis and S. J. Knell, eds., The age of the Earth: from 4004 BC to AD 2002: Geological Society of London Special Publication 190, p. 237–252.

Carter, R. M., 2010, Climate: The counter consensus, Stacey International, London, 315 p.

Cartwright, J. A., Haddock, R. C., and Pinheiro, L. M., 1993, The lateral extent of sequence boundaries, in Williams, G. D., and Dobb, A., eds., Tectonics and seismic sequence stratigraphy: Geological Society, London, Special Publication 71, p. 15–34.

Castradori, D., 2002, A complete standard chronostratigraphic scale: how to turn a dream into reality? Episodes, v. 25, p. 107–110.

Catuneanu, O., 2006, Principles of sequence stratigraphy: Elsevier, Amsterdam, 375 p.

Catuneanu, O., Abreu, V., Bhattacharya, J. P., Blum, M. D., Dalrymple, R. W., Eriksson, P. G., Fielding, C. R., Fisher, W. L., Galloway, W. E., Gibling, M. R., Giles, K. A., Holbrook, J. M., Jordan, R., Kendall, C. G. St. C., Macurda, B., Martinsen, O. J., Miall, A. D., Neal, J. E., Nummedal, D., Pomar, L., Posamentier, H. W., Pratt, B. R., Sarg, J. F., Shanley, K. W., Steel, R. J., Strasser, A., Tucker, M. E., and Winker, C., 2009, Toward the Standardization of Sequence Stratigraphy: Earth Science Reviews, v. 92, p. 1–33.

Catuneanu, O., Bhattacharya, J. P., Blum, M. D., Dalrymple, R. W., Eriksson, P. G., Fielding, C. R., Fisher, W. L., Galloway, W. E., Gianolla, P., Gibling, M. R., Giles, K. A., Holbrook, J. M., Jordan, R., Kendall, C. G. St. C., Macurda, B., Martinsen, O. J., Miall, A. D., Nummedal, D., Posamentier, H. W., Pratt, B. R., Shanley, K. W., Steel, R. J., Strasser, A., and Tucker, M. E., 2010, Sequence stratigraphy: common ground after three decades of development: First Break, v. 28, p. 21–34.

Catuneanu, O., Galloway, W.E., Kendall, C.G.St.C., Miall, A.D., Posamentier, H.W., Strasser A., and Tucker M.E., 2011, Sequence Stratigraphy: Methodology and Nomenclature: Report to ISSC: Newsletters on Stratigraphy, v. 4 (3), p. 173–245.

Chlupáč, I., 1972, The Silurian-Devonian boundary in the Barrandian: Bulletin of Canadian Petroleum Geology, v. 20, p. 104–174.

Christie-Blick, N., Mountain, G. S., and Miller, K. G., 1990, Seismic stratigraphy: record of sea-level change, in Revelle, R., ed., Sea-level change: National Research Council, Studies in Geophysics, Washington, National Academy Press, p. 116–140.

Cloud, P. E., 1948, Some problems and patterns of evolution exemplified by fossil invertebrates: Evolution, v. 2, p. 322–350.

Cloud, P. E., 1968, Atmospheric and hydrospheric evolution on the primitive Earth: Science, v. 160, p. 729–736.

Cody, R. M., Levy, R. H., Harwood, D. M., and Sadler, P. M., 2008, Thinking outside the zone: high-resolution quantitative biochronology for the Antarctic Neogene: Palaeogeography, Palaeoclimatology, Palaeoecology, v. 260, p. 92–121.

Cohee, G. V., Glaessner, M. F., and Hedberg, H. D., eds., 1978, Contributions to the geologic time scale: American Association of Petroleum Geologists Studies in Geology No. 6.

Cohen, K.M., Finney, S., and Gibbard, P.L., 2012, International Chronostratigraphic Chart: International Commission on Stratigraphy, www.stratigraphy.org

Conkin, B. M., and Conkin, J. E., eds., 1984, Stratigraphy: foundations and concepts: Benchmark Papers in Geology, New York: Van Nostrand Reinhold, 363 p.

Cooper, R. A., Nowlan, G. S., and Williams, S. H., 2001, Global Stratotype Section and Point for base of the Ordovician System: Episodes, v. 24, p. 19–28.

Cowie, J. W., 1986, Guidelines for boundary stratotypes: Episodes, v. 9, p. 78–82.

Cox, A., 1969, Geomagnetic reversals: Science, v. 163, p. 237–245.

Cramer, B. D., and Jarvis, I., 2020, Carbon isotope stratigraphy, in Gradstein, F. M., Ogg, J. G., Schmitz, M. D., and Ogg, G. M., eds., Geologic time scale 2020: Elsevier, Amsterdam, p. 309–343.

Cramer, B. D., Loydell, D. K., Samtleben, C., Munnecke, A., Kaljo, D., Männik, P., Martma, T., Jeppsson, L., Kleffner, M. A., Barrick, J. E., Johnson, C. A., Emsbo, P., Joachimski, M. M., Bickert, T., and Saltzman, R., 2010, testing the limits of Paleozoic chronostratigraphic correlation via high-resolution (<500 k.y.) integrated conodont, graptolite, and carbon isotope ($\delta^{13}C_{carb}$) biochemostratigraphy across the Llandovery-Wenlock (Silurian) boundary: is a unified Phanerozoic time scale achievable? Geological Society of America Bulletin, v. 122, p. 1700–1716.

Cramer, B. D., Vandenbroucke, T. R. A., and Ludvigson, G. A., 2015, High-resolution event stratigraphy (HiRES) and the quantification of stratigraphic uncertainty: Silurian examples of the quest for precision in stratigraphy: Earth Science Reviews, v. 141, p. 136–153.

Croll, J., 1864, On the physical cause of the change of climate during geological epochs: Philosophical Magazine, v. 28, p. 435–436.

Crowell, J. C., 1978, Gondwanan glaciation, cyclothems, continental positioning, and climate change: American Journal of Science, v. 278, p. 1345–1372.

Crutzen, P.J., 2002, Geology of mankind: Nature, v. 415, p. 23, doi: 10.1038/ 415023a.

Davies, N. S., and Shillito, A. P., 2018, Incomplete but intricately detailed: The inevitable preservation of true substrates in a time-deficient stratigraphic record: Geology, v. 46, p. 679–682.

Davis, W. J., 2017, The relationship between atmospheric carbon dioxide carbon dioxide concentrations and global temperature for the last 425 million years: MDPI Journals, Climate, v. 5, 76, 35 p.

de Boer, P. L., and Smith, D. G., 1994, Orbital forcing and cyclic sequences, in de Boer, P. L., and Smith, D. G., eds., Orbital forcing and cyclic sequences: International Association of Sedimentologists Special Publication 19, p. 1–14

Derby, J. R., Fritz, R. D., Longacre, S. A., Morgan, W. A., and Sternbach, C. A., eds., 2012, The great American carbonate bank: The geology and economic resources of the Cambrian–Ordovician Sauk megasequence of Laurentia: AAPG Memoir 98.

Doyle, J. A., 1977: Spores and pollen: the Potomac Group (Cretaceous) Angiosperm sequence; in Kauffman, E. G., and Hazel, J. E., eds., Concepts and methods of biostratigraphy: Dowden, Hutchinson and Ross Inc., Stroudsburg, Pennsylvania, p. 339–364.

Edwards, L. E., 1984, Insights on why graphic correlation (Shaw's method) works: Journal of Geology, v. 92, p. 583–597.

Edwards, L. E., 1985, Insights on why graphic correlation (Shaw's method) works: A reply [to discussion]: Journal of Geology, v. 93, p. 507–509.

Einsele, G., and Seilacher, A., eds., 1982, Cyclic and event stratification: Springer-Verlag Inc., Berlin, 536 p.

Elder, W. P., Gustason, E. R., and Sageman, B. B., 1994, Correlation of basinal carbonate cycles to nearshore parasequences in the Late Cretaceous Greenhorn seaway, Western Interior, U.S.A.: Geological Society of America Bulletin, v. 106, p. 892–902.

Eldredge, N., and Gould, S. J., 1972, Punctuated equilibrium: an alternative to phyletic gradualism, in Schopf, T. J. M., ed., Models in paleobiology: San Francisco, Freeman, Cooper and Company, p. 82–115.

Eldredge, N., and Gould, S. J., 1977, Evolutionary models and biostratigraphic strategies; in Kauffman, E. G., and Hazel, J. E., eds., Concepts and methods of biostratigraphy: Dowden, Hutchinson and Ross Inc., Stroudsburg, Pennsylvania, p. 25–40.

El Meknassi, S., Dera, G., Cardone, T., De Rafélis, M., Brahmi. C., and Chavagnac, V., 2018, Sr isotope ratios of modern carbonate shells: good and bad news for chemostratigraphy: Geology, v. 46, p. 1003–1006.

Embry, A. F., 1995, Sequence boundaries and sequence hierarchies: problems and proposals, in Steel, R. J., Felt, V. L., Johannessen, E. P., and Mathieu, C., eds., Sequence stratigraphy on the Northwest European margin: Norsk Petroleumsforening Special Publication 5, Elsevier, Amsterdam, p. 1–11.

Embry, A. F., and Johannessen, E. P., 1992, T-R sequence stratigraphy, facies analysis and reservoir distribution in the uppermost Triassic-Lower Jurassic succession, western Sverdrup Basin, Arctic Canada, in Vorren, T. O., Bergsager, E., Dahl-Stamnes, O. A., Holter, E., Johansen, B., Lie, E., and Lund, T. B., eds., Arctic geology and petroleum potential: Norwegian Petroleum Society Special Publication 2, p. 121–146.

Emery, D., and Myers, K. J., 1996, Sequence stratigraphy: Blackwell, Oxford, 297 p.

Emiliani, C., 1955, Pleistocene temperatures: Journal of Geology, v. 63, p. 538–578.

Eriksson, P. G., Banerjee, S., Catuneanu, O, Corcoran, P. L., Eriksson, K. A., Hiatt, E. E, Laflamme, M., Lenhardt, N., Long, D. G. F., Miall, A. D., Mints, M. V., Pufahl, P. K., Sarkar, S., Simpson, E. L., Williams, G. E., 2013, Secular changes in sedimentation systems and sequence stratigraphy, in Kusky, T., Stern, R., and Dewey, J., eds., Secular Changes in Geologic and Tectonic Processes: Gondwana Research, Special issue, v. 24, issue #2, p. 468–489.

Eriksson, P. G., Condie, K. C., Tirsgaard, H., Mueller, W. U., Alterman, W., Miall, A. D., Aspler, L. B., Catuneanu, O., and Chiarenzelli, J. R., 1998, Precambrian clastic sedimentation systems, in Eriksson, P. G., ed., Precambrian clastic sedimentation systems: Sedimentary Geology, Special Issue, v. 120, p. 5–53.

Eriksson, P. G., and Mazumder, R., eds., 2020, Editorial Preface, Special Issue: Archean Earth Processes: Earth Science Reviews, v. 202, #103058, 6p.

Eyles, N., 1993, Earth's glacial record and its tectonic setting: Earth Science Reviews, v. 35, p. 1–248.

Eyles, N., 2008, Glacio-epochs and the supercontinent cycle after ∼3.0 Ga: tectonic boundary conditions for glaciation: Palaeogeography, Palaeoclimatology, Palaeoecology, v. 258, p. 89–129.

Eyles, N., and Januszczak, N., 2004, "Zipper-rift": A tectonic model for Neoproterozoic glaciations during the breakup of Rodinia after 750 Ma: Earth Science Reviews, v. 65, p. 1–73.

Fejfar, O., and Heinrich, W. D., 1989, Muroid rodent biochronology of the Neogene and Quaternary, in Lindsay, E. H., Fahlbusch, V., and Mein, P., eds., European mammal chronology: NATO Advanced Research Workshop, p. 91–118.

Fischer, A. G., 1981, Climatic oscillations in the biosphere, in Nitecki, M. H., ed., Biotic crises in ecological and evolutionary time: Academic Press, New York, p. 102–131.

Fischer, A. G., 1984, The two Phanerozoic cycles, in Berggren, W. A., and Van Couvering, J. A., eds., Catastrophes and Earth History: The New Uniformitarianism: Princeton University Press, New Jersey, p. 129–150.

Fischer, A. G., and & Arthur, M. A. 1977, Secular variations in the pelagic realm: Society of Economic Paleontologists and Mineralogists Special Publication 25, p. 19–50.

Frazier, D. E., 1974, Depositional episodes: their relationship to the Quaternary stratigraphic framework in the northwestern portion of the Gulf Basin: Bureau of Economic Geology, University of Texas, Geological Circular 74–1, 26 p.

Friend, P. F. and Williams, B. P. J., eds., 2000, New Perspectives on the Old Red Sandstone. Geological Society, London, Special Publication 180.

Galloway, W. E., 1989, Genetic stratigraphic sequences in basin analysis I: Architecture and genesis of flooding-surface bounded depositional units: American Association of Petroleum Geologists Bulletin, v. 73, p. 125–142.

Gibling, M. R., and Bird, D. J., 1994, Late Carboniferous cyclothems and alluvial paleovalleys in the Sydney Basin, Nova Scotia: Geological Society of America Bulletin, v. 106, p. 105–117.

Gibling, M. R., 2018, River systems and the Anthropocene: a Late Pleistocene and Holocene timeline for human influence: Quaternary, v. 1, no. 21, https://doi.org/10.3390/quat1030021.

Gilbert, G. K., 1895, Sedimentary measurement of geologic time: Journal of Geology, v. 3, p. 121–127.

Gradstein, F. M., Ogg, J. G., and Smith, A. G., eds., 2004a, A geologic time scale: Cambridge University Press, Cambridge, 610 p.

Gradstein, F. M., Cooper, R. A., and Sadler, P. M., 2004b, Biostratigraphy: time scales from graphic and quantitative methods, in Gradstein, F. M., Ogg, J. G., and Smith, A. G., eds., A geologic time scale: Cambridge University Press, Cambridge, p. 49–54.

Gradstein, F. M., Ogg, J. G., Schmitz, M. D., and Ogg, G. M., 2012, The Geologic time scale 2012: Elsevier, Amsterdam, 2 vols., 1176 p.

Gradstein, F. M., Ogg, J. G., Schmitz, M. D., and Ogg, G. M., eds., 2020, Geologic time scale 2020: Elsevier, Amsterdam, 1357 p.

Grant, B. R., and Grant, P. R., 2008, Fission and fusion of Darwin's finches populations: Philosophical Transactions of the Royal Society, London Series B, Biological Sciences, v. 363 (1505), p. 2821–2829.

Hancock, J. M., 1977, The historic development of biostratigraphic correlation, in Kauffman, E. G. and Hazel, J. F., eds., Concepts and methods of biostratigraphy: Dowden, Hutchinson and Ross Inc., Stroudsburg, Pennsylvania, p. 3–22.

Harland, W. B., 1978, Geochronologic scales, in Cohee, G. V., Glaessner, M. F. and Hedberg, H. D., eds., Contributions to the Geologic time scale: American Association of Petroleum Geologists Studies in Geology 6, p. 9–32.

Harland, W. B., 1993, Stratigraphic regulation and guidance: a critique of current tendencies in stratigraphic codes and guides: Discussion: Geological Society of America Bulletin, v. 105, p. 1135–1136.

Harland, W. B., Armstrong, R. L., Cox, A. V., Craig, L. E., Smith, A. G., and Smith, D. G., 1990, A geologic time scale, 1989: Cambridge Earth Science Series, Cambridge University Press, Cambridge, 263 p.

Harland, W. B., Cox, A. V., Llewellyn, P. G., Pickton, C. A. G., Smith, A. G., and Walters, R., 1982, A geologic time scale: Cambridge Earth Science Series, Cambridge University Press, Cambridge, 131 p.

Harland, W. B., and Francis, H., eds., 1964, The Phanerozoic time scale (A symposium dedicated to Professor Arthur Holmes): Quarterly Journal of the Geological Society of London, v. 120s, 458 p.

Harland, W. B., and Francis, H., eds., 1971, The Phanerozoic time scale — A Supplement. Geological Society of London Special Publication 5, 356 p.

Harper, C. W., Jr., and Crowley, K. D., 1985, Insights on why graphic correlation (Shaw's method) works: A discussion: Journal of Geology, v. 93, p. 503–506.

Harrison, C. G. A., and Funnell, B. M., 1964, Relationship of palaeomagnetic reversals and micropalaeontology in two Late Cenozoic cores from the Pacific Ocean: Nature, v. 204, p. 566.

Hawkesworth, C. J., Cawood, P. A., Dhuime, B., and Kemp, T. I., 2017, Earth's continental lithosphere through time: Annual Review of Earth and Planetary Sciences, v. 45, p. 169–198,

Hay, W. W., and Southam, J. R., 1978, Quantifying biostratigraphic correlation: Annual Review of Earth and Planetary Sciences, v. 6, p. 353–375.

Hays, J. D., Imbrie, J., and Shackleton, N. J., 1976, Variations in the earth's orbit: pacemaker of the ice ages: Science, v. 194, p. 1121–1132.

Heckel, P. H., 1986, Sea-level curve for Pennsylvanian eustatic marine transgressive-regressive depositional cycles along midcontinent outcrop belt, North America: Geology, v. 14, p. 330–334.

Heckert, A. B., and Lucas, S. G., 2004, Simplifying the stratigraphy of time: Comments and Reply: Geology, v. 32, p. e58.

Hedberg, H. D., ed., 1976, International Stratigraphic Guide: Wiley, New York, 200 p.

Herbert, T. D., Premoli Silva, P, Erba, E., and Fischer, A. G., 1995. Orbital chronology of Cretaceous-Paleocene marine sediments, in Berggren, W. A., Kent, D. V., Aubry, M.-P., and Hardenbol, J., eds., Geochronology, time scales and global stratigraphic correlation: Society for Sedimentary Geology Special Publication 54, p. 81–93.

Hildebrand, A., Penfield, G., Kring, D., Pilkington, M., and Camargo Z., 1991, Chicxulub crater: a possible Cretaceous/Tertiary boundary impact crater on the Yucatan Peninsula, Mexico: Geology, v. 19, p. 867–871.

Hilgen, F. J., 1991. Extension of the astronomically calibrated (polarity) time scale to the Miocene/Pliocene boundary: Earth and Planetary Sciences Letters, v. 107, p. 349–368.

Hilgen, F. J., Brinkhuis, H., and Zachariasse, W. J., 2006, Unit stratotypes for global stages. The Neogene perspective: Earth Science Reviews v. 74, p. 113–125.

Hilgen, F. J., Hinnov, L. A., Aziz, H. A., Abels, H. A., Batenburg, S., Bosmans, J. H. C., de Boer, B., Hüsings, S. K., Kuiper, K. F., and Lourens, L. J., 2015, Stratigraphic continuity and fragmentary sedimentation: the success of cyclostratigraphy as part of integrated stratigraphy in Smith, D. G., Bailey, R., J., Burgess, P., and Fraser, A., eds., Strata and time: Geological Society, London, Special Publication 404, p. 157–197.

Hilgen, F. J., Lourens, L. J., Pälike, H., and research support team, 2020, Should Unit-Stratotypes and Astrochronozones be formally defined? A dual proposal (including postscriptum). Newsletters on Stratigraphy, v. 53, p. 19–39.

Hinnov, L. A., and Ogg, J. G., 2007, Cyclostratigraphy and the astronomical time scale: Stratigraphy, v. 4, p. 239–251.

Hoffman, P. F., Schrag, D. P., 2002, The snowball Earth hypothesis: testing the limits of global change. Terra Nova, v. 14, p. 129–155.

Holbrook, J. M., and Bhattacharya, J. P., 2012, Reappraisal of the sequence boundary in time and space: Case and considerations for an SU (subaerial unconformity) that is not a sediment bypass surface, a time barrier, or an unconformity: Earth Science Reviews, v 113, p. 271–302.

Holland, C. H., 1986, Does the golden spike still glitter? Journal of the Geological Society, London, v. 143, p. 3–21.

Holmes, A., 1960, A revised geological time-scale: Transactions of the Edinburgh Geological Society, v. 17, p. 183–216.

Holmes, A., 1965· Principles of Physical Geology, second edition, Nelson, London, 1288 p.

House, M. R., 1985, A new approach to an absolute timescale from measurements of orbital cycles and sedimentary microrhythms: Nature, v. 315, p. 721–725.

House. M. R., and Gale, A. S., eds., 1995, Orbital forcing timescales and cyclostratigraphy: Geological Society, London, Special Publication 85, 210 p.

Hunt, D., and Tucker, M. E., 1992, Stranded parasequences and the forced regressive wedge systems tract: deposition during base-level fall: Sedimentary Geology, v. 81, p. 1–9.

Imbrie, J., 1985, A theoretical framework for the Pleistocene ice age: Journal of the Geological Society, London, v. 142, p. 417–432.

Imbrie, J., Hays, J. D., Martinson, D. G., McIntyre, A., Mix, A. C., Morley, J. J., Pisias, N. G., Prell, W. L., Shackleton, N. J., 1984. The orbital theory of Pleistocene climate: support from a revised chronology of the marine $\delta^{18}O$ record, in: Berger, A. L., Imbrie, J., Hays, J., Kukla, G., and Saltzman, B., eds., Milankovitch and Climate. D. Reidel, Norwell, Mass, p. 269–305.

Imbrie, J., and Imbrie, K. P., 1979, Ice ages: solving the mystery: Enslow, Hillside, New Jersey, 224 p.

International Subcommission on Stratigraphic Classification, 1987, Unconformity-bounded stratigraphic units: Geological Society of America Bulletin, v. 98, p. 232–237.

Irving, E., 1966: Paleomagnetism of some Carboniferous rocks from New South Wales and its relation to geological events: Journal of Geophysical Research, v. 71, p. 6025–6051.

Jarvis, I., Gale, A. S., Jenkyns, H. C. and Pearce, M. A., 2006, Secular variation in Late Cretaceous carbon isotopes and sea-level change: evidence from a new ?13C carbonate reference curve for the Cenomanian - Campanian (99.6 -70.6 Ma). Geological Magazine, v. 143, p. 561–608.

Jarvis, I., Trabucho-Alexandre, J., Gröcke, D. R., Ulicny, D., and Laurin, J., 2015, Intercontinental correlation of organic carbon and carbonate stable isotope records: evidence of climate and sea-level change during the Turonian (Cretaceous): The Depositional Record, v. 1, p. 53–90.

Jenkins, D. G., and Gamson, P., 1993, The late Cenozoic *Globorotalia truncatulinoides* datum plane in the Atlantic, Pacific and Indian Oceans, in Hailwood, E. A., and Kidd, R. B., eds., High resolution stratigraphy: Geological Society, London, Special Publication 70, p. 127–130.

Johnson, G. D., Johnson, N. M., Opdyke, N. D., and Tahirkheli, R. A. K., 1979, Magnetic reversal stratigraphy and sedimentary tectonic history of the Upper Siwalik Group, eastern Salt Range and southwestern Kashmir, in Farah, A., and DeJong, K. A., eds., Geodynamics of Pakistan: Geological Survey of Pakistan, Quetta, Pakistan, p. 149–165.

Johnson, N. M., Stix, J., Tauxe, L., Cerveny, P. F., and Tahirkheli, R. A. K., 1985, Paleomagnetic chronology, fluvial processes, and tectonic implications of the Siwalik deposits near Chinji Village, Pakistan: Journal of Geology, v. 93, p. 27–40.

Kauffmann, E. G., 1977, Evolutionary rates and biostratigraphy; in Kauffman, E. G., and Hazel, J. E., eds., Concepts and methods of biostratigraphy: Dowden, Hutchinson and Ross Inc., Stroudsburg, Pennsylvania, p. 109–142.

Kauffman, E. G., 1988, Concepts and methods of high-resolution event stratigraphy: Annual Reviews of Earth and Planetary Sciences, v. 16, p. 605–654.

Kelley, P. H., Fastovsky, D. E., Wilson, M. A., Laws, R. A., and Raymond, A., 2013, From paleontology to paleobiology: A half-century of progress in understanding life history, in Bickford, M. E., ed., The web of geological sciences: Advances, impacts and interactions: Geological Society of America Special Paper 500, p. 191–232.

Keller, H. M., Tahirkheli, R. A. K., Mirza, M. A., Johnson, G. D., and Johnson, N. M., 1977, Magnetic polarity stratigraphy of the Upper Siwalik deposits, Pabbi Hills, Pakistan: Earth and Planetary Science Letters, v. 36, p. 187–201.

Kemple, W. G., Sadler, P. M., and Strauss, D. J., 1995, Extending graphic correlation to many dimensions: stratigraphic correlation as constrained optimization, in Mann, K. O., and Lane, H. R., eds., Graphic correlation: Society for Sedimentary Geology, Special Publication 53, p. 65–82.

Kennedy, K. Eyles, N., and Broughton, D., 2019, Basinal setting and origin of thick (1.8 km) mass-flow dominated Grand Conglomérat, Kamoa, Democratic Republic of Congo: resolving climate and tectonic controls during Neoproterozoic glaciations: Sedimentology, v. 66, p. 556–589.

Kennedy, W. J., and Cobban, W. A., 1977, The role of ammonites in biostratigraphy; in Kauffman, E. G. and Hazel, J. E., eds., Concepts and methods of biostratigraphy: Dowden, Hutchinson and Ross Inc., Stroudsburg, Pennsylvania, p. 309–320.

Kennedy, W. J., Walaszczyk, I., and Cobban, W. A., 2005, The Global Boundary Stratotype Section and Point for the base of the Turonian

Stage of the Cretaceous, Pueblo, Colorado, U.S.A.: Episodes, v. 28, p. 93–104.

Kennett, J. P., 1977, Cenozoic evolution of Antarctic glaciation, the circum-Antarctic Ocean, and their impact on global paleoceanography: Journal of Geophysical Research, v. 82, p. 3843–3860.

Kennett, J. P., ed., 1980, Magnetic stratigraphy of sediments: Dowden, Hutchinson and Ross Inc., Stroudsburg, Pennsylvania, Benchmark Papers in Geology 54, 438 p.

Knoll, A., Walter, M. R., Narbonne, G. M., and Christie-Blick, N., 2006, The Ediacaran Period: a new addition to the geologic time scale: Lethaia, v. 39, p. 13–30.

Kolla, V., Posamentier, H. W., and Eichenseer, H., 1995, Stranded parasequences and the forced regressive wedge systems tract: deposition during base-level fall—discussion: Sedimentary Geology, v. 95, p. 139–145.

Koster, E., 2020, Anthropocene: Transdisciplinary shorthand for human disruption of the earth system: Geoscience Canada, v. 47, p. 59–64.

Lamb, H. H., 1972, Climate: Present, past and future: Routledge, Abingdon, Oxford, 624 p.

Landing, E., Geyer, G., Brasier, M. D., and Bowring, 2013, S. A., Cambrian evolutionary radiation: context, correlation and chronostratigraphy—Overcoming deficiencies of the first appearance datum (FAD) concept: Earth Science Reviews, v. 123, p. 133–172.

Langereis, C. G., Krijgsman, W., Muttoni, G., and Menning, M., 2010, Magnetostratigraphy—concepts, definitions, and applications: Newsletters on Stratigraphy, v. 43, p. 207–233.

Laskar, J., 2020, Astrochronology, in Gradstein, F. M., Ogg, J. G., Schmitz, M. D., and Ogg, G. M., eds., Geologic time scale 2020: Elsevier, Amsterdam, p. 139–158.

Laskar, J., Robutel, P., Joutel, F., Gastineau, M., Correia, A. C. M., and Levrard, B., 2004, A long-term numerical solution for the insolation quantities of the earth: Astronomy & Astrophysics, v. 428, p. 261–285.

Le Heron, D. P., Eyles, N., and Busfield, M. E., 2020, The Laurentian Neoproterozoic glacial interval: reappraising the extent and timing of glaciation: Austrian Journal of Earth Sciences, v. 113, p. 59–70.

Lewis, S. L. and Maslin, M. A., 2015, Defining the Anthropocene: Nature, v. 519, p. 171–180.lewis

Lin, W., Bhattacharya, J. P., Jicha, J. P., Singer, B. S., and Matthews, W., 2021, Has Earth ever been ice-free? Implications for glacio-eustasy in the Cretaceous greenhouse age using high-resolution sequence stratigraphy: Geological Society of America Bulletin, v. 133, p. 243–252.

Loutit, T. S., Hardenbol, J., Vail, P. R., and Baum, G. R., 1988, Condensed sections: the key to age dating and correlation of continental margin sequences, in Wilgus, C. K., Hastings, B. S., Kendall, C. G. St. C., Posamentier, H. W., Ross, C. A., and Van Wagoner, J. C., eds., Sea-level Changes: an integrated approach: Society of Economic Paleontologists and Mineralogists Special Publication 42, p. 183–213.

MacLeod, N., and Keller, G., 1991, How complete are Cretaceous/Tertiary boundary sections? A chronostratigraphic estimate based on graphic correlation: Geological Society of America Bulletin, v. 103, p. 1439–1457.

Mann, K. O., and Lane, H. R., eds., 1995, Graphic correlation: Society for Sedimentary Geology, Special Publication 53, 263 p.

Martinsen, O. J., 1993, Namurian (Late Carboniferous) depositional systems of the Craven-Askrigg area, northern England: implications for sequence-stratigraphic models, in Posamentier, H. W., Summerhayes, C. P., Haq, B. U., and Allen, G. P., eds., Sequence stratigraphy and facies associations: International Association of Sedimentologists Special Publication 18, p. 247–281.

Matthews, R. K., 1984, Oxygen-isotope record of ice-volume history: 100 million years of glacio-isostatic sea-level fluctuation, in Schlee, J. S., ed., Interregional unconformities and hydrocarbon accumulation: American Association of Petroleum Geologists Memoir 36, p. 97–107.

Matthews, R. K., 1988, Sea level history: Science, v. 241, p. 597–599.

Matthews, S. C., and Missarzhevsky, V. V., 1975, Small shelly fossils of late Precambrian and early Cambrian age: a review of recent work: Journal of the Geological Society, London, v. 131, p. 289–304.

Mattinson, J. M., 2013, The geochronology revolution, in Bickford, M. E., The web of geological sciences: advances, impacts and interactions: Geological Society of America Special paper 500, p. 303–320.

McArthur, J. M., 1994, Recent trends in strontium isotope stratigraphy, Terra Nova, v. 6, p. 331–358.

McArthur, J. M., 1998, Strontium isotope stratigraphy, in Doyle, P. and Bennett, M. R., eds., Unlocking the stratigraphical record: John Wiley and Sons, Chichester, p. 221–241.

McArthur, J. M., and Howarth, R. J., 2004, Strontium isotope stratigraphy, in Gradstein, F. M., Ogg, J. G., and Smith, A. G., eds., A geologic time scale: Cambridge University Press, Cambridge, p. 96–105.

McGowran, B., 2005, Biostratigraphy: Microfossils and Geological Time: Cambridge University Press, Cambridge, 459 p.

McLaren, D. J., 1970, Presidential address: time, life and boundaries: Journal of Paleontology, v. 44, p. 801–813.

Miall, A. D., 1994, Sequence stratigraphy and chronostratigraphy: problems of definition and precision in correlation, and their implications for global eustasy: Geoscience Canada, v. 21, p. 1–26.

Miall, A. D., 1995, Whither stratigraphy? Sedimentary Geology, v. 100, p. 5–20.

Miall, A. D., 2004, Empiricism and model building in stratigraphy: the historical roots of present-day practices. Stratigraphy: American Museum of Natural History, v. 1, p. 3–25.

Miall, A. D., 2010, The geology of stratigraphic sequences, second edition: Springer-Verlag, Berlin, 522 p.

Miall, A. D., 2013, Sophisticated stratigraphy, in Bickford, M. E., ed., The web of geological sciences: Advances, impacts and interactions: Geological Society of America Special Paper 500, p. 169–190.

Miall, A. D., 2015, Updating uniformitarianism: stratigraphy as just a set of "frozen accidents", in Smith, D. G., Bailey, R., J., Burgess, P., and Fraser, A., eds., Strata and time: Geological Society, London, Special Publication 404, p. 11–36.

Miall, A. D., 2016, The valuation of unconformities: Earth Science Reviews, v. 163, p. 22–71.

Miall, A. D., and Blakey, R. C., 2019, The Phanerozoic tectonic and sedimentary evolution of North America, in Miall, A. D., ed., The Sedimentary Basins of the United States and Canada, Second edition: Sedimentary basins of the World, v. 5, K. J. Hsü, Series Editor, Elsevier Science, Amsterdam, p. 2–38.

Miall, A. D., and Miall, C. E., 2001, Sequence stratigraphy as a scientific enterprise: the evolution and persistence of conflicting paradigms: Earth Science Reviews, v. 54, #4, p. 321–348.

Milankovitch, M., 1930, Mathematische klimalehre und astronomische theorie der klimaschwankungen, in Koppen, W., and Geiger, R., eds., Handbuch der klimatologie, I (A); Gebruder Borntraeger, Berlin.

Milankovitch, M., 1941, Kanon der Erdbestrahlung und seine Anwendung auf das Eiszeitenproblem: Akad. Royale Serbe, 133, 633 p.

Miller, F. X., 1977, The graphic correlation method in biostratigraphy, in Kauffman, E. G., and Hazel, J. E., eds., Concepts and methods in biostratigraphy: Dowden, Hutchinson and Ross, Inc., Stroudsburg, Pennsylvania, p. 165–186.

Miller, K. G., Wright, J. D., and Browning, J. V., 2005, Visions of ice sheets in a greenhouse world: Marine Geology, v. 217, p. 215–231.

Mitchum, R. M., Jr., Vail, P. R., and Sangree, J. B., 1977, Seismic stratigraphy and global changes of sea level, Part six: Stratigraphic interpretation of seismic reflection patterns in depositional sequences, in Payton, C. E., ed., Seismic stratigraphy—applications to hydrocarbon exploration; American Association of Petroleum Geologists Memoir 26, p. 117–133.

Mitchum, R. M., Jr., and Van Wagoner, J. C., 1991, High-frequency sequences and their stacking patterns: sequence-stratigraphic evidence of high-frequency eustatic cycles: Sedimentary Geology, v. 70, 131–160.

Mungall, C., and McLaren, D. J., eds., 1990, Planet under Stress: Oxford University Press, Toronto, 344 p.

Naish, T. R, Field, B. D., Zhu, H., Melhuish, A., Carter, R. M., Abbott, S. T., Edwards, S., Alloway, B. V., Wilson, G. S., Niessen, F., Barker, A., Browne, G. H., and Maslen, G., 2005, integrated outcrop, drill core, borehole and seismic stratigraphic architecture of a cyclothemic, shallow-marine depositional system, Wanganui Basin, New Zealand: Journal of the Royal Society of New Zealand, v. 35, p, 91–122.

North American Commission on Stratigraphic Nomenclature, 1983, North American Stratigraphic Code: American Association of Petroleum Geologists Bulletin, v. 67, p. 841–875.

North American Commission on Stratigraphic Nomenclature (NACSN), 2005, North American Stratigraphic Code: American Association of Petroleum Geologists Bulletin, v. 89, p. 1547–1591.

Nummedal, D., and Swift, D. J. P., 1987, Transgressive stratigraphy at sequence-bounding unconformities: some principles derived from Holocene and Cretaceous examples, in Nummedal, D., Pilkey, O. H., and Howard, J. D., eds., Sea-level fluctuation and coastal evolution; Society of Economic Paleontologists and Mineralogists Special Publication 41, p. 241–260.

Odin, G. S., 1982, Numerical dating in Stratigraphy, v. 1 and 2, Chichester: Wiley-Interscience.

Ogg, J. G., 2019, Integrated global stratigraphy and geologic timescales, with some future directions for stratigraphy in China, Earth Science Reviews, v. 189, p. 6–20.

Ogg, J. G., and Smith, A. G., 2004, The geomagnetic polarity time scale, in Gradstein, F. M., Ogg, J. G., and Smith, A. G., eds., A geologic time scale: Cambridge University Press, Cambridge, p. 63–86.

Opdyke, N. D., 1972, Paleomagnetism of deep-sea cores; Review of Geophysics and Space Physics, v. 10, p. 213.

Opdyke, N. D., Glass, B., Hays, J. D., and Foster, J., 1966, Paleomagnetic study of Antarctic deep-sea cores: Science, v. 154, p. 349–357.

Penck, A. and E. Bruckner, 1909, Die Alpen im Eiszeitalter Bd 1-3. Chr.-Herm. Tauchnitz. Leipzig.

Picard, N. D., 1964, Paleomagnetic correlation of units within the Chugwater (Triassic) Formation, west-central Wyoming: American Association of Petroleum Geologists Bulletin, v. 48, p. 269–291.

Pipiringos, G. N., and O'Sullivan, R. B., 1978, Principal unconformities in Triassic and Jurassic rocks, western interior United States: a preliminary survey: U.S. Geological Survey, Professional Paper, 1035-A, p. 1–29.

Plimer, I., 2009, Heaven and earth: Global warming, the missing science, Taylor Trade Publishing, New York, 504 p.

Plint, A. G., 1988, Sharp-based shoreface sequences and "offshore bars" in the Cardium Formation of Alberta: their relationship to relative changes in sea level, in Wilgus, C. K., Hastings, B. S., Kendall, C. G. St. C., Posamentier, H. W., Ross, C. A., and Van Wagoner, J. C., eds., Sea-level Changes: an integrated approach: Society of Economic Paleontologists and Mineralogists Special Publication 42, p. 357–370.

Plint, A. G., 1990, An allostratigraphic correlation of the Muskiki and Marshybank Formations (Coniacian-Santonian) in the foothills and subsurface of the Alberta Basin: Bulletin of Canadian Petroleum Geology, v. 38, p. 288–306.

Plint, A. G., 1991, High-frequency relative sea-level oscillations in Upper Cretaceous shelf clastics of the Alberta foreland basin: possible evidence for a glacio-eustatic control? in Macdonald, D. I. M., ed., Sedimentation, tectonics and eustasy: sea-level changes at active margins: International Association of Sedimentologists Special Publication 12, p. 409–428.

Plint, A. G., and Kreitner, M. A., 2007, Extensive thin sequences spanning Cretaceous foredeep suggest high-frequency eustatic control: Late Cenomanian Western Canada foreland basin: Geology, v. 35, p. 735–738.

Plint, A. G., Walker, R. G., and Bergman, K. M., 1986, Cardium Formation 6. Stratigraphic framework of the Cardium in subsurface: Bulletin of Canadian Petroleum Geology, v. 34, p. 213–225.

Posamentier, H. W., Jervey, M. T., and Vail, P. R., 1988, Eustatic controls on clastic deposition I—Conceptual framework, in Wilgus, C. K., Hastings, B. S., Kendall, C. G. St. C., Posamentier, H. W., Ross, C. A., and Van Wagoner, J. C., eds., Sea level Changes - an integrated approach: Society of Economic Paleontologists and Mineralogists Special Publication 42, p. 109–124.

Posamentier, H., and Kolla, V., 2003, Seismic geomorphology and stratigraphy of depositional elements in deep-water settings: Journal of Sedimentary Research, v. 73, p. 367–388.

Pratt, B. R., and Holmden, C., eds., 2008, Dynamics of epeiric seas: Geological Association of Canada Special Paper 48.

Raikes, R., 1967, Water, weather and prehistory: London.

Raymo, M. E., and Ruddiman, W. F., 1992, Tectonic forcing of late Cenozoic climate: Chicken or egg? Nature, v. 346, p. 29–34.

Remane, J., 2000a, Explanatory note and international stratigraphic chart: UNESCO, Division of Earth Sciences, Paris.

Remane, J., 2000b, Should the golden spike glitter?—Comments to the paper of M.-P. Aubry et al.: Episodes, v. 23, p. 211–213.

Ruddiman, W. F., ed., 1997, Tectonic uplift and climate change: Springer, Boston, 535 p.

Ruddiman, W. F., 2008, Earth's climate, past and future, second edition: W. H. Freeman and Co., New York, 388 p.

Ruddiman, W. F., 2013, The Anthropocene: Annual Review of Earth and Planetary Science, v. 41, p. 45–68.

Rutter, N. Coppold, M., and Rokosh, D., 2006, Climate Change and Landscape in the Canadian Rocky Mountains: The Burgess Shale Geoscience Foundation, Field, British Columbia, 137 p.

Sadler, P. M., 1999, Constrained optimization approaches to stratigraphic correlation and seriation problems. A user's guide and reference manuals to the CONOP program family: University of California, Riverside, 142 p.

Sadler, P. M., Cooper, R. A., and Crampton, J. S., 2014, High-resolution geobiologic time-lines: progress and potential, fifty years after the advent of graphic correlation: The Sedimentary Record, v. 12, #3, p. 4–9.

Sadler, P. M., Cooper, R. A., and Melchin, M., 2009: High-resolution, early Paleozoic (Ordovician-Silurian) time scales: Geological Society of America Bulletin, v. 121, p. 887–906.

Sageman, B. B., Myers, S. R., and Arthur, M. A., 2006, Orbital time scale and new C-isotope record for Cenomanian-Turonian boundary stratotype: Geology, v. 34, p. 125–128.

Sageman, B. B., Singer, B. S., Meyers, S. R., Siewert, S. E., Walaszczyk, I., Condon, D. J., Jicha, B. R., Obradovich, J. D., and Sawyer, D. A., 2014, Integrating $^{40}Ar/^{39}Ar$e, U-Pb and astronomical clocks in the Cretaceous Niobrara Formation, Western Interior Basin, USA: Geological Society of America Bulletin, v. 126, p. 956–973.

Salvador, A., ed., 1994, International Stratigraphic Guide, Second edition: International Union of Geological Sciences, Trondheim, Norway, and Geological Society of America, Boulder, Colorado, 214 p.

Schaen, A. E., and 42 co-authors, 2021, Interpreting and reporting $^{40}Ar/^{39}Ar$ geochronologic data: Geological Society of America Bulletin, v. 133, p. 461–487.

Schlager, W., 1989, Drowning unconformities on carbonate platforms, in Crevello, P. D., Wilson, J. L., Sarg, J. F., and Read, J. F., eds., Controls on carbonate platforms and basin development: Society of Economic Paleontologists and Mineralogists Special Publication 44, p. 15–25.

Schlager, W., 2005, Carbonate sedimentology and sequence stratigraphy: SEPM Concepts in Sedimentology and Paleontology #8, 200p.

Schlanger, S. O., Arthur, M. A., Jenkyns, H. C., and Scholle, P. A., 1987, The Cenomanian-Turonian Anoxic Event, I. Stratigraphy and distribution of organic carbon-rich beds and the marine $\delta^{13}C$ excursion, in Brooks, J., and Fleet, A. J., eds., Marine petroleum source rocks: Geological Society, London, Special Publication 26, p. 371–399.

Schlanger, S. O., and Jenkyns, H. C., 1976, Cretaceous oceanic anoxic events: causes and consequences. Geol. Mijnbouw, v. 55, p. 179–184.

Scholle, P. A. and Arthur, M. A., 1980, Carbon isotope fluctuation in Cretaceous pelagic limestones: potential stratigraphic and petroleum exploration tool. American Association of Petroleum Geologists Bulletin, v. 64, p. 67–87.

Schwarzacher, W., 1993, Cyclostratigraphy and the Milankovitch theory: Elsevier, Amsterdam, Developments in Sedimentology 52, 225 p.

Scotese, C. R., 2001, Atlas of Earth History, v. 1. Paleogeography, Paleomap Project, Arlington, Texas, 52 p.

Shackleton, N. J., and Opdyke, N. D., 1976, Oxygen isotope and paleomagnetic stratigraphy of equatorial Pacific core V28-239, Late Pliocene to latest Pliocene: Geological Society of America Memoir 145, p. 449–464.

Shackleton, N. J., McCave, I. N., and Weedon, G. P., eds., 1999, Astronomical (Milankovitch) calibration of the geological time-scale: Philosophical Transactions of the Royal Society, London, Series A, v. 357, p 1731–2007.

Shank, J. A., and Plint, A. G., 2013, Allostratigraphy of the Upper Cretaceous Cardium Formation in subsurface and outcrop in southern Alberta, and correlation to equivalent strata in northwestern Montana: Bulletin of Canadian Petroleum Geology, v. 61, p. 1–40.

Shaw, A. B., 1964, Time in stratigraphy: McGraw Hill, New York, 365 p.

Shepard, F. P., and Wanless, H. R., 1935, Permo-Carboniferous coal series related to southern hemisphere glaciation: Science, v. 81, p. 521–522.

Sheriff, R. E., 1976, Inferring stratigraphy from seismic data; American Association of Petroleum Geologists Bulletin: v. 60, p. 528–542.

Sloss, L. L., 1963, Sequences in the cratonic interior of North America: Geological Society of America Bulletin, v. 74, p. 93–113.

Sloss, L. L., Krumbein, W. C., and Dapples, E. C., 1949, Integrated facies analysis; in Longwell, C. R., ed., Sedimentary facies in geologic history: Geological Society of America Memoir 39, p. 91–124.

Smith, A. G., Barry, T., Bown, P., Cope, J., Gale, A., Gibbard, P., Gregory, J., Hounslow, M., Kemp, D., Knox, R., Marshall, J., Oates, M., Rawson, P., Powell, J., and Waters, C., 2015, GSSPs, global stratigraphy and correlation: in Smith, D. G., Bailey, R., J., Burgess, P., and Fraser, A., eds., Strata and time: Geological Society, London, Special Publication 404, p. 37–67.

Smith, M. P., and Harper, D. A. T., 2013, Causes of the Cambrian Explosion: Science v. 341, p. 1355–1356.

Stanley, S. M., 2005, Earth system history: W. H. Freeman and Company, New York, 567 p.

Strong, N., and Paola, C., 2008, Valleys that never were: time surfaces versus stratigraphic surfaces: Journal of Sedimentary Research, v. 78, p. 579–593.

Summerhayes, C. P., 2015, Earth's climate evolution, Wiley-Blackwell, Chichester, 394 p.

Sylvester-Bradley, P.C., 1977: Biostratigraphical tests of evolutionary theory; in Kauffman, E. G., and Hazel, J. E., eds., Concepts and methods of biostratigraphy: Dowden, Hutchinson and Ross Inc., Stroudsburg, Pennsylvania, p. 41–64.

Tarling, D. H., 1982, Land bridges and plate tectonics: Geobios, v. 15, Supplement 1, p. 361–374.

Torrens, H. S., 2002, Some personal thoughts on stratigraphic precision in the twentieth century, in Oldroyd, D. R., ed., The Earth inside and out: some major contributions to geology in the twentieth century: Geological Society, London, Special Publication 192, p. 251–272.

Twidale, C. R., 1997. The great age of some Australian landforms: examples of, and possible explanations for, landscape longevity. In: Widdowson, M. ed., Palaeosurfaces: Recognition, Reconstruction and Palaeoenvironmental Interpretation. Geological Society, London, Special Publication v. 120, p. 13–23.

Underhill, J R., and Partington, M. A., 1993, Jurassic thermal doming and deflation in the North Sea: implications of the sequence stratigraphy evidence, in Parker, J. R., ed., Petroleum geology of northwest Europe: Proceedings of the 4th Conference, Bath, Geological Society, London, v. 1, p. 337–346.

Vai, G. B., 2001, GSSP, IUGS and IGC: an endless story toward a common language in the Earth Sciences: Episodes: v. 24, p. 29–31.

Vail, P. R., Mitchum, R. M., Jr., Todd, R. G., Widmier, J. M., Thompson, S., III, Sangree, J. B., Bubb, J. N., and Hatlelid, W. G., 1977, Seismic stratigraphy and global changes of sea-level, in Payton, C. E., ed., Seismic stratigraphy - applications to hydrocarbon exploration: American Association of Petroleum Geologists Memoir 26, p. 49–212.

Vail, P. R., and Todd, R. G., 1981, Northern North Sea Jurassic unconformities, chronostratigraphy and sea-level changes from seismic stratigraphy, in Illing, L. V., and Hobson, G. D., eds., Petroleum Geology of the continental shelf of northwest Europe: Institute of Petroleum, London, p. 216 235.

Van Hinte, J. E., 1976a, A Jurassic time scale: American Association of Petroleum Geologists Bulletin, v. 60, p. 489–497.

Van Hinte, J. E., 1976b, A Cretaceous time scale: American Association of Petroleum Geologists Bulletin, v. 60, p. 498–516.

Van Wagoner, J. C., Mitchum, R. M., Campion, K. M. and Rahmanian, V. D. 1990, Siliciclastic sequence stratigraphy in well logs, cores, and outcrops: American Association of Petroleum Geologists Methods in Exploration Series 7, 55 p.

Van Wagoner, J. C., Mitchum, R. M., Jr., Posamentier, H. W., and Vail, P. R., 1987, Seismic stratigraphy interpretation using sequence stratigraphy, Part 2: key definitions of sequence stratigraphy, in Bally, A. W., ed., Atlas of seismic stratigraphy: American Association of Petroleum Geologists Studies in Geology 27, v. 1, p. 11–14.

Van Wagoner, J. C., Posamentier, H. W., Mitchum, R. M., Jr., Vail, P. R., Sarg, J. F., Loutit, T. S., and Hardenbol, J., 1988, An overview of the fundamentals of sequence stratigraphy and key definitions, in Wilgus, C. K., Hastings, B. S., Kendall, C. G. St. C., Posamentier, H. W., Ross, C. A., and Van Wagoner, J. C., eds., Sea level Changes - an integrated approach: Society of Economic Paleontologists and Mineralogists Special Publication 42, p. 39–45.

Veeken, 2007, Seismic stratigraphy, basin analysis and reservoir characterization: Elsevier, Amsterdam, Seismic Exploration, v. 37, 509 p.

Veizer, J. 1989, Strontium isotopes in seawater through time: Annual Review of Earth and Planetary Science Letters, v. 17, p. 141–167.

Villeneuve, M., 2004, Radiogenic isotope geochronology, in Gradstein, F. M., Ogg, J. G., and Smith, A. G., eds., A geologic time scale: Cambridge University Press, Cambridge, p. 87–95.

Visser, C. F., and Johnson, G. D., 1978, Tectonic control of Late Pliocene molasse sedimentation in a portion of the Jhelum re-entrant, Pakistan: Geologische Rundschau, v. 67, p. 15–37.

Walker, R. G., 1990, Facies modeling and sequence stratigraphy: Journal of Sedimentary Petrology, v. 60, p. 777–786.

Walker, R. G., 1992, Facies, facies models and modern stratigraphic concepts, in Walker, R. G. and James, N. P., eds., Facies models: response to sea-level change: Geological Association of Canada, p. 1–14.

Walsh, S. L., 2004, Solutions in chronostratigraphy: the Paleocene/Eocene boundary debate, and Aubry vs. Hedberg on chronostratigraphic principles: Earth Science Reviews, v. 64, p. 119–155.

Wanless, H. R., and Weller, J. M., 1932, Correlation and extent of Pennsylvanian cyclothems: Geological Society of America Bulletin, v. 43, p. 1003–1016.

Westphal, H., Munnecke, A., and Brandano, M., 2008, Effects of diagenesis on the astrochronological approach of defining stratigraphic boundaries in calcareous rhythmites: The Tortonian GSSP: Lethaia, v. 41, p. 461–476.

Wignall, P. B., 1991, Ostracod and foraminifera micropaleontology and its bearing on biostratigraphy: a case study from the Kimmeridgian (Late Jurassic) of north west Europe: Palaios, v. 5, p. 219–226.

Wilson, J. T., 1966, Did the Atlantic close and then re-open? Nature, v. 211, p. 676–681.

Withjack, M. O., Schlische, R. W., and Olsen, P. E., 1998, Diachronous rifting, drifting and inversion on the passive margin of Central Eastern North America: an analog for other passive margins: American Association of Petroleum Geologists, v. 82, #5A, p. 817–835.

Worsley, T. R., Nance, D., and Moody, J. B., 1984, Global tectonics and eustasy for the past 2 billion years: Marine Geology, v. 58, p. 373–400.

Zalasiewicz, J., Smith, A., Brenchley, P., Evans, J., Knox, R., Riley, N., Gale, A., Gregory, F. J., Rushton, A., Gibbard, P., Hesselbo, S., Marshall, J., Oates, M., Rawson, P., and Trewin, N., 2004, Simplifying the stratigraphy of time: Geology, v. 32, p. 1–4.

Zalasiewicz, J., Williams, M., Smith, A., Barry, T. L., Coe, A. L., Bown, P. R., Brenchley, P., Cantrill, D., Gale, A., Gibbard, P., Gregory, F. J., Hounslow, M. W., Kerr, A. C., Pearson, P., Knox, R., Powell, J., Waters, C., Marshall, J., Oates, M., Rawson, P., and Stone, P., 2008, Are we now living in the Anthropocene?: GSA Today, v. 18, p. 4–8, doi: https://doi.org/10.1130/GSAT01802A.1.

Zecchin, M., and Catuneanu, O., 2013, High-resolution sequence stratigraphy of clastic shelves I: units and bounding surfaces: Marine and Petroleum Geology, v. 39, p. 1–25.

Ziegler, A. M., Cocks, L. R. M., and McKerrow, W. S., 1968, The Llandovery transgression of the Welsh borderland: Paleontology, v. 11, p. 736–782.

Ziegler, P. A., 1988, Evolution of the Arctic-North Atlantic and the Western Tethys: American Association of Petroleum Geologists, Memoir 43, 198 p.

Zuchuat, V., Midtkandal, I., Poyatos-Moré, M., Da Costa, S., Brooks, H. L., Halvorsen, K., Cote, N., Sundal, A. and Braathen, A., 2019, Composite and diachronous surfaces in low-gradient, transitional settings: the J-3 'unconformity' and the Curtis Formation, east-central Utah, U.S.A., Journal of Sedimentary research, v. 89, p. 1075-1095

The Future of Time

8

Contents

© The Author(s), under exclusive license to Springer Nature Switzerland AG 2022
A. D. Miall, *Stratigraphy: A Modern Synthesis*, Springer Textbooks in Earth Sciences, Geography and Environment,
https://doi.org/10.1007/978-3-030-87536-7_8

Abstract

The stratigraphic record is hierarchical and three dimensional. It is composed of complex lithologic units formed over time scales ranging from seconds to hundreds of millions of years that may be defined and described on all scales from the thin-section to the basin fill. Preservation is fragmentary, yet stratigraphic order may be perceived at many levels, from the cyclicity observed in vertical profile, to the sequences mappable across and between basins. Modern work on some specific field examples is described in order to illustrate recent concepts concerning the "stratigraphy machine," which is described and explained in this chapter. Current developments in high-resolution stratigraphy include the recognition of orbitally forced cyclicity at many levels in the Phanerozoic record, and this is facilitating the development of an astrochronology for the Cenozoic and parts of earlier systems.

8.1 Introduction

The purpose of this chapter is to assess the current status of the measurement of geologic time as stored in the stratigraphic record and to present some arguments regarding present pitfalls and future potential regarding the measurement and interpretation of geological time. As Holbrook and Miall (2020, p. 1) asked: "what does the stratigraphic record actually record?" We provide some examples of what it records, based on very detailed stratigraphic data bases, and the results are fascinating.

Modern methods of relative and "absolute" age dating of the stratigraphic record, and the current status of the Geological Time Scale, are discussed in the previous chapter. What these modern methods reveal is that the stratigraphic record is far more fragmentary than most geologists are accustomed to thinking. Many key concepts in sedimentary geology carry an implication of continuity in the sedimentary record, including the practices of stratigraphic classification and correlation, Walther's law, cyclic sedimentation, facies models, and sequence stratigraphy. However, as argued in this chapter, it is becoming increasingly clear that we need to carefully re-evaluate these assumptions of continuity. A key criterion to keep in mind is that of time scale. We tend to assume that ancient sedimentary records representing very long time intervals on the human time scale ($>10^4$ a) may be reliably compared with observations made over the much shorter time scales accessible to human observation ($\leq 10^2$ a). This is the basis of the Hutton-Lyellian aphorism "the present is the key to the past" (and the reverse). But a question that persists is that

concerning the relevance and significance of transient processes and ephemeral modern deposits to the interpretation of the rock record, given questions about the highly variable preservability of different sedimentary facies.

8.2 Where We Are Now and How We Got Here

Barrell (1917), in a review and discussion that was about seven decades ahead of its time, was probably the first to fully recognize the incompleteness of the stratigraphic record. Barrell (1917, Fig. 8.7) constructed a diagram showing the "Sedimentary Record made by Harmonic Oscillation in Baselevel" (Fig. 5.2 in this book). This is remarkably similar to diagrams that have appeared in some of the Exxon sequence model publications since the 1980s (e.g., Van Wagoner et al. 1990, Fig. 8.59; see Fig. 5.5 of this book), and represents a thoroughly modern deductive model of the way in which "time" is stored in the rock record. Curve A-A simulates the record of long-term subsidence and the corresponding rise of the sea. Curve B-B simulates an oscillation of sea levels brought about by other causes. Barrell discussed diastrophic and climatic causes, including glacial causes, and applied these ideas to the rhythmic stratigraphic record of the "upper Paleozoic formation of the Appalachian geosyncline" in a discussion that would appear to have provided the foundation for the interpretations of "cyclothems" that appeared in the 1930s. Barrell showed that when the long-term and short-term curves of sea-level change are combined, the oscillations of base level provide only limited time periods when sea level is rising and sediments can accumulate. "Only one-sixth of time is recorded" by sediments (Barrell 1917, p. 797).

Wheeler (1958, 1959) developed the concept of the chronostratigraphic cross-section, in which the vertical dimension in a stratigraphic cross-section is drawn with a time scale instead of a thickness scale (Fig. 8.1). In this way, time gaps (unconformities) become readily apparent, and the nature of time correlation may be accurately indicated. Such graphs have come to be termed "Wheeler diagrams." They are commonly constructed for use as stratigraphic tables in regional reports, but typically the time scale is arbitrary and variable, which means that information about missing time is distorted and usually not considered beyond qualitative statements about the significance of regional unconformities. In fact, this quote from Grabau (1960, p. 1097) would seem to imply that stratigraphic continuity is the norm:

> It is of first importance to the chronographer of earth history that he should find a continuous record, in order that he may have a measure by which to judge the partial records of any given region and to discover the breaks and imperfections in the local records thus presented. The question then arises: under what

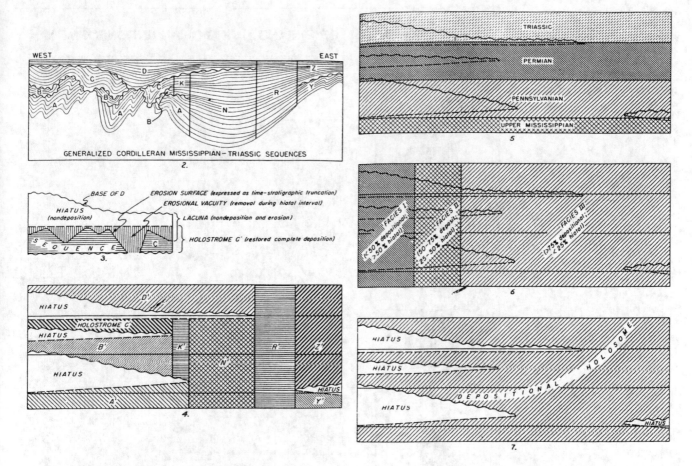

Fig. 8.1 Wheeler's development of chronostratigraphic diagrams. Diagram 2 shows a complex stratigraphic cross-section containing numerous unconformities. Diagram 3 explains some of the terminology used, and the remaining diagrams are "Wheeler plots" of the section in Diagram 2, ornamented in various ways to highlight different geological features (Wheeler 1958). AAPG © 1958, reprinted by permission of the AAPG whose permission is required for further use

conditions may we expect to obtain a continuous record and how are we to guard against the introduction of errors?

We return to the use of Wheeler plots later in Sect. 8.10.2.

Ager's (1981) famously remarked that the sedimentary record is "more gap than record." In a later book he expanded on the theme of gaps. Following a description of the major unconformities in the record at the Grand Canyon, he said, Ager (1993, p. 14):

> We talk about such obvious breaks, but there are also gaps on a much smaller scale, which may add up to vastly more unrecorded time. Every bedding plane is, in effect, an unconformity. It may seem paradoxical, but to me the gaps probably cover most of earth history, not the dirt that happened to accumulate in the moments between. It was during the breaks that most events probably occurred.

Much has now been learned about the significance of unconformities (Miall 2016; see Sect. 7.6). Sequence boundaries are unconformities, but assigning an age to an unconformity surface is not necessarily a simple matter, as illustrated in the useful theoretical discussion by Aubry (1991). An unconformity represents a finite time span at any one location, it may have a complex genesis, representing amalgamation of more than one event (Fig. 8.2). It may also be markedly diachronous, because the transgressions and regressions that occur during the genesis of a stratigraphic sequence could span the entire duration of the sequence. Ravinement surfaces, which commonly form sequence boundaries, are the product of diachronous erosion during transgression (Nummedal and Swift 1987; Zecchin et al. 2019). Kidwell (1988) demonstrated that transgression results in an offset in sequence boundary unconformities by as much as one half of a cycle between basin center and basin margin. Christie-Blick et al. (1990, 2007) pointed out several geological situations where unconformities may be diachronous. On a small scale, it has been demonstrated by flume experiments and observations of the rock record that laminated mudstones may preserve less than 10% of elapsed time as a result of rapid deposition and frequent reworking (Trabucho-Alexandre 2015).

Fig. 8.2 The interpretation of
unconformities: The generation of
a single major unconformity
could represent a single long-term
sedimentary break, as in the
column labeled Section A, or it
could have been generated by one
or more events, T1, T2 and/or T3,
that have been identified in
Section B (adapted from Aubry
1991)

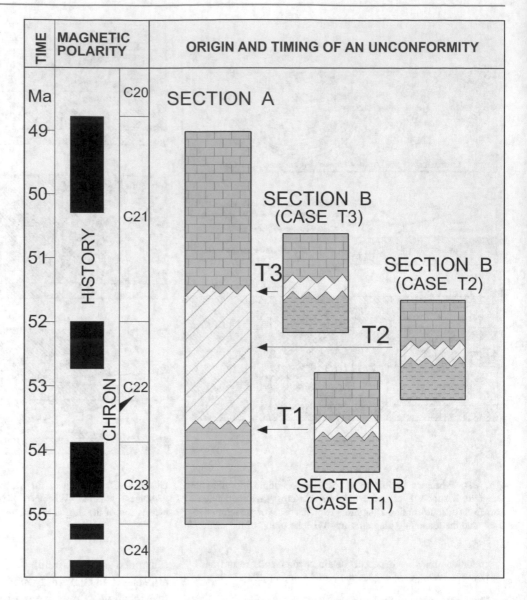

The subaerial erosion surfaces that underlie incised valley
fills at continental margins, and which commonly provide
important sequence boundaries, are good examples of
unconformities that evolve over time and are not everywhere
the same age (Holbrook and Bhattacharya 2012; see Fig. 5.8).
Strong and Paola (2008) pointed out an important distinction
between a *topographic surface* and a *stratigraphic surface*.
The topographic surface that corresponds to the subaerial
erosion surface undergoes continual change until it is finally
buried and preserved. It is the stratigraphic surface that we
map in the rock record, but this is a surface that never actually
existed in its entirety as a topographic surface in its final
preserved form, because it undergoes continuous modification
by erosion or sedimentation until final burial. The subaerial
erosion surface may also violate one of the fundamental
principles of an unconformity, which is that all the deposits
below the unconformity surface are older than all the beds

above the surface. Deepening and widening of an alluvial
valley may continue during the final stages of the evolution of
a subaerial surface, even while a turn-around in the base-level
cycle has begun to transgress and bury the surface during the
beginning of a transgression. Channel or overbank deposits
that are preserved as terrace remnants, resting on the basal
erosion surface, could therefore predate the coastal deposits
formed during the final stages of base-level fall, and would
therefore be older than the sequence boundary at the coast,
although resting on it. Holbrook and Bhattacharya (2012)
demonstrated that subaerial erosion surfaces are typically
highly complex and diachronous surfaces (Fig. 5.8).

One of the most significant achievements of modern
stratigraphic work is the increasing accuracy and precision
of the Geological Time Scale and our ability to provide ever
more refined dating of the geological record (Sect. 7.8.6).
A particularly instructive example of this advance (although

dependent on detailed paleontological work that goes back to the early nineteenth century!) is the biostratigraphic subdivision of parts of the European Jurassic section. In Chap. 7 the work of Callomon (1995) is highlighted (see Fig. 7.41 of this book). The shallow-marine Inferior Oolite, which in the study area of southern Britain is only some 5 m thick, may be subdivided into as many as 56 faunal zones spanning about 5 m.y. The biozones there average about 90,000 years in duration. Note that none of the thirteen sections examined for this project contains a complete record of the zones, and the preserved record is different in every section. This formation is replete with local diastems that record local areas of negative accommodation (Fig. 8.3). Almost certainly there are many additional breaks present that are below the resolution of the ammonite chronostratigraphy. The numerous breaks suggest that there may have been continuous adjustments in accommodation, or sediment supply, or the

current regime. Buckman (1910, p. 90) concluded, after many years of meticulous analysis of this unit that the Inferior Oolite of Dorset, an extremely condensed shelf sequence, "might be defined as a series of gaps united by thin bands of deposits .... the deposits are so local, the deposits of one place correspond to the gaps of another."

Overall, it is impossible to distinguish any ordered pattern to the record of sedimentation and erosion in these sections. How typical is this of shallow-marine sedimentation in general? Does the availability of an unusually detailed ammonite biostratigraphy in the Inferior Oolite enable us to develop a much more detailed record of local change than would otherwise be available? And should this section therefore be regarded as a model for the interpretation of other shallow-marine carbonate sections? Would it be correct to conclude that many other shallow-marine (and non-marine?) successions should similarly be regarded as

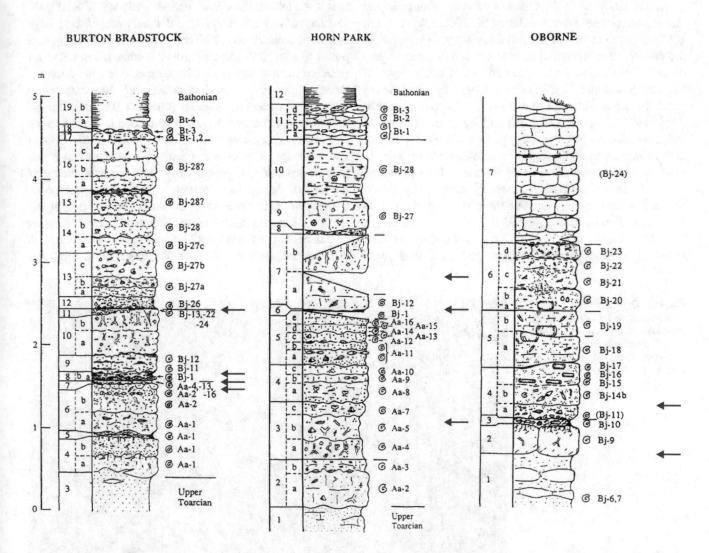

Fig. 8.3 Three of the stratigraphic sections through the Inferior Oolite studied by Callomon (1995) showing (at left) the series of faunal zones recorded in the rocks and (at right) red arrows indicating the likely position of diastems where one or more zones are missing. This unit spans the Aalenian to lower Bathonian, a duration of about 6 m.y

containing numerous local diastems? If so, what does this tell us about short- to long-term sedimentary processes? We return to this example later (Sect. 8.11.1).

An examination of a typical shallow-marine succession (Fig. 8.4) suggests the likelihood of many additional breaks in sedimentation at a still finer scale. This outcrop exposes a succession of sandstones, siltstone and mudstones, with clearly defined lithologic differentiation defined by sharp contacts. Each of the sharp contacts highlighted by the red arrows represents a break in sedimentation. This could be a shift in tidal currents, an erosion surface, a seasonal change, or a temporary cessation in sediment accumulation. We have no idea whether any sediment has been removed at the breaks or what elapsed time each break represents. The breaks may be termed cryptic diastems, and the missing time at each could range from minutes to years. Additional cryptic breaks may be present in the mudstone units.

As the study of modern sedimentary environments and facies interpretations evolved during the 1960s (Sects. 1.2.6, 1.2.7), it began to be realized that there is a wide variation in the energy, magnitude and time scales of sedimentary processes. Some facies, such as turbidites and storm deposits, are clearly formed rapidly over geologically insignificant periods of time. Although he was not the first to discuss this, Ager (1973) is credited with the coining of the term **event sedimentation** to encompass this pattern of rapid deposition. One of his stratigraphic studies was carried out in order to raise the question of the time significance of the preserved record. He re-evaluated the sedimentary history of the Sutton Stone, a beach conglomerate that forms the base of the Blue Lias succession in Glamorgan, South Wales, where it ranges from 10 to 13.5 m in thickness (Fig. 8.5). The conglomerate rests on a transgressive surface, and had long been interpreted as the product of slow sedimentation during a protracted transgression. According to Wobber (1965) it contains the fossils representing at least four ammonite zones of Hettangian and Lower Sinemurian age which, according to the data provided by Gradstein et al. (2004a, Fig. 18.1) span a total of about 4 m.y. However, Ager (1986) suggested that the deposit was formed in a single major tropical storm or hurricane. He stated (p. 35) "I do not think it took the three or four million years or so of three or four or five ammonite chronozones." His one-line conclusion: "It all happened one Tuesday afternoon."

The "Tuesday afternoon" remark is, of course deliberately provocative (one might ask, why not Wednesday?), but Ager's point was to emphasize the rapidity by which certain geological processes may accomplish spectacular results. Storms do, indeed, accomplish most of their erosional and depositional work within a space of a few hours or days, at most. A four-million-year period includes 208,000,000 Tuesdays. Are we to understand that only one of these days left a sedimentary record? The question is fascinating and points to a nagging issue regarding sedimentation rates and preservation that has gradually emerged with the accumulation of an ever-increasing volume of data concerning modern and ancient sedimentary processes. The central point is that when an ancient deposit is compared to the sediments accumulating in an equivalent environment at the present day, the rates of modern sedimentary processes and stratigraphic accumulation would allow for the accumulation of the ancient deposit in a fraction of the time that chronostratigraphic data indicate is available. It is not at all uncommon to find that the accumulation of a given thickness of sediment in any given environment could take place in as little as one tenth of the available time. What happened

Fig. 8.4 Outcrop of the shallow-marine Dunlevy/Gething Formation at the W.A.C. Bennett Dam in northeastern British Columbia. Each red arrow marks a minor break in sedimentation

Fig. 8.5 An exposure of the Hettangian (basal Lias) marginal marine Sutton Stone conglomerate and bioclastic beds onlapping palaeo relief developed by denudation on the Variscan unconformity, at this location developed atop folded Lower Carboniferous carbonate strata. The Hettangian strata have been interpreted as products of palaeo sea cliff erosion during still stands of relative sea-level during an overall transgressive interval that progressed from a late Triassic lacustrine environment to Early Jurassic marine conditions (Fletcher 1988). The location is Southerndown on the Bristol Channel coast, near Ogmore-by-Sea in the Vale of Glamorgan, South Wales, UK. (photograph and caption courtesy of P. Burgess)

during the rest of the time that passed, according to the chronostratigraphy of the unit? Callomon's demonstration of numerous diastems in the Inferior Oolite might be pointing the way to a more general conclusion.

It does not materially undermine Ager's (1986) broader arguments about event sedimentation to note that later workers have disputed his claim that the Sutton Stone was deposited in a single event. Evidence of encrusting organisms at more than one level within the conglomerate, and the observations of varying conglomerate facies through the unit suggested an alternative model of cliff collapse on an exposed rocky shoreline, perhaps subsequently influenced by several or many storms (Fletcher et al. 1986; Johnson and McKerrow 1995).

Sadler (1981) compiled thousands of records concerning sedimentation in deposits ranging in age from ancient to modern and demonstrated that the relationship between sedimentation rate and elapsed time (the time period over which the sedimentation rate is measured) is linear and inverse on a log–log scale (Fig. 8.6). His data base consisted of 25,000 records of accumulation rates. Measured sedimentation rates vary by eleven orders of magnitude, from 10^{-4} to 10^7 m/ky.

This huge range of values would seem to suggest the presence of an increasing number and duration of intervals of non-deposition or erosion that become factored into the measurements as the length of the measured stratigraphic record increases. A comparable range of sedimentation rates was revealed in his later compilation of records focusing on shallow-marine carbonate deposits (Kemp and Sadler 2014).

Although Sadler's (1981) synthesis has been in the public domain for more than thirty years, few stratigraphers have attempted to wrestle with its significance for the geological interpretations that we make based on the Hutton-Lyell Principle of Uniformitarianism: "The present is the key to the past." Although Gould (1965) argued, on logical grounds, that this Principle is no longer necessary, given the assumption of the invariance of natural laws, it nonetheless exerts a powerful influence on the methods of modern sedimentology, which are very firmly based on the practice of seeking modern analogues for an ancient deposit of interest. As argued in this chapter, it is becoming clear that there are some important provisos and limitations that must now be inserted into this practice.

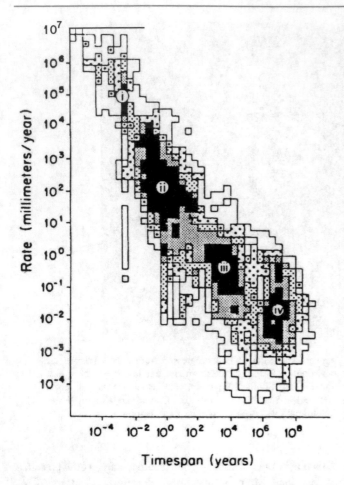

Fig. 8.6 The relationship between sedimentation rate and elapsed time in the stratigraphic record (Sadler 1981)

8.3 A Natural Hierarchy of Sedimentary Processes

Bailey and Smith (2010, p. 57–58) pointed out the ephemeral nature of most sedimentary processes:

> There would seem to be a very small chance of the preservation in 'stratigraphic snapshots' of, say, one particular ripple-marked shoreface out of the thousands or millions, created and destroyed diurnally through geologic time. Such instances suggest that such stratigraphic records are better viewed as the outcome of temporary cessation of the erosion and redistribution of sediment: 'frozen accidents' of accumulation.

It is now widely recognized that not only the durations of the gaps, but also the distribution of layer thicknesses and sedimentation rates in stratigraphic successions have fractal-like properties (Plotnick 1986; Sadler 1999; Schlager 2004; Bailey and Smith 2005; Smith et al. 2015). However, the practical development of this concept is hampered by the current methods of stratigraphic documentation. Bailey and Smith (2010, p. 58) noted that current classifications of

stratigraphic units, based as they are on "a human-scale observer" (e.g., lithostratigraphy, sequence stratigraphy) constitute hierarchies that are somewhat arbitrary, and make statistical analysis of bedding and its contained gaps difficult. In addition, it is extremely difficult to operationalize a lithologically, petrologically and statistically reliable practical field definition of what constitutes a "layer," at all scales from the lamina to the basin fill.

To circumvent this problem Bailey and Smith (2010) and Bailey and Schumer (2012) developed a method of analyzing the stratigraphic record termed the "Layer Thickness Inventory." The analysis is carried out on continuous, digitized records, such as wireline logs. The gamma-ray log is particularly suitable for this purpose, because it is readily interpreted in lithologic terms. A computer routine works its way through the digital data records one at a time, searching the data string above and below for records in which the GR reading is higher and lower than that at each sample point, and records the calculated thicknesses. All lithologically defined layers are thereby recorded, ignoring hiatuses. The procedure records layers within layers, which therefore overlap, and in this way "it recognizes that the various sedimentary influences on lithology operate in overlapping time frames, rather than as the succession of discrete process-response effects suggested by conventional hierarchical stratigraphic subdivisions." (Bailey and Smith 2010, p. 59).

Their analysis of a range of geological examples demonstrated that log–log plots of bed thickness against number of records always generate linear distributions, regardless of the scale of the stratigraphic section and the nature of the lithology, suggesting "that there is a universal relationship between layer thickness and frequency of occurrence in the record." (Bailey and Smith 2010, p. 62).

Bailey and Smith (2010) raised the question of the degree to which the evidently fractal record is representative of past surface processes, and made the following points:

- The notion of continuous deposition, on which the historicity of the record depends, has no theoretical or evidential basis. In relation to the accumulation of particulate solids it is, in fact, an impossibility. At best, it is a scale-dependent descriptive convenience.
- If there is no continuity in accumulation, the sequential preservation of laterally contiguous facies, according to Walther's Law, becomes questionable.
- Stratigraphic hierarchies are constructs, commonly tailored to human-scale analysis of the fractal record. They are a practical, convenient, but incomplete, representation of this record.
- Currently observable sedimentary processes and facies underpin uniformitarian stratigraphic interpretations. Yet there is no way of determining whether a present-day

deposit will be preserved millions of years hence. Equally, if there is "more gap" there is the question of the degree to which the preserved record is representative of the continuous operations of past sedimentary systems. Specifically, are the snapshot "frozen accidents of preservation" representative?

- As Sadler (1999) has shown, local calculations of accumulation rate are time scale dependent.

Is the stratigraphic record fundamentally unrepresentative of the geological past? These conclusions would appear to invalidate virtually the whole of the last two centuries of stratigraphic progress!

However, all is not lost!

Miall (1991) suggested that the sedimentary time scale constitutes a natural hierarchy corresponding to the natural hierarchy of temporal processes (diurnal, lunar, seasonal, geomorphic threshold, tectonic, etc.) and the main purpose of this chapter is to develop this idea further, making extensive use of modern quantitative data dealing with sedimentation rates and accumulation rates. Most of the discussion that follows relates to clastic sedimentation in shallow-marine and nonmarine environments. I return to the above discussion points in a later section of this chapter, where it is argued that most may be managed within the context of the appropriate time frame.

In an attempt to understand the log–log sedimentation rate: duration plot of Sadler (1981), Miall (1991; 2010, Chap. 13) undertook an analysis of the relationship between sedimentation rate and sedimentary process in the published record of shallow-marine stratigraphy. It emerged that there does, in fact, appear to be a natural hierarchy of process and preservation based on the natural time scales of sedimentary processes. The model which emerged from this looks remarkably like a fractal analysis, although it was generated entirely in the absence of any guidance from fractal theory (Fig. 8.7). Two cycles with frequencies in the million-year range are plotted on a chronostratigraphic scale (column MC), and successively broken down into components that reflect an increasingly fine scale of chronostratigraphic subdivision. The second column shows hundred-thousand-year cycles (HC), followed by depositional systems (DS) and individual lithosomes (L), such as channels, deltas and beaches. At this scale chronostratigraphic subdivision is at the limit of line thickness, and is therefore generalized, but does not represent the limit of subdivision that should be indicated, based on the control of deposition by events of shorter duration and recurrence interval (e.g., infrequent hurricanes, seasonal dynamic events, etc.).

The record shows that when measurements of sedimentation rate are calculated at the appropriate scale, they are internally consistent and can be related to natural processes occurring within that time scale. Miall (1991) identified ten

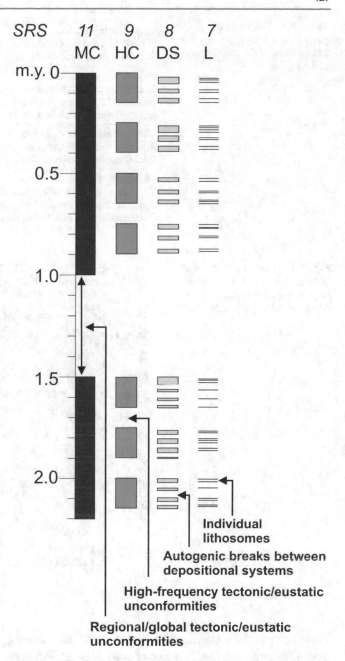

Fig. 8.7 A demonstration of the predominance of missing time in the sedimentary record. MC = million-year scale cycles; HC = hundred-thousand-year cycles; DS = depositional systems; L = lithosomes

informal groupings of sediment packages, based on sedimentation rate. This classification, updated by Miall (2010, Table 8.1; Miall 2015), is reproduced here as column 1 in Table 3.3. It constitutes a natural range of *Sedimentation Rate Scales* (*SRS*), now expanded to twelve groupings (Fig. 8.8). Column 2 in this table indicates the time scale of measurement, and column 3 provides the instantaneous sedimentation rate for deposits formed over that time scale. Stratigraphic and sedimentologic studies ranging from the

Table 8.1 The range of geological processes involved in the creation of the sedimentary record

• Seconds to months
Diurnal to Tidal flow variations
Turbulent flow (bedforms)
• Years to thousands of years
Long-term geomorphic processes
(seasonal to 100-year flood, paleosol development, channel migration, delta switching, channel avulsion, sediment gravity flow and storm events)
• Tens to hundreds of thousands of years
Regional tectonism (tectonic cyclothems, paleoslope tilts, regional changes in sea level)
Orbital cyclicity (climate change, eustasy, changes in sediment supply)
• Millions of years
Continental-scale tectonism (angular unconformities)
Dynamic topography (cratonic unconformities)
Eustasy related to sea-floor spreading rates (major sequence boundaries)

Fig. 8.8 Rates and durations of sedimentary processes. Numerals refer to the Sedimentation Rate Scale (see also Table 3.3)

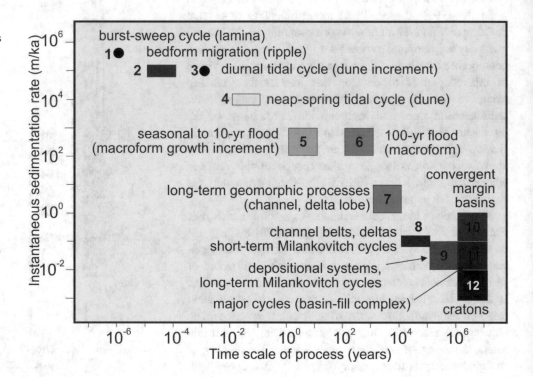

micro scale to the regional, and based on time scales ranging from the short term (e.g., studies of processes in laboratory models or modern settings) to the long term (e.g., the evolution of major sedimentary basins), are best carried out at the appropriate *SRS*, much as photography uses lenses of different focal length, from macro to telephoto to wide-angle, to focus in on features at the desired scale.

Many detailed chronostratigraphic compilations have shown that marine stratigraphic successions commonly consist of intervals of "continuous" section representing up to a few million years of sedimentation, separated by disconformities spanning a few hundred thousand years to more than one million years (e.g., MacLeod and Keller 1991, Fig. 15; Aubry 1991, Fig. 8.8). The first column of Fig. 8.7, labeled MC (for cycles in the million-year range), illustrates an example of such a succession. Detailed studies of such

cycles demonstrate that only a fraction of elapsed time is represented by sediment. For example, Crampton et al. (2006) demonstrated that an average of 24% of time is recorded in a suite of Upper Cretaceous sections in New Zealand, when measured at a 10^6-year time scale, whereas in a suite of drill cores through the Lower Cretaceous to Miocene stratigraphic record of New Jersey, the plots of Browning et al. (2008) show that about 82% of elapsed time is represented by sediments, although some sections are more complete than others. Each million-year cycle may be composed of a suite of high-frequency cycles, such as those in the hundred-thousand-year range, labeled HC in Fig. 8.7.

Chronostratigraphic analyses of many cyclic successions demonstrate that the hiatuses between the cycles represent as much or more missing time than is recorded by actual sediment (e.g., Ramsbottom 1979; Heckel 1986; Kamp and

Turner 1990). Sedimentation rates calculated for such sequences (Table 3.3) confirm this, and the second column of Fig. 8.7 indicates a possible chronostratigraphic breakdown of the third-order cycles into component Milankovitch-band cycles (labeled HC), which may similarly represent incomplete preservation. For example, a detailed chronostratigraphic correlation of the coastal Wanganui Basin sequences in New Zealand ("5th-order Milankovitch cycles of *SRS 8* in Table 3.3) shows that at the 10^5-year time scale only 47% of elapsed time is represented by sediments (Kamp and Turner 1990). Each Milankovitch cycle consists of superimposed depositional systems (column DS) such as delta or barrier-strand plain complexes, and each of these, in turn, is made up of individual lithosomes (column L), including fluvial and tidal channels, beaches, delta lobes, etc.

Devine's (1991) lithostratigraphic and chronostratigraphic model of a typical marginal marine sequence demonstrates the importance of missing time at the sequence boundary (his subaerial hiatus). Shorter breaks in his model, such as the estuarine scours, correspond to breaks between depositional systems (the DS column in Fig. 8.7), but it is suggested that more are present in such a succession than Devine (1991) has indicated. Additional discontinuities at the lithosome level (L in Fig. 8.7) correspond to the types of breaks in the record introduced by switches in depositional systems, channel avulsions, storms and hurricanes.

According to the hierarchical breakdown of Table 3.3, the four columns in Fig. 8.7 correspond to sediment *SRSs 11, 9, 8,* and *7,* in order from left to right. In each case, moving (from left to right) to a smaller scale of depositional unit focuses attention on a finer scale of depositional subdivision, including contained discontinuities. The evidence clearly confirms Ager's (1973, 1981) assertion that the sedimentary record consists of "more gap than record."

Guidance for the interpretation of fragmentary records, as we now recognize them, was offered by Holbrook and Miall (2020), including the following:

- Each "order" in a hierarchy of stratigraphic elements is an amalgam of elements sampled from lower orders.
- The sedimentation rates and processes for strata must be scaled to the appropriate order in the stratigraphic hierarchy.
- Stratigraphic completeness is scaled to the order in the hierarchy at which the sample is categorized.
- Allogenic processes do not generate stratigraphy but are recorded by the changes they cause to autogenic processes.
- Instantaneous sedimentation rates are typically much higher than the local rate of accommodation generation. This is compensated by lateral shifts in the locus of deposition.

- Clastic sediment supply from a basin margin is typically much more regular and continuous than the basin fill that it feeds. This is explained by lateral shifts in depositional axes, temporary sediment storage and sediment bypass, resulting in the characteristic fragmentary nature of the record at any given location.

Very detailed studied by Davies et al. (2019) and Davies and Shillito (2021) have demonstrated how bedding surfaces may be preserved as "true substrates." These are surfaces that accumulate significant evidence of biogenic activity, and other processes that modify but that add or subtract little from a bedding surface but reveal the passage of significant intervals of time, thus corresponding to what Tipper (2015) referred to as surfaces of "stasis."

8.4   Sedimentation Rates

Examples of measured and calculated sedimentation rate from modern and ancient environments provide the basis for the ranges of values indicated in column 3 in Table 3.3 and in Fig. 8.8.

SRS 2: small-scale ripples typically migrate a distance equivalent to their own wavelength in 20 to 60 min (Southard et al. 1980). A 5-cm-high ripple that forms in 30 min is equivalent to an instantaneous sedimentation rate of 876,000 m/ky. Clearly, this number is meaningless, but it will serve to emphasize the extremes of sedimentation rate, to compare with more geologically typical rates discussed later.

SRS 3: Tidal sand waves have similarly very high instantaneous rates. In the Bay of Fundy Dalrymple (1984) demonstrated that in one tidal cycle sand waves migrate a distance about equivalent to their average height, which is 0.8 m. Bay of Fundy tides are semi-diurnal, and so this migration is equivalent to a sedimentation rate of 584,000 m/ky.

SRS 5: The deposits formed by seasonal or more irregular runoff events have extremely variable instantaneous sedimentation rates. The flood deposit in Bijou Creek, Colorado, described by McKee and others (1965) was formed by the most violent flood in 30 years. It formed 1 to 4 m of sediment in about 12 h, an instantaneous sedimentation rate of 730,000–2,920,000 m/ky. Assuming no erosion, and a repeat of such floods every 30 years this translates into a rate of 33–133 m/ky averaged over a few hundred years. In fact, scour depths during the flood ranged from 1.5 to 3 m, and true net preservation of any one flood deposit over periods of hundreds or thousands of years may be negligible. Long-term rates measured over hundreds to thousands of years are likely to be an order of magnitude less, in the range of 10^{-1} m/ky. Leclair (2011) demonstrated that the large-scale dunes that may characterize seasonal to longer term floods do not necessarily have a higher preservation potential than the deposits formed during non-peak flood periods.

In the Rio Grande valley in Texas, Dean et al. (2011) used tree-ring studies to establish the recent sedimentation history. Over a 13-year period they determined that overbank floods occurred with a recurrence interval of 1.5–7 years, at accretion rates of 16–35 cm/year, an instantaneous sedimentation rate of 10^2 m/ky.

It is simple to observe the inconsistency and incompleteness of the sedimentary record in shallow-marine and non-marine environments over human time scales ($\leq 10^2$ yrs). For example, if carefully documented by surveying and photography, repeated visits to active rivers and tidal flats will show subtly different lithofacies successions and different assemblages of sedimentary structures each time, indicating the constant repetition of erosion and resedimentation events, modulated by the variability and episodicity of seasonal, tidal and weather-related sedimentary processes.

SRS 6: The point bars in the Wabash River are on average about 5 m thick. The active area of each bar is about 200 m wide. If the bars migrate at a maximum rate of 2 km in 50 years (Jackson 1976) they would take 5 years to migrate one point-bar width, which is equivalent to an instantaneous sedimentation rate of 1,000 m/ky.

Comparable rates may be calculated from the migration of distributary mouth bars. The Southwest Pass of the Mississippi Delta migrated a distance of 9 km in 100 years (Gould 1970, Fig. 20). The mouth-bar deposits, from the mouth of the channel to the toe of the distal bar, are about 4 km wide (in a dip direction) and about 70 m thick. This lateral migration is equivalent to an instantaneous sedimentation rate of 730 m/ky. Oomkens (1970) quoted sedimentation rates of 35 cm/year (350 m/ky) for the delta front of the modem Rhone River. A short-term rate of 4.4 cm/month (528 m/ky) was determined close to the mouth of the Yangtze River, China, by McKee and others (1983), who studied the decay of short-lived radionucleides in the uppermost 15 cm of recent deposits, representing about 100 days of accumulation *(SRS 5)*. The uppermost 200 cm of section, representing about 100 years of accumulation *(SRS 6)*, yielded a rate an order of magnitude lower, 5.4 cm/a (54 m/ky).

Bhattacharya et al. (2020) stated:

> Bhattacharya et al. (2019) showed that the regressive, prograding, coarsening-upwards components of prodeltaic to delta-front facies successions (i.e., parasequences) may include a remarkably high-fidelity record of sedimentation over the period of time it takes for a delta to vertically fill any given point as it progrades. This may be a thousand years or less, depending upon the depth of the water and the rate of sediment supplied.

Bhattacharya et al. (2019) also showed that much of the time associated with deltaic systems is wrapped up in various bypass and transgressive surfaces, and especially the slowly deposited and commonly pervasively bioturbated transgressive facies that cap para-sequences, versus the more rapidly deposited regressive phase.

Bhattacharya et al. (2020) calculated sedimentation rates of several centimetere to tens of centimeters per year (*SRS 5–6*) in some modern and ancient delta front deposits (distributary channels, splays, regressive beaches) in aggradational and progradational (clinoform) settings.

Data from modem Dutch tidal deposits summarized by Yang and Nio (1989) showed that ebb-tidal deltas accumulate at rates of 100 to 450 m/ky for periods of about 20 years, before abandonment occurs. Van den Bergh et al. (2007) used ^{210}Pb methods, which provide age information on the 100-a scale (*SRS 6*), to assess sedimentation rates on the prodelta of the Red River off Vietnam. They range from high values of 330 to 940 m/ky on the proximal prodelta slope, to less than 10 m/ky on the distal margin. The distal sedimentation rate is at the lower limit for *SRS 6* sedimentation rates when measured at the appropriate 100-year scale.

Rates calculated for the upper Bengal submarine fan, the world's largest and most active depositional system are consistent with the *SRS 5–6* time scales. Using ^{14}C age determinations for the last 10,000 years of sedimentation, Weber et al. (1997, p. 317) calculated that "on the shelf, sedimentation rates are currently extremely high in the foreset region of the recent delta (as much as 8 cm/a)" where water depth is 30–70 m, "and especially in the head of the canyon" where sedimentation rates reach as much as 1 m/a (10^3 m/ky). A sedimentation rate of 1 m/a places this system in the range of *SRS 5* and *6*, for which such sedimentation rates occur within time spans of 10^0–10^3 years. In this case the preservation machine at work is the removal of sediment by slope failure and slumping, with huge volumes of sediment moving downslope as sediment gravity flows, Weber et al. (1997, p. 317) noted that "This sediment load would fill the entire canyon in less than 1000 years. Therefore, we conclude that, because of the steep gradients at the head of the canyon, frequent slumping and formation of turbidity currents occur even during the current sea-level highstand." The frequency of turbidite occurrence ranges between 500 and 10,000 years (Stow et al. 1983, p. 58), which is well within the 10^0–10^3-year time range.

SRS 7: Most studies of post-glacial sedimentation are carried out on a "long-term" (10^3–10^4 year) time scale that is assigned to *SRS 7*. Rates of post-glacial sea-level rise (and accommodation generation) reached as high as 18 m/ky (e.g., East China Sea: Wellner and Bartek 2003), although rates of 1–6 m/ky were more characteristic.

In the coastal river valleys of Texas, Blum (1993) demonstrated that late Pleistocene-Holocene cycles of degradation and aggradation depended on climatically controlled variations in discharge and sediment supply, not on sea-level change. Humid periods corresponded to episodes of aggradation. During two such periods, lasting 6 ky, as much as 10 m of valley fill sediment accumulated in the upper Colorado drainage, indicating sedimentation rates of 1.7 m/ky. Long-term (10^4 a) floodplain aggradation rates reported by Bridge and Leeder (1979) ranged between 0.035 and 0.2 cm/a (0.35–2 m/ky).

Turning to the deltaic environment, if we assume that a distributary will only build out across a given area of the delta front once during the migration of one major delta lobe, we can calculate the sedimentation rate of the mouth-bar averaged over the life of the lobe. In the post-glacial Mississippi Delta, major lobes are formed and abandoned in about 1000 years (Kolb and Van Lopik 1966; Frazier 1967), giving an average sedimentation rate for that period of 70 m/ky. Lobe sedimentation took place during the post-glacial period of sea-level rise.

The recurrence interval of delta lobes themselves depends on subsidence rates, sediment supply and the configuration of the continental shelf. In the case of the Mississippi complex the river is attempting to switch discharge (and delta construction) to the Atchafalaya River, where one of the earliest lobes developed about 6 to 8 ka BP. On this scale deposition of a 45 m thick mouth-bar deposit represents a sedimentation rate of 5.6 to 7.5 m/ky, although this calculation does not take into account the bay-fill and other facies interbedded with the mouth-bar deposit. This compares with typical values for Holocene sedimentation rates of 6 to 12 m/ky that are commonly quoted for the Mississippi Delta complex (e.g., Weimer 1970), and the Rhone Delta, which has a thickness of only 50 m, accumulated since about 8.2 ka, indicating an average sedimentation rate of 6.1 m/ky (Oomkens 1970).

Dating of peat layers and other units in the Rhine–Meuse Delta indicated that aggradation of ribbon-like (anastomosed) channel belts kept pace with the average sea-level rise of 1.5 mm/a, at sedimentation rates in the order of 10^0 m/ky (Blum and Törnqvist 2000). Highly detailed studies of the development of the Rhine–Meuse system using multiple ^{14}C dates have provided a wealth of detail regarding the aggradation history (Stouthamer et al. 2011). Regional flood basin accumulation rates for the upper delta are in the range of 0.3 to 1 mm/a (0.3–1 m/ky). Some channel belts indicate local rates as high as 2 8 mm/a (2.8 m/ky). The ratio of local to regional aggradation rate ranges between 0.4 and 4.0, but all these values, measured over time periods of 10^3–10^4 years, are within the *SRS 7* range.

A few sedimentation rates can be calculated for tidal-inlet and barrier deposits. The Galveston Island barrier is 12 m thick and 3.5 ka old at its base (Bernard and others 1962,

Fig. 8.60), indicating an average sedimentation rate of 3.4 m/ky. A tidal-inlet at Fire Island, New York has migrated 8 km in 115 years (Kumar and Sanders 1974). The depositional slope from spit crest to channel floor is about 500 m wide, suggesting that at any one point the entire tidal-inlet fill could form by lateral accretion in about 7 years. The sequence is 12 m thick, indicating an instantaneous sedimentation rate of 1,714 m/ky. This migration rate is unusually rapid. Tidal inlets at Sapelo Island, Georgia appear to have migrated only about 2.5 km since the post-glacial sea-level rise (Hoyt and Henry 1967; Fig. 8.6C), indicating a sedimentation rate of 4.5 m/ky. Sommerfield (2006) calculated accumulation rates and stratigraphic completeness for modern oceanic continental margins. His research was based on measures of mass per unit area with time and translates into sedimentation rates in the *SRS 7* to *8* range.

Coal seams are estimated to represent 4,000–12,000 years of peat accumulation, at accumulation rates of 1–3 m/ky (Nemec 1988; Phillips and Bustin 1996). Allowing for a gradual 3:1 compaction during accumulation, Nemec (1988, p. 163) calculated an accommodation generation rate in a mire of 0.4–1.1 m/ky, which, as he noted, is one to two orders of magnitude greater than the long-term subsidence typical of the cratonic basins he studied (Illinois, South Wales, SW Poland).

Rates of sedimentation on modern alluvial fans and fluvial floodplains have been measured using ^{14}C dates on plant material, and tephrochronology. Available data were summarized by Miall (1978a) and shown to encompass a wide range, from 0.08 to 50 m/ky. However, these measurements have not been correlated to specific scales of architectural units, such as the depositional groups defined here.

Plio-Pleistocene slope and basin deposits in the Gulf of Mexico are characterized by sedimentation rates ranging from 0.16 to 6.45 m/ky (10^1–10^0 m/ky). The lower values are from areas characterized by slow subsidence rates and a high proportion of hemipelagic sedimentation; the high values were derived from areas with a high proportion of sediment gravity flow deposits and accumulation in salt-withdrawal basins. These values were calculated by Fiduk and Behrens (1993) for tectono-stratigraphic sequences representing between 0.3 and 1.25 Ma. These are unusually rapid rates of accumulation, the rates corresponding to *SRS 7* or *8*, over time scales of *SRS 9*, confirming the unique setting of the Gulf Coast basins or, alternatively, suggesting that the geological preservation machine (see below) may not yet have completed its work.

Scarponi et al. (2013) used amino-acid racemization data to demonstrate varying accumulation rates at the *SRS 7–8* scale, within the topmost 100-ka Holocene coastal sequence of the Po river coastal plain, Italy. Lowest rates (0.2–0.7 m/ky) were recorded within the nonmarine lowstand systems tract, increasing to 1.4–2.5 m/ky in the transgressive systems tract and to as much as 10 m/ky in the highstand. This variability was interpreted as a response to the systematic change in the probability of preservation of depositional events as accommodation increased with rising base level, versus changing sediment supply in the shallow-marine realm as the depositional systems shifted from retrogradational to progradational. Higher values of net accumulation rate were recorded in the transgressive and highstand systems tracts, and lower values (with concomitantly more diastems) in the condensed section formed around the time of the maximum flooding surface. The highest values up to 19.6 m/ky were recorded in homogeneous sandy sediment that may incorporate fewer diastems, in accord with Sadler's (1981) analysis that high accumulation rates scale inversely with time span.

The compilation of sedimentation rates in shallow-marine carbonate sediments by Kemp and Sadler (2014) indicated that when normalized to a 5-ka time scale—"the typical duration of Holocene and Pleistocene sections lacking major exposure surfaces" (Kemp and Sadler 2014, p. 1290) maximum accumulation rates range between 10 and 60 m/ky, depending on latitude, with rates dropping off in the higher latitudes.

SRS 8 and 9: At time scales of 10^4–10^5 years, measured sedimentation rates are in the range of 0.1 to 1.0 m/ky (*SRS 8*). At time scales of 10^5–10^6 years, rates are 0.01 to 0.1 m/ky (*SRS 9*).

Stratigraphic studies based on magnetostratigraphic dating and correlation, and studies of Late Cenozoic deposits using such techniques as high-resolution reflection-seismic data provide the appropriate focus. Cyclic successions developed by orbital forcing typically fall into this category. Examples of such deposits include the Quaternary shelf-margin sequences of Suter and others (1987), the classic 41-ky cycles of the Wanganui Basin, New Zealand (Pillans et al. 2005), and the minor and major cyclothems of Heckel (1986).

The sequences described from the Gulf Coast by Suter and others (1987) averaged 25,000 years in duration and range in thickness from about 25 to 160 m, indicating average accumulation rates of 1–6.4 m/ky. Heckel (1986) documented the chronology of 55 cycles of Westphalian–Stephanian age in the US Mid-continent. Estimates of the length of this time span range from 8 to 12 Ma The thickness of the succession varies from 260 m in Iowa to 550 m in Kansas. These values indicate an average accumulation rate of between 0.02 and 0.07 m/ky (*SRS 12*). Many of the cycles contain substantial fluvial-deltaic sandstone units and, according to Ramsbottom (1979), who studied similar cyclothems in Europe, rates of lateral deltaic growth must have been about as rapid as that of the modem Mississippi; yet the average sedimentation (vertical aggradation) rate is

two orders of magnitude less than that of the Holocene Mississippi Delta complex and its Pleistocene shelf-margin precursors on the Louisiana Gulf Coast. Part of the explanation for this marked contrast is that the Carboniferous cyclothems that were the subject of Heckel's study are located in a cratonic region, where subsidence rates would be expected to be substantially lower than on the continental margin of the Gulf Coast. As demonstrated by Runkel et al. (2007, 2008) cratonic sequences may develop by very low-angle lateral accretion.

Oxygen isotope and magnetostratigraphic data for the Wanganui Basin sections confirm the predominance of the 41-ky orbital cycle. In several composite sections it can be demonstrated that about 1 km of section accumulated in 1.2 Ma between the Olduvai and Brunhes paleomagnetic stages, indicating an average sedimentation rate of 0.8 m/ky (Pillans et al. 2005, Fig. 11).

A well-known ancient example of interpreted orbitally forced cyclic sedimentation comprises the Newark-type lacustrine cycles of eastern North America. Olsen et al. (1990) reconstructed characteristic orbital cyclic frequencies based on available chronostratigraphic information, and this yielded average sedimentation rates of 0.27 m/ky at the 10^4–10^5-year time scale.

A magnetostratigraphic study of the Siwalik fluvial deposits of Pakistan provides some control data for an ancient fluvial system (Johnson et al. 1985, 1988). In the Miocene Chinji formation, 400–500 m thick, and deposited over a time span of approximately 3.5 Ma, fluvial cycles representing channel belts up to several kilometers wide (Johnson et al. 1988, Fig. 9.3) range from 12 to 50 m in thickness. At an average sedimentation rate of 0.12 m/ky, these represent cycle return periods of between 10^4 and 10^5 years. Detailed studies reported in their second paper indicate considerable variation in local sedimentation rate, with evidence that specific magnetostratigraphic reversals are missing, indicating gaps in sedimentation on a 10^4-year scale. The formation as a whole corresponds to a *SRS-9* assemblage, but it is likely that the fluvial cycles, containing missing intervals, represent channel belts of *SRS-8*.

Jones et al. (2004) explored sedimentation rates and sediment transport rates in a foreland basin in Spain. The long-term sedimentation rate for their complete section averaged 0.075 m/ky over 12 Ma (*SRS-9*). Magnetostratigraphic dating of short intervals within this section indicated sedimentation rates varying between 0.03 and 0.2 m/ky (*SRS 8–9*), with much of the local variability being attributed to syndepositional folding affecting accommodation rates.

Fluvial cycles representing similar long-term avulsion processes were described by Hofmann et al. (2011) from the Cretaceous Piceance Basin of Colorado. The cycles average 120 m in thickness and are estimated to represent about 400 ky, accumulating at an average sedimentation rate of

0.305 m/ky (*SRS 8*). Clusters of channels develop an alluvial belt, the topographic elevation of which eventually leads to avulsive switching to lower areas on the floodplain, a process termed *compensational stacking*. Magnetostratigraphic studies of the Eocene Escanilla Formation, a braided-stream deposit, in the Spanish Pyrenees yield similar *SRS* values, at 0.17–0.57 m/ky over time periods of 10^5 years (Bentham et al. 1993).

Cores through interpreted precessional cycles of Cretaceous age on the floor of the Atlantic Ocean off tropical west Africa (66 22-ky cycles totalling 37 m of core: Beckmann et al. 2005) yield sedimentation rates of 0.025 m/ky. This is an order of magnitude slower than the *SRS-8* range characteristic of high-frequency orbital cycles.

The long-term sedimentation rates of *SRSs 8* to *12* depend largely on long-term rates of generation of sedimentary-accommodation space. This depends both on basin subsidence, which is controlled by tectonic setting, and by changes in base level, such as eustasy. Miall (1978a) showed that most nonmarine basins, in various tectonic settings, have sedimentation rates averaged over millions of years of 0.03 to 1.5 m/ky.

SRS 10: Basins in convergent margins are provided their own category in Table 3.3 and Fig. 8.8, because of the exceptionally high rates of subsidence and sedimentation that have been recorded in this tectonic setting. Miall (2010, Table 8.2, p. 280–281) summarized data from settings such as the Banda Arc, the Himalayan foreland basin, the Cretaceous forearc basin of Baja California and a forearc basin in Japan where sedimentation rates of 10^{-1} to 10^0 m/ky have been measured over durations in the order of 10^6 years. Magnetostratigraphic calibration of several sections in Andean foreland basin strata of Argentina indicated sedimentation rates ranging between 0.22 and 1.71 m/ky over intervals ranging between approximately 0.5 and 5 Ma (Echavarria et al. 2003). Hiatuses lasting up to 2 Ma, bring the average sedimentation rates, measured over total sections representing between 7 and 13.5 Ma, down to between 0.183 and 0.571 m/ky.

Growth strata that develop adjacent to active structures, such as basin-margin thrust faults are typically deposited at *SRS-10* rates. Burbank et al. (1996, Fig. 8.8) provided an example where accumulation rates averaged 0.117 m/ky over 1.7 Ma. Data provided by Medwedeff (1989) indicate a growth rate of 0.305 m/ky over 8 Ma. At the margins of the Tarim Basin, in western China, Sun et al. (2010) used magnetostratigraphic data to determine the rates of accumulation of "growth strata" in proximity to a growing anticline. Sedimentation rates increased from 0.325 m/ky prior to the syndepositional movement of the growth structure, to 0.403 m/ky during the period of active tectonism, over a total time span of about 10 Ma.

Rapid subsidence is indicated in basins developed along the San Andreas Fault system. Dorsey et al. (2011) calculated subsidence rates of between 0.4 and 2.1 mm/a (10^{-1}–10^0 m/ky) measured over periods of a few million years.

Not all convergent-margin basins are characterized by high sedimentation rates. In deep-water basins, where sediment supply is low, sedimentation rates may be much less. Finney et al. (1996) compiled data indicating that Paleozoic graptolitic shales in the Taconic foreland basin of the southeast USA accumulate at average rates of 0.01–0.03 m/ky measured over time periods of a few million years (comparable to *SRS 9*).

SRS 11: This is the rate characteristic of long-term geological processes. Aschoff and Steel (2011) calculated sedimentation rates for the Upper Cretaceous clastic wedge of the Book Cliffs (Utah-Colorado) in order to explore relationships between sedimentation and tectonism in the Sevier foreland basin. The range of rates is 0.047–0.14 m/ky, calculated over stratigraphic times spans of between 2.1 and 6.5 Ma. These are within the range for *SRS 11* but are low relative to those recorded in some Andean basins (*SRS-10*, above). The Catskill Delta of New York-Pennsylvania accumulated at comparable rates. Data provided by Ettensohn (2008) indicate a maximum rate for the proximal part of the "delta" (in reality a major clastic wedge deposited in a range of nonmarine to shallow-marine environments) of 0.096 m/ky (maximum thickness of 3 km accumulated over about 9 Ma between the Givetian and the Famennian).

Aschoff and Steel (2011) speculated about the possible influence of basement uplift within the Sevier foreland basin, which would tend to cancel out some of the subsidence due to flexural loading. This seems particularly likely for the middle portion of the clastic wedge, that characterized by the Castlegate sandstone, the sheet-like nature of which has, for some time, been attributed to a slow rate of regional subsidence (Yoshida et al. 1996). The incipient activation of Laramide structures within the basin, as suggested by Aschoff and Steel (2011), would be consistent with these characteristics of the clastic wedge. However, no such special influence on rates of accommodation has been suggested for the Appalachian basin.

SRS 12: Long-term sedimentation rates in cratonic environments include the lowest that have been recorded. Runkel et al. (2008) demonstrated that the deposits flanking the Trans-continental Arch in Wisconsin-Minnesota accumulated at an average rate of 0.013 m/ky, measured over a 15-Ma Upper Cambrian depositional record. Detailed biostratigraphic studies indicate that the section is relatively complete, and accumulated as a series of thin, offlapping shingles. The thickness of each shingle, as indicated by the biostratigraphic zonation, varies from 50 to as little as 5 m, indicating an order of magnitude variation in sedimentation

rate within this average rate. The Lower Mannville formation of southern Alberta consists of a series of
high-frequency sequences with multiple phases of incised
valley cut-and-fill. Long-term sedimentation rates range
between 0.0013 and 0.02 m/ky (Zaitlin et al. 2002).

As an example of our new ability to quantify sedimentation at
all time scales, Fig. 8.9 applies our new knowledge of rates and
time scales to the alluvial hierarchy illustrated in Fig. 3.67.
Fluvial deposits have long been known to consist of a hierarchy
of depositional units accumulated over a wide range of time
scales and sedimentation rates (Miall 1978b). An architectural
approach to field studies was introduced by Allen (1983),
developed by Miall (1985, 1988a, b, c, 1996), and has now been
fully documented in a number of detailed field studies (e.g.,
Holbrook 2001). However, owing to the difficulty of dating
nonmarine deposits, assigning ages and calculating sedimentation rates for these deposits is usually not possible. However,
order of magnitude estimates are now possible, as detailed here.

First, the sample of drill core at the lower right shows the
accumulation of the basic lithofacies of which fluvial systems are constructed. Sedimentary processes consist primarily of the generation and preservation of bedforms,
driven by turbulent flow and the autogenic changes in flow

patterns over minutes to hours. First-order bounding surfaces
between bedforms are minor scour surfaces, constituting the
smallest-scale elements in the range of natural sedimentary
breaks. Second-order surfaces record minor changes in flow
conditions, as indicated by changes in bedform type. Mud
drapes record the record of waning floods that drape bars and
bedforms. Inter-flood periods may be marked by footprints
and mud-rip-up horizons. The presence of bioturbation or
trace fossils on a first- or second-order surface may indicate
a period of stasis longer than the time suggested by bedform
architecture (Davies and Shillito 2018, 2021).

Next, the construction of channels and bars by channel
changes acting over time periods up to thousands of years.
At these time scales autogenic processes of channel migration and avulsion dominate. Third-order surfaces are the
pause surfaces within macroforms generated by seasonal or
longer term flood events. Fourth-order surfaces represent the
upper bounding surfaces of macroform architectural elements, such as lateral-or downstream-accretion units. They
are generated when channels switch in position, leaving the
macroform abandoned. Fifth-order surfaces are the bases of
major channels. These vary in erosional relief and lateral
extent (10^0–10^3 m) according to the scale of the river.

Fig. 8.9 The construction of an
alluvial stratigraphic unit showing
the contribution of the full range
of sedimentary processes over
time scales ranging from minutes
to millions of years. The purple
numerals in circles refer to the
system of bounding surfaces
identified by Miall (1996). Rates,
time scales and *SRS* values
(Sedimentation Rate Scale) are
summarized in Table 3.3

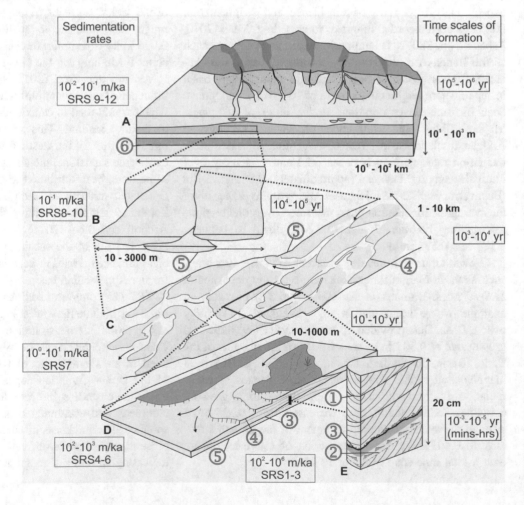

Mudstones are deposited during episodes of low-energy sedimentation, during seasonal low-flow periods, in abandoned channels, or in backwater areas (at 2nd–5th-order surfaces) and in floodplain settings. The presence of roots, coals or paleosoils provides some indication of the time represented by the beds. Roots may develop as a result of plant colonization on abandoned areas after a few years, whereas well-developed paleosoils indicate periods of stasis lasting in the order of 10^3 years (Wright 1990).

At the time scale of tens of thousands of years, larger channels and bar complexes are constructed, leading to the accumulation of complex channel fills. Extensive surfaces of non-deposition of sixth-order rank may be generated where alluvial systems switch position following periods of sediment build-up (e.g., as a result of compensational stacking, as described by Straub et al. (2009), Wang et al. (2011).

Whole fluvial systems represent the product of longer term tectonic or climatic changes, over time periods up to hundreds of thousands of years. Alluvial paleogeography is controlled by tectonic slope changes or climatically driven changes in runoff and sediment yield.

Finally, the basin fill is shown, which represents accumulation of the surviving deposits in the accommodation created by long-term geological processes. Sixth-order surfaces here represent the bounding surfaces of alluvial complexes, which could be defined as nonmarine sequences.

Other examples of stratigraphic hierarchies (Table 3.3) are discussed in Sect. 8.7.

8.5 The Fractal-Like Character of Sedimentary Accumulation

The log-linear relationship between sedimentation rates and time span was addressed by Plotnick (1986), Middleton et al. (1995), and Sadler (1999), who demonstrated that it could be interpreted using the fractal "Cantor bar" model of Mandelbrot (1983). Using the process described by Mandlebrot (1983) as "curdling," Plotnick (1986, p. 885) developed a Cantor bar for a hypothetical stratigraphic section by successively emplacing hiatuses (within portions of the section, at ever-increasing levels of detail (the result is illustrated in Fig. 8.10):

Assume a sedimentary pile 1000 m thick, deposited over a total interval of 1,000,000 years. The measured sedimentation rate for the entire pile is, therefore, 1 m/1000 years. Now assume that a recognizable hiatus exists exactly in the middle of the section, corresponding to a third of the total time (i.e., 333,333 years). The subpiles above and below the hiatus each contain 500 m of sediment, each deposited over 333,333 years, so that the measured sedimentation rate for each subpile is 1.5 m/1000 years. We now repeat the process, introducing hiatuses of 111,111 years in each of the two subpiles. This produces 4 subpiles, each 250 m thick, each with a duration of 111,111 years. The measured sedimentation rate is now 2.25 m/1000 years. The process

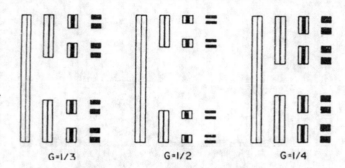

Fig. 8.10 Cantor bars generated using three different gap sizes (G) (Plotnick 1986, Fig. 1)

can be reiterated endlessly, producing subpiles representing progressively shorter periods of time with higher sedimentation rates (Fig. 8.7 of this book). Nevertheless, because the total sediment thickness is conserved at each step, the sedimentation rate of the entire pile remains 1 m/1000 years.

The selection of one third as the length of the hiatus, or gap (G in Fig. 8.10) is arbitrary. Other gap lengths generate Cantor bars that differ only in detail. Figure 8.10 is remarkably similar to Fig. 8.7, which was constructed by Miall (1997) based on the hierarchies of sedimentation rates compiled by Miall (1991), but with no knowledge of fractals.

The dependence of sedimentary accumulation on the availability of accommodation was understood by Barrell (1917; Fig. 5.2 of this book) and is the basis of modern sequence stratigraphy (Jervey 1988; Van Wagoner et al. 1990). The fractal model provides an elegant basis for integrating this knowledge with the data on varying sedimentation rates and varying scales of hiatuses discussed in the paragraphs above. Mandelbrot (1983) and Plotnick (1986) provided a version of an accumulation graph, called a Devil's staircase (Fig. 8.11). This shows how sediments accumulate as a series of clusters of varying lengths. Vertical increments of the graph correspond to intervals of sedimentation; horizontal plateaus represent periods of non-accumulation (or sedimentation removed by erosion). Sequence stratigraphy is essentially a study of the repetitive cycle of accumulation followed by the next gap, at various scales. The larger, more obvious gaps (the longer plateaus in Fig. 8.11) define for us the major sequences, over a range of time scales. The prominence of particular ranges of "accumulation + gap" length in the first data sets compiled by Vail et al. (1977) was what led to the establishment of the sequence hierarchy of first-, second-order, and so on. That this has now been shown to be an incomplete representation of nature (Miall 1997, 2010; Schlager 2005) does not alter the fact that there is a limited range of processes that control accumulation, and these have fairly well-defined rates which, nevertheless, overlap in time to some extent.

As Plotnick (1986, Table 1) and Sadler (1999) demonstrated, the incompleteness of the stratigraphic record

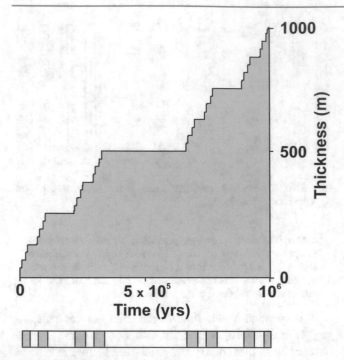

Fig. 8.11 Stratigraphic thickness accumulation viewed as a Cantor function—what has been termed a Devils's staircase—constructed with $G = \frac{1}{2}$. The corresponding Cantor bar is shown at a lower resolution below the graph (adapted from Plotnick 1986, Fig. 8.7)

depends on the scale at which that record is examined. Sections spanning several million years may only represent as little as 10% of elapsed time at the 1000-year measurement scale, although this is below the resolution normally obtainable in geological data. Sequences, as we know them, each essentially consist of clusters of the shorter "accumulation + gap" intervals separated by the longer gaps—those more readily recognizable from geological data.

Schlager (2005) illustrated the relationship between sedimentation rate and time span in a different way (Fig. 8.12). This diagram highlights the central point that the length and frequency of the gaps determine the calculated sedimentation rate. Almost all actual geological data sets yield correlation lines similar to that indicated by the black line of correlation. This line suggests a relatively slow rate of sedimentation with few, widely dispersed hiatuses. Such an interpretation inevitably follows from 1) the limited ability of geological methods to provide numerous tightly constrained age dates, and 2) the cryptic nature of most sedimentary hiatuses. Reality might be much closer to one or other of the red or orange correlation lines, which incorporate closer spacings of hiatuses and higher short-term sedimentation rates.

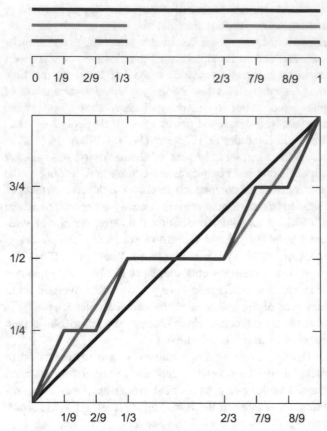

Fig. 8.12 A fractal plot of elapsed time (X-axis) versus sediment thickness (Y-axis) showing how the choice of gap length determines the distribution of hiatuses and sedimentation rate (from Schlager 2005)

8.6 Apparent Anomalies of High Sedimentation Rate Versus Slow Rate of Accommodation Generation

Bailey (2011) highlighted the preservation of fossil tree trunks in some coal-bearing strata as examples of the apparent dilemma posed by what appear to be exceptional modes of stratigraphic preservation. Tree trunks would be expected to decay rapidly, probably within decades (But were decay rates this rapid in the Carboniferous? Was there the same range of microorganisms that we observe at the present?), so the preservation of tree trunks up to 12 m high, in good condition, appears to argue for an unusually rapid rate of burial. Bailey (2011) suggested a rate of ~ 100 m/ky, well in excess of the 0.005 to 0.1 m/ky rates of accommodation generation indicated by the setting of the fossil trees within orbital cycles accumulating within tectonically active basins (the *SRS-9* time scale). The trees would seem to qualify as "frozen accidents," to use Bailey and Smith's (2010) term. But how unusual is this? And does it require a special explanation, as Bailey (2011) suggested? He argued for episodic, rapid (indeed, instantaneous) seismogenic subsidence to create the necessary accommodation.

Another example of an apparent stratigraphic puzzle is shown in Fig. 8.13. This is a carefully calibrated and dated stratigraphic record from the Guadalquivir foreland basin in SW Spain (Salvaney et al. 2011). High-precision dating for this Pliocene to Recent succession has been provided by the magnetostratigraphic record. For the purpose of this discussion, the interesting point is the increase in sedimentation rate following a hiatus at 1.6 Ma (calculated sedimentation rates are indicated next to the line of correlation). Prior to the hiatus, sedimentation rates were in the order of 10^{-2} m/ky (*SRS 9*). After the break they rose to 0.5 m/ky (10^{-1} m/ky: *SRS 8*) and then to 3.5 m/ky (10^{0} m/ky: *SRS 7*). Why? What happened?

Surface sediments in this basin are not yet as compacted as in the ancient record, but this is unlikely to account for more than a few percent of the total thickness of the post-1.6 Ma section. I suggest that what we are seeing is typical sediment accumulations that have yet to be completely processed by the geological preservation machine.

The post-11.5 ka section represents sedimentation in accommodation generated by the post-glacial sea-level rise

which averaged 11 m/ky between 12 and 8 ka (10^{1} m/ky: *SRS 7*). The calculated sedimentation rate is entirely in accord with sedimentation rates calculated over a 1000-year time scale (*SRS 7*). Future events could include a fall in sea level over a 10^{4}-year period, if the orbital cycles that have characterized Earth history for the last 2.5 Ma continue. That could potentially remove most or all of the top 95 m of the sedimentary record, as rivers grade themselves to the lower base level, thereby completing the work of the geological preservation machine at the *SRS 8* or *9* time scale. It is suggested that the pre-1.6 Ma section consists of short intervals of stratigraphy which accumulated at rates comparable to that calculated for the top of the section, but separated from each other by numerous unrecognized hiatuses, a pattern comparable to the red line of correlation in Fig. 8.12. Sadler (1999) had explained a similar pattern of apparently accelerating accumulation in younger sediments, based on his study of sedimentation rates and time scales.

Whereas most published stratigraphic data sets contain the generalization exemplified by the black line of correlation in Fig. 8.12, the reality for the top of the Guadalquivir foreland basin section (Fig. 8.13), as measured at appropriate times scales, would be closer to the red line of correlation in this diagram, where short intervals of time characterized by high rates of sedimentation are separated from each other by hiatuses. For example, this interpretation can explain the intervals of rapid sedimentation (*SRS 5* or *6*) required to

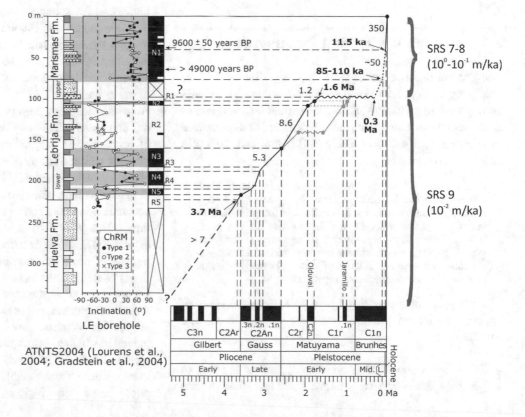

Fig. 8.13 Drill core from the modern Guadalquivir foreland basin in SW Spain. This is the Lebrija borehole, drilled from the surface of the modern, active floodplain, down through post-glacial estuarine deposits and into the older stratigraphic record. Sedimentation rates calculated from paleomagnetic calibration of the section are indicated by the values, in cm/ka, along the diagonal line of correlation. Adapted from Salvaney et al. (2011), with permission

preserve the tree trunks described by Bailey (2011). The rapid rate can then be seen as part of a predictable spectrum of sedimentation rates, when measured at the appropriate time scale. As noted above, coal seams represent *SRS-7* deposits. Valley fills and deltas lobes, where many coal seams (and fossil trees) accumulate, are *SRS-7* and *8* deposits.

In a specific attempt to "disentangle time" in the preserved rock record, de Natris (2012) and de Natris and Helland-Hansen (2012) used rates of sedimentation derived from modern shallow-marine environments to calculate elapsed time in the Tarbert Formation (Mid-Late Jurassic) of the northern North Sea. They applied these rates to facies successions interpreted to have been deposited in these environments. A summation of rate versus thickness for each facies explained only 7% of the 2.8 Ma elapsed time span of the Tarbert Formation as measured at geological time scales (*SRS 9* and higher). However, these calculations were carried out using rates in the *SRS 6–7* range, and therefore did not account for the longer term events recorded in the system.

Scott and Stephens (2015) addressed the issue of missing time through detailed calculations of sedimentation rates of Carboniferous coal-bearing successions in Britain. They cited sedimentation rates of 10^0–10^{-2} m/ky, based on studies of estuarine and deltaic environments. These are equivalent to *SRS* 7, 8 or 9. Their results indicate a representation of elapsed time ranging between 4.9 and 33%, depending on the interval selected and such factors as corrections for compaction of the peat to coal.

8.7   Accommodation and Preservation

The general question arises from the cases discussed in the preceding section: how could sediments accumulate at rates an order of magnitude or more greater than the local rate of accommodation generation (these processes are summarized in columns 4, 5 and 6 of Table 3.3). Blum and Törnqvist (2000, p. 20) noted:

> It … seems that accommodation, as it is commonly used, somewhat imprecisely mixes processes that operate over a range of rates and temporal scales; it is difficult to reconcile the time scales over which sediments are deposited in the first place, whether or not those deposits will be preserved in the stratigraphic record, and the manner in which ancient alluvial successions are interpreted in terms of changes in accommodation or an accommodation/sediment supply ratio.

Six important points listed help to explain the types of apparent anomalies described above (see also Holbrook and Miall 2020):

- While accommodation is typically quantified in terms of vertical space relative to sea level (base level) and the rate

at which it is created or removed, many important sedimentary processes are dominated by lateral sedimentary accretion. Sediments accumulate on mid-channel bars and on meander bends by lateral (cross-channel) and downstream accretion. Deltas and continental margins accumulate by oceanward progradation, typically with the development of clinoforms.

- Fluvial, tidal and other channels, and valleys, ranging up in scale to major incised valley systems, are locations where accommodation is not controlled by base level but are best understood with reference to the buffer concept of Holbrook et al. (2006). Accommodation generation on geomorphic time scales is therefore not dependent on tectonic subsidence rates and may be substantially higher (at the appropriate *SRS*).

- In two other major settings, accommodation is not restricted by base level: Deposition landward of the shoreline and in inland nonmarine basins is constrained by depositional slopes that are dependent on upstream controls, such as rates of tectonic uplift, river discharge and sediment load (Holbrook et al. 2006). Also, deep-marine sediments are not in any way constrained by rates of accommodation generation but are largely dependent on sediment supply and slope.

- Allogenic and autogenic sedimentary processes may generate predictable, ordered stratigraphic patterns at all time scales. The order and predictability may include erosional processes as well as processes of accumulation. This has always been the basis for Walther's Law and, more recently, sequence stratigraphy. Therefore, contrary to the random or chaotic processes of accumulation implied by Bailey and Smith (2010) stratigraphic order, including cyclicity, may be preserved in the rock record and may be understood and interpreted within the focus of the appropriate *SRS*.

- Although sediment preservation is extremely discontinuous and spasmodic at any one location, the shifting locus of accumulation (aggrading channels, delta lobes, prograding clinoforms, etc.) means that substantially more elapsed time is represented by preserved sediment in three dimensions than the percentages relating to vertical accumulation noted earlier in this section. At intermediate time scales (*SRS 5–8*) (and in the experiments of Sheets et al. 2002), sedimentation is continuous for lengthy periods of time, but distributed across an entire depositional system. Our tools for reconstructing these processes in the ancient record are quite limited.

- Actual sedimentation rates in most geological settings are always likely to be much higher—typically orders of magnitude higher—than those calculated from the rock record, based on observable geological data, such as extrapolations from datable ash beds or biohorizons, or rates based on regional rates of accommodation generation. There is no conflict between the rapid sedimentation that can commonly be observed in modern settings, and the rates that prevailed in the past. In that limited sense, traditional uniformitarianism ("the present is the key to the past") is correct, but with the additional proviso that analyses of the past must take into account the ubiquitous hiatuses, many quite cryptic, that occur at all time scales.

There is not necessarily a simple relationship between forcing processes and a stratigraphic result. Many surface processes are characterized by a **geomorphic threshold**, whereby a steady or unsteady process may operate without significant effect until a particular critical level is reached, whereupon sudden and dramatic change may take place (Schumm 1973). Theoretical and experimental studies of nonmarine and coastal environments have confirmed the non-linear relationship between such forcing functions as tectonism and climate change and the resulting effects on erosion patterns and sediment delivery (e.g., Allen and Densmore 2000; Kim and Paola 2007; Allen 2008). In fact, Jerolmack and Paola (2010) refer to the "shredding of environmental signals by sediment transport." Allen (2008, p. 20) suggested that "Large alluvial systems with extensive floodplains should therefore strongly buffer any variations in sediment supply with frequencies of less than 10^5–10^6 years. This has strong implications for the detection of high-frequency driving mechanisms in the stratigraphy of sedimentary basins." Jones et al. (2004) demonstrated that in a foreland basin in Spain, the effects of basin-margin thrust faulting and erosional sediment unroofing would take in the order of 1 Ma to be recorded in petrographic changes 20 km into the basin.

Jerolmack and Sadler (2007) examined the relationship between the transience of autogenic processes and the persistence of the "nested hierarchy of beds and bedding planes" constituting "the patchwork record of former landscape surfaces" that are ultimately preserved over the longer term. The purpose of their paper was to develop a quantitative stochastic diffusion model to simulate the multitude of overlapping processes.

Sadler supplemented his data compilation on vertical aggradation rates (Sadler 1981) with a similar compilation on lateral accumulation rates (the lateral growth of ripples, bars, delta lobes, continental margin clinoforms, etc.) (Sadler and Jerolmack 2012, 2015). By expressing downstream accretion in terms of dip-parallel cross-sectional area, they showed that growth rates of fluvial, coastal and shelf clastics are fairly constant at around 1 m^2/a, at all time scales, as measured over time spans from months to hundreds of millions of years. This continuity of sediment flux is, of course, not evident from the geologic record but, as noted

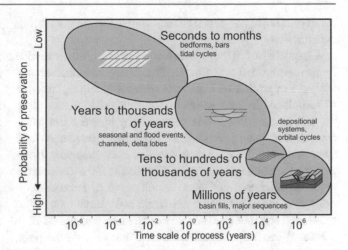

Fig. 8.14 The processes that constitute the mechanics of the stratigraphy machine (Sect. 8.8). Geological processes are grouped for the purposes of discussion into four major groups, based on sedimentation rates and durations (Table 8.1). The preservation of any unit depends initially on processes operating at a given time scale, but longer term preservation depends on the set of processes operating at the next longer term time scale (to the right and down)

above, non-deposition or erosion in one location is expected to be contemporaneous with deposition elsewhere,

In the next sections I discuss how specific sedimentary processes operating over a range of rates can lead to the generation of "frozen accidents" over a range of time scales. Rates and time scales may be conveniently classified into four broad (and partly overlapping groups) as summarized in Table 8.1 and Fig. 8.14.

8.7.1 Preservation at a Scale of Seconds to Months

Ripples and dunes form and migrate continuously under running water. Typically, trains of bedforms migrate down channels and across bar flanks, with one bedform replacing another, resulting in no net sedimentation. Temporary accumulations may form by lateral accretion where bedform trains build into areas of increasing water depth, such as scour pools or the flanks of bars. Under conditions of high bedload, ripple sets may become superimposed, to form climbing sets. Sedimentation on tidal flats can be affected by the lunar cycle from neap- to spring-tide conditions, which has been observed in some cases to be recorded as rhythmicity in lamina thickness (e.g., the Dutch tidal flats: Visser 1980). Mud in marine systems of all types is constantly undergoing deposition, erosion and transportation as tides, storms and other processes affect the sea floor (Trabouche-Alexandre 2015). How do such ephemeral deposits become preserved? Some must be so preserved, because we see them in ancient deposits in settings that clearly indicate the types of environmental processes just described (e.g., Archer et al. 1991).

Channels in flowing systems are ephemeral over a wide range of physical and time scales. Bank erosion, bar and meander migration are a result of the ever-changing structure of turbulence in the system. Minor changes in discharge or the direction of flow in one channel may trigger a cascading set of changes downstream. Sediment movement and deposition are therefore dynamic and ever changing. However, these autogenic processes will lead to the abandonment of channel reaches and bars, which serve as areas of temporary sediment storage, even while sedimentation may be relatively continuous when the system is considered as a whole. When we walk across a tidal flat or a fluvial point bar, it is these temporary deposits that we see. Returning to the same location days or years later, the deposits may have the same appearance, but there is a high degree of probability that the specific deposits we took note of have been replaced by others. What happens next is again a matter of chance. Long-term preservation depends on the events at the next time scale.

8.7.2 Preservation at a Scale of Years to Thousands of Years

The time scale of years to thousands of years is what Sheets et al. (2002) termed the stratigraphic "mesoscale." Within this time frame, "the depositional pattern shifts from reflecting the short-term flow pattern to reflecting long-term basinal accommodation. Individual events are averaged to produce large-scale stratal patterns" (op. cit., p. 288). This process is what Duller et al. (2012) termed the transition from "noisiness" to "drift."

Walther's Law is based on the concept of shifting depositional environments that are represented by deposits stacked in a vertical succession. Many of the environments to which this law has been applied generate the characteristic vertical profiles (e.g., fluvial fining-upward cycles, deltaic mouth bars and crevasse splays) over time spans of 10^1–10^3 years, and may, therefore be interpreted within the framework of *SRS* 5–7. Examples are summarized below. Longer term processes are described in the next section.

Most changes in fluvial systems take place during episodes of maximum discharge, which may be regular spring floods, or rarer flood events. Tidal systems are most affected during spring tides and during storms. These can result in large-scale changes in channel and bar location and orientation. In braided fluvial systems, entire minor channel systems may be abandoned. In meandering systems, meander chute and neck cutoff can leave earlier deposits behind. The deposits that fill scours, such as those which form at channel confluences, have a particularly high preservation

potential at this time scale. Davies and Shillito (2018, 2021) pointed out how the preservation of bioturbation or trace fossils at first-or second-order surfaces may indicate the passage of more time than would be suggested by basic bedform architecture. In fact, many sedimentary surfaces may remain in a state of stasis for very lengthy periods, being regularly modified by minor sedimentation, erosion and bioturbation (Davies et al. 2019). Holbrook et al. (2006) defined what they termed the buffer zone, the zone of instantaneous preservation space for fluvial systems. For graded rivers, this is the space between the deepest level of scour and the highest level to which levees and floodplains can aggrade during normal year-to-year flow conditions.

At a somewhat longer time scale, from hundreds to a few thousand years, nodal avulsion of river channels is an important process for deposit abandonment (Schumm 1977), which considerably increases the preservability of the deposits so affected.

Deposits that form by flow expansion, including alluvial fans, crevasse splays and delta lobes are abandoned by switching of the distributary system as a result of slope advantages. This is a well-known process based on the example of the Mississippi Delta, from the scale of the interdistributary bay-fill delta (Coleman and Gagliano 1964) up to the scale of the major delta lobes (Kolb and Van Lopik 1966; Frazier 1967). Once abandoned, the larger delta lobes undergo natural compaction and subsidence, which gradually takes them below the level of active scour and increases their chance of long-term preservation.

In tidal systems, channels, bars and tidal deltas evolve in the same way. Long-shore drift can displace inlet mouths, with resultant abandonment of inlet fill and bar-flank deposits. Scarponi et al. (2013), as noted earlier, demonstrated how accumulation rates and the probability of preservation of individual depositional events changed through a 100-ky base-level cycle.

In the experimental braid-delta constructed by Sheets et al. (2002), localized episodes of rapid aggradation occurred by avulsive channel switching across the experimental tank, eventually evening out "regional" deposition to the point that average aggradation equaled subsidence. Sheets et al. (2002, p. 300) scaled this up to a scenario whereby at a long-term aggradation rate of 1 m/ky the "depositional transition from flow control to subsidence control would occur on a time scale of the order of 15,000–30,000 years." This corresponds exactly to the conditions prevailing within the *SRS* 7 and 8 scale of depositional units. Sheets et al. (2002) demonstrated that in their model, after the deposition of a sediment layer equivalent to between five and ten channel-depths, which required an equivalent number of avulsion events to occur, the resultant layer had evolved a relatively consistent thickness and that the regional variation in this thickness could be related to the pattern of subsidence. What this means from the perspective of this chapter is that sedimentation rates for individual channels in a fluvial or deltaic system differ from the sedimentation rate for the entire depositional system by one half to about one order of magnitude (*SRS 5–6* versus *SRS 7–8*).

Fans and deltas illustrate the important point that (as expected following the arguments of Sadler and Jerolmack 2012, 2015) a continuous sediment flux still yields a discontinuous record because of the patterns of channel, channel belt and lobe switching that take place as a result of natural avulsion processes. Sediment bypass in one area is contemporaneous with sediment accumulation elsewhere.

At these time scales, several processes are repetitive and can result in deposits that preserve an element of internal repetition or cyclicity. Channel aggradation and bar accretion will record the range of flow conditions from higher velocity flow at the base of the channel, the bar toe or the base of the bar flank, to low-flow at the bar crest or the channel bank. Seasonal flooding and flash floods can impose a crude cyclicity of minor erosion followed by sedimentation that decreases in grain size as flow energy dissipates. All these processes tend to generate upward-fining successions which have a chance of being preserved at the decadal to millennial scale. Progradational deposits develop an upward-coarsening profile as deeper water environments are gradually filled with sediment.

All the processes described here include erosional episodes operating at the same time scale, and constituting integral components of the geological preservation machine.

8.7.3 Preservation at the Scale of Tens of Thousands to Hundreds of Thousands of years

At this time scale, sedimentary processes (the generation and removal of accommodation, sediment accumulation and erosion) may be dominated by high-frequency tectonic processes, or by Milankovitch processes (including sea-level change), or by both. Walther's Law may be applicable to processes operating at *SRS 7–9*. The section illustrated in Fig. 8.13 illustrates the geological preservation machine in action at this scale, the topmost 100 m of the section representing preservation at the *SRS 7–8* scale, but with long-term geological processes, that would most likely remove much of this section, still to come (Fig. 8.14).

As the discussion in this chapter has demonstrated, in most settings, local sedimentation rates are more rapid than the rate of accommodation generation and complete preservation of any succession at any time scale is unlikely. For example, the detailed Jurassic ammonite studies of Callomon (1995) reveal an extremely variable and fragmentary record of shallow-marine preservation over the 10^5-year time scale in the Jurassic deposits of Dorset (see Sect. 8.11.1 and discussions of Fig. 7.41 and 8.3).

Clastic wedges or *tectonic cyclothems* (Blair and Bilodeau 1988) are cycles developed under tectonic control over time scales of 10^4–10^7 years. Accommodation is generated by differential movement at basin margins. Episodic thrust loading within a foreland basin setting may generate regional basement adjustments at this time scale and has been suggested as one of the generating mechanisms for tectonic cyclothems (Peper et al. 1992). The angle or direction of tilt of depositional slopes may be changed by changes in intraplate stress, by extensional subsidence or faulting, or by changes in the supracrustal load in the case of contractional settings (e.g., foreland basins). Heller et al. (1993) modeled localized crustal flexure (uplift and subsidence) at rates of up to 0.16 m/ky over time periods of 10^5 years caused by tectonic reactivation of lines of crustal weakness in response to far-field intraplate stresses. Zecchin et al. (2010) compared the architecture of Milankovitch cycles generated under conditions of high-frequency sea-level change and tectonism, with accommodation changing at rates of 1–10 m/ky. Plint et al. (2012) employed detailed isopach maps to demonstrate along-strike shifts in depocentres on a time scale of about 200 ky in the Western Interior Basin (we return to this study in Sect. 8.12).

Entire depositional systems may be affected by these processes, leading to formation, and then abandonment and preservation of previously formed deposits. Kim and Paola (2007) modeled a coastal fluvial-deltaic system and demonstrated that autogenic cycles of delta and channel switching may, under the influence of fault movement, develop cyclothem-like cycles over time periods of 10^5 years (*SRS 8–9*). Allen (2008) suggested that the response time of fluvial systems to tectonic perturbations of an alluvial landscape would be in the order of 10^{5-6} years. Tectonic cyclothems are discussed further in the next section.

Autogenic switching of alluvial channel belts at a 10^3–10^5-year time scale has been described by Hajek et al. (2010) and Hofmann et al. (2011). Clusters of channels generate an alluvial ridge leading to instability and avulsion into neighboring low areas on the alluvial valley. A greater degree of compaction of the adjacent floodplain units relative to the channel deposits is a factor in creating the additional accommodation. This process, termed compensational stacking, generates cycles of about 120 m in thickness, and clearly requires the switching mechanism to be superimposed on a long-term process of tectonic subsidence.

In nonmarine successions, major surfaces of non-deposition may be difficult to distinguish from autogenic scour surfaces. Miall and Arush (2001b) labeled such unconformities "cryptic sequence boundaries." They may be recognized by detailed ground mapping, by careful petrographic work, and also by the application of special petrophysical methods that reflect the subtle diagenetic signatures of such surfaces (Filomena and Stollhofen 2011).

Sequence models for nonmarine systems have long incorporated the concept of variable stacking patterns of channelized sand bodies. Following the modeling experiments of Bridge and Leeder (1979), the standard interpretation has been that the architecture of channels and channel belts depends largely on the balance between the rate of avulsion and the rate of accommodation (Wright and Marriott 1993; Shanley and McCabe 1994). The models assert that rapid accommodation generation may increase the likelihood of a given channel deposit being buried as an isolated sandbody by floodplain deposits before channel migration and scour removes it from the record. Slow accommodation generation favors the accumulation of channel bodies that erode laterally into each other, as a fluvial system slowly migrates across a floodplain, developing laterally amalgamated sand bodies. Such interpretations must, however, be consistent in terms of time scale. Alluvial architecture—the preserved complex of amalgamated macroforms—is determined by sedimentary processes that occur within the SRS-6 to SRS 8 range, that is, on time scales of 10^2 to 10^5 years and sedimentation rates of 10^{-1} to 10^2 m/ky. These are the rates used by Bridge and Leeder (1979) based on information they compiled from modern rivers. However, this means that the formation of the elements of alluvial stratigraphy occurs within time scales that are several orders of magnitude more rapid than the rate at which accommodation is typically generated by regional geological subsidence (SRS 11). Therefore, it seems likely that in the ancient record there would be a genetic relationship between fluvial architecture and rates of accommodation only in the case of high-frequency sequences, those formed within the SRS-8 range, e.g., orbital cycles. Nonmarine sequences formed over longer time periods (many have been documented on a 10^6-year time scale) require different interpretations, in which changes in alluvial architecture are related to the migration of facies belts or to changes in fluvial style in response to tectonic or climatic forcing.

Orbital cycles are superbly exposed in pelagic sedimentary records in southern Italy and have been studied as a basis for erecting an astrochronologic time scale (Hilgen 1991; Hilgen et al. 2006, 2007, 2015). In this setting, accommodation is not an issue. Sea level, water chemistry, sediment supply, and every other important aspect of the environment, including even the regularity of processes generating hiatuses, is controlled by orbital forcing, and a clear orbital signal and a representative sedimentary record can be expected to be preserved. Calibration of the cyclicity against a retrodicted astronomical record confirms this. Lacustrine settings may also provide the necessary

environment of total control by orbitally forced parameters for development and preservation of a cyclostratigraphic signature (e.g., Aziz et al. 2003), and the classic cyclothems of the US Mid-continent, controlled as they were by substantial climatic and glacio eustatic sea-level change, also provide excellent examples of preservation at this time scale (other examples include the Triassic cycles of the eastern US: Olsen 1990; Green River Formation of Wyoming: Fischer and Roberts 1991). Some ancient carbonate-platform margins also offer convincing cases. In other settings, however, questions of preservability of the cyclostratigraphic record arise, particularly because of the likelihood of overprinting by autogenic processes (Miall and Miall 2004, 2010; Bailey 2009). As Kemp (2012) demonstrated by numerical modeling, in the case of marine cycles, if sea-level changes occurring at a higher than Milankovitch frequency are embedded in the preserved record, the evidence of orbital control may be masked. Even in deep-marine sections, the sedimentary record may not be complete (Trabucho-Alexandre 2015), and so caution must be applied. Each putative example needs to be examined on its merits. Strasser et al. (2006, p. 81) said:

> It is clear that astronomical climate forcing is most accurately recorded in depositional settings where the preservation potential is highest (deep marine basins, rapidly subsiding shelves, long-lasting deep lakes).

Recent developments in the field of cyclostratigraphy and astrochronology are discussed in Sect. 8.13.

8.7.4 Preservation at the Scale of Millions of years

On extensional continental margins, subsidence rates range from 0.2 m/ky at the initiation of rifting, decreasing to less than 0.05 m/ky during the flexural subsidence phase, at time scales of 10^6–10^7 years. Foreland basins subside at rates of 0.2–0.5 m/ky, and cratonic basins at 0.01–0.04 m/ky (rates from Allen and Allen 2005, p. 364–365). Intraplate stress changes can generate regional accommodation changes at 0.01 to 0.1 m/ky at time scales of 10^6 years (Cloetingh (1988, p. 216). Detailed studies of offsets on growth faults in the Niger delta, based on measuring the displacement of maximum flooding surfaces, indicated rates of accommodation generation on the downthrown side of 0.01–0.12 m/ky, over time scales of 10^5–10^6 years (Pochat et al. 2009). These rates are all within the range of SRS 9–11.

Where accommodation generation is rapid, sedimentation rates comparable to the "long-term geomorphic" rates of SRS 7 (10^0 m/ky) have been recorded over intervals of several millions of years. Such settings include forearc and foreland basins and some basins associated with strike-slip faults

(Miall 2010, p. 280–281). These are categorized as convergent-margin basins (including transgressive settings) in Fig. 8.8 and assigned to *SRS-10*. As discussed above, variations on these long-term rates may provide clues concerning subsidence and uplift mechanisms.

Sedimentation and long-term preservation by lateral accretion may be the key to understanding the thin sedimentary succession present in cratonic-interior settings (*SRS 12*). It required the highly detailed chronostratigraphic reconstructions of Runkel et al. (2007, 2008) to demonstrate that the Upper Cambrian-Lower Ordovician succession on the flank of the Trans-Continental Arch in Wisconsin and Minnesota developed by gradual offlapping of successive sedimentary shingles on a very gently dipping ramp.

On continental margins, given the importance of lateral progradation, total accumulation is not limited by vertical accommodation generation, and sedimentation is supply dominated. The following generalizations are adapted from Stow et al. (1983, p. 58): Typical long-term accumulation rates on carbonate or clastic shelves are from 10 to 40 m/ky

(10^1 m/ky). Pelagic ooze sedimentation will normally not exceed 0.03 m/ky, although under upwelling areas it may reach 0.1 m/ky (10^{-1} m/ky). Resedimentation of material to deeper water results in accumulation rates on modern deep-sea fans from 0.1 to 2 m/ky, and up to 10 m/ky (10^{-1}–10^1 m/ky) in small tectonically active basins.

8.8 The Stratigraphy Machine

The generation of the stratigraphic record by multiple processes overlapping in time, as summarized in the preceding section, can be encapsulated in the form of a machine, as suggested in Fig. 8.15. The term "Stratigraphy Machine" is borrowed from Smith (1994). To read and interpret this diagram, start with the red horizontal arrow at the left. Where accommodation is created by subsidence or base-level change, sedimentation is initiated by short-term processes, such as marine or nonmarine movement of sediment to generate bedforms, bars and channels. These may be

Fig. 8.15 The Stratigraphy Machine (from Holbrook and Miall 2020)

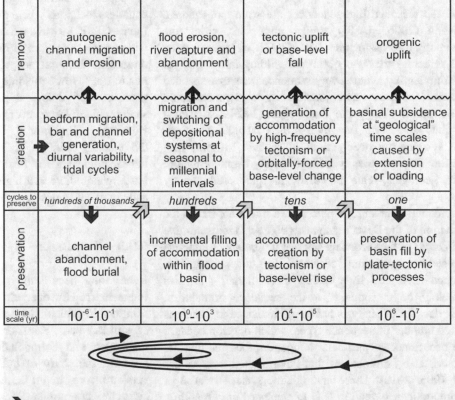

replaced many times by local autogenic processes, as suggested by the row labeled "cycles to preserve." Ultimately a few centimeters or meters of deposit are either preserved by channel abandonment and burial, as shown by the downward-directed red arrow, or are uplifted and removed, as indicated by the upward-directed black arrow. Where preservation occurs, over longer term time periods, this temporary deposit will then be subjected to longer term geological processes, as shown by the diagonal yellow arrow, and so this continues through the full extent of geological time, with further preservation, or removal, subject to the action of geological processes over progressively longer time periods. The final fragments, containing all the preserved sedimentary breaks, are conserved in whatever accommodation has been generated by long-term subsidence or are subject to orogenic uplift and permanent removal.

8.9 Implications of Missing Time for Modern Stratigraphic Methods

8.9.1 Sequence Stratigraphy

It has been reasoned that sequences are scale independent. Catuneanu (2006, p. 10) argued this point, citing the observations by Posamentier et al. (1992) on a small delta only 1 m across that was observed building into a pool of water from a gully, where every sequence process associated with base-level change and systems tract development could be observed on a tiny scale. Paola et al. (2009) also argued this point, suggesting that their scaled laboratory experimental systems run over periods of seconds to hours in a tank a few meters across could legitimately be interpreted to explain geological-scale processes in major sedimentary basins.

Fragmentary the stratigraphic record might be, but as the construction of the Devil's staircase (Fig. 8.11) indicates, the fractal nature of the record means that it consists of intervals of succession fragments separated by larger gaps that developed at higher time scales. These larger gaps can legitimately be considered as the sequence boundaries. Several decades of analysis have now indicated that there is a limited number of sequence types, which develop because of the occurrence of particular allogenic processes that are characterized by time scales that have limited time ranges (Miall 1995, 2010). These natural time scales, because of their predominance, tend to lead to enhanced preservability, and it is for this reason that sequence stratigraphy "works." The original concept of a sequence hierarchy—the five or six "orders" of Vail et al. (1977) has been shown to be unworkable (Schlager, 2004), but a crude hierarchy does

exist, based on the nature of the processes that develop sequences, which range over time scales from 10^4 to 10^8 years (Table 3.3; Miall 1995, 2010; Table 4.1).

8.9.2 Implications for Stratigraphic Continuity, the Concept of Correlation and the Principal of the GSSP

The discipline of stratigraphy is dependent on the principle that the sedimentary record is amenable to dating and correlation. The thrust of this chapter has been to demonstrate the fragmentary nature of the stratigraphic record, whereas the existence of the property of correlatability—which has been amply demonstrated by two hundred years of stratigraphic practice, implies continuity. There is no contradiction here. Consider the evolution of time at *SRS 1*, the present moment. At this time more than half of the Earth's surface is under water and accumulating water-laid deposits in fluvial and marine environments, or otherwise in a condition which is favorable to instantaneous sediment accumulation, e.g., the migration of an eolian dune, or alluvial fan deposits banking up against a fault. Therefore, there are countless locations where the deposits that could constitute a future GSSP for "Now" are accumulating. Many of these will survive the rigors of *SRS 2* to *10*, *11* or *12* (depending on tectonic location) to provide the framework for the correlation of "Now" at some distant point in the future. This work of the geological preservation machine is all that stratigraphers have ever assumed and depended upon. However, as Smith et al. (2015) have pointed out, most of the officially ratified GSSPs that currently form the basis of the Geological Time Scale are defined in shallow-marine strata, where there is a high probability of multiple hiatuses and significant missing time.

8.9.3 Discussion

It has long been known that the sedimentary record is fragmentary. However, this has not stopped stratigraphers from making calculations about sedimentation rates and the ages of key beds, based on assumptions of continuous sedimentation and extrapolation from known horizons. For example, the study of cyclostratigraphy requires the conversion of the "depth domain" to the "time domain" using age calibration points (Strasser et al. 2006, p. 82). This type of analysis should always be treated with considerable caution, and the arguments presented here only serve to emphasize this point. This could be argued before on an ad hoc basis, but not until the advent of the fractal concept has it been possible to systematize these observations and place them into a framework that suggests a continuity of process over all time scales. The fractal framework constitutes a useful method of statistical description of the geological preservation machine, but because the time frames of geological processes are not genetically related, there is no reason to expect that the framework will constitute anything other than a mathematical approximation. Accordingly, it may be more appropriate to describe the relationships as "fractal-like."

The significant differences highlighted in this chapter between (1) the preservation of the products of modern sedimentary processes, (2) those preserved in the recent (post-glacial) record, and (3) those preserved in the more ancient record, indicate the need for a modified use in geological work of the concepts of *uniformitarianism*, hence the title of the paper from which this section has been adapted (Miall 2015). The same applies to the comparable term "*actualism*," which is "the principle that the same processes and natural laws applied in the past as those active today" (Donaldson et al. 2002). With the exception of some unique conditions in the Precambrian relating to marine and atmospheric chemistry (Eriksson et al. 1998; see Sect. 7.9), the *processes* of sediment creation have been comparable throughout geological time. It is the issue of *preservation* that has required a re-evaluation. Interpretations of the geological record that use modern, active, post-glacial depositional systems as analogues (e.g., deltas, valley fills, prograding continental margins) need to take into account that these deposits can only illustrate the working of the geological preservation machine up to the time scale of *SRS 7* or *8* (Fig. 8.14). Application of uniformitarianist concepts to the longer term geological time scale (*SRS 9–12*) needs to be carried out with these cautions in mind. In this sense, the concept of uniformitarianism, as applied in practice, is incomplete.

These concepts likely hold a key to an improved way of studying and interpreting the sedimentary record, requiring us to go back and look at that record again, ironically, to document what is not there in greater detail: the record of missing time. Sedimentary units at all scales need to be evaluated in terms of the sedimentation rates they indicate over the full range of scales, at the appropriate *SRS*, in order to unravel the complexity of preservation and removal. De Natris (2012) and de Natris and Helland-Hansen (2012) have made a start at this form of analysis. Amongst other consequences, this research will re-emphasize the value of the "mesoscale" experiments described by Paola et al. (2009). For example, he and his colleagues have demonstrated that incised valleys formed by shoreline incision and fill during cycles of base-level change do not represent single chronostratigraphic surfaces (as commonly assumed for the purpose of sequence definition) but represent amalgamated fragments of surfaces and deposits that evolve both during falling and rising stages of the base-level cycle (Strong and Paola 2008; Holbrook and Bhattacharya 2012). The experiments of Sheets et al. (2002) provided a quantitative basis for the transition from an autogenic to an allogenic time scale in the gradual filling of an alluvial-deltaic basin by avulsive switching of channel belts. All of these findings will help to clarify the concepts of time and the behavior of the geological preservation machine.

Quantitative stratigraphic studies (e.g., those based in time-series analysis) are becoming increasingly popular. However, lest the new fractal concepts tempt geologists to focus in future on quantitative studies based on fractal theory, the warning of the field sedimentologist needs to be heard. Quantitative analyses too frequently ignore the field reality of the rocks under study (e.g., see Sect. 8.13). Without careful analyses of facies details, a careful search for grain size and lithologic changes, and a focus on the nature of facies contacts (sharp versus transitional), researchers can mistake mathematical or statistical rigor for geological reality.

The examples used here to illustrate the *Sedimentation Rate Scales* merely brush the surface of a potentially instructive form of deductive investigation in which tectonic and geomorphic setting, sedimentary processes and preservation mechanisms can be evaluated against each other both qualitatively and quantitatively, leading to more complete quantitative understanding of the geological preservation machine, and a more grounded approach than earlier treatments of stratigraphic completeness. For example, wide variations in sedimentation rate in foreland basins, and in continental margin sedimentation have been touched on here, and may help to refine future geological interpretations. Once tectonic setting is taken into account, the variability in the data in Sadler's (1981, 1999) linear log–log plot become comprehensible. When accommodation generation is particularly rapid, as in many convergent-margin settings, and where accommodation is essentially limitless, as on continental margins, and sedimentation is supply dominated, long-term sedimentation rates may be one to two orders of magnitude greater than Sadler's compilation would suggest. Sediment flux may be relatively constant over a wide range of time scales (Sadler and Jerolmack 2012, 2015), but processes of sediment distribution, sedimentation and preservation "shred" the resulting record, as explained in this chapter. We return to this debate in Sect. 8.10.2, which is a discussion of the construction and interpretation of chronostratigraphic charts (Wheeler diagrams) for stratigraphic successions.

It is now clear that the stratigraphic record is more than just incomplete. To extend Ager's famous thought: there are gaps within the gaps, and the record is permeated with them, at every scale. The frozen accidents that the gaps enclose can still tell us a great deal, but only if we get the time scales right.

We are confronted with an apparent paradox: on the one hand, the recent studies of sedimentation rates and time scales have revealed the fragmentary nature of sedimentary preservation; yet detailed stratigraphic studies (some of which we discuss in Sects. 8.11 and 8.12) are increasingly confirming the record of stratigraphic order, such as the evidence of orbital cyclicity in the Phanerozoic rock record, while those other aspects of sedimentology that depend on similar assumptions of continuity, including vertical profile analysis, Walther's Law and sequence stratigraphy continue to "work." How can this be? The explanation would seem to be that throughout geologic time multiple processes with rates that vary over orders of magnitude have been at work simultaneously. This is what has yielded the fragmentary record. Yet, the power of particular processes to simultaneously influence depositional systems over regional, continental, even global areas, results in recognizable signals, such as orbital cyclicity, that may be identified through the confusing overprint of other autogenic and allogenic processes. It requires very detailed stratigraphic studies to simultaneously document the disorder while mapping out the significant process indicators. We come back to this point in Sect. 8.11.

8.10 The Future of Conventional Chronostratigraphy

8.10.1 Current Examples of Outstanding Work

The modern geological time scale (Gradstein et al. 2004, 2012, 2020, and the updated scale maintained at www.stratigraphy.org; see Fig. 7.33 of this book) takes into account the types of potential error discussed in Sect. 7.8.5 by collating data from multiple sources. The construction of Composite Standard Reference Sections using graphic correlation methods (Sect. 7.5.4) is part of this process. Currently finalized global stratotypes for systems, series and stages are identified in Gradstein et al. (2004, 2012, 2020) and on the website, with references to published documentation, most of which consists of reports in the journal *Episodes* by representatives from boundary working groups. Realistic error estimates are provided for Phanerozoic stages, and range from very small values (10^4-year range) for most of the Cenozoic, the time scale for which is increasingly linked to an astrochronological record, to as much as $\pm$ 4 m.y. for several stages between the Middle Jurassic and Early Cretaceous. Figure 8.16 illustrates the expected error in age estimation through the Phanerozoic, based on information available in 2007. The refinement of the scale expected to accrue from the integration of astrochronology into the data

base (Fig. 8.16E) may be regarded as optimistic, but the key workers in this field make a good case for such a development, as discussed in Sect. 8.13.

Miall (2010, Chap. 14) provided a discussion of the global scale as applied to the issue of the global correlation of sequences, particularly as this relates to the possible global extent of sequences based on their presumed origin as a result of eustatic changes of sea level. An example of this type of work is illustrated in Figs. 8.17 and 8.18. The stratigraphy of the New Jersey continental margins has been extensively studied by K. Miller and his colleagues, in part to provide a chronostratigraphic baseline for global sequence studies. Figure 8.17 shows the sequence subdivision and generalized lithologies in relevant boreholes along a transect across a portion of the continental margin which consists of a series of prograding Oligocene clinoform sets. Figure 8.18 is an example of the highly detailed chronostratigraphic documentation of the Oligocene–Miocene stratigraphy that can now be developed, based on integrated biostratigraphy, oxygen isotope stratigraphy, strontium isotope stratigraphy and magnetostratigraphy. The details of this work are presented in several major regional studies (Miller et al. 1998, 2008; Browning et al. 2008; Kominz et al. 2008) and are discussed in detail by Miall (2010, Sect. 14.6.1).

Figure 8.19 reproduces a correlation diagram developed by Betzler et al. (2000) comparing the Miocene-Pliocene 10^6-year sequence record on the Bahamas Bank and the Queensland plateau, to which I have added the sequence boundaries from the same interval of the New Jersey continental margin succession (from Kominz et al. 2008). The Queensland-Bahamas study is based on correlations of calcareous nannoplankton and planktonic foraminifera. The density of the biostratigraphic data ranges from one to three biohorizons per million years, calibrated to the Berggren et al. (1995) time scale. Gradstein et al. (2004, Fig. 1.7) indicated that stage-boundary ages have been adjusted by between about 0.1 and 0.8 m.y. from the Berggren et al. (1995) scale, not enough to significantly affect the level of correlation with the more recently dated New Jersey section.

Figure 8.19 indicates a high degree of correlation between Queensland and the Bahamas. Betzler et al. (2000, p. 727) stated that throughout the Miocene-Pliocene "The isochroneity of sea level low stands in two tectonically unrelated carbonate platforms is strong evidence for eustatic sea level changes as the controlling factor on large to medium-scale (1–5 Ma) stratigraphic packaging." Betzler et al. (2000) went on to note that the number of sequence boundaries in their data does not correspond to the number of boundaries in the Hardenbol et al. (1998) chart—an issue that, as discussed by Miall and Miall (2001) constitutes one of the criteria for sequence correlation that has consistently

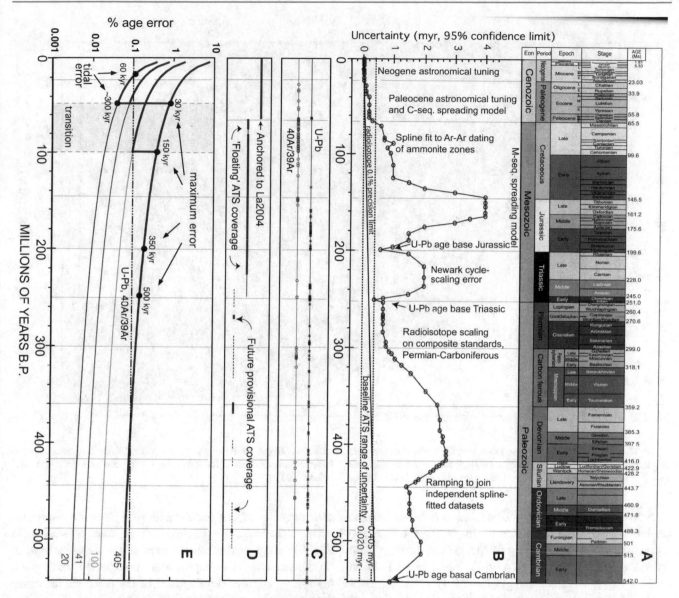

Fig. 8.16 **A** The standard divisions of the Phanerozoic International Geologic Time Scale (Gradstein et al. 2004). **B** Estimated uncertainty (95% confidence level) in the ages of stage boundaries. **C.** Distribution of radiometric ages used in the construction of the time scale. **D** Documented and potential astrochronological time series. **E** Estimated error to be expected when the astrochronological time scale through the Phanerozoic is completed and integrated into the International Geologic Time Scale. (Hinnov and Ogg, 2007)

misled and confused stratigraphers. We do not consider the lack of correlation with the Hardenbol chart further, for reasons discussed elsewhere (Miall 2010). Betzler et al. (2000) suggested that the accuracy of their sequence boundary correlation might be improved by more intensive biostratigraphic analysis. However, for our purposes, it is much more significant to evaluate the Queensland and Bahaman correlations with the sequence boundary record from New Jersey (Fig. 8.19; accurate sequence boundary ages for the succession younger than 10 Ma are not available from this area). This is a comparison between two independent studies and therefore fulfills one of the most important criteria for a meaningful test of the global-eustasy paradigm. In this case, the evidence indicates a strong degree of correlation between New Jersey, The Bahamas, and Queensland. Despite the small differences in the time scales used in these two different interpretations, the comparisons between the data sets are striking. The adjustments that would be required to recalibrate the Betzler et al. (2000) study to GTS2004 would not significantly change the pattern visible in Fig. 8.19. The case for eustatic control of Miocene stratigraphy is considerably strengthened by this comparison.

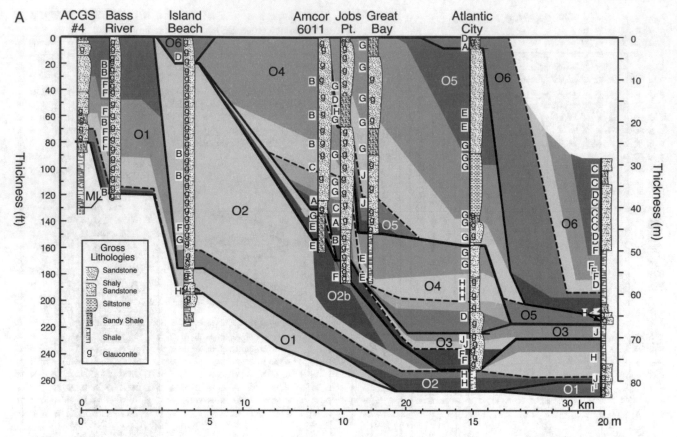

Fig. 8.17 Details of the Oligocene sequences O1 to O6 forming a set of seaward-dipping clinoforms along a NW–SE transect across the New Jersey continental margin. Gross lithologies for each borehole are indicated. Upper case letters to the left of each borehole indicate foraminiferal biofacies, identified by factor analysis (Kominz and Pekar, 2001)

Also shown in Fig. 8.19, for reference purposes, are the sequence boundaries documented in the "great Neogene sedimentary wedge" (wording of McGowran, 2005, p. 190) reconstructed by Vail et al. (1991), based in part on an analysis of the Antarctic margin (Bartek et al. 1991). Their model of the Neogene wedge claims to illustrate a global pattern of sea-level change and accompanying stratigraphic architecture that can be recognized worldwide. However, there is very little correspondence between these sequence boundaries and the New Jersey-Queensland-Bahamas events documented in Fig. 8.19, which calls into question the chronostratigraphic basis on which the wedge model was drawn. McGowran (2005, p. 191) indicated that the number and ages of the sequence boundaries has been modified by Hardenbol et al. (1998), but the boundaries and ages indicated in his redrawn version of the Vail et al. (1991, Fig. 12) diagram are the same as in that diagram, which were, in turn, based on the Haq et al. (1987, 1988) global cycle chart.

This discussion of the New Jersey margin and what the chronostratigraphy tells us provides an excellent example of the capabilities of modern chronostratigraphic methods. The

demonstration of the correlatability of Neogene sequences between tectonically unrelated areas that is shown in Fig. 8.19 supports the interpretation of glacioeustatic control for the sequence stratigraphy, a not unexpected result given the likely extensive development of the Antarctic ice cover during the Neogene (Miller et al. 2005).

The second example discussed here is the detailed study of late Cenozoic sedimentation of the Brazos-Trinity depositional system in the Gulf Coast margin by Pirmez et al. (2012) and Prather et al. (2012). This study was discussed in Sect. 6.3.2 as an example of the construction of a regional stratigraphy using 2-D seismic data integrated with well and core data. Here I focus on the chronostratigraphic methods used in the study. Figure 8.20 provides the basis for chronostratigraphic dating and correlation used in this study. Biostratigraphic zonation of planktonic foraminifera and nannofossils obtained from core was supplemented by oxygen isotope measurements. Additional age and correlation information was provided by volcanic ash geochemistry (correlated to Caribbean volcanic events). For deposits younger than 40 ka, radiocarbon dating was used on

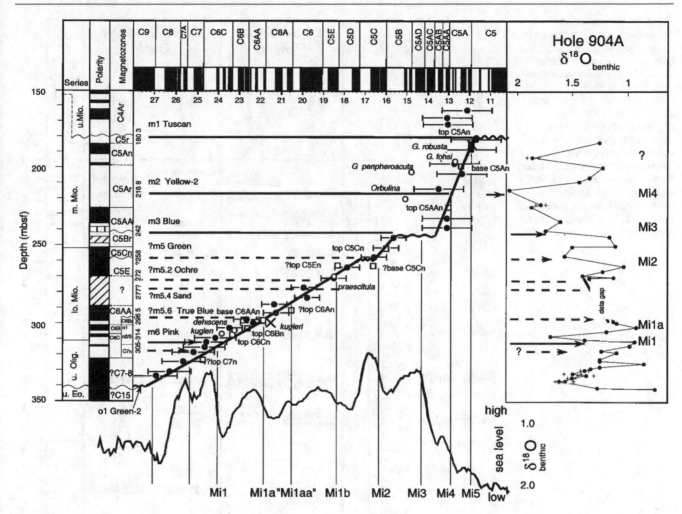

Fig. 8.18 Age-depth diagram for Hole 904, New Jersey continental slope. The core record, with magnetozones, is indicated at left. Sequence boundaries, recognized from seismic and facies studies, are shown by the horizontal lines extending from the core log across to the right, to the sloping line of correlation. Sequences are named, from the top down, "m1 Tuscan", etc. The standard magnetostratigraphic time scale is shown across the top of the figure and the δ¹⁸O record is shown along the bottom. Dated sample points are indicated by symbols along the line of correlation: strontium isotope values, with error bars (black circles), planktonic foraminifer lowest occurrences (open circles) and highest occurrences (crosses), and magnetostratigraphic reversal boundaries (squares with chron number). δ¹⁸O from hole 904A are indicated in the box at right. Unconformities are indicated by wavy lines in the line of correlation (Miller et al. 1998)

planktonic foraminifera, reworked plant debris and shell material. Magnetostratigraphy proved not to be useable because of the presence of sand and silt layers, which did not provide magnetic information, and because of highly variable sedimentation rates, making the "bar-code" matching of the magnetic signal to the standard record difficult to impossible. The absolute ages of first and last appearance datums provided key datable and mappable horizons, as shown in Fig. 8.21. Their ages are based on standard studies referenced in Pirmez et al. (2012, p. 119).

The average sedimentation rate for the entire succession examined in this project averages 2.1 m/ky (a maximum of 300 m of section deposited in about 140 ky). This corresponds in time range and sedimentation rate to *SRS-7* (Table 3.3). However, the detailed chronostratigraphy reveals that there are significant sedimentation breaks and condensed intervals, and intervals within which sedimentation rates are substantially higher (Fig. 8.22). Hemipelagic units reveal sedimentation rates between 0.1 and 9 m/ky, which is a similar order of magnitude to the long-term rate

Fig. 8.19 Comparison of Neogene sequence boundaries in the platform carbonate margins of Queensland and the Bahamas Bank. The sequence boundaries from the Hardenbol et al. (1998) scale are shown at right. From Betzler et al. (2000). Two additional sets of sequence boundaries have been added, at left, the sequence boundaries from the Miocene portion of the New Jersey record (from Kominz et al. 2008), and the Neogene record of the Antarctic (from Vail et al. 1991, with boundary ages revised by Hardenbol et al. 1998)

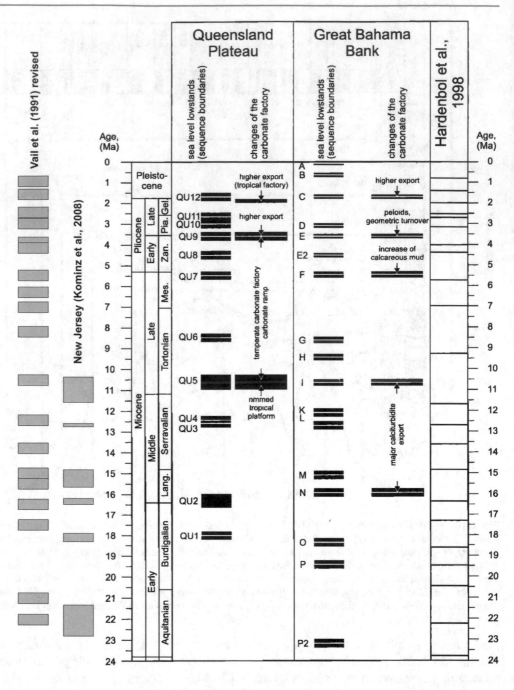

but in at least one instance more than four times the long-term rate, which emphasizes the importance of knowing the time scale over which the measurement is made. Sand-rich ponded apron deposits yield sedimentation rates an order of magnitude greater, ranging between 20 and 64 m/ky, which corresponds to the higher-rate end of *SRS-7*.

Many additional details may be gleaned from detailed study of these data, including the partitioning of the sediment flux between different basins within the project area. This, in turn, throws light on the evolution of the dispersal

system as it crossed the shelf and fed several separate basins arranged along a downstream trend through the Sigsbee Knolls.

The first example discussed in this section, the Neogene stratigraphy of the New Jersey continental margin, indicates that modern conventional chronostratigraphic techniques may be able to provide ages and regional to global correlations for the late Paleogene (Oligocene) to within about ± 0.5 m.y. In the case of the second example, much younger rocks were under discussion, and accuracy appears

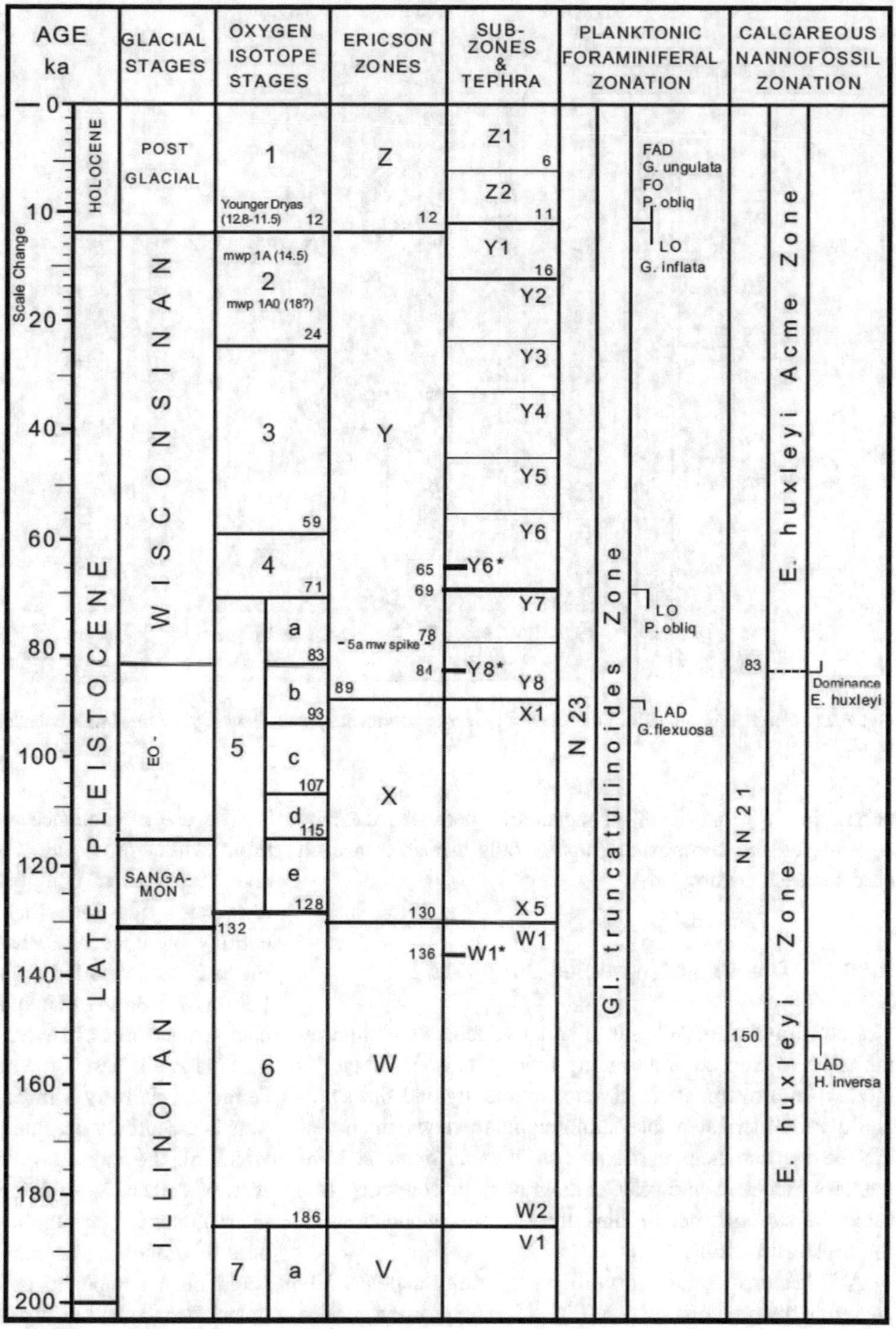

Fig. 8.20 Chronostratigraphic chart used for the Brazos-Trinity depositional system. Mwp = meltwater pulse. LO/FO = last/first occurrence locally observed in the area. FAD/LAD = first/last appearance datum. Ash layers are identified by an asterisk (Pirmez et al. 2012, Fig. 7, p. 119)

to be in the 10^5-year range, with at least an order of magnitude greater accuracy attainable for the last 40 ky of stratigraphic time through the use of radiocarbon dating. Each of these examples was approached by their authors as studies of typical stratigraphic problems, making use of all available chronostratigraphic data. This is different from the objective of perfecting the geological time scale, when field locations are selected specifically to maximize the chronostratigraphic information available, rather than to answer some sedimentological or tectonic question. This is an important distinction because, as noted earlier (see Fig. 8.16), accuracy and precision of the Geological Time Scale in the 10^5-year range is expected to be attainable for the Mesozoic when cyclostratigraphic studies are completed. Even if this proves to be correct, this does not mean that every Mesozoic section will then be capable of dating and

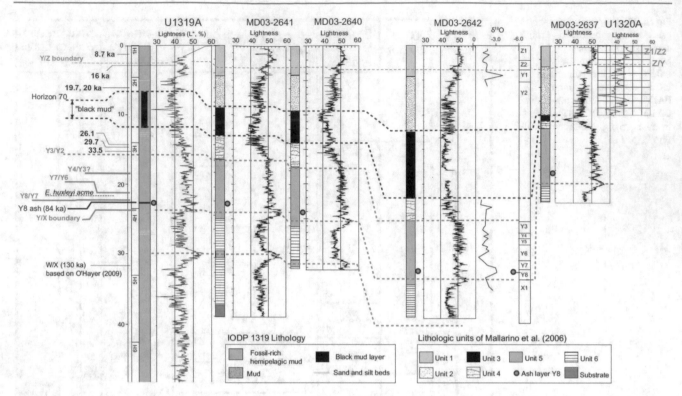

Fig. 8.21 Correlation of wells in the Brazos-Trinity depositional system, showing identified unit boundaries at left (Pirmez et al. 2012, Fig. 10, p. 122)

correlation at that level of accuracy, because accuracy depends on the chronostratigraphic tools available in each case for each section.

8.10.2   The Use of Wheeler Diagrams

The construction of Wheeler diagrams focuses attention on the issue of elapsed and missing time (Fig. 8.1). Given the modern improvements in chronostratigraphy and the availability of modern sequence-stratigraphic methods, the use of this device can help to throw considerable light on sedimentary processes and provide diagrammatic constructs that force the stratigrapher to think through the implications of their reconstructions.

A Wheeler diagram for the Brazos-Trinity deposits discussed in the previous section (Fig. 8.23) helps to emphasize the substantial variability in sedimentation rates that characterized this region. Most of the elapsed time between 140 and 25 ka is represented by hemipelagic or condensed intervals. Sand-rich intervals (seismic units 30–70) were deposited rapidly during a relatively very short interval (25–15 ka), corresponding to the Last Glacial Maximum, when sea-levels were low, and immediately before a meltwater pulse from the Mississippi system corresponding to the commencement of rapid post-glacial sea-level rise.

The use of sequence methods for mapping and correlation, while providing much more logical architectural reconstructions of complex sedimentary units (Sect. 6.2.3), may raise questions that can take interpretations to an entirely new level (Bhattacharya 2011). For example, Fig. 8.24 is a detailed sequence correlation of a delta in the Ferron Sandstone of Utah. At first sight the incised valley fill complexes (units at several levels colored red in the figure) that mark the base of sequences 1 and 2 would appear to define a relatively simple, almost layer-cake stratigraphy. This is essentially the interpretation shown in Fig. 8.25A, in which all the incised valleys are assigned to Sequence 2. However, there is no definitive proof that this is the correct interpretation. The regional erosion surface that marks the base of sequence 2 across the left-hand end of the section could be a composite of erosion surfaces cut at different times after the end of deposition of parasequence 12. This is the basis for the interpretation shown in Fig. 8.25B, in which it is suggested that the incised valley fills that comprise the first valley fill complex were formed at different times following the deposition of parasequence 9, each one correlated in time to a different shoreface tongue. The alternative interpretations may be followed through with questions about sedimentation rates and missing time, of the type raised in Sect. 8.9. What differences in fluvial style, if any, may be suspected from the different interpretations shown in

Fig. 8.22 Age-depth curve for two wells in the Brazos-Trinity depositional system. Numbers 10–79, at right, refer to seismic units (see Fig. 6.20) (Pirmez et al. 2012, Fig. 12, p. 125)

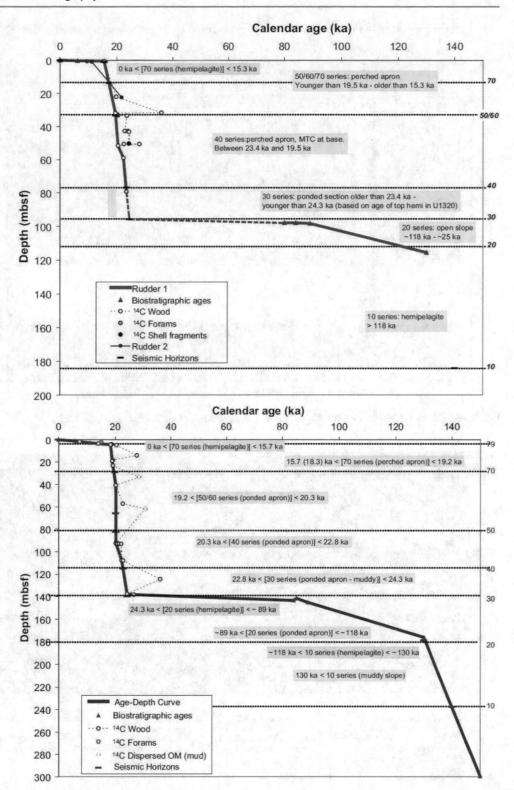

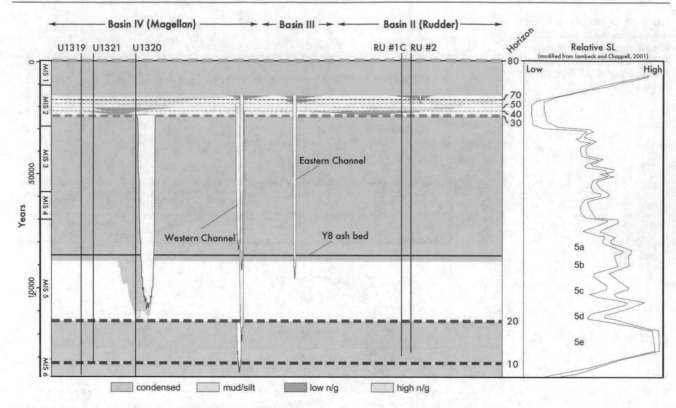

Fig. 8.23 Wheeler diagram for the Brazos-Trinity system (Pirmez et al. 2012, Fig. 14, p. 128)

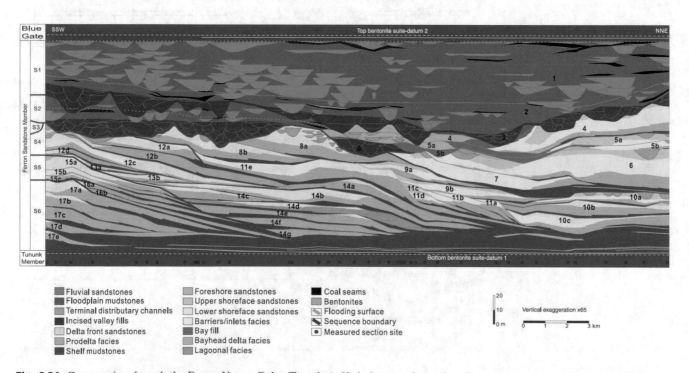

Fig. 8.24 Cross-section through the Ferron Notom Delta (Turonian), Utah, hung on bentonites. Parasequences are numbered 1 to 17 and are grouped into six sequences. Two alternative Wheeler diagrams for this cross-section are shown in Fig. 8.32 (Bhattacharya 2011, Fig. 11, p. 135)

Fig. 8.25 Two alternative Wheeler diagrams for the cross-section shown in Fig. 8.31. The differences between them focus primarily on the chronostratigraphic interpretation of the incised valley fills (Bhattacharya 2011, Fig. 14, p. 138)

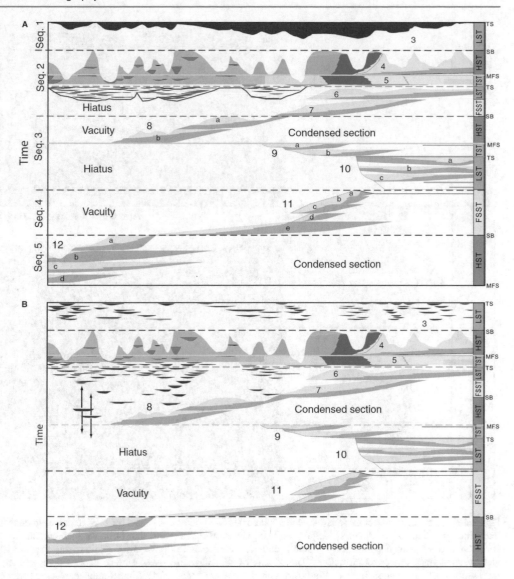

Fig. 8.25? Was there a single rapid episode of changing accommodation or sediment supply that gave rise to a single, amalgamated incised valley complex (Fig. 8.25A)? Or was this a more drawn-out process, reflecting longer term processes, or a more distant signal that became distorted ("shredded"?) down the transport direction, which is what could be implied by the interpretation indicated in Fig. 8.25B?

Developments in the acquisition and processing of 3-D seismic data have provided the possibility for the automated generation of Wheeler diagrams for subsurface sections, including four-dimensional visualizations that illustrate paleogeographic evolution in a temporal framework. Stark et al. (2013) described the evolution of a technique for extracting time information from seismic data called "computational seismic chronostratigraphy." Qayyum et al. (2015) discussed the method and provided worked examples. Other worked examples are available in the special

issue of "The Leading Edge" edited and introduced by Stark et al. (2013).

The procedure is as follows: automatic tracking software is used to generate multiple mapped horizons within the seismic volume. Seismic terminations, corresponding to such real structural features as unconformities or downlap surfaces, can be recognized and mapped by the software by the convergence of reflections to within a pre-set spacing value. A Wheeler diagram may then be constructed by flattening each surface to the horizontal. In this way thickness, the Z dimension (the y-axis of a 2-D plot), is converted to relative time. Structural and thickness information are lost in this process, and the resulting diagram emphasizes non-depositional and erosional hiatuses by the presence of blank space. The horizontal extent of each surface reflects either (1) its erosional extent, for example, the technique provides a clear time–space map of subaerial unconformities, or (2) the original depositional extent, such as that of a

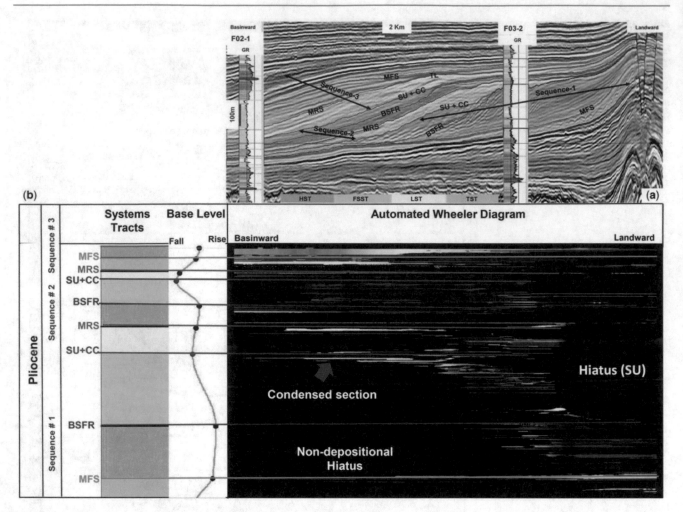

Fig. 8.26 a Seismic transect through two wells with overlain systems tract interpretation of a Pliocene deltaic unit in the North Sea. **b** Automated Wheeler diagram of the studied interval, with sequence and systems tract interpretation at left. The Y-axis of the diagram represents relative geological time. The color-coded lines are autotracked events, and the colors represent relative thickness within each clinoform unit, extending from its coastal origins at the systems tract. Note that Sequence 1 shows a low rate of sedimentation in the basinward direction. In seismic Wheeler diagrams, such condensed sections show up as hiatuses. TL, transgressive lag; BSFR, basal surface of forced regression; MFS, maximum flooding surface; MRS, maximum regressive surface; SU, subaerial unconformity; CC, correlative conformity (Qayyum et al. 2015, Fig. 5)

clinoform unit, extending from its coastal origins at the mouth of a delta, down the continental margin to its lapout at the foot of the continental slope (this is, of course, within the limit of seismic resolution. In fact, distal clinoform sets may extend for considerable distances into a deep basin as thin pelagic units). The vertical dimension of the plot depends on the number of autotracked horizons, which means that it varies in scale depending on the heterogeneity of the section and is also dependent on seismic resolution. The time scale to be derived from the vertical axis is therefore relative, and variable. Nonetheless, the resulting ability to visualize missing section, reflecting non-depositional or erosional hiatuses may provide a valuable addition to the tools for sequence interpretation and the development of paleogeographic models.

An example of the use of this technique is illustrated in Figs. 8.26 and 8.27. The gradual stepping out of the axis of deposition as the clinoforms prograde across the continental margin is graphically seen in the Wheeler plot, and the substantial age span of the hiatus at right representing a subaerial unconformity is particularly striking. This hiatus spans essentially the entire interval from one basal surface of forced regression to the next such surface above.

One problem with this method of developing Wheeler plots is the distortion of time that results from variable sedimentation rates and would also result from long-continued deposition of homogeneous sediment, resulting in reflection-free records. For example, in Fig. 8.26 note the two pale-colored reflections labeled "condensed section." The labeling arises from the seismic interpretation

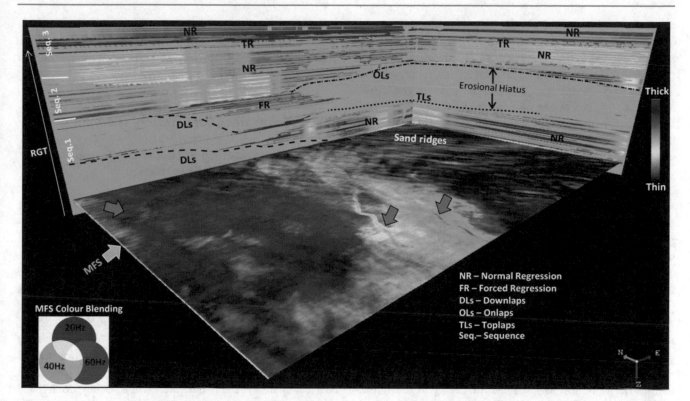

Fig. 8.27 4-D Wheeler diagram for the Pliocene interval shown in Fig. 8.26. The three-dimensional space (two horizontal dimensions plus relative time in the vertical dimension) is filled with information from the 4th dimension, systems tracts thickness in this case. The bottom slice is a color-blended spectral decomposition attribute slice for a particular reflection event (in this case, a MFS, maximum flooding surface). Along the surface, several geomorphological features are identifiable—NE- to SW-direction flowing deep-water channels (blue arrows) and NW–SE oriented elongated features that are interpreted as sand ridges comparable to those existing in the North Sea at the present day (Qayyum et al. 2015, Fig. 6)

of the section, which suggests a thin unit that blankets sequence 1. The term "condensed section" typically indicates condensation of time, such as is commonly seen in the thin units that form offshore during rapid transgression (Loutit et al. 1988). However, in Fig. 8.26 there is no suggestion of the condensation of time because time, here, is in fact a distorted representation of seismic heterogeneity. Compare this with the lengthy intervals of condensation shown in Fig. 8.23, which are based on careful chronostratigraphic calibration of the seismic and well records. The Wheeler plot shown here was drawn by the seismic interpretation algorithms, but it is possible to imagine a further step, in which information on sedimentation rates and chronostratigraphic tie points is used to adjust the vertical scale (y-axis) of the plot to more accurately represent elapsed time.

8.10.3 Improving Accuracy and Precision

Several developments are underway that promise to improve chronostratigraphic accuracy and precision, both for the purposes of local to regional correlation (for which, relative dating would suffice) and also for the purpose of improving the Geological Time Scale. Developments in cyclostratigraphy and astrochronology are discussed in Sect. 8.13. In this section we discuss briefly the work to maximize the information available from the biostratigraphic record.

The rapid increase in computer power, in addition to enormously increasing the utility of 3-D seismic data, is also providing the opportunity for many different approaches to the consolidation, sorting, selection and correlation of all the different types of chronostratigraphic information that are increasingly available from the rock record. The concept of the **composite standard reference section** (Sect.7.5.4) can be vastly expanded by adding such observational data as isotopic excursions, key seismic reflections, paleomagnetic reversals, chemostratigraphic determinations, dated volcanic ash beds and bentonites to the data on taxon ranges. Taxon ranges, as observed in any given section, may be incomplete, for reasons explained in Sect. 7.5, and where subsurface information is used, borehole caving may introduce inaccuracies into the record, which is why the use of the composite standard reference section method is so powerful, because it constitutes a method that includes the opportunity for continuous correction and refinement. There are now numerous algorithms available for sorting and standardizing the various types of data now being assembled in digital data bases. Sadler et al. (2014) provided a useful summary.

Figure 8.28A provides an example of the problems that may arise with the plotting of raw chronostratigraphic data points. Because of ecological selection, selective preservation or incomplete sampling, the indicated correlations of taxa between sample points may include crossing correlations. First-appearance datums may appear too high in some sections because these problems, whereas last appearance datums might be too low because of caving in the subsurface, or because of reworking. By contrast, unless borehole caving is a problem, such data points as dated ash beds can be regarded as fixed points that may be used as "nailed" datums (Sadler et al.'s 2014 term; see Fig. 8.29), around which bioevents may be adjusted. Corrections to local taxon ranges may involve expansion to fit the composite standard ("jacked" apart in Fig. 8.29) or reductions, where corrections from other data, including fixed or "nailed" datums are present ("clamped": Fig. 8.29). Figure 8.28C shows the result of a fully automated correlation exercise that has adjusted and corrected the raw chronostratigraphic data to provide a best-fit solution. It may be compared with Fig. 8.28B, which represents a lithostratigraphic correlation guided by older, conventional biostratigraphy.

An insight into the accuracy and precision that is now available for the chronostratigraphic calibration of the distant geologic past was provided by Sadler et al. (2009), who reported on the use of a global graptolite data base for the Ordovician–Silurian interval comprising 17,861 locally observed range-end events for 1983 taxa, and 131 local observations of 57 other events (23 dated events and 34 marker beds) in 446 sections worldwide. This paper describes the details of how data points are ordered, culled, corrected and calibrated to generate a time scale. The resolving power of the resulting scale is mostly in the 10^4- to 10^5-year time range (Figs. 8.30, 8.31), which is at least an order of magnitude greater precision than that obtainable by conventional biostratigraphically based methods for this time interval (Fig. 8.16). To repeat a caution made elsewhere in this book, the availability of such a precise scale does not automatically mean that any graptolite-bearing Ordovician–Silurian section may be dated with a comparable precision. The application of the scale, in practice, depends on the locally available suite of datable materials.

8.11   Accounting for Missing Time

8.11.1   Constructing Wheeler Diagrams for Selected Examples

Ager's (1986) provocative paper on the Sutton Stone, which we introduced in Sect. 8.2 (Fig. 8.5) served to alert geologists to the issue of rapid sedimentation and potential missing time in the sedimentary record nearly 40 years ago. An examination of an ordinary outcrop of shallow-marine sandstones, siltstones and mudstones, introduced here as Fig. 8.4, reveals the probability of multiple short breaks in sedimentation. We expand on these issues in this section.

Callomon's (1995) documentation of the remarkable ammonite biostratigraphy of the British Inferior Oolite (Figs. 7.41, 8.3) reveals the potential for detailed geological analysis to answer many questions while posing problematic new ones. We can now argue that the overall rate of sedimentation of the Inferior Oolite was extremely slow. Five meters of sedimentation in 5 m.y. is a sedimentation rate of 0.001 m/ky (10^{-3} m/ky), which is slower than the rates measured on most cratons (see Table 3.3; Fig. 8.8). However, further analysis reveals some potentially interesting contrasts. Examining the Burton Bradstock section in Fig. 8.3 we see 25 beds/biozones, comprising the 5 m of section; that is an average of 0.2 in thickness for each bed. Given the average 90,000 years duration for the biozones (56 biozones in a section deposited in 5 m.y.), this means that only 2.25 m.y. of the total 5 m.y. duration of the Inferior Oolite is represented by sediment at this location. As shown in Fig. 7.41, all the Inferior Oolite sections studied by Callomon show similar lengthy gaps and substantial "missing" time. It gets more interesting! Units of 90,000 duration are typically in the *SRS* 7 or 8 category (Fig. 8.8; Table 3.3), with sedimentation rates of 10^{-1} to 10^1 m/ky. At these rates, the 5 m of beds could have been deposited at between 0.5 and 50 ky, as little as 0.01 to 1% of the total elapsed time of about 5 m.y. It seems likely that the carbonate shelf environment where the Inferior Oolite was deposited was subjected to variations in sedimentary controls over periods of 10^5 years that left parts of the shelf in a condition of stasis for tens to hundreds of thousands of years. However, when sedimentation did occur, it was rapid. It is possible that the 25 beds preserved at Burton Bradstock represent only a fraction of the sediment that was transported into the area, with most of it deposited temporarily, only to be moved on out of the area before permanent burial.

Fig. 8.28 Correlation of
foraminifera, nannofossils,
dinoflagellates, spores and pollen
from 8 wells in New Zealand's
Taranaki Basin. **A** Raw
correlation of range tops of the 87
most reliable of 351 taxa. Range
bases were potentially
compromised by borehole caving.
B Lithostratigraphy and
traditional biostratigraphic
correlation. **C** Outcome of
automated correlation to a fully
resolved, composite, scaled
time-line by minimal adjustment
of observed ranges and insertion
of missing taxa (Sadler et al.
2014, Fig. 2, p. 6)

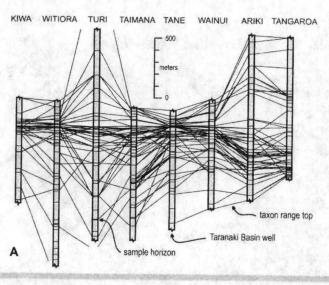

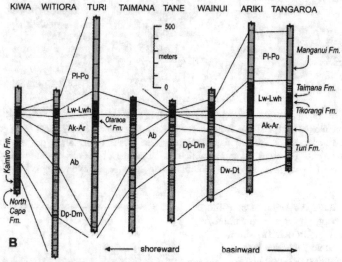

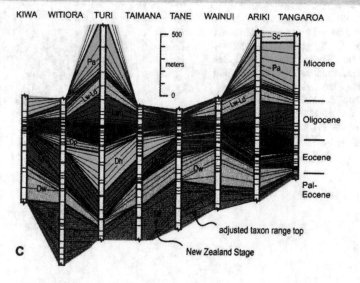

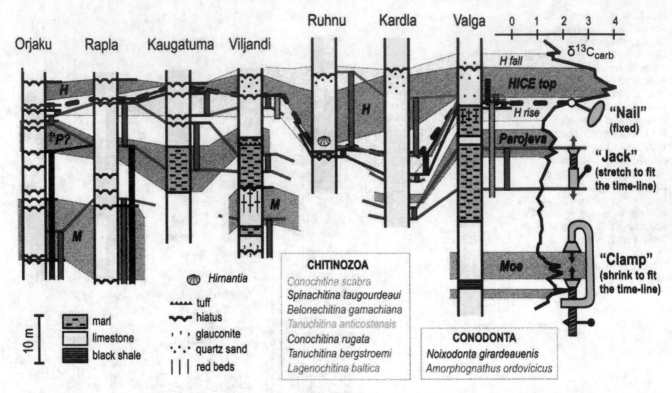

Fig. 8.29 A sample of end-Ordovician (Hirnantian) information: 9 taxon ranges and 3 stable isotope excursions constrained by 92 local event observations from 7 Estonian cores (see Sadler et al. 2014 for references). To align all cores with one sequence of events, some observed event pairs may be jacked farther apart (e.g., taxon range ends); others may be clamped closer together (e.g., uncertainty intervals on parts of stable isotope excursions); and some must be left in-place—the nailed horizons of steepest onset (thick dashed line) of the Hirnantian Isotopic Carbon Excursion (HICE) (Sadler et al. 2014, Fig. 3, p. 8)

Fig. 8.30 Variation of potential resolving power as a function of age in an Ordovician–Silurian time scale generated by the computer-based optimization process described by Sadler et al. (2009). Differences in interpolated age were determined for all 3904 pairs of adjacent events. These highly variable data are summarized by 10-point (thin gray line) and 200-point moving averages (thick gray line), plotted at the center point of the moving window (Sadler et al. 2009, Fig. 8, p. 900)

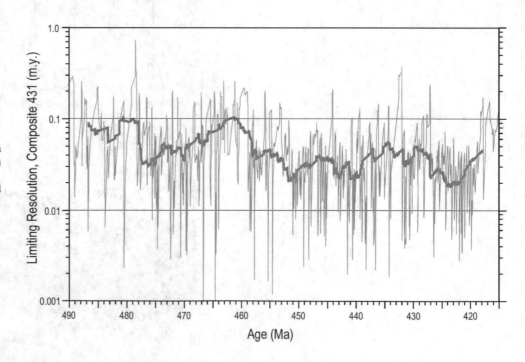

Fig. 8.31 An Ordovician–Silurian time scale developed by the automated processing of a chronostratigraphic data base consisting primarily of global graptolite occurrences. Gray intervals in the "uncertainty" column show the quantified potential error associated with each zone boundary (Sadler et al. 2009, Fig. 9, p. 901)

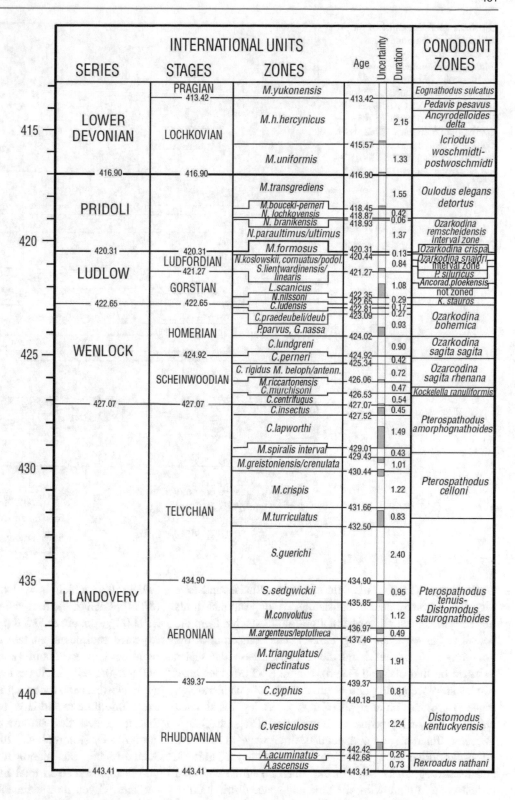

Davies et al. (2019) described a comparable example, the Pliocene–Pleistocene Red Crag Formation of Eastern England, where bedding surfaces provide evidence of deposition, erosion and bioturbation, including multiple episodes of all three processes interacting over very short intervals.

They suggested that many such surfaces would have existed in the condition of stasis "for the majority of the time they were active" (Davies et al. 2019, p. 1156).

Turning to a clastic example, Fig. 8.32 is adapted from the work of Holbrook (1996, 2001) and Holbrook and

Fig. 8.32 A 3-km panel of the fluvial Mesa Rica Sandstone (Middle Cretaceous), southeastern Colorado, showing the hierarchy of bounding surfaces from sequence boundary to channel belt. **B, C, D** Wheeler diagrams at three levels of the stratigraphic hierarchy. Time scales are arbitrary. Based on the work of Holbrook (1996, 2001) and Holbrook and Bhattacharya (2012)

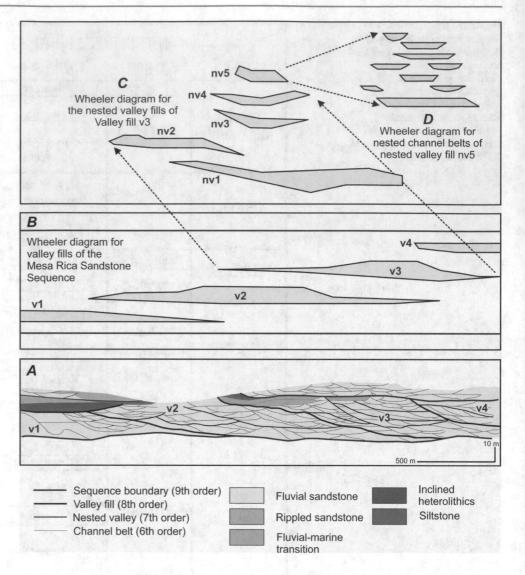

Bhattacharya (2012). This field study of a fluvial sandstone revealed an architecture consisting of depositional units nested within each other over a nine-fold hierarchy, from the bedset to the sequence. (Fig. 8.32A). The numbering of the bounding surfaces that define the units over these nine scales is based on that of Miall (1996). Figure 8.32 provides a set of nested Wheeler diagram at three levels of this hierarchy. Panel B provides an analysis of the four valley fills situated between the major sequence boundaries; Diagram C is a Wheeler diagram at an expanded time scale of the five 7th-order nested valley fills that constitute valley fill unit #3, and inset diagram D is a Wheeler diagram of the youngest nested valley fill showing the time and space distribution of channel belts (bounding-surface order 6). The time scales for all these diagrams are arbitrary. However, the Mesa Rica Sandstone is 38 m thick and represents approximately 1 m. y., and so some order of magnitude estimates of the time represented at the various orders of bounding surface may be attempted. The overall sedimentation rate for the sequence is

0.038 m/ky, which is typical of major fluvial complexes (*SRS 9*). Each of the four valley fills represents an average of 250,000 years, but at *SRS 8* (typical rate for valley fills) at an expected sedimentation rate of 10^{-1} m/ky they could each represent between about 42 and 380 ky (sedimentation rate of 0.9 to 0.1 m/ky). The valley fills offlapped each other, and at the slowest rate, reflecting autogenic shifts in depositional axes, little time would have been represented by the erosion that generated the 8th-order bounding surfaces (as diagrammed by Holbrook and Bhattacharya 2012, their Fig. 9). At the higher rate, significantly more time (about 800,000 yrs) is represented in total by the four 8th order bounding surfaces. Given that sediment supply rates may have been relatively constant through the deposition of the sequence, it is likely that this "missing" time is represented by other valley fills out of the line of section, and that each valley fill represents significantly less than 250,000 years.

Assuming each of the valley fills represents between 42 and 380 ky, the two to five nested valley fills in each of the valley fills (Fig. 8.32C shows the five units in v3) each represent between 8400 and 190,000 years, which is within the range of *SRS 7*. Similar speculations can be carried out for the channel belts (Fig. 8.32D), which assign them to *SRS 6* or *SRS 7*, typical of large macroforms, channels, and (at the slower end of the rate range), channel belts.

Another example, which illustrates the importance of stasis, is an analysis of the modern Canterbury Plains (Fig. 8.33). Bhattacharya et al. (2019) constructed arguments based on the avulsive behaviors of modern rivers, together with the known recent stratigraphic history of the Plains to argue that, at any given location, the surface was undergoing stasis for 95% of elapsed time during the last 60 ky. Active sedimentation in one of the major braided rivers crossing the plains was occurring at any location for only 5% of the time, as they switched across the plains by avulsion. Portions of the alluvial plain may have been left in the condition of stasis for as long as 20,000 years. The Wheeler diagram in Fig. 8.33 represents a speculative model constructed for one of the rivers. Along the coast, where coastal cliffs reveal a strike-parallel outcrop transect of the plains (Leckie 1994; Browne and Naish 2003), the deposits consist

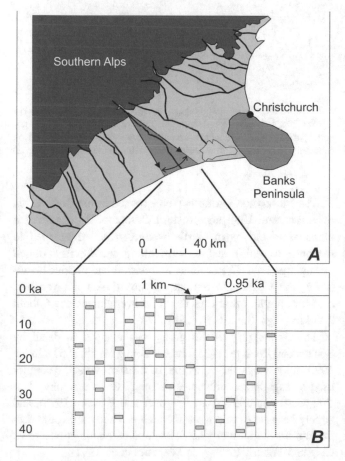

Fig. 8.33 Speculative Wheeler diagram for the glacial to post-glacial Canterbury Plains, New Zealand (from Bhattacharya et al. 2019)

of a complex of fluvial gravels and sands with numerous major and minor scour surfaces generated by bedform and channel erosion. Outcrop characteristics provide little or no evidence of the lengthy periods of erosion or stasis that must be represented by many of these surfaces.

The reader is also referred to two studies of the preservation of eolian dune systems, which includes speculative Wheeler diagrams, by Hassan et al. (2018) and Kocurek and Day (2018).

8.11.2 A Well-Documented Quaternary Example

A study of the Quaternary fluvial plain and delta of the Po River, Italy, by Amorosi et al. (2017) supported by numerous ^{14}C dates of drill core samples provided an excellent illustration of the wide range of sedimentation rates and the extent of "missing" time in a well-preserved coastal succession (Fig. 8.34). The overall sedimentation rate of the approximately 30-m thick section of delta front and prodelta sediments was 3 m/ky (10^0 m/ky; *SRS 7*). This compares with the average rate of post-glacial sea-level rise of 11 m/ky between 16 and 8 ka, after which sea-level rise slowed (from IPCC Assessment #4, WG-1 Report, Fig. 6.8). As seen in Fig. 8.34, the initial post-glacial transgression took about 3 ky, until 8 ka ago, after which highstand sedimentation changed from aggradation to markedly progradational, indicating that sediment supply outpaced accommodation generation.

The interval of condensed sedimentation was an essentially non-depositional hiatus lasting most of the last 9 ky, with 90% of the elapsed time unrepresented by strata. Dating of the preserved condensed deposits in drill core indicates a sedimentation rate of 0.6 m/ky (10^{-1} m/ky; *SRS 8*).

High sedimentation rates were recorded in the delta-plain fluvial and delta front deposits, at 8–10 m/ky (10^0 m/ky; *SRS 7*). Higher rates were recorded from fine-grained delta lobes, where mud accumulation rates reached 40 m/ky (10^1 m/ky: *SRS 7*). Beyond the landward extent of marine influence, vertically stacked shoaling-upward successions of fresh- to brackish-water swamp deposits accumulated at typical rates of 2.3 to 3 m/ky (10^0 m/ky; *SRS 7*), with one interval indicating that 6 m of beds accumulated in 5 years (1200 m/ky; 10^3 m/ky; *SRS 5*). Peat layers exhibited lower rates, averaging 0.7 m/ky (10^{-1} m/ky; *SRS 8*).

In conclusion, typical sedimentation rates for individual successions (parasequences) in the Po system varied over three orders of magnitude, from 10^{-1} to 10^1 m/ky, with one exceptionally high rate probably representing an unusual flood event. These rates are in accordance with the *SRS* tabulation of Miall (2015), as summarized in Table 3.3. The accumulation of channel systems and delta lobes typically occurs at *SRS 7* rates. Individual units accumulated at rates typically more than double the overall rate for the entire system, at 3 m/ky.

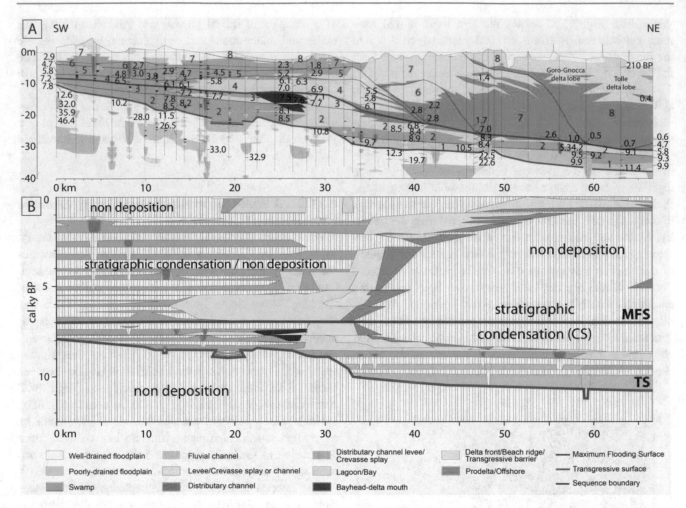

Fig. 8.34 Regional cross-section of the late Quaternary Po Plain system, Italy, and accompanying Wheeler diagram representing the hierarchy of stratigraphic hiatuses identified at systems tract and parasequence time scales, based upon 83 radiocarbon dates. The transgressive surface and the condensed section (CS) encompassing the MFS record major stratigraphic gaps, though the CS does not represent a true hiatal break. Short-duration depositional hiatuses, mostly corresponding to parasequence boundaries, pervade the succession on centennial to millennial time scales. From Amorosi et al. (2017, Fig. 2)

8.11.3 The Example of the Mesaverde Group, Utah

Compared to well-documented and well-dated recent deposits, such as the Po Delta system, examined here, and the Rhine-Meuse system, the sedimentological analysis of pre-Holocene deposits is made much more difficult by the typically sparse and imprecise chronostratigraphic control. Miall (2014b) made an attempt at an analysis of sedimentation rates and "missing time" in the Upper Cretaceous Mesaverde group of the Book Cliffs, Utah, and much of that study was included in the first edition of this book. The detailed mapping of this unit by Pattison (2018, 2019, 2020a, b) has required a re-evaluation of some of the conclusions of that study, and a revised and more concise version of the analysis is included here.

The Mesaverde Group is well exposed in the Book Cliffs of east-central Utah and western Colorado. It constitutes a classic foreland basin clastic wedge, which developed in response to uplift and erosion along the Sevier Orogen during the Late Cretaceous. A resistant nonmarine sandstone, the Castlegate Sandstone, constitutes a prominent cliff-forming unit that caps the cliffs over a distance of about 200 km (Figs. 8.35, 8.36 and 8.37).

The stratigraphy of the group was mapped in detail by Spieker and Reeside (1925) and by Young (1955, 1957). In the 1990s these rocks served as one of a set of worked examples used to explain the principles of sequence stratigraphy (Van Wagoner et al. 1990; Van Wagoner and Bertram 1995), and largely because of this work the area has been a favorite destination for field trips by academic and corporate groups (e.g., Cole and Friberg 1989; Van Wagoner et al. 1991).

Fig. 8.35 Location of the Book Cliffs study area, showing the locations of detailed sections and other illustrations

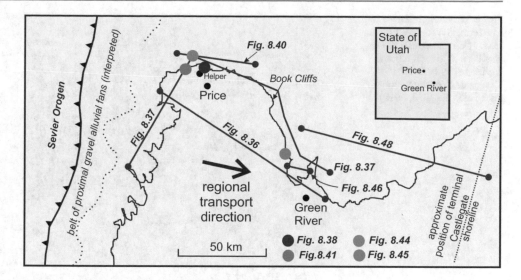

Fig. 8.36 Regional cross-section of the Mesaverde Group. Adapted from Seymour and Fielding (2013), after the detailed mapping of Young (1955)

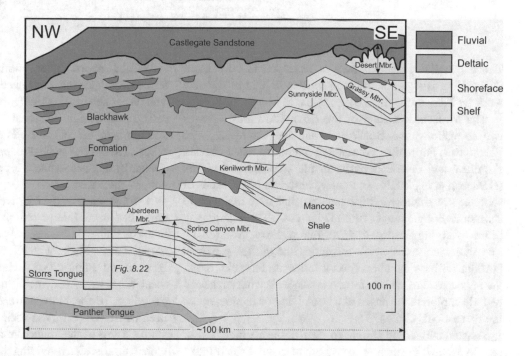

In more recent years the Mesaverde Group has become the focus for examinations of many concepts about stratigraphy and sedimentation. A detailed treatment, in which the author explored many quantitative models concerning sedimentation in an ancient foreland basin, was provided by Hampson (2010). More recently, the extremely detailed ground mapping of the Mesaverde Group by Pattison (2018, 2019, 2020a, b), particularly its distal portion to the north and east of Green River Utah, has demonstrated the inadequacy of earlier sequence models and indicated the need for a wholesale re-evaluation of the evolution of this group. However, the issue of fragmentary preservation of the record has not been discussed. This is not an inconsequential

issue because, as the work of Hampson and Pattison exemplifies, the stratigraphic and sedimentologic detail that are now available for the Mesaverde Group are amongst the most extensive available for clastic wedges of this type, and the unit should, therefore, provide an ideal test bed for the exploration of advanced sedimentological concepts. Amongst the few workers who have addressed the issue of missing time in the Mesaverde Group, and speculated about its implications for correlations and sequence stratigraphy, are Howell and Flint (2003), who reviewed the chronostratigraphy of the succession and the presence of gaps at the parasequence scale, and Bhattacharya (2011), who discussed alternative sequence models for the Castlegate Sandstone.

Fig. 8.37 Stratigraphy of the Castlegate Sandstone. Adapted from Miall and Arush (2001a). Based on Willis (2000) and Yoshida (2000). Base of Castlegate Sandstone modified after Pattison (2018, 2019)

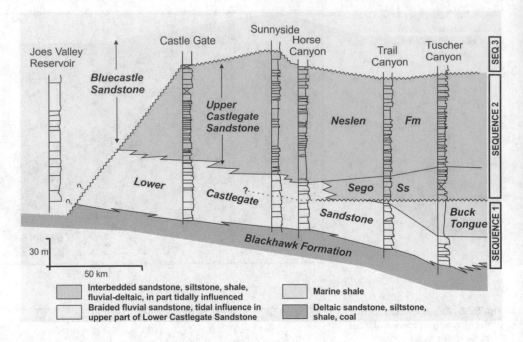

The discussion presented here is based on that by Miall (2014b) but is significantly revised here in light of the new work by Pattison (2018, 2019; 2020a, b).

The Mesaverde Group of the Book Cliffs encompasses two formations, the Blackhawk and the Castlegate. The Blackhawk Formation consists of undifferentiated fluvial, shoreface and deltaic deposits in the western Book Cliffs (Hampson et al., 2013), and passes eastward into a series of shoreface to shallow-marine tongues that were assigned to six members by Young (1955) (Fig. 8.36). The stratigraphy of the overlying Castlegate Sandstone is shown in Fig. 8.37. The shoreface sandstones constituting the thicker, proximal portions of these members form a series of cliffs (Fig. 8.38), the appearance of which, from a distance, is what suggested to early explorers the pages of a book lying on its side, hence the name Book Cliffs.

As summarized by Hampson (2010), the six members of the Blackhawk Formation have now been subdivided into a total of 23 submembers. The first sequence-stratigraphic studies of these rocks interpreted them as the product of eustatic sea-level change (Van Wagoner et al., 1990), but these units have subsequently been much discussed as examples of a type of high-frequency sequence stratigraphy controlled by regional tectonism and its effect on accommodation (e.g., Krystinik and de Jarnett 1995; Hampson 2010; Aschoff and Steel 2011). Other interpretations are discussed below.

The Castlegate Sandstone, named after the Castle Gate, a prominent landmark in the Price River Canyon north of Helper, was interpreted as a third-order sequence by Olsen et al. (1995), but was subdivided into two sequences by Miall and Arush (2001a; see Fig. 8.37). Bhattacharya (2011) speculated about other stratigraphic models. All these sequence studies have followed the original interpretation of Van Wagoner et al. (1990) that the Castlegate Sandstone rests on a major erosional unconformity extending for 200 km from the Wasatch Plateau eastward to the Colorado border. However, detailed mapping by Pattison (2018, 2019, 2020a, b) has necessitated a complete re-evaluation of this unit, as discussed later.

The age range of the Mesaverde Group and its constituent units has been interpreted primarily from the ammonite fauna contained in the marine portions of the Blackhawk and in the Mancos Shale, with which it is interbedded to the east. Howell and Flint (2003), Hampson (2010), Aschoff and Steel (2011), and Seymour and Fielding (2013) provided overviews of the biostratigraphic and other work that has been carried out on these rocks since mapping began in the 1920s. Three of the ammonite zones can be correlated to the global time scale, providing numerical-age tie points. The following data are taken from Aschoff and Steel (2011).

The base of the Blackhawk Formation is Lower Campanian in age and is dated at 81.86 Ma. The top is placed at 79 Ma, and the top of the Castlegate is Middle Campanian, at 77 Ma. The time span of the Mesaverde Group is therefore 4.86 m.y. Near Price the section is up to 700 m thick, which indicates an average sedimentation rate of 0.14 m/ky. This rate (10^{-1} m/ky) is at the upper end (more rapid) of rates characteristic of long-term geological subsidence measured over periods of 10^5–10^7 yr, including thermal subsidence and flexural loading of the crust (*SRS 10–11*), and is comparable to rates associated with low-frequency orbital cycles (*SRS 9*) (Miall 2013). The Blackhawk Formation is about 500 m thick. Each of the members has now

Fig. 8.38 View from Helper of a portion of the Mesaverde Group

been subdivided into submembers, totalling 23 for the Blackhawk Formation (the individual shoreface tongues in Fig. 8.36), averaging 124 ky in duration. The overall sedimentation rate for the formation is 0.17 m/ky (500 m/2.86 m.y.). That is 10^{-1} m/ky (*SRS 8, 9, 10,* or *11*).

A comparison of Book Cliffs stratigraphy with that of the Henry Mountains area, some 200 km to the south of Prince (Seymour and Fielding, 2013) indicates that there is no detailed correlation between the Blackhawk members and comparable units in that section. This, and the lack of spatial regularity of the Blackhawk members, argues against orbital cyclicity as a major mechanism for the generation of the members and submembers of the Blackhawk Formation.

Units of 124 ky are *SRS 8* or *9*. An average mid-range sedimentation rate for units in this *SRS* class is 10^{0} m/ky (1 m/ky). At this rate all the submembers could have been deposited in 500 ka (17% of elapsed time of 2.86 m.y.).

A speculative chronostratigraphic chart for the Blackhawk Formation that includes the suggestion of significant gaps at the sequence boundaries is presented here as Fig. 8.39. This follows the approximate timing of the sequence boundaries shown by Hampson (2010, his Fig. 7), but otherwise differs substantially from that chart in the following principal ways: (1) in the updip coastal-plain region represented by the Blackhawk Formation, most of the elapsed time is completely unrepresented. Some of this unit consists of fluvial deposits formed above marine base level (Hampson et al. 2013) and here, accommodation can be explained by the buffer concept of Holbrook et al. (2006). These deposits are shown schematically in Fig. 8.39. (2) The arrangement of sequence boundaries in Hampson's (2010, Fig. 7) suggests that the completeness of the Blackhawk and the contemporary shallow-marine record (that of the Blackhawk members) decreases basinward, whereas the opposite is more likely to be the case. Sedimentation of the Mancos Shale toward the basin center is likely to be much more continuous and therefore more complete at the 10^4–10^6 time scale than the proximal sediments of the coastal region, but may contain substantial disconformities, as shown in Fig. 8.39. The pattern shown here in Fig. 8.39 is more like that developed by Krystinik and DeJarnett (1995, Fig. 3), in which it was suggested that the proximal region to the west of the basin was a region of uplift and erosion. Units of condensed (slow) sedimentation may be expected to develop offshore at times of regional transgression, and there is also evidence that sediment gravity flows occurred on distal clinoform slopes during the falling stage of some of the sequence cycles (Pattison, 2005; as shown by Hampson (2010), his Fig. 7, but not included in Fig. 8.39 of this book). (3) Following from the previous point, the time span

Fig. 8.39 An interpretation of the chronostratigraphy of the Blackhawk Formation. Age information is from Hampson (2010) and Seymour and Fielding (2013)

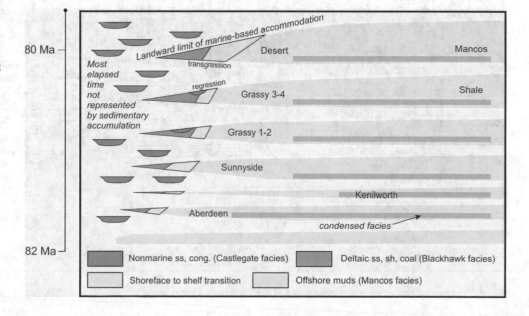

represented by the member (sequence) boundaries increases landward, and the undifferentiated proximal, deltaic Blackhawk formation represents, in total, much less than one half of the available elapsed time, with substantial erosional breaks embedded in this coastal succession.

The next step in this analysis is to examine sedimentation rates and preservation at shorter time scales. *SRS 7* is the time scale for long-term geomorphic processes, those occurring over time periods of 10^3–10^4 years, such as the members and submembers of the Blackhawk Formation. Such processes include the development of major delta lobes and alluvial channel belts, regressive shoreface complexes, major coal seams, etc. Sedimentation rates for many post-glacial successions in the *SRS 7* category were provided in Sect. 8.4 and are typically in the range of 10^0–10^1 m/ky, which is up to an order of magnitude greater than the *SRS 8* range, which applies to high-frequency orbital cycles, and up to two orders of magnitude greater than long-term geological rates (*SRS 11*).

To study sedimentation at this scale, a more detailed examination is presented here of the Spring Canyon section, located west of Helper (Figs. 8.40, 8.41; locations shown in Fig. 8.35). The stratigraphic framework of this section is shown in Fig. 8.40, and the section is illustrated in Fig. 8.41. The section is shown in Fig. 8.41 in the traditional form, as a continuous succession of facies units, just as it appears in actuality in the field. However, as first pointed out by Barrell

(1917), this form of plot obscures the numerous cryptic hiatuses that permeate all stratigraphic sections.

In Fig. 8.42 units 1–15 of the same section are plotted against an *SRS 8* time scale. At the mid-range *SRS 8* sedimentation rate of 0.29 m/ky used by Miall (2014b), the 79.2 m of Spring Canyon section (units 2–15) shown in this figure would represent 273 ky of elapsed time. The same fifteen units are shown at the right re-plotted using *SRS 7* rates. As suggested in Fig. 8.23, and following all previous interpretations, the section can be interpreted in terms of four progradational successions, or parasequences, averaging 19.8 m in thickness. Using the same lithofacies unit numbers as in the original section, these are displayed at the right in this figure. Three regressive shoreface successions, the last capped by a coal swamp (unit 13 is the Hiawatha coal), are followed by a progradational delta. At an *SRS 7* sedimentation rate range of 1.5–6 m/ky, this 79.2 m of section would have accumulated over a time span of between 52.8 ky and 13.2 ky (13.2 to 3.3 ky for each of the four component cycles). A mid-range of 27 ky used here (arbitrarily) as an average for illustration is one tenth the time assumed for the sequences timed at the longer term *SRS 8* rate in the discussion above (and comparable to the 7% of elapsed time represented by sediment that was calculated by de Natris (2012). How is preservation and non-preservation distributed through the estimated 273 ky represented by section (Fig. 8.43)?

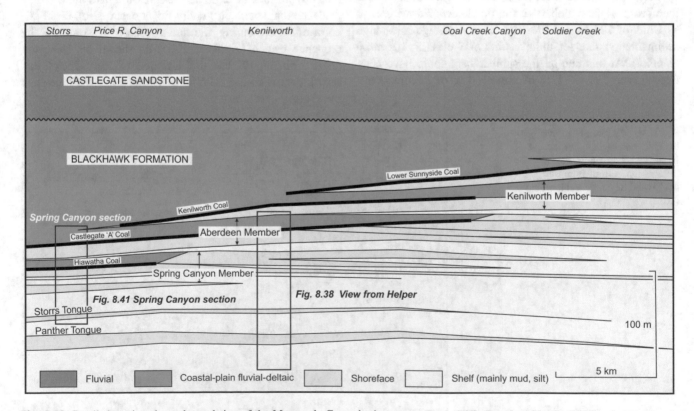

Fig. 8.40 Detailed stratigraphy and correlation of the Mesaverde Group in the western Book Cliffs. Based on Young (1955)

Fig. 8.41 The section at Spring Canyon. Based on Cole and Friberg (1989) and Kamola and Van Wagoner (1995). The section is subdivided into 22 lithofacies units, as numbered at left

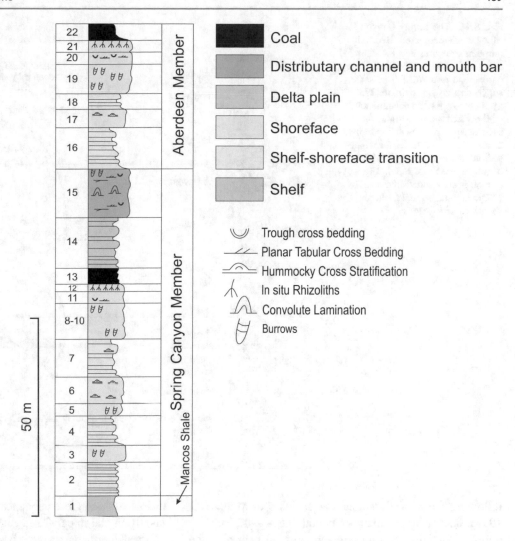

Figure 8.42 provides an illustration of the problem that sedimentologists have yet to address fully, that of the lengthy periods of empty time, time for which there is no sedimentary record within apparently continuous successions. For example, Hampson's (2010) detailed examination of the clastic wedge includes an analysis of the trajectory of shorelines in time and space through the evolution of the Mesaverde Group. The outcome of the analysis (Hampson 2010, his Fig. 14) is displayed in the form of a height-versus-distance diagram that provides quantitative estimates of forestepping (progradation) and backstepping (flooding) and is interpreted in terms of changing bathymetry and sediment supply in the basin as the succession gradually prograded eastward across the foreland basin. The analysis is keyed to the evolution of the Blackhawk and Castlegate members, but timing is not discussed beyond the labeling of each eastward movement in the shoreline with the name of the appropriate prograding submember. However, the plotting of the trajectory as a continuous zig-zag line implies continuity and does not include the possibility that the shoreline may have regressed entirely across the basin

(leading to lengthy periods of exposure and erosion), even though this possibility is hinted at by the way his chronostratigraphic diagram is constructed (Hampson 2010, his Fig. 7), nor does the trajectory diagram provide any indication of the point argued in this section, that there were lengthy periods of time for which we have no preserved record from which to construct such a diagram. This problem is addressed further in the next section.

The Castlegate Sandstone has been a subject of interest for many years. It is one of the best and most well-exposed (and accessible) examples of a regional sheet-braided deposit in the North American geological record. The first sequence-stratigraphic analysis of the Castlegate Sandstone was carried out by Van Wagoner et al. (1990), who introduced the concept of the basal sequence boundary, as a subaerial unconformity extending for 200 km along the Book Cliffs. A later study by Olsen et al. (1995) divided the formation near the type section (Castle Gate, at the mouth of the Price River Canyon, north of Helper; Fig. 8.44; location shown in Fig. 8.44) into a lower, sandstone-dominated member, deposited in a braided-stream environment,

Fig. 8.42 The Spring Canyon Member dismembered. Bed unit numbers are as in Fig. 8.41. At left, the section is plotted to correspond to a *SRS 8* time scale, suggesting an approximate 273 ky timespan for the accumulation of the succession. At right, the same section is evaluated in terms of an *SRS 7* time scale, for which sedimentation rates are an order of magnitude more rapid. The section is subdivided into a set of progradational coastal-plain and shoreface successions

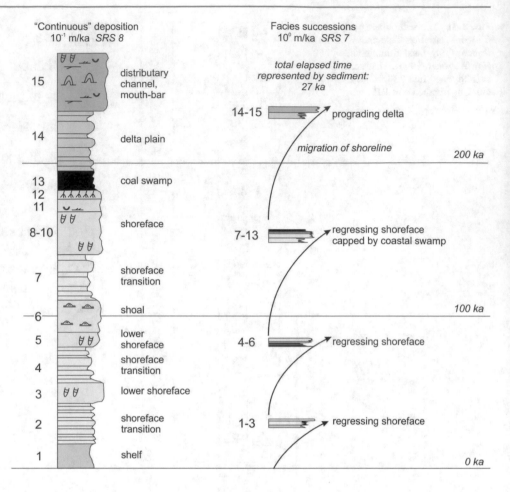

following the fluvial architectural analyses of Miall (1993, 1994), and an upper member containing a significant proportion of interbedded mudstones, units of inclined hetero-lithic strata (terminology of Thomas et al. 1987), and evidence of marine, tidal influence in the form of flaser bedding and *Skolithos* burrows. They interpreted the formation thus subdivided as a "third-order" sequence. To explain the subdivision into the two members, Olsen et al. (1995) turned to the fluvial models of Wright and Marriott (1993) and Shanley and McCabe (1994). The sandstone-rich lower member and the more heterogeneous upper member were interpreted, respectively, as low- and high-accommodation systems tracts. However, Miall (2014a, Sect. 6.2) has argued that such interpretations are untenable, given the issue of dramatically different sedimentation rates for the models and for the Castlegate Sandstone.

In a further analysis (McLaurin and Steel 2000) subdivided the upper member into five higher order (fourth-order) sequences and mapped a transition within these sequences between the fluvial deposits in the west into barrier, deltaic and estuarine deposits near Green River, and ultimately into the offshore mudstones of the Buck Tongue. However, Willis (2000), who recognized a high-frequency sequence stratigraphy in the Sego Sandstone east of Green River, was

unable to trace these sequences westward into the predominantly fluvial upper Castlegate Sandstone.

In an alternative analysis based on detailed mapping north of Green River, Yoshida et al. (1996), Willis (2000), and Yoshida (2000) argued that the Buck Tongue is truncated by the upper member of the Castlegate, at an angular unconformity that cuts gradually down section along the Book Cliffs to the northwest. According to this interpretation (shown in Fig. 8.37) the beds overlying the unconformity above the Buck Tongue in the east (Sego Sandstone) are stratigraphically equivalent to the upper part of the lower Castlegate Sandstone at the type section. As noted by Miall and Arush (2001a), based on this interpretation, the truncation of the Buck Tongue implies that updip from the pinch-out of this unit, approximately 1 m.y. of section is missing in proximal parts of the Book Cliffs, including at the type section of the Castlegate Sandstone.

The evidence to enable a choice to be made for any of these interpretations depends on the ability to trace ("walk out") key surfaces between sections. Even in the case of the Book Cliffs, where exposure is much better than average, it is not possible reliably to trace key surfaces based on facies and outcrop characteristics for long distances within what is a very heterogeneous succession. Accordingly, Miall and

Fig. 8.43 Outcrops in Spring Canyon (Fig. 8.41): **a** The top of a distributary-channel and mouth-bar complex that forms the top of the Spring Canyon Member. Unit 15 is 12 m thick. **b** The top of the Aberdeen member, consisting of a shoreface to foreshore complex. Person for scale. **c** The Storrs Tongue, a shoreface succession 9 m thick exposed at the base of the Spring Canyon section. It is capped by a ravinement surface. **d** A shoreface-transition to lower shoreface succession that forms the base of the Aberdeen Member. Person for scale. Unit numbers for a, b and d are as in Fig. 8.41

Arush (2001a) sought to develop other means to analyze the stratigraphy and determined, on the basis of petrographic evidence, that the best evidence for missing time at the type section consists of changes in detrital composition and evidence for early cementation at surface "D" in the type section (Fig. 8.44). According to this interpretation, the lower Castlegate at this location comprises parts of two sequences (sequences 1 and 2 in Fig. 8.37), and the upper part of this unit at the type section passes laterally (downdip) through a facies transition into the more heterogeneous beds of the upper Castlegate and the Sego Sandstone to the east and southeast (Willis 2000). This correlation of the Buck Tongue truncation surface into the Castlegate type section is now regarded as problematic.

Yet another interpretation of Castlegate sequence stratigraphy was offered by Bhattacharya (2011, his Fig. 17), in which he speculated about the relationship between the lower Castlegate Sandstone and the underlying Desert Member in the area east of Green River. The original interpretation of Van Wagoner et al. (1990) and Van

Wagoner (1995) was that the Desert member is entirely older than the Castlegate, and that they are separated by a major regional unconformity (as shown in Figs. 8.36 and 8.40). However, Bhattacharya (2011), referring to a discussion by Van Wagoner (1995) about the whereabouts of the coastal marine equivalents of the Castlegate fluvial sandstones (i.e., where are the mouths of the Castlegate rivers?), suggested that the Castlegate may in fact comprise a suite of high-frequency sequences, each with its own attached "Desert" shoreface (Bhattacharya 2011, his Fig. 18). This has now been confirmed by the detailed mapping of Pattison (2018, 2019, 2020a, b). He made two significant discoveries: (A) the base of the braided fluvial sand sheets comprising the lowermost Castlegate Sandstone is not at a constant stratigraphic level, but can be observed in cliff sections to migrate up and down by as much as several channel stories within a few hundred yards, (e.g., Fig. 8.45), indicating that the contact is not a regional unconformity but consists of a series of basal fluvial channel scour surfaces and represents a rapid but conformable environmental transition from the coastal

Fig. 8.44 The type section of the Castlegate Sandstone, with key bounding surfaces. Surface D of Miall and Arush (2001a) was tentatively identified as a major intraformational unconformity and sequence boundary. Width of field of view is about 250 m

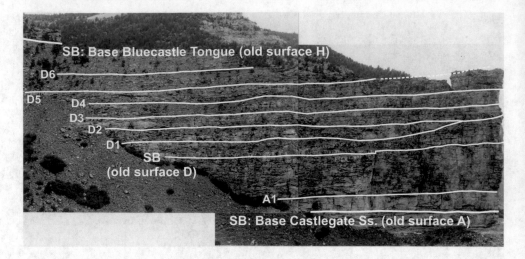

Fig. 8.45 The base of the Lower Castlegate Sandstone north of Price River Canyon, showing multiple options for the base of the Sandstone. Photo courtesy of S. Pattison

deltaic facies assemblages of the underlying Desert Member both vertically and laterally into the braided sheets of the classic Castlegate facies. The revised configuration of this contact is shown schematically in Fig. 8.37. The second discovery, (B), is that the Lower Castlegate Sandstone does indeed pass laterally into the shallow-marine facies of the Desert member. Examples of this transition are shown in Figs. 8.46 and 8.47, and the overall stratigraphy of this transition is shown in Fig. 8.48. These revisions to the stratigraphy have fundamental implications for the interpretation of the chronostratigraphy and evolution of the Mesaverde Group, as discussed below.

A detailed examination of the age spectra and provenance of the detrital zircon suites in the Mesaverde Group by Petit et al. (2019) led to several conclusions that help to confirm the work of Pattison (2018). Petit et al. (2019, p. 289) stated:

> The basal Castlegate surface does not represent significant missing time. The lack of missing time combined with the lack of provenance change at any locality are consistent with formation of this surface by autogenic scour rather than as a significant unconformity that reflects external forcing.

Pattison's (2018) new model solves the long-standing problem of the nature of the transition from the fluvial Castlegate environment into the marine environment of the

Fig. 8.46 The interfingering of the Desert Member and the Castlegate Sandstone north of Green River (Pattison 2020b, Fig. 8)

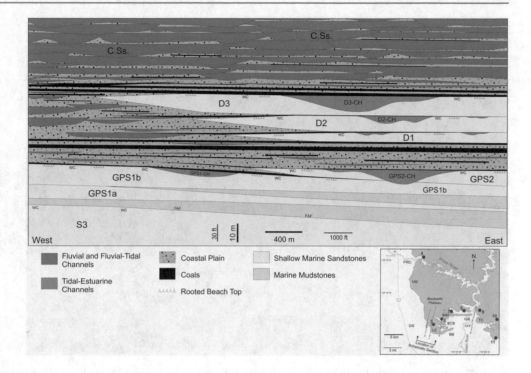

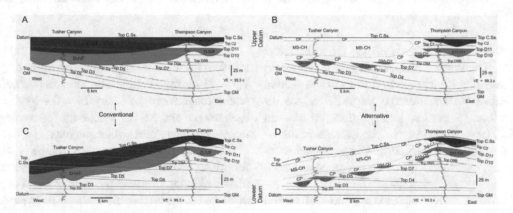

Fig. 8.47 The relationship between the Desert Member and the Castlegate Sandstone east of Green River. **A** and **C** illustrate the conventional interpretation, based originally on the work of Van Wagoner et al. (1990), plotted using two different datums. **B** and **D** illustrate the re-interpretation of the stratigraphy by Pattison (2019).

In the original reconnaissance mapping all the basal Castlegate channel sandstones were assumed to correlate above a regional unconformity. The new, more detailed mapping shows that this is incorrect. Each channel sandstone represents a different position of the fluvial-marine transition, that was gradually, and conformably, prograding eastward

Mancos Shale. As is clear from Figs. 8.46, 8.47 and 8.48, the transition was geologically rapid, continuous, regionally conformable, and stratigraphically traces out a major, and relatively sudden change through time in depositional conditions, probably caused by a rapid, tectonically driven upstream increase in coarse sediment supply.

According to information summarized by Miall (2014b), the Castlegate Sandstone spans about 2 m.y. But this estimate applies to the combined Lower and Upper Castlegate in the type area. As a first approximation, a working hypothesis is that the Lower and Upper part of the Castlegate each represents about 1 m.y. at the type section. Pattison's new

work indicates that between the Castle Gate and the Green River the Lower Castlegate interfingers with the upper Blackhawk member. Two or three fluvial channels, of typical Castlegate facies, may be stacked at the base of the Castlegate, and pass laterally into typical Blackhawk facies (floodplain sandstone, shale, coal). The uppermost Blackhawk north of Green River, which is transitional into Desert member tongues D1 to D3 (Fig. 8.46), is probably equivalent to the lowermost part of the Lower Castlegate at the type section. If this is the case, then the entire D1 to D11 and C1 to C5 succession is equivalent to the Lower Castlegate at the type section and similarly spans about 1 m.y. In this new

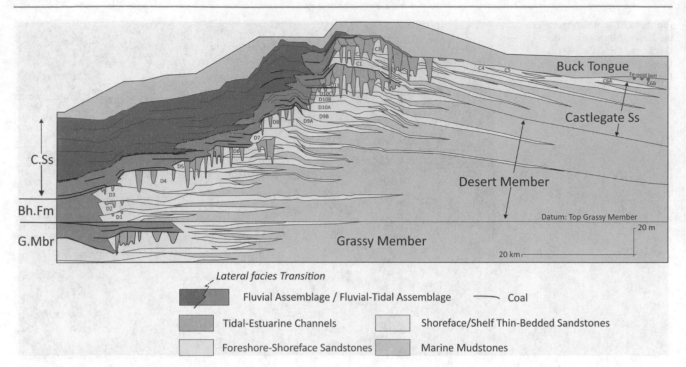

Fig. 8.48 Stratigraphy of the Desert Member and Castlegate Sandstone from north of Green River to the seaward limit of the sandstone. Adapted from Pattison (2018)

reconstruction, there are 19 tongues named D1 to D11 and C1-5 (Fig. 8.48). The entire Lower Castlegate (from D1 to C5) reaches a maximum thickness of 150 m in the section illustrated in Fig. 8.48. The average thickness of the 19 tongues is therefore 7.9 m. The average duration of each tongue is 1 m.y./19 = 53,000 years. The average sedimentation rate for this 150 m of section in 1 m.y. is 0.15 m/ky (10^{-1} m/ky), which puts the tongues into the range of SRS 8 units.

SRS 8 and *9* units have frequencies typical of orbital cycles. The time scale and sedimentation rate refer to the entire time span of the unit, including internal surfaces of erosion/stasis, and any longer term gap between the tongues due to withdrawal of the sea.

Turning to the internal sedimentary processes within each tongue, the sedimentation rate for sedimentation units (the actual preserved sediment) that span 10^4 years (53,000 years) is typically *SRS 7* or *8*. That is 10^1–10^{-1} m/ky, extreme range. On average 10^0 m/ky is a range of 1–9 m/ky. Considering only the accumulation of sedimentary units (the actual sediment) within the entire succession at this rate and ignoring gaps, diastems, etc., for a range of 1 to 9 m/ky the total thickness of 150 m would be deposited in between 150 and 17 ky. That is, between 15% and 1.7% of total elapsed time of 1 m.y., which suggests considerable missing elapsed time between and probably within each tongue.

For the range of sedimentation rates: Low: 150 ky/19 = 7.9 ky for deposition of each tongue, and 53–7.9 = 45 ky for missing time at each tongue boundary and within tongues. High: 17 ky/19 = 0.9 ky for deposition of each tongue and 53–0.9 = 52.1 ky for missing time at each tongue boundary and within tongues.

The 53 ky time span for each tongue is within the range for orbital cyclicity, an episodicity that is increasingly being identified in the Cretaceous record of the Western Interior, as discussed in the next section. It may be hypothesized that the tongues are generated by modest glacioeustasy or by other changes related to climatic cycles, including cyclic changes in runoff and/or sedimentation rate.

Are the tongues bounded by surfaces of erosion or stasis? This could be demonstrated by following tongue boundaries updip into the Castlegate fluvial succession to seek evidence of erosion, pedogenesis, early cementation, etc.; a project for the future.

An alternative to orbital forcing could be simple autogenic switching of fluvial axes similar to the process of compensational stacking or deltaic lobe switching. But the lateral persistence and extent of the numbered tongues would seem to argue against this. Could some additional tongues be missing from the composite line of section, but present elsewhere, e.g., further north in the subsurface, or removed by erosion to the south? Patterns of mixing detrital zircon populations in this area suggested to Petit et al. (2019) that river systems downdip were distributive in nature, derived from several different updip sources, and subject to frequent avulsion. This would be consistent with the possibility of additional Castlegate tongues present out of the line of the available outcrop section. These are the kinds of research questions that are now being raised by the new high-resolution stratigraphy (Miall et al. 2021). Some further examples, with some specific, and exciting outcomes, are discussed in the next section.

8.11.4 High-Resolution Stratigraphy

Stratigraphy has achieved much through the recognition of patterns and the development of models, but patterns and models are forms of generalization, and we are now at a point that generalizations can no longer add much to our level of understanding of stratigraphic processes. This section presents two examples of advanced stratigraphic studies that are based on very detailed documentation. The first, based on a largely subsurface data base of tens of thousands of wells, demonstrates how careful synthesis can add considerable detail to an interpretation of the tectonic control of a foreland basin. The second example shows how careful quantitative work, based on detailed outcrop mapping, provides a convincing demonstration of the processes of glacioeustasy at work in the Late Cretaceous, a period that has traditionally been thought of as characterized by a long-term greenhouse climate.

A. G. Plint and his students and colleagues have, over a period of decades, worked to systematically map the Cretaceous stratigraphy of Alberta using allostratigraphic principles. Most of this work has been based on the use of petrophysical logs for regional mapping, with ties to outcrops in the Rocky Mountain Foothills (e.g., see Fig. 6.4). A synthesis of much of this work, dealing with the mid-Cretaceous (Albanian-Campanian) stratigraphy of central and northern Alberta demonstrates how the architecture of the basin fill elucidates the tectonic evolution of Western Canada on a regional scale, while also throwing fascinating light on the details of the thrust-loading process controlling foreland basin subsidence (Plint et al. 2012). This synthesis exploits a highly detailed regional stratigraphy in which units (allomembers) representing time periods as short as a few tens of thousands of years, have been traced for hundreds of kilometers.

The dynamic relationship between thrust-sheet loading and foreland basin development was recognized by Price (1973) and explored quantitatively by Beaumont (1981) and Jordan (1981). Cant and Stockmal (1989) argued that the accretion of successive terranes to the outboard margin of western North America could be correlated to the successive development of the clastic wedges in the foreland basin to the east, and many subsequent workers have attempted to refine these correlations based on the increasingly detailed knowledge of the trajectories of plate movements in the marginal seas of eastern Panthalassa. The data base regarding the timing of magmatism, tectonism and terrane accretion events, and the linking of sedimentation to the uplift of provenance terranes by zircon studies, has been much enlarged since this first attempt, and this has confirmed some correlations and raised additional questions (Miall and Catuneanu 2019). Plint et al. (2012) developed tectonic syntheses for four time periods during the Cretaceous which help to explain the evolving architecture of the foreland basin (Fig. 8.49). From these maps is can be seen that the development of the main locus of accommodation in the basin, as revealed by isopach patterns, is located close to that part of the orogen that was under the greatest contractional stress at the time. Areas under limited contractional stress are associated with modest sediment thicknesses in the basin, as for example, in southwestern Alberta during the Barremian-Albian, when the motion of the accreting terranes was sinistral strike-slip (Fig. 8.49A). During the Albian-Cenomanian, the major point of contraction was located in northwestern British Columbia, and the foreland basin became rotated such that the forebulge was oriented northeast-southwest. The Dunvegan Delta was derived from the contemporary orogen and prograded transversely across the foreland basin in a southeasterly direction (Figs. 8.49C).

In detail, the locus of crustal loading may move along strike as individual thrusts respond separately to contractional movement (Fig. 8.50). Regional mapping demonstrates that, in plan view, the hingeline at the edge of the forebulge has an arcuate trend, the point at the center of the arc corresponding to the main locus of loading during each thrusting episode. This locus moved northward several hundred kilometers between the Campanian and the Paleocene, which is consistent with the right-lateral transcurrent deformation that the orogen was undergoing at this time (Monger 1993). Isopach patterns, similarly, can be used to point to the likely location of active crustal loads. The isopach of the entire Puskwasku Alloformation (Fig. 8.50A), spanning 2.5 m.y. of the Campanian-Santonian, illustrates the classic pattern of crust-sheet loading. When isopachs for the three individual allomembers are plotted separately (Figs. 8.50B, C, D), it can be seen that the locus of the load shifts laterally by more than 200 km over periods in the order of 800 ka, suggesting that subsidence is responding to movements on individual thrusts within an advancing structural complex. Plint et al. (2012, p. 503) stated:

> Drawing on a homely analogy, the notion of "trampoline tectonics" might summarize the essence of our findings: Viewed from below, children jumping on a trampoline produce a pattern of temporally and spatially discrete "loading centers" that

Fig. 8.49 Tectono-stratigraphic reconstructions of the Canadian Cordillera for four intervals during the Cretaceous showing regional isopachs for the basin fill deposited during each interval. Arrows show the direction of convergence of the Panthalassa plate (from Plint et al. 2012)

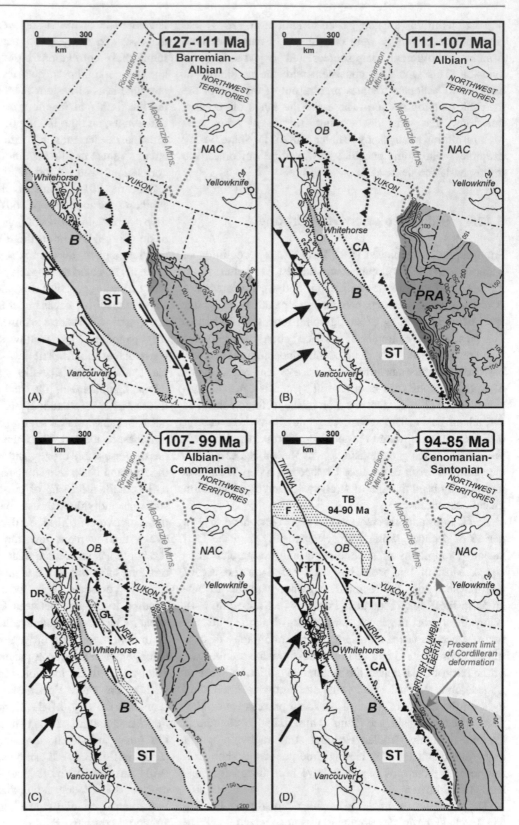

Fig. 8.50 Isopachs of the Santonian Puskwasku Alloformation in the Alberta Basin (**A**) and its individual allomembers (**B**, **C**, **D**) (Plint et al. 2012)

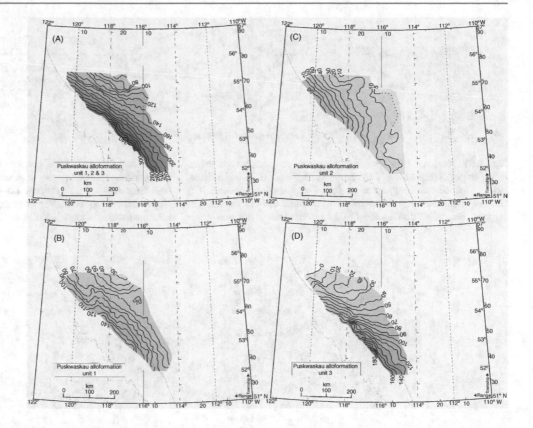

resemble the shifting pattern of subsidence recorded by Colorado allogroup rocks, and by implication, the shifting pattern of loading by the adjacent Cordillera.

The second example discussed briefly here is the work of Lin et al. (2021) to reconstruct stratigraphic evidence for high-frequency eustasy from the Upper Cretaceous Gallup Sandstone of New Mexico. A distinction between tectonic and eustatic control of stratigraphic architecture can be made based on the architecture of the units. As Ryer (1993) suggested, thin, sheet-like units that extend for considerable distances are likely to be of eustatic origin, whereas units that display wedge- or lozenge-shaped isopach patterns are more likely to be the product of differential subsidence. Plint et al. (2012) made extensive use of this feature in the study discussed above. As Lin et al. (2021) noted, there have been a number of recent studies of the cyclostratigraphy of Cretaceous sections, including several in the Western Interior basin, that have focused on facies or geochemical cyclicity in distal-shelf and basinal settings (e.g., see Figs. 7.43, 7.44, 8.58); but while these have provided good evidence of cyclicity with orbital frequency, direct evidence of eustatic sea-level change cannot be derived from such studies. Cyclic changes in distal and basinal facies, including geochemical cycles, may be the product of other features of orbitally forced climate change, including productivity, redox balance, runoff, sediment supply and other physical changes attributable to climate change. To assess sea-level change,

units deposited in coastal environments are required, within which shoreline migration can be tracked directly. Quantification of the amplitude and rate of sea-level change requires well-exposed sediments with good chronostratigraphic control, where shoreline migration in successive units can be directly traced in the field. Lin et al. (2021) based their study on the Gallup Sandstone of New Mexico, where chronostratigraphic control was provided by a series of datable bentonite horizons. Adjustments to allow for compaction and subsidence were carried out using backstripping methods (Watts and Steckler 1979; Liu and Nummedal 2004).

The stratigraphic synthesis on which the analysis is based is shown in Fig. 8.51. This was constructed from 68 measured sections with an average spacing between them of 880 m. Backstripping and geochronologic methods are not discussed here. The detailed architectures of a transgressive and a falling-stage parasequence set are shown in Fig. 8.52.

Sea-level changes between individual parasequences were estimated by tracking the vertical distance of boundaries of shoreline facies on the backstripped stratigraphic cross-section. The results show high-frequency low-amplitude eustatic sea-level fluctuations throughout the deposition of the Gallup system. The maximum estimated sea-level fall and rise are −28 m and +22 m, respectively, and are associated with parasequences 6 and 5a in the upper Gallup sequences, whereas the average sea-level fall and rise are −11 m and +7 m (Fig. 8.53). The magnitudes of sea-level fall and rise are generally similar, but the falls are

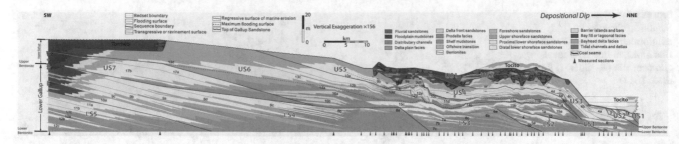

Fig. 8.51 Regional high-resolution sequence stratigraphy correlation in depositional dip of the Cretaceous Gallup system, New Mexico, USA. Two bentonite layers were found and used as datums. The "Lower Bentonite" is used as the bottom datum for correlation. The "Upper Bentonite" divides the Gallup system into lower and upper parts. Sixty-five individual parasequences, composed of twenty-nine parasequence sets and twelve sequences are identified that demonstrate high-frequency stratal cyclicity. LS—Lower Gallup sequence; US—Upper Gallup sequence. Vertical exaggeration is × 156. Locations of measured sections are labeled at the bottom (Lin et al. 2021, Fig. 2)

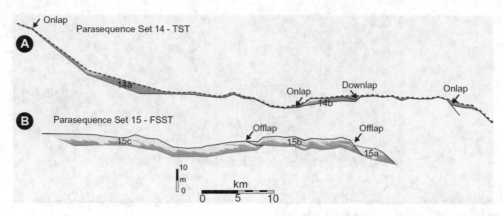

Fig. 8.52 Parasequence stacking patterns and lapout terminations of the Late Cretaceous Gallup system, New Mexico, USA. **A** Parasequence Set 14, interpreted as a transgressive systems tract (TST), shows non-accretionary backstepping retrogradational stacking. **B** Parasequence Set 15 shows offlapping termination and downstepping degradation, indicating a falling-stage systems tract (FSST). See Fig. 8.51 for legend (Lin et al. 2021, Fig. 6)

usually a few meters greater than rises (Fig. 8.53). The cumulative sea-level curve shows a 108 m relative rise, which is smaller than the maximum subsidence of 123 m. This difference likely indicates 15 m of net eustatic sea-level fall which surpasses the tectonic subsidence during the Gallup system's time interval. The overall eustatic sea-level fall may explain the overall progradation of the Gallup system into the basin (Fig. 8.51) (Lin et al. 2021, p. 246).

It was determined that the average depositional duration of a sequence, parasequence set, and parasequence is 100 ky, 41 ky, and 19 ky, respectively. These depositional time scales are remarkably consistent with eccentricity, obliquity, and precession cycles, even considering 15% of parasequences may have formed by autogenic processes. The results imply an astronomical control.

8.11.5 Cyclostratigraphy and Astrochronology

Cyclostratigraphy and astrochronology are amongst the most active areas of stratigraphic research at the time of writing. While there are many success stories, including the development of reliable astrochronological time scales, with associated GSSPs, for much of the Cenozoic, and many individual studies of "floating" scales for parts of the older Phanerozoic (Hilgen et al. 2015), most of the successes relate to studies of deep-marine and lacustrine successions, where it could be expected that orbital control may act essentially undisturbed over millions of years within environments where autogenic influences are weak and other allogenic controls are longer term and readily recognized. A study of a shallow-marine clastic succession is discussed in the previous section. The cautions expressed by Miall and Miall (2004) should continue to provide guidance for researchers in this field. Amongst these are the following issues, which relate to all current research into high-resolution stratigraphic definitions and correlation, whether concerned with cyclostratigraphy or not:

The continuity of the stratigraphic record: The discussion in this chapter has focused on the issue of sedimentation rates and the ubiquity of "missing" time in the stratigraphic record. Hilgen et al. (2015) asserted that when an integrated approach is used to build a high-resolution time scale it can

Fig. 8.53 Estimated sea-level change curve of parasequences and cumulative sea-level curve of the Late Cretaceous Gallup system, New Mexico, USA. The sea-level change curve shows generally symmetrical high-frequency, low-amplitude sea-level fluctuations. Sea-level changes range between −8 m and + 22 m. The absolute amplitudes of sea-level fall are in general greater than those of sea-level rise. The cumulative sea-level curve shows the overall accommodation increase resulting from tectonic subsidence with the high-frequency signals of eustatic sea-level changes. The magnitude of the overall subsidence-generated accommodation is consistent with the estimates of subsidence from the backstripping analysis (Lin et al. 2021, Fig. 4)

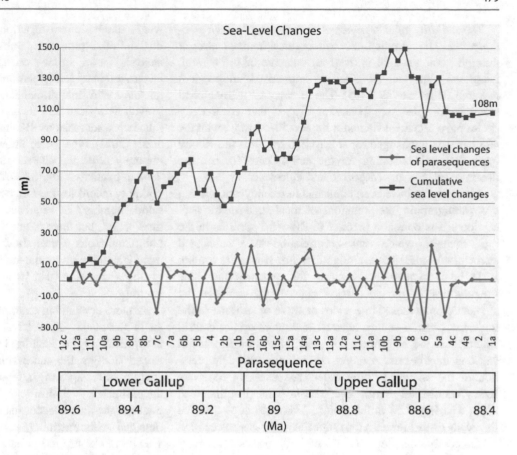

be demonstrated that many successions are complete at the Milankovitch time scale (10^4–10^5 years), although it is conceded that not all cycles, especially those of short period, may be recorded. It is conceivable that in some settings orbital forcing may be the dominant allogenic mechanism at orbital time scales, overwhelming or at least modifying autogenic processes, such as the control of fluvial style through a climatic control on fluvial runoff and sediment supply. In such cases, even the distribution of hiatuses in the record might illustrate orbital control. A case study of the Coniacian–Santonian (Upper Cretaceous) interval in the Western Interior of the United States (Locklair and Sageman 2008; Sageman et al. 2014) provides an excellent example of the use of multiple chronostratigraphic criteria in the establishment of more accurately dated stage boundaries and biozones, and the documentation of a 400-ky cyclicity in the Upper Cretaceous of the Western Interior Basin (see Fig. 7.43). They commented that "The use of accurate radioisotope data sets to determine how time is distributed in the strata of interest becomes ever more challenging as the level of temporal resolution increases" (Sageman et al. 2014, p. 966).

Tuning-induced Milankovitch spectra: "The potential for Milankovitch frequencies to be introduced into records through astronomical tuning is well known and is a serious drawback" (Hilgen et al. 2015). Another issue is that autogenic and other allogenic processes can generate a cyclicity that may mimic an orbital cycle thickness spectrum (Algeo and Wilkinson 1988). The use of various statistical tests and a careful attention to the real cyclicity in actual stratigraphic data should alleviate this problem.

Potential error in the assignment of numerical ages to orbital successions: Until relatively recently, the ability of radiogenic dating systems to provide sufficiently accurate age determinations was a major stumbling block in the research on orbital control. As noted in Sect. 7.8.2, it is now possible, under some circumstances, to develop numerical ages with uncertainties of ± 0.1% (e.g., 100,000 years at 100 Ma). This is at the scale of the long eccentricity cycle (405 ky) for Mesozoic rocks, which means that the extension of an astrochronological scale back into the Mesozoic may be possible. Systematic uncertainties in numerical dating, such as possible errors in a decay constant, are not a concern in the calculation of orbital frequencies and relative timing for floating sections, so long as the same systematics are used for all relevant age determinations. However, such potential error will still impact the development of a global time scale so long as these imprecisions remain of an order of magnitude comparable to that of orbital frequencies. Current work to develop a High-Resolution Event Stratigraphy (HiRES) is discussed below.

The stability of astronomical frequencies through geologic time. The retrodiction of astronomical frequencies back through geologic time is itself an outcome of the careful, integrated stratigraphic studies emphasized in modern research (Hilgen et al. 2015). The most recent astronomical solutions of planetary behavior suggest that eccentricity cycles have remained stable for the last 50–55 m.y., with the 405 ky cycle unchanged for at least 250 m.y. It is the 405-ky cycle that guides tuning for the early Cenozoic, and the floating scales of the Mesozoic and Paleozoic.

Many papers have been published in recent years offering cyclostratigraphic interpretations of local stratigraphic successions. It is common practice to show the sections in the time domain, which makes correlation to a calculated astrochronological scale straightforward. However, caution is to be recommended, because this practice may hide irregularities in the succession caused, for example, by autogenic processes. Time series analysis of sections in the thickness domain cannot be used to explore orbital control where there are significant autogenic effects on lithofacies and unit thicknesses. A recent cyclostratigraphic interpretation of the so-called "third-order" sequences of the New Jersey continental margin is a case in point (Boulila et al. 2011; e.g., Fig. 8.54 of this book). A time-depth plot for the five wells drilled through the Oligocene–Miocene succession on the continental slope (ODP Leg 150) reveals very irregular sedimentation rates (Miller and Mountain 1994, Fig. 8.55 of this book). Facies analyses of these deposits (Browning et al. 2008) indicate that they were deposited under a range of coastal, deltaic and shelf environments where autogenic redistribution of clastic sediment by wave, tide and other processes is ubiquitous (Fig. 8.56). Sequence analysis based strictly on empirical seismic, facies and biostratigraphic interpretations, indicates that, both in the thickness and time domains, sequence boundaries are quite irregularly spaced and there is no obvious cyclicity in the lithofacies successions (Kominz et al. 2008). Yet Boulila et al. (2011, Figs. 4, 6) show correlations to a calculated obliquity sequence characterized by 1.2-Ma cycles, or by 2.4-Ma cyclicity (Fig. 8.54). The sequence record and the obliquity record for the Oligocene–Miocene record show the same "number" of sequence boundaries for the 5–34 Ma time period, but there is not the same exact "match" for the Late Cretaceous cycles shown in Fig. 8.54, and otherwise there is nothing about the stratigraphy or the sedimentology of the succession that would suggest cyclostratigraphic control.

A more convincing example of cyclostratigraphic analysis is that of the Green River Formation, by Aswasereelert et al. (2013). Long identified as an example of an orbitally forced stratigraphic succession (Bradley 1929), this recent study used the ages of six tephra layers interbedded with the approximately 350 m of section to generate a thickness-to-time transformation. Ages of the tephras were obtained with maximum errors of ± 0.21 m.y., and provided a model of modest, smooth changes in sedimentation rate. Detailed stratigraphic analysis confirmed the importance of the 100-ky astronomical cycle in sedimentological forcing that generated the distinctive layering in this well-known formation.

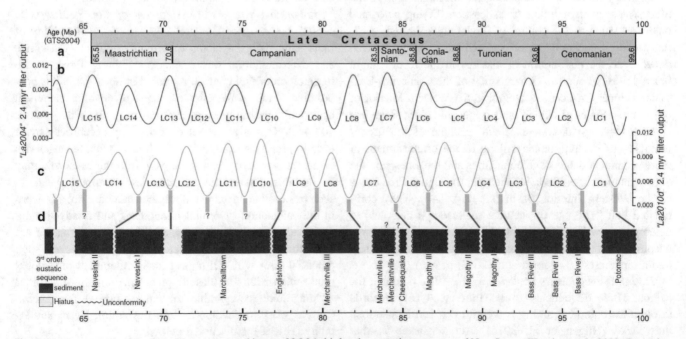

Fig. 8.54 Comparison of the Late Cretaceous (∼ 64 to ∼ 99 Ma) third-order eustatic sequences of New Jersey (Kominz et al., 2008; Browning et al., 2008) with ∼ 2.4 myr orbital eccentricity cycles (Boulila et al. 2011, Fig. 8, p. 104)

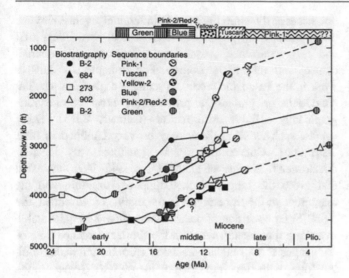

Fig. 8.55 Subsidence plots for drill holes on the New Jersey continental margin (Miller and Mountain 1994)

Where the evidence of autogenic activity, or of tectonic influence acting within a similar time scale to that of cyclostratigraphy (10^4–10^5-yr) is strong, it is clearly incumbent on proponents of cyclostratigraphic control (especially for hanging sections representing the distant geological past) to do more than provide statistical "proof" of their reality, such as from amplitude spectra of time-series studies. Stratigraphic sections should be shown in the depth domain, and a detailed facies analysis performed. Statistical analysis cannot take account of changes in facies or sedimentation rate, the presence of cryptic hiatuses, etc., without the use of special methods. The facies successions should be clearly cyclic, with regularity of facies and of unit thicknesses. Bailey (2009) discussed these and other problems, pointing out the difficulty of designing sufficiently rigorous statistical and other tests for the detection of orbital control. Meyers (2012) discussed the issue of "red noise" in stratigraphic sections, proposing statistical tests to evaluate the suitability of given data sets for cyclostratigraphic analysis. As he pointed out, it is possible for "noise to look like signal." Even random variation can generate "cycles" (e.g., Hiscott 1981).

Fig. 8.56 A depositional model for the Niger Delta, used as an illustration of the depositional environments and autogenic processes that, it is interpreted, were active during the deposition of the Early Cretaceous and Paleogene (from Browning et al. 2008, Fig. 77, p. 235)

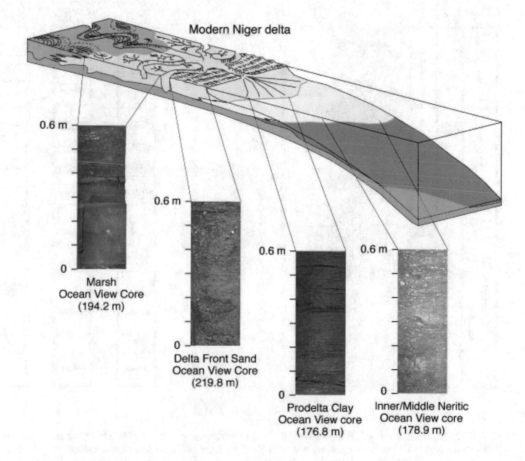

The current "best practices" for chronostratigraphic analysis make use of the techniques of High-Resolution Event Stratigraphy, or HiRES. "The cornerstone of HiRES is the integration of every available piece of stratigraphic information into a single set of data that can be cross-correlated and internally calibrated" (Cramer et al. 2015, p. 138). Originally conceived by Kauffman (1986, 1988), the guiding principle is that there are many features of a sedimentary section, beyond biostratigraphy, that may be used to develop tools for correlation and dating. Stratigraphic events, such as storm beds, tephras, and flooding events, may have limited regional extents, but when used in combination across multiple stratigraphic sections and integrated with chemostratigraphic, magnetostratigraphic and other indicators, they may permit highly detailed stratigraphic syntheses to be developed. Figure 8.57 illustrates the principles involved in the development of an "integrated event stratigraphy." The application of this methodology is

not necessarily straightforward. Individual events may be diachronous to a greater or lesser degree, particularly first and last occurrence taxon data, reflecting environmental control and migration patterns. For example, for HiRES work in the Paleozoic record, it remains a question whether conodonts or graptolites provide the most chronostratigraphically reliable information (Cramer, 2015). Other events, such as storm beds may be very confined in their distribution. Chemostratigraphic signatures vary in their diachroneity. Cramer et al. (2015, p. 148–149) suggested that given the long residence time of strontium and the magnitude of the Sr reservoir in the marine environment, the $^{87}Sr/^{86}Sr$ composition of ocean waters should be quite stable over long time scales, whereas the shorter residence time of ^{13}C makes $\delta^{13}C$ a higher precision chronostratigraphic tool, but one with an imprecision in the 10^{3-4}-year range reflecting a mixing time of a few thousand years.

High-Resolution Event Stratigraphy (HiRES)

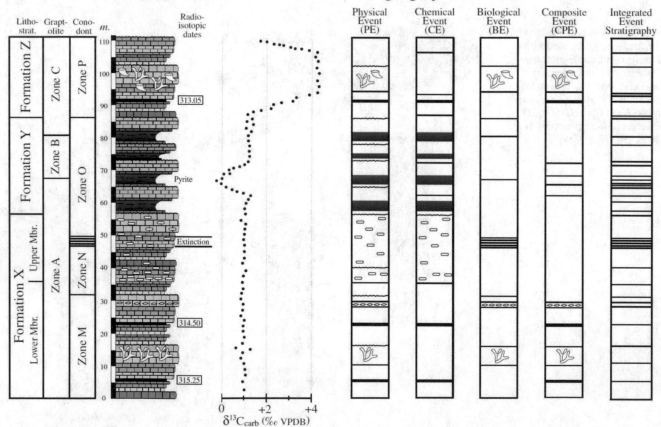

Fig. 8.57 Demonstration of High-Resolution Event Stratigraphy (HiRES) concepts and methods by using a hypothetical stratigraphic section and data (modified from Kauffman 1988). Lithostratigraphic nomenclature, biostratigraphy, lithostratigraphy, biotic events, radioisotopic age determinations, and stable carbon isotope stratigraphy shown at the left. The right five columns illustrate the principles of HiRES in which all stratigraphic information is included and a series of "events"

is delimited within the section. All of the chronostratigraphically useful horizons are combined into the integrated event stratigraphy at the far right. In principle, the lithostratigraphic names and biozones at the far left provide a total of 12 discreet horizons for correlation while the integrated event stratigraphy at the far right provides many more potential discreet horizons for correlation and an improved chronostratigraphic resolution (Cramer et al. 2015, Fig. 2, p. 139)

Cramer et al. (2015, p. 144) suggested that the process of establishing and refining a HiRES chronostratigraphic record could be clarified by the use of three terms:

Geochronometry refers to the numerical dating of stratigraphic events by the use of radioisotopes.

Geochronology is the synthetic art of reconciling all available geochronologic and chronostratigraphic data to provide temporal calibration of the stratigraphic record.

Chronostratigraphy refers to the documentation, testing and calibration of a regional to global framework of age-dated stratigraphy.

The use of these three terms should help to illuminate what exactly is being done in each specific research endeavor (see also Sect. 7.8). Thus, for example, a discussion of the potential error in correlation, in stratigraphic interval (expressed in meters) of a GSSP is an exercise in chronostratigraphy. The numerical calibration of a specific section is an exercise in geochronology. The laboratory processing of a sample and assignment of an age and range of error is geochronometry. In this last case, consideration of the specifics of precision and accuracy and the difference between them may become significant.

Locklair and Sageman (2008) and Sageman et al. (2014) reported on a study to investigate the cyclostratigraphy of the Niobrara Formation (Santonian-Coniacian) and laterally equivalent units in the Western Interior Basin of the United States. The chalk-marl cycles of the Niobrara were digitized using microscanner resistivity data from drill core. The 85-m core was divided into eight segments, corresponding to the lithologically defined members of the formation. Ages were assigned using the available time scale for this unit, which is characterized by one of the best established chronostratigraphic records in the Phanerozoic time scale. The depth-to-time conversion and the analysis for orbital cyclostratigraphy were carried out using a succession of statistical methods. An essential step is a technique termed evolutive harmonic analysis (Locklair and Sageman 2008, Fig. 5), which is designed to take into account gradual changes in average sedimentation rate reflecting allogenic forcing at a time scale longer than that of orbital forcing. Calculated rates vary between 0.0075 and 0.0235 m/ky with an average of about 0.07 m/ky, which corresponds to *SRS*-9 and is a typical accumulation rate for orbital cycles (Miall 2015). The cyclostratigraphic analysis, carried out over successive segments of the core, suggested that sedimentation of this section at this locality was continuous at the 10^5-yr scale, a conclusion supported by biostratigraphic analysis and bentonite correlations. Tuning to the temporally stable 400-yr eccentricity period revealed obliquity and precessional frequencies in the record. The later study (Sageman et al. 2014) extended the cyclic analysis to other cores and to a series of outcrop sections in New Mexico, Utah and Montana (Fig. 8.58), in which the cyclostratigraphic

analysis was combined with new radioisotope data with the objective of refining laboratory methods, recalibrating decay constants and developing more precise correlations between methods in order to increase the accuracy and precision of the Coniacian–Santonian time scale. The potential presence of hiatuses in the sampled sections was recognized to be a potential problem. Breaks with a length equivalent to a whole cycle or multiple of whole cycles may not be detected by spectral analysis (Meyers and Sageman 2004). The duration of several gaps that were suggested by biostratigraphic zonation were within the potential error of radioisotopic calibrations ($\sim 10^5$ yrs), which means that they could not be ruled out in the cyclostratigraphic analysis and require further study by means of regional correlations and additional data. This means that correlation to specific cycles in the 400-ky record could be in error by one or more cycles, but the results, nevertheless, represent a substantial increase in the accuracy and precision of this part of the Cretaceous time scale, and constitute a substantial contribution in the project to extend an astrochronological time scale back through the Mesozoic.

8.11.6 Conclusions

Two hundred years ago, William Smith gave us the first complete geological map (Smith 1815) and started us on the road to an understanding of Earth's geologic history. Lyell (1830–1833) gave us the necessary tools for interpretation, based on the principle of uniformitarianism, and Holmes, beginning a little more than one hundred years ago, began the development of the modern geological time scale (culminating in his first book: Holmes 1913). Barrell (1917) was the first to attempt to synthesize these critical developments, but most of his ideas were forgotten or ignored for decades. The development of the formal principles of stratigraphy, the evolution of sedimentology as a mature discipline, the stimulus provided by seismic stratigraphy, all have been necessary developments in the evolution of the modern synthesis that constitutes the science described in this book.

So where have we arrived at today, and what may we predict as possible future developments, and outcomes from the application of this science?

Analysis of depositional and erosional processes over the full range of time scales provides insights into long-term geological preservation. At time scales of days to months (10^{-1}-10^{-6} yrs), preservation of individual lithofacies units is essentially random, reflecting autogenic processes, such as diurnal changes in current speed and tidal activity. Packages of strata that survive long enough are subject to the next cycle of preservational processes, such as autogenic channel switching, the "100-year flood," or storm activity at the 10^0-10^3-yr time scale. At the 10^4-10^5-yr time scale, so-called

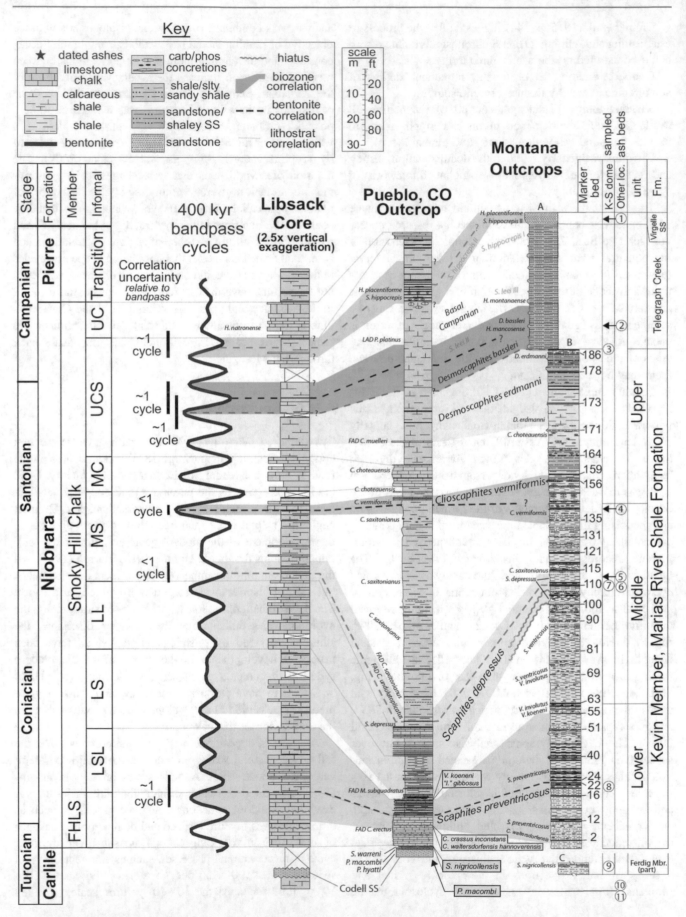

Fig. 8.58 Lithostratigraphic and biostratigraphic correlation of Coniacian-Campanian sections from Colorado to Montana, showing the 400-ka cycle record derived by time-series analysis (Sageman et al. 2014, Fig. 2, p. 959)

"high-frequency" geological processes come into play, including orbital forcing of climate and sea level, and local tectonic episodicity. Ultimately, all remaining stratigraphic accumulations are subject to the long-term (10^6–10^7-yr) geological (largely tectonic) controls on basin accommodation (Miall, 2014b). The interplay of these processes constitutes what we have called here the "stratigraphy machine" (Miall et al. 2021).

An important insight emerging from this analysis has critical implications for the application of uniformitarianist principles. Geological processes interpreted from successions accumulated over the post-glacial period, such as those of the Mississippi and Rhine–Meuse Deltas (descriptions and analyses of which comprise substantial contributions to sedimentological literature on fluvial and deltaic systems) can only be used as analogs for interpretations of geological processes up to the 10^4-year time scale. The geological record contains many examples of coastal fluvial-deltaic successions spanning millions of years, but a 10^6–10^7-year record cannot be interpreted simply by "scaling-up" an analysis carried out on a 10^4–10^5-year time scale. Firstly, coastal succession such as those on present-day continental margins could be largely eliminated by subaerial erosion during the next glacial cycle of lowered sea level, as the geological preservation machine begins to operate over the next longer time scale. Secondly, the time scales implied for sedimentary processes would be

wrong. For example, sequence models for fluvial systems, which relate channel-stacking behavior to rates of sedimentary accommodation, are largely based on measurements of rates of sedimentation and channel switching in modern rivers and in post-glacial alluvial valley fills, at time scales no greater than 10^3 years. Applications of these models to the ancient record deal mostly with so-called "third-order" sequences (durations in the 10^6-yr range) for which calculated accumulation rates are one to three orders of magnitude slower than those on which the sequence models are based (Miall 2014a). Colombera et al. (2015) confirmed, by a detailed study of twenty ancient fluvial systems, that there is no relationship between channel-stacking pattern and aggradation rate. It is suggested that the observed changes in channel-stacking patterns that have been observed in the rock record are the product of longer term processes, such as tectonically controlled changes in paleoslope or sediment supply (Miall 2014a).

Modern stratigraphic techniques are at the very core of current methods of petroleum exploration and development. Figure 8.59 is an excellent example of a continuing major thrust in the petroleum industry, the development of ever more sophisticated techniques of seismic data acquisition, processing and visualization. The image shown here is from an advertisement appearing in *"AAPG Explorer,"* the monthly news magazine of the American Association of

Fig. 8.59 An image generated from three-dimensional seismic data showing the upper surface of a salt horizon in the Gulf Coast, intersected by vertical and horizontal 2-D sections (Petroleum Geo-Services, Houston)

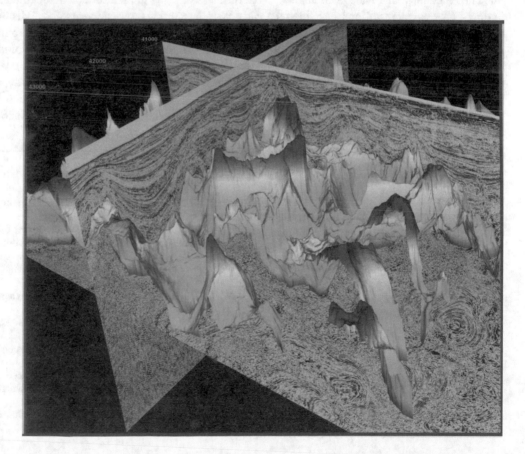

Petroleum Geologists. It shows an example of the type of seismic imagery now being developed in the highly competitive field of seismic exploration. It is from an area of the world that has received considerable attention from the industry in recent years because geophysicists have developed techniques for "seeing" through the thick layers of evaporites that occur near the bottom of the sedimentary layer on the floor of the Gulf of Mexico. The so-called "sub-salt play" has been the hottest area of exploration in North America for several years, and is finding increasing application in other comparable basinal settings worldwide, such as the deep Atlantic Ocean, offshore from Brazil and Angola. Other examples of advanced seismic techniques are illustrated and discussed in Chap. 6. The increasingly precise information now obtainable from seismic data, together with the modern techniques of directional drilling, and geosteering, backed up by sequence-stratigraphic mapping and detailed chronostratigraphic control, have substantially improved the odds of successful drilling in frontier areas. On a smaller scale, such detail is also an essential component of projects to efficiently exploit heavy oil using in-situ methods. Issues such as shale distribution and bed length, and reservoir cap integrity, are key to the successful use of the steam-assisted gravity drainage (SAG-D) method for oil extraction in the Alberta Oil Sands. It is one of the cruel ironies of life that the enormous intellectual energy and immense ingenuity that have been applied to the business of finding and extracting oil and gas since the early twentieth century have been so easily dismissed by the followers of the anthropogenic model of climate change. Oil and gas are just "dirty" fuels that apparently, we don't need any more.

Fresh water is a critical natural resource. About one-quarter of the world's supply is stored in the ground (most of the rest is in major ice caps), and its health is very much a concern of sedimentary geologists. Groundwater movement depends on the porosity and permeability architecture of the rocks (mostly sediments) through which water moves. In some deep basins, groundwater bodies at depth contain enough heat to offer considerable potential for geothermal power generation. Hydrogeology involves many of the same stratigraphic approaches utilized by petroleum geologists. The tracking and treatment of pollutants involves adaptations of traditional geological methods of exploration and remote sensing, including down-hole petrophysics, high-resolution seismic exploration and ground-penetrating radar.

One of the lines of evidence that modern climate scientists point to as an indication of the anthropogenic influence on climate is the supposed rapidity with which global warming has occurred (essentially all of which has taken place since about 1850) compared with the supposedly much slower rates of change that can be observed in the geological record. In fact, earth scientists have not been able to either prove or disprove this assertion because the accuracy and precision of the geological time scale for deep time has not

permitted refinements in the dating of geological events down to the 10^2-year time scale. The development of the astrochronological time scale is getting is closer, with a time scale in a 10^{4-5}-year accuracy range, but this gap is unlikely ever to be completely closed. However, refinements in the paleoclimatic reconstructions from the stratigraphic record, in terms of the scope, type and rate of past climate change, should be able to provide much better comparisons, and analogues for testing climate models and understanding Earth's behavior under different climatic conditions than have been available in the recent past.

High-resolution stratigraphic studies are increasingly pointing to a much more complex climatic history during the Cretaceous than the simple, long-term "greenhouse" that has long been assumed. Studies in both the American and Canadian portions of the Western Interior Basin, some of which are referred to in Sects. 8.11 and 8.12, are providing ever more convincing evidence for what Miller et al. (2005) referred to as "cold snaps," in the isotopic record, in the recognition of cyclostratigraphic signatures, and in the record of stratigraphic sequences and erosion surfaces that can be traced for hundreds, if not thousands of kilometers, indicating likely glacioeustatic control on sedimentation.

In a more general sense, as the well-known sedimentologist Harvey Blatt said in 1982: "Sediments and sedimentary rocks are the only source of knowledge about conditions on the Earth's surface before the invention of written language a few thousand years ago." Historical geology depends largely on the study of the stratigraphic record. It is from the rocks that we know about the evolution of life, the plate-tectonic development of the earth's crust, and ancient climate changes. Continued detailed study, and the development of new techniques, can therefore be expected in the future. Ask a stratigrapher about Earth History. It's in our files.

References

Ager, D. V., 1973, The nature of the stratigraphical record: New York, John Wiley, 114 p.

Ager, D. V., 1981, The nature of the stratigraphical record (second edition): John Wiley, New York, 122 p.

Ager, D., 1986, A reinterpretation of the basal 'Littoral Lias' of the Vale of Glamorgan, Proceedings of the Geologists Association, v. 97, p. 29-35.

Ager, D. V., 1993, The new catastrophism, Cambridge University Press, 231 p.

Algeo, T. J., and Wilkinson, B. H., 1988, Periodicity of mesoscale Phanerozoic sedimentary cycles and the role of Milankovitch orbital modulation: Journal of Geology, v. 96, p. 313–322.

Allen, J. R. L., 1983, Studies in fluviatile sedimentation: bars, bar complexes and sandstone sheets (low-sinuosity braided streams) in the Brownstones (L. Devonian), Welsh Borders: Sedimentary Geology, v. 33, p. 237-293.

Allen, P. A., 2008, Time scales of tectonic landscapes and their sediment routing systems, in Landscape Evolution: Denudation,

Climate and Tectonics Over Different Time and Space Scales, edited by K. Gallagher, S. J. Jones, and J. Wainwright, Spec. Publ. Geol. Soc., 296, 7–28,

Allen, P. A., and Allen, J. R., 2005, Basin analysis, principles and applications, second edition, Wiley-Blackwell, Hoboken, New Jersey, 500 p.

Allen, P. A., and Densmore, A. L., 2000, Sediment flux from an uplifting fault block: processes and controls in the stratigraphic development of extensional basins: Basin Research, v. 12, p. 367–380.

Amorosi A., Bohacs K.M., Bruno L., Campo B., and Drexler T.M., 2017. How close is geological thought to reality? The concept of time as revealed by the sequence stratigraphy of the late Quaternary record, in Hart, B., Rosen, N.C., West, D., D'Agostino, A., Messina, C., Hoffman, M. & Wild, R., eds., Sequence Stratigraphy: The Future Defined: 36th Annual Gulf Coast Section SEPM Foundation Perkins-Rosen Research Conference, Marathon Conference Center, Houston, December 4–5, 2017, p. 47–86.

Archer, A. W., Kvale, E. P., and Johnson, H. R., 1991, Analysis of modern equatorial tidal periodicities as a test of information encoded in ancient tidal rhythmites, in Clastic tidal sedimentology: Canadian Society of Petroleum Geologists, Memoir 16, p. 189-196.

Aschoff, J. L., and Steel, R. J., 2011, Anomalous clastic wedge development during the Sevier-Laramide transition, North American Cordilleran foreland basin, USA: Geological Society of America Bulletin, v. 123, p. 1822-1835.

Aswasereelert, W., Mayers, S. R., Carroll, A. R., Peters, S. E., Smith, M. E., and Feigl, K. L., 2013, basin-scale cyclostratigraphy of the green River Formation, Wyoming: Geological Society of America Bulletin, v. 125, p. 216–228.

Aubry, M.-P., 1991, Sequence stratigraphy: eustasy or tectonic imprint: Journal of Geophysical Research, v. 96B, p. 6641–6679.

Aziz, A., Krijgsman, W., Hilgen, F. J., Wilson, D. S., and Calvo, J. P., 2003, An astronomical polarity time scale for the late middle Miocene based on cyclic continental sequences: Journal of Geophysical Research, v. 108B, doi:https://doi.org/10.1029/2002JB001818.

Bailey, R. J., 2009, Cyclostratigraphic reasoning and orbital time calibration: Terra Nova, v. 21, #5, p. 340–351.

Bailey, R. J., 2011, Buried trees and basin tectonics: a Discussion: Stratigraphy, v. 8, p. 1–6.

Bailey, R. J., and Smith, D. G., 2005, Quantitative evidence for the fractal nature of the stratigraphic record: results and implications: Proceedings of the Geologists' Association, v. 116, p. 129–138.

Bailey, R. J., and Smith, D. G., 2010, Scaling in stratigraphic data series: implications for practical stratigraphy: First Break, v. 28, p. 57–66.

Bailey, R. J., and Schumer, R., 2012, The statistical properties of stratigraphic layering and their possible significance: The Geological Society, London, William Smith Meeting 2012, Strata and Time: Probing the gaps in our understanding,

Barrell, Joseph, 1917, Rhythms and the measurement of geologic time: Geological Society of America Bulletin, v. 28, p. 745–904.

Beaumont, C, 1981, Foreland basins: Geophysical Journal of the Royal Astronomical Society, v. 65, p. 291–329.

Beckmann, B., Flögel, S., Hofmann, P., Schulz, M., and Wagner, T., 2005, Orbital forcing of Cretaceous river discharge in tropical Africa and ocean response: Nature, v. 437, p. 241–244.

Bentham, P. A., Talling, P. J., and Burbank, D. W., 1993, Braided stream and flood-plain deposition in a rapidly aggrading basin: the Escanilla formation, Spanish Pyrenees, in Best, J. L., and Bristow, C. S., eds., Braided rivers: Geological Society, London, Special Publication 75, p. 177–194.

Berggren, W. A., Kent, D. V., Aubry, M.-P., and Hardenbol, J., eds., 1995, Geochronology, time scales and global stratigraphic correlation: Society for Sedimentary Geology Special Publication 54, 386 p.

Bernard, H. A., Leblanc, R. J., and Major, C. J., 1962, Recent and Pleistocene geology of southeast Texas, in Rainwater, E. H., and Zingula, R. P., eds., Geology of the Gulf Coast and central Texas: Geological Society of America, Guidebook for 1962 Annual Meeting., p. 175–224.

Betzler, C., Kroon, D., and Reijmer, J. G. J., 2000, Sychroneity of major late Neogene sea level fluctuations and paleoceanographically controlled changes as recorded by two carbonate platforms: Paleoceanography, v. 15, p. 722–730.

Bhattacharya, J. P., 2011, Practical problems in the application of the sequence stratigraphic method and key surfaces: integrating observations from ancient fluvial-deltaic wedges with Quaternary and modelling studies: Sedimentology, v. 58, p. 120–169.

Bhattacharya, N. P., Howell, C. D., MacEachern, J. A., and Walsh, J. P., 2020, Bioturbarion, sedimentation rates, and preservation of flood events in deltas: Palaeogeograpphy, Palaeoclimatology, Palaeoecology, v. 560, #110049. 22 p.

Bhattacharya, J.P., Miall, A.D., Ferron, C., Gabriel, J., Randazzo, N., Kynaston, D., Jicha, B. R., and Singer, S., 2019, Balancing sediment budgets in deep time and the nature of the stratigraphic record: Earth-Science Reviews. v. 199, 102985, 25 p.

Blair, T. C., and Bilodeau, W. L., 1988, Development of tectonic cyclothems in rift, pull-apart, and foreland basins: sedimentary response to episodic tectonism: Geology, v. 16, p. 517–520.

Blum, M. D., 1993, Genesis and architecture of incised valley fill sequences: a late Quaternary example from the Colorado River, Gulf Coastal Plain of Texas, in Weimer, P., and Posamentier, H. W., eds., Siliciclastic sequence stratigraphy: Recent developments and applications: American Association of Petroleum Geologists Memoir 58, p. 259–283.

Blum, M. D., and Törnqvist, T. E., 2000, Fluvial responses to climate and sea-level change: a review and look forward: Sedimentology, v. 47, p. 2–48.

Boulila, S., Galbrun, B., Miller, K. G., Pekar, S. F., Browning, J. V., Laskar, J., and Wright, J. D., 2011, On the origin of Cenozoic and Mesozoic "third-order" eustatic sequences: Earth Science Reviews, v. 109, p. 94–112.

Bradley, W. H., 1929, The Varves and Climate of the Green River Epoch: U.S. Geological Survey Professional Paper 158-E, 110 p.

Bridge, J. S. and Leeder, M. R., 1979, A simulation model of alluvial stratigraphy: Sedimentology, v. 26, p. 617–644.

Browne, G.H., and Naish, T.R., 2003. Facies development and sequence architecture of a late Quaternary fluvial-marine transition, Canterbury Plains and shelf, New Zealand: implications for forced regressive deposits: Sedimentary Geology, v. 158(1-2), p. 57-86.

Browning, J. V., Miller, K. G., Sugarman, P. J., Kominz, M. A., McLaughlin, P. P. Kulpecz, A. A., and Feigenson, M. D., 2008, 100 Myr record of sequences, sedimentary facies and sea level changes from Ocean Drilling Program onshore coreholes, US Mid-Atlantic coastal plain: Basin Research, v. 20, p. 227-248.

Buckman, S. S. 1910. Certain Jurassic ('Inferior Oolite') species of ammonites and brachiopoda. Quarterly Journal of the Geological Society, London, v. 66, p. 90-108.

Burbank, D. W., Meigs, A., and Brozovic, N., 1996, Interactions of growing folds and coeval depositional systems: Basin Research, v. 8, p. 199–223.

Callomon, J. H., 1995, Time from fossils: S. S. Buckman and Jurassic high-resolution geochronology, in Le Bas, M. J., ed., Milestones in Geology: Geological Society of London Memoir 16, p. 127-150.

Cant, D. J., and Stockmal, G. S., 1989, The Alberta foreland basin: relationship between stratigraphy and terrane-accretion events: Canadian Journal of Earth Sciences, v. 26, p. 1964–1975.

Catuneanu, O., 2006, Principles of sequence stratigraphy: Elsevier, Amsterdam, 375 p.

Christie-Blick, N., Mountain, G. S., and Miller, K. G., 1990, Seismic stratigraphy: record of sea-level change, in Revelle, R., ed., Sea-level change: National Research Council, Studies in Geophysics, Washington, National Academy Press, p. 116-140.

Christie-Blick, N., Pekar, S. F., and Madof, A. S., 2007, Is there a role for sequence stratigraphy in chronostratigraphy? Stratigraphy, v. 4, p. 217-229.

Cloetingh, S., 1988, Intraplate stress: a new element in basin analysis, in Kleinspehn, K., and Paola, C., eds., New Perspectives in basin analysis: Springer-Verlag, New York, p. 205–230.

Cole. R.D., and Friberg, J.F., 1989, Stratigraphy and sedimentation of the Book Cliffs, Utah, in Nummedal, D. and Wright, R., eds., Cretaceous Shelf Sandstone and Shelf Depositional Sequences, Western Interior Basin, Utah, Colorado, and New Mexico: 28th International Geological Congress, Field Trip Guidebook T119.

Coleman, J. M., and Gagliano, S. W., 1964, Cyclic sedimentation in the Mississippi River deltaic plain: Transactions of the Gulf Coast Association of Geological Societies, v. 14, p. 67–80.

Colombera, L., Mountney, N. P., and McCaffrey, W. D., 2015, A meta-study of relationships between fluvial channel-body stacking pattern and aggradation rate: implications for sequence stratigraphy: Geology, v. 43, p. 283–286.

Cramer, B. D., Vandenbroucke, T. R. A., and Ludvigson, G. A., 2015, High-resolution event stratigraphy (HiRES) and the quantification of stratigraphic uncertainty: Silurian examples of the quest for precision in stratigraphy: Earth Science Reviews, v. 141, p. 136–153.

Crampton, J. S., Schiøler, P., and Roncaglia, L., 2006, Detection of Late Cretaceous eustatic signatures using quantitative biostratigraphy: Geological Society of America Bulletin, v. 118, p. 975–990.

Dalrymple, R. W., 1984, Morphology and internal structure of sandwaves in the Bay of Fundy: Sedimentology, v. 31, p. 365–382.

Davies, N. S., and Shillito, A. P., 2018, Incomplete but intricately detailed: The inevitable preservation of true substrates in a time-deficient stratigraphic record: Geology, v. 46, p. 679–682.

Davies, N. S., and Shillito, A. P., 2021, True substrates: the exceptional resolution and unexceptional preservation of deep-time snapshots on bedding surfaces: Sedimentology, in press.

Davies, N. S., Shillito, A. P., and McMahon, W. J., 2019, Where does the time go? Assessing the chronostratigraphic fidelity of sedimentary geological outcrops in the Pliocene-Pleistocene red Crag Formation, eastern England: Journal of the Geological Society, v. 176, p. 1154-1168.

de Natris, M. F., 2012, Facies- and time-analysis of the upper part of the Brent Group (mid-upper Jurassic) in the Greater Oseberg area, northern North Sea: M.Sc., thesis, University of Bergen, 90 p.

de Natris, M., and Helland-Hansen, W., 2012, Where has all the time gone? Disentangling time in the Mid-Late Jurassic Tarbert Formation, northern North Sea: The Geological Society, London, William Smith Meeting 2012, Strata and Time: Probing the gaps in our understanding, Abstract, p. 40.

Devine, P. E., 1991, Transgressive origin of channeled estuarine deposits in the Point Lookout Sandstone, northwestern New Mexico: a model for Upper Cretaceous, cyclic regressive parasequences of the U. S. Western Interior: American Association of Petroleum Geologists Bulletin, v. 75, p. 1039-1063.

Donaldson, J. A., Eriksson, P. G. and Altermann, W., 2002, Actualistic versus non-actualistic conditions in the Precambrian: a reappraisal of an enduring discussion, in: Altermann, W. and Corcoran, P. L., eds., Precambrian Sedimentary Environments: a Modern Approach to Ancient Depositional Systems. International Association of Sedimentologists Special Publication 33, p. 3–13.

Dorsey, R. J., Housen, B. A., Janecke, S. U., Fanning, C. M., and Spears, A. L. F., 2011, Stratigraphic record of basin development within the San Andreas fault system: Late Cenozoic Fish Creek-Vallecito basin, southern California: Geological Society of America Bulletin, v. 123, p. 771–793.

Duller, R. A., Kougioumtzoglou, I., Dunning, S., and Fedele, J., 2012, Vertical and lateral patterns of grain size in the stratigraphic record, The Geological Society, London, William Smith Meeting 2012, Strata and Time: Probing the gaps in our understanding, Abstract, p. 10.

Echavarria, L., Hernández, R., Allmendinger, R., and Reynolds, J., 2003, Subandean thrust and fold belt of northwestern Argentina: Geometry and timing of the Andean evolution: American Association of Petroleum Geologists Bulletin, v. 87, p. 965–985.

Eriksson, P. G., Condie, K. C., Tirsgaard, H., Mueller, W. U., Alterman, W., Miall, A. D., Aspler, L. B., Catuneanu, O., and Chiarenzelli, J. R., 1998, Precambrian clastic sedimentation systems, in Eriksson, P. G., ed., Precambrian clastic sedimentation systems: Sedimentary Geology, Special Issue, v. 120, p. 5-53.

Ettensohn, F. R., 2008, The Appalachian foreland basin in eastern United States, in Miall, A. D., ed., The Sedimentary Basins of the United States and Canada: Sedimentary basins of the World, v. 5, K. J. Hsü, Series Editor, Elsevier Science, Amsterdam, p. 105–179.

Fiduk, J. C., and Behrens, E. W., 1993, A comparison of Plio-Pleistocene to recent sediment accumulation rates in the East breaks area, northwestern Gulf of Mexico, in Armentrout, J. M., Bloch, R., Olson, H. C., and Perkins, B. E., eds., Rates of Geologic Processes: Gulf Coast Section, Society of Economic Paleontologists and Mineralogists, Fourteenth Annual Research Conference, Houston, Texas, p. 41–55.

Filomena, C. M., Stollhofen, H., 2011, Ultrasonic logging across unconformities — outcrop and core logger sonic patterns of the Early Triassic Middle Buntsandstein Hardegsen unconformity, southern Germany, Sedimentary Geology, v. 236, p. 185-196.

Finney, S. C., Grubb, B. J., and Hatcher, R. D., Jr, 1996, Graphic correlation of Middle Ordovician graptolite shale, southern Appalachian: an approach for examining the subsidence and migration of a Taconic foreland basin: Geological Society of America Bulletin, v. 108, p. 355–371.

Fischer, A. G., and Roberts, L. T., 1991, Cyclicity in the Green River Formation (lacustrine Eocene) of Wyoming: Journal of Sedimentary Petrology, v. 61, p. 1146–1154.

Fletcher, C. J. N., 1988, Tidal erosion, solution cavities and exhalative mineralisation associated with the Jurassic unconformity at Ogmore, South Glamorgan: Proceedings of the Geologists Association, v. 99, p. 1-14.

Fletcher, C. J. N., Davies, J. R., Wilson, D., and Smith, M., 1986, The depositional environment of the basal 'Littoral Lias' in the Vale of Glamorgan – a discussion of the reinterpretation by Ager (1986): Proceedings of the Geologists' Association, v. 97, p. 383-384.

Frazier, D. E., 1967; Recent deltaic deposits of the Mississippi River—their development and chronology: Transactions of the Gulf Coast Association of Geological Societies, v. 17, p. 287–315.

Gould, H. R., 1970, The Mississippi Delta complex, in Morgan, J. P., ed., Deltaic sedimentation: modern and ancient: Society of Economic Paleontologists and Mineralogists Special Publication 15, p. 3-30.

Gould, S. J., 1965, Is uniformitarianism necessary? American Journal of Science, v. 263, p. 223–228.

Grabau, A. W., 1906, Types of sedimentary overlap. Geological Society of America Bulletin, v. 17, p. 567-636.

Gradstein, F. M., Ogg, J. G., and Smith, A. G., eds., 2004a, A geologic time scale: Cambridge University Press, Cambridge, 610 p.

Gradstein, F. M., Ogg, J. G., Schmitz, M. D., and Ogg, G. M., 2012, The Geologic time scale 2012: Elsevier, Amsterdam, 2 vols., 1176 p.

Gradstein, F. M., Ogg, J. G., Schmitz, M. D., and Ogg, G. M., eds., 2020, Geologic time scale 2020: Elsevier, Amsterdam, 1357 p.

Hajek, E. A., Heller, P. L., and Sheets, B. A., 2010, Significance of channel-belt clustering in alluvial basins: Geology, v. 38, p. 535–538.

Hampson, G.J., 2010, Sediment dispersal and quantitative stratigraphic architecture across and ancient shelf: Sedimentology, v. 57, p. 96–141.

Hampson, G.J., Jewell, T.O., Irfan, N., Gani, M.R. and Bracken, B., 2013, Modest changes in fluvial style with varying accommodation in regressive alluvial-to-coastal-plain wedge: Upper Cretaceous Blackhawk Formation, Wasatch Plateau, central Utah, U.S.A.: Journal of Sedimentary Research, v. 83, p. 145–169.

Haq, B. U., Hardenbol, J., and Vail, P. R., 1987, Chronology of fluctuating sea levels since the Triassic (250 million years ago to present): Science, v. 235, p. 1156–1167.

Haq, B. U., Hardenbol, J., and Vail, P. R., 1988, Mesozoic and Cenozoic chronostratigraphy and cycles of sea-level change, in Wilgus, C. K., Hastings, B. S., Kendall, C. G. St. C., Posamentier, H. W., Ross, C. A., and Van Wagoner, J. C., eds., Sea-level Changes: an integrated approach: Society of Economic Paleontologists and Mineralogists Special Publication 42, p. 71–108.

Hardenbol, J., Thierry, J., Farley, M. B., Jacquin, T., Graciansky, P.-C., and Vailo, P. R., 1998. Mesozoic and Cenozoic sequence chronostratigraphic framework of European basins, in Graciansky, P.-C. de, Hardenbol, J., Jacquin, T., and Vail, P. R., eds., Mesozoic and Cenozoic sequence stratigraphy of European basins, Society for Sedimentary Geology (SEPM) Special Publication 60, p. 3–13.

Hassan, M. S., Venetikidis, A., Bryant, G., and Miall, A. D., 2018, The sedimentology of an erg margin: the Kayenta-Navajo transition (Lower Jurassic), Kanab, Utah: Journal of Sedimentary Research, v. 88, p. 613-640.

Heckel, P. H., 1986, Sea-level curve for Pennsylvanian eustatic marine transgressive-regressive depositional cycles along midcontinent outcrop belt, North America: Geology, v. 14, p. 330-334.

Heller, P. L., Beekman, F., Angevine, C. L., and Cloetingh, S. A. P. L., 1993, Cause of tectonic reactivation and subtle uplifts in the Rocky Mountain region and its effect on the stratigraphic record: Geology, v. 21, p. 1003–1006.

Hilgen, F. J., 1991. Extension of the astronomically calibrated (polarity) time scale to the Miocene/Pliocene boundary: Earth and Planetary Sciences Letters, v. 107, p. 349–368.

Hilgen, F. J., Brinkhuis, H., and Zachariasse, W. J., 2006, Unit stratotypes for global stages. The Neogene perspective: Earth Science Reviews v. 74, p. 113-125.

Hilgen, F. J., Hinnov, L. A., Aziz, H. A., Abels, H. A., Batenburg, S., Bosmans, J. H. C., de Boer, B., Hüsings, S. K., Kuiper, K. F., and Lourens, L. J., 2015, Stratigraphic continuity and fragmentary sedimentation: the success of cyclostratigraphy as part of integrated stratigraphy in Smith, D. G., Bailey, R., J., Burgess, P., and Fraser, A., eds., Strata and time: Geological Society, London, Special Publication 404, p. 157–197.

Hilgen, F. J., Kuiper, K., Krijgsman, W., Snel, E., and van der Laan, E., 2007, Astronomical tuning as the basis for high resolution chronostratigraphy: the intricate history of the Messinian Salinity Crisis: Stratigraphy, v. 4, p. 231–238.

Hiscott, R. N., 1981, Deep-sea fan deposits in the Macigno Formation (Middle-Upper Oligocene) of the Gordana valley, northern Apennines, Italy — Discussion: Journal of Sedimentary Petrology, v. 51, p. 1015–1033.

Hofmann, M. H., Wroblewski, A., and Boyd, R., 2011, Mechanisms controlling the clustering of fluvial channels and the compensational stacking of cluster belts: Journal of Sedimentary Research, v. 81, p. 670–685.

Holbrook, J.M., 1996, Complex fluvial response to low gradients at maximum regression: a genetic link between smooth sequence-boundary morphology and architecture of overlying sheet sandstone: Journal of Sedimentary Research, v. 66, p. 713–722.

Holbrook, J. M., 2001, Origin, genetic interrelationships, and stratigraphy over the continuum of fluvial channel-form bounding surfaces: an illustration from middle Cretaceous strata, southeastern Colorado: Sedimentary Geology, v. 144, p. 179–222.

Holbrook, J. M., and Bhattacharya, J. P., 2012, Reappraisal of the sequence boundary in time and space: Case and considerations for an SU (subaerial unconformity) that is not a sediment bypass surface, a time barrier, or an unconformity: Earth Science Reviews, v 113, p. 271–302.

Holbrook, J. M., and Miall, A. D., 2020, Time in the Rock: A field guide to interpreting past events and processes from siliciclastic stratigraphy: Earth Science Reviews, v. 203, 103121, 23 p.

Holbrook, J., Scott, R. W., and Oboh-Ikuenobe, F. E., 2006, Base-level buffers and buttresses: a model for upstream versus downstream control on fluvial geometry and architecture within sequences: Journal of Sedimentary Research, v. 76, p. 162–174.

Holmes, A., 1913, The age of the Earth: Harper, London

Howell, J.A., and Flint, S.S., 2003, Siliciclastics case study: The Book Cliffs, in Coe, A., ed., The Sedimentary Record of Sea-level Change: Cambridge, U.K., Open University and Cambridge University Press, p. 135–208.

Hoyt, J. H., and Henry, V. J., Jr., 1967, Influence of island migration on barrier island sedimentation: Geological Society of America Bulletin, v. 78, p. 77–86.

Jackson, R. G. II, 1976, Depositional model of point bars in the lower Wabash River: Journal of Sedimentary Petrology, v. 46, p. 579–594.

Jerolmack, D. J., and Paola, C., 2010, Shredding of environmental signals by sediment transport: Geophysical Research Letters, v. 37, L10401, 5 p.

Jerolmack, D., and Sadler, P., 2007, Transience and persistence in the depositional record of continental margins: Journal of Geophysical Research, v. 112, F03S13, 14 p.

Jervey, M. T., 1988, Quantitative geological modeling of siliciclastic rock sequences and their seismic expression, in Wilgus, C. K., Hastings, B. S., Kendall, C. G. St. C., Posamentier, H. W., Ross, C. A., and Van Wagoner, J. C., eds., Sea level Changes - an integrated approach: Society of Economic Paleontologists and Mineralogists Special Publication 42, p. 47–69.

Johnson, M. E., and McKerrow, W. S., 1995, The Sutton Stone: an early Jurassic rocky shore deposit in South Wales: Palaeontology, v. 38, p. 529–541.

Johnson, N. M., Sheikh, K. A., Dawson-Saunders, E., and McRae, L. E., 1988, The use of magnetic-reversal time lines in stratigraphic analysis: a case study in measuring variability in sedimentation rates, in Kleinspehn, K. L., and Paola, C., eds., New perspectives in basin analysis: Springer-Verlag Inc., Berlin and New York, p. 189–200.

Johnson, N. M., Stix, J., Tauxe, L., Cerveny, P. F., and Tahirkheli, R. A. K., 1985, Paleomagnetic chronology, fluvial processes, and tectonic implications of the Siwalik deposits near Chinji Village, Pakistan: Journal of Geology, v. 93, p. 27–40.

Jones, M. A., Heller, P. L., Roca, E., Garcés, M., and Cabrera, L., 2004, Time lag of syntectonic sedimentation across an alluvial basin: theory and example from the Ebro basin, Spain, Basin Research, v. 16, p. 467-488.

Jordan, T. E., 1981, Thrust loads and foreland basin evolution, Cretaceous, western United States: American Association of Petroleum Geologists Bulletin, v. 65, p. 2506-2520.

Kamp, P. J. J., and Turner, G. M., 1990, Pleistocene unconformity-bounded shelf sequences (Wanganui Basin, New Zealand) correlated with global isotope record: Sedimentary Geology, v. 68, p. 155–161.

Kauffman, E. G., 1986, High-resolution event stratigraphy: regional and global bio-events, in Walliser, O. H. ed., Global bioevents:

Lecture Notes on Earth History: Springer-Verlag, Berlin, p. 279–335.

Kauffman, E. G., 1988, Concepts and methods of high-resolution event stratigraphy: Annual Reviews of Earth and Planetary Sciences, v. 16, p. 605–654.

Kemp, D. B., 2012, Stochastic and deterministic controls on stratigraphic completeness and fidelity: International Journal of Earth Sciences (Geol. Rundschau), v. 101, p. 2225–2238. DOI https://doi.org/10.1007/s00531-012-0784-1.

Kemp, D. B., and Sadler, P. M., 2014, Climatic and eustatic signals in a global compilation of shallow marine carbonate accumulation rates: Sedimentology, v. 61, p. 1286–1297.

Kidwell, S. M., 1988, Reciprocal sedimentation and noncorrelative hiatuses in marine-paralic siliciclastics: Miocene outcrop evidence: Geology, v. 16, p. 609–612.

Kim, W., and Paola, C., 2007, Long-period cyclic sedimentation with constant tectonic forcing in an experimental relay ramp: Geology, v. 35, p. 331–334.

Kocurek, G., and Day, M., 2018, What is preserved in the aeolian ock record? A Jurassic Entrada Sandstone case study at the Utah-Arizona border: Sedimentology, v. 65, p. 1301-1321.

Kolb, C. R., and Van Lopik, J. R., 1966, Depositional environment of the Mississippi River deltaic plain - southeastern Louisiana, in Shirley, M. L., ed., Deltas in their geologic framework: Houston Geological Society, p. 17–61.

Kominz, M. A., Browning, J. V., Miller, K. G., Sugarman, P. J., Mizintserva, S., and Scotese, C. R., 2008, late Cretaceous to Miocene sea-level estimates from the New Jersey and Delaware coastal plain coreholes: an error analysis: Basin Research, v. 20, p. 211–226.

Krystinik, L.F., DeJarnett, B.B., 1995, Lateral variability of sequence stratigraphic framework in the Campanian and Lower Maastrichtian of the Western Interior Seaway, in Van Wagoner, J.C., and Bertram, G.T., eds., Sequence Stratigraphy of Foreland Basin Deposits: American Association of Petroleum Geologists, Memoir 64, p. 11–25.

Kumar, N., and Sanders, J. E., 1974, Inlet sequence: a vertical succession of sedimentary structures and textures created by the lateral migration of tidal inlets: Sedimentology, v. 21, p. 491–532.

Leckie, D.A., 1994, Canterbury Plains, New Zealand–implications for sequence stratigraphic models. American Association of Petroleum Geologists, Bulletin, v. 78(8), p. 1240-1256.

Leclair, S. F., 2011, Interpreting fluvial hydromorphology from the rock record: large-river peak flows leave no clear signature, in Davidson, S. K., Leleu, S., and North, C. P., eds., From river to rock record: Society for Sedimentary Geology (SEPM) Special Publication 97, p. 113–123.

Lin, W., Bhattacharya, N. P., Jicha, B. R., Singer, B. S., and Matthews, W., 2021, Has Earth ever been ice-free? Implications for glacio-eustasy in the Cretaceous greenhouse age using high-resolution sequence stratigraphy: Geological Society of America Bulletin, v. 133, p. 243-252.

Liu, S., and Nummedal, D., 2004, Late Cretaceous subsidence in Wyoming: Quantifying the dynamic component: Geology, v. 32, p. 397–400.

Locklair, R. E., and Sageman, B. B., 2008, Cyclostratigraphy of the Upper Cretaceous Niobrara Formation, Western Interior, U.S.A.: A Coniacian-Santonian timescale: Earth and Planetary Science Letters, v. 269, p. 539–552.

Loutit, T. S., Hardenbol, J., Vail, P. R., and Baum, G. R., 1988, Condensed sections: the key to age dating and correlation of continental margin sequences, in Wilgus, C. K., Hastings, B. S., Kendall, C. G. St. C., Posamentier, H. W., Ross, C. A., and Van Wagoner, J. C., eds., Sea-level Changes: an integrated approach: Society of Economic Paleontologists and Mineralogists Special Publication 42, p. 183–213.

Lyell, C., 1830–1833, Principles of Geology, 3 vols.: John Murray, London (reprinted by Johnson Reprint Corp., New York, 1969).

MacLeod, N., and Keller, G., 1991, How complete are Cretaceous/Tertiary boundary sections? A chronostratigraphic estimate based on graphic correlation: Geological Society of America Bulletin, v. 103, p. 1439-1457.

Mandelbrot, B. B., 1983, The fractal geometry of nature: Freeman, New York, 468 p.

McGowran, B., 2005, Biostratigraphy: Microfossils and Geological Time: Cambridge University Press, Cambridge, 459 p.

McKee, B. A., Nittrouer, C. A., and Demaster, D. J., 1983, Concepts of sediment deposition and accumulation applied to the continental shelf near the mouth of the Yangtze River: Geology, v. 11, p. 631–633.

McKee, E. D, Crosby, E. J., and Berryhill, H. L., Jr., 1967, Flood deposits, Bijou Creek, Colorado, June 1965: Journal of Sedimentary Petrology, v. 37, p. 829-851.

Medwedeff, D. A., 1989, Growth fault-bend folding at Southeast Lost Hills, San Joaquin valley, California: American Association of Petroleum Geologists Bulletin, v. 73, p. 54-67.

Meyers, S., 2012, Seeing red in cyclic stratigraphy, The Geological Society, London, William Smith Meeting 2012, Strata and Time: Probing the gaps in our understanding, Abstract, p. 18.

Meyers, S. R., and Sageman, B. B., 2004, detection, quantification, and significance of hiatuses in pelagic and hemipelagic strata: Earth and Planetary Sciences Letters, v. 224, p. 55–72.

Miall, A. D., 1978a, Tectonic setting and syndepositional deformation of molasse and other nonmarine-paralic sedimentary basins; Canadian Journal of Earth Sciences, v. 15, p. 1613-1632.

Miall, A. D., 1978b, Fluvial sedimentology: an historical review, in Miall, A. D., ed., Fluvial Sedimentology: Canadian Society of Petroleum Geologists Memoir 5, p. 1-47.

Miall, A. D., 1985, Architectural-element analysis: A new method of facies analysis applied to fluvial deposits: Earth Science Reviews, v. 22, p. 261–308.

Miall, A. D., 1988a, Reservoir heterogeneities in fluvial sandstones: lessons from outcrop studies: American Association of Petroleum Geologists Bulletin, v. 72, p. 682–697.

Miall, A. D., 1988b, Facies architecture in clastic sedimentary basins, in Kleinspehn, K., and Paola, C., eds., New perspectives in basin analysis: Springer-Verlag Inc., New York, p. 67–81.

Miall, A. D., 1988c, Architectural elements and bounding surfaces in channelized clastic deposits: notes on comparisons between fluvial and turbidite systems, in Taira, A., and Masuda, F., eds., Sedimentary facies in the active plate margin: Terra Scientific Publishing Company, Tokyo, Japan, p. 3–15.

Miall, A. D., 1991b, Hierarchies of architectural units in terrigenous clastic rocks, and their relationship to sedimentation rate, in A. D. Miall and N. Tyler, eds., The three-dimensional facies architecture of terrigenous clastic sediments and its implications for hydrocarbon discovery and recovery: Society of Economic Paleontologists and Mineralogists, Concepts in Sedimentology and Paleontology, v. 3, p. 6–12.

Miall, A. D., 1995, Whither stratigraphy? Sedimentary Geology, v. 100, p. 5-20.

Miall, A. D., 1996, The geology of fluvial deposits: sedimentary facies, basin analysis and petroleum geology: Springer-Verlag Inc., Heidelberg, 582 p.

Miall, A. D., 1997, The geology of stratigraphic sequences, First edition: Springer-Verlag, Berlin, 433 p.

Miall, A. D., 2010, The geology of stratigraphic sequences, second edition: Springer-Verlag, Berlin, 522 p.

Miall, A. D., 2014a, Fluvial depositional systems: Springer-Verlag, Berlin 316 p.

Miall, A. D., 2014b, The emptiness of the stratigraphic record: A preliminary evaluation of missing time in the Mesaverde Group, Book Cliffs, Utah: Journal of Sedimentary Research, v. 84, p. 457-469.

Miall, A. D., 2015, Updating uniformitarianism: stratigraphy as just a set of "frozen accidents", in Smith, D. G., Bailey, R., J., Burgess, P., and Fraser, A., eds., Strata and time: Geological Society, London, Special Publication 404, p. 11–36.

Miall, A. D., 2016, The valuation of unconformities: Earth Science Reviews, v. 163, p. 22–71.

Miall, A. D., and Arush, M., 2001a, The Castlegate Sandstone of the Book Cliffs, Utah: sequence stratigraphy, paleogeography, and tectonic controls: Journal of Sedimentary Research, v. 71, p. 536–547.

Miall, A.D., and Arush, M., 2001b, Cryptic sequence boundaries in braided fluvial successions: Sedimentology, v. 48, p. 971–985.

Miall, A. D., Catuneanu, O, 2019, The Western Interior Basin, in Miall, A. D., ed., The Sedimentary Basins of the United States and Canada, Second edition: Sedimentary basins of the World, v. 5, K. J. Hsü, Series Editor, Elsevier Science, Amsterdam, p. 401–443.

Miall, A. D., Holbrook, J. M., and Bhattacharya, J. P., 2021, The stratigraphy machine: Journal of Sedimentary Research, v. 91, p. 595–610.

Miall, A. D., and Miall, C. E., 2001, Sequence stratigraphy as a scientific enterprise: the evolution and persistence of conflicting paradigms: Earth Science Reviews, v. 54, #4, p. 321–348.

Miall, A. D., and Miall, C. E., 2004, Empiricism and Model-Building in stratigraphy: Around the Hermeneutic Circle in the Pursuit of Stratigraphic Correlation. Stratigraphy: American Museum of Natural History, v. 1, p. 27–46.

Middleton, G. V., Plotnick, R. E., and Rubin, D. M., 1995, Nonlinear dynamics and fractals; New numerical techniques for sedimentary data sets: Society for Sedimentary Geology, Tulsa, Oklahoma, Short Course No. 36, 174 p.

Miller, K. G., Browning, J. V., Aubry, M.-P., Wade, B., Katz, M. E., Kulpecz, A. A., and Wright, J. D., 2008, Eocene-Oligocene global climate and sea-level changes: St. Stephens Quarry, Alabama: Geological Society of America Bulletin, v. 120, p. 34-53.

Miller, K. G., and Mountain, G. S., 1994, Global sea-level change and the New Jersey margin, in Mountain, G. S., Miller, K. G., Blum, P., et al., eds., Proceedings of the Ocean Drilling Program, Initial Reports, v. 150, p. 11–20.

Miller, K. G., Mountain, G. S., Browning, J. V., Kominz, M., Sugarman, P. J., Christi-Blick, N., Katz, M. E., and Wright, J. D., 1998, Cenozoic global sea level, sequences, and the New Jersey transect: results from coastal plain and continental slope drilling: Reviews of Geophysics, v. 36, p. 569–601.

Miller, K. G., Wright, J. D., and Browning, J. V., 2005, Visions of ice sheets in a greenhouse world: Marine Geology, v. 217, p. 215–231.

Monger, J. W. H., 1993, Cretaceous tectonics of the North American Cordillera, in Caldwell, W. G. E., and Kauffman, E. G., eds., Evolution of the Western Interior Basin: Geological Association of Canada Special Paper 39, 31–47.

Nemec, W., 1988, Coal correlations and intrabasinal subsidence: a new analytical perspective, in Kleinspehn, K. L., and Paola, C., eds., New perspectives in basin analysis: Springer-Verlag Inc., Berlin and New York, p. 161–188.

Nummedal, D., and Swift, D. J. P., 1987, Transgressive stratigraphy at sequence-bounding unconformities: some principles derived from Holocene and Cretaceous examples, in Nummedal, D., Pilkey, O. H., and Howard, J. D., eds., Sea-level fluctuation and coastal evolution; Society of Economic Paleontologists and Mineralogists Special Publication 41, p. 241–260.

Olsen, P. E., 1990, Tectonic, climatic, and biotic modulation of lacustrine ecosystems—examples from Newark Supergroup of eastern North America, in Katz, B. J., ed., Lacustrine basin exploration: case studies and modern analogs: American Association of Petroleum Geologists Memoir 50, p. 209-224.

Olsen, T., Steel, R. J., Høgseth, K., Skar, T., and Røe, S.-L., 1995, Sequential architecture in a fluvial succession: sequence stratigraphy in the Upper Cretaceous Mesaverde Group, Price Canyon, Utah: Journal of Sedimentary Research, v. B65, p. 265-280.

Oomkens, E., 1970, Depositional sequences and sand distribution in the postglacial Rhône delta complex, in Morgan, J. P., ed., Deltaic sedimentation: modern and ancient: Society of Economic Paleontologists and Mineralogists Special Publication 15, p. 198-212.

Paola, C., Straub, K., Mohrig, D., and Reinhardt, L., 2009, The unreasonable effectiveness of stratigraphic and geomorphic experiments: Earth Science Reviews, v. 97, p. 1–43.

Pattison, S.A.J., 2018, Using classic outcrops to revise sequence stratigraphic models: Reevaluating the Campanian Desert Member (Blackhawk Formation) to lower Castlegate Sandstone interval, Book Cliffs, Utah and Colorado, USA: Geology, v. 47, p. 11-14.

Pattison, S.A.J., 2019, Re-evaluating the sedimentology and sequence stratigraphy of classic Book Cliffs outcrops at Tusher and Thompson canyons, eastern Utah, USA: applications to correlation, modelling, and prediction in similar nearshore terrestrial to shallow marine subsurface settings worldwide. Marine and Petroleum Geology, v. 102, p. 202–230.

Pattison, S.A.J., 2020a, Sediment-supply-dominated stratal architecture in a regressively stacked succession of shoreline sand bodies, Campanian Desert Member to Lower Castlegate Sandstone interval, Book Cliffs, Utah-Colorado, USA: Sedimentology, v. 67, p. 390-430.

Pattison, S.A.J., 2020b, No evidence for an unconformity at the base of the lower Castlegate Sandstone in the Campanian Book Cliffs, Utah-Colorado, United States: Implications for sequence models: American Association of Petroleum Geologists Bulletin, v. 104, p. 595-628.

Peper, T., Beekman, F., and Cloetingh, S., 1992, Consequences of thrusting and intraplate stress fluctuations for vertical motions in foreland basins and peripheral areas: Geophysical Journal International, v. 111, p. 104–126.

Petit, B. S., Blum, M., Pecha, M., McLean, N., Bartschi, N. C. and Saylor, J. E., 2019, Detrital-zircon U-Pb paleodrainage reconstruction and geochronology of the Campanian Blackhawk-Castlegate succession, Wasatch Plateau and Book Cliffs, Uta, U.S.A., Journal of Sedimentary Research, v. 89, p. 273-292.

Phillips, S., and Bustin, R. M., 1996, Sedimentology of the Changuinola peat deposit: organic and clastic sedimentary response to punctuated coastal subsidence: Geological Society of America Bulletin, v. 108, p. 794–814.

Pillans, B., Alloway, B., Naish, T., Westgate, J., Abbott, S., and Palmer, A., 2005, Silicic tephras in Pleistocene shallow-marine sediments of Wanganui Basin, New Zealand: Journal of the Royal Society of New Zealand, v. 35, p. 43-90.

Pirmez, C., Prather, B. E., Mallarino, G., O'Hayer, W. W., Droxler, A. W, and Winker, C. D., 2012, Chronostratigraphy of the Brazos-Trinity depositional system, western Gulf of Mexico: implications for deepwater depositional models, in Applications of the Principles of seismic geomorphology to continental slope and base-of-slope systems: case studies from seafloor and near-seafloor analogues: Society for Sedimentary Geology Special Publication 99, p. 111-143.

Plint, A.G., Tyagi, A.A., McCausland, P.J.A., Krawetz, J.R., Zhang, H., Roca, X., Varban, B.L., Hu, Y.G., Kreitner, M.A., and Hay, M. J., 2012, Dynamic relationship between subsidence, sedimentation, and unconformities in mid-Cretaceous, shallow-marine strata of the Western Interior Foreland Basin: Links to Cordilleran tectonics, in Busby, C., and Azor, A., eds., Tectonics of Sedimentary Basins: Recent Advances: Wiley-Blackwell, Chichester, p. 480–507.

Plotnick, R. E., 1986, A fractal model for the distribution of stratigraphic hiatuses: Journal of Geology, v. 94, p. 885–890.

Pochat. S., Castelltort, S., Choblet, G., and Driessche, J. V. Den, 2009, High-resolution record of tectonic and sedimentary processes in growth strata: Marine and Petroleum Geology, v. 26, p. 1350–1364.

Posamentier, H. W., Allan, G. P., and James, D. P., 1992, High-resolution sequence stratigraphy – the East Coulee Delta, Alberta: Journal of Sedimentary Petrology, v. 62, p. 310–317.

Prather, B. E., Pirmez, C., and Winker, C. D., 2012, Stratigraphy of linked intraslope basins: Brazos-Trinity System, western Gulf of Mexico, in Application of the Principles of Seismic Geomorphology to Continental-Slope and Base-of-Slope Systems: Case Studies from Seafloor and Near-Seafloor Analogues: SEPM Special Publication 99, p. 83–109.

Price, R. A., 1973, large-scale gravitational flow of supracrustal rocks, southern Canadian Rockies, in De Jong, K. A., and Scholten, R., eds., Gravity and Tectonics, Wiley-Interscience, New York, p. 491-502.

Qayyum, F., de Groot, P., Hemstra. N., and Catuneanu, O., 2015, 4D Wheeler diagrams: concept and applications, in Smith, D. G., Bailey, R., J., Burgess, P., and Fraser, A., eds., Strata and time: Geological Society, London, Special Publication 404, p. 223–232.

Ramsbottom, W. H. C., 1979, Rates of transgression and regression in the Carboniferous of NW Europe: Journal of the Geological Society, London, v. 136, p. 147-153.

Runkel, A. C., Miller, J. F., McKay, R. M., Palmer, A. R., and Taylor, J. F., 2007, High-resolution sequence stratigraphy of lower Paleozoic sheet sandstones in central North America: the role of special conditions of cratonic interiors in development of stratal architecture: Geological Society of America Bulletin, v. 119, p. 860–881.

Runkel, A. C., Miller, J. F., McKay, R. M., Palmer, A. R., and Taylor, J. F., 2008, The record of time in cratonic interior strata: does exceptionally slow subsidence necessarily result in exceptionally poor stratigraphic completeness? in Pratt, B. R., and Holmden, C., eds., Dynamics of epeiric seas: Geological Association of Canada Special Paper 48, p. 341–362.

Ryer, T. A., 1993, Speculations on the origins of mid-Cretaceous clastic wedges, central Rocky Mountain region, United States, in Caldwell, W. G. E., and Kauffman, E. G., eds., Evolution of the Western Interior Basin: Geological Association of Canada Special Paper 39, p. 189–198.

Sadler, P. M., 1981, Sedimentation rates and the completeness of stratigraphic sections: Journal of Geology, v. 89, p. 569–584.

Sadler, P. M., 1999, The influence of hiatuses on sediment accumulation rates: GeoResearch Forum, v. 5, p. 15–40.

Sadler, P. M., Cooper, R. A., and Crampton, J. S., 2014, High-resolution geobiologic time-lines: progress and potential, fifty years after the advent of graphic correlation: The Sedimentary Record, v. 12, #3, p. 4–9.

Sadler, P. M., Cooper, R. A., and Melchin, M., 2009: High-resolution, early Paleozoic (Ordovician-Silurian) time scales: Geological Society of America Bulletin, v. 121, p. 887–906.

Sadler, P. M., and Jerolmack, D., 2012, Scaling laws for aggradation, denudation and progradation rates: the case for time-scale invariance at sedimentary sources and sinks: The Geological Society, London, William Smith Meeting 2012, Strata and Time: Probing the gaps in our understanding, Abstract, p. 32.

Sadler, P. M., and Jerolmack, D., 2015, Scaling laws for aggradation, denudation and progradation rates: the case for time-scale invariance at sediment sources and sinks, in Smith, D. G., Bailey, R., J., Burgess, P., and Fraser, A., eds., Strata and time: Geological Society, London, Special Publication 404, p. 69–88.

Sageman, B. B., Singer, B. S., Meyers, S. R., Siewert, S. E., Walaszczyk, I., Condon, D. J., Jicha, B. R., Obradovich, J. D., and Sawyer, D. A., 2014, Integrating $^{40}Ar/^{39}Ar$, U-Pb and astronomical clocks in the Cretaceous Niobrara Formation, Western Interior Basin, USA: Geological Society of America Bulletin, v. 126, p. 956-973.

Salvany, J. M., Larrasoaña, J. G., Mediavilla, G., and Rebollo, A., 2011, Chronology and tecto-sedimentary evolution of the Late Pliocene to Quaternary deposits from the Lower Guadilquivir foreland basin, SW Spain: Sedimentary Geology, v. 241, p. 22-39.

Scarponi, D., Kaufman, D., Amarosi, A. and Kowalewski, M., 2013, Sequence stratigraphy and the resolution of the fossil record: Geology, v. 41, p. 239–242.

Schlager, W., 2004, Fractal nature of stratigraphic sequences, Geology, v. 32, p. 185-188.

Schlager, W., 2005, Carbonate sedimentology and sequence stratigraphy: SEPM Concepts in Sedimentology and Paleontology #8, 200p.

Schumm S. A., 1973, Geomorphic thresholds and complex response of drainage systems, in Morisawa, M. ed., Fluvial geomorphology, Publications in Geomorphology, State University of New York, Binghamton, N.Y., p. 299-310.

Schumm, S. A., 1977, The fluvial system, John Wiley and Sons, New York, 338 p.

Scott, A. C., and Stephens, R. S., 2015, British Pennsylvanian (Carboniferous) coal-bearing sequences: where is the time? in Smith, D. G., Bailey, R., J., Burgess, P., and Fraser, A., eds., Strata and time: Geological Society, London, Special Publication 404, p. 283–302.

Shanley, K. W., and McCabe, P. J., 1994, Perspectives on the sequence stratigraphy of continental strata: American Association of Petroleum Geologists Bulletin, v. 78, p. 544–568.

Sheets, B. A., Hickson, T. A., and Paola, C., 2002, Assembling the stratigraphic record: depositional patterns and time-scales in an experimental alluvial basin: Basin Research, v. 14, p. 287–301.

Smith, A. G., Barry, T., Bown, P., Cope, J., Gale, A., Gibbard, P., Gregory, J., Hounslow, M., Kemp, D., Knox, R., Marshall, J., Oates, M., Rawson, P., Powell, J., and Waters, C., 2015, GSSPs, global stratigraphy and correlation: in Smith, D. G., Bailey, R., J., Burgess, P., and Fraser, A., eds., Strata and time: Geological Society, London, Special Publication 404, p. 37–67.

Smith, D.G., 1994, Cyclicity or chaos? in de Boer, P.L.,and Smith, D. G., eds., Orbital Forcing and Cyclic Sequences. International Association of Sedimentologists, Special Publication 19, p. 531–544.

Smith, W., 1815, A memoir to the map and delineation of the strata of England and Wales, with part of Scotland: London: John Carey, 51 p.

Sommerfield, C. K., 2006, On sediment accumulation rates and stratigraphic completeness: lessons from Holocene ocean margins: Continental Shelf Research, v. 26, p. 2225–2240.

Southard, J. B., Bohuchwal, L. A., and Romea, R. D., 1980, Test of scale modelling of sediment transport in steady unidirectional flow: Earth Surface Processes, v. 5, p. 17–23.

Spieker, E.M., and Reeside, J.B., Jr., 1925, Cretaceous and Tertiary formations of the Wasatch Plateau, Utah: Geological Society of America, Bulletin, v. 36, p. 429–454.

Stark, T. J., Zeng, H., and Jackson, A., 2013, An introduction to this special section: Chronostratigraphy: The Leading Edge, v. 32, p. 132–138.

Stouthamer, E., Cohen, K. M., and Gouw, M. J. P., 2011, Avulsion and its implication for fluvial-deltaic architecture: insights from the Holocene Rhine-Meuse delta, in Davidson, S. K., Leleu, S., and North, C. P., eds., From river to rock record: Society for Sedimentary Geology (SEPM) Special Publication 97, p. 215–231.

Stow, D. A. V., Howell, D. G., and Nelson, C. H., 1983, Sedimentary, tectonic, and sea-level controls on submarine fan and slope-apron turbidite systems: Geo-marine Letters, v. 3, p. 57–64.

Strasser, A., Hilgen, F. J., and Heckel, P. H., 2006, Cyclostratigraphy — Concepts, definitions, and applications: Newsletters in Stratigraphy, v. 42, p. 75–114.

Straub, K. M., Paola, C., Mohrig, D., Wolinsky, M. A., and George, T, 2009, Compensational stacking of channelized sedimentary deposits: Journal of Sedimentary Research, v. 79, 673-688.

Strong, N., and Paola, C., 2008, Valleys that never were: time surfaces versus stratigraphic surfaces: Journal of Sedimentary Research, v. 78, p. 579–593.

Sun, J., Li, Y., Zhang, Z., and Fu, B., 2010, Magnetostratigraphic data on Neogene growth folding in the foreland basin of the southern Tianshan Mountains: Geology, v. 37, p. 1051–1054.

Suter, J. R., Berryhill, H. L., Jr., and Penland, S., 1987, Late Quaternary sea-level fluctuations and depositional sequences, southwest Louisiana continental shelf, in Nummedal, D., Pilkey, O. H., and Howard, J. D., eds., 1987, Sea-level fluctuation and coastal evolution: Society of Economic Paleontologists and Mineralogists Special Publication 41, p. 199–219.

Tipper, J. C., 2015, The importance of doing nothing: stasis in sedimentation systems and its stratigraphic effects: in Smith, D. G., Bailey, R., J., Burgess, P., and Fraser, A., eds., Strata and time: Probing the Gaps in Our Understanding: Geological Society, London, Special Publication 404, p. 105–122.

Trabucho-Alexandre, J. 2015. More gaps than shale: erosion of mud and its effect on preserved geochemical and palaeobiological signals. in: Smith, D. G., Bailey, R. J., Burgess, P. M., and Fraser, A.J., eds., Strata and Time: Probing the Gaps in Our Understanding: Geological Society, London, Special Publications, 404, p. 251–270.

Vail, P. R., Audemard, F., Bowman, S. A., Eisner, P. N., and Perez-Crus, C., 1991, The stratigraphic signatures of tectonics, eustasy and sedimentology—an overview, in Einsele, G., Ricken, W., and Seilacher, A., eds., Cycles and events in stratigraphy: Springer-Verlag, Berlin, p. 617–659.

Vail, P. R., Mitchum, R. M., Jr., Todd, R. G., Widmier, J. M., Thompson, S., III, Sangree, J. B., Bubb, J. N., and Hatlelid, W. G., 1977, Seismic stratigraphy and global changes of sea-level, in Payton, C. E., ed., Seismic stratigraphy - applications to hydrocarbon exploration: American Association of Petroleum Geologists Memoir 26, p. 49-212.

Van den Bergh, G. D., Boer, W., Schaapveld, M.A. S., Duc, D. M., and van Weering, T. C. E., 2007, Recent sedimentation and sediment accumulation rates of the ba Lat prodelta (Red River, Vietnam), Journal of Asian Earth Sciences, v. 29, p. 545-557.

Van Wagoner, J.C., 1995, Sequence stratigraphy and marine to nonmarine facies architecture of foreland basin strata, Book Cliffs, Utah, USA, in Van Wagoner, J.C., and Bertram, G.T., eds., Sequence Stratigraphy of Foreland Basin Deposits: American Association of Petroleum Geologists, Memoir 64, p. 137–223.

Van Wagoner, J.C., and Bertram, G.T., eds., 1995, Sequence Stratigraphy of Foreland Basin Deposits - Outcrop and Subsurface Examples from the Cretaceous of North America: American Association of Petroleum Geologists, Memoir 64, 489 p.

Van Wagoner, J.C., Jones, C.R., Tayler, D.R., Nummedal, D., Jennette, D.C., and Riley, G.W., 1991, Sequence stratigraphy: applications to shelf sandstone reservoirs: American Association of Petroleum Geologists, Field Conference, September 1991.

Van Wagoner, J. C., Mitchum, R. M., Campion, K. M. and Rahmanian, V. D. 1990, Siliciclastic sequence stratigraphy in well logs, cores, and outcrops: American Association of Petroleum Geologists Methods in Exploration Series 7, 55 p.

Visser, M. J., 1980, Neap-spring cycles reflected in Holocene subtidal large-scale bedform deposits: a preliminary note: Geology, v. 8, p. 543–546.

Wang, Y., Straub, K. M., and Hajek, E. A., 2011, Scale-dependent compensational stacking: an estimate of autogenic time scales in channelized sedimentary deposits: Geology, v. 39, p. 811–814.

Watts, A.B., and Steckler, M.S., 1979, Subsidence and eustasy at the continental margin of eastern North America: American Geophysical Union, Maurice Ewing Series, v. 3, p. 218–234.

Weber, M.E., Wiedicke, M.H., Kudrass, H.R., Huebscher, C., and Erlenkeuser, H., 1997, Active growth of the Bengal Fan during sea-level rise and highstand: Geology, v. 25, p. 315–318.

Weimer, R. J., 1970, Rates of deltaic sedimentation and intrabasin deformation, Upper Cretaceous of Rocky Mountain region, in Morgan, J. P., ed., Deltaic sedimentation modern and ancient: Society of Economic Paleontologists and Mineralogists Special Publication 15, p. 270-292.

Wellner, R. W., and Bartek, L. R., 2003, The effect of sea level, climate, and shelf physiography on the development of incised-valley complexes: a modern example from the East China Sea: Journal of Sedimentary Research, v. 73, p. 926–940.

Wheeler, H. E., 1958, Time-stratigraphy: American Association of Petroleum Geologists Bulletin, v. 42, p. 1047–1063.

Wheeler, H. E., 1959, Stratigraphic units in time and space: American Journal of Science, vol. 257, p. 692–706.

Willis, A. J., 2000, Tectonic control of nested sequence architecture in the Sego Sandstone, Neslen Formation, and Upper Castlegate Sandstone (Upper Cretaceous), Sevier Foreland Basin, Utah, U.S. A.: Sedimentary Geology, vol. 136, p. 277–318.

Wobber, F. J., 1965, Sedimentology of the Lias (Lower Jurassic) of South Wales: Journal of Sedimentary Petrology, v. 35, p. 683–703.

Wright, V. P., and Marriott, S. B., 1993, The sequence stratigraphy of fluvial depositional systems: the role of floodplain sediment storage: Sedimentary Geology, v. 86, p. 203–210.

Yang Chang-shu, and Nio, S.-D., 1989, An ebb-tide delta depositional model - a comparison between the modern Eastern Scheldt tidal basin (southwest Netherlands) and the Lower Eocene Roda Sandstone in the southern Pyrenees (Spain): Sedimentary Geology, v. 64, p. 175–196.

Yoshida, S., Willis, A., and Miall, A., 1996, Tectonic control of nested sequence architecture in the Castlegate Sandstone (Upper Cretaceous), Book Cliffs, Utah: Journal of Sedimentary Research, v. 66, p. 737-748.

Young, R.G., 1955, Sedimentary faces and intertonguing in the Upper Cretaceous of the Book Cliffs, Utah-Colorado: Geological Society of America, Bulletin, v. 66, p. 177–202.

Young, R.G., 1957, Late Cretaceous cyclic deposits, Book Cliffs, eastern Utah, American Association of Petroleum Geologists, Bulletin, v. 41, p. 1760-1774.

Zaitlin, B. A., Warren, M. J., Potocki, D., Rosenthal, L., and Boyd, R., 2002, depositional styles in a low accommodation foreland basin setting: an example from the Basal Quartz (Lower Cretaceous), southern Alberta: Bulletin of Canadian Petroleum Geology, v. 50, p. 31–72.

Zecchin, M., Caffau, M., Tosi, L., Civile, D., Brancolini, G., Rizzetto, F., and Roda, C., 2010, The impact of Late Quaternary glacio-eustasy and tectonics on sequence development: evidence from both uplifting and subsiding settings in Italy: Terra Nova, v. 22, p. 324–329.

Zecchin, M., Catuneanu, O., and Caffau, M., 2019, Wave-ravinement surfaces: classification and key characteristics: Earth Science Reviews, v. 188, p. 210–239.

Author Index

Subject Index

A

Accommodation, 6, 14, 16, 17, 22, 23, 107, 123, 149, 159, 219, 232, 233, 299, 322, 372, 373, 423, 429, 430–437, 445, 455, 463, 466, 467, 470, 475, 479, 485
 and supply, 234
 and sequent models, 243–260
 and preservation, 438–444
Acoustic anisotropy, 57
Acoustic impedance, 285
Actualism, 4, 6, 9, 445
Aeolian. *See* Eolian environments and facies models
Aggradation, 6, 7, 106, 149, 178, 218, 236, 243, 253, 256, 257, 261, 306, 375, 377, 430, 431, 432, 439, 440, 441, 463, 485
Alberta basin/deposits, 15, 16, 21, 67, 87, 116, 142, 152, 190, 199, 202, 214, 216, 217, 222, 245, 246, 274, 275, 277, 282, 288, 302, 305, 309, 317, 319, 320, 321, 322, 325, 326, 327, 328, 408, 434, 475, 477, 486
Algal mat, 148, 216, 217
Alizarin red-S, 49
Allochthonous carbonates, 152, 219, 293
Allocyclic, 5, 149 *See also* Allogenic
Allogenic, 5, 6, 19, 22, 23, 107, 110, 149, 161, 165, 184, 232, 234, 242, 255, 262–265, 274, 343, 372, 398, 429, 438, 444, 445, 446, 478, 479, 483
Allopatric speciation, 354
Allostratigraphy, 22, 110, 157, 233, 344, 368, 408
Alluvial environment/facies models. *See* Fluvial
Alluvial fan, 112, 113, 150, 157, 159
Amazon fan, 200, 202, 203, 205, 206, 253
Amazon River, 190, 203, 249, 253
American Association of Petroleum Geologists, 3, 19
American Stratigraphic Company, 75, 78
Anastomosed river/facies model, 106, 178, 180, 256, 431
Anisotropy, acoustic, 57
Anisotropy, magnetic susceptibility, 57, 186, 327
Anoxic environment/deposit/event, 117, 118, 261, 279, 391, 402, 406
Antarctic continent/ice cap, 106, 186, 187, 266, 292, 383, 385, 405, 407–409, 448, 450
Anthropocene, 402, 410
Antidune, 123, 334
Architectural element, 22, 101, 109, 158, 161–164
Arenicolites, 69
Ash, volcanic, 81, 139, 195, 345, 346, 380, 382, 384, 391, 392, 397, 438, 448, 458
Astrochronology, 26, 395, 396–403, 446, 478–483
Astronomic forcing. *See* orbital forcing
Atchafalaya River/delta, 194, 431
Athabasca Oil Sands, 190

Autocyclic, 5, 149 *See also* Autogenic
Autogenic, 5, 7, 19, 22, 23, 101, 110, 149, 153, 161, 165, 242, 246, 249, 253, 257, 258, 261, 262, 263, 264, 274, 308, 366, 372, 373, 398, 429, 434, 438, 439–444, 445, 446, 462, 472, 474, 478–481
Avalanche face, 53, 120
Avulsion, 107, 149, 193, 253, 255, 428, 432, 434, 440, 441, 442, 463, 475

B

Backstepping, 149, 190, 261, 377, 469, 478
Bahamas Bank/Platform, 11, 12, 101, 102, 106, 115, 210, 212, 213, 216, 219, 243, 258, 446, 447, 448, 450
Bahamite, 12
Bakken Formation, 277, 278, 305, 322
Ball structures, 61
Barchan dune, 131, 182, 183
Barrier coastline, environment/facies model, 11, 105, 106, 118, 120, 149, 157, 159, 181, 183, 187–191, 197, 210, 211, 212, 218, 245, 279, 284, 330, 373, 429, 431, 470
Barrier reef, 210–212, 257, 260, 261
Basal surface of forced regression, 233, 246, 252, 369, 370, 376, 456
Base level, 17, 21, 22, 23, 178, 204, 232, 233, 234, 235, 236, 237, 243–259, 274, 308, 366, 367, 368, 369, 370–376, 406, 420, 422, 432, 433, 437, 438, 440, 443, 444, 445, 467
Baselevel [sic], 16, 17, 420 *See also* Base level
Basin models, 14
Beach Deposit/Environment, 106, 114, 119, 120, 122, 126, 128, 149, 150, 181, 183, 187–194, 197, 245, 284, 333, 424
Beach rock, 205
Bedding, classification of, 50–52
Bedform hierarchy, 158, 330–331
Bedforms, classification of, 52–57
Bell-shaped log, 153, 155, 156, 275
Bengal, Bay of, 193
Bengal fan, 200, 203, 207, 250, 430
Bingham plastic, 134
Biofacies, 92–94, 95, 96, 101, 142–145, 331, 351–353, 390, 448
Biogenic
 carbonate, 187, 207, 234
 sedimentary structures, 48, 68, 78, 138, 429
Biogeography, 140, 345, 351, 353, 357
Bioherm, 27, 78, 212, 215, 257
Biostratigraphic sampling, 68, 72
Biostratinomy, 142
Bioturbation, 2, 68, 98, 119, 141, 142, 147, 152, 201, 375, 383, 434, 440, 461
Bioturbation index, 147